Cell Calcium Metabolism

Physiology, Biochemistry,
Pharmacology, and
Clinical Implications

GWUMC Department of Biochemistry
Annual Spring Symposia

Series Editors:
Allan L. Goldstein, Ajit Kumar, and J. Martyn Bailey
The George Washington University Medical Center

CARDIOVASCULAR DISEASE
Molecular and Cellular Mechanisms, Prevention, and Treatment
Edited by Linda L. Gallo

CELL CALCIUM METABOLISM
Physiology, Biochemistry, Pharmacology, and Clinical Implications
Edited by Gary Fiskum

DIETARY FIBER IN HEALTH AND DISEASE
Edited by George V. Vahouny and David Kritchevsky

EUKARYOTIC GENE EXPRESSION
Edited by Ajit Kumar

NEURAL AND ENDOCRINE PEPTIDES AND RECEPTORS
Edited by Terry W. Moody

PROSTAGLANDINS, LEUKOTRIENES, AND LIPOXINS
Biochemistry, Mechanism of Action, and Clinical Applications
Edited by J. Martyn Bailey

THYMIC HORMONES AND LYMPHOKINES
Basic Chemistry and Clinical Applications
Edited by Allan L. Goldstein

Cell Calcium Metabolism

Physiology, Biochemistry,
Pharmacology, and
Clinical Implications

Edited by
Gary Fiskum

The George Washington University Medical Center
Washington, D.C.

Plenum Press · New York and London

Library of Congress Cataloging in Publication Data

Cell calcium metabolism: physiology, biochemistry, pharmacology, and clinical implications / edited by Gary Fiskum.
    p.   cm.—(GWUMC Department of Biochemistry annual spring symposia)
    Based on papers presented at the VIIth International Spring Symposium on Health Sciences held in Washington, D.C., May 1987.
    Includes bibliographies and index.
    ISBN-13: 978-1-4684-5600-4   e-ISBN-13: 978-1-4684-5598-4
    DOI: 10.1007/ 978-1-4684-5598-4
    1. Calcium—Metabolism—Congresses. 2. Calcium—Physiological transport—Congresses. 3. Calcium—Pathophysiology. 4. Calcium channels—Congresses. 5. Calcium channels—Effect of drugs on—Congresses. 6. Pathology, Molecular—Congresses. I. Fiskum, Gary. II. International Spring Symposium on Health Sciences (7th: 1987: Washington, D.C.) III. Series.
    [DNLM: 1. Calcium—metabolism—congresses. 2. Cells—physiology—congresses. QV 276 C3933 1987]
QP535.C2C44 1989
574.87'6—dc19
DNLM/DLC   89-3815
for Library of Congress   CIP

Cover photograph of the U.S. Capitol by James P. Kendrick, R.B.P.,
The George Washington University Medical Center

© 1989 Plenum Press, New York
Softcover reprint of the hardcover 1st edition 1989
A Division of Plenum Publishing Corporation
233 Spring Street, New York, N.Y. 10013

Preface

A widespread appreciation for the role that calcium plays in cell physiology and pathophysiology has now been achieved due to the pioneering studies of many of the scientists who attended the VIIth International Spring Symposium on Health Sciences at George Washington University in Washington, D.C. The participants in this unique meeting represented diverse fields of basic and clinical research, such as molecular physiology, oncology, molecular genetics, cardiology, bioenergetics, pathology, and endocrinology. The content of the proceedings of the symposium represents work in these and other areas of biomedical research. Organization of the book is aimed at striking a balance between the biochemistry and physiology of normal cell Ca^{2+} metabolism and the pathological consequences of alterations in cell Ca^{2+} homeostasis.

The first section of the book is devoted to the transport mechanisms responsible for regulating intracellular Ca^{2+} and the pharmacological modalities for controlling cell Ca^{2+}. Particular attention is given to the molecular basis for plasma membrane transport activities, including the ATP-driven Ca^{2+} pump, the Na^+-Ca^{2+} exchange system, and voltage sensitive Ca^{2+} channels.

The second section covers the exciting relationships between phosphoinositide metabolism, signal transduction, and cell Ca^{2+} metabolism. This section begins with an eloquent overview by Professor Michael Berridge, who was the keynote speaker at the symposium and the recipient of the scientific merit award. Chapters in this section deal with studies performed with a wide variety of tissues, including cardiac and skeletal muscle, liver, platelets, and cells of the immune system.

The involvement of Ca^{2+} in cell proliferation and differentiation was another important field covered at the symposium and is the topic of the third section of this book. The relationship of Ca^{2+} movements to the action of cell growth factors and the expression of oncogenes is highlighted in these chapters.

The fourth section provides in-depth coverage of the regulation of metabolism by Ca^{2+}. Energy metabolism, stimulus–secretion coupling, hormonal activities, and cholesterol and prostanoid metabolism are examples of the topics covered in this part of the book.

The last two sections are primarily concerned with the involvement of Ca^{2+} in cell pathophysiology. The chapters in these sections examine topics such as ischemic tissue damage, anesthetic actions on cardiac and skeletal muscle contractility, and various aspects of cardiovascular pathophysiology.

The studies reviewed in this book describe many new advances in the area of cell Ca^{2+} metabolism that are profoundly influencing today's approaches to many problems in biomedical research. Discoveries such as sustained oscillations in intracellular Ca^{2+}; the presence of new, biologically active phosphoinositide metabolites; and the interrelationships among cell surface receptors, GTP binding proteins, protein phosphorylation, and Ca^{2+} fluxes are discussed. The molecular basis for Ca^{2+}-mediated tissue damage during cardiac, renal, and cerebral ischemia as well as the role Ca^{2+} plays in diseases such as muscular dystrophy and hypertension are also examined in the sections concerned with pathophysiology. Most importantly, many thought-provoking questions are raised throughout the book that will heighten the curiosity of those interested in cell Ca^{2+} metabolism.

Gary Fiskum

Washington, D.C.

Contents

PART II—PHOSPHOINOSITIDE METABOLISM

PART III—CELL PROLIFERATION AND DIFFERENTIATION

PART IV—REGULATION OF METABOLISM

PART V—MECHANISMS OF CELL INJURY

PART VI—CARDIOVASCULAR PATHOPHYSIOLOGY

I

Transport and Pharmacology

Hormonal Inhibition of the Liver Plasma Membrane $(Ca^{2+}-Mg^{2+})$ATPase Is Mediated by a G_s-like Protein

SOPHIE LOTERSZTAJN, CATHERINE PAVOINE,
ARIANE MALLAT, DOMINIQUE STENGEL, PAUL INSEL,
and FRANÇOISE PECKER

1. INTRODUCTION

The calcium pump in liver plasma membrane, which is supported by a high-affinity calcium-activated ATPase (Lotersztajn *et al.*, 1981, 1982; Pavoine *et al.*, 1987), is responsible for extrusion of calcium out of the cell. We have shown that the liver Ca^{2+} pump is specifically inhibited by pharmacological concentrations of glucagon (Lotersztajn *et al.*, 1984, 1985). This inhibition is independent of adenylate cyclase activation (Lotersztajn *et al.*, 1984) and is probably unrelated to activation of phospholipase C by nanomolar concentrations of glucagon (Wakelam *et al.*, 1986). The liver Ca^{2+} pump is also regulated by two proteins, an activator that directly stimulates activity of the purified $(Ca^{2+}-Mg^{2+})$ATPase (Lotersztajn *et al.*, 1981), and a 30,000-Da inhibitor that interacts with the purified enzyme only in the presence of the activator and Mg^{2+} (Lotersztajn and Pecker, 1982) and that mediates enzyme inhibition by glucagon (Lotersztajn *et al.*, 1985). The question arose as to the possible analogy between these activator and inhibitor proteins of $(Ca^{2+}-Mg^{2+})$ATPase and guanine nucleotide-binding (G) proteins.

G proteins are a family of homologous, membrane-associated proteins that act as transducers of receptor-mediated signals (for a review, see Gilman, 1984). Originally described as regulatory components of adenylate cyclase (G_i and G_s) and cGMP phosphodiesterase (transducin), G proteins have now been involved in the control of multiple other cellular events, such as K^+ channel activity, Na^+/H^+ exchange, excitation–contraction coupling, phosphoinositide metabolism, receptor-mediated Ca^{2+} mobilization and Ca^{2+} gating, insulin action, and agonist-promoted inhibition of Mg^{2+} influx (for reviews, see Bourne, 1986;

SOPHIE LOTERSZTAJN, CATHERINE PAVOINE, ARIANE MALLAT, DOMINIQUE STENGEL, PAUL INSEL, and FRANÇOISE PECKER • INSERM U-99, Hôpital Henri Mondor, 94010 Créteil, France.

Taylor and Merrit, 1986; Spiegel, 1987). Purification of G_i, G_s, G_o, and transducin has shown that they all are composed of three subunits: α, β, and γ (for a review, see Birnbaumer et al., 1985). The α subunits bind guanine nucleotides and some are specifically ADP-ribosylated by either cholera toxin (α_s), or pertussis toxin (α_i, α_o), or both toxins (α subunit of transducin). The sensitivity of α components to both toxins and/or their guanine nucleotide-binding activity has been used as a means to infer the role of G proteins in the mediation of hormonal responses. The purpose of the present report is to make the point on our recent results concerning the hormonal control of the liver Ca^{2+} pump and the role of G proteins in this regulation.

2. EXPERIMENTAL PROCEDURES

Female Wistar rats weighing 120–150 g were used. Highly purified glucagon was purchased from Novo (Copenhagen). Cholera toxin was obtained from Sigma (Saint Louis, MO). Pertussis toxin was from List Biological (Campbell, CA). ^{14}C-Methylated markers and [α-^{32}P]ATP (22 Ci/mmol) were from Amersham (Radiochemical Centre, UK). Cyclic [8-^{3}H]AMP was from CEA (Saclay, France), and [^{32}P]NAD (nicotinamide-adenine dinucle-otide) (10–50 Ci/mmol) was from New England Nuclear. Glucagon (19–29) was prepared as described in Mallat et al. (1987).

Liver plasma membranes were prepared according to the procedure of Neville (1968) up to step 11. The (Ca^{2+}–Mg^{2+})ATPase activity was assayed as described in Lotersztajn et al. (1981). Liver plasma membrane proteins (3–10 μg) were preincubated with glucagon or glucagon (19–29) for 10 min at 4°C. The reaction was then initiated by the addition of the assay medium, which contained in a final volume of 100 μl, 0.25 mM ATP, 50 mM Tris-HCl (pH 8), 2 mM EGTA, and 1.95 mM total $CaCl_2$ (0.1 μM free Ca^{2+}) unless oth-erwise indicated. After 10 min at 30°C, aliquots were assayed for Pi by colorimetric deter-mination using malachite green according to Kallner (1975). The (Ca^{2+}–Mg^{2+})ATPase activity was calculated by subtracting values obtained in the presence of EGTA alone from those obtained in the presence of chelator plus calcium.

Treatment of rats with cholera toxin was performed as described in Lotersztajn et al. (1987). Prior to use, cholera toxin was dialyzed for one night at 4°C against 3 liters of distilled water. Rats were injected intraperitoneally at a dose of 120 μg/100 g body weight. The rats were killed 26 hr after injection and liver plasma membranes were prepared as described above. Treatment of liver plasma membranes with pertussis toxin or cholera toxin was performed as described in Lotersztajn et al. (1987). Prior to use, toxins were activated by incubation with 20 mM DTT (dithiothreitol) for 30 min at 37°C. Rat liver plasma mem-branes (0.5–1 mg protein) were incubated with either 10 μg/ml activated pertussis toxin or vehicle, or 300–600 μg/ml cholera toxin or vehicle, in a final volume of 1.5 ml containing 50 mM Tris-HCl (pH 7.6), 1 mM ATP, 3 mM DTT, 0.1 mM GTP, and 10 μM NAD, for 30 min at 30°C. Samples were then washed twice by 15-fold dilution in ice-cold 50 mM Tris-HCl, pH 8, and centrifugation at 15,000g for 10 min. The final pellet was resuspended in 50 mM Tris-HCl, pH 8.

As a control experiment, in order to ensure that the effect of cholera toxin was not due to the production of cAMP, liver plasma membranes were preincubated with 1 mM cAMP for 30 min at 30°C under the conditions described for cholera toxin treatment, i.e., with 50 mM Tris-HCl (pH 7.6), 1 mM ATP, 3 mM DTT, and 0.1 mM GTP, but NAD was omitted. Samples were then washed as described above.

Treatment of plasma membranes with pertussis or cholera toxin was performed as described above with 10 μM [^{32}P]NAD (40 $\times$ 10^6 cpm). Radiolabeled membranes were solubilized in Laemmli sample buffer and subjected to sodium dodecyl sulfate (SDS) polyacrylamide gel electrophoresis in 12% acrylamide slab gels. After electrophoresis, the gels were fixed in 10% trichloracetic acid, dried, and exposed to Kodak XAR5 film for 12–36 hr at $-80°$C.

3. RESULTS

3.1. Hormonal Inhibition of the Liver Plasma Membrane Ca^{2+} Pump: Effect of Glucagon and Glucagon (19–29)

We have shown that glucagon specifically inhibits the Ca^{2+} pump in liver plasma membranes independently of adenylate cyclase activation (Lotersztajn et al., 1984). However, as shown in Fig. 1, this inhibition is only observed at high concentrations of glucagon (K_i = 0.7 μM). Moreover, in the presence of bacitracin, an inhibitor of glucagon degradation, the Ca^{2+} pump is no longer sensitive to glucagon (Mallat et al., 1985), which suggests that a fragment of glucagon might be the true effector of the liver Ca^{2+} pump. Pairs of basic amino acids are recognized as potential cleavage sites in posttranslational processing of peptide hormones (Steiner et al., 1980; Patzelt et al., 1984; Georges et al., 1985), and the glucagon molecule includes a dibasic doublet (Arg^{17}–Arg^{18}). Therefore, we have examined the action of glucagon (19–29) on the liver Ca^{2+} pump. This peptide was obtained from glucagon by tryptic cleavage and separated by reverse phase high-performance liquid chromatography. As shown in Fig. 1, glucagon (19–29), which is totally ineffective in activating adenylate cyclase (Mallat et al., 1987), inhibited (Ca^{2+}–Mg^{2+})ATPase activity in liver plasma membranes with an efficiency 1000-fold higher than that of glucagon. Maximal

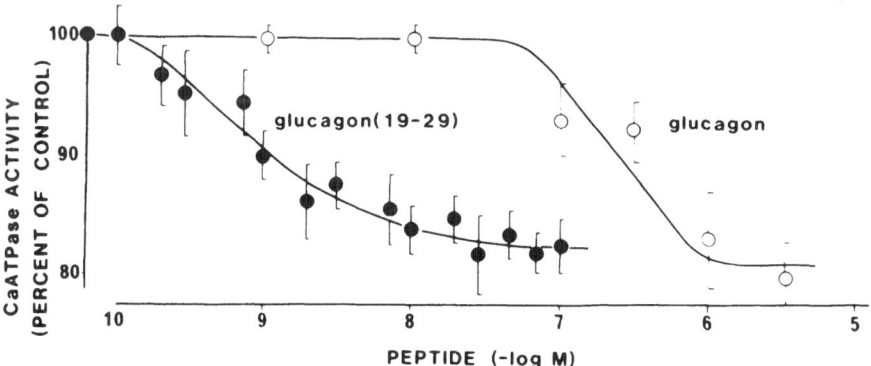

FIGURE 1. Inhibition of (Ca^{2+}–Mg^{2+})ATPase activity in purified liver plasma membranes by glucagon and glucagon (19–29). Plasma membranes (4–8 μg protein) were preincubated at 4°C either with glucagon (19–29) (0.1 nM–0.1 μM) (●) diluted in 50 mM Tris-HCl (pH 8), 0.01% bovine serum albumin, and 0.87 μM lactose or glucagon (○). After 10 min, (Ca^{2+}–Mg^{2+})ATPase was assayed as described in Sec. 2 in the presence of 0.4 mM EGTA and 393 μM total CaCl$_2$ (0.1 μM free Ca^{2+}).

inhibition was 15–20% at 10 nM glucagon (19–29) and half-maximal inhibition occured at 0.75 ± 0.25 nM glucagon (19–29). The specificity of the effect of glucagon (19–29) was demonstrated by the study of related peptides. Glucagon (1–21) was completely inactive; glucagon (18–29) and glucagon (22–29) acted only as partial agonists of glucagon (19–29) (Mallat *et al.*, 1987). These results indicated that glucagon (19–29), obtained by proteolytic cleavage of glucagon, is likely to be the active peptide involved in the inhibition of the liver Ca^{2+} pump.

3.2. Role of a G_s-like Protein in the Hormonal Inhibition of the Liver Ca^{2+} Pump

We have used cholera toxin and pertussis toxin as a way to test the possible involvement of a G protein in the hormonal regulation of $(Ca^{2+}-Mg^{2+})$ATPase in liver plasma membrane.

3.2.1. Effect of Cholera Toxin

In a first series of experiments, we studied the response of the $(Ca^{2+}-Mg^{2+})$ATPase to glucagon after treatment of the liver plasma membranes with cholera toxin. In control membranes, addition of 4 μM glucagon resulted in a 14 ± 2% maximal decrease in $(Ca^{2+}-Mg^{2+})$ATPase activity, with half-maximal inhibition occurring in the presence of 2 μM glucagon (Fig. 2A). By contrast, in cholera toxin-treated membranes no inhibition by glucagon was observed (Fig. 2A). It is known that cholera toxin treatment, which inhibits the GTPase activity of the α subunit of G_s, leads to a persistent activation of adenylate cyclase activity (for a review, see Birnbaumer *et al.*, 1985). In order to ensure that the effect of cholera toxin on $(Ca^{2+}-Mg^{2+})$ATPase was not due to the production of cAMP, we preincubated liver plasma membranes with 1 mM cAMP instead of cholera toxin. Unlike results obtained with cholera toxin, treatment of membranes with cAMP did not suppress the inhibition of $(Ca^{2+}-Mg^{2+})$ATPase by glucagon (Fig. 2A).

We next examined the effect of *in vivo* treatment of rats with an intraperitoneal injection of cholera toxin. Liver plasma membranes were prepared from rats injected with cholera toxin and killed 26 hr later. As shown in Fig. 2B, addition of 4 μM glucagon to control membranes resulted in a 18 ± 3% inhibition of $(Ca^{2+}-Mg^{2+})$ATPase activity, half-maximal inhibition being observed with 2 μM glucagon. In plasma membranes obtained from cholera toxin-treated rats, the inhibition of $(Ca^{2+}-Mg^{2+})$ATPase by glucagon (0.5–5 μM) did not occur.

The effectiveness of cholera toxin treatment was estimated by measuring adenylate cyclase activity in membranes obtained after *in vitro* or *in vivo* treatment. As previously observed by several authors (for a review, see Birnbaumer *et al.*, 1985), cholera toxin treatment induced a 2 ± 0.3-fold increase in basal adenylate cyclase activity as compared to its activity in control membranes (not shown). In addition, activation of adenylate cyclase activity by 10 mM NaF, which is known to stimulate G_s, was reduced to 4 ± 0.5-fold after *in vitro* or *in vivo* treatment with cholera toxin as compared to 8 ± 2-fold in control membranes, due to the increase in "basal" activity (not shown). Also, we used SDS polyacrylamide gel electrophoresis to verify that, in the conditions we used, cholera toxin effectively stimulated ADP ribosylation of two specific proteins of 52,000 and 42,000 Da (Fig. 2C), identified as two forms of the α subunit of G_s (see Birnbaumer *et al.*, 1985). In

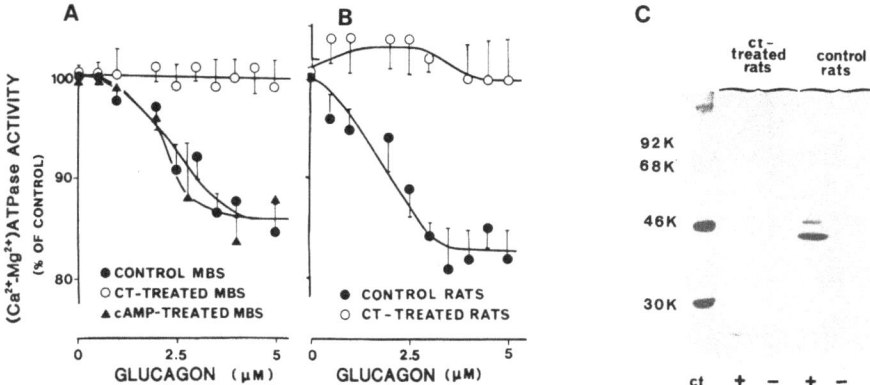

FIGURE 2. Effect of *in vitro* treatment of liver plasma membranes or *in vivo* treatment of rats with cholera toxin. (A) Liver plasma membranes were subjected to an *in vitro* treatment with cholera toxin (○), cAMP (▲), or vehicle as control (●). (B) Rats were injected with cholera toxin and killed 26 hr after treatment. Liver plasma membranes were then prepared from control (●) or treated rats (○). (Ca²⁺–Mg²⁺)ATPase activity was assayed as a function of increasing concentrations of glucagon diluted in 50 mM Tris-HCl, pH 8, containing 0.01% bovine serum albumin. Results are expressed as a percentage of control activity in the absence of glucagon. (A) 3.0 ± 0.14, 2.8 ± 0.14, and 2.9 ± 0.3 μmol Pi/mg/10 min, for cholera toxin-treated, cAMP-treated, and control membranes, respectively. (B) 1.4 ± 0.4 and 1.42 ± 0.3 μmol Pi/mg/10 min for control rats or cholera toxin-treated rats. Results are the mean ± SEM of four experiments. (C) Liver plasma membranes obtained from control rats or cholera toxin-treated rats were treated with cholera toxin (ct) or vehicle (O) in the presence of [³²P]NAD and subjected to 12% SDS polyacrylamide gel electrophoresis. The molecular weight ($\times 10^{3}$) of protein standards are shown on the left.

addition, an unidentified 39,000-Da protein was also specifically labeled. In contrast, these bands were only faintly visible in membranes obtained from cholera toxin-treated rats and then ADP-ribosylated with cholera toxin *in vitro* (Fig. 2C). However, no ADP ribosylation of proteins, which would be the (Ca²⁺–Mg²⁺)ATPase (110,000 Da; Lotersztajn *et al.*, 1984) or its inhibitor (30,000 Da; Lotersztajn *et al.*, 1982, 1985), respectively, was detected (Fig. 2C).

3.2.2. Effect of Pertussis Toxin

In a second series of experiments, liver plasma membranes were subjected to treatment with pertussis toxin in the presence of 10 μM NAD under conditions that induce ADP ribosylation of a 41,000-Da protein (Fig. 3B), identified as the α subunit of G_i (for a review, see Birnbaumer *et al.*, 1985). In pertussis toxin-treated membranes, 4 μM glucagon caused a 21 ± 2% maximal inhibition of (Ca²⁺–Mg²⁺)ATPase, which was comparable to the 19 ± 1.5% inhibition observed in control membranes (Fig. 3A). The apparent affinity of glucagon for the enzyme was similar under both conditions (2.5 and 1.5 μM, respectively; Fig. 3A).

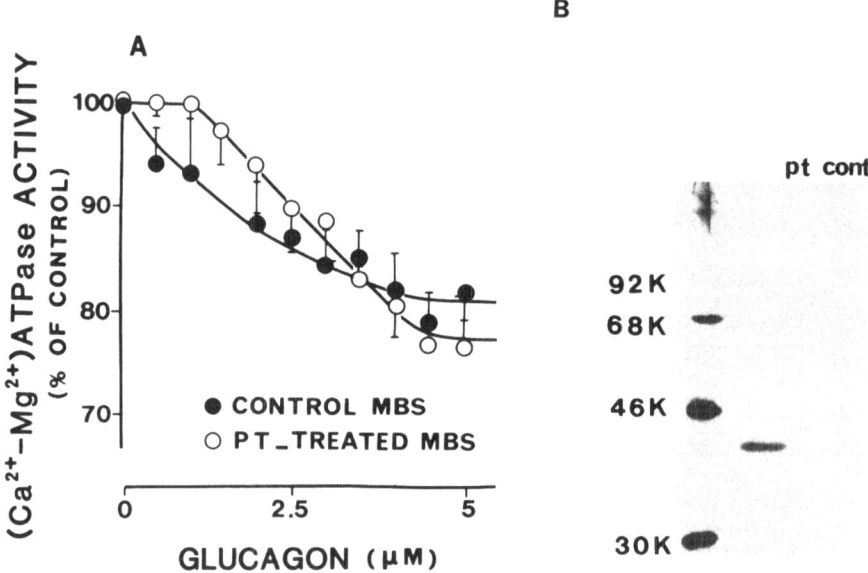

FIGURE 3. Effect of *in vitro* treatment of liver plasma membrane with pertussis toxin. Liver plasma membranes were subjected to a treatment with pertussis toxin or vehicle as control, as described under Sec. 2. (A) $(Ca^{2+}-Mg^{2+})$ATPase activity was assayed in control membranes (●) or pertussis-treated membranes (○) as a function of increasing concentrations of glucagon diluted in 50 mM Tris-HCl, pH 8, containing 0.01% bovine serum albumin. Results are expressed as a percentage of control activity in the absence of effectors (3.3 and 3.7 μmol Pi/mg/10 min for pertussis toxin-treated and control membranes, respectively). Results are the mean ± SEM of four experiments. (B) Liver plasma membrane proteins were treated with pertussis toxin (pt) or vehicle (cont) in the presence of [^{32}P]NAD and submitted to 12% SDS gel electrophoresis as described under Sec. 2.

4. DISCUSSION

Our results strongly suggest that the inhibition of the liver $(Ca^{2+}-Mg^{2+})$ATPase by glucagon or glucagon (19–29) depends on a G protein that is a specific substrate for cholera toxin (G_a in Fig. 4). Nevertheless, it is not mediated by activation of adenylate cyclase, since the action of the toxin is not mimicked by cAMP. Since cholera toxin treatment does not enhance $(Ca^{2+}-Mg^{2+})$ATPase activity, we hypothesize that the active form of the enzyme in liver plasma membranes is the one associated with the α_a subunit ("activated state" in Fig. 4). In fact, such a model for regulation of adenylate cyclase has recently been proposed by Levitski (1987). Under the influence of inhibitory hormones, e.g., glucagon, glucagon (19–29), and/or guanyl nucleotides, dissociation of the α_a subunit from enzyme would then lead to a decrease in $(Ca^{2+}-Mg^{2+})$ATPase activity ("deactivated state," Fig. 4). Our recent data indicate that nonhydrolyzable analogs of GTP [GTPγ_s, G_{pp}(NH)$_p$] mimic the action of glucagon/glucagon (19–29) and inhibit the liver Ca^{2+} pump (Lotersztajn, Pavoine, and Pecker, manuscript in preparation). ADP ribosylation of G_a by cholera toxin would prevent dissociation of enzyme-α_a complex and thereby would lead to a loss of sensitivity of $(Ca^{2+}-Mg^{2+})$ATPase to hormonal inhibition.

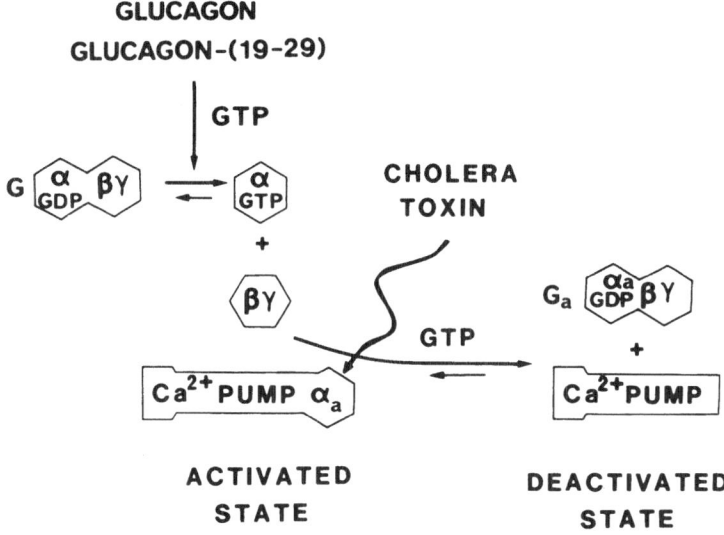

GLUCAGON
GLUCAGON-(19-29)

FIGURE 4. Hypothesis is for the inhibition mechanism of the liver Ca^{2+} pump by hormones. In the active state, the liver Ca^{2+} pump is associated with the α_a subunit of a G_a protein, sensitive to cholera toxin. Glucagon/glucagon (19–29) in the presence of guanyl nucleotides promotes dissociation of G proteins to α and $\beta\gamma$ subunits. The released $\beta\gamma$ subunits would interact with α_a and induce reassociation of G_a subunits. Dissociation of α_a from the Ca^{2+} pump will result in the deactivation of the system ("inhibition"). Cholera toxin, by causing ADP ribosylation of α_a, will prevent reassociation of α_a with $\beta\gamma$ and thus hinder hormonal inhibition of the Ca^{2+} pump.

The activator of the liver plasma membrane $(Ca^{2+}-Mg^{2+})$ATPase might be a candidate for the α subunit of this G_s-like protein, since it activates the enzyme by direct interaction (Lotersztajn et al., 1981). In this hypothesis, interaction of the activator with a $\beta\gamma$-like subunit would result in the deactivation of $(Ca^{2+}-Mg^{2+})$ATPase. One could postulate that the inhibitor, which reverses the stimulation of the $(Ca^{2+}-Mg^{2+})$ATPase induced by the activator (Lotersztajn et al., 1982) and mediates the inhibitory signal produced by glucagon (Lotersztajn et al., 1985), plays the role of a $\beta\gamma$-like subunit (see Gilman, 1984; Cerione et al., 1985; Bockaert et al., 1985).

The present results constitute evidence for the involvement of a G_s-like protein that we call G_a, sensitive to cholera toxin, in the regulation of plasma membrane $(Ca^{2+}-Mg^{2+})$ATPase. Several reports have established the multiple role of G_i-like proteins in hormonal action in cAMP-dependent (see Birnbaumer et al., 1985) as well as in cAMP-independent pathways (for reviews, see Bourne, 1986; Taylor and Merrit, 1986; Spiegel, 1987). Implication of G_s-like proteins in metabolic responses that are independent of adenylate cyclase have also been evoked in chemotaxis (Askamit et al., 1985), insulin action (Heyworth et al., 1985), regulation of the T-cell antigen receptor (Imboden et al., 1986), agonist-promoted Mg^{2+} influx inhibition (Maguire et al., 1980), and inhibition of glucagon-stimulated phospholipase C (Wakelam et al., 1986). However, in none of these systems could the precise role of G_s-like proteins in regulation of specific enzymes be defined. The present data suggest that the plasma membrane $(Ca^{2+}-Mg^{2+})$ATPase provides a useful system to examine enzymatic regulation by G_s-like proteins.

ACKNOWLEDGMENTS. The authors wish to thank Drs. Jacques Hanoune and Michele Good-hardt for their most helpful comments and Ms. Lydie Rosario for her skillful secretarial assistance. This work was supported by the Institut National de la Santé et de la Récherche Médicale, the Délégation Générale de la Récherche Scientifique et Technique, the Université Paris-Val de Marne, the National Institutes of Health grants GM 31987 and HL 35847, and the National Science Foundation grant DCB85-02168.

REFERENCES

Aksamit, R. P., Backlund, P. S. Jr., and Cantoni, G. L., 1985, Cholera toxin inhibits chemotaxis by a cAMP-independent mechanism, *Proc. Natl. Acad. Sci. USA* **82:**7475–7479.

Birnbaumer, L., Codina, J., Mattera, R., Cerione, R. A., Hildebrandt, J. D., Sunyer, T., Rojas, F. J., Caron, M. G., Lefkowitz, R. J., and Iyengar, R., 1985, Regulation of hormone receptors and adenylyl cyclases by guanine nucleotide binding N proteins, *Recent Prog. Hormone Res.* **41:** 41–99.

Bockaert, J., Deterre, P., Pfister, C., Guillon, G., and Chabre, M., 1985, Inhibition of hormonally-regulated adenylate cyclase by the beta gamma subunit of transducin, *EMBO J.* **4:**1413–1417.

Bourne, H. R., 1986, One molecular machine can transduce diverse signals, *Nature* **321:**814–816.

Cerione, R. A., Codina, J., Kilpatrick, B. F., Staniszewski, C., Gierschik, P., Somers, R. L., Spiegel, A. M., Birnbaumer, L., Caron, M. G., and Lefkowitz, R. J., 1985, Transducin and inhibitory nucleotide regulatory protein inhibit the stimulatory nucleotide regulatory protein mediated stimulation of adenylate cyclase in phospholipid vesicle systems, *Biochemistry* **24:**4499–4503.

Georges, S. K., Uttenthal, L. O., Ghiglione, M., and Bloom, S. R., 1985, Molecular forms of glucagon-like peptides in man, *FEBS Lett.* **192:**275–278.

Gilman, A. G., 1984, G Protein and dual control of adenylate cyclase, *Cell* **36:**577–579.

Heyworth, C. M., Whetton, A. D., Wong, S., Martin, B. R., and Houslay, M. D., 1985, Insulin inhibits the cholera-toxin-catalysed ribosylation of a Mr-25000 protein in rat liver plasma membranes, *Biochem. J.* **228:**593–603.

Imboden, J. B., Shoback, D. M., Pattison, G., and Stobo, J. D., 1986, Cholera toxin inhibits the T-cell antigen receptor mediated increase in inositol trisphosphate and cytoplasmic free calcium, *Proc. Natl. Acad. Sci. USA* **83:**5673–5677.

Kallner, A., 1975, Determination of phosphate in serum and urine by a single step malachite-green method, *Clin. Chim. Acta.* **59:**35–39.

Levitski, A., 1987, Regulation of adenylate cyclase by hormones and G proteins, *FEBS Lett.* **211:** 113–118.

Lotersztajn, S., and Pecker, F., 1982, A membrane-bound protein inhibitor of high affinity Ca ATPase in rat liver plasma membranes, *J. Biol. Chem.* **257:**6638–6641.

Lotersztajn, S., Hanoune, J., and Pecker, F., 1981, A high affinity calcium-stimulated magnesium dependent ATPase in rat liver plasma membranes, *J. Biol. Chem.* **256:**11209–11215.

Lotersztajn, S., Mavier, P., Clergue, J., Dhumeaux, D., and Pecker, F., 1982, Human liver plasma membrane Ca ATPase: Identification and sensitivity to calcium antagonists, *Hepatology* **2:**843–848.

Lotersztajn, S., Epand, R., Mallat, A., and Pecker, F., 1984, Inhibition by glucagon of the calcium pump in liver plasma membranes, *J. Biol. Chem.* **259:**8195–8201.

Lotersztajn, S., Mallat, A., Pavoine, C., and Pecker, F., 1985, The inhibition of liver plasma membrane $(Ca^{2+}-Mg^{2+})$ATPase, *J. Biol. Chem.* **260:**9692–9698.

Lotersztajn, S., Pavoine, C., Mallat, A., Stengel, D., Insel, P. A., and Pecker, F., 1987, Cholera toxin blocks glucagon-mediated inhibition of the liver plasma membrane $(Ca^{2+}-Mg^{2+})$ATPase, *J. Biol. Chem.* **262:**3114–3117.

Maguire, M. E., and Erdos, J. J., 1980, Inhibition of magnesium uptake by β-adrenergic agonists and prostaglandin E_1 is not mediated by cAMP, *J. Biol. Chem.* **255:**1030–1035.

Mallat, A., Pavoine, C., Lotersztajn, S., and Pecker, F., 1985, Inhibition of the Ca pump in liver plasma membranes by glucagon is due to a metabolite of the hormone, *Fed. Proc.* **44:**1392.

Mallat, A., Pavoine, C., Dufour, M., Lotersztajn, S., Bataille, D., and Pecker, F., 1987, A glucagon fragment is responsible for inhibition of the liver Ca^{2+} pump by glucagon, *Nature* **325:**620–622.

Neville, D. M., 1968, Isolation of an organ specific protein antigen from cell-surface membrane of rat liver, *Biochim. Biophys. Acta.* **154:**540–552.

Patzelt, C., and Schiltz, E., 1984, Conversion of proglucagon in pancreatic alpha cells: The major end products are glucagon and a single peptide, the major proglucagon fragment that contains two glucagon sequences, *Proc. Natl. Acad. Sci. USA* **81:**5007–5011.

Pavoine, C., Lotersztajn, S., Mallat, A., and Pecker, F., 1987, The high affinity (Ca^{2+}–Mg^{2+})ATPase in liver plasma membranes is a Ca^{2+} pump, *J. Biol. Chem.* **262:**5113–5118.

Spiegel, A. M., 1987, Signal transduction by guanine nucleotide binding proteins, *Mol. Cell Endocrinol.* **49:**1–16.

Steiner, D. F., Quinn, P. S., Chan, S. J., Marsh, J., and Tager, H. S., 1980, Processing mechanism in the biosynthesis of proteins, *Ann. N.Y. Acad. Sci.* **343:**1–16.

Taylor, C. W., and Merrit, J. E., 1986, Receptor coupling to polyphosphoinositine turnover: A parallel with the adenylate cyclase system, *Trends Pharmacol. Sci.* **7:**238–242.

Wakelam, M. J. O., Murphy, G. J., Hruby, G. J., and Houslay, M. D., 1986, Activation of two signal transduction systems in hepatocytes by glucagon, *Nature* **323:**68–71.

2

The High-Affinity $(Ca^{2+}-Mg^{2+})$ATPase of Rat Liver Plasma Membrane Hydrolyzes Extracellular ATP

SUE-HWA LIN

The intracellular free Ca^{2+} concentration is very low (in the region of 0.1 μM) in contrast to extracellular $[Ca^{2+}]$, which is usually in the millimolar range. An energy-dependent Ca^{2+} transporter in the plasma membrane is therefore important in maintaining the concentration gradient. Studies on the biochemical properties of a Ca^{2+} pump were first carried out with sarcoplasmic reticulum membrane that is enriched with a Ca^{2+} pump (for review, see Tada *et al.*, 1978). These studies established several properties of the sarcoplasmic reticulum Ca^{2+} pump: (1) The Ca^{2+} pump utilizes both Ca^{2+}- and Mg^{2+}-ATP as substrates. (2) It exhibits a high affinity for Ca^{2+} (in the submicromolar range). (3) It has a Ca^{2+}-stimulated ATPase activity. When the enzyme is incorporated into phospholipid vesicles, a coupled Ca^{2+}-stimulated ATP hydrolysis and Ca^{2+} transport can be demonstrated. (4) The mechanism of Ca^{2+}-stimulated ATP hydrolysis involves an enzyme phosphate intermediate. As a result of such a property, both the Ca^{2+}-stimulated ATP hydrolysis and Ca^{2+} transport can be inhibited by vanadate, a transition state analog of phosphate.

Are the properties of plasma membrane Ca^{2+} pumps different from those of the sarcoplasmic reticulum Ca^{2+} pump? Most studies on a plasma membrane Ca^{2+} pump have utilized the enzyme from the erythrocyte membrane. These studies showed that although the erythrocyte Ca^{2+} pump may be different from the sarcoplasmic reticulum Ca^{2+} pump, they share several properties: the mechanism of ATP hydrolysis, vanadate inhibition of activity, and formation of a phosphorylated intermediate (for review, see Schatzmann, 1983). Studies on the plasma membrane Ca^{2+}-pumping activities of several other tissues indicated that they are more or less similar to the erythrocyte Ca^{2+} pump.

As a result of these studies, it is likely that all Ca^{2+} pumps should have a Ca^{2+}-stimulated ATPase activity. However, is it correct to assume that any plasma membrane protein that has a Ca^{2+}-stimulated ATPase activity is a Ca^{2+} pump? A functional correlation between a Ca^{2+}-stimulated ATPase activity and Ca^{2+}-pumping activity requires the recon-

SUE-HWA LIN • Department of Biochemistry and Molecular Biology, Harvard University, Cambridge, Massachusetts 02138.

stitution of the purified Ca^{2+}-ATPase protein into artificial liposomes and the demonstration that it can transport Ca^{2+} against a Ca^{2+} concentration gradient. Due to the low abundance of plasma membrane Ca^{2+} pumps, few have been purified. I have studied the liver plasma membrane Ca^{2+} pump. This report summarizes the central observations.

By measuring the Ca^{2+}-stimulated ATPase activity, we (Lin et al., 1983) and others (Lotersztajn et al., 1981) found a high-affinity $(Ca^{2+}-Mg^{2+})$ATPase localized in the plasma membrane of liver parenchymal cells. The enzyme had a distribution similar to that of the plasma membrane-specific markers alkaline phosphodiesterase and 5'-nucleotidase (Lin et al., 1983) and a high affinity for Ca^{2+}, with an apparent half-saturation constant ($K_{0.5}$) of 75 nM for ionized Ca^{2+}. We developed a procedure for the purification of $(Ca^{2+}-Mg^{2+})$ATPase from rat liver plasma membrane with about 20% recovery of total activity (Lin and Fain, 1984). The enzyme was purified about 330-fold from the plasma membrane. By sodium dodecyl sulfate (SDS) polyacrylamide gel electrophoresis, the purified $(Ca^{2+}-Mg^{2+})$ATPase has a molecular weight of 70,000 in nonreducing conditions. In the presence of reducing agent or by heating, the molecular weight is 95,000. The affinity for Ca^{2+} of the detergent-solubilized, purified protein is 87 nM. However, in the purified state, the protein possesses Mg^{2+}-ATPase activity (Lin, 1985b). In the absence of Ca^{2+}, the enzyme was activated by Mg^{2+} with half-maximum activation in the range of 0.16 μM free Mg^{2+}. The effect of Ca^{2+} and Mg^{2+} on the ATPase activity is not additive indicating that these two activities reside in the same protein (Lin, 1985b). In the presence of Ca^{2+} or Mg^{2+}, the enzyme can hydrolyze GTP, UTP, CTP, ADP, and GDP in addition to its normal substrate ATP. The Ca^{2+}-ATPase is also insensitive to vanadate, oligomycin, NBD-Cl (7-chloro-4-nitrobenzo-2-oxa-1,3-diazole), N-ethylmaleimide, and p-chloromercuribenzoate treatments. These properties of the Ca^{2+}-stimulated ATPase activity are quite different from those of sarcoplasmic reticulum Ca^{2+}-ATPase and human erythrocyte Ca^{2+}-ATPase.

Is this high-affinity $(Ca^{2+}-Mg^{2+})$ATPase a Ca^{2+} pump? To address this question, we reconstituted the purified $(Ca^{2+}-Mg^{2+})$ATPase into artificial liposomes. No ATP-stimulated Ca^{2+} transport could be detected in the reconstituted system. However, an ATP-stimulated Ca^{2+} transport activity could be detected by reconstituting the liver plasma membrane proteins into artificial liposomes. This indicates that in the rat liver plasma membranes there is an ATP-dependent Ca^{2+} transporter. The purified protein that possesses Ca^{2+}-stimulated ATPase activity, however, is not the Ca^{2+} pump. We characterized the properties of the ATP-dependent Ca^{2+}-pumping activity by using the reconstituted system (Lin, 1985a), and the differences between the properties of the Ca^{2+}-stimulated ATP hydrolysis activities of the

TABLE I. Comparison of $(Ca^{2+}-Mg^{2+})$ATPase and Ca^{2+} Pump

Feature	$(Ca^{2+}-Mg^{2+})$ATPase	Ca^{2+} pump
Ca^{2+}-pumping activity	No	Yes
Vanadate inhibition	No	Yes
Formation of E-P	No	Yes
Mg^{2+} requirement	0.01 μM	23 μM
$K_{0.5}$ for ATP	100–200 μM	6 μM
Substrate specificity	Nonspecific	ATP
Protein mol wt	70,000	118,000
Specific activity in liver membrane	100–200 nmol/mg/min	2–6 nmol/mg/min
Localization	Canalicular membrane	Sinusoidal membrane

purified enzyme and those of the ATP-dependent Ca^{2+} transport activities are listed in Table I. The Ca^{2+}-pumping activity is sensitive to vanadate while the $(Ca^{2+}-Mg^{2+})$ATPase activity is not. The Ca^{2+} pump can only use ATP as a substrate while the $(Ca^{2+}-Mg^{2+})$ATPase has a broad substrate specificity. The Ca^{2+} pump requires Mg^{2+} in order to function, while the $(Ca^{2+}-Mg^{2+})$ATPase does not require Mg^{2+}. In fact, since the latter is also a Mg^{2+}-ATPase, no Ca^{2+}-stimulated activity was observed in the presence of saturating amounts of Mg^{2+} (Lin, 1985b). Since the Ca^{2+} pump activity is inhibited by vanadate, a property associated with a phosphorylated protein as an obligatory intermediate, we can determine the molecular weight of the Ca^{2+} pump by trapping the Ca^{2+}-stimulated phosphorylated intermediate of the Ca^{2+} pump protein. The molecular weight of the plasma membrane Ca^{2+} pump is 118,000 (Lin, 1985a and Fig. 1). In the same experiment, we also showed that the

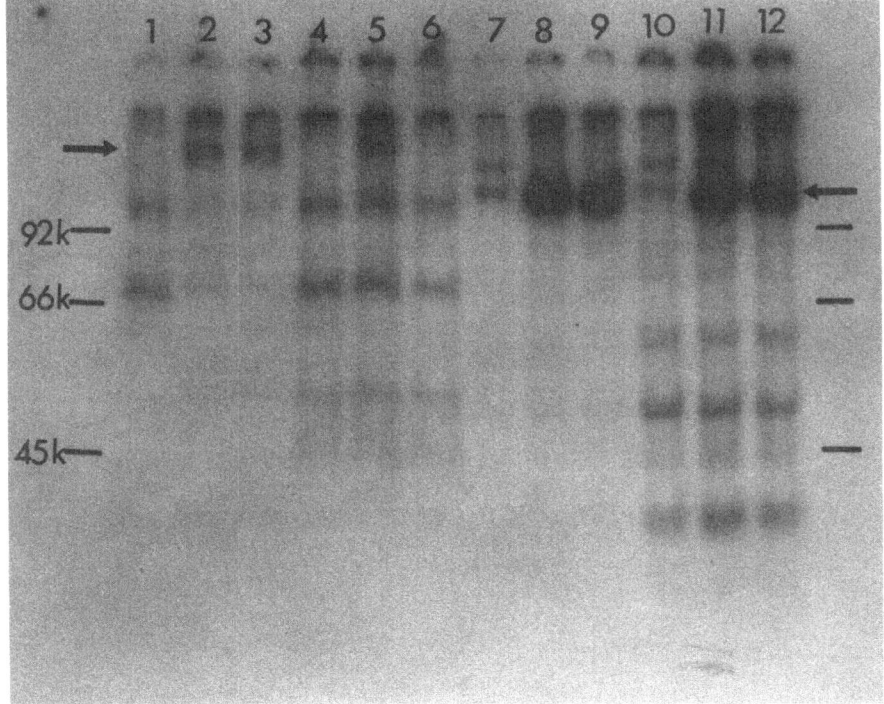

FIGURE 1. Electrophoretic pattern of phosphoproteins from the phospholipid vesicles reconstituted with liver plasma membrane proteins and from the liver microsomes. Phosphorylation of reconstituted vesicles (lanes 1–6, 50 µg protein each lane) and liver microsomes (lanes 7–12, 20 µg protein each lane) was performed as described in Lin (1985a). Lanes 1–3 and 7–9, phosphorylation in the absence of added Mg^{2+}; lanes 4–6 and 10–12, phosphorylation in the presence of 10 mM Mg^{2+}. Lanes 1, 4, 7, and 10, phosphorylation in the presence of 2 mM EGTA; lanes 2, 5, 8, and 11, phosphorylation in the presence of 20 µM Ca^{2+}; lanes 3, 6, 9, and 12, phosphorylation in the presence of 20 µM Ca^{2+} and 50 µM vanadate.

liver endoplasmic reticulum Ca^{2+} pump has a molecular weight of 100,000 and has a different sensitivity to vanadate compared to that of the Ca^{2+} pump of the plasma membrane. This excludes the possibility that the 118,000 Ca^{2+}-stimulated phosphoprotein may be a contaminant of the endoplasmic reticulum Ca^{2+} pump. In the liver plasma membrane, the ATPase activity of the Ca^{2+} pump is about 20 times lower than that of the $(Ca^{2+}-Mg^{2+})$ATPase. Therefore, the ATPase activity of the Ca^{2+} pump cannot be detected easily in the plasma membrane. With antibodies against the purified $(Ca^{2+}-Mg^{2+})$ATPase as a probe, immunofluorescence on frozen sections of rat liver showed that the $(Ca^{2+}-Mg^{2+})$ATPase is localized in the canalicular region of the liver membrane. The formation of vanadate-sensitive Ca^{2+}-stimulated phosphoenzyme formation, on the other hand, indicated that the Ca^{2+} pump is located in the sinusoidal membrane fraction. From these observations, we concluded that the previously purified $(Ca^{2+}-Mg^{2+})$ATPase and the Ca^{2+} pump are two different proteins. As a result, we are also interested in the physiological function of the $(Ca^{2+}-Mg^{2+})$ATPase

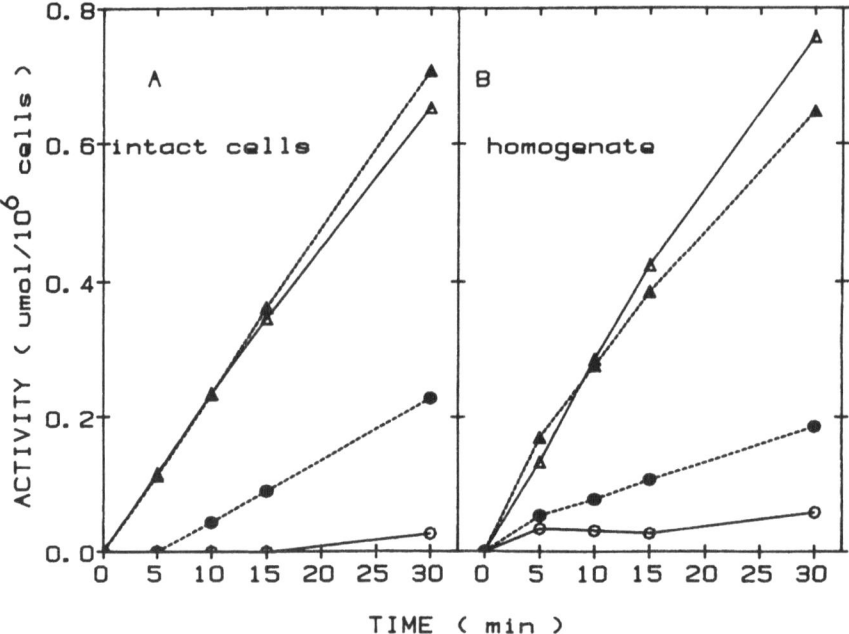

FIGURE 2. Time course of Ca^{2+}- and Mg^{2+}-stimulated ATPase activities of hepatocytes in primary culture and of disrupted hepatocytes. (A) The hepatocytes in primary culture were incubated in the presence of 120 mM NaCl, 5 mM KCl, 20 mM Hepes/Tris (pH 7.4), 2 mM EGTA without added Ca^{2+} or Mg^{2+} (○), or with 2 mM Ca^{2+} (●), or with 1 mM Mg^{2+} (△), or with 2 mM Ca^{2+} plus Mg^{2+} (▲), respectively. The reactions were started by the addition of ATP to a final concentration of 2 mM. After incubating the cells with ATP at room temperature for different periods of time as indicated, aliquots of cell medium were withdrawn and the amount of inorganic phosphate determined. (B) Hepatocytes in primary culture were removed from dishes and homogenized with buffer A, which contains 120 mM NaCl, 5 mM KCl, 20 mM Hepes/Tris (pH 7.4), and 2 mM EGTA (○), or with buffer A plus 2 mM Ca^{2+} (●), or with buffer A plus 1 mM Mg^{2+} (△), or with buffer A plus 2 mM Ca^{2+} and 1 mM Mg^{2+} (▲), respectively. ATPase activities were measured as described above.

that is not a calcium pump. It has been observed that several tissues, including liver, contain plasma membrane ecto-ATPases, i.e., enzymes with their ATP-hydrolyzing sites on the outside of the cells. We first tested whether this $(Ca^{2+}-Mg^{2+})$ATPase of the liver plasma membrane is an ecto-ATPase by comparing the properties of the purified $(Ca^{2+}-Mg^{2+})$ATPase with those of liver ecto-ATPase.

Figure 2A shows the time course of Ca^{2+}- and Mg^{2+}-stimulated ATPase activities of hepatocytes in primary culture. Disruption of the hepatocytes by scraping the cells off the collagen-coated dishes followed by sonication for 1 min did not increase either the Ca^{2+}-ATPase or Mg^{2+}-ATPase activities (Fig. 2B). At the end of a 30-min incubation, aliquots of cell medium from the intact cells and the disrupted cell preparations were taken for the determination of lactate dehydrogenase activity. The lactate dehydrogenase activity of the cell medium from the intact cells was 14% of that of the disrupted cells (0.27 U/10^6 cells versus 1.92 U/10^6 cells). Addition of saponin (40 μg/ml) to the intact cell preparation after a 30-min incubation with ATP caused release of all the lactate dehydrogenase activity to the cell medium (2.03 U/10^6 cells). These results indicate that the hepatocytes in primary culture were still intact after 30 min of incubation with 2 mM ATP, and the Ca^{2+}-ATPase and Mg^{2+}-ATPase activities measured in the whole-cell preparation were due to cell surface ATPase activities. Addition of both Ca^{2+} and Mg^{2+} to the ATPase assay medium gave the same ATPase activity as Mg^{2+} alone in both whole cells and disrupted cell preparations (Fig. 2A,B). The nonadditive effect of Ca^{2+} and Mg^{2+} indicates that both activities are from the same enzyme.

In order to test the possibility that the ecto-$(Ca^{2+}-Mg^{2+})$ATPase is the same protein we purified previously, we compared the properties of the ecto-$(Ca^{2+}-Mg^{2+})$ATPase with those of the previously purified $(Ca^{2+}-Mg^{2+})$ATPase. The nucleotide specificity of the ecto-ATPase activities is shown in Table II. The ecto-$(Ca^{2+}-Mg^{2+})$ATPase activity has broad

TABLE II. Substrate Specificity of Ecto-$(Ca^{2+}-Mg^{2+})$ATPase and $(Ca^{2+}-Mg^{2+})$ATPase[a]

Substrate	Ecto-$(Ca^{2+}-Mg^{2+})$ATPase (%)	$(Ca^{2+}-Mg^{2+})$ATPase (%)
ATP	100	100
ADP	32	33
AMP	27	0
AMP-PNP	20	0
GTP	93	92
GDP	39	52
UTP	76	81
CTP	78	52
p-Nitrophenyl phosphate	6	0

[a] The ecto-$(Ca^{2+}-Mg^{2+})$ATPase activity was measured by incubating hepatocytes in primary culture with buffer A (120 mM NaCl, 5 mM KCl, 2 mM EGTA, 20 mM Hepes/Tris, pH 7.4), or buffer A plus 2 mM Ca^{2+}. The reactions were started by adding different nucleotides to a final concentration of 2 mM and incubating at room temperature for 30 min.

The $(Ca^{2+}-Mg^{2+})$ATPase activity was assayed by incubating the purified enzyme for 30 min at 37°C in an assay medium containing, in a final volume of 500 μl, 20 mM sodium azide, 50 mM Hepes/Tris (pH 7.4), 2 mM EGTA, 2 mM nucleotide, with or without 2 mM Ca^{2+}. $C_{12}E_9$ (polyoxyethylene-9-laurylether) at a concentration of 0.2 mg/ml was included in the assay medium.

The ATPase activity was determined by measuring the inorganic phosphate released as described by Ames (1966), except that the time for color development was 20 min at 37°C instead of 1 hr at 37°C. Ca^{2+}-stimulated ATPase activity was determined at the end of a 30 min incubation period by subtracting values obtained with EGTA alone from those with calcium plus chelator.

substrate specificities. The relative nucleotide-hydrolyzing rates are about the same as that of the purified $(Ca^{2+}-Mg^{2+})ATPase$ except with AMP and AMPPNP. The hydrolysis of AMP may be due to the presence of 5′-nucleotidase, which is known to be a hepatocyte ectoenzyme, in the intact cell. The hydrolysis of AMPPNP in the presence of Ca^{2+} may be due to the sequential hydrolysis of AMPPNP by the Ca^{2+}-dependent adenosine triphosphate pyrophosphohydrolase activity (Flodgaard and Torp-Pedersen, 1978) and the 5′-nucleotidase.

One of the distinct properties of the previously purified $(Ca^{2+}-Mg^{2+})ATPase$ is its insensitivity to several known ATPase inhibitors (Lin, 1985b). The ecto-$(Ca^{2+}-Mg^{2+})ATPase$ activity, like the previously purified $(Ca^{2+}-Mg^{2+})ATPase$, was not inhibited by 0.5 mM vanadate, 5 mM N-ethylmaleimide, 100 μM p-chloromercuribenzoate, 1 mM ouabain, and 50 μM NBD-Cl.

Since the ATP-hydrolyzing site of the ecto-$(Ca^{2+}-Mg^{2+})ATPase$ is extracellular, it was interesting to see whether proteolysis would destroy the ATPase activity from the outside of the cells. The ecto-$(Ca^{2+}-Mg^{2+})ATPase$ activity of hepatocytes in primary culture is not sensitive to trypsin, chymotrypsin, or papain treatment. Although treatment of the hepatocytes in primary culture with proteases causes dissociation of the cells from the collagen-coated culture dishes, less than 25% of the ecto-$(Ca^{2+}-Mg^{2+})ATPase$ activity was lost with 50 $\mu g/ml$ of trypsin, chymotrypsin, or papain for 90 min at room temperature. When the purified high-affinity $(Ca^{2+}-Mg^{2+})ATPase$ was treated with proteases under the same conditions, the purified $(Ca^{2+}-Mg^{2+})ATPase$ activity was similarly insensitive to protease treatment.

From these studies, it is concluded that the previously purified $(Ca^{2+}-Mg^{2+})ATPase$ is a membrane ecto-ATPase. Recently, Charest *et al.* (1985) reported that stimulation of isolated hepatocytes with ATP or ADP induced a rapid but transient increase of the cytosolic Ca^{2+} concentration and the transient response was probably due to rapid hydrolysis of the ATP or ADP. Our finding that the high-affinity $(Ca^{2+}-Mg^{2+})ATPase$ hydrolyzes extracellular ATP and ADP suggests that it may play a role in terminating the effect of ATP and ADP on stimulating hepatocyte Ca^{2+} mobilization.

SUMMARY

Rat liver plasma membrane contains a high-affinity $(Ca^{2+}-Mg^{2+})ATPase$ which we have purified and characterized (Lin and Fain, 1984). The properties of this enzyme are different from those of the plasma membrane Ca^{2+} pump studied by reconstitution (Lin, 1985a, 1985b) indicating that the high-affinity $(Ca^{2+}-Mg^{2+})ATPase$ and the plasma membrane Ca^{2+} pump are different proteins. In this study, I examine the active site arrangement and the inhibitor and substrate specificities of the $(Ca^{2+}-Mg^{2+})ATPase$ in intact hepatocytes.

Ca^{2+}-ATPase and Mg^{2+}-ATPase activities were detected by the addition of ATP to hepatocytes in primary culture. The localization of the active site of the ATPase was determined by measuring the Ca^{2+}-ATPase and Mg^{2+}-ATPase activities in intact cells and cell homogenates. The result is that ATPase activities are identical in the two preparations, indicating that the active site of the plasma membrane $(Ca^{2+}-Mg^{2+})ATPase$ is extracellular. The effect of Ca^{2+} and Mg^{2+} on the ecto-ATPase activity is not additive, indicating that both Ca^{2+}- and Mg^{2+}-ATPase activities are part of the same enzyme.

The possibility that the ecto-$(Ca^{2+}-Mg^{2+})ATPase$ may be the same protein as the previously purified high-affinity $(Ca^{2+}-Mg^{2+})ATPase$ was tested by comparing the properties of the ecto-$(Ca^{2+}-Mg^{2+})ATPase$ with those of $(Ca^{2+}-Mg^{2+})ATPase$. The nucleotide specificity of the ecto-$(Ca^{2+}-Mg^{2+})ATPase$ is the same as that of the purified enzyme. The ecto-$(Ca^{2+}-Mg^{2+})$ATPase activity, like the $(Ca^{2+}-Mg^{2+})$ATPase, is not sensitive to oligomycin, vanadate, N-ethylmaleimide, or p-chloromercuribenzoate; and both the ecto-$(Ca^{2+}-Mg^{2+})$ATPase and purified $(Ca^{2+}-Mg^{2+})$ATPase are insensitive to protease treatments. These properties indicate that the previously purified high-affinity $(Ca^{2+}-Mg^{2+})$ATPase is the ecto-$(Ca^{2+}-Mg^{2+})$ATPase and may function in terminating the effect of ATP and ADP on stimulating hepatocyte Ca^{2+} mobilization.

ACKNOWLEDGMENT. I thank professor Guido Guidotti, in whose laboratory this work was done, for his support and advice during the course of this work.

REFERENCES

Ames, B. N., 1966, Assay of inorganic phosphate, total phosphate and phosphatases, *Methods Enzymol.* **8:**115–117.

Charest, R., Blackmore, P. F., and Exton, J. H., 1985, Characterization of responses of isolated rat hepatocytes to ATP and ADP, *J. Biol. Chem.* **260:**15789–15794.

Flodgaard, H., and Torp-Pedersen, C., 1978, A calcium ion-dependent adenosine triphosphate pyrophosphohydrolase in plasma membrane from rat liver, *Biochem. J.* **171:**817–820.

Lin, S. H., 1985a, Novel ATP-dependent calcium transport component from rat liver plasma membranes: The transporter and the previously reported $(Ca^{2+}-Mg^{2+})$ATPase are different proteins, *J. Biol. Chem.* **260:**7850–7856.

Lin, S. H., 1985b, The rat liver plasma membrane high affinity $(Ca^{2+}-Mg^{2+})$ATPase is not a calcium pump: Comparison with ATP-dependent calcium transporter, *J. Biol. Chem.* **260:**10976–10980.

Lin, S. H., and Fain, J. N., 1984, Purification of $(Ca^{2+}-Mg^{2+})$ATPase from rat liver membranes, *J. Biol. Chem.* **259:**3016–3020.

Lin, S. H., Wallace, M. A., and Fain, J. N., 1983, Regulation of $(Ca^{2+}-Mg^{2+})$ATPase activity in hepatocyte plasma membranes by vasopressin and phenylephrine, *Endocrinology* **113:**2268–2275.

Lotersztajn, S., Hanoune, J., and Pecker, F., 1981, A high affinity calcium-stimulated magnesium-dependent ATPase in rat liver plasma membranes, *J. Biol. Chem.* **256:**11209–11215.

Schatzmann, H. J., 1983, The red cell calcium pump, *Ann. Rev. Physiol.* **45:**303–312.

Tada, M., Yamamoto, T., and Tonomura, Y., 1978, Molecular mechanism of active calcium transport by sarcoplasmic reticulum, *Physiol. Rev.* **58:**1–79.

3

The Plasma Membrane Calcium Pump
Some Molecular Properties of the Purified Enzyme, with Emphasis on the Calmodulin-Binding Domain

ERNESTO CARAFOLI

The Ca pump of plasma membranes is an ATPase that has a high affinity for Ca (K_m 0.5 μM or less) and thus it interacts with it even at the normal low cytosolic concentration (about 0.2 μM). The ATPase can therefore be considered as a fine regulator of cell Ca, as opposed to the other Ca-exporting system, the Na/Ca exchanger, which is particularly active in excitable tissues, and handles much larger amounts of Ca with lower affinity. A number of regulators of potential physiological significance modulate (i.e., activate) the ATPase, among them calmodulin (Jarret and Penniston, 1987; Gopinath and Vincenzi, 1987), acidic phospholipids and polyunsaturated fatty acids (Niggli *et al.*, 1981), and a cAMP-dependent phosphorylation reaction (Neyses *et al.*, 1985). Of interest, although probably not significant physiologically, is the activation of the ATPase by a controlled exposure to proteases like trypsin (Taverna and Hanahan, 1980; Sarkadi *et al.*, 1980).

Calmodulin activates the ATPase by direct interaction, a finding that has permitted the isolation of the enzyme to essential homogeneity using calmodulin affinity columns (Niggli *et al.*, 1979); the purified ATPase repeats the properties of the enzyme *in situ*, including the stimulation by calmodulin and by controlled proteolysis with trypsin (Niggli *et al.*, 1981).

TRYPSIN PROTEOLYSIS OF THE PURIFIED ATPase

The original work on the trypsin proteolysis of the purified enzyme (Zurini *et al.*, 1984) described a number of products, of which some, having M_r between 90 and 76 kDa, appear of particular interest because they contain the active site of the enzyme (ATP binding and phosphoenzyme formation). However, calmodulin binding, as determined by interacting the enzyme with azido-modified, iodinated calmodulin, or by gel overlay experiments with ^{125}I-labeled calmodulin, while still present in the 90-kDa fragment, disappears on degradation of the latter to a fragment of 81 kDa. The implication of this finding is that the calmodulin-binding domain of the ATPase is contained in a stretch of about 9 kDa, removed by trypsin

ERNESTO CARAFOLI ● Laboratory of Biochemistry, Swiss Federal Institute of Technology (ETH), 8092 Zurich, Switzerland.

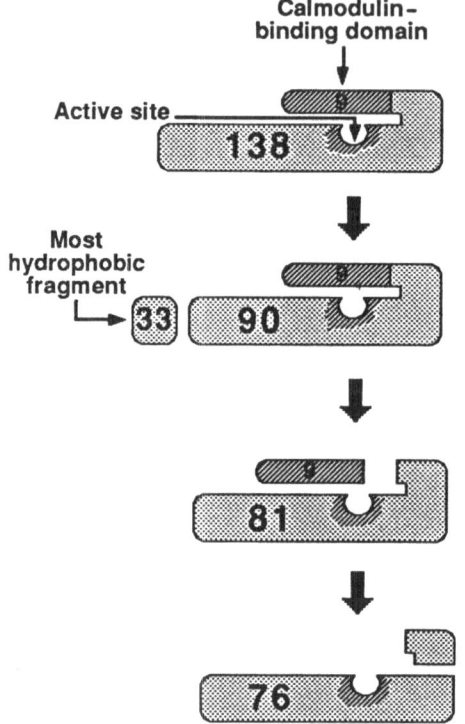

Calmodulin–
binding domain

Active site

Most
hydrophobic
fragment

FIGURE 1. Pattern of degradation of the purified ATPase by trypsin under the conditions described by Zurini et al. (1984). Only the portion of the pattern spanning the M_r region 138–76 kDa is shown. Details are given in the text and in Zurini et al. (1984).

in the transition between 90 and 81 kDa. It is of interest that the fragments of M_r 90 and 81 kDa, when reconstituted in liposomes (Zurini et al., 1984; Benaim et al., 1986), behave like functional ATPases, i.e., they carry out ATP-dependent transport of Ca. A summary of the original fractionation work using trypsin, with additional information on the properties of some of the fragments produced, is presented in Fig. 1. It is necessary to point out that the location of the calmodulin-binding domain in the C-terminal region is only presumptive, since the work with trypsin offers no indications for it; however, very recent work in which carboxypeptidase digestion was found to eliminate calmodulin binding (Sarkadi et al., 1986), as well as work to be described below on chymotrypsin digestion, make the assignment plausible.

More recent work with trypsin, in which the protease was applied under milder conditions and in the presence of different effectors (Benaim et al., 1984), revealed additional details of the digestion pattern in the region 90–76 kDa. Specifically, a fragment of 85 kDa becomes visible and accumulates if the digestion is carried out in the presence of Ca and calmodulin, whereas the transition between the 81- and 76-kDa fragments is inhibited in the presence of vanadate and Mg, leading to the accumulation of the 81-kDa product. Functional tests have shown that the 85-kDa fragment has limited ATPase activity and is still able to bind calmodulin, but is no longer stimulated by it. The 81-kDa fragment, on the other hand, has fully expressed ATPase activity. These findings have led to the suggestion that the 9-kDa calmodulin interacting domain consists of a 5-kDa peripheral calmodulin-binding site proper, and of a 4-kDa ''inhibitory'' stretch that must be removed from the active site for full expression of the ATPase activity (Fig. 2).

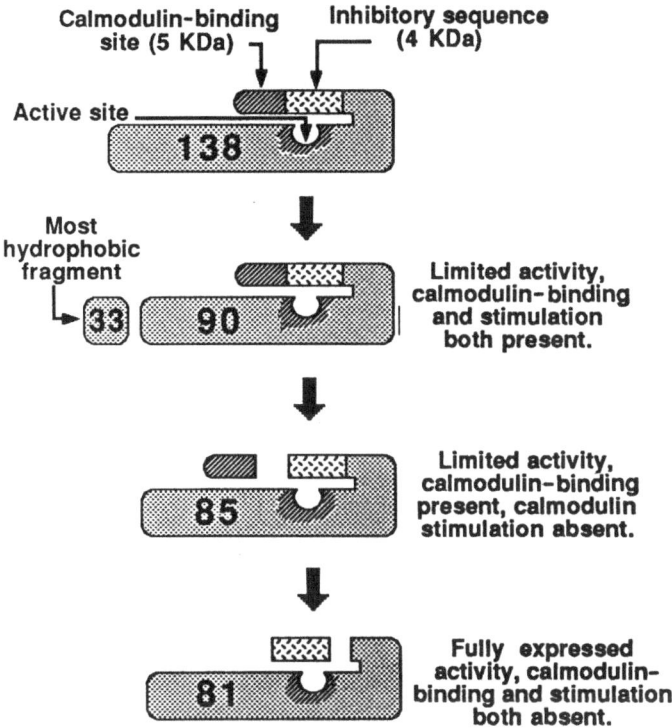

FIGURE 2. Pattern of degradation of the purified ATPase by trypsin under the conditions described by Benaim *et al.* (1984). Only the portion of the pattern spanning the M_r region between 138 and 81 kDa is shown. Details are given in the text and in Benaim *et al.* (1984).

CHYMOTRYPSIN PROTEOLYSIS OF THE PURIFIED ATPase

Chymotrypsin activates the ATPase in the absence of calmodulin more rapidly than trypsin. In addition, while in the presence of trypsin the activity tends to remain in the maximally stimulated state for a rather prolonged period, in the presence of chymotrypsin it drops significantly after a burst of activation. Gel electrophoresis analysis of the digested ATPase shows a rather complex fragmentation pattern (Carafoli *et al.*, 1987), characterized by the early removal from the main body of the enzyme of a fragment of M_r about 12 kDa that is the only product, among the many formed, still able to interact with calmodulin in gel overlay experiments with [125I]calmodulin. Since even at the earliest times of digestion investigated (1 or 2 min) the numerous high M_r fragments formed (i.e., 120, 105, 85 kDa, etc.) fail to bind calmodulin, the previously drawn conclusion that the calmodulin-binding domain is peripheral seems tenable. However, it is not necessary to assume that the 12 kDa fragment is located at the very C-terminus of the ATPase, i.e., it could be located somewhat more internally, next to the C-terminus domain. Figure 3 summarizes the degradation pattern of the enzyme during the initial phases of chymotrypsin attack.

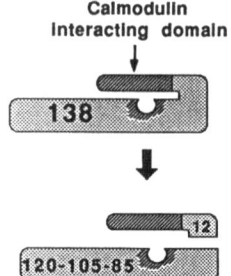

FIGURE 3. Pattern of degradation of the purified ATPase by chymotrypsin under the conditions described by Carafoli *et al.* (1987). The scheme only indicates the fragments of higher M_r and refers to very short proteolysis times (1–2 min). Details are given in the text and in Carafoli *et al.* (1987).

CALPAIN PROTEOLYSIS OF THE PURIFIED ATPase

Calpain is a Ca-activated protease that attacks several membrane proteins. When applied to the purified ATPase, it produces a modest activation of its basal activity and leaves the activation by calmodulin evident even after very prolonged times of incubation (up to 4 hr under the experimental conditions employed in the experiments described by Carafoli *et al.*, 1987). The pattern of degradation by calpain is rather unusual: a fragment of about 14 kDa is removed from the ATPase almost immediately, leaving behind a fragment of M_r 124 kDa. No further attack occurs on the latter, even after extended incubations (2 hr). As expected from the results of experiments on the effects of calpain on the ATPase activity and on its stimulation by calmodulin, gel overlay experiments with [^{125}I]calmodulin have shown that the calmodulin-binding domain remains associated with the 124-kDa fragment. The scheme shown in Fig. 4 interprets the degradation pattern by calpain as emerging from the results summarized here. However, it is also possible that the fragment of about 14 kDa is removed from the C-terminus, without reaching into the calmodulin binding domain, or reaching only minimally into it.

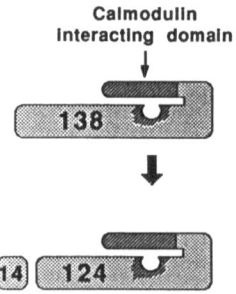

FIGURE 4. Pattern of degradation of the purified ATPase by calpain under the conditions described by Carafoli *et al.* (1987). Details are given in the text and in Carafoli *et al.* (1987).

CONCLUSIONS

The work described in this report offers strong indications that the calmodulin-interacting domain of the Ca-ATPase resides at one of the ends of the enzyme, probably the C end. It apparently consists of two subdomains: one that binds calmodulin, and one that does not bind it but is essential for the expression of stimulation. When only the binding domain is in place, the ATPase activity becomes repressed, suggesting that the binding domain functions as an inhibitor of the activity.

Chymotrypsin separates from the enzyme the calmodulin-interacting domain as a low M_r (12-kDa) fragment. Pilot experiments have shown that the fragment can be purified on calmodulin affinity columns: work on this fragment is currently under way in this laboratory, with the aim of determining the primary structure of the calmodulin-interacting domain. Further work now in progress is aimed at deducting the entire primary structure of the ATPase by cloning and sequencing the cDNA encoding it.

ACKNOWLEDGMENT. The experimental work described in this report has been supported by the financial contribution of the Swiss Nationalfonds (grants 3.658-0.84 and 3.531-0.86).

REFERENCES

Benaim, G., Zurini, M., and Carafoli, E., 1984, Different conformational states of the purified Ca^{2+}-ATPase of the erythrocyte plasma membrane revealed by controlled trypsin proteolysis. *J. Biol. Chem.* **259**:8471.

Benaim, G., Clark, A., and Carafoli, E., 1986, ATPase activity and Ca^{2+} transport by reconstituted tryptic fragments of the Ca^{2+} pump of the erythrocyte plasma membrane. *Cell Calcium* **7**:175.

Carafoli, E. Fischer, R., James, P., Krebs, J., Maeda, M., Enyedi, A., Morelli, A., and de Flora, A. 1987, The calcium pump of the plasma membrane. Recent studies on the purified enzyme and on its proteolytic fragments, with particular attention to the calmodulin binding domain, *in: Calcium Binding Proteins and Calcium Function in Health and Disease* (A. W. Norman, T. C. Vanaman, and A. R. Means, eds.), pp. 78–91, Academic Press, San Diego.

Gopinath, R. M., and Vincenzi, F. F., 1977, Phosphodiesterase protein activator mimics red blood cell cytoplasmic activator of $(Ca^{2+}-Mg^{2+})$ATPase. *Biochem. Biophys. Res. Commun.* **77**:1203.

Jarret, H. W., and Penniston, J. T., 1977, Partial purification of the $(Ca^{2+}-Mg^{2+})$ATPase activator from human erythrocytes: Its similarity to the activator of $3':5'$-cyclic nucleotide phosphodiesterase. *Biochem. Biophys. Res. Commun.* **77**:468.

Neyses, L., Reinlib, L., and Carafoli, E., 1985, Phosphorylation of the Ca^{2+}-pumping ATPase of heart sarcolemma and erythrocyte plasma membrane by the cAMP-dependent protein kinase, *J. Biol. Chem.* **260**:10283.

Niggli, V., Penniston, J. T., and Carafoli, E., 1979, Purification of the calcium-magnesium ATPase from human erythrocyte membranes using a calmodulin affinity column. *J. Biol. Chem.* **254**:9955.

Niggli, V., Adunyah, E. S., and Carafoli, E., 1981, Acidic phospholipids, unsaturated fatty acids, and limited proteolysis mimic the effect of calmodulin on the purified erythrocyte Ca^{2+} ATPase. *J. Biol. Chem.* **256**:8588.

Niggli, V., Adunyah, E. S., Penniston, J. T., and Carafoli, E., 1981, Purified $(Ca^{2+}-Mg^{2+})$ATPase of the erythrocyte membrane; reconstitution and effect of calmodulin and phospholipids. *J. Biol. Chem.* **256**:395.

Sarkadi, B., Enyedi, A., and Gardos, G., 1980, Molecular properties of the red cell calcium pump. Effects of calmodulin, proteolytic digestion and drugs on the kinetics of active calcium uptake in inside-out red cell membrane vesicles. *Cell Calcium* **1**:287.

Sarkadi, B., Enyedi, A., Földes-Papp, Z., and Gardos, G., 1986, Molecular characterization of the in situ red cell membrane calcium pump by limited proteolysis. *J. Biol. Chem.* **261:**9552.

Taverna, R. D., and Hanahan, D. J., 1980, Modulation of human erythrocyte (Ca^{2+}/Mg^{2+}) ATPase activity by phospholipase A2 and protease. A comparison with calmodulin. *Biochem. Biophys. Res. Commun.* **94:**652.

Zurini, M., Krebs, J., Penniston, J. T., and Carafoli, E., 1984, Controlled proteolysis of the purified Ca^{2+} ATPase of the erythrocyte membrane. *J. Biol. Chem.* **259:**618.

4

The Cardiac Sodium–Calcium Exchange System

JOHN P. REEVES and JOO CHEON

1. INTRODUCTION

The strength of contraction of cardiac muscle is controlled in part by regulating the amount of Ca^{2+} that enters and leaves the myocardial cells with each beat. This involves a very complex interplay between Ca^{2+} channels and pumps located in both the plasma membrane of the heart cell (sarcolemma) and intracellular organelles, particularly the sarcoplasmic reticulum (SR). There are two principal sources for the Ca^{2+} that activates contraction: the extracellular medium and Ca^{2+} stored within the SR. The rate of delivery of Ca^{2+} to the myofilaments from each source is determined by the activity of various Ca^{2+} channels located within the SR membrane and the sarcolemma. The distribution of Ca^{2+} between the two compartments is determined primarily by the relative activities of Ca^{2+} pumps located within the same membrane systems. One such pump is the Ca-ATPase of the SR membrane, an enzyme that couples the hydrolysis of ATP to the accumulation of Ca^{2+} within the lumen of the SR. In effect, the SR Ca-ATPase "competes" for Ca^{2+} with two different types of Ca^{2+} pumps in the sarcolemma: a Ca-ATPase (which is a different molecular entity than the SR Ca-ATPase) and the Na–Ca exchange system.

The latter is a carrier-mediated transport process that couples the movement of Ca^{2+} in one direction to the movement for Na^+ in the opposite direction. Its stoichiometry is three Na^+ per Ca^{2+} (Reeves and Hale, 1984). Under resting conditions in mammalian cardiac cells, it appears to function primarily as a Ca^{2+} efflux process, utilizing the energy of the inwardly directed Na gradient (maintained by the activity of the Na–K–ATPase) to pump Ca^{2+} out of the cell. Its activity as a Ca^{2+} pump is determined not only by kinetic factors but also by the thermodynamic gradient of Na^+ ions. Thus, the Na–Ca exchange system is a mechanism for translating changes in intracellular Na^+ activity to changes in the amount of Ca^{2+} stored in the SR. This is an important mechanism for regulating cardiac contractility and appears to be involved in mediating at least part of the inotropic effect of cardiac glycosides.

JOHN P. REEVES and JOO CHEON • Roche Institute of Molecular Biology, Roche Research Center, Nutley, New Jersey 07110.

Work in our laboratory has been directed toward studying the Na–Ca exchange system in a subcellular preparation consisting of osmotically sealed cardiac sarcolemmal vesicles (Reeves and Sutko, 1979). These are prepared by homogenizing cardiac ventricular tissue and isolating an enriched preparation of sarcolemmal membranes by differential and density gradient centrifugation. The vesicles are devoid of cytoplasmic constituents and internal organelles but retain many of the transport characteristics of the intact cell. We measure Na–Ca exchange activity by loading the vesicles internally with 160 mM NaCl and then diluting them 50- to 100-fold into an isosmotic Na-free medium (e.g., 160 mM KCl) containing $^{45}Ca^{2+}$. The outwardly directed concentration gradient of Na^+ drives the accumulation of $^{45}Ca^{2+}$ into the vesicles via the Na–Ca exchange system. Appropriate control experiments have demonstrated that the $^{45}Ca^{2+}$ is indeed transported into the vesicle interior and is not simply bound to sites on the external surface.

2. CHARACTERISTICS OF Na–Ca EXCHANGE

The experimental use of membrane vesicles has contributed in an important way to delineating the fundamental properties of the exchange process. Table I summarizes some of the information resulting from work in our laboratory on the various modes of operation of the exchange system in sarcolemmal vesicles. The stoichiometry determinations referred to in the table involve a thermodynamic approach in which the imposition of a membrane potential as the driving force for exchange activity is balanced against an offsetting Na^+ gradient such that no net Ca^{2+} movement occurs. The theoretical balance point in such an experiment differs for different assumed Na^+/Ca^{2+} stoichiometries and the results of nine separate determinations yielded an average value of 2.97 ± 0.03 (Reeves and Hale, 1984).

Although vesicles would seem to be the ideal system for the precise measurement of the kinetics of the exchange process, the results have been somewhat confusing. As depicted in Table I, reported values for the K_m for Ca^{2+} vary from 1.5 μM to >200 μM. Part of this variability is probably a reflection of the large number of factors that affect the K_m for Ca^{2+} in the vesicle system. A partial list of agents or treatments that stimulate exchange activity by reducing the K_m for Ca^{2+} includes proteolysis (Philipson and Nishimoto, 1982), anionic amphiphiles (Philipson, 1984), membrane phosphorylation (Caroni and Carafoli, 1983), and certain redox reagents (Reeves et al., 1986).

Another factor that was not appreciated until recently is the effect of intravesicular Ca^{2+} itself, which appears to lower the apparent K_m for external Ca^{2+} (Reeves and Poronnik, 1987). The mechanism of this effect is uncertain; the most appealing interpretation is that the binding of Ca^{2+} to sites on the intravesicular membrane surface accelerates a rate-limiting step in the exchange cycle, perhaps by influencing the electrostatic field within the membrane. Whatever the explanation, this effect has disturbing implications for kinetic measurements in vesicles. The levels of intravesicular Ca^{2+} attained during 1 sec initial rate measurements of activity (2–10 nmol/mg protein) are sufficient to alter the kinetics of the exchange process itself; thus, true initial rate conditions are unlikely to prevail during these measurements, so that the kinetic parameters obtained may not accurately reflect the true properties of the system. Furthermore, variations in the amount of endogenous Ca^{2+} present among different membrane preparations or storage conditions may produce corresponding variations in the kinetic properties of the vesicles; this may account for at least a portion of the variability in K_m values in the literature.

TABLE I. Sarcolemmal Vesicles and Sodium–Calcium Exchange

Mode of operation	Reference
Ca^{2+} uptake	
$\quad K_{m.Ca} = 1.5$ to >200 μM	Reeves, 1985
$\quad K_{m.Na} = 20$–30 mM; $n = 2$–3	
$\quad V_{max} = 20$–40 nmol/mg	
$\quad\quad$ protein/sec	
$\quad$ Competitively inhibited by Na_0	Reeves and Sutko, 1983
$\quad\quad K_i = 15$ mM; $n = 1$–2	
Ca^{2+} efflux	Kadoma et al., 1982
$\quad K_{m.Na} = 20$–30 mM; $n = 2$–3	
Ca–Ca exchange	Slaughter et al., 1983
$\quad$ Stimulated by Li^+, Na^+, K^+, Rb^+	
Na–Na exchange	Reeves and Sutko, 1979b
$\quad$ Sigmoidal $[Na]_0$ dependence	
$\quad$ Inhibited by $Ca^{2+}{}_0$	
Stoichiometry	Reeves and Hale, 1984
$\quad$ Na/Ca $= 2.97 \pm 0.03$ $(N = 9)$	

3. IDENTIFICATION OF THE EXCHANGE CARRIER

Establishing the molecular identity of the exchange carrier is a central goal of current research efforts in several different laboratories. Progress in this area has been hampered by the lack of a specific high-affinity probe or inhibitor that could be used to label the carrier. Consequently, one must rely on protein separation techniques with detergent-solubilized membranes and subsequent assays of exchange activity in reconstituted proteoliposomes to provide insight into the carrier's identity.

Earlier work from our laboratory had indicated that the exchange carrier might be an 82-kDa glycoprotein that was highly resistant to proteolysis (Hale et al., 1984). Subsequent work, however, showed that when cholate extracts of the membranes were passed over a column of immobilized concanavalin A, the 82-kDa band bound strongly to the column but activity did not. These and other observations indicate that the 82-kDa glycoprotein is not the exchange carrier. Other workers have suggested polypeptides at 70 kDa (Barzilai et al., 1984) and 33 kDa (Soldati et al., 1985) as possible candidates for the exchange carrier. Proof that either of these proteins (or any other candidate) is indeed the exchange carrier must await their complete purification in an active state and/or the development of specific antibodies against them.

A report that initially seemed to be a major breakthrough claimed that monoclonal antibodies developed against a particular surface antigen of lymphocytic cell lines inhibited Na–Ca exchange activity in cardiac sarcolemmal vesicles (Michalak et al., 1986). However, subsequent studies in several different laboratories, including ours, failed to confirm these findings. Thus, it seems unlikely that these antibodies will serve as useful probes for the Na–Ca exchange carrier or its activity.

4. SITE DENSITY OF THE EXCHANGE CARRIER

An important step in identifying the exchange carrier is to determine the number of carrier molecules in the membrane. The absence of specific, high-affinity probes for the exchanger precludes a direct measurement of this number. We have approached this issue by trying to estimate the fraction of proteoliposomes in a reconstituted preparation that exhibit exchange activity. For this purpose, vesicles were solubilized in 1% Triton X-100 and reconstituted into proteoliposomes after adding soybean phospholipids and removing the detergent using Biobeads SM-2 (Bio-Rad). This procedure results in proteoliposomes with a highly uniform size distribution and an average radius of 50 nm.

To estimate the proportion of vesicles carrying the exchange carrier, we measured the uptake of $^{45}Ca^{2+}$ by the vesicles under equilibrium conditions, with both internal and external media having the following composition: 40 mM NaCl, 120 mM KCl, 0.1 mM CaCl$_2$. The measured uptake was corrected for the uptake observed in proteoliposomes in which the Na–Ca exchanger had been inactivated and compared to the total uptake observed in the presence of the Ca^{2+} ionophore A23187. The results, which will be published in detail elsewhere, suggest that only 2–5% of the proteoliposomes exhibit exchange activity.

We make the assumption that the distribution of proteins among the proteoliposomes is random and that the proteoliposomes exhibiting exchange activity contain just one molecule of exchange carrier. With this assumption, one can calculate that the density of exchange carrier in the reconstituted proteoliposomes is approximately 10–20 pmol/mg protein. If the carrier had a molecular weight of 100 kDa, this would correspond to 0.1–0.2% of the protein. Ca^{2+} flux measurements in the reconstituted system suggest that, with this site density, the turnover number of the exchanger is 1000–2000 sec^{-1}. This is similar to the turnover number recently estimated for the Na–H exchanger in kidney brush border membranes (Vigne *et al.*, 1985) and rat thymocytes (Dixon *et al.*, 1987). Thus, the Na–Ca exchange carrier appears to be a protein with a relatively low site density and high turnover number, an assessment that emphasizes the need for caution in trying to identify the carrier on the basis of sodium dodecyl sulfate gels of partially purified preparations.

REFERENCES

Barzilai, A., Spanier, R., and Rahamimoff, H., 1984, Isolation, purification, and reconstitution of the Na$^+$ gradient-dependent Ca^{2+} transporter (Na$^+$–Ca^{2+} exchanger) from brain synaptic plasma membranes, *Proc. Natl. Acad. Sci. USA* **81**:6521–6525.

Caroni, P., and Carafoli, E., 1983, The regulation of the Na–Ca exchanger of heart sarcolemma, *Eur. J. Biochem.* **132**:451–460.

Dixon, S. J., Cohen, S., Cragoe, E. J., Jr., and Grinstein, S., 1987, Estimation of the number and turnover rate of Na$^+$/H$^+$ exchangers in lymphocytes, *J. Biol. Chem.* **262**:3626–3632.

Hale, C. C., Slaughter, R. S., Ahrens, D. C., and Reeves, J. P., 1984, Identification and partial purification of the cardiac sodium-calcium exchange carrier, *Proc. Natl. Acad. Sci. USA* **81**: 6569–6573.

Michalak, M., Quackenbush, E., and Letarte, M., 1986, Inhibition of Na$^+$/Ca^{2+} exchanger activity in cardiac and skeletal muscle sarcolemmal vesicles by monoclonal antibody 44D7, *J. Biol. Chem.* **261**:92–95.

Philipson, K. D., 1984, Interaction of charged amphiphiles with Na$^+$–Ca^{2+} exchange in cardiac sarcolemmal vesicles, *J. Biol. Chem.* **259**:13999–14002.

Philipson, K. D., and Nishimoto, A. Y., 1982, Stimulation of Na–Ca exchange in cardiac sarcolemmal vesicles by proteinase pretreatment, *Am. J. Physiol.* **243:**C191–C195.

Reeves, J. P., and Hale, C. C., 1984, The stoichiometry of the cardiac sodium-calcium exchange system, *J. Biol. Chem.* **259:**7733–7739.

Reeves, J. P., and Poronnik, P., 1987, Modulation of Na^+–Ca^{2+} exchange in sarcolemmal vesicles by intravesicular Ca^{2+}, *Am. J. Physiol.* **252:**C17–C23.

Reeves, J. P., and Sutko, J. L., 1979, Sodium-calcium ion exchange in cardiac membrane vesicles, *Proc. Natl. Acad. Sci. USA* **76:**590–594.

Reeves, J. P., Bailey, C. A., and Hale, C. C., 1986, Redox modification of sodium-calcium exchange activity in cardiac sarcolemmal vesicles, *J. Biol. Chem.* **261:**4948–4955.

Soldati, L., Longoni, S., and Carafoli, E., 1985, Solubilization and reconstitution of the Na^+/Ca^{2+} exchanger of cardiac sarcolemma, *J. Biol. Chem.* **260:**13221–13227.

Vigne, P., Jean, T., Barbry, P., Frelin, C., Fine, L. G., and Lazdunski, M., 1985, [^{3}H]Ethylpropylamiloride, a ligand to analyze the properties of the Na^+/H^+ exchange system in the membranes of normal and hypertrophied kidneys, *J. Biol. Chem.* **260:**14120–14125.

Molecular Properties of Dihydropyridine-Sensitive Calcium Channels

WILLIAM A. CATTERALL, MICHAEL J. SEAGAR, MASAMI TAKAHASHI, and BENSON M. CURTIS

1. INTRODUCTION

In muscle tissues, voltage-sensitive calcium channels mediate calcium influx during cellular depolarization and play an important role in excitation-contraction coupling (reviewed by Reuter, 1979; Hagiwara and Byerly, 1981). In neurons, they produce action potentials in dendrites (Schwartzkroin and Slawsky, 1977; Llinas et al., 1981) and couple changes in membrane potential at nerve terminals to the release of neurotransmitter (Katz and Miledi, 1969). Multiple classes of calcium channels have been distinguished in neurons (Carbone and Lux, 1984; Armstrong and Matteson, 1985; Nowycky et al., 1985) and in cardiac muscle cells (Nilius et al., 1985; Bean, 1985). This article focuses on molecular properties of calcium channels that are blocked by dihydropyridine calcium antagonists. These are the most prominent calcium channels in smooth, cardiac, and skeletal muscle and they are also present in neurons and neurosecretory cells.

2. ORGANIC CALCIUM ANTAGONISTS AS PROBES OF CALCIUM CHANNEL STRUCTURE AND FUNCTION

Calcium channels are inhibited by three different classes of organic antagonists: the dihydropyridines including nifedipine and nitrendipine, the phenylalkylamines and diphenylalkylamines including verapamil, and the thiobenzazepine diltiazem (Triggle, 1982). ^{3}H-labeled derivatives of nitrendipine and other dihydropyridines have been prepared and their binding to high-affinity receptor sites in homogenates and membrane preparations from excitable tissues has been extensively examined (Janis and Scriabine, 1983). A single class of high-affinity binding sites is observed with a K_D for nitrendipine in the range of 0.1–1.0 nM.

WILLIAM A. CATTERALL, MICHAEL J. SEAGAR, and BENSON M. CURTIS • Department of Pharmacology, University of Washington, Seattle, Washington 98195.   MASAMI TAKAHASHI • Department of Pharmacology, University of Washington, Seattle, Washington 98195; and Mitsubishi, Kasei Institute of Life Sciences, Machida-Shi, Japan.   Present address for B.M.C.: Division of Biology, California Institute for Technology, Pasadena, California 91125.

The structure-activity relationships for occupancy of these high-affinity binding sites correlate closely with the concentrations of various dihydropyridines required to block contraction of smooth muscle over a wide range of K_D values (Bolger *et al.*, 1982).

The binding of nitrendipine to its receptor site is modulated via an allosteric mechanism by diltiazem and various mono- and diphenylalkylamines, such as verapamil (Murphy *et al.*, 1983), which interact with two separate receptor sites (Garcia *et al.*, 1986). Verapamil decreases nitrendipine binding and enhances the rate of dissociation of the nitrendipine-receptor complex, whereas diltiazem increases binding and slows the rate of dissociation of the nitrendipine-receptor complex. Thus the calcium antagonist receptor of the calcium channel contains at least three drug-binding sites, one specific for dihydropyridines and one each for diltiazem and phenylalkylamines, which interact allosterically in modulating calcium channel function and have opposite effects on binding of dihydropyridine calcium channel blockers.

The dihydropyridine receptor site on the calcium channel can also mediate enhanced activation of the channel when occupied by dihydropyridine calcium channel activators, exemplified by BAY K 8644 (Schramm *et al.*, 1983). These agents cause smooth-muscle contraction and stimulation of the heart by favoring activation of calcium channels at more negative membrane potentials (Kokubun and Reuter, 1984). Thus, occupancy of the dihydropyridine receptor sites by agents of closely related structure can cause either channel block or channel activation. These results provide firm evidence that the dihydropyridine receptor modulates the gating of the calcium channel as its principal mode of action.

The development of radioactively labeled derivatives of high affinity and specificity now allows the use of these drugs as probes to identify and isolate the protein components of calcium channels and understand their molecular properties.

3. PURIFICATION OF THE CALCIUM ANTAGONIST RECEPTOR FROM SKELETAL MUSCLE

Since it is likely to be an intrinsic membrane protein, the first essential step in purification and biochemical characterization of the calcium antagonist receptor is solubilization from an appropriate membrane source and characterization of the solubilized protein. After a survey of several detergents, we concluded that digitonin is the most effective detergent for solubilization of a specific [^{3}H]nitrendipine-receptor complex from brain and skeletal muscle transverse tubule membranes (Curtis and Catterall, 1983, 1984). Up to 40% of the receptor-ligand complex is solubilized. The dissociation of bound [^{3}H]nitrendipine from the complex is accelerated by verapamil and slowed by diltiazem through allosteric interactions between the nitrendipine-binding site and the binding sites for those ligands as previously observed in intact membranes. These results show that the three different binding sites for calcium antagonist drugs remain associated as a complex after detergent solubilization of the calcium antagonist receptor and provide further support for the conclusion that a specific receptor complex has been solubilized under conditions that allow retention of the functional allosteric regulation of dihydropyridine binding.

Sedimentation of the solubilized [^{3}H]nitrendipine-receptor-digitonin complex through sucrose gradients gives a single peak of specifically bound nitrendipine with a sedimentation coefficient of 19–20S. Comparison of the sedimentation behavior of the solubilized complex from brain, heart, and skeletal muscle indicates that they have identical size. These results

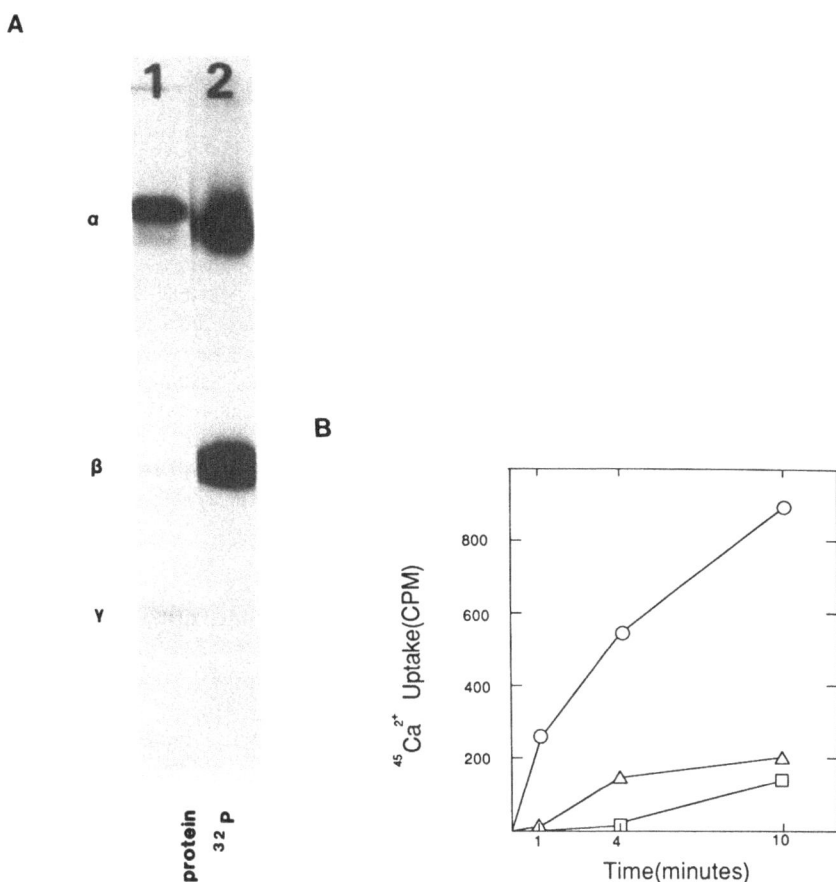

FIGURE 1. Purification and reconstitution of the calcium channel. (A) Noncovalently associated protein components. The dihydropyridine-sensitive calcium channel from skeletal muscle transverse tubules was purified as described by Curtis and Catterall (1984). Lane 1: Purified calcium channels were analyzed by SDS-PAGE without reduction of disulfide bonds and proteins were visualized by silver staining. Three size classes of polypeptide components are recognized under these conditions: α, 167 kDa; β, 50 kDa; and γ, 30 kDa. Lane 2: The purified calcium channel was phosphorylated by reaction with cAMP-dependent protein kinase and [γ-^{32}P]ATP as described by Curtis and Catterall (1985), analyzed by SDS-PAGE as for lane 1, and visualized by autoradiography. The α- and β-protein bands are labeled by ^{32}P. (B) Calcium flux mediated by the purified calcium channel. The calcium antagonist receptor from skeletal muscle transverse tubules was purified and incorporated into phosphatidylcholine vesicles as described by Curtis and Catterall (1986). Initial rates of ^{45}Ca^{2+} influx into reconstituted vesicles were measured in the presence of 1 μM BAY K 8644 (○), 1 μM BAY K 8644 and 100 μM verapamil (△), or no added drug (□).

provide support for the view that the calcium antagonist receptor in different tissues is quite similar.

Many plasma membrane proteins are glycosylated during their synthesis and transport to the cell surface. The solubilized calcium antagonist receptor from brain or skeletal muscle is specifically adsorbed to and eluted from affinity columns with immobilized wheat germ agglutinin or other lectins (Glossmann and Ferry, 1983; Curtis and Catterall, 1983, 1984; Borsotto et al., 1984). Evidently, one or more of the subunits of the calcium antagonist receptor is a glycoprotein.

We have purified the calcium antagonist receptor solubilized from T-tubule membranes by digitonin 330-fold by affinity chromatography on wheat germ agglutinin-Sepharose, ion exchange chromatography on DEAE-Sephadex, and velocity sedimentation through sucrose gradients (Curtis and Catterall, 1984). The purified preparation contained 1950 pmol calcium antagonist receptor per mg protein, 81% of the value expected for a protein of 416,000 daltons (see below), which binds one calcium antagonist per molecule. Analysis of the purified preparation by polyacrylamide gel electrophoresis after denaturation in sodium dodecyl sulfate (SDS) without reduction of disulfide bonds revealed three major protein bands as illustrated in Fig. 1. Their apparent molecular weights are α, 167,000; β, 50,000; and γ, 33,000. All three of these polypeptides quantitatively comigrate with the [^{3}H]nitrendipine-receptor complex during velocity sedimentation in sucrose gradients and therefore are likely to be components of the calcium antagonist receptor.

4. PHOSPHORYLATION AND cAMP-DEPENDENT REGULATION OF CALCIUM CHANNELS

The classic work of Reuter, Tsien, and their colleagues (Tsien et al., 1972; Reuter, 1974, 1983) established that the positive inotropic effect of epinephrine and norepinephrine on the heart is mediated by a cAMP-dependent increase in inward calcium current. This effect is due to an increase in the number of calcium channels that are active during the cardiac action potential (Cachelin et al., 1983). These effects are mimicked by intracellular injection of the catalytic subunit of cAMP-dependent protein kinase suggesting that protein phosphorylation mediates the effects of cAMP (Brum et al., 1983). It is uncertain whether the substrate for cAMP-dependent phosphorylation is the calcium channel itself or another regulatory component. In addition to these studies on mammalian heart, calcium channels in mammalian dorsal root ganglion neurons and both calcium channels and calcium-activated K$^+$ channels in molluscan neurons are regulated by cAMP-dependent protein phosphorylation. Recent results indicate that calcium channels in cultured skeletal muscle cells are also regulated in this manner (Schmid et al., 1985). Thus, modulation of calcium channel properties by cellular regulatory processes may be a general phenomenon.

Although regulation of calcium channels via a pathway involving cAMP and protein phosphorylation is now well established, it is not known with certainty whether the site of phosphorylation is the calcium channel itself or another intracellular protein that in turn regulates channel function. As a step toward resolving this question, we have investigated whether the subunits of the purified calcium antagonist receptor can serve as substrates for cAMP-dependent protein kinase in the purified state and in intact T-tubule membranes (Curtis and Catterall, 1985). When incubated with physiological concentrations of cAMP-dependent protein kinase, proteins in the α- and β-subunit bands are phosphorylated as illustrated in Fig. 1. Up to 0.85–0.9 mol ^{32}P per calcium antagonist receptor can be incorporated into each subunit band at rates consistent with a physiologically significant phosphorylation reaction. In contrast, the γ subunit is not a substrate for cAMP-dependent protein kinase, even under forcing reaction conditions (Curtis and Catterall, 1985). These results identify proteins in the α- and β-subunit bands of the calcium antagonist receptor as potential sites of regulation of the voltage-sensitive calcium channel by cAMP-dependent phosphorylation.

5. FUNCTIONAL PROPERTIES OF THE PURIFIED CALCIUM ANTAGONIST RECEPTOR IN PHOSPHOLIPID VESICLES

We purify the calcium antagonist receptor as a preformed complex with the calcium antagonists [^{3}H]nitrendipine or [^{3}H]PN 200-110. After solubilization, it retains allosteric interactions among the separate binding sites for verapamil, diltiazem, and dihydropyridines (Curtis and Catterall, 1983). However, other aspects of calcium channel function cannot be assessed in detergent solution. It is important, therefore, to return the purified calcium antagonist receptor to a membrane environment and to determine whether the purified protein is capable of mediating voltage-dependent calcium flux.

In order to maximize the probability of purification of the calcium channel in an active state, the calcium antagonist receptors in T-tubule membranes were incubated with sufficient [^{3}H]PN 200-110 to label approximately 1% of the binding sites. This label was used to identify the calcium antagonist receptor during purification as in previous studies. The T-tubule membranes were then incubated in an excess of the specific calcium channel activator BAY K 8644 so that the remaining 99% of dihydropyridine sites would be occupied by this agent. The calcium antagonist receptors were then solubilized in digitonin and purified in the continued presence of BAY K 8644 using previously described procedures (Curtis and Catterall, 1986). Phosphatidylcholine for reconstitution was solubilized in the zwitterionic detergent CHAPS because it is poorly soluble in digitonin. Purified calcium antagonist receptor dispersed in digitonin was then mixed with phosphatidylcholine dispersed in CHAPS and single-walled phospholipid vesicles were formed by removal of the detergents by molecular sieve chromatography. Analysis of the resulting vesicle preparations by sucrose density gradient sedimentation show that the purified calcium antagonist receptors are quantitatively incorporated into phosphatidylcholine vesicles and 15–25% of vesicles contain at least one calcium antagonist receptor. These preparations therefore provide a suitable system for analysis of the ion transport properties of this purified protein.

Initial rates of influx of ^{45}Ca^{2+} or ^{133}Ba^{2+} into reconstituted phosphatidylcholine vesicles were measured under countertransport conditions in which the intravesicular compartment contains a high concentration of unlabeled Ca^{2+} or Ba^{2+}. These conditions greatly increase the amount of ^{45}Ca^{2+} or ^{133}Ba^{2+} uptake required to achieve isotopic equilibrium and greatly slow the approach to isotopic equilibrium as described previously for reconstituted sodium channels (Talvenheimo et al., 1982). They therefore maximize any ion flux mediated by reconstituted channels. Under these conditions, calcium influx into phosphatidylcholine vesicles containing reconstituted calcium antagonist receptors was two- to threefold greater than influx into protein-free phosphatidylcholine vesicles in the presence of BAY K 8644 (Fig. 2). This increase is completely blocked by verapamil at a concentration of 100 μM. If the calcium channel activator BAY K 8644 is removed from the vesicle preparation by molecular sieve chromatography, ^{45}Ca^{2+} influx is markedly reduced. These results show that at least a fraction of the purified calcium antagonist receptors can function as calcium channels when incorporated into phosphatidylcholine vesicles.

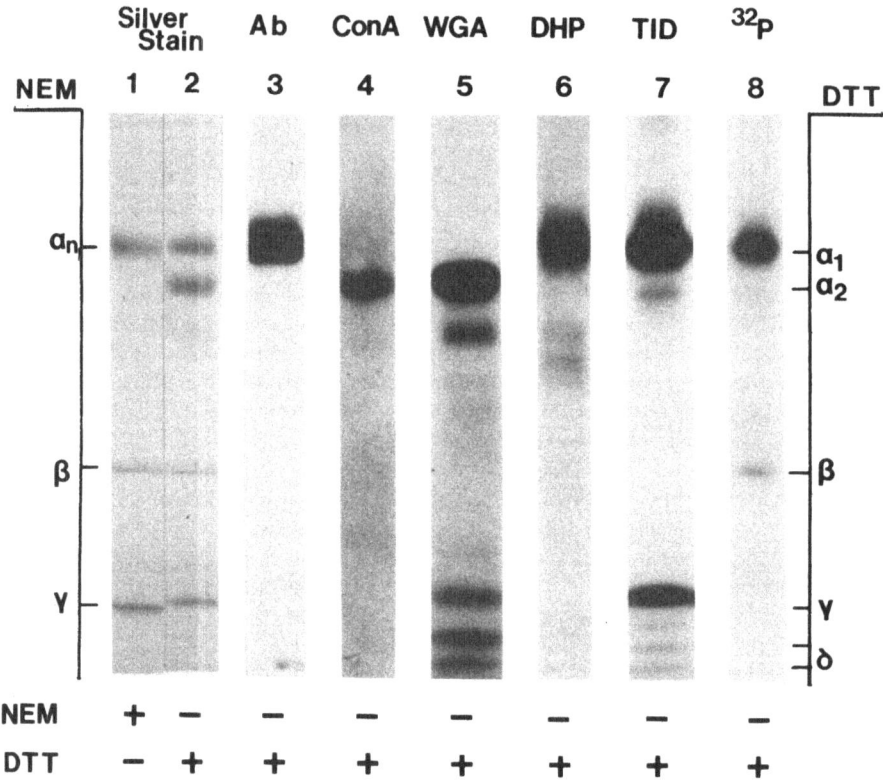

FIGURE 2. Differential labeling of calcium channel subunits. Lanes 1 and 2: Purified calcium channels were analyzed by SDS-PAGE and silver staining with or without reduction of disulfide bonds as indicated beneath each lane. Lane 3: Polypeptides separated by SDS-PAGE with reduction of disulfide bonds were electrophoretically transferred to nitrocellulose strips and immunolabeled by incubation with the antibody PAC-10, or followed by incubation with [^{125}I]protein A, washing, and autoradiography. Lane 4: Calcium channel subunits were transferred to a nitrocellulose sheet and labeled with [^{125}I]ConA. Lane 5: Calcium channel subunits were labeled directly in the polyacrylamide gel with [^{125}I]WGA. Lane 6: Photoaffinity labeling. T-tubule membranes (0.4 mg/ml) in 25 mM Hepes, 1 mM $CaCl_2$ adjusted to pH 7.5 with Tris base were incubated with 6 nM [^{3}H]azidopine and were irradiated for 15 min at 4°C with a 30-watt UV source (λ_{max} 356 nM). The membranes were solubilized in 1% digitonin, 10 mM Hepes, 185 mM NaCl, 0.5 mM $CaCl_2$, 0.1 mM PMSF (phenylmethanesulfonyl fluoride), 1 μM pepstatin A adjusted to pH 7.5 with Tris base, and calcium channels were partially purified by chromatography on WGA-Sepharose and analyzed by SDS-PAGE and fluorography. Lane 7: Hydrophobic labeling. [^{125}I]TID (15 Ci/mmol) was prepared and purified calcium channel was labeled with 100 μCi/ml [^{125}I]TID in a buffer containing 0.1% digitonin as previously described (Reber and Catterall, 1987). Lane 8: Phosphorylation. Purified calcium channel was incubated with 0.3 μM cAMP-dependent kinase catalytic subunit and 0.12 μM carrier free [γ-^{32}P]ATP for 15 min at 37°C as previously described (Curtis and Catterall, 1985).

The inhibition of the reconstituted calcium channels by different concentrations of organic calcium channel blockers from 10^{-9} to 10^{-4} M was examined. Half-maximal inhibition was observed with approximately 1.5 μM verapamil, 1.0 μM D600, or 0.2 μM PN 200-110. These concentrations are similar to those that give half-maximal inhibition of voltage-activated calcium currents in intact skeletal muscle fibers consistent with the conclusion that the calcium flux in reconstituted vesicles is mediated by functional purified calcium channels.

To determine whether the calcium influx stimulated by BAY K 8644 and blocked by PN 200-110 and verapamil required the presence of the subunits of the calcium antagonist receptor and not other detectable proteins, purified preparations were sedimented through sucrose gradients and each fraction was examined for bound [³H]PN 200-110, polypeptide composition, and ability to mediate $^{133}Ba^{2+}$ influx when incorporated into phosphatidylcholine vesicles (Curtis and Catterall, 1986). A close quantitative correlation was observed between the presence of the α-, β-, and γ-subunit bands of the calcium antagonist receptor, as analyzed by SDS-polyacrylamide gel electrophoresis (PAGE) without reduction of disulfide bonds, and the ability to mediate $^{133}Ba^{2+}$ flux. No other polypeptides were noted whose presence correlated with ion flux activity. Thus, these results are also consistent with the conclusion that the purified calcium antagonist receptor is capable of mediating ion flux with the pharmacological characteristics expected of voltage-sensitive calcium channels.

6. SUBUNIT STRUCTURE OF DIHYDROPYRIDINE-SENSITIVE CALCIUM CHANNEL

6.1. Polypeptide Components

As shown above (Fig. 1), analysis of the purified protein by SDS-PAGE under alkylating conditions and silver staining revealed three classes of polypeptides that we have designated α (167 kDa), β (54 kDa), and γ (30 kDa) (Fig. 3, lane 1). When disulfide bonds were cleaved with dithiothreitol, the α-band split into two clearly resolved protein populations with molecular weights of 175 and 143 kDa (Fig. 3, lane 2). In the initial studies from this laboratory, the anomalous behavior of the α polypeptide was ascribed to partial cleavage and/or reformation of intrachain disulfide bonds resulting in a variable fraction of the protein with smaller apparent size (Curtis and Catterall, 1984). The more recent use of a battery of specific labeling methods has now shown that the 175- and 143-kDa polypeptides are two distinct calcium channel subunits, α_1 and α_2, which have similar size but clearly different properties (Takahashi et al., 1987).

A polyclonal antibody (PAC-10), obtained from the ascites fluids of a SJL/J mouse immunized with purified calcium channel, selectively labeled only the 175-kDa polypeptide after reduction (Fig. 3, lane 3). No immunolabeling was observed with preimmune serum or with PAC-10 that had been preadsorbed with purified calcium channel. These observations indicate that the 175 and 143-kDa components are distinct polypeptides.

6.2. Subunit Glycosylation

Solubilized [³H]dihydropyridine receptors specifically bind to various immobilized lectins and affinity chromatography on wheat germ agglutinin-Sepharose is the most efficient purification step (Glossmann and Ferry, 1983; Borsotto et al., 1984; Curtis and Catterall,

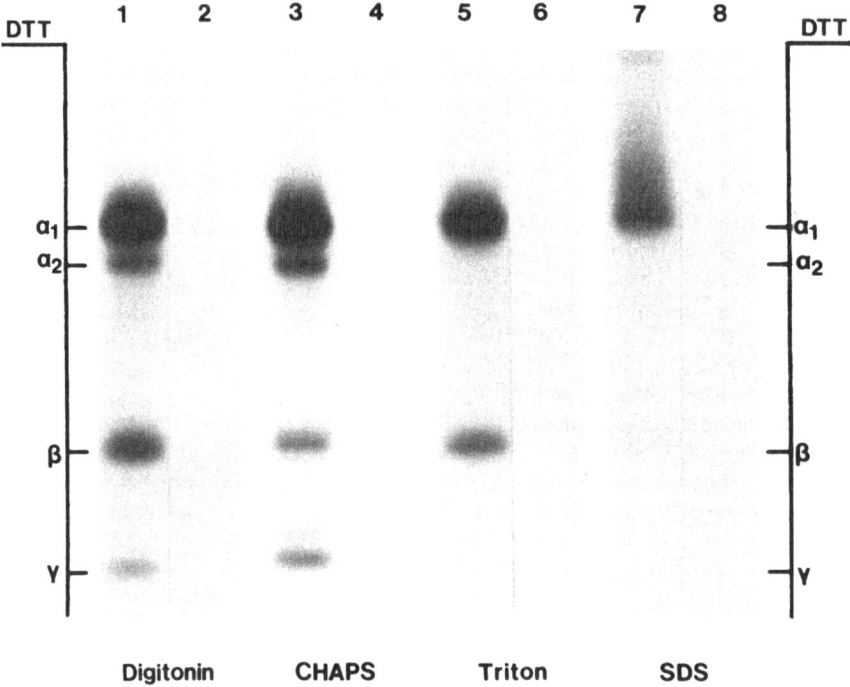

FIGURE 3. Immunoprecipitation of calcium channel subunits by anti-α_1 antibodies. ^{125}I-labeled calcium channel was immunoprecipitated as described previously (Takahashi *et al.*, 1987) using affinity-purified anti-α_1 antibodies (lanes 1, 3, 5, and 7) or a control preparation (lanes 2, 4, 6, and 8) in an immunoassay buffer containing the indicated detergents: 0.5% digitonin, 0.1% CHAPS, and 1% Triton X-100, or 1% SDS for 2 min at 100°C followed by immunoassay in 0.5% digitonin detergent after exchange by gel filtration on a 2-ml Sephadex G-50 column.

1984). These results imply that at least one subunit is glycosylated. The oligosaccharide chains of the purified calcium channel were detected by separating subunits of SDS-PAGE and probing the resolved polypeptides with [^{125}I]ConA or [^{125}I]WGA (wheat germ agglutinin). After disulfide reduction [^{125}I]ConA labeled only the α_2 subunit (Fig. 3, lane 4). [^{125}I]WGA bound to both the α_2 and γ subunits after dithiothreitol treatment (Fig. 3, lane 5). In addition, disulfide reduction leads to the appearance of two new [^{125}I]WGA-labeled components at 24–27 kDa that are clearly distinct from the γ subunit. These polypeptides were also detected, but much less distinctly, by silver staining (Fig. 3, lane 2). They appear to be disulfide-linked to the α_2 subunit under nonreducing conditions. Previous immunochemical evidence (Schmid *et al.*, 1986) suggests that the 24-kDa peptide may be proteolytically derived from the 27-kDa peptide so we refer to them collectively as the δ subunit. No labeling of α_1 or β was detected with either lectin. [^{125}I]ConA and [^{125}I]WGA binding to calcium channel subunits was blocked in the presence of 100 mM α-methylmannoside or N-acetylglucosamine, respectively (not shown).

To determine the extent of glycosylation and the core polypeptide size of the calcium channel subunits, purified channel preparations were labeled with ^{125}I, incubated with glycosidases to remove oligosaccharide chains, and analyzed by SDS-PAGE and autoradiography. Sequential deglycosylation with neuraminidase and endoglycosidase F caused a reduction in the apparent sizes of the α_2 and γ subunits reaching core polypeptide sizes of 105 and 20 kDa, respectively. Poor iodination of the δ subunit prevented estimation of its carbohydrate content by this method. No shift in the mobility of the α_1 and β subunits was noted confirming the absence of N-linked carbohydrate in these two subunits.

6.3. Covalent Labeling of Calcium Channel Subunits

[³H]PN200-110 and [³H]azidopine have been shown to covalently label a 145- to 170-kDa polypeptide in T-tubule membranes (Ferry et al., 1984, 1985; Galizzi et al., 1986) and purified calcium channels (Striessnig et al., 1986) that presumably corresponds to one of the two α subunits. In our preparations, [³H]azidopine was incorporated by UV photolysis into a polypeptide that migrated as a band of 175 kDa after dithiothreitol treatment (Fig. 3, lane 6). The electrophoretic behavior of this polypeptide identifies it as the α_1 subunit. The α_2 subunit is not labeled and no labeling of α_1 was observed in the presence of 2 μM PN 200-110.

Ion channel-forming polypeptides should contain transmembrane segments that may be detected using the hydrophobic probe [¹²⁵I]3-(trifluoromethyl)-3-(m-iodophenyl)diazirine (TID). This photoreactive compound partitions into free detergent micelles and detergent associated with the major hydrophobic domains of integral membrane proteins, and is specifically incorporated into these regions by photolysis (Brunner and Semenza, 1981). The α_1 and γ subunits were prominently labeled by TID, with a much lower level of incorporation into α_2 and δ (Fig. 3, lane 7). The β subunit was not detectably labeled. Quantitation of [¹²⁵I]TID in excised protein bands showed that the α_1 and γ subunits incorporated 10-fold more TID per unit mass than the α_2 or δ subunits, even though, as shown below, nearly all α_1 and γ subunits are associated with an α_2 subunit. These results indicate that the α_1 and γ subunits are the principal transmembrane components of the purified calcium channel complex.

Previous work showed that protein components of the α- and β-subunit bands were good substrates for cAMP-dependent protein kinase and were therefore likely to be the sites of cAMP-dependent regulation of the calcium channel (Curtis and Catterall, 1985, figure 1). Comparison of the electrophoretic mobility of the phosphorylated bands after reduction of disulfide bonds (Fig. 3, lane 8) showed that the α_1 subunit is a good substrate for this enzyme while the α_2 and δ polypeptides are not labeled. The β subunit was more weakly labeled at the low ATP concentration used in this experiment (see legend). These results identify the α_1 and β subunits of the calcium channel as the probable sites of regulation of the voltage-sensitive calcium channel by cAMP-dependent phosphorylation.

6.4. Analysis of Noncovalent Subunit Interactions

By the use of several labeling techniques, we established that α_1 has the properties expected of the calcium channel including a binding site for dihydropyridine calcium antagonists, at least one cAMP-dependent phosphorylation site, and extensive hydrophobic domains. It is important to determine whether other polypeptides present in the purified preparation are persistent impurities or specifically associated components of the oligomeric calcium channel complex.

The data presented in Fig. 3, lane 4 demonstrate that PAC-10 antibodies recognize only the α_1 subunit of NaDodSO$_4$-denatured calcium channel. However, this polyclonal serum produced by immunization with native calcium channel may contain antibodies that bind only to native conformations of the $\alpha_2\delta$, β, or γ subunits. To eliminate any antibodies with this specificity, a nitrocellulose strip containing α_1 subunit was used as an affinity matrix to purify anti-α_1 antibodies. Purified anti-α_1 antibodies specifically precipitated [³H]PN 200-110-labeled calcium channel, while proteins from PAC 10 that were nonspecifically adsorbed to a bare nitrocellulose strip did not.

Immunoprecipitation of ¹²⁵I-labeled calcium channel (Fig. 3) showed that only α_1 was precipitated after NaDodSO₄ denaturation (lane 7). In contrast, α_1, α_2, β, and γ were precipitated as a complex in 0.5% digitonin (lane 1) or 0.1% CHAPS (lane 3), detergent conditions that are known to stabilize dihydropyridine binding, allosteric coupling of the three calcium antagonist receptor sites, and ion conductance activity. A higher concentration of CHAPS (1%) caused dissociation of α_2 from the complex (data not shown). In addition, experiments in 1% Triton X-100 (lane 5) showed complete dissociation of the α_2 subunit. The β subunit and a small fraction of the γ subunit (not easily seen in Fig. 3) were coimmunoprecipitated with α_1 in Triton X-100. The results of these immunoprecipitation experiments have been confirmed by selective elution of α_2 subunits from immune complexes. Purified calcium channels were immunoprecipitated in 0.5% digitonin as in Fig. 3, lane 1. Resuspension in a buffer containing 1% Triton X-100 caused complete release of the $\alpha_2\delta$ dimer and partial release of γ without loss of β from the precipitate (data not shown). Complementary data supporting these observations have also been obtained using lentil lectin agarose, which has the same specificity as ConA (see Fig. 1, lane 8) and is a selective probe for the α_2 subunit (Takahashi *et al.*, 1987).

6.5. An Oligomeric Model for the Dihydropyridine-Sensitive Calcium Channel

On the basis of present knowledge of the structure of the dihydropyridine-sensitive calcium channel, and in analogy with current models of the structure of voltage-sensitive sodium channels (Catterall, 1986), we propose a model (Fig. 4) based on a central ion channel-forming element interacting with three other noncovalently associated subunits. The α_1 subunit, which contains the calcium antagonist-binding sites, cAMP-dependent phosphorylation sites, and the largest hydrophobic domains, is proposed to be the central ion channel-forming component of the complex. Its apparent molecular weight of 175 kDa from SDS-PAGE is likely to be a reasonable approximation of the true polypeptide molecular weight since no N-glycosylation was detected. This calcium channel subunit is therefore large enough to contain four homologous transmembrane domains analogous to those of the rat brain sodium channel α subunit, whose mRNA alone encodes a functional ion channel (Goldin *et al.*, 1986; Noda *et al.*, 1986). Like α_1, the sodium channel α subunit also contains cAMP-

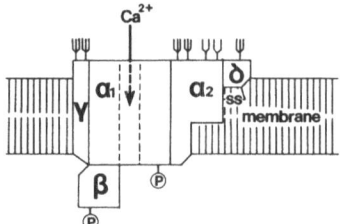

FIGURE 4. Proposed model for calcium channel structure. Sites of cAMP-dependent phosphorylation (P), glycosylation, and interaction with the membrane are illustrated.

dependent phosphorylation sites (Catterall, 1986) and extensive hydrophobic domains that are efficiently labeled by TID (Reber and Catterall, 1987).

The β subunit is also a substrate for cAMP-dependent kinase (Curtis and Catterall, 1985), but hydrophobic labeling indicates that it does not interact with the membrane phase and it is not a glycoprotein. Since it remains associated with α_1 in Triton X-100 while α_2 is dissociated, it is probably tightly associated with an intracellular domain of α_1 (Takahashi *et al.*, 1987).

The γ subunit of 30 kDa interacts independently with α_1, contains at least one transmembrane segment, and consists of approximately 30% carbohydrate. All these properties are similar to those of the β_1 subunit of the rat brain and skeletal muscle Na^+ channels (Catterall, 1986). A polypeptide of similar size appears to be associated with the apamin-sensitive, calcium-activated potassium channel (Seagar *et al.*, 1986) and it is interesting to speculate that this subunit may be a conserved constituent of voltage- or calcium-dependent ion channels.

The $\alpha_2\delta$ dimer appears to interact more weakly with α_1, although the conditions necessary to achieve dissociation result in a loss of dihydropyridine-binding activity. The 105-kDa core polypeptide of α_2 contains a heavily glycosylated extracellular domain but displays weak hydrophobic labeling, indicating a limited intramembrane domain. For this reason it seems unlikely that the ion channel is formed jointly by α_1 and α_2 at their zone of interaction.

The proposed model assumes a complex containing one of each subunit type. Our present results and previous data showing quantitative binding of solubilized calcium channels to ConA (Glossmann and Ferry, 1983) suggest that each complex contains at least one α_1 and one α_2 subunit, but do not specify the stoichiometry of any subunits. The α_1 and α_2 appear to be present in approximately equal amounts on silver-stained gels, and the α_1 and β subunits incorporate approximately 1 mol of ^{32}P per mole of complex. A complete hydrodynamic analysis of the skeletal muscle calcium channel has not been reported. However, a size of 370 kDa determined for the rat ventricular muscle dihydropyridine receptor (Horne *et al.*, 1986) is within reasonable range of the predicted size of the complex represented in Fig. 4 (416 kDa). Thus, an assumption of 1 mol of each subunit in the complex is plausible but requires direct experimental verification.

7. CONCLUSION

The molecular properties of dihydropyridine-sensitive calcium channels are now being elucidated by following a general strategy that was used previously in studies of voltage-sensitive sodium channels in this laboratory: identification by specific ligand binding and covalent labeling, solubilization and isolation by conventional protein purification methods, and reconstitution of channel function *in vitro*. The work reviewed here illustrates the substantial progress achieved with this approach to date. Future directions include further definition of the functional properties of the purified calcium channel complex in reconstituted phospholipid vesicles; analysis of the mechanism of regulation of the channel by protein phosphorylation; determination of the primary structure of the channel subunits; extension of these molecular studies to calcium channels in heart, brain, and smooth muscle; and comparison of the structural features of calcium channels with those of voltage-sensitive sodium channels in order to define common structural themes underlying the function of voltage-sensitive ion channels in general.

REFERENCES

Armstrong, C. M., and Matteson, D. R., 1985, Two distinct populations of calcium channels in a clonal line of pituitary cells, *Science* **227**:65–67.

Bean, B. P., 1985, Two populations of calcium channels in canine atrial cells, *J. Gen. Physiol.* **86**: 1–30.

Bolger, G. T., Gengo, P. J., Luchowki, E. M., Seigel, H., Triggle, D. J., and Janis, R. A., 1982, High affinity binding of a calcium channel antagonist to smooth and cardiac muscle, *Biochem. Biophys. Res. Commun.* **104**:1604–1609.

Borsotto, M., Barhanin, J., Norman, R. I., and Lazdunski, M., 1984, Purification of the dihydropyridine receptor of the voltage-dependent Ca^{2+} channel from skeletal muscle transverse tubules using (+) [^{3}H]PN 200-110, *Biochem. Biophys. Res. Commun.* **122**:1357–1365.

Brum, G., Flockerzi, V., Hofmann, F., Osterrieder, W., and Trautwein, W., 1983, Injection of catalytic subunit of cAMP-dependent protein kinase into isolated cardiac myocytes, *Pflugers Archiv.* **398**: 147–154.

Brunner, J., and Semenza, G., 1981, Selective labeling of the hydrophobic core of membranes with 3-(trifluoromethyl)-3-(m-[^{125}I]iodophenyl)diazirine, a carbene generating reagent, *Biochemistry* **20**: 7174–7182.

Cachelin, A. B., de Peyer, J. E., Kukubun, S., and Reuter, H., 1983, Ca^{2+} channel modulation by 8-bromocyclic AMP in cultured cells, *Nature* **304**:462–464.

Carbone, E., and Lux, H. D., 1984, A low voltage-activated fully inactivating Ca channel in vertebrate sensory neurones, *Nature* **310**:501–502.

Catterall, W. A., 1986, Molecular properties of voltage-sensitive sodium channels, *Ann. Rev. Biochem.* **55**:953–985.

Curtis, B. M., and Catterall, W. A., 1983, Solubilization of the calcium antagonist receptor from rat brain, *J. Biol. Chem.* **258**:7280–7283.

Curtis, B. M., and Catterall, W. A., 1984, Purification of the calcium antagonist receptor of the voltage-sensitive calcium channel from skeletal muscle transverse tubules, *Biochemistry* **23**:2113–2118.

Curtis, B. M., and Catterall, W. A., 1985, Phosphorylation of the calcium antagonist receptor of the voltage-sensitive calcium channel by cAMP-dependent protein kinase, *Proc. Natl. Acad. Sci. USA* **82**:2528–2532.

Curtis, B. M., and Catterall, W. A., 1986, Reconstitution of the voltage-sensitive calcium channel purified from skeletal muscle transverse tubules, *Biochemistry* **25**:3077–3083.

Ferry, D. R., Rombusch, M., Goll, A., and Glossmann, H., 1984, Photoaffinity labelling of Ca^{2+} channels with [^{3}H]azidopine, *FEBS Lett.* **169**:112–167.

Ferry, D. R., Kampf, K., Goll, A., and Glossmann, H., 1985, Subunit composition of skeletal muscle transverse tubule calcium channels evaluated with the 1,4-dihydropyridine photoaffinity probe, [^{3}H]azidopine, *EMBO J.* **4**:1933–1940.

Galizzi, J. P., Borsotto, M., Barhanin, J., Fosset, M., and Lazdunski, M., 1986, Characterization and photoaffinity labeling of receptor sites for the Ca^{2+} channel inhibitors d-cis-diltiazem, (±)-bepridil, desmethoxyverapamil and (+) PN 200-110 in skeletal muscle transverse tubule membranes, *J. Biol. Chem.* **261**:1393–1397.

Garcia, M. L., King, V. F., Siegl, P. K. S., Reuben, J. P., and Kaczorowski, G. J., 1986, Binding of calcium entry blockers to cardiac sarcolemmal membrane vesicles. Characterization of diltiazem-binding sites and their interaction with dihydropyridine and aralkylamine receptors, *J. Biol. Chem.* **261**:8146–8157.

Glossmann, H., and Ferry, D. R., 1983, Solubilization and partial purification of putative calcium channels labelled with [^{3}H]nimodipine, *Naunyn-Schmiedeberg's Arch. Pharmacol.* **323**:279–291.

Goldin, A. L., Snutch, T., Lubbert, H., Dowsett, A., Marshall, J., Auld, V., Downey, W., Fritz, L. C., Lester, H. A., Dunn, R., Catterall, W. A., and Davidson, N., 1986, Messenger RNA coding for only the α subunit of the rat brain Na channel is sufficient for expression of functional channels in *Xenopus* oocytes, *Proc. Natl. Acad. Sci. USA* **83**:7503–7509.

Hagawara, S., and Byerly, L., 1981, Calcium channel, *Ann. Rev. Neurosci.* **4**:69–125.

Horne, W. A., Weiland, G. A., and Oswald, R. E., 1986, Solubilization and hydrodynamic characterization of the dihydropyridine receptor from rat ventricular muscle, *J. Biol. Chem.* **261**:3588–3594.

Janis, R. A., and Scriabine, A., 1983, Sites of action of Ca^{2+} channel inhibitors, *Biochem. Pharmacol.* **32**:3499–3507.

Katz, B., and Miledi, R., 1969, Tetrodotoxin-resistant electric activity in presynaptic terminals, *J. Physiol.* **203**:459–487.

Kokubun, S., and Reuter, H., 1984, Dihydropyridine derivatives prolong the open state of Ca channels in cultured cardiac cells, *Proc. Natl. Acad. Sci. USA* **81**:4824–4827.

Llinas, R., Yarom, Y., and Sugimori, M., 1981, Isolated mammalian brain in vitro: New technique for analysis of electrical activity of neuronal circuit function, *Fed. Proc.* **40**:2240–2245.

Murphy, K. M. M., Gould, R. J., Largent, B. L., and Snyder, S. H., 1983, A unitary mechanism of calcium antagonist drug action, *Proc. Natl. Acad. Sci. USA* **80**:860–864.

Nilius, B., Hess, P., Lansman, J. B., and Tsien, R. W., 1985, A novel type of cardiac calcium channel in ventricular cells, *Nature* **316**:443–446.

Noda, M., Ikeda, T., Suzuki, H., Takeshima, H., Takahashi, T., Kuno, M., and Numa, S., 1986, Expression of functional sodium channels from cloned cDNA, *Nature* **322**:826–828.

Nowycky, M. C., Fox, A. P., and Tsien, R. W., 1985, Three types of neuronal calcium channel with different calcium agonist sensitivity, *Nature* **316**:440–443.

Reber, B. F. X., and Catterall, W. A., 1987, Hydrophobic properties of the β_1 and β_2 subunits of the rat brain sodium channel, *J. Biol. Chem.* **262**:11369–11374.

Reuter, H., 1974, Localization of beta adrenergic receptors, and effects of noradrenaline and cyclic nucleotides on action potentials, ionic currents and tension in mammalian cardiac muscle, *J. Physiol. (London)* **242**:429–451.

Reuter, H., 1979, Properties of two inward membrane currents in the heart, *Ann. Rev. Physiol.* **41**:413–424.

Reuter, H., 1983, Calcium channel modulation by neurotransmitters, enzymes and drugs, *Nature* **301**:569–574.

Schmid, A., Renaud, J.-F., Lazdunski, M., 1985, Short term and long term effects of β-adrenergic effectors and cyclic AMP on nitrendipine-sensitive voltage-dependent Ca^{2+} channels of skeletal muscle, *J. Biol. Chem.* **260**:13041–13046.

Schmid, A., Barhanin, J., Coppola, T., Borsotto, M., and Lazdunski, M., 1986, Immunochemical analysis of subunit structures of 1,4-dihydropyridine receptors associated with voltage dependent Ca^{++} channels in skeletal, cardiac and smooth muscle, *Biochemistry* **25**:3492–3495.

Schramm, M., Thomas, G., Towart, R., Franckowiak, G., 1983, Novel dihydropyridines with a positive inotropic action through activation of Ca^{2+} channels, *Nature* **303**:535–537.

Schwartzkroin, P. A., and Slawsky, M., 1977, Probable calcium spikes in hippocampal neurons, *Brain Res.* **135**:157–161.

Seagar, M. J., Labbe-Julle, C., Granier, C., Goll, A., Glossmann, H., Van Reitschoten, J., and Couraud, F., 1986, Molecular structure of the rat brain apamin receptor: Differential photoaffinity labeling of putative K^+ channel subunits and target size analysis, *Biochemistry* **25**:4051–4057.

Striessnig, J., Moosburger, K., Goll, A., Ferry, D. R., and Glossmann, H., 1986, Stereoselective photoaffinity labeling of the purified 1,4-dihydropyridine receptor of the voltage dependent calcium channel, *Eur. J. Biochem.* **161**:603–609.

Takahashi, M., Seagar, M. J., Jones, J. F., Reber, B. F. X., and Catterall, W. A., 1987, Subunit structure of dihydropyridine-sensitive calcium channels from skeletal muscle, *Proc. Natl. Acad. Sci. USA* **84**:5478–5482.

Talvenheimo, J. A., Tamkun, M. M., and Catterall, W. A., 1982, Reconstitution of neurotoxin-stimulated sodium transport by the voltage-sensitive sodium channel purified from rat brain, *J. Biol. Chem.* **257**:11868–11871.

Triggle, D. J., 1982, Biochemical pharmacology of calcium blockers, in: *Calcium Blockers: Mechanism of Actions and Clinical Applications* (S. F. Flaim and R. Zelis, eds.), Urban and Schwarzenberg, Baltimore, pp. 121–134.

Tsien, R. W., Giles, W., and Greengard, P., 1972, Cyclic-AMP mediates the action of epinephrine on the action potential plateau of cardiac Purkinje fibers, *Nature New Biol.* **240**:181–183.

Monoclonal Antibodies That Coimmunoprecipitate the 1,4-Dihydropyridine and Phenylalkylamine Receptors and Reveal the Ca^{2+} Channel Structure

MICHEL FOSSET, SYLVIE VANDAELE, JEAN-PIERRE GALIZZI, JACQUES BARHANIN, and MICHEL LAZDUNSKI

Monoclonal hybridoma cell lines secreting antibodies against the (+)PN 200-110 and the (−)desmethoxyverapamil binding components of the voltage-dependent calcium channel from rabbit transverse-tubule membranes have been isolated. The specificity of these monoclonal antibodies was established by their ability to coimmunoprecipitate (+)[^{3}H]PN 200-110 and (−)[^{3}H]desmethoxyverapamil receptors. Monoclonal antibodies described in this work cross-reacted with rat, mouse, chicken, and frog skeletal muscle Ca^{2+} channels, but not with crayfish muscle Ca^{2+} channels. Cross-reactivity was also detected with membranes prepared from rabbit heart, brain, and intestinal smooth muscle.

These antibodies were used in immunoprecipitation experiments with ^{125}I-labeled detergent (CHAPS and digitonin)-solubilized membranes. They revealed a unique immunoprecipitating component of molecular weight 170,000 in nonreducing conditions. After disulfide bridge reduction, the CHAPS solubilized (+)PN 200-110-(−)desmethoxyverapamil-binding component gave rise to a large peptide of M_r 140,000 and to smaller polypeptides of M_r 30,000 and 26,000, whereas the digitonin-solubilized receptor appeared with subunits at M_r 170,000, 140,000, 30,000, and 26,000.

All these results taken together are interpreted as showing that both the 1,4-dihydropyridine and the phenylalkylamine receptors are part of a single polypeptide chain of M_r 170,000.

Organic Ca^{2+} channel inhibitors have proved to be of great importance (Janis and Triggle, 1984) in studies of the mechanism and molecular structure of the slow type of Ca^{2+}

MICHEL FOSSET, SYLVIE VANDAELE, JEAN-PIERRE GALIZZI, JACQUES BARHANIN, and MICHEL LAZDUNSKI • Center for Biochemistry of the CNRS, Parc Valrose, 06034 Nice Cedex, France.

channel. The best known Ca^{2+} channel inhibitors include (1) 1,4-dihydropyridines such as nitrendipine and (+)PN 200-110, (2) verapamil-like compounds, and (3) other types of molecules such as diltiazem and bepridil.

Skeletal muscle transverse tubule (T-tubule) membranes are the best source to study the properties of the 1,4-dihydropyridine-sensitive Ca^{2+} channel (Fosset et al., 1983; Ferry et al., 1984). The (+)[^{3}H]PN 200-110 binding component of the channel protein has been detergent-solubilized and purified. However, different purification procedures in different laboratories have provided different evaluations of the subunit structure of the putative Ca^{2+} channel. It was found in this laboratory that the 1,4-dihydropyridine receptor is a protein of 170,000 assembled from a large subunit of M_r 140,000 and a smaller subunit of 33,000–29,000, the two subunits being covalently linked by disulfide bridge(s) (Borsotto et al., 1984a, 1985; Schmid et al., 1986). It was also found that the large subunit is the target of cAMP-dependent phosphorylation (Hosey et al., 1986). Conversely, it was reported by another laboratory that the 1,4-dihydropyridine receptor is made by the assembly of three noncovalently linked polypeptides of M_r 160,000/130,000–50,000–33,000 (Curtis and Catterall, 1984), the peptide of M_r 50,000 being the one that is phosphorylated by a cAMP-dependent kinase (Curtis and Catterall, 1985). The first purpose of this work is to approach the problem of the subunit structure of the Ca^{2+} channel by a different route that does not require purification procedures and uses monoclonal antibodies. The second purpose is to show that the two distinct receptors (Janis and Triggle, 1984) for 1,4-dihydropyridines and verapamil-like compounds are present in the same protein.

RESULTS AND DISCUSSION

Biochemical and electrophysiological experiments previously showed that (+)PN 200-110 is a very active 1,4-dihydropyridine calcium channel inhibitor of the mammalian skeletal muscle calcium channel (Cognard et al., 1986). Monoclonal antibodies were prepared by intraperitoneal injection in mice of purified rabbit T-tubule membranes.

The putative Ca^{2+} channel protein from T-tubule membranes was previously detergent-solubilized with CHAPS (Borsotto et al., 1984b, 1985) and purified using (+)[^{3}H]PN 200-110, which with its solubilized receptor form a stable complex with a half-life of dissociation

$\longrightarrow$

FIGURE 1. Immunoprecipitation of the (+)[^{3}H]PN 200-110 and (−)[^{3}H]D888-binding components by monoclonal antibody 3007 and equilibrium binding assay of (−)[^{3}H]D888 to T-tubule membranes in CHAPS-solubilized form. Experiments were performed in duplicate. (A) The CHAPS-solubilized T-tubule membranes from rabbit skeletal muscle were incubated with (+)[^{3}H]PN 200-110 at increasing concentrations of ascites fluid. In this experiment, rabbit anti-mouse IgG antibodies were omitted. Immunoprecipitation with 3007 in the absence (■) or presence (□) of 1 μM unlabeled (+)PN 200-110. Nonrelevant ascites fluid in the absence (●) or presence (○) of 1 μM (+)PN 200-110. Antibody 3007 without CHAPS-solubilized membranes in the absence (▲) or presence (△) of 1 μM (+)PN 200-110. (B) Inset A: T-tubule membranes at 20 μg/ml in 20 mM Hepes/NaOH buffer (pH 7.5 and 0°C) were incubated for 60 min (time sufficient to attain equilibrium) with increasing concentrations of (−)[^{3}H]D888 in the absence (●) or presence (■) of 1 μM (−)D888. Each sample was then CHAPS-solubilized. Bound (−)[^{3}H]D888 was separated from the free ligand by filtration through Sephadex G-50 columns. Inset B: Scatchard plot for the specific (−)[^{3}H]D888-binding component. Main panel: T-tubule membranes from rabbit skeletal muscle labeled with (−)[^{3}H]D888 and CHAPS-solubilized membranes were incubated with increasing concentrations of ascites fluid (Vandaele et al., 1987). Immunoprecipitation with 3007 in the absence (☆) or presence (★) of 1 μM (−)D888.

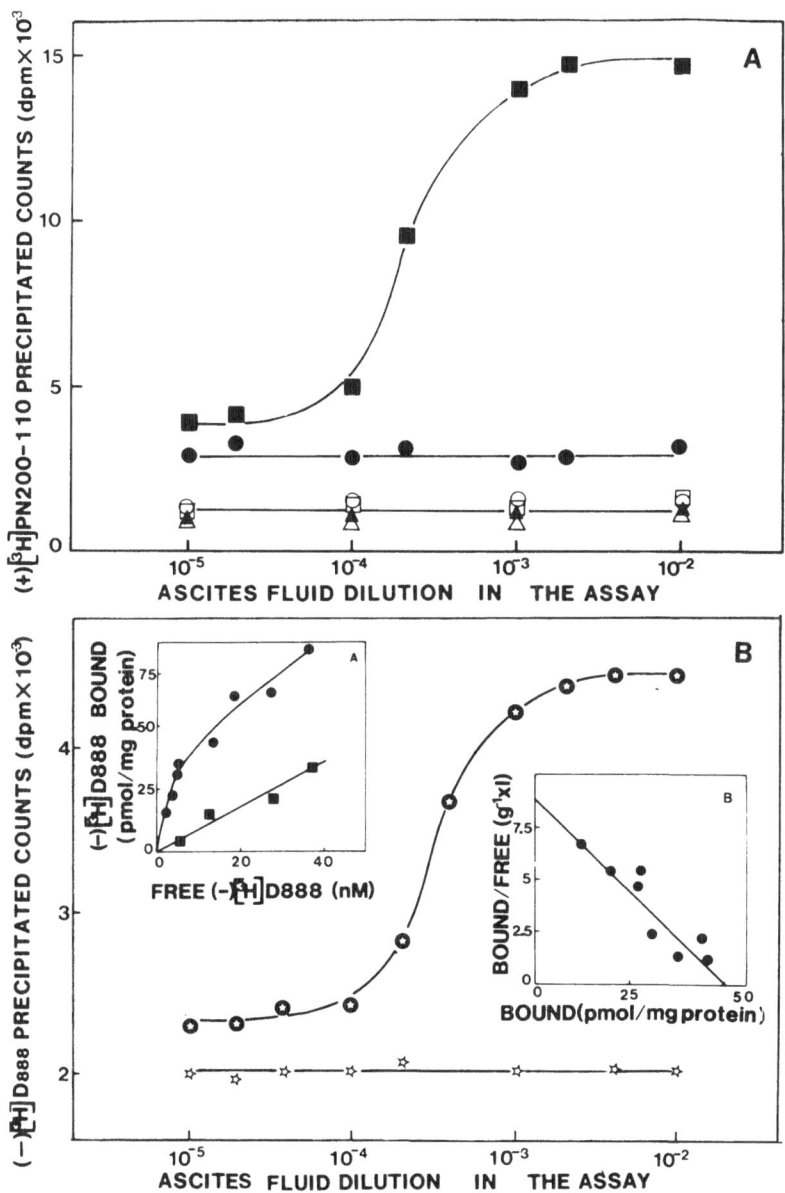

of 34 hr at 4°C and an equilibrium dissociation constant K_D = 0.2–0.4 nM (Borsotto et al., 1984a,b). We took advantage of these properties in screening experiments with supernatants of hybridoma cultures in which we looked for supernatants able to precipitate the CHAPS-solubilized (+)[³H]PN 200-110–receptor complex. Hybridoma supernatants were considered to be positive when the amount of counts precipitated was at least twice the background value. Figure 1A shows a typical immunoprecipitation curve of the (+)[³H]PN 200-110 binding component as a function of ascites concentration for the 3007 hybridoma secreting monoclonal antibody. Precipitation of the nonspecific binding component was independent of the ascites concentration. The concentration of IgG necessary to give a half-maximum precipitation of the specific counts in Fig. 1A was estimated to be 2–6 nM. Another type of purification using digitonin solubilization has been used by other authors (Curtis and Catterall, 1984). We have verified by using the same ascites that an identical type of immunoprecipitation curve was obtained for the digitonin extract of the (+)PN 200-110 receptor (Fig. 3A).

One important problem is to determine whether monoclonal antibodies that immuno-precipitate the 1,4-dihydropyridine receptor also immunoprecipitate the phenylalkylamine (verapamil-like) receptor. It is known from binding studies that the receptors for those two categories of Ca^{2+} channel inhibitors are distinct and allosterically linked (Janis and Triggle, 1984). The first step was to analyze the properties of binding of (–)[³H]D888, a potent verapamil analog (Galizzi et al., 1986), to solubilized T-tubule membranes.

Figure 1B, inset A shows a typical equilibrium binding experiment of (–)[³H]D888 to CHAPS-solubilized skeletal muscle T-tubule membranes in the absence (total binding) and in the presence of 1 μM of unlabeled (–)D888. The Scatchard plot for the specific binding component (Fig. 1B, inset B) was linear, indicating the presence of a single class of sites. The equilibrium dissociation constant was K_d = 5.7 ± 0.5 nM. The affinity of (–)[³H]D888 for the solubilized receptor was only slightly altered by the solubilization procedure (K_D = 1.5 ± 0.5 nM for the membrane-bound receptor (Galizzi et al., 1986)). The maximum binding capacity was B_{max} = 50 ± 5 pmol/mg of protein as for the membrane-bound receptor. Figure 1B, main panel, shows a typical immunoprecipitation curve of the solubilized (–)[³H]D888-binding component as a function of ascites fluid concentration for the 3007 hybridoma secreting monoclonal antibody. Precipitation of the nonspecific binding component was independent of the ascites concentration. The concentration of IgG necessary to give a half-maximum precipitation of the specific counts (Fig. 1B) was estimated to be 2–6 nM, the same concentration as that found for half-maximal precipitation of (+)[³H]PN 200-110 binding. We have also verified that prelabeling of the 1,4-dihydropyridine receptor with (+)[³H]PN 200-110 prior to CHAPS solubilization gave the same immunoprecipitation curve as that observed when labeling of the receptor with (+)[³H]PN 200-110 was posterior to solubilization.

Seven relevant clones were obtained out of a general screening of several thousand clones coming from different fusions between myeloma and spleen cells. All seven mono-clonal antibodies had a high titer of 5000. The seven monoclonal antibodies were protein A-sensitive. They belonged to the IgG class: IgG₃ (3007, 3106); IgG₁ (3329, 7447, 6322, 6332, 6308).

Dot immunobinding assays against either native T-tubule or microsome membranes have indicated (not shown) that monoclonal antibodies against the rabbit skeletal muscle 1,4-dihydropyridine receptor cross-reacted with rat, mouse, chicken, and frog muscle mem-branes. No cross-reaction was found with crayfish muscle. Cross-reactivity was also found with membranes prepared from rabbit heart, brain, and intestinal smooth muscle.

In order to analyze the subunit structure of the immunoprecipitated protein(s) containing both the 1,4-dihydropyridine and the phenylalkylamine receptor, CHAPS-solubilized transverse tubule (T-tubule) and wheat germ agglutinin (WGA) extracts (Fig. 2) and digitonin-WGA extract (Fig. 3B) were iodinated, and the ^{125}I-labeled preparations were submitted to immunoprecipitation by monoclonal antibodies. SDS-polyacrylamide gel electrophoresis was performed either in non-disulfide-reducing conditions (iodoacetamide, Fig. 2, lanes 1–6; Fig. 3B, lanes 1–3) or in disulfide-reducing conditions (β-mercaptoethanol, Fig. 2, lanes 7–12; Fig. 3B, lanes 4–6).

The typical monoclonal antibody 3007 was found to precipitate both from CHAPS and digitonin extracts a single polypeptide chain of apparent M_r 170,000 in nonreducing conditions (iodoacetamide; Fig. 2, lanes 2 and 5; Fig. 3B, lane 2).

In disulfide bridge-reducing conditions, two different patterns were obtained depending on the type of detergent used for solubilization: (1) Polypeptides of apparent M_r of 140,000, 30,000 and 26,000 were observed (Fig. 2, lanes 8 and 11) when CHAPS was used for solubilization. (2) When digitonin was used, polypeptides of M_r 170,000, 140,000, 30,000, and 26,000 were observed (Fig. 3B, lane 5). Peptide maps of proteins of M_r 170,000 and 140,000 using *Staphylococcus aureus* V8 protease showed that they were nearly indistinguishable (Fig. 3C).

The seven selected monoclonal antibodies gave the same pattern of precipitation as antibody 3007 (not shown). Neither mouse IgG nor irrelevant monoclonal antibodies were able to specifically precipitate any identifiable component (Fig. 2, lanes 3, 6, 9, and 12; Fig. 3B, lanes 3 and 6).

Immunoprecipitation of the CHAPS-solubilized 1,4-dihydropyridine/phenylalkylamine receptor gave results similar to those of our previous purifications (Borsotto *et al.*, 1984a, 1985). The intact receptor has a total M_r of 170,000; after disulfide reduction it is split into several pieces. One subunit, the large one, has a M_r of 140,000, identical to that found previously (Borsotto *et al.*, 1984a, 1985; Schmid *et al.*, 1986), the other ones have a M_r of 30,000 and 26,000 (Schmid *et al.*, 1986). The M_r 26,000 subunit is either a degradation product of the 30,000 subunit or a very similar but distinct peptide. In the latter case, the unique 140,000 subunit could be covalently linked to either a 30,000 or a 26,000 subunit to give isochannel proteins of M_r 170,000.

A subunit structure with a large chain of M_r 140,000 and a smaller chain of M_r 30,000 would be radically different from that found by others using a digitonin solubilization procedure (Curtis and Catterall, 1984) prior to purification. It is for that reason that monoclonal antibodies have also been used with the digitonin solubilized material. It is clear from Fig. 3B that in these solubilization conditions, too, the intact 1,4-dihydropyridine/phenylalkylamine receptor has a M_r of 170,000. Therefore no differences are found between CHAPS and digitonin solubilizations for the nonreduced 1,4-dihydropyridine/phenylalkylamine receptor. However, in disulfide-reducing conditions the gel pattern is different. The M_r 170,000 digitonin-solubilized receptor, unlike the CHAPS-solubilized receptor, is not completely converted into a M_r 140,000 protein. The easiest interpretation of this result is the following: (1) The 1,4-dihydropyridine/phenylalkylamine receptor is a single polypeptide chain of M_r 170,000 with internal disulfide bridges. (2) This large polypeptide is vulnerable to proteolysis that is known, from Na$^+$ channel work (Casadei *et al.*, 1984), to be very efficient in skeletal muscle. There is a "hot spot" for peptide cleavage capable of leading to the transformation of the M_r 170,000 unit in two pieces of M_r 140,000 and 30,000 that remain covalently linked by disulfide bridges in nonreducing conditions. (3) This cleavage is complete (in spite of the cocktail of protease inhibitors) in the CHAPS-solubilized material (M_r 170,000 → M_r

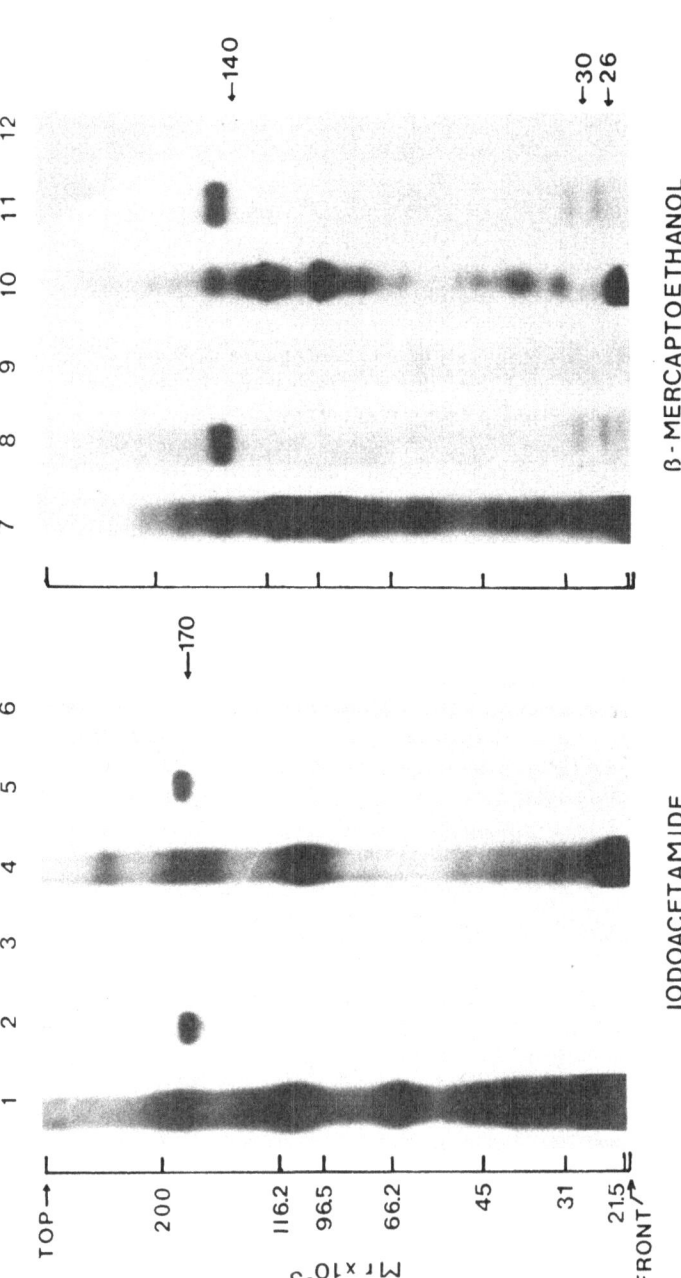

FIGURE 2. Immunoprecipitation by monoclonal antibody 3007 from the radioiodinated CHAPS-solubilized membranes and from the CHAPS-WGA extract of rabbit T-tubule membranes. Detergent extracts were iodinated and subjected to immunoprecipitation. Autoradiogram: SDS-polyacrylamide gel electrophoreses were performed in the presence of iodoacetamide (lanes 1–6) or β-mercaptoethanol (lanes 7–12). Iodinated proteins from detergent-solubilized membranes (lanes 1 and 7; 10-μl aliquots from starting volume of 100 μl) or after immunoprecipitation by monoclonal antibody 3007 (lanes 2 and 8) and control mouse IgG (lanes 3 and 9). Iodinated proteins from WGA extract before (lanes 4 and 10) or after immunoprecipitation by monoclonal antibody 3007 (lanes 5 and 11) and control mouse IgG (lanes 6 and 12). Lanes 1, 4, 7, and 10 were exposed to the film for 5 hr, and the other lanes were exposed for 24 hr. Molecular weight markers are presented at the left side of the figure.

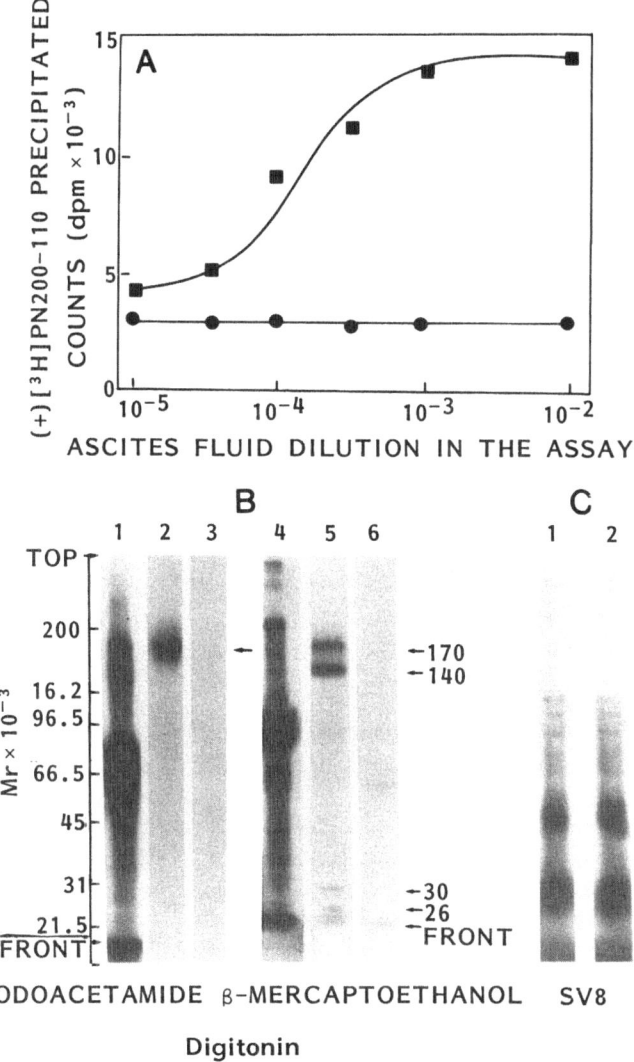

FIGURE 3. Immunoprecipitation by monoclonal antibody 3007 from digitonin-WGA extract and peptide maps. (A) Immunoprecipitation of the $(+)[^3H]PN$ 200-110 binding component. The digitonin-WGA extract labeled with $(+)[^3H]PN$ 200-110 was incubated at increasing concentrations of ascites fluids. Immunoprecipitation with 3007 in the absence (■) or presence (●) of 1 μM unlabeled $(+)PN$ 200-110. (B) Immunoprecipitation of the radioiodinated $(+)PN$ 200-110 receptor. The digitonin-WGA extract was iodinated and subjected to immunoprecipitation. Autoradiogram: SDS-polyacrylamide gel electrophoresis were performed in the presence of iodoacetamide (lanes 1–3) or β-mercaptoethanol (lanes 4–6). Iodinated proteins from digitonin-WGA extract (lanes 1 and 4) or after immunoprecipitation by monoclonal 3007 (lanes 2 and 5) and control mouse IgG (lanes 3 and 6). (C) Peptide maps. The immunoprecipitated proteins obtained on nitrocellulose were iodinated and submitted to limited proteolytic digestion. Autoradiogram: 140-kDa (lane 1) and 170-kDa (lane 2) peptides were submitted to limited proteolytic digestion with *S. aureus* V8 protease (0.3 μg of SV8, lanes 1 and 2).

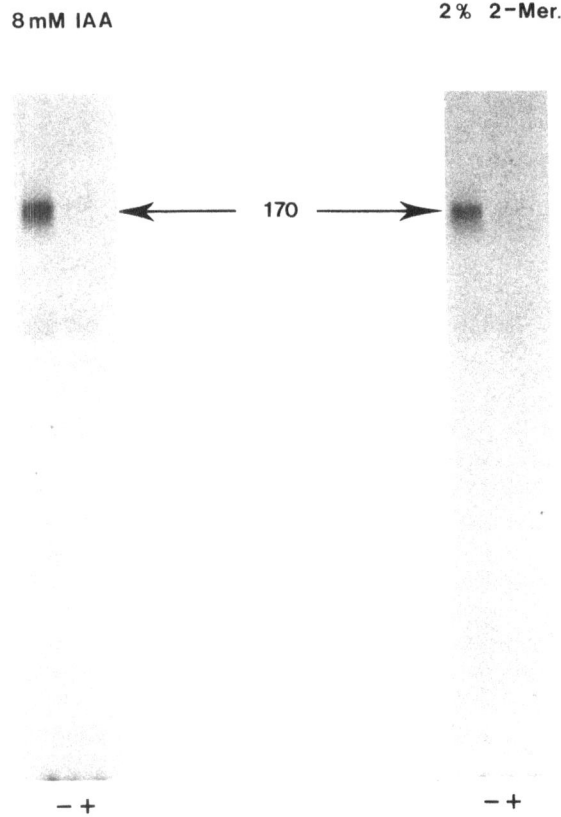

FIGURE 4. Photoaffinity labeling of the phenylalkylamine receptor with [³H]azidopamil. T-tubule membranes [50 pmol of (−)[³H]D888 binding sites per mg of protein] were incubated at a final concentration of 960 μg of protein per ml with 22 nM [³H]azidopamil ([³H]LU-47781) in 20 mM Hepes-NaOH pH 7.4 under dim light at 4°C. After 1 hr incubation, incubation samples were washed by centrifugation and irradiated with UV light (40 watts, 257.3 nm) for 2 min in the same buffer at 4°C. Membranes were pelleted, washed again, and denaturated in 2% SDS, 9% glycerol, 75 nM Tris-HCl at pH 6.8, and 2% mercaptoethanol (disulfide-reducing conditions) or 8 mM idoacetamide (nonreducing conditions). After electrophoresis on 4–14% linear polyacrylamide gradient gels (100 μg protein per lane), gels were stained with Coomassie blue, impregnated with Amplify (Amersham) for fluorography, and exposed to Kodak X-OMAT AR films for 20 days at − 70°C. IAA: iodoacetamide; 2-Mer.: 2-mercaptoethanol; − : incubation in the absence of unlabeled (±)verapamil; + : incubation in the presence of 1 μM unlabeled (±)verapamil (protection).

140,000 + 30,000 in reducing conditions) and only partial (M_r 170,000 → M_r 170,000 + M_r 140,000 + 30,000) in the digitonin-solubilized material.

In conclusion, whatever the interpretation that one can make on differences of gel patterns in disulfide-reducing conditions for the CHAPS- and digitonin-solubilized materials, this work clearly indicates that (1) the same protein is the receptor for both 1,4-dihydro-pyridines and phenylalkylamines, (2) this protein has a M_r of 170,000 and other subunits of M_r 50,000 and 33,000 found by others (Curtis and Catterall, 1984) in nonreducing conditions are unlikely to belong to the receptor.

Affinity-labeling experiments with (+)PN 200-110 have previously shown that the 170,000 M_r protein is the receptor of 1,4-dihydropyridine (Galizzi et al., 1986). Figure 4 shows that this protein is also the receptor of phenylalkylamines. These results are in good agreement with those provided by the monoclonal antibody approach. These monoclonal antibodies, because of their immunoprecipitating properties, will be very useful in future studies of the biosynthesis, integration, and histochemical localization of the Ca^{2+} channel protein at the surface membrane of excitable cells. Another group has confirmed our results concerning the immunochemical analysis of the subunit studies of the Ca^{2+} channel (Norman et al., 1987) and developed monoclonal antibodies against the Ca^{2+} channel protein that do not immunoprecipitate but that are useful probes to immunoblot the peptide components of the Ca^{2+} channel.

ACKNOWLEDGMENTS. This work was supported by the Association des Myopathes de France, the Centre National de la Recherche Scientifique, the Fondation sur les Maladies Vasculaires, and the Ministère de la Recherche et de la Technologie (grant 85C1137). We thank Dr. R. P. Hof for the gift of (+)PN 200-110 and Drs. Hollmann and Traut for the gift of (−)desmethoxyverapamil. Thanks are due to C. Widmann, C. Roulinat-Bettelheim, M. Valetti, M. Tomkowiak, and M. T. Bohn for expert technical assistance. We also thank the American Chemical Society, copyright owner, for permission to reproduce Figs. 1–3 from *Biochemistry* **26**, 1987, 5–9, with their legends.

REFERENCES

Borsotto, M., Barhanin, J., Norman, R. I., and Lazdunski, M., 1984a, Purification of the dihydro-pyridine receptor of the voltage-dependent Ca^{2+} channel from skeletal muscle transverse tubules using (+)[^{3}H]PN 200-110, *Biochem. Biophys. Res. Commun.* **122**:1357–1366.

Borsotto, M., Norman, R. I., Fosset, M., and Lazdunski, M., 1984b, Solubilization of the nitrendipine receptor from muscle transverse tubule membranes: Interactions with specific inhibitors of the voltage-dependent Ca^{2+} channel, *Eur. J. Biochem.* **142**:449–455.

Borsotto, M., Barhanin, J., Fosset, M., and Lazdunski, M., 1985, The 1,4-dihydropyridine receptor associated with the skeletal muscle voltage-dependent Ca^{2+} channel. Purification and subunit composition, *J. Biol. Chem.* **260**:14255–14263.

Casadei, J. M., Gordon, R. D., Lampson, L. A., Schotland, D. L., and Barchi, R. L., 1984, Monoclonal antibodies against the voltage-sensitive Na^+ channel from mammalian skeletal muscle, *Proc. Natl. Acad. Sci. USA* **81**:6227–6231.

Cognard, C., Romey, G., Galizzi, J. P., Fosset, M., and Lazdunski, M., 1986, Dihydropyridine-sensitive Ca^{2+} channels in mammalian skeletal muscle cells in culture: Electrophysiological properties and interactions with Ca^{2+} channel activator (Bay K8644) and inhibitor (PN 200-110), *Proc. Natl. Acad. Sci. USA* **83**:1518–1522.

Curtis, B. M., and Catterall, W. A., 1984, Purification of the calcium antagonist receptor of the voltage-sensitive calcium channel from skeletal muscle transverse tubules, *Biochemistry* **23:**2113–2118.

Curtis, B. M., and Catterall, W. A., 1985, Phosphorylation of the calcium antagonist receptor of the voltage-sensitive calcium channel by cAMP-dependent protein kinase, *Proc. Natl. Acad. Sci. USA* **82:**2528–2532.

Ferry, D. R., Rombush, M., Goll, A., and Glossmann, H., 1984, Photoaffinity labeling of Ca^{2+} channels with [^{3}H]azidopine, *FEBS Lett.* **169:**112–118.

Fosset, M., Jaimovich, E., Delpont, E., and Lazdunski, M., 1983, [^{3}H]nitrendipine receptors in skeletal muscle. Properties and preferential localization in transverse tubules, *J. Biol. Chem.* **258:**6086–6092.

Galizzi, J. P., Borsotto, M., Barhanin, J., Fosset, M., and Lazdunski, M., 1986, Characterization and photoaffinity labeling of receptor sites for the Ca^{2+} channel inhibitors d-cis-diltiazem, ($\pm$)bepridil, ($-$)desmethoxyverapamil and ($+$)PN 200-110 in skeletal muscle transverse tubule membranes, *J. Biol. Chem.* **261:**1393–1397.

Hosey, M. M., Borsotto, M., and Lazdunski, M., 1986, Phosphorylation and dephosphorylation of the major component of the voltage-dependent Ca^{2+} channel in skeletal muscle membranes by cyclic AMP- and Ca^{2+}-dependent processes, *Proc. Natl. Acad. Sci. USA* **83:**3733–3737.

Janis, R. A., and Triggle, D. J., 1984, *Calcium Channel Antagonists: New Perspectives from the Radioligand Binding Assay, Vol. 2, Modern Methods Pharmacol.* Alan R. Liss, New York, pp. 1–28.

Norman, R. I., Burgess, A. J., Allen, E., and Harrison, T. M., 1987, Monoclonal antibodies against the 1,4-dihydropyridine receptor associated with voltage-sensitive Ca^{2+} channels, *FEBS Lett.* **212:** 127–132.

Schmid, A., Barhanin, J., Coppola, T., Borsotto, M., and Lazdunski, M., 1986, Immunochemical analysis of subunit structure of 1,4-dihydropyridine receptors associated with voltage-dependent Ca^{2+} channels in skeletal, cardiac and smooth muscles, *Biochemistry* **25:**3492–3495.

Vandaele, S., Fosset, M., Galizzi, J. P., and Lazdunski, M., 1987, Monoclonal antibodies that coimmunoprecipitate the 1,4-dihydropyridine and phenylalkylamine receptors and reveal the Ca^{2+} channel structure, *Biochemistry* **26:**5–9.

Criteria for Identification of α_1 and α_2 Subunits in a Purified Skeletal Muscle Calcium Channel Preparation

PAL L. VAGHY, KUNIHISA MIWA, KIYOSHI ITAGAKI, FERENC GUBA, Jr., EDWARD McKENNA, and ARNOLD SCHWARTZ

Skeletal and cardiac muscle Ca channels have been isolated in several different laboratories (Curtis and Catterall, 1984; Borsotto et al., 1985; Striessnig et al., 1986a; Cooper et al., 1987a; Nakayama et al., 1987). The procedures are modified versions of one of the two originally described methods, i.e., the Curtis and Catterall (1984) and the Lazdunski procedure (Borsotto et al., 1985). Curtis and Catterall (1984) isolated T-tubular membranes from fresh (not frozen) rabbit skeletal muscle, solubilized the membranes with 1% digitonin, and employed a four-step purification procedure, which included the combined use of wheat germ agglutinin affinity chromatography and ion exchange chromatography followed by sucrose density gradient centrifugation. This procedure resulted in the enrichment of the preparation in three different polypeptides termed α (135-kDa), β (55-kDa), and γ (33-kDa) subunits. The α subunit had a characteristic behavior under sodium dodecyl sulfate polyacrylamide gel electrophoresis (SDS-PAGE); its apparent molecular weight was higher (165 kDa) under nonreducing conditions than under disulfide-reducing conditions (135 kDa) that were in the presence of dithiothreitol or β-mercaptoethanol. The Lazdunski group isolated T-tubular membranes from frozen rabbit skeletal muscle thawed at room temperature, solubilized the T-tubular membranes with 1% CHAPS, and employed a combination of chromatographic steps utilizing anion exchange, lectin affinity, and gel filtration (Borsotto et al.,

Since the submission of this paper for publication, the primary structure of both α_1 and α_2 subunits have been determined (Tanabe et al., 1987, Nature 328: 313–318; Ellis et al., 1988, Science 241:1661–1664). The biochemical properties of these polypeptides were extensively reviewed (Hofmann et al., 1987, TIPS 8:393–398; Campbell et al., 1988, TINS 11: 425–430; Catterall et al., 1988, J. Biol. Chem. 263:3535–3538; Froehner, 1988, TINS 11:90–92; Glossmann and Striessnig, 1988, ISI Atlas Sci. Pharmocol.: 202–210; Vaghy et al., 1988, TIPS 9:398–402).

PAL L. VAGHY, KUNIHISA MIWA, KIYOSHI ITAGAKI, FERENC GUBA, JR., EDWARD McKENNA, and ARNOLD SCHWARTZ • Department of Pharmacology and Cell Biophysics, University of Cincinnati College of Medicine, Cincinnati, Ohio 45267-0575.

1985). With this latter procedure only one large molecular weight polypeptide (170 kDa) was purified, which was dissociated to two smaller polypeptides, 140 and 30 kDa, upon treatment with disulfide-reducing agents (Lazdunski *et al.*, 1986). In this respect this protein is identical with the α subunit first described by Curtis and Catterall (1984). Both of these purification procedures were established by prelabeling the channel with a radioactive dihydropyridine and following the enrichment in the specific radioactivity throughout the purification. Since enrichment with the radioactivity paralleled the enrichment in the 165- or 170-kDa polypeptide, the conclusion was that the drug-binding sites were associated with this polypeptide. Direct identification methods such as photoaffinity labeling were not used to prove this conclusion. Striessing *et al.* (1986a) developed a rapid two-step purification procedure that consisted of wheat germ agglutinin affinity chromatography and sucrose density gradient centrifugation. With this method the preparation was very similar to the one obtained by Curtis and Catterall (1984). Photoaffinity labeling of the DHP receptor preparations has been done in several laboratories but the data lead to contradictory results. Both low (Campbell *et al.*, 1984; Sarmiento *et al.*, 1986) and high (Ferry *et al.*, 1984, 1985; Galizzi *et al.*, 1986; Streissnig *et al.*, 1986b) molecular weight proteins were found to be photoaffinity-labeled in partially purified Ca channel preparations and the identity of the drug-binding subunit remained a controversial issue.

We employed a rapid two-step procedure to purify Ca channels from digitonin-solubilized T-tubular membranes isolated from fresh (not frozen) rabbit skeletal muscle (Vaghy *et al.*, 1987). Solubilized membranes were equilibrated with wheat germ agglutinin–Sepharose, washed with solubilization buffer, and glycoproteins absorbed to the column were eluted with N-acetyl-D-glucosamine. The upper panel of Fig. 1 shows the elution profile of total protein and $(+)[^3H]PN200$-110 binding sites in fractions collected from the wheat germ agglutinin–Sepharose column. The peak $(+)[^3H]PN$ 200-110 binding fractions were pooled and layered on the top of a 5–20% linear sucrose density gradient. The gradient was centrifuged, fractionated from the bottom of the tube, and the protein concentration, and the $(+)[^3H]PN$ 200-110 binding was measured in each fraction. These results are shown on the lower part of Fig. 1. Two to three peak $(+)[^3H]PN$ 200-110 binding fractions were pooled and termed purified 1,4-dihydropyridine (DHP) receptor preparation.

The polypeptide composition of this preparation was analyzed by SDS-PAGE. The DHP receptor preparation was solubilized either in the presence of N-ethylmaleimide (nonreducing conditions) or dithiothreitol (disulfide-reducing conditions). Two large molecular weight proteins in the α region described by Curtis and Catterall (1984) were identified with SDS-PAGE under both conditions (Fig. 2). Under nonreducing conditions the apparent molecular weights were 160–190 kDa for the larger and 155–170 kDa for the smaller polypeptide. On the other hand, the larger was 155–170 kDa (α_1) and the smaller polypeptide was 130–150 kDa (α_2) under disulfide-reducing conditions. Three possible reasons were considered to explain this doublet. The first possibility is that the doublet is not real but represents products of a single polypeptide as has been suggested by Vandaele *et al.* (1987). The second possibility is that two distinct polypeptides exist and both have smaller apparent molecular weight under reducing conditions than under nonreducing conditions. The third possibility is that the 160–190 kDa polypeptide migrates with an apparent molecular weight of 130–150 kDa under reducing conditions and the 155–170 kDa polypeptide does not change its electrophoretic mobility upon treatment with disulfide-reducing agents.

The first apparent difference between the two high molecular weight polypeptides is their differential staining with silver. While the 160–190 kDa polypeptide present under nonreducing conditions and the 130–150 kDa polypeptide present under reducing conditions

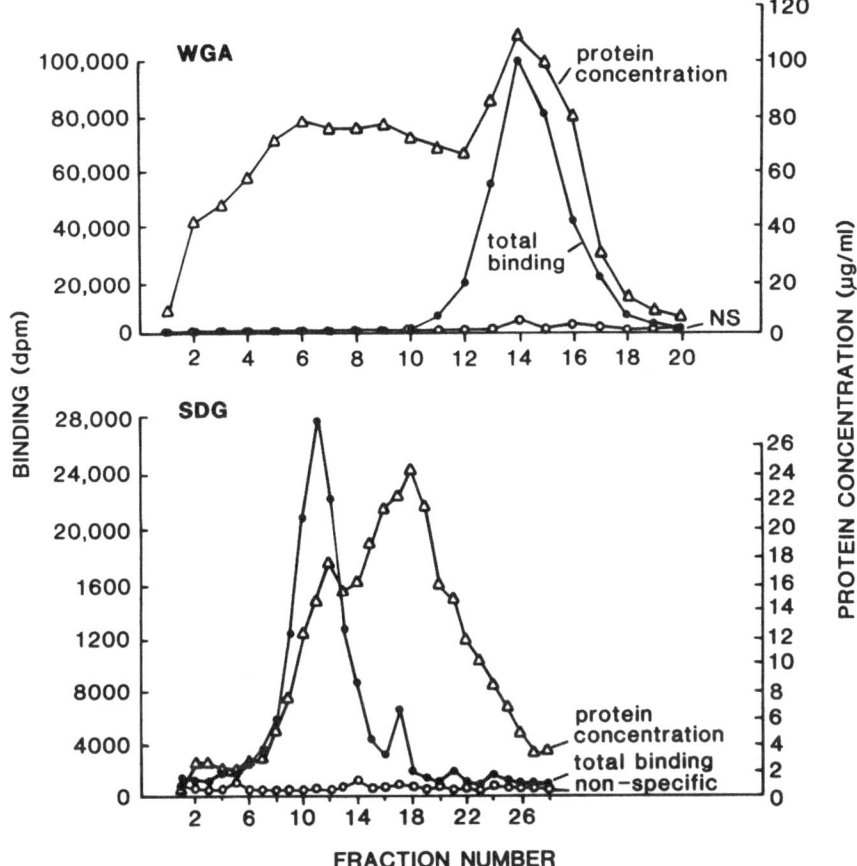

FIGURE 1. Two-step purification of 1,4-dihydropyridine receptor/calcium channel complex from skeletal muscle. Aliquots were removed from each fraction eluted from the wheat germ agglutinin (WGA) column or fractionated from the sucrose density gradient (SDG). Total (●) and non-specific (○) (+)[³H]PN200-110 binding was measured as described by Vaghy et al. (1987) using 5 nM (+)[³H]PN200-110. Protein concentration (△) was determined by the Bio-Rad microassay with bovine serum albumin as the standard.

are heavily stained with silver (dark gray), the 155–170 kDa (α_1) polypeptide present in both conditions is only weakly stained with this dye (light brown). Beside this simple visual examination of the doublet, more sophisticated techniques were also employed for determining similarities and differences between the two polypeptides. Two arylazide photoaffinity probes have recently been developed for identifying drug-binding subunit(s) of the voltage-dependent calcium channels (Fig. 3). One is a 1,4-dihydropyridine derivative, (−)[³H]azidopine (Glossmann et al., 1987), and the other is a phenylalkylamine, [N-methyl-³H]LU49888 (Striessnig et al., 1987). We used both of these ligands to identify the drug-binding subunit of the purified Ca channel. Total and nonspecific photoaffinity labeling were carried out with

PAL L. VAGHY et al.

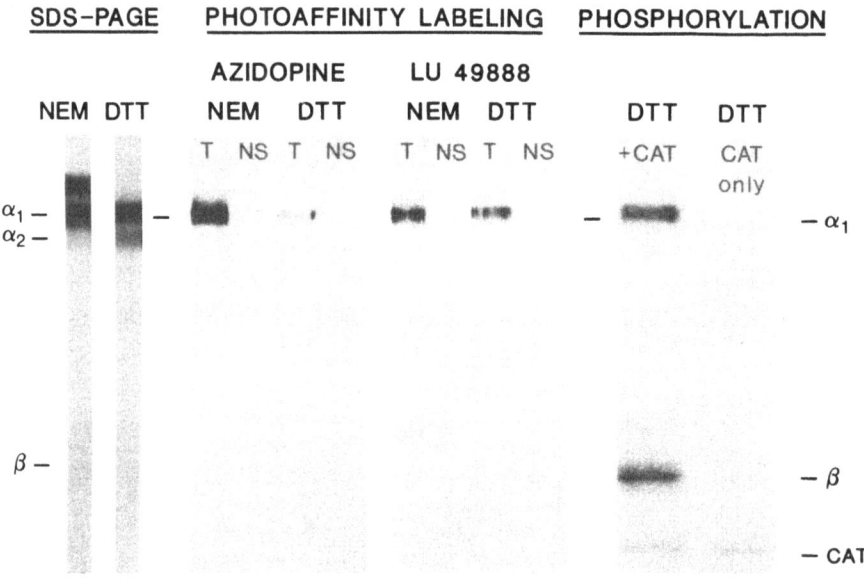

FIGURE 2. Photoaffinity labeling and phosphorylation of the purified skeletal muscle 1,4-di-hydropyridine receptor/calcium channel complex. The peak (+)[³H]PN200-110 binding fraction from the sucrose density gradient was subjected to 5–15% SDS-PAGE (Laemmli, 1970) under nonreducing (NEM) and reducing (DTT) conditions and silver stained as described by Ansorge (1985) (lanes 1 and 2). Photoaffinity labeling (lanes 3–6; fluorograms) was performed in the absence (TOT) and the presence (NS) of 1 μM (+)PN200-110 using (−)[³H]azidopine or 3 μM (−)D888 using LU49888. Purified protein was phosphorylated (lane 7: autoradiogram) with the catalytic subunit (CAT) of cAMP-dependent protein kinase. Autophosphorylation of the catalytic subunit is shown in lane 8.

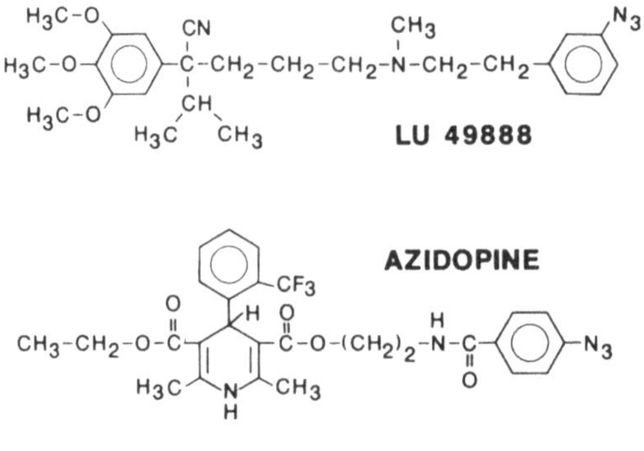

LU 49888

AZIDOPINE

FIGURE 3. Chemical structure of azi-dopine and LU49888.

both ligands and the gels were run under both nonreducing and disulfide-reducing conditions. The results are shown in Fig. 2.

Our data indicate that only one of the two polypeptides, the 155–170 kDa (α_1) polypeptide, was specifically photoaffinity-labeled by the two ligands under both nonreducing and disulfide-reducing conditions. These data, taken together with the silver staining, suggest first that the two components of the doublet are distinct polypeptides because only one of them, the 155–170 kDa (α_1) polypeptide, carries the receptors for 1,4-dihydropyridines and phenylalkylamines, and second that the 155–170 kDa (α_1) polypeptide, which is weakly stained with silver, does not change its electrophoretic mobility upon treatment with disulfide-reducing agents, and third that the 160–190 kDa polypeptide present only under nonreducing conditions is converted to a 130–150 kDa polypeptide upon treatment with disulfide-reducing agents. The 160–190/130–150 kDa protein described here seems to be identical with the α subunit (165/135 kDa) present in the Curtis and Catterall (1984) preparation, and with the 170/140 kDa protein of the Lazdunski *et al.* (1986) preparation. Reports by Curtis and Catterall (1985) and by Hosey *et al.* (1986) suggested that the 165/135 kDa protein in the case of Curtis and Catterall (1985) and the 160/140 kDa protein in the case of Hosey *et al.* (1986) were phosphorylated with the catalytic subunit of the cAMP-dependent protein kinase. Therefore, we carried out protein phosphorylation experiments using the *purified* DHP receptor/Ca channel complex and the catalytic subunit of the cAMP-dependent protein kinase. The data are shown in Fig. 2. We found that the 155–170 kDa polypeptide and *not the 160–190/130–150 kDa polypeptide is phosphorylated* by the catalytic subunit of the cAMP-dependent protein kinase. In addition, a 55-kDa polypeptide is also phosphorylated by the catalytic subunit of the cAMP-dependent protein kinase. These *in vitro* data are not sufficient to determine the physiological significance of the protein phosphorylation. However, they clearly show that a 155–170 kDa polypeptide that does not change its electrophoretic mobility upon treatment with disulfide-reducing agents, i.e., the same that is labeled with both photoaffinity probes, is phosphorylated with the catalytic subunit of the cAMP-dependent protein kinase. Most recent data by Flockerzi *et al.* (1986), Catterall *et al.* (this volume), and Hosey (this volume), appear to agree with this conclusion.

The 155–170 kDa (α_1) polypeptide, which we think is the main component of skeletal muscle voltage-dependent calcium channels, was not present in previously purified Ca channel preparations in sufficiently high concentration and therefore its existence has been overlooked. The use of fresh (not frozen) muscle and a rapid, two-step purification procedure as well as employment of highly specific photoaffinity probes together with phosphorylation of the purified protein and the use of highly sensitive silver staining (Ansorge, 1985), as well as using the appropriate type of gel, were all required for discovering this novel subunit of the skeletal muscle voltage-dependent Ca channel.

Considering the molecular weights of the subunits determined under reducing conditions, the two components of the doublet are termed α_1 and α_2 where α_1 is the larger, 155–170 kDa, drug-binding subunit that is phosphorylated with the catalytic subunit of the cAMP-dependent protein kinase. The α_2 is a 130–150 kDa polypeptide which is a component of a 160–190 kDa polypeptide present only under nonreducing conditions. It is likely that α_2 is linked to a smaller (approximately 30 kDa) polypeptide with disulfide bridge(s) (Schmid *et al.*, 1986). Assuming the existence of a 50–65 kDa β and a 30–33 kDa γ subunit (Curtis and Catterall, 1984), it is reasonable to term the smallest, 30-kDa polypeptide, which is disulfide-linked to the 130–150 kDa α_2 subunit, δ subunit.

The criteria for identification of α_1 and α_2 subunits are listed in Table I. Besides the differential silver staining, M_r, photoaffinity labeling, and phosphorylation discussed above,

TABLE I. Criteria for Identification of Two Large Molecular Weight Putative Subunits of the Skeletal Muscle Voltage-Dependent Calcium Channel

Criteria	α_1	α_2
Silver staining	Weak	Strong
Apparent M_r (SDS-PAGE[a]),		
10 mM NEM[b]	155–170 kDa	160–190 kDa ($\alpha_2 + \delta$)
10 mM DTT[c] or β-ME[d]	155–170 kDa	130–150 kDa
Photoaffinity labeling		
[³H]azidopine	Yes	No
[³H]LU49888	Yes	No
Phosphorylation	Yes	No
Glycosylation	Weak	Strong
Antigenic regions	Different from α_2	Different from α_1
Primary structure	Different from α_2	Different from α_1
Hydrophobic character[e]	Yes	No

[a] Sodium dodecyl sulfate–polyacrylamide gel electrophoresis.
[b] N-ethylmaleimide.
[c] Dithiothreitol.
[d] β-mercaptoethanol.
[e] Catterall, this volume.

our unpublished data suggest that glycoprotein staining and the use of antibodies are also appropriate for identification of α_1 and α_2 subunits. Preliminary molecular biological studies also suggest that the primary structure of the two subunits is different. Furthermore, Catterall (this volume) presented evidence that the α_1 subunit contains hydrophobic domains that may represent transmembrane regions of the Ca channel. Similar hydrophobic domains have not been found in the α_2 subunit.

SUMMARY

Our view on the subunit composition of the purified skeletal muscle voltage-dependent Ca channel has changed significantly. Evidence has been provided indicating that the disulfide-linked 140- and 30-kDa polypeptides do not bind to Ca-antagonist drugs and cannot be phosphorylated. A 155–170 kDa polypeptide that is termed α_1 subunit does not change its electrophoretic mobility upon treatment with disulfide-reducing agents, contains both 1,4-dihydropyridine- and phenylalkylamine-binding sites, can be phosphorylated, is weakly glycosylated, and contains hydrophobic domains. It is suggested that the α_1 subunit is the major component of the voltage-dependent Ca channel. The structural and functional role of the α_2, β, γ, and δ subunits remains to be determined.

ACKNOWLEDGMENTS. The excellent technical assistance of Ms. Susan Koelliker and Mr. Ross Agnor is greatly appreciated. We thank Drs. Hartmut Glossmann and Jorg Striessnig for the ($-$)[³H]azidopine and [N-methyl-³H]LU 49888. Mr. Kiyoshi Itagaki is a visiting scientist from Otsuka Pharmaceutical Co., Ltd., Japan and Dr. Ferenc Guba, Jr., is a visiting scientist from Medical University of Szeged, Hungary. Supported by NIH grant PO1 HL-22619.

REFERENCES

Ansorge, W., 1985, Fast and sensitive detection of protein and DNA bands by treatment with potassium permanganate, *J. Biochem. Biophys. Methods* **11**:13–20.

Borsotto, M., Barhanin, J., Fosset, M., and Lazdunski, M., 1985, The 1,4-dihydropyridine receptor associated with the skeletal muscle voltage-dependent Ca^{2+} channel, *J. Biol. Chem.* **260**:14255–14263.

Campbell, K. P., Lipshutz, G. M., and Denney, G. H., 1984, Direct photoaffinity labeling of the high affinity nitrendipine-binding site in subcellular membrane fractions isolated from canine myocardium, *J. Biol. Chem.* **259**:5384–5387.

Catterall, W.A., Seagar, M. J., Musami, T., and Reber, B. F. X., 1987, Molecular properties of voltage-sensitive calcium channels, in this volume.

Cooper, C. L., Vandaele, S., Barhanin, J., Fosset, M., Lazdunski, M., and Hosey, M. M., 1987a, Purification and characterization of the dihydropyridine-sensitive voltage-dependent calcium channel from cardiac tissue, *J. Biol. Chem.* **262**:509–512.

Cooper, C. L., O'Callahan, C. M., and Hosey, M. M., 1987b, Phosphorylation of dihydropyridine-sensitive calcium channels from cardiac and skeletal muscle, in this volume.

Curtis, B. M., and Catterall, W. A., 1984, Purification of the calcium antagonist receptor of the voltage-sensitive calcium channel from skeletal muscle transverse tubules, *Biochemistry* **23**:2113–2118.

Curtis, B. M., and Catterall, W. A., 1985, Phosphorylation of the calcium antagonist receptor of the voltage-sensitive calcium channel by cAMP-dependent protein kinase, *Proc. Natl. Acad. Sci. USA* **82**:2528–2532.

Ferry, D. R., Rombusch, M., Goll, A., and Glossmann, H., 1984, Photoaffinity labeling of Ca^{2+} channels with [³H]azidopine, *FEBS Lett.* **169**:112–118.

Ferry, D. R., Kampf, K., Goll, A., and Glossmann, H., 1985, Subunit composition of skeletal muscle transverse tubule calcium channels evaluated with the 1,4-dihydropyridine affinity probe, [³H]azidopine, *EMBO J.* **4**:1933–1940.

Flockerzi, V., Oeken, H-J., Hofmann, F., Pelzer, D., Cavalie, A., and Trautwein, W., 1986, Purified dihydropyridine-binding site from skeletal muscle t-tubules is a functional calcium channel. *Nature* **323**:66–68.

Galizzi, J-P., Borsotto, M., Barhanin, J., Fosset, M., and Lazdunski, M., 1986, Characterization and photoaffinity labeling of receptor sites for the Ca^{2+} channel inhibitors d-cis-diltiazaem, ($\pm$) bepridil, desmethoky verapamil, and (+)PN200-110 in skeletal muscle transverse tubule membranes, *J. Biol. Chem.* **261**:1393–1397.

Glossmann, H., Ferry, D. R., Striessnig, J., Goll, A., and Moosburger, K., 1987, Resolving the structure of the Ca^{2+} channel by photoaffinity labeling. *Trends Pharmacol. Sci.* **8**:95–100.

Hosey, M. M., Borsotto, M., and Lazdunski, M., 1986, Phosphorylation and dephosphorylation of dihydropyridine-sensitive voltage-dependent Ca^{2+} channel in skeletal muscle membranes by cAMP- and Ca^{2+}-dependent processes, *Proc. Natl. Acad. Sci. USA* **83**:3733–3737.

Laemmli, U. K., 1970, Cleavage of structural proteins during the assembly of the head of bacteriophage T4, *Nature* **227**:680–685.

Lazdunski, M., Barhanin, J., Borsotto, M., Fosset, M., Galizzi, J-P., Renaud, J. F., Romey, G., and Schmid, A., 1986, Dihydropyridine-sensitive Ca^{2+} channels: Molecular properties of interaction with Ca^{2+} channel blockers: Purification, subunit-structure, and differentiation, *J. Cardiovasc. Pharmacol.* **8**(Suppl. 8):513–519.

Nakayama, N., Kirley, T. L., Vaghy, P. L., McKenna, E., and Schwartz, A., 1987, Purification of a putative Ca^{2+} channel protein from rabbit skeletal muscle: Determination of the amino-terminal sequence, *J. Biol. Chem.* **262**:6572–6576.

Sarmiento, J. G., Epstein, P. M., Rowe, W. A., Chester, D. W., Smilowitz, H., Wehinger, E., and Janis, R. A., 1986, Photoaffinity labeling of a 33–35,000 dalton protein in cardiac, skeletal and smooth muscle membrane using a new ^{125}I-labeled 1,4-dihydropyridine calcium channel antagonist, *Life Sci.* **39**:2401–2409.

Schmid, A., Barhanin, J., Coppola, T., Borsotto, M., and Lazdunski, M., 1986, Immunochemical analysis of subunit structures of 1,4-dihydropyridine receptors associated with voltage-dependent Ca^{2+} channels in skeletal, cardiac and smooth muscles, *Biochemistry* **25**:3492–3495.

Striessnig, J., Goll, A., Moosburger, K., and Glossmann, H., 1986a, Purified calcium channel have three allosterically coupled drug receptors, *FEBS Lett.* **197**:204–210.

Striessnig, J., Moosburger, K., Goll, A., Ferry, D. R., and Glossmann, H., 1986b, Stereoselective photoaffinity labelling of the purified 1,4-dihydropyridine receptor of the voltage-dependent calcium channel, *Eur. J. Biochem.* **161**:603–609.

Striessnig, J., Knaus, H-G., Grabner, M., Moosburger, K., Seitz, W., Leitz, H., and Glossmann, H., 1987, Photoaffinity labelling of the phenylalkylamine receptor of the skeletal muscle transverse-tubule calcium channel, *FEBS Lett.* **212**(2):247–253.

Vaghy, P. L., Williams, J. S., and Schwartz, A., 1987a, Receptor pharmacology of calcium entry blocking agents, *Am. J. Cardiol.* **59**:9A–17A.

Vaghy, P. L., Striessnig, J., Miwa, K., Knaus, H.-G., Itagaki, K., McKenna, E., Glossmann, H., and Schwartz, A., 1987b, Identification of a novel 1,4-dihydropyridine- and phenylalkylamine-binding polypeptide in calcium channel preparations, *J. Biol. Chem.* **262**:14337–14342.

Vandaele, S., Fosset, M., Galizzi, J-P., and Lazdunski, M., 1987, Monoclonal antibodies that coimmunoprecipitate the 1,4-dihydropyridine and phenylalkylamine receptors and reveal the Ca^{2+} channel structure, *Biochemistry* **26**:5–9.

8

Phosphorylation of Dihydropyridine-Sensitive Calcium Channels from Cardiac and Skeletal Muscle

CHRISTY L. COOPER, CLIFF M. O'CALLAHAN, and
M. MARLENE HOSEY

1. INTRODUCTION

The entry of Ca into many excitable cells occurs through voltage-dependent Ca channels that are activated in response to depolarization. An interesting aspect of some Ca channels is that in addition to their voltage dependence, they can be regulated by neurotransmitters. This has been most thoroughly characterized in electrophysiological studies of cardiac cells (Reuter, 1983; Tsien *et al.*, 1986). The probability of opening of the predominant type of cardiac Ca channel, that which is sensitive to dihydropyridines (Nilius *et al.*, 1985), is increased by norepinephrine and decreased by acetylcholine (Reuter, 1983; Heschler *et al.*, 1986; Tsien *et al.*, 1986). This regulation is believed to be mediated by cyclic AMP (cAMP) and to involve a cAMP-dependent phosphorylation of the channel itself or an associated regulatory protein (Reuter, 1983). A similar type of regulation by cAMP has also been observed with Ca channels in skeletal muscle and some neuronal preparations (Schmid *et al.*, 1985; Chad and Eckert, 1986). The regulation of Ca channels by mechanisms implicating protein kinases other than cAMP-dependent protein kinase also has been reported (Di Virgilio *et al.*, 1986; Paupardin-Tritsch *et al.*, 1986; Rane and Dunlap, 1986; Galizzi *et al.*, 1987; Strong *et al.*, 1987). A goal of our laboratory is to elucidate the biochemical events involved in the regulation of Ca channels by phosphorylation-dependent processes. To this end we initiated studies with dihydropyridine-sensitive Ca channels from cardiac and skeletal muscle to determine if any of the peptide components of these channels serve as a substrate for cAMP-dependent, as well as Ca-dependent, protein kinases.

CHRISTY L. COOPER, CLIFF M. O'CALLAHAN, and M. MARLENE HOSEY • Department of Biological Chemistry and Structure, University of Health Sciences/The Chicago Medical School, North Chicago, Illinois 60064.

2. MATERIALS AND METHODS

Dihydropyridine-sensitive Ca channels were purified from rabbit skeletal and chick cardiac muscle as previously described (Hosey *et al.*, 1986; Cooper *et al.*, 1987). Phosphorylation of Ca channels was performed using either the purified Ca channel preparations or transverse tubule membranes from rabbit skeletal muscle as substrate. The phosphorylation reactions and analyses of phosphopeptide products were performed as described by Hosey *et al.* (1986). The catalytic subunit of cAMP-dependent protein kinase was purified from bovine heart according to the method of Sugden *et al.* (1976). The Ca/calmodulin-dependent protein kinase was purified from rat brain according to the method of Klee (C. Klee, personal communication). Sodium dodecyl sulfate (NaDodSO$_4$) gel electrophoresis was performed using a gradient of 5–15% polyacrylamide according to Laemmli (1977). Phosphopeptide mapping was performed according to the method of Hunter and Sefton (1980).

3. RESULTS

3.1. Phosphorylation of Skeletal Muscle Ca Channels

Preparations of dihydropyridine-sensitive Ca channels were partially purified from transverse tubule membranes isolated from rabbit skeletal muscle. These membranes contained 50–100 pmol of dihydropyridine-binding sites/mg protein as assessed with the ligand [^{3}H]PN 200-110. Ca channels were purified using this ligand to follow the purification. The purified material was enriched in two large peptides. One migrated on NaDodSO$_4$ gels with an apparent M_r of 170,000 when electrophoresed under nonreducing conditions, or as a peptide of 140,000 when electrophoresed under reducing conditions (Fig. 1). This peptide had properties similar to that found in highly purified preparations obtained from rabbit muscle by others (Borsotto *et al.*, 1984, 1985; Curtis and Catterall, 1984). Previous studies (Schmid *et al.*, 1986; Cooper *et al.*, 1987) showed that the conversion of the 170-kDa peptide to the 140-kDa peptide is accompanied by the appearance of two small peptides of 30–32 kDa that are not visible in Fig. 1. The second peptide had a M_r of 160 kDa under reducing and nonreducing conditions (Fig. 1). When the purified Ca channel preparation from skeletal muscle was phosphorylated with the catalytic subunit of cAMP-dependent protein kinase and [γ-^{32}P]ATP and analyzed by electrophoresis under reducing and nonreducing conditions, the major phosphoprotein observed was a peptide with an apparent M_r of 160,000 (Fig. 1). The electrophoretic mobility of this phosphopeptide was only slightly affected by reducing versus nonreducing conditions. No evidence for ^{32}P incorporation into the 170/140-kDa peptide was observed. The 160-kDa phosphopeptide corresponded to a stained peptide that was present in a somewhat smaller amount than the 170/140-kDa peptide. That the 160-kDa peptide is a component of dihydropyridine-sensitive Ca channels was suggested by recent studies with a photoaffinity-labeling reagent of the phenylalkylamine series of Ca channel inhibitors (Striessnig *et al.*, 1987). These studies showed that a peptide of 155 kDa whose M_r was little affected by reducing conditions was specifically labeled, whereas the 170/140-kDa peptide did not appear to be significantly labeled (Striessnig *et al.*, 1987).

3.2. Phosphorylation of Cardiac Ca Channels

Cardiac membranes contain significantly fewer binding sites for dihydropyridines than skeletal muscle. Consequently, it has been more difficult to purify the cardiac Ca channel.

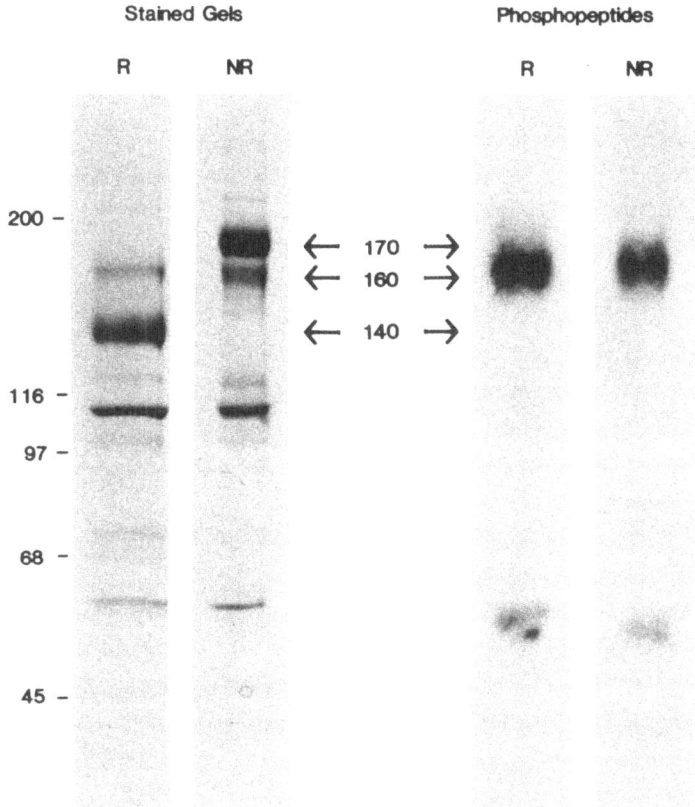

FIGURE 1. Calcium channel preparations purified from skeletal muscle. Calcium channels were purified as dihydropyridine receptors using the radioligand [³H]PN 200-110 to follow the purification. Transverse tubule membranes isolated from rabbit skeletal muscle were solubilized with the detergent CHAPS and the solubilized material was subjected to a one-step purification using wheat germ agglutinin–Sepharose (Hosey et al., 1986). The purified material was phosphorylated with the catalytic subunit of cAMP-dependent protein kinase and subjected to electrophoresis under reducing (R) and nonreducing (NR) conditions on 5–15% NaDodSO₄ gels. The left part of the figure shows the Coomassie blue-stained polypeptides present in the partially purified preparation, while the right part of the figure is an autoradiogram of the gel that shows the phosphorylated peptides.

However, we have recently achieved a 900-fold purification of the cardiac protein from chick heart (Cooper et al., 1987). The preparation purified from cardiac muscle, like that purified from skeletal muscle, was enriched in a peptide that migrated as a species of 170 kDa when electrophoresed on NaDodSO₄ gels under nonreducing conditions, or as a peptide of 140 kDa when electrophoresed under reducing conditions (Fig. 2). When the purified cardiac Ca channel preparation was phosphorylated with the catalytic subunit of cAMP-dependent protein kinase and [γ-³²P]ATP, ³²P incorporation into a peptide of 160 kDa was observed (Fig. 2). The mobility of the phosphopeptide was only slightly affected by reducing versus nonreducing

Stained Gels   Phosphopeptides

R   NR   R   NR

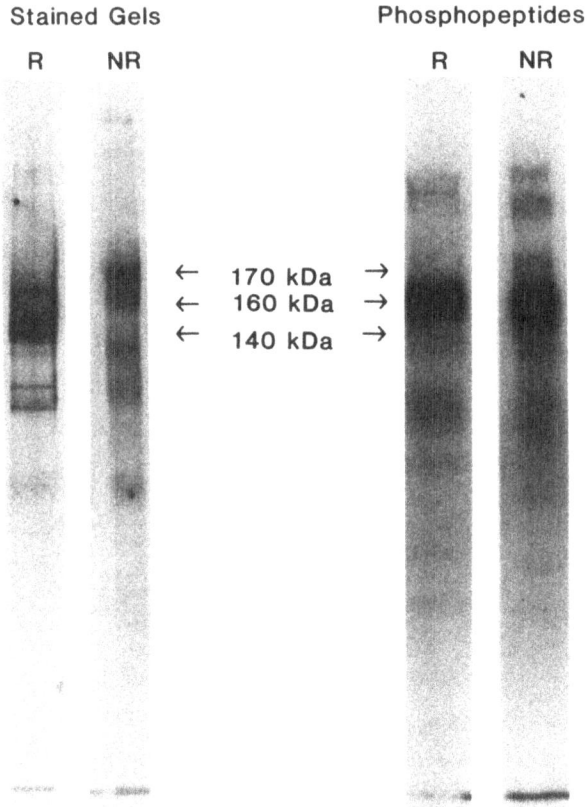

← 170 kDa →
← 160 kDa →
← 140 kDa →

FIGURE 2. Calcium channel preparations purified from chick cardiac muscle. Chick cardiac membranes were prelabeled with [³H]PN 200-110 and solubilized with digitonin. Ca channels were purified as described by Cooper *et al.* (1987), and the purified preparation was phosphorylated with cAMP-dependent protein kinase. The purified fraction was subjected to Na-DodSO₄ gel electrophoresis on a 5–15% linear gradient gel under reducing (R) and nonreducing (NR) conditions. The left part of the figure shows the peptide composition of the purified fraction as revealed by silver staining, while the right part of the figure shows an autoradiogram depicting the phosphopeptides formed by the cAMP-dependent protein kinase.

conditions. As was the case with the skeletal muscle Ca channel, the 160-kDa phosphopeptide observed in the cardiac preparations corresponded to a stained peptide of 160 kDa that was present in smaller quantities than the 170/140-kDa peptide. In no instance did we observe phosphorylation of the cardiac 170/140 peptide. However, the 160 kDa cardiac phosphopeptide was not a component of cardiac Ca channels, as it was separated from the channels by a sucrose density gradient step (Chang and Hosey, unpublished observations).

3.3. Multiple Phosphorylation Sites in the 160-kDa Peptide

In a previous study we demonstrated that a multifunctional Ca/calmodulin-dependent protein kinase (McGuiness *et al.*, 1983) also phosphorylated the 160-kDa peptide in skeletal muscle membranes (Hosey et al., 1986). [In that study different types of NaDodSO$_4$ gels were used in which the electrophoretic behavior of the 160-kDa peptide was similar to, and could not be easily distinguished from, that of the 170/140-kDa peptide. In contrast, in the present studies the 160-kDa peptide was clearly resolved from the 170-kDa peptide (Figs. 1 and 2).] In order to determine if there are multiple sites for phosphorylation in the 160-kDa peptide, we performed experiments with optimal concentrations of cAMP-dependent protein kinase and the multifunctional Ca/calmodulin-dependent protein kinase, alone and together. These experiments were performed by phosphorylating the peptide in its native state in transverse tubule membranes, and then purifying Ca channels in order to assess the state of phosphorylation of the 160-kDa peptide. The results of these experiments showed that the phosphorylation of the 160-kDa peptide by the two kinases was more than that achieved by either kinase alone (Fig. 3). Direct counting of the gel pieces containing the 160-kDa peptide showed that the level of phosphorylation achieved by both kinases together was equivalent to 90% of that expected if the phosphorylation by each kinase was strictly additive (these values were calculated by subtracting the amount of ^{32}P incorporated in the absence of exogenous protein kinases from that incorporated in the presence of each kinase alone and together). This result suggested that there may be distinct sites phosphorylated by the two protein kinases. This finding was directly demonstrated by analyzing the phosphorylated peptide by two-dimensional peptide mapping (Fig. 4). The results of this analysis showed that there were several phosphopeptides that were phosphorylated by both kinases; however, there was one distinctive phosphopeptide that was only observed in the presence of the cAMP-dependent protein kinase (Fig. 4A, indicated by arrow). In addition, there were other phosphopeptides that appeared to be more heavily phosphorylated by the Ca/calmodulin-dependent protein kinase than by the cAMP-dependent protein kinase (Fig. 4B, upper left and lower left). The phosphopeptide map obtained for the peptide phosphorylated by the combination of the two kinases confirmed the conclusion that the phosphorylation observed under these conditions was additive. These results demonstrated that the 160-kDa peptide can be multiply phosphorylated in vitro by cAMP-dependent and Ca/calmodulin-dependent protein kinases.

4. DISCUSSION

The results show that preparations of dihydropyridine-sensitive Ca channels purified from rabbit skeletal muscle contain two distinct peptides of similar molecular weight. One peptide migrates on NaDodSO$_4$ gels with an apparent M_r of 170,000 but is converted by reduction with sulfhydryl reagents to a peptide of M_r 140,000. The other peptide has an apparent M_r of 160,000, and its electrophoretic mobility is less affected by reducing versus non-reducing conditions. The 160 kDa peptide corresponds to the receptor for dihydropyridines and phenylalkylamines. The 160-kDa peptide isolated from either cardiac or skeletal muscle, but not the 170-kDa peptide, was phosphorylated in vitro by cAMP-dependent protein kinase. These results suggest that the cAMP-dependent regulation of the probability of Ca channel opening may occur through the phosphorylation of the peptide bearing the receptors for Ca-channel effectors. This may be true in both cardiac and skeletal muscle cells, since it has been clearly demonstrated in both cell types that agents that elevate cAMP in either cell cause increased Ca flux through the dihydropyridine-sensitive Ca channels (Reuter, 1983; Schmid et al., 1985; Tsien et al., 1986). In additional studies we found that the 160-kDa peptide could be multiply phosphorylated in skeletal muscle transverse tubule membranes by cAMP-dependent and Ca/calmodulin-dependent protein kinases. A possible role for the

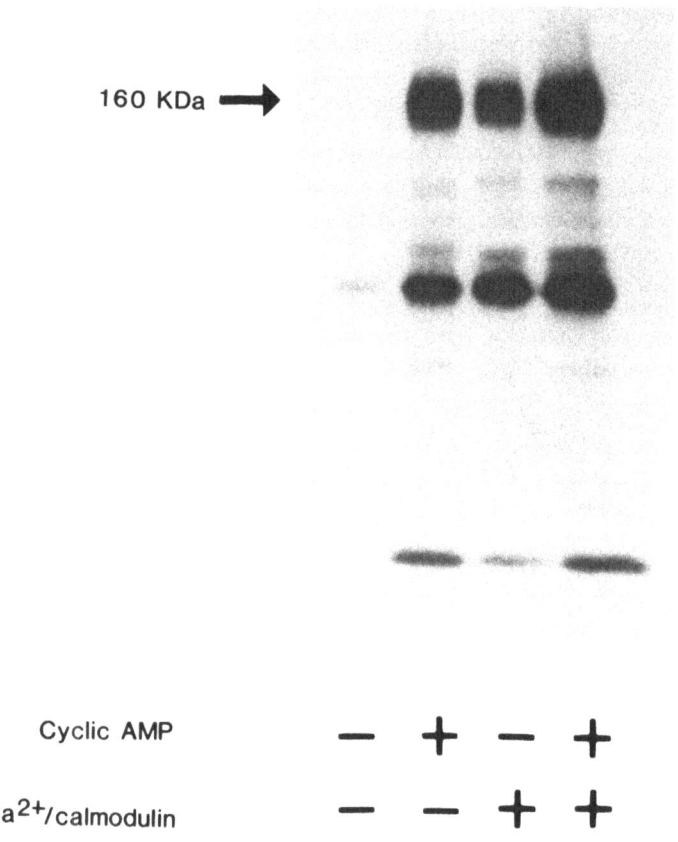

160 KDa ➡

Cyclic AMP   — + — +

Ca^{2+}/calmodulin   — — + +

FIGURE 3. Phosphorylation of the 160-kDa peptide in skeletal muscle transverse tubule membranes by cAMP-dependent- and Ca/calmodulin-dependent protein kinases. Transverse tubule membranes were phosphorylated with optimal concentrations of the purified catalytic subunit of cAMP-dependent protein kinase and/or with partially purified preparations of a Ca/calmodulin-dependent protein kinase. Ca channels were purified from the phosphorylated membranes in the presence of phosphatase inhibitors as described (Hosey *et al.*, 1986). In order to quantitate the phosphorylation, the bands corresponding to the 160-kDa peptide were excised and ^{32}P-quantified by liquid scintillation counting. The amounts of ^{32}P incorporated into the 160-kDa peptide under the different conditions were 70 cpm (lane 1, no addition); 414 cpm (lane 2, cAMP-dependent protein kinase alone); 250 cpm (lane 3, Ca/calmodulin-dependent protein kinase alone); 540 cpm (lane 4, both kinases together).

phosphorylation of the 160-kDa peptide by the Ca/calmodulin-dependent protein kinase may be in the regulation of Ca channels by Ca-dependent events; however, further evidence is necessary to establish whether or not this is so. Future studies will be directed at elucidating the mechanism whereby phosphorylation of the 160-kDa peptide results in regulation of Ca channel activity, as well as establishing the relationship between the 170/140- and the 160-kDa peptides.

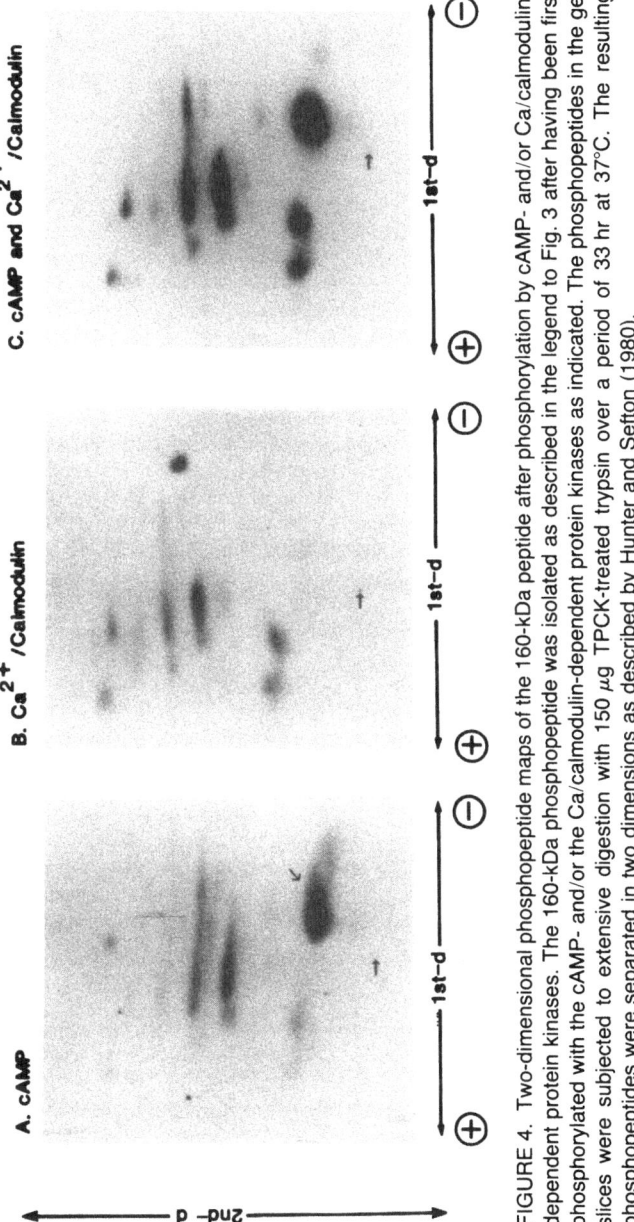

FIGURE 4. Two-dimensional phosphopeptide maps of the 160-kDa peptide after phosphorylation by cAMP- and/or Ca/calmodulin-dependent protein kinases. The 160-kDa phosphopeptide was isolated as described in the legend to Fig. 3 after having been first phosphorylated with the cAMP- and/or the Ca/calmodulin-dependent protein kinases as indicated. The phosphopeptides in the gel slices were subjected to extensive digestion with 150 µg TPCK-treated trypsin over a period of 33 hr at 37°C. The resulting phosphopeptides were separated in two dimensions as described by Hunter and Sefton (1980).

ACKNOWLEDGMENTS. This work was supported by NIH grant HL23306 and a grant-in-aid from the Chicago Heart Association. C.L.C. was a Junior Fellow of the Chicago Heart Association. M.M.H. was an Established Investigator of the American Heart Association. We thank Mrs. Cheryll Johnson for assistance in the preparation of the manuscript.

REFERENCES

Borsotto, M., Barhanin, J., Norman, R. I., and Lazdunski, M., 1984, Purification of the dihydropyridine receptor of the voltage-sensitive calcium channel from skeletal muscle transverse tubules using (+)[^{3}H]PN 200–110, Biochem. Biophys. Res. Commun. 122:1357–1366.

Borsotto, M., Barhanin, J., Fosset, M., and Lazdunski, M., 1985, The 1,4-dihydropyridine receptor associated with skeletal muscle voltage-dependent Ca channels, J. Biol. Chem. 260:14255–14263.

Chad, J. E., and Eckert, R., 1986, An enzymatic mechanism for calcium current inactivation in dialysed Helix neurones, J. Physiol. 378:31–51.

Cooper, C. L., Vandaele, S., Barhanin, J., Fosset, M., Lazdunski, M., and Hosey, M. M., 1987, Purification and characterization of the dihydropyridine-sensitive voltage-dependent calcium channel from cardiac tissue, J. Biol. Chem. 262:509–512.

Curtis, B. M., and Catterall, W. A., 1984, Purification of the calcium antagonist receptor of voltage-sensitive calcium channels from skeletal muscle transverse tubules, Biochemistry 23:2113–2118.

Di Virgilio, F., Pozzan, T., Wolheim, C. B., Vicentini, L. M., and Meldolesi, J., 1986, Tumor promoter phorbol myristate acetate inhibits Ca^{2+} influx through voltage-gated Ca^{2+} channels in two secretory cell lines, PC12 and RINm5F, J. Biol. Chem. 2611:32–35.

Galizzi, J.-P., Qar, J., and Fosset, M., VanRenterghem, C., and Lazdunski, M., 1987, Regulation of calcium channels in aortic smooth muscle cells by protein kinase C activators (diacylglycerol and phorbol esters) and by peptides (vasopressin and bombesin) that stimulate phosphoinositide breakdown, J. Biol. Chem. 262:6947–6950.

Heschler, J., Kameyama, M., and Trautwein, W., 1986, On the mechanism of muscarinic inhibition of the cardiac Ca current, Pflugers Arch. 407:182–189.

Hosey, M. M., Borsotto, M., and Lazdunski, M., 1986, Phosphorylation and dephosphorylation of the dihydropyridine-sensitive voltage-dependent Ca channel in skeletal muscle membranes by cAMP- and Ca-dependent processes, Proc. Natl. Acad. Sci. USA 83:3733–3737.

Hunter, T., and Sefton, B. M., 1980, Transforming gene product of Rous sarcoma virus phosphorylates tyrosine, Proc. Natl. Acad. Sci. USA 77:1311–1315.

Laemmli, U. K., 1970, Cleavage of structural proteins during the assembly of the head of the bacteriophage T4, Nature 227:680–685.

McGuiness, T. L., Lai, Y., Greengard, P., Woodgett, J. R., and Cohen, P., 1983, A multifunctional calmodulin-dependent protein kinase, FEBS Lett. 163:329–334.

Nilius, B., Hess, P., Lansman, J. B., and Tsien, R. W., 1985, A novel type of cardiac calcium channel in ventricular cells, Nature 316:443–446.

Paupardin-Tritsch, D., Hammond, C., Gerschenfeld, H. M., Nairn, A. C., and Greengard, P., 1986, cGMP-dependent protein kinase enhances Ca current and potentiates the serotonin-induced Ca current increase in snail neurones, Nature 323:812–814.

Rane, S. G., and Dunlap, K., 1986, Kinase C activator 1,2-oleoylacetylglycerol attenuates voltage-dependent calcium current in sensory neurons, Proc. Natl. Acad. Sci. USA 83:184–188.

Reuter, H., 1983, Calcium channel modulation by neurotransmitters, enzymes and drugs, Nature 301: 569–574.

Schmid, A., Renaud, J.-F., and Lazdunski, M., 1985, Short term and long term effects of β-adrenergic effectors and cyclic AMP on nitrendipine-sensitive voltage-dependent Ca^{2+} channels of skeletal muscle, J. Biol. Chem. 260:13041–13046.

Schmid, A., Barhanin, J., Coppola, T., Borsotto, M., and Lazdunski, M., 1986, Immunochemical analysis of subunit structures of 1,4-dihydropyridine receptors associated with voltage-dependent Ca^{2+} channels in skeletal, cardiac, and smooth muscle, *Biochemistry* **25:**3492–3495.

Streissnig, J., Knaus, H.-G., Grabner, M., Moosburger, K., Seitz, W., Leitz, H., and Glossman, H., 1987, Photoaffinity labeling of the phenylalkylamine receptor of the skeletal muscle transverse-tubule calcium channel, *FEBS Lett.* **212:**247–253.

Strong, J. A., Fox, A. P., Tsien, R. W., and Kaczmarek, L. K., 1987, Stimulation of protein kinase C recruits covert calcium channels in *Aplysia* bag cell neurons, *Nature* **325:**714–717.

Sugden, P. H., Holladay, L. A., Reimann, E. M., and Corbin, J. D., 1976, Purification and characterization of the catalytic subunit of adenosine, 3′:5′-cyclic monophosphate-dependent protein kinase from bovine liver, *Biochem. J.* **159:**409–422.

Tsien, R. W., Bean, B. P., Hess, P., Lansman, J. B., Nilius, B., and Nowycky, M. C., 1986, Mechanisms of calcium channel modulation by beta-adrenergic agents and dihydropyridine calcium agonists, *J. Mol. Cell. Cardiol.* **18:**691–710.

9

Phorbol Esters Stimulate Calcium Influx via Voltage-Dependent Channels in A$_7$r$_5$ Vascular Smooth-Muscle Cells

WILSON S. COLUCCI and GIOVANNI SPERTI

1. INTRODUCTION

Phorbol esters have been shown to cause a slowly developing and sustained contraction of vascular smooth muscle from a variety of species (Rasmussen *et al.*, 1984; Danthuluri and Deth, 1984; Forder *et al.*, 1985; Gleason and Flaim, 1986; Sybertz *et al.*, 1986; Chatterjee and Tejada, 1986). This action of phorbol esters is presumed to be due to stimulation of protein kinase C, a Ca^{2+}-phospholipid-dependent protein kinase (Nishizuka, 1984; Castagna *et al.*, 1982). Phorbol ester-stimulated vascular smooth muscle contraction in many systems is dependent on extracellular Ca^{2+} (Rasmussen *et al.*, 1984; Danthuluri and Deth, 1984; Forder *et al.*, 1985; Gleason and Flaim, 1986). In rabbit aorta, phorbol-stimulated contraction is inhibited by organic Ca^{2+} channel antagonists, and is facilitated by the Ca^{2+} channel agonist BAY K 8644 or by an increase in extracellular Ca^{2+} concentration (Forder *et al.*, 1985).

Recently, electrophysiological studies in *Aplysia* neurons and *Hermissenda* photoreceptor cells showed that phorbol esters can stimulate inward Ca^{2+} conductance through voltage-dependent Ca^{2+} channels (DeRiemer *et al.*, 1985; Farley and Auerbach, 1985). In view of these observations, we considered the possibility that phorbol ester-stimulated contraction of vascular smooth muscle might involve an increase in Ca^{2+} influx by way of voltage-dependent channels. As an initial approach to this hypothesis, we evaluated the effect of the phorbol ester 12-O-tetradecanoylphorbol-13-acetate (TPA) on $^{45}Ca^{2+}$ uptake in the A$_7$r$_5$ cell line derived from vascular smooth muscle of fetal rat aorta (Kimes and Brandt, 1976; Ruegg *et al.*, 1985).

WILSON S. COLUCCI and GIOVANNI SPERTI ● Cardiovascular Division, Department of Medicine, Brigham and Women's Hospital and Harvard Medical School, Boston, Massachusetts 02115.

2. MATERIALS AND METHODS

2.1. Cell Culture

A$_7$r$_5$ cells obtained from the American Type Culture Collection, Bethesda, MD were plated in 35-mm plastic culture dishes at an initial density of 5000–9000 cells/cm^2 and grown in Dulbecco's modified Eagle medium containing antibiotics and 10% fetal calf serum in a humidified atmosphere at 37°C under 5% CO_2/95% air. Experiments were conducted 5–8 days later, when cells had reached confluence and numbered approximately 35,000 cells/cm^2.

2.2. Ca^{2+} Uptake

^{45}Ca^{2+} uptake experiments were performed in balanced salt solution (BSS) consisting of 145 mM NaCl, 5 mM KCl, 1 mM $MgCl_2$, 1 mM $CaCl_2$, 10 mM glucose, and 5 mM Hepes at pH 7.40. Depolarizing solution was of the same composition except that NaCl was 95 mM and KCl 55 mM. The experiment was initiated by washing the cells three times with BSS at 37°C, and adding 1 ml of BSS or depolarizing solution that contained approximately 1.5 μCi ^{45}Ca^{2+}. ^{45}Ca^{2+} uptake was terminated by washing each plate five times with 2 ml of ice-cold BSS that was modified to contain 10 mM $LaCl_3$ and no $CaCl_2$. Cells remained an additional 10 min in the last wash (4°C), and subsequently the solution was aspirated and the cells were dried at room temperature for 20 min. ^{45}Ca^{2+} was determined by dissolving the cells in 0.1 N HNO_3 for 20 min at 4°C and transferring the solution to vials to which 13 ml of scintillation cocktail were added.

^{45}Ca^{2+} was obtained from ICN (Irvine, CA). Nitrendipine, BAY K 8644, and verapamil were generously supplied by Miles Pharmaceuticals, Bayer EG, and Searle Pharmaceuticals, respectively. All other drugs were obtained from Sigma.

3. RESULTS

3.1. Effect of TPA on Initial Ca^{2+} Uptake Rate

^{45}Ca^{2+} uptake was linear for at least 5 min ($r \geq 0.98$) and therefore reflected primarily unidirectional Ca^{2+} influx. Under basal conditions, the ^{45}Ca^{2+} influx rate was 0.006 ± 0.037 nmol/10^6 cells/min ($n = 3$; Fig. 1A). In preliminary experiments it was observed that the effect of TPA on ^{45}Ca^{2+} uptake required 2–5 min. Therefore, in order to evaluate the initial ^{45}Ca^{2+} uptake rate in response to TPA, cells were preincubated with 10 nM TPA for 30 min prior to the initiation of the ^{45}Ca^{2+} uptake experiment. Under these conditions, TPA caused a marked increase in unidirectional Ca^{2+} influx to 0.335 ± 0.010 nmol/10^6 cells/min ($p < 0.025$ versus basal; $n = 3$; Fig. 1A).

3.2. Effects of KCl and TPA on ^{45}Ca^{2+} Uptake

In uptake experiments up to 60 min, the effect of TPA was similar to that of 55 mM KCl (Fig. 1B). Net ^{45}Ca^{2+} uptake was maximal at 30 min, the time point used in subsequent experiments. By 24 hr, ^{45}Ca^{2+} content in TPA-stimulated cells was similar to that in control

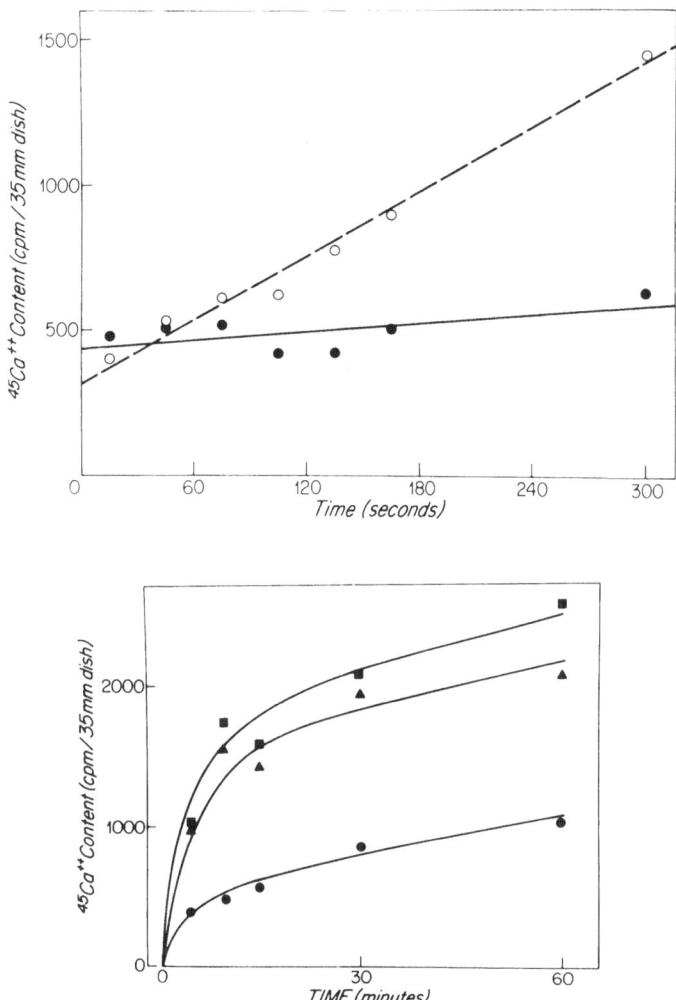

FIGURE 1. (A) Effect of TPA on initial ^{45}Ca^{2+} influx rate. Under basal conditions, the ^{45}Ca^{2+} influx rate in the depicted experiment was 0.031 nmol/10^6 cells/min, and increased to 0.240 nmol/10^6 cells/min following a 30-min preincubation with 10 nM TPA (● = basal; ○ = 10 mM TPA). (B) Time course of the effects of 55 mM KCl (▲) and 10 nM TPA (■) on net ^{45}Ca^{2+} uptake (● = basal uptake). (Reproduced with permission from Sperti and Colucci, 1987b.)

cells, indicating that the effect of TPA was to increase ^{45}Ca^{2+} influx rate rather than total cellular Ca^{2+} content. The effect of TPA was maximal by 10 nM, with a half-maximal effect between 1 and 3 nM.

The effects of maximal KCl (55 mM) and TPA (10 nM) stimulation of ^{45}Ca^{2+} uptake were similar, and the effect of the two agents together was no greater than that of either agent alone (Fig. 2). Likewise, maximal TPA-stimulated ^{45}Ca^{2+} uptake was fully inhibited

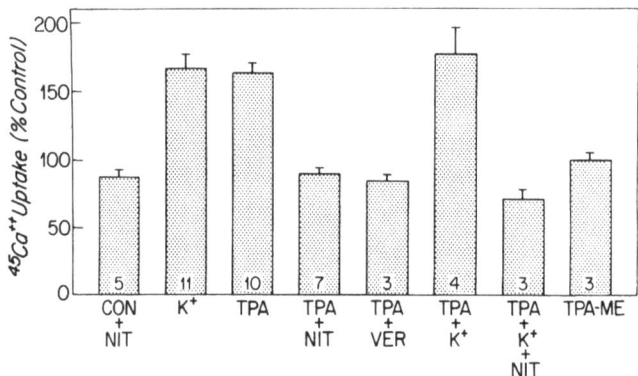

FIGURE 2. Effects of TPA (10 nM) and KCl (55 mM) alone and in combination on net 30-min $^{45}Ca^{2+}$ uptake. The effects of TPA alone or of TPA plus KCl were fully blocked by 1 μM nitrendipine (NIT). The inactive phorbol ester TPA–methyl ether (TPA-ME) had no effect. (Reproduced with permission from Sperti and Colucci, 1987b.)

by 1 μM nitrendipine or 100 μM verapamil, and the effect of TPA and KCl together was fully inhibited by 1 μM nitrendipine (Fig. 2). The biologically inactive phorbol ester TPA-methyl ether had no effect on $^{45}Ca^{2+}$ uptake (Fig. 2). The concentration response curves for nitrendipine inhibition of maximal KCl or TPA-stimulated $^{45}Ca^{2+}$ uptake were similar, with complete inhibition at 1/μM and half-maximal inhibition of approximately 3 nM (Fig. 3).

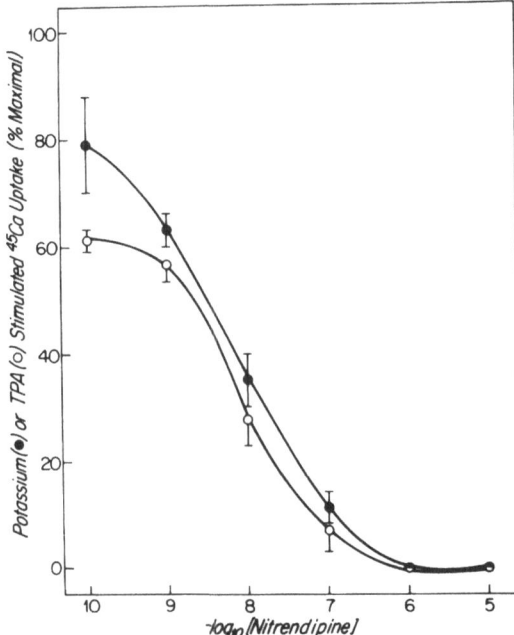

FIGURE 3. Concentration-response curves for nitrendipine inhibition of maximal TPA (○) or KCl (●) stimulated $^{45}Ca^{2+}$ uptake.

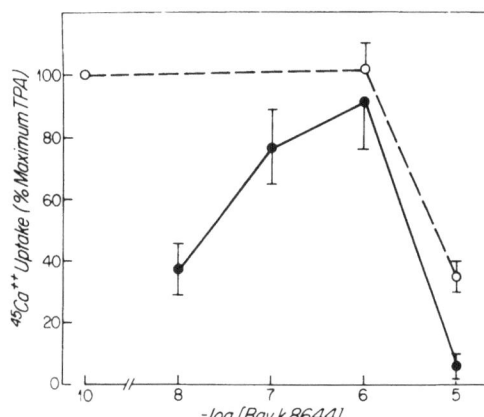

FIGURE 4. Relative effects of 10 nM TPA and various concentrations of BAY K 8644 on net 30-min ^{45}Ca^{2+} uptake. (Reproduced with permission from Sperti and Colucci, 1987b.)

3.3. Effect of BAY K 8644 on TPA-Stimulated ^{45}Ca^{2+} Uptake

The dihydropyridine Ca^{2+} channel agonist BAY K 8644 caused a concentration-related stimulation of ^{45}Ca^{2+} uptake that was maximal at approximately 1 μM (Fig. 4). At BAY K 8644 concentrations of $\geq$10 μM, ^{45}Ca^{2+} uptake was not stimulated. The effect of maximal TPA (10 nM) and BAY K 8644 (1 μM) together on ^{45}Ca^{2+} uptake was no greater than the effect of either agent alone (Fig. 4). In addition, 10 μM BAY K 8644 caused an approximately 65% inhibition of maximal TPA-stimulated ^{45}Ca^{2+} uptake (Fig. 4).

4. DISCUSSION

In A$_7$r$_5$ cells, the phorbol ester TPA caused a marked increase in both the unidirectional Ca^{2+} influx rate and net 30-min Ca^{2+} uptake. This phorbol-stimulated Ca^{2+} uptake was fully inhibited by two classes of Ca^{2+} channel antagonists, was not additive with either depolarizing solution or a maximally effective concentration of the Ca^{2+} channel agonist BAY K 8644, and was inhibited by high concentrations of BAY K 8644. These pharmacological observations strongly suggest that phorbol-stimulated Ca^{2+} uptake occurs by way of dihydropyradine-sensitive, voltage-dependent channels. Both T-type and L-type Ca^{2+} channels have been identified in A$_7$r$_5$ cells (D. Fish *et al.*, 1988) and A$_{10}$ cells (Friedman *et al.*, 1986), a closely related sister clone of the A$_7$r$_5$ line. Our data suggest that the effect of phorbol esters is primarily mediated by L-type channels.

From these data we cannot determine the mechanism by which phorbol esters cause an increase in ^{45}Ca^{2+} influx. Phorbol esters, presumably acting through activation of protein kinase C, are well known to cause phosphorylation of several membrane components, and therefore might act directly by phosphorylating the voltage-dependent Ca^{2+} channel. Alternatively, phorbol esters might cause membrane depolarization and hence Ca^{2+} influx through voltage-dependent channels. It was shown that phorbol esters can inhibit a voltage-dependent K$^+$ current (Higashida and Brown, 1986). Recently, Fish *et al.* (1988) found that 100 nM TPA caused a 32% increase in the L-type Ca^{2+} current in A$_7$r$_5$ cells. Since membrane potential is held constant in this technique, at least part of the effect of TPA is not likely due to membrane depolarization.

This effect of phorbol esters on Ca^{2+} influx does not appear unique to A_7r_5 cells. It has been shown by patch-clamp techniques that phorbol esters also cause inward Ca^{2+} conductance in neurons and photoreceptor cells (DeRiemer et al., 1985; Farley and Auerbach, 1985). Likewise, in cultured juxtaglomerular cells from rat, phorbol esters stimulate $^{45}Ca^{2+}$ (Kurtz et al., 1986). However, phorbol esters do not stimulate influx in all cells: in embryonic chicken dorsal ganglion neurons, phorbol esters were shown to cause a decrease in Ca^{2+} conductance (Rane and Dunlap, 1986). Although our findings in A_7r_5 cells are supported by preliminary data from Ruegg (1986), they appear to differ from those of Galizzi et al. (1987), who recently reported that TPA inhibits KCl or BAY K 8644-stimulated $^{45}Ca^{2+}$ uptake in A_7r_5 cells. However, this effect required substantially higher concentrations of TPA (EC_{50} = 25 nM, maximal effect = 1 μM) than those found to stimulate uptake in our study or that of Ruegg (1986), since in both cases stimulation was half-maximal by 1 nM and maximal by 3 nM. Thus, TPA may stimulate Ca^{2+} influx in non-depolarized cells (which have a low basal influx rate), and inhibit influx in depolarized cells which have a high influx rate. The different EC_{50}'s for these effects of TPA (stimulation, approximately 1 nM, inhibition, approximately 25 nM) raises the possibility that two different mechanisms are involved. In view of these observations, it may be important to note that basal (i.e., unstimulated) dihydropyridine-sensitive $^{45}Ca^{2+}$ influx varies several fold between quiescent and growing A_7r_5 cells (Sperti and Colucci, 1987a).

Potentially, Ca^{2+} influx through voltage-dependent channels may play a modulatory role during the sustained phase of the cellular response to physiological stimulation. Although the precise mechanism by which this Ca^{2+} influx would act to modulate the tonic phase of cellular response is not clear, one possibility is that it would provide a source of transmembrane Ca^{2+} flux in the vicinity of protein kinase C, thereby allowing continued activation of the kinase despite a return of $[Ca^{2+}]_i$ back toward basal levels during the tonic phase of the cellular response (Rasmussen, 1986). Recently, we found that TPA-stimulated Ca^{2+} influx is not associated with significant increases in $[Ca^{2+}]_i$ or net cellular Ca^{2+}, thereby suggesting that TPA may directly stimulate bidirectional transmembrane Ca^{2+} flux (Sperti and Colucci, 1987b). It will be important to determine whether this phenomenon is generalized to other vascular smooth muscle cells and whether a similar effect is evoked by stimulation of vasoconstrictor receptors. The A_7r_5 cell line provides a convenient model system for evaluating the relationship between protein kinase C and voltage-dependent channels.

5. SUMMARY

The purpose of these studies was to determine whether phorbol esters cause an increase in Ca^{2+} influx in vascular smooth-muscle cells. In the A_7r_5 clonal cell line derived from smooth muscle of fetal rat aorta, $^{45}Ca^{2+}$ uptake was linear over the first 5 min, presumably reflecting unidirectional influx, and showed a marked increase in influx rate in response to 10 nM 12-O-tetradecanylphorbol-13-acetate (TPA). Likewise, net 30-min $^{45}Ca^{2+}$ uptake was increased 63 $\pm$ 9% by 10 nM TPA. The mean effective concentration for TPA-stimulated $^{45}Ca^{2+}$ uptake was between 1 and 3 nM, and a maximal effect was achieved between 3 and 10 nM. Maximal TPA-stimulated $^{45}Ca^{2+}$ uptake was equivalent to that stimulated by 55 mM KCl. TPA-stimulated $^{45}Ca^{2+}$ uptake was not additive to that caused by 55 mM KCl and was fully inhibited by 100 μM verapamil or 1 μM nitrendipine. The biologically inactive phorbol ester TPA–methyl ether (10 nM) had no effect on $^{45}Ca^{2+}$ uptake. The calcium channel agonist BAY K 8644 increased ^{45}Ca^{2+} uptake in a concentration-related manner with a maximal effect at approximately 1 μM. Maximally effective concentrations of TPA (10 nM) and BAY K 8644 (1 μM) were not additive, and at high concentrations ($\geq$10 μM), BAY K 8644 inhibited TPA-stimulated Ca^{2+} influx. The concentration-response curves for nitrendipine inhibition of TPA- or KCl-stimulated ^{45}Ca^{2+} uptake were similar. Thus, phorbol esters cause an increase in ^{45}Ca^{2+} influx in A$_7$r$_5$ cells via channels with the pharmacological characteristics of L-type voltage-dependent Ca^{2+} channels. This effect of phorbol esters may be due to a direct action on the Ca^{2+} channel or may be secondary to membrane depolarization.

ACKNOWLEDGMENT. This work was supported in part by NIH grant R01-HL-34874-02 and a Clinician-Scientist Award of the American Heart Association to WSC.

REFERENCES

Castagna, M., Takai, Y., Taibuchi, K., Sano, K., Kikkawa, U., and Nishizuka, Y., 1982, Direct activation of calcium-activated, phospholipid-dependent protein kinase by tumor-promoting phorbol esters, *J. Biol. Chem.* **257:**1847–1851.

Chatterjee, M. and Tejada, M., 1986, Phorbol ester-induced contraction in chemically-skinned vascular smooth muscle, *Am. J. Physiol.* **251:**C356–C361.

Danthuluri, N. R., and Deth, R. C., 1984, Phorbol ester-induced contraction of arterial smooth muscle and inhibition of alpha-adrenergic response, *Biochem. Biophys. Res. Commun.* **125:**1103–1109.

DeRiemer, S. A., Strong, J. A., Albert, K. A., Greengard, D., and Kaczmarek, L. K., 1985, Enhancement of calcium current in *Aplysia* neurones by phorbol ester and protein kinase C, *Nature* **313:**313–316.

Forder, J., Scriabine, A., and Rasmussen, H., 1985, Plasma membrane calcium flux, protein kinase C activation and smooth muscle contraction, *J. Pharmacol. Exp. Ther.* **235:**267–273.

Farley, J., and Auerbach, S., 1985, Protein kinase C activation induces conductance changes in *Hermissenda* photoreceptors like those seen in associative learning, *Nature* **319:**220–223.

Fish, R. D., Sperti, G., Colucci, W. S., and Clapham, D. E., 1988, Phorbol ester increases the dihydropyridine-sensitive calcium conductance in a vascular smooth muscle cell line, *Circ. Res.* **62:**1049–1054.

Friedman, M. E., Suarez-Kurtz, G., Kaczorowski, G. J., Katz, G. M., and Reuben, J. P., 1986, Two calcium currents in a smooth muscle cell line, *Am. J. Physiol.* **250:**H699–H703.

Galazzi, J. P., Qar, J., Fosset, M., Van Renterghem, C., and Lazdunski, M., 1987, Regulation of calcium channels in aortic muscle cells by protein kinase-C activators and by peptides that stimulate phosphoinositide breakdown, *J. Biol. Chem.* **262:**6947–6950.

Gleason, M. M., and Flaim, S. F., 1986, Phorbol ester contracts rabbit thoracic aorta by increasing intracellular calcium and by activating calcium influx, *Biochem. Biophys. Res. Commun.* **138:** 1362–1369.

Higashida, H., and Brown, D. A., 1986, Two polysphatidylinositide metabolites control two K$^+$ currents in a neuronal cell, *Nature* **323:**333–335.

Kimes, B. W., and Brandt, B. L., 1976, Characterization of two putative smooth muscle cell lines from rat thoracic aorta, *Exp. Cell Res.* **98:**349–366.

Kurtz, A., Pfeilschifter, J., Hutter, A., Buhrle, C., Nobiling, R., Taugner, R., Hachenthal, E., and Bauer, C., 1986, Role of protein kinase-C in inhibition in renin release caused by vasoconstrictors, *Am. J. Physiol.* **250**(Cell Physiol. **19**):C563–C571.

Nishizuka, Y., 1984, The role of protein kinase C in cell surface signal transduction and tumour promotion, *Nature* **308:**693–698.

Rane, S. G., and Dunlap, K., 1986, Kinase C activator 1,2-oleoylacetyl-glycerol attenuates voltage-dependent calcium current in sensory neurons, *Proc. Natl. Acad. Sci. USA* **83:**184–188.

Rasmussen, H., 1986, The calcium messenger system, *N. Engl. J. Med.* **314:**1094–11011, 1164–1170.

Rasmussen, H., Forder, J., Kojima, and Scriabine, A., 1984, TPA-induced contraction of isolated rabbit vascular smooth muscle, *Biochem. Biophys. Res. Commun.* **122:**776–784.

Ruegg, U. T., 1986, Activation of potential operated calcium channels by activation of protein kinase-C. 11th Scientific Meeting of the International Society of Hypertension, August 31–September 6, 1986, Heidelberg, Federal Republic of Germany (abstract).

Ruegg, U. T., Doyle, V. M., Zuber, J. F., and Hof, R. P., 1985, A smooth muscle cell line suitable for the study of voltage sensitive calcium channels, *Biochem. Biophys. Res. Commun.* **130:**447–453.

Sperti, G., and Colucci, W. S., 1987a, Ca^{2+} channel antagonists inhibit DNA synthesis and proliferation in A_7r_5 vascular smooth muscle cells, *Fed. Proc.* **46:**974 (abstract).

Sperti, G., and Colucci, W. S., 1987b, Phorbol-stimulated bidirectional transmembrane Ca^{2+} flux in A_7r_5 vascular smooth muscle cells, *Mol. Pharmacol.* **32:**37–42.

Sybertz, E. J., Desiderio, D. M., Tetzloff, G., and Chiu, P. J. S., 1986, Phorbol dibutyrate contractions in rabbit aorta: Calcium dependence and sensitivity to nitrovasodilators and 8-BR-cyclic GMP, *J. Pharmacol. Exp. Ther.* **239:**78–83.

10

Effects of cAMP-Dependent, Calmodulin-Dependent, and C-Type Protein Kinases on Platelet Calcium Transport

WILLIAM L. DEAN and SAMUEL EVANS ADUNYAH

1. INTRODUCTION

Cyclic AMP (cAMP) and Ca^{2+} are interrelated intracellular messengers known to modify cellular function through specific protein kinases (Cohen, 1982). In platelets, increases in cAMP inhibit platelet activation, an effect that is reversed by inhibitors of adenylate cyclase (Salzman, 1972; Chiang et al., 1975), while Ca^{2+} is involved in activation rather than inhibition of platelet function (Detwiler et al., 1978). In addition to cAMP- and calmodulin-dependent protein kinases, a third kinase is intimately involved in platelet activation—protein kinase C (Kishimoto et al., 1980). This kinase is activated by diacylglycerol, a product of phospholipase C–mediated hydrolysis of phosphatidylinositol.

Internal membranes termed the dense tubular system possess one mechanism by which cytoplasmic Ca^{2+} levels are controlled in the platelet (Menashi et al., 1982; Kaser-Glanzmann et al., 1977). These membranes actively sequester Ca^{2+} and have been shown to release this stored Ca^{2+} upon exposure to inositol trisphosphate, a soluble second messenger formed by hydrolysis of phosphatidylinositol-4,5-bisphosphate (O'Rourke et al., 1985; Adunyah and Dean, 1986a). We have characterized highly purified internal platelet membranes with regard to Ca^{2+} uptake and release activities (Dean and Sullivan, 1982; Dean, 1984; Adunyah and Dean, 1985; Adunyah and Dean, 1986a,b) and have shown these membranes to be highly enriched in endoplasmic reticulum markers (Adunyah and Dean, 1986a). The purified membranes possess a Ca^{2+}-ATPase that is quite similar in functional characteristics to the Ca^{2+} pump of muscle sarcoplasmic reticulum (Dean, 1984; Adunyah and Dean, 1986a) and an inositol trisphosphate-activated Ca^{2+} release channel (Adunyah and Dean, 1985, 1986b). Since two other groups have published conflicting reports on the effects of cAMP-dependent protein kinase on the platelet Ca^{2+} pump (Kaser-Glanzmann et al., 1977; Le Peuch et al., 1983), in the present report we have examined the ability of cAMP-dependent, calmodulin-

WILLIAM L. DEAN and SAMUEL EVANS ADUNYAH • Department of Biochemistry, University of Louisville School of Medicine, Louisville, Kentucky 40292.

dependent, and C-type protein kinases to modify Ca^{2+} sequestration by purified dense tubular system membranes.

2. MATERIALS AND METHODS

Internal platelet membranes were obtained from outdated human platelets as described previously (Dean, 1984). Calmodulin was purified from bovine brain according to Gopalakrishna and Anderson (1982). The catalytic subunit of cAMP-dependent protein kinase was purchased from Sigma (rabbit muscle) as was the cAMP-dependent protein kinase inhibitor (bovine heart). Protein kinase C was purified from rabbit kidney by the method of Kikkawa *et al.* (1982).

Ca^{2+}-ATPase activity was determined by a coupled assay (Dean and Sullivan, 1982) and Ca^{2+} uptake was measured by membrane filtration with Millipore filters (Adunyah and Dean, 1985). Phosphorylation of platelet membrane proteins was carried out as described by Le Peuch *et al.* (1979) except that the phosphorylation medium consisted of 10 mM TES buffer, pH 7.5, containing 100 mM KCl, 10 mM NaF, 50 mM KPO_4, 5 mM EGTA, 30–100 μg of purified membranes, and 2–10 μg/ml of the catalytic subunit of cAMP-dependent protein kinase. The reaction was initiated by the addition of 10 mM Mg^{2+} and 100 μM ATP containing 20 μCi of [^{32}P]ATP. For phosphorylation by endogenous calmodulin-dependent protein kinase, membranes were suspended in 50 μM chlorpromazine and 5 mM EGTA, and incubated on ice for 30 min. The calmodulin-stripped membranes were then pelleted by centrifugation in a Beckman Airfuge for 30 min at 100,000g. The membranes were then washed twice with 10 mM TES buffer, pH 7.4, containing 100 mM KCl, 0.5 mM dithiothreitol, and 6% glycerol. Phosphorylation was carried out as described for cAMP-dependent protein kinase except that 60 μM Ca^{2+} and 10 μg/ml calmodulin were added prior to the addition of ATP. Protein kinase C-dependent phosphorylation was performed in Tris buffer, pH 7.4, containing 1 mM Ca^{2+}, 5 mM $MgCl_2$, 50 μg/ml phosphatidylserine, 1 μg/ml diolein, 80 μg/ml platelet membranes, and 40 μM ATP containing 5 μCi of [^{32}P]ATP (Kikkawa *et al.*, 1982). In some experiments diolein was replaced with 10 nM phorbol-12-myristate-13-acetate. Reactions were carried out at 30°C for various times and were terminated by addition of sodium dodecyl sulfate (SDS) sample buffer (Laemmli, 1970) to a final concentration of 1.7% SDS, 8.5% glycerol, and 4.3% mercaptoethanol. Sodium dodecyl sulfate polyacrylamide gel electrophoresis on 15% gels was carried out according to Laemmli (1970). Phosphorylated bands were identified by autoradiography using Kodak X-Omat AR film.

3. RESULTS

As shown in Table I, all three protein kinases caused significant changes in the activity of the platelet Ca^{2+} pump located in internal membranes. The catalytic subunit of cAMP-dependent protein kinase stimulated both Ca^{2+}-ATPase activity and Ca^{2+} pumping by approximately twofold at a concentration of 10 μg/ml. Stimulation of both activities could be totally inhibited by 10 μg/ml of cAMP-dependent protein kinase inhibitor added prior to the addition of the catalytic subunit. After stripping of calmodulin from the membranes as described in Sec. 2, addition of 60 μM Ca^{2+} and 10 μg/ml calmodulin stimulated Ca^{2+}-ATPase activity by a factor of 1.4. This effect could be totally inhibited by the addition of 50 μM chlorpromazine. The addition of purified protein kinase C had the opposite effect;

TABLE I. Effects of cAMP-Dependent, Calmodulin-Dependent, and
C-Type Kinases on Ca^{2+}-ATPase and Ca^{2+} Uptake by Purified
Platelet Membranes[a]

Kinase	Ca^{2+}-ATPase	Rate of Ca^{2+} uptake
cAMP-dependent	(+) 1.7-fold	(+) 1.9-fold
Calmodulin-dependent	(+) 1.4-fold	ND
C-Type	(−) 0.53-fold	ND

[a] Ca^{2+}-ATPase activity and Ca^{2+} uptake rates were determined as described in Sec. 2. Ten micrograms of the catalytic subunit of cAMP-dependent kinase was used for stimulating ATPase activity and uptake in a 1-ml assay. Ten micrograms of calmodulin was used to stimulate calmodulin-stripped membranes prepared as described in Sec. 2. Protein kinase C (60 μl of a purified preparation with an undetectable protein concentration) was used in combination with diolein as described in Sec. 2. (+) and (−) refer to stimulation and inhibition, respectively. ND indicates that the experiment was not carried out.

Ca^{2+}-ATPase activity was inhibited by one-half. The effects of the latter two kinases on Ca^{2+} transport were not determined.

In order to correlate the effects of kinases with membrane phosphorylation, autoradiography of platelet membrane proteins separated on SDS polyacrylamide gels was carried out after phosphorylation with γ-labeled ATP. As shown in Fig. 1, time-dependent phosphorylation of a 22-kDa protein occurred upon addition of the catalytic subunit of cAMP-dependent protein kinase. This phosphorylation was linear with time for 2 min and considerable phosphorylation had occurred by 20 sec. Since stimulation of ATPase activity and Ca^{2+} uptake was apparent by 20 sec, the phosphorylation of this protein could account for stimulation of the platelet Ca^{2+} pump. Rapid phosphorylation of 50- and 100-kDa proteins also was observed. However, these phosphorylations were not specifically inhibited by the cAMP-dependent protein kinase inhibitor while phosphorylation of the 22-kDa protein was inhibited. This result is similar to that observed for phospholamban in cardiac sarcoplasmic reticulum, which also results in stimulation of the Ca^{2+} pump (Tada et al., 1975; Wegener et al., 1986).

Calmodulin-dependent phosphorylation of the 22-kDa protein in calmodulin-stripped membranes is shown in Fig. 2. This phosphorylation depends on the presence of both Ca^{2+} and calmodulin as demonstrated by comparing lanes 2 and 4, and is inhibited by chlorpromazine as shown in lane 3. In addition, concerted phosphorylation of the 22-kDa protein is observed in the presence of calmodulin and the catalytic subunit of cAMP-dependent protein kinase as shown in lane 5. Again, this is the same behavior exhibited by phospholamban in the cardiac membrane system (Tada et al., 1975; Wegener et al., 1986).

Addition of exogenous protein kinase C purified from rabbit kidney caused phosphorylation of a 40-kDa protein as shown in Fig. 3. This phosphorylation was enhanced by the addition of phosphatidylserine and diolein (lane 3) or phorbol ester and phosphatidylserine (lane 4) as would be expected for this kinase (Nishizuka, 1984). Protein kinase C catalyzes phosphorylation of a 40-kDa protein in intact platelets upon addition of either the natural activator thrombin or phorbol ester (Sano et al., 1983). Thus it appears that addition of protein kinase C to internal membranes may result in phosphorylation of the same protein that is acted upon in vivo.

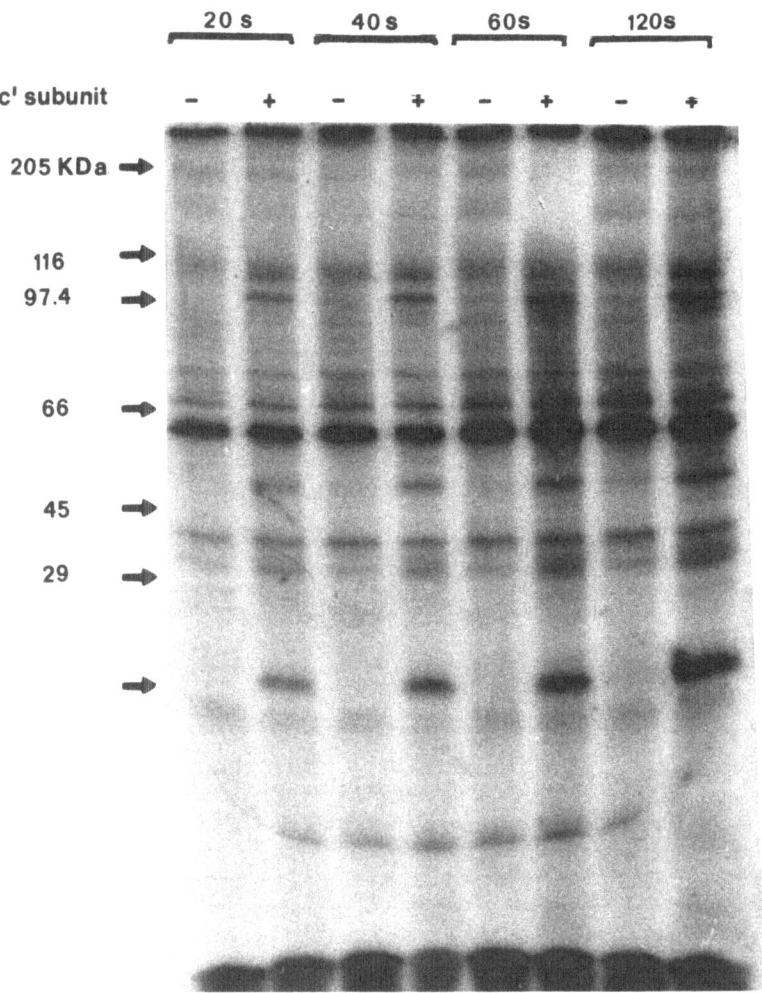

FIGURE 1. Time course of cAMP-dependent protein phosphorylation. Membranes (100 μg) prewashed with chlorpromazine and EGTA as described in Sec. 2 were phosphorylated in the presence of 2 μg/ml catalytic subunit. At the indicated times, phosphorylation was quenched by the addition of SDS sample buffer followed immediately by boiling for 1 min. The unlabeled arrow indicates the position of the 22-kDa polypeptide.

4. DISCUSSION

We have shown that cAMP-dependent protein kinase stimulates Ca^{2+}-ATPase activity and Ca^{2+} uptake in purified internal membranes to the same extent and with the same concentration dependence. This stimulation was inhibited by a specific inhibitor of cAMP-dependent protein kinase. Our results are in agreement with Kaser-Glanzmann *et al.* (1977) but in disagreement with the published work of Le Peuch *et al.* (1983). However, the latter

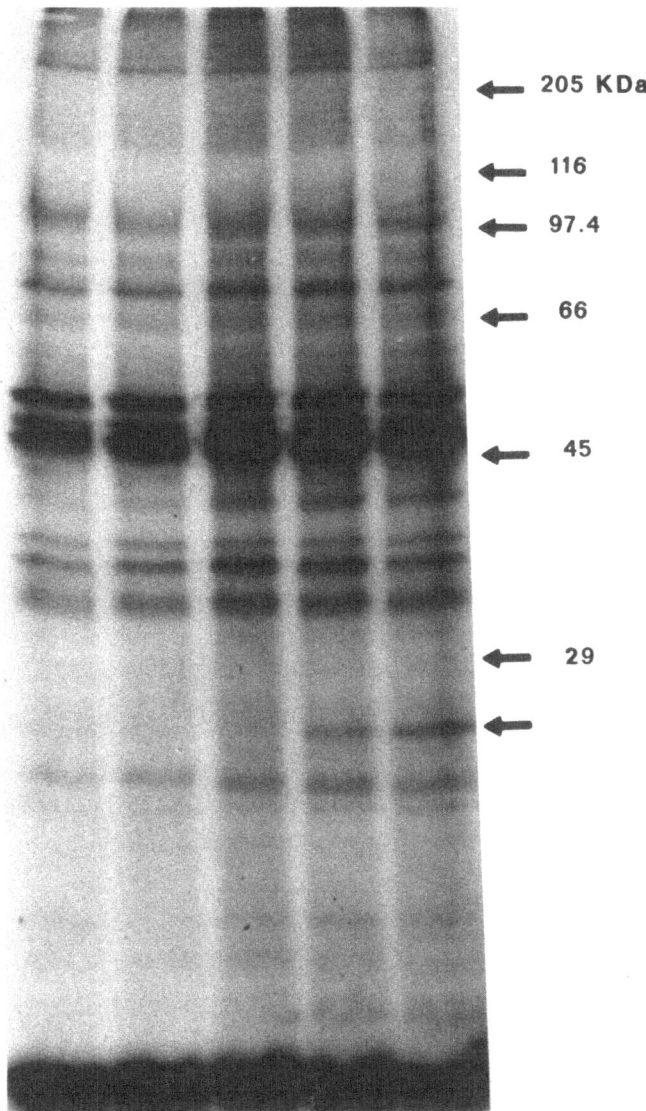

FIGURE 2. Concerted phosphorylation of purified human platelet membranes by exogenous cAMP- and endogenous calmodulin-dependent protein kinases. Membranes (100 μg) pre-washed with chlorpromazine and EGTA were phosphorylated as described in Sec. 2. Three micrograms of protein was applied to a 12% polyacrylamide gel. The conditions used for phosphorylation were: lane 1, 1 mM EGTA; lane 2, 60 μM Ca^{2+}; lane 3, 60 μM Ca^{2+} + 50 μM chlorpromazine + 10 μg/ml calmodulin; lane 4, 60 μM Ca^{2+} + 10 μg/ml calmodulin; and lane 5, 60 μM Ca^{2+} + 10 μg/ml calmodulin + 2 μg/ml catalytic subunit of cAMP-dependent protein kinase. The arrow indicates the position of the 22-kDa polypeptide.

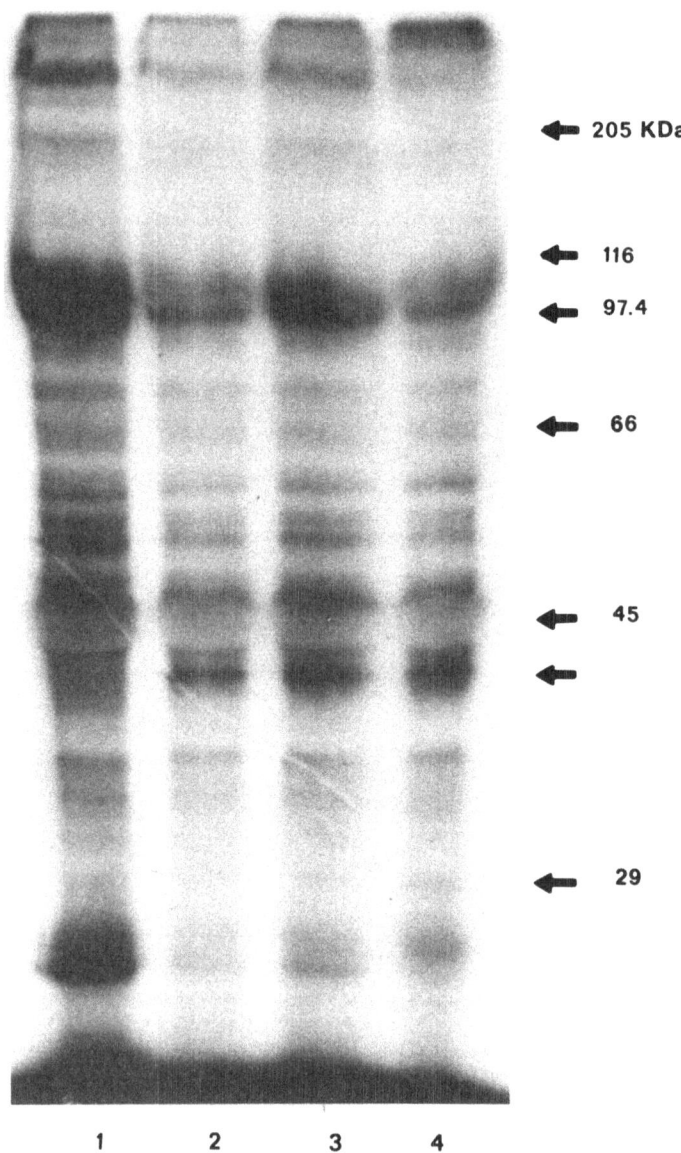

FIGURE 3. Phosphorylation of platelet membranes by purified protein kinase C. After addition of 50 μl of the purified protein kinase C preparation described in Table I, phosphorylation of purified platelet membranes (100 μg) was carried out as described in Sec. 2. Lane 1, phosphorylation in the presence of 1 mM Ca^{2+} and 40 μM ATP but no added protein kinase C; lane 2, protein kinase C + 1 mM Ca^{2+}; lane 3, protein kinase C + 1 mM Ca^{2+} + 50 μg/ml phosphatidylserine + 1 μg/ml diolein; lane 5, same as lane 4 except diolein was replaced with 10 nM phorbol-12-myristate-13-acetate. The unlabeled arrow indicates the position of the 40-kDa protein.

group reported at this meeting that they now observe stimulation of Ca^{2+} uptake in platelet membranes by the catalytic subunit of cAMP-dependent protein kinase. Thus there is now total agreement that cAMP-dependent protein kinase stimulates the Ca^{2+} pump in internal platelet membranes and that this action may be a factor in inhibition of platelet activation by agents that elevate cAMP.

Calmodulin-dependent protein kinase also stimulated Ca^{2+}-ATPase activity in calmodulin-stripped membranes. Furthermore, the addition of both cAMP-dependent protein kinase and calmodulin resulted in concerted activation of the Ca^{2+}-ATPase. This is the same result observed in cardiac sarcoplasmic reticulum (Le Peuch *et al.*, 1979; Gasser *et al.*, 1986). Thus it appears that the platelet Ca^{2+} pump may be under the same type of control as cardiac sarcoplasmic reticulum, suggesting that a phospholamban-like protein may be present in internal platelet membranes.

Addition of protein kinase C to purified membranes resulted in inhibition of the Ca^{2+} pump. Since there are conflicting reports of the action of protein kinase C on cytoplasmic Ca^{2+} in intact platelets (Zavoico *et al.*, 1985; Ware *et al.*, 1985), the physiological significance of this observation is not clear. Furthermore, protein kinase C becomes associated with the plasma membrane and not internal membranes when it is activated (Nishizuka, 1984). Thus it is possible that protein kinase C never exerts a direct effect on internal membranes *in vivo*, although our results indicate that protein kinase C can inhibit the pump which could lead to an increase in cytoplasmic Ca^{2+}.

The results of phosphorylation experiments indicate that the phosphorylation events in platelet membranes are similar to those observed in cardiac membranes. Both cAMP-dependent and calmodulin-dependent kinases phosphorylate a 22-kDa protein in internal membranes. This phosphorylation is concerted and the kinetics of phosphorylation are consistent with the rate of Ca^{2+} pump stimulation. Addition of the cAMP-dependent protein kinase inhibitor and chlorpromazine, a calmodulin antagonist, both inhibited phosphorylation of this protein. These results indicate a functional similarity to cardiac phospholamban. However, preliminary experiments indicate that the platelet protein is not immunochemically related to phospholamban and that dissociation of the platelet protein upon heating or in the presence of urea does not occur. Thus the platelet 22-kDa protein is functionally related but not identical to cardiac phospholamban.

Protein kinase C phosphorylated a 40-kDa protein in purified internal membranes that may be the same polypeptide labeled *in vivo*. The identity of this protein is unknown, although it has been suggested to be either lipocortin or inositol phosphate phosphatase (Touqui *et al.*, 1986; Connolly *et al.*, 1986). Neither of these activities should affect the Ca^{2+}-ATPase. It is not clear whether it is phosphorylation of this 40-kDa protein or some other component of the platelet membranes that affects the Ca^{2+}-ATPase.

In conclusion, we have shown that phosphorylation by protein kinases can have a profound effect on the activity of the internal membrane Ca^{2+} pump in platelets. Further study is necessary to determine the possible role of protein kinases in controlling the cytoplasmic level of Ca^{2+} *in vivo*.

ACKNOWLEDGMENTS. We wish to acknowledge Paul W. Eichenberger for preparing purified platelet membranes and assisting in purification of protein kinase C. This work was supported by U.S. Public Health Service grant HL36303.

REFERENCES

Adunyah, S. E., and Dean, W. L., 1985, Inositol trisphosphate-induced Ca^{2+} release from human platelet membranes, *Biochem. Biophys. Res. Commun.* **128**:1274–1280.

Adunyah, S. E., and Dean, W. L., 1986a, Ca^{2+} transport in human platelet membranes: kinetics of active transport and passive release, *J. Biol. Chem.* **261**:3122–3127.

Adunyah, S. E., and Dean, W. L., 1986b, Effects of sulfhydryl reagents and other inhibitors on Ca^{2+} transport and inositol trisphosphate-induced Ca^{2+} release from human platelet membranes, *J. Biol. Chem.* **261**:13071–13075.

Chiang, T. M., Beachy, E. H., and Kang, A., 1975, Interaction of a chick skin collagen fragment with human platelets, *J. Biol. Chem.* **250**:6916–6922.

Cohen, P., 1982, The role of protein phosphorylation in neural and hormonal control of cellular activity, *Nature* **296**:613–620.

Connolly, T. M., Lawing, Jr., W. J., and Majerus, P. W., 1986, Protein kinase C phosphorylates human platelet inositol trisphosphate 5'-phosphomonoesterase, increasing the phosphatase activity, *Cell* **46**:951–958.

Dean, W. L., 1984, Purification and reconstitution of a Ca^{2+} pump from human platelets, *J. Biol. Chem.* **259**:7343–7348.

Dean, W. L., and Sullivan, D. M., 1982, Structural and functional properties of a Ca^{2+}-ATPase from human platelets, *J. Biol. Chem.* **257**:14390–14394.

Detwiler, T. C., Charo, I. F., and Feinman, R. D., 1978, Evidence that calcium regulates platelet function, *Thrombos. Haemostas.* **40**:207–212.

Gasser, J. Th., Chesi, M., and Carafoli, E., 1986, Concerted phosphorylation of the 26 kDa phospholamban oligomer and of the low molecular weight phospholamban subunits, *Biochemistry* **25**:7615–7623.

Gopalakrishna, R., and Anderson, W. B., 1982, Ca^{2+}-induced hydrophobic site on calmodulin: application for purification of calmodulin by phenyl-Sepharose affinity chromatography, *Biochem. Biophys. Res. Commun.* **104**:830–836.

Kaser-Glanzmann, R., Jakabova, M., George, J. N., and Luscher, E. F., 1977, Stimulation of calcium uptake in platelet membrane vesicles by cAMP and protein kinase, *Biochim. Biophys. Acta* **466**:429–440.

Kikkawa, U., Takai, T., Minakuchi, R., Inohara, S., and Nishizuka, Y., 1982, Calcium-stimulated phospholipid-dependent protein kinase from rat brain, *J. Biol. Chem.* **257**:13341–13348.

Kishimoto, A., Takai, Y., Mori, T., Kikkawa, U., and Nishizuka, Y., 1980, Activation of calcium and phospholipid dependent protein kinase by diacylglycerol, its possible relation to phosphatidylinositol turnover, *J. Biol. Chem.* **255**:2273–2276.

Laemmli, U. K., 1970, Synthesis of T_4 phage proteins, *Nature* **227**:680–685.

Le Peuch, C. J., Haiech, J., and Demaille, J. G., 1979, Concerted regulation of cardiac sarcoplasmic reticulum calcium transport by cAMP-dependent and calcium-calmodulin-dependent phosphorylations, *Biochemistry* **18**:5150–5157.

Le Peuch, C. J., Le Peuch, D. A. M., Katz, S., DeMaille, J., Hinkle, M. T., Bredoux, R., Enouf, J., Levy-Toledano, S., and Caen, J., 1983, Regulation of calcium accumulation and efflux from platelet membrane vesicles: possible role for cAMP-dependent phosphorylation and calmodulin, *Biochim. Biophys. Acta* **731**:456–464.

Menashi, S., Davis, C., and Crawford, N., 1982, Calcium uptake associated with an intracellular membrane fraction prepared from human blood platelets by high-voltage, free-flow electrophoresis, *FEBS Lett.* **140**:298–302.

Nishizuka, Y., 1984, The role of protein kinase C in cell surface signal transduction and tumor promotion, *Nature* **308**:693–698.

O'Rourke, F. A., Halenda, S. P., Zavoico, G. B., and Feinstein, M. B., 1985, Inositol 1,4,5-trisphosphate releases Ca^{2+} from a Ca^{2+}-transporting membrane vesicle fraction derived from human platelets, *J. Biol. Chem.* **260**:956–962.

Salzman, E. W., 1972, Cyclic AMP and platelet function, *N. Engl. J. Med.* **286**:358–363.

Sano, K., Takai, Y., Yamanishi, J., and Nishizuka, Y., 1983, A role of Ca^{2+}-activated phospholipid-dependent protein kinase in human platelet function, *J. Biol. Chem.* **258**:2010–2013.

Tada, M., Kirchberger, M. A., and Katz, A. M., 1975, Phosphorylation of a 22,000 Dalton component of the cardiac sarcoplasmic reticulum by cAMP-dependent protein kinase, *J. Biol. Chem.* **250**:2640–2647.

Touqui, L., Rothhut, B., Shaw, A. M., Fradin, A., Vargaftig, B., and Russo-Marie, F., 1986, Platelet activation-a role for a 40K anti-phospholipase A$_2$ protein indistinguishable from lipocortin, *Nature* **321**:177–179.

Ware, J. A., Johnson, P. C., Smith, M., and Salzman, E. W., 1985, Aequorin detects increases in cytoplasmic calcium in platelets stimulated with phorbol esters or diacylglycerol, *Biochem. Biophys. Res. Commun.* **133**:98–104.

Wegener, A. D., Simmerman, H. K. B., Liepnieks, J., and Jones, L., 1986, Proteolytic cleavage of phospholamban purified from canine cardiac sarcoplasmic reticulum vesicles, *J. Biol. Chem.* **261**:5154–5159.

Zavoico, G. B., Halenda, S. P., Sha'afi, R. I., and Feinstein, M. B., 1985, Phorbol myristate acetate inhibits thrombin-stimulated Ca^{2+} mobilization and phosphatidylinositol 4,5-bisphosphate hydrolysis in human platelets, *Proc. Natl. Acad. Sci. USA* **82**:3859–3862.

11

Molecular Structure of Canine Cardiac Phospholamban, the Regulatory Protein of Ca Pump ATPase of Sarcoplasmic Reticulum

MICHIHIKO TADA, MASAAKI KADOMA, and JUNICHI FUJII

1. INTRODUCTION

The excitation-contraction coupling of the myocardium represents a three-part process, involving three kinds of subcellular systems. These are sarcolemma, sarcoplasmic reticulum (SR), and myofibrillar proteins. Information transfer among these systems is exclusively carried out by Ca ions (Tada et al., 1978), in that both membranes of sarcolemma and SR exhibit bidirectional Ca fluxes and the myofibrillar system contains the Ca receptor protein troponin. It is important to note that all of these three subcellular systems provide phosphorylation sites for protein kinases and, in addition, such phosphorylation reactions are thought to accompany profound alterations in Ca-related events in these systems. Among these, phosphorylation of phospholamban and its functional consequences are extensively defined (Tada and Katz, 1982; Tada et al., 1982) in that phospholamban presumably serves to modulate the Ca pump ATPase of SR by augmenting the key elementary steps of ATPase (Tada et al., 1979, 1980). Based on steady-state and presteady-state kinetic analysis of the Ca pump ATPase, phospholamban was proposed to function as a regulatory cofactor of the Ca pump ATPase type I, found in SR of cardiac and slow-contracting skeletal muscles (Kirchberger and Tada, 1976; Jorgensen and Jones, 1986). This is in contrast to the finding that SR of fast-contracting skeletal muscle, having type II Ca pump ATPase, is devoid of phospholamban. Phospholamban of cardiac SR was purified to near homogeneity and was sequenced by amino acid sequencing and by cDNA sequencing, demonstrating unique molecular properties. This paper defines the structural characteristics of phospholamban in its purified form and attempts to propose a molecular model for the functioning unit of phos-

MICHIHIKO TADA, MASAAKI KADOMA, and JUNICHI FUJII • Division of Cardiology, First Department of Medicine, and Department of Pathophysiology, Osaka University School of Medicine, Fukushima-ku, Osaka 553, Japan.

pholamban to understand, at least partly, the molecular mechanism by which the Ca pump ATPase is controlled by the phosphorylation of phospholamban.

2. STRUCTURAL CHARACTERISTICS OF PHOSPHOLAMBAN

2.1. Purification of Phospholamban

Several attempts have been made to purify phospholamban. These procedures employed organic solvents, sodium dodecyl sulfate (SDS), and nonionic detergents like $C_{12}E_8$ and Zwittergent® (*N*-tetradecyl-*N*,*N*'-dimethyl-3-amino-1-propanesulfonate) for fractionating hydrophobic membrane proteins. We overcame several difficulties by using $C_{12}E_8$ (octaethyleneglycol-*n*-dodecylether) and obtained purified phospholamban with its inherent properties reasonably preserved (Inui *et al.*, 1985). Judging from SDS-polyacrylamide gel electrophoresis and the extent of phosphorylation, our procedures yielded phospholamban with more than 99% purity. Approximately 0.06 mg of phospholamban was purified from 80 mg of canine cardiac SR. When amounts of cAMP-dependent phosphorylation were determined by incubating with the $[\gamma\text{-}^{32}P]ATP$ and the catalytic subunit of cAMP-dependent protein kinase, purified phospholamban incorporated about 125 nmol of phosphate/mg protein, in contrast to the original SR vesicles which incorporated about 1.55 nmol of phosphate/mg of SR protein. These findings indicate an 80-fold purification with overall recovery of 6% from cardiac SR.

2.2. Molecular Assembly of Phospholamban

The molecular weight of phospholamban was originally reported to be 22,000 (Tada *et al.*, 1975) based on electrophoretic mobility of ^{32}P-labeled phospholamban on Weber and Osborn gel system. A number of reports indicated that the molecular weight, determined by similar procedures, was in accord with our original report (Tada *et al.*, 1975). Employing purified phospholamban in unphosphorylated form, we demonstrated that the apparent molecular weight of phospholamban varied by varying the gel system for electrophoresis. In the Weber and Osborn neutral gel system, the molecular weight of phospholamban was 22,000 as originally reported, against 27,000 in the Laemmli alkaline gel system, when either gel system consisted of 15% polyacrylamide and 0.1% SDS. A similar shift of electrophoretic mobility on SDS-gel is reported in another SR protein, calsequestrin, in which the molecular weight is 44,000 in neutral system against 55,000 in the alkaline system. In the subsequent process to identify the phospholamban molecule, Laemmli gel system was largely employed.

The purified phospholamban preparation exhibited unusual electrophoretic behavior (Fig. 1) (Fujii *et al.*, 1986). While phospholamban migrated as a 27,000-Da component in the presence of SDS above the critical micelle concentration, the heat treatment (90°C or above) of phospholamban preparation lowered the apparent molecular weight quite significantly. Thus, SDS-polyacrylamide gel electrophoresis of nonheated preparation gave a 27,000-Da band and a trace band at the 6000-Da component. Upon heat treatment, the 6000-Da component was predominant. This temperature-dependent conversion was reversible because the 27,000-Da component was predominant when the heat-treated preparation was incubated at $-20°C$ overnight. We tentatively designated the 27,000-Da form of phospholamban as PN_H and the 6000-Da form as PN_L.

More precise examination of heat modifiability, performed by changing the temperature

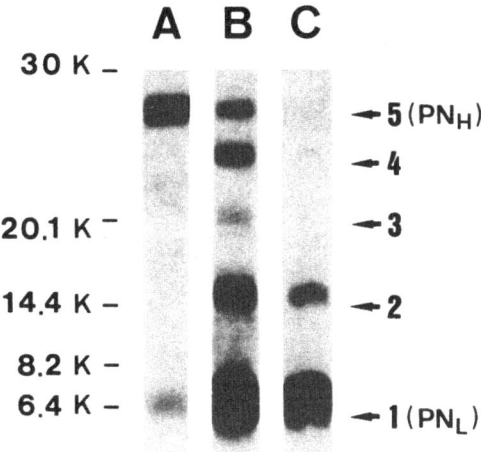

FIGURE 1. Effect of heat treatment on purified phospholamban in SDS-polyacrylamide gel electrophoresis. Purified phospholamban was solubilized in 2% SDS and subjected to various heat treatments for 1 min prior to electrophoresis. Heat treatments were performed at low (lane A: 30°C), moderate (lane B: 60°C), and high (lane C: 100°C) temperatures. Lane B represents the typical electrophoretic pattern exhibiting five bands, which were usually observed at the temperature ranging between 50 and 70°C. PN_H and PN_L designate the high and low M_r forms of phospholamban, respectively. (From Fujii *et al.*, 1986.)

between 30 and 100°C, demonstrated the existence of three intermediate electrophoretic bands between PN_H and PN_L, resulting in a total of five bands (Fig. 1) (Fujii *et al.*, 1986). Five bands were seen when the SDS concentration in the heat treatment was lower or the temperature was mild (50–70°C). An autoradiogram of phosphorylated phospholamban under these conditions indicated that all of the five bands contained phosphorylation sites. Estimation by electrophoretic mobility suggested the possibility that PN_H and PN_L represent a pentamer and a protomer, respectively. The electrophoretic mobility of phospholamban also changed by changing the extent of the phosphorylation. While nonphosphorylated phospholamban exhibited 27,000 molecular weight, phospholamban fully phosphorylated by cAMP-dependent protein kinase exhibited 29,000 molecular weight (Inui *et al.*, 1985). More peculiarly, such a shift of apparent molecular weight occurred in stepwise fashion, in that four intermediary bands were identifiable when the extent of phosphorylation was graded by altering incubation time (Wegener and Jones, 1984; Imagawa *et al.*, 1986). In PN_L, the phosphorylation-induced shift in electrophoretic mobility of phospholamban showed only one step. This phenomenon also supported the view that holoprotein of phospholamban consists of five identical monomers.

3. PRIMARY STRUCTURE OF PHOSPHOLAMBAN MONOMER

3.1. Partial Amino Acid Sequence of Phospholamban

Purified phospholamban was subjected to amino acid sequencing (Fujii *et al.*, 1986). By direct Edman degradation, we could not detect significant PTH (phenylthiohydan-

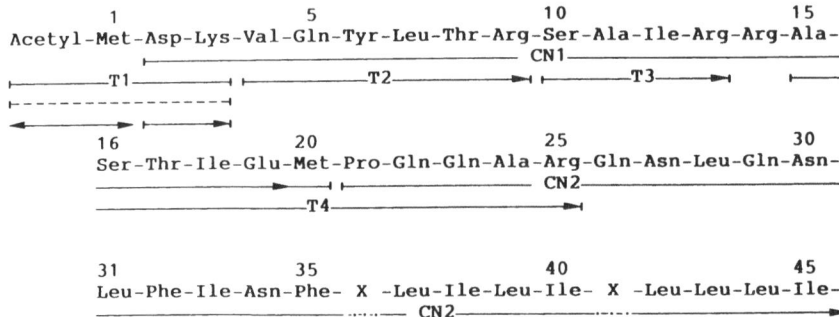

FIGURE 2. Partial amino acid sequence of phospholamban from canine cardiac SR. Designations are: CN, cyanogen bromide-cleaved peptide; T, tryptic peptide; X, unidentified residue; ⊢⊣, amino acid analysis; →, automatic Edman degradation; ⊦---⊣, fast-atom bombardment mass spectrometry; ↔, reverse phase HPLC. (From Fujii *et al.*, 1986.)

toin)–amino acid derivative released from intact and S-aminoethylated phospholamban, indicating that the amino terminus of the protein is blocked. When the intact protein was cleaved by cyanogen bromide, two peptide fragments, CN1 and CN2, were obtained by fractionation through HPLC on a phenyl 5PW-RP column. Edman degradation determined the first 18 amino acids of CN1, which are finally placed from Asp 2 to Glu 19 (Fig. 2). Although the C-terminal homoserine remained undetermined, Met 20 was placed at the C terminus of CN1 from the sequence of T4 as mentioned later. Analysis of CN2 gave the unambiguous determination of the N-terminal 25 amino acids of which two of the residues were not identified.

The S-aminoethylated protein was digested with TPCK (*N*-tosyl-*L*-phenylalanyl chloromethyl ketone)-trypsin and fractionated by Cosmosil columns, resulting in the separation of four tryptic fragments, T1–T4 (Fig. 2). T1 was a tripeptide composed of Met, Asp, and Lys and could not generate the N-terminal PTH-amino acid, suggesting that T1 was originated from the blocked amino terminus of the protein. Fast-atom bombardment mass spectrometry gave a major peak at M/Z 435.0 (acetyl–Met–Asp–Lys–OH, mol wt = 434.5), indicating that T1 was N$^\alpha$-acetylated. Cyanogen bromide-cleaved T1 showed sequence Asp–Lys and acetylhomoserine was determined by reverse phase high pressure liquid chromatography (HPLC). The N terminus of T1 was concluded to be acetyl–Met. T4 provided sequential overlap for CN1 and CN2. It is likely that phospholamban with the free N terminus as reported by Simmerman *et al.* (1986) may represent a partially proteolyzed polypeptide. We could not obtain any peptide different from the sequence, indicating that the preparation contains homologous polypeptides. The observed amino acid sequence would represent the N termini of the homooligomer of phospholamban.

3.2. Complete cDNA-Derived Amino Acid Sequence of Phospholamban

To isolate cDNA for phospholamban, a canine cardiac cDNA library was screened by hybridization with a mixture of 32 synthetic oligodeoxyribonucleotide probes predicted from a partial amino acid sequence (Glu 19–Met–Pro–Gln–Gln–Ala 24) of phospholamban (Fujii *et al.*, 1987). Three hybridization-positive clones were isolated from about 3000 transformants. All these plasmids contained the same size insert of about 800 bases with the same restriction maps, suggesting that they were derived from the same mRNA.

```
                                              5'-AGAAAACTTTCTAACTAAACAC -159

CGATAAGACTTCATACAACTCACAATACTTTATATTGTAATCATCACAAGAGCCAAGGCTACCTAAAAGAAGAGAGTGG  -80

TTGAGCTCACATTTGGCCGCCAGCTTTTTACCTTTCTCTTCACCATTTAAAACTTGAGACTTCCTGCTTTCCTGGGGTC   -1

1                              10                                    20
Met Asp Lys Val Gln Tyr Leu Thr Arg Ser Ala Ile Arg Arg Ala Ser Thr Ile Glu Met
ATG GAT AAA GTC CAA TAC CTC ACT CGC TCT GCT ATT AGA AGA GCT TCA ACC ATT GAA ATG   60

21                             30                                    40
Pro Gln Gln Ala Arg Gln Asn Leu Gln Asn Leu Phe Ile Asn Phe Cys Leu Ile Leu Ile
CCT CAA CAA GCA CGT CAA AAT CTT CAG AAC CTA TTT ATA AAT TTC TGT CTC ATT TTA ATA  120

41                           50       52
Cys Leu Leu Leu Ile Cys Ile Ile Val Met Leu Leu End
TGT CTC TTG TTG ATC TGC ATC ATT GTG ATG CTT CTC TGA AGTTCTGCTGCAATCTCCAGTGATGCA  187


ACTTGTCACCATCAACTTAATATCTGCCATCCCATGAAGAGGGGAAAATAATACTATATAACAGACCACTTCTAAGTAG  266

AAGATTTTACTTGTGAAAAGGTCAAGATTCAGAACAAAAGAAATTATTAACAAATGTCTTCATCTGTGGGATTTTGTAA  345

ACATGAAAAGAGCTTTATTTTCAAAAATTAACTTCAAAATGACTATAGGTGCGCATAATGTAATTGCTGAATTCCTCAA  424

CAAAGCTTGTAAAAGTTTCTATGCCAAATTTTTTCTGAGGGTAAAGTAGGAGTTTAGTTTTAAAACTGCTCTGCTAACC  503

AGTTCACTTCACATATAAAGCATTAGCTTCACTATTTGAGCTAAATATTTATATTGTACTGTAAATGCCTATGTAATGT  582

TTATTAAGATTTTTCAAGTCTCCGCTAAGTACGAAAATAATCATCCA AATGAA GTCATCATTTGAAATAGC-3'        652
```

FIGURE 3. Nucleotide sequence of pPLB1 cDNA and deduced amino acid sequence of canine cardiac phospholamban. Nucleotide residues are numbered in the 5' to 3' direction, beginning with the first residue of ATG triplet encoding the initiator methionine. Nucleotides on the 5' side of residue 1 are indicated by negative numbers. A poly(A) tail on the 3' end is not shown. The predicted amino acid sequence of phospholamban is displayed above the nucleotide sequence with its residue number beginning with the initiator methionine. A *box* indicates the presumptive polyadenylation signal. The in-frame stop codon preceding the initiator codon is *underlined*. (From Fujii *et al.*, 1987.)

Figure 3 depicts the complete nucleotide sequence of the cDNA insert of the plasmid, with 832 base pairs. Analysis of the cloned cDNA showed an open reading frame of 156 nucleotides (52 codons) starting with the ATG codon (position 1) and ending with the TGA stop codon (position 157). Figure 3 also shows the amino acid sequence deduced from the 52 codons. In this reading frame, the sequence of nucleotide residues 1–105, encoding amino acids 1–35, corresponded precisely to the 35 N-terminal sequence from acetyl–Met 1 to Phe 35 (Fig. 2). The two residues unidentified by Edman degradation for purified phospholamban, positions 36 and 41 (Fig. 2), were both deduced to be cysteine residues. Seven amino acids following Ile 45 would represent C-terminal residues ending at Leu 52, although protein sequencing is necessary to confirm these findings.

The cloned cDNA extended for 180 nucleotides to the 5' end of the sequence coding for the N terminus. Because this sequence contained an in-frame stop codon (TGA, position −78) and did not contain an ATG codon between the TGA and the Met 1, we conclude that phospholamban is synthesized without an N-terminal signal sequence. Therefore, it is likely that the information required for integration of phospholamban into the membrane is encoded in an uncleavable signal sequence in the C terminus.

3.3. Molecular Characteristics of Phospholamban

The molecular weight of encoded protein was estimated to be 6080 in good agreement with the apparent molecular weight of phospholamban monomer (6000) (Fujii *et al.*, 1986). The hydropathic profile of phospholamban indicates that the protein is amphipathic in nature. The protein is divided into two domains. The N-terminal domain from Met 1 to Asn 30 (domain I) was composed largely of hydrophilic amino acids, whereas the other domain representing 22 amino acids from Leu 31 to Leu 52 (domain II) was highly hydrophobic. Domain I, but not domain II, contained phosphorylatable serine and threonine residues, indicating that domain I faces cytoplasmic surface of SR membrane. Quite recently, Simmerman *et al.* (1986) reported a partial amino acid sequence and phosphorylation sites in phospholamban. They indicated 36 amino acid residues, which correspond precisely to those in the peptide from Ser 10 to Ile 45 in Fig. 2. Phosphorylation sites at Ser 7 and Thr 8 in their sequence would correspond to Ser 16 and Thr 17 in our sequence, respectively.

Secondary structure prediction of the protein suggested that domain I consists of an α helix. In view of an unusual behavior of phospholamban molecule (Wegener and Jones, 1984; Inui *et al.*, 1985; Imagawa *et al.*, 1986), it is intriguing to ask whether this helix breaks into two portions (domains IA and IB), possibly at around Pro 21, allowing side chains of each to express hydrophobic interactions (Fig. 4). Although alternative molecular models are possible, such an assumption may explain phosphorylation-induced structural changes of the phospholamban molecule, leading to profound functional consequences. Figure 4 illustrates the pentameric model of phospholamban, in which each phosphorylation site faces the cytoplasmic milieu with Pro 21 serving as a hinge (Tada *et al.*, 1987). In this

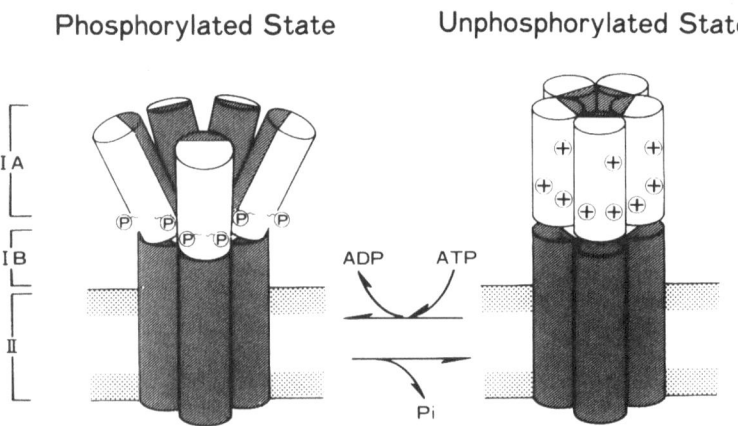

FIGURE 4. Pentameric model of phospholamban. When each monomer is assembled into a pentamer, the hydrophilic residues including phosphorylatable Ser 16 and Thr 17 face the cytoplasmic milieu. The positive charge of the arginine residue in unphosphorylated phospholamban is indicated by ⊕, whereas the phosphate group covalently bound to the side chain of Ser 16 and/or Thr 17 is indicated by ℗~. The cross-hatched area indicates the hydrophobic region of the α helix. In domain IA, the helical wheel analysis indicates that the hydrophobic residues face the core of pentamer. (From Tada *et al.*, 1988.)

model, phosphorylation of Ser 16 and/or Thr 17 would induce an alteration in charge distribution of the helical unit (domains IA and IB) and would result in a conformational change of domain I. This hypothesis would presumably account for an electrophoretic mobility shift induced by phosphorylation (Wegener and Jones, 1984; Imagawa *et al.*, 1986). In this sense, the hydrophobic pocket, formed in the core of the pentamer at domain IA level, is assumed to play a critical role, in which a key moiety of the Ca pump ATPase may be trapped and released, depending on the unphosphorylated and phosphorylated states, respectively. Domain II, consisting largely of hydrophobic amino acids, would also form an α helix that allows a hydrophobic interaction with membrane lipids. Three cysteine residues are reported to exist as free sulfhydryl groups (Simmerman *et al.*, 1986). Domain II coding 21 amino acids is clearly immersed in the lipid bilayer of SR membrane and could anchor phospholamban in the membrane.

It is interesting to speculate as to which portion of the phospholamban protomer would permit intermolecular interactions to form the pentameric assembly. A tryptic fragment of phospholamban holoprotein that was devoid of the phosphorylation site remained pentameric (Wegener *et al.*, 1986). Since trypsinization under these conditions results in the removal of N-terminal residues up to Arg 25, the ability to maintain an oligomeric organization would reside in the C-terminal residues, possibly in the carboxyl end of domain I (domain IB) or in domain II (Fig. 4).

Phospholamban was previously proposed to exert control over the Ca pump ATPase of cardiac SR (Tada and Katz, 1982). However, no structural evidence has yet been presented to support such a molecular interaction with a Ca pump ATPase. Suzuki and Wang (1986) observed that a monoclonal antibody directed against phospholamban blocked the phosphorylation and dephosphorylation of phospholamban but markedly increased the ATP-dependent Ca^+ pump activity by cardiac SR. Possibly the residues near phosphorylation sites in domain IA are responsible for exerting an influence over the Ca pump ATPase. It is intriguing to assume that the hydrophobic residues inside the pocket by the pentameric assembly in the model (Fig. 4) could exhibit a direct protein-protein interaction with the certain key residues of the Ca^+ pump ATPase (Tada *et al.*, 1988).

4. PHYSIOLOGICAL RELEVANCE OF PHOSPHOLAMBAN-ATPase SYSTEM

The two mechanical effects of catecholamines on the myocardium are the increased contractility (i.e., positive inotropic effect) and the abbreviation of systole (i.e., accelerated rates of contraction and relaxation). Such effects of catecholamines are considered to be produced during the excitation-contraction coupling by altering Ca fluxes across the two principal membrane systems, SR and sarcolemma, of the myocardial cells (Tada and Katz, 1982; Tada and Inui, 1983). The cAMP and phospholamban-ATPase system in SR could alter the rate of Ca uptake, subsequently changing the rate of Ca release (see below). Ca influx across the sarcolemmal membrane was also found to increase during β-adrenergic stimulation of the myocardial cells (Osterrieder *et al.*, 1982), possibly due to the phosphorylation of a channel protein by cAMP-dependent protein kinase. In the presence of cAMP, the acceleration of Ca uptake by the ATPase-phospholamban system may explain the acceleration of relaxation because the increased rate of Ca uptake by SR would increase the rate at which Ca^{2+} is removed from troponin. This effect could eventually increase the amount of Ca^{2+} stored within the SR, for some of the Ca^{2+} remaining within the SR might otherwise be lost during diastole. Catecholamine-induced enhancement of Ca influx across sarcolemma would produce the following two effects on the SR: (1) enhancement of Ca-induced Ca release from the SR and (2) increased amounts of Ca loading on SR. Increased amounts of Ca accumulation into the SR, brought out by the increases in Ca uptake in SR and Ca influx in sarcolemma, could add to the amounts of Ca^{2+} available for delivery to the myofibrillar proteins in subsequent contractions, thus promoting myocardial contractility. Increased Ca influx across sarcolemma could increase the Ca release due to the Ca-induced Ca release mechanism, thus partly contributing to the latter effect. Increased Ca release from the SR could increase both the rate and extent of myofibrillar contractions.

The *in vivo* evidence supporting these intracellular mechanisms was obtained by several investigators (Iwasa and Hosey, 1983; Le Peuch *et al.*, 1980; Lindemann *et al.*, 1983), who documented that the addition of isoproterenol to the isolated heart or the sliced heart tissue perfused with $[^{32}P]Pi$ resulted in increased ^{32}P incorporation into phospholamban *in situ*, with the simultaneous increase in the rates of contraction and relaxation. Interestingly, cholinergic agonists are found to antagonize the isoproterenol-induced augmentation of phospholamban phosphorylation. Calmodulin inhibitor (fluphenazine) significantly reduced *in vivo* phosphorylation of phospholamban (Le Peuch *et al.*, 1980), although the physiological relevance of such an effect is not entirely clear.

There is other evidence consistent with the mechanism by which Ca^{2+} fluxes of SR are controlled by the cAMP-phospholamban system. Employing a skinned cardiac cell, which exhibits cycles of phasic contractions upon addition of Ca^{2+}, Fabiato and Fabiato (1975) demonstrated that a brief preincubation with cAMP results in an increased amplitude of contraction and faster rates of tension development and relaxation. The more direct evidence was obtained by Allen and Blinks (1978), who measured intracellular Ca^{2+} by aequorin, a Ca^{2+}-sensitive bioluminescent protein. They found that isoproterenol augments the initial rate of Ca^{2+} release from SR during the early phase of contraction, with the simultaneous enhancement in the rate of Ca^{2+} reduction at the onset of relaxation.

ACKNOWLEDGMENTS. This work was supported by research grants from the Ministries of Education, Science and Culture, and of Health and Welfare of Japan, and a grant-in-aid from the Muscular Dystrophy Association of America.

REFERENCES

Allen, D. G., and Blinks, J. R., 1978, Calcium transients in aequorin-injected frog cardiac muscle, *Nature* **273**:509–513.

Fabiato, A., and Fabiato, F., 1975, Relaxing and inotropic effects of cyclic AMP on skinned cardiac cells, *Nature* **253**:556–558.

Fujii, J., Kadoma, M., Tada, M., Toda, H., and Sakiyama, F., 1986, Characterization of structural unit of phospholamban by amino acid sequencing and electrophoretic analysis, *Biochem. Biophys. Res. Commun.* **138**:1044–1050.

Fujii, J., Ueno, A., Kitano, K., Tanaka, S., Kadoma, M., and Tada, M., 1987, Complete complementary DNA-derived amino acid sequence of canine cardiac phospholamban, *J. Clin. Invest.* **79**:301–304.

Imagawa, T., Watanabe, T., and Nakamura, T., 1986, Subunit structure and multiple phosphorylation sites of phospholamban, *J. Biochem. (Tokyo)* **99**:41–53.

Inui, M., Kadoma, M., and Tada, M., 1985, Purification and characterization of phospholamban from canine cardiac sarcoplasmic reticulum, *J. Biol. Chem.* **260**:3708–3715.

Iwasa, Y., and Hosey, M. M., 1983, Cholinergic antagonism of β-adrenergic stimulation of cardiac membrane protein phosphorylation in situ, *J. Biol. Chem.* **258**:4571–4575.

Jorgensen, A. O., and Jones, L. R., 1986, Localization of phospholamban in slow but not fast canine skeletal muscle fibers: An immunocytochemical and biochemical study, *J. Biol. Chem.* **261**:3775–3781.

Kirchberger, M. A., and Tada, M., 1976, Effects of adenosine 3′:5′-monophosphate-dependent protein kinase on sarcoplasmic reticulum isolated from cardiac and slow and fast contracting skeletal muscle, *J. Biol. Chem.* **251**:725–729.

Le Peuch, C. J., Guilleux, J.-C., and Demaille, J. G., 1980, Phospholamban phosphorylation in the perfused rat heart is not solely dependent on β-adrenergic stimulation, *FEBS Lett.* **114**:165–168.

Lindemann, J. P., Jones, L. R., Hathaway, D. R., Henry, B. G., and Watanabe, A. M., 1983, β-Adrenergic stimulation of phospholamban phosphorylation and Ca^{2+}-ATPase activity in guinea pig ventricles, *J. Biol. Chem.* **258**:464–471.

Osterrieder, W., Brum, G., Hescheler, J., Trautwein, W., Flockerzi, V., and Hofmann, F., 1982, Injection of subunits of cyclic AMP-dependent protein kinase into cardiac myocytes modulates Ca^{2+} current, *Nature* **298**:576–578.

Simmerman, H. K. B., Collins, J. H., Theibert, J. L., Wegener, A. D., and Jones, L. R., 1986, Sequence analysis of phospholamban: Identification of phosphorylation sites and two major structural domains, *J. Biol. Chem.* **261**:13333–13341.

Suzuki, T., and Wang, J. H., 1986, Stimulation of bovine cardiac sarcoplasmic reticulum Ca^{2+} pump and blocking of phospholamban phosphorylation and dephosphorylation by a phospholamban monoclonal antibody, *J. Biol. Chem.* **261**:7018–7023.

Tada, M., and Inui, M., 1983, Regulation of calcium transport by the ATPase-phospholamban system, *J. Molec. Cell. Cardiol.* **15**:565–575.

Tada, M., and Katz, A. M., 1982, Phosphorylation of the sarcoplasmic reticulum and sarcolemma, *Annu. Rev. Physiol.* **44**:401–423.

Tada, M., Kirchberger, M. A., and Katz, A. M., 1975, Phosphorylation of a 22,000-dalton component of the cardiac sarcoplasmic reticulum by adenosine 3′:5′-monophosphate-dependent protein kinase, *J. Biol. Chem.* **250**:2640–2647.

Tada, M., Yamamoto, T., and Tonomura, Y., 1978, Molecular mechanism of active calcium transport by sarcoplasmic reticulum, *Physiol. Rev.* **58**:1–79.

Tada, M., Ohmori, F., Yamada, M., and Abe, H., 1979, Mechanism of the stimulation of Ca^{2+}-dependent ATPase of cardiac sarcoplasmic reticulum by adenosine 3′:5′-monophosphate-dependent protein kinase, *J. Biol. Chem.* **254**:319–326.

Tada, M., Yamada, M., Ohmori, F., Kuzuya, T., Inui, M., and Abe, H., 1980, Transient state kinetic studies of Ca^{2+}-dependent ATPase and calcium transport by cardiac sarcoplasmic reticulum, *J. Biol. Chem.* **255**:1985–1992.

Tada, M., Yamada, M., Kadoma, M., Inui, M., and Ohmori, F., 1982, Calcium transport by cardiac sarcoplasmic reticulum and phosphorylation of phospholamban, *Molec. Cell. Biochem.* **46**:73–95.

Tada, M., Kadoma, M., Inui, M., and Fujii, J., 1988, Regulation of Ca^{2+} pump from cardiac sarcoplasmic reticulum, *Methods Enzymol.* **157**:107–154.

Wegener, A. D., and Jones, L. R., 1984, Phosphorylation-induced mobility shift in phospholamban in sodium dodecyl sulfate-polyacrylamide gels, *J. Biol. Chem.* **259**:1834–1841.

Wegener, A. D., Simmerman, H. K. B., Liepnieks, J., and Jones, L. R., 1986, Proteolytic cleavage of phospholamban purified from canine cardiac sarcoplasmic reticulum vesicles, *J. Biol. Chem.* **261**:5154–5159.

12

Ca²⁺ Transport by Liver and Plant Mitochondria
Aspects Linked to the Biological Role

ANIBAL E. VERCESI, LUCIA PEREIRA-DA-SILVA,
IONE S. MARTINS, EVA G. S. CARNIERI,
CELENE F. BERNARDES, and MARCIA M. FAGIAN

1. INTRODUCTION

Mitochondria isolated from vertebrate tissues possess a very active Ca^{2+} transport system that is believed to participate in the intracellular Ca^{2+} homeostasis. In respiring mitochondria the Ca^{2+} distribution between the matrix and extramitochondrial compartments, in steady state, is kinetically regulated by the simultaneous operation of two distinct pathways for Ca^{2+} influx and efflux. Ca^{2+} uptake takes place by a uniport mechanism driven electrophoretically by the negative-inside membrane potential and the efflux pathway appears to promote the electroneutral exchange of matrix Ca^{2+} by external Na^+ or H^+ (Fiskum and Lehninger, 1981; Hansford, 1985; McCormack and Denton, 1986).

The role of this Ca^{2+} transport system in the cell physiology has been a matter of intensive study from which three major propositions have emerged:

1. The ability of mitochondria, particularly in the presence of Mg^{2+} and ATP, to buffer external free Ca^{2+} at concentrations approaching those of the cytosol *in vivo*, has led many investigators to the conclusion that mitochondria may act as a Ca^{2+} sink, regulating the cytoplasmic Ca^{2+} concentration (Fiskum and Lehninger, 1981).
2. Recent determination of mitochondrial Ca^{2+} content in vivo, using electron probe X-ray microanalysis or even careful conventional homogenization and fractionation procedures in the presence of ruthenium red or EGTA to minimize Ca^{2+} accumulation, has indicated values of 1–5 nmol Ca^{2+}/mg mitochondrial protein (Hansford, 1985). This Ca^{2+} content is incompatible with the kinetic characteristics of the influx-efflux Ca^{2+} pathways that allow an accurate buffering of cytoplasmic Ca^{2+}

ANIBAL E. VERCESI, LUCIA PEREIRA-DA-SILVA, IONE S. MARTINS, EVA G. S. CARNIERI, CELENE F. BERNARDES, and MARCIA M. FAGIAN • Department of Biochemistry, I.B. The State University of Campinas, C.P. 6109, CEP. 13081, Brazil.

concentration (Hansford, 1985). Instead, these values are compatible with a hypothesis by which the mitochondrial Ca^{2+} transport system regulates the free Ca^{2+} concentration in the matrix in a range that allows the regulation of three intramitochondrial dehydrogenases (pyruvate dehydrogenase, PDH; NAD^+-isocitrate dehydrogenase, NAD-ICDH; and 2-oxoglutarate dehydrogenase, OGDH) considered to play a regulatory role in oxidative metabolism (Hansford, 1985; McCormack and Denton, 1986).

3. Several reports (Vallières et al., 1975; Sordahl, 1974; Malmström and Carafoli, 1977; Villalobo and Lehninger, 1980; Abou-Khalil et al., 1981; Roman et al., 1981; Moreno-Sánchez, 1983) have also indicated that intramitochondrial Ca^{2+} inhibits oxidative phosphorylation leading to the proposition of a possible role of the mitochondrial Ca^{2+} transport system in the regulation of oxidative phosphorylation.

The present paper summarizes recent results from our laboratory approaching some aspects of Ca^{2+} transport by liver and plant mitochondria related to each of the above propositions.

2. Ca^{2+} EFFLUX INDUCED BY THE OXIDIZED STATE OF MITOCHONDRIAL PYRIDINE NUCLEOTIDES

It was first shown by Lehninger et al. (1978) that Ca^{2+} efflux from isolated mitochondria could be stimulated by the oxidized state of mitochondrial pyridine nucleotides. This was subsequently confirmed by other laboratories, not only in isolated mitochondria but also in intact cells (Bellomo et al., 1982) and perfused liver (Sies et al., 1981). On the basis of these results, it was proposed that Ca^{2+} release from mitochondria, in response to an oxidized state of mitochondrial NAD(P)H associated to a low cytosolic phosphorylation potential (ΔGp), could function as a feedback mechanism to increase the cytosolic phosphorylation potential and the $NAD(P)H/NAD(P)^+$ ratio through the stimulation of cytoplasmic catabolism by Ca^{2+}.

Despite much study, the mechanism of $NAD(P)^+$-stimulated Ca^{2+} efflux from mitochondria remains poorly understood and controversial. Some authors claimed that this efflux is due mainly to a nonspecific increase in membrane permeability and it is preceded by collapse in membrane potential ($\Delta\psi$) (Nicholls and Brand, 1980; Broekemeier et al., 1985). Others conclude that this release mechanism is independent of gross alterations in membrane permeability and decrease in $\Delta\psi$, and could be physiologically relevant (Fiskum and Lehninger, 1979; Frei et al., 1985). With respect to the molecular mechanism it has been proposed that this Ca^{2+} release is mediated by an NAD-binding membrane protein (Panfili et al., 1980) or by a membrane protein that is ADP-ribosylated after the hydrolysis of intramitochondrial NAD^+ by an ATP-sensitive NADase to ADP-ribose, nicotinamide, and 5'-AMP (Frei et al., 1985). Other results (Broekemeier et al., 1985) implicated the activity of the intramitochondrial phospholipase A_2 causing an increase in membrane permeability due to accumulation of lysophospholipids and free fatty acids. According to these results, in the presence of NAD(P)H oxidants the reacylation of lysophospholipids, which depends on the action of sulfhydryl-sensitive enzymes, would be inhibited due to a decreased availability of reduced glutathione.

Our recent work (Bernardes et al., 1986) with isolated liver mitochondria (Fig. 1) incubated in reaction medium containing intracellular concentrations of ATP, Mg^{2+}, and t-

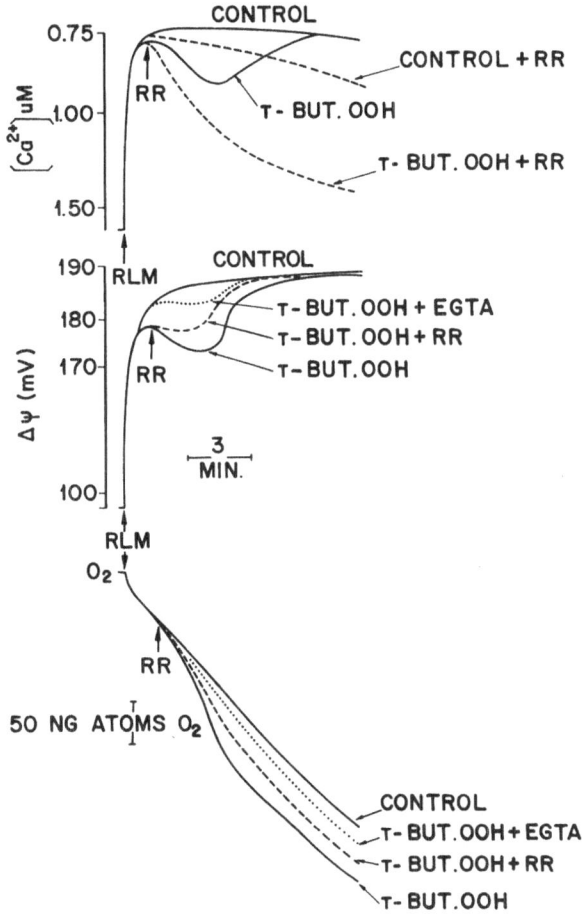

FIGURE 1. Changes in $\Delta\psi$ and rate of O_2 consumption during transient t-butylhydroperoxide-induced Ca²⁺ release from liver mitochondria.

Mitochondria (1 mg) were suspended in 1.0 ml of medium containing 125 mM sucrose, 65 mM KCl, 3.0 mM Hepes buffer (pH 7.2), 4.0 mM Mg²⁺, 3.0 mM ATP, 0.05% bovine serum albumin, 4.0 μM rotenone, 0.5 μg oligomycin/mg, 10 mM K⁺-acetate and 2.0 mM succinate, in the presence or absence of 50 μM t-butylhydroperoxide. Ruthenium red (RR) (0.7 μM) was added where indicated (dashed lines). The dotted lines indicated an experiment carried out in the presence of 0.5 mM EGTA added prior to addition of mitochondria. RLM, rat liver mitochondria. (From Bernardes et al. (1986), by permission of the copyright holder.)

butylhydroperoxide as NAD(P)H oxidant have shown that (1) the transient NAD(P)H oxidation that occurs during the metabolism of t-butylhydroperoxide is paralleled by an increase in the steady-state concentration of extramitochondrial Ca²⁺, decrease in $\Delta\psi$, and increase in the rate of respiration and mitochondrial swelling; (2) the decrease in $\Delta\psi$ precedes the beginning of Ca²⁺ efflux and is only partially blocked by EGTA or ruthenium red; (3) with the exception of mitochondrial swelling all other events were found to be reversible; and (4) $NADP^+$ rather than NAD^+ is linked to Ca^{2+} efflux, although not directly.

The decrease in $\Delta\psi$ and the parallel mitochondrial swelling, which is not the cause of Ca^{2+} efflux per se (Vercesi, 1984a), do not support the idea that Ca^{2+} release induced by $NADP^+$ represents a regulatory mechanism. It is possible, however, that this process is involved in the perturbation of intracellular Ca^{2+} homeostasis and loss of cell viability that occurs during oxidative stress (Bellomo et al., 1982).

In regard to the molecular mechanism of stimulated Ca^{2+} efflux, our data (Vercesi, 1984b) indicate that Ca^{2+} release induced by the oxidized state of mitochondrial pyridine nucleotides is a consequence of ultrastructural alterations of the inner mitochondrial membrane due to oxidation of protein thiol groups similar to those alterations that cause hemolysis due to disulfide formation in the membranes of erythrocytes (Kosower et al., 1982). In fact, there is considerable evidence that sulfhydryl groups are involved in the control of the selective permeability of the inner mitochondrial membrane (Lê-Quôc and Lê-Quôc, 1985) and that NAD(P)H maintains glutathione and membrane thiols in the reduced state (Bellomo et al., 1982).

3. INHIBITION OF OXIDATIVE PHOSPHORYLATION BY INTRAMITOCHONDRIAL Ca^{2+}

It has been shown that intramitochondrial Ca^{2+} inhibits oxidative phosphorylation in mitochondria isolated from different sources such as tumor (Abou-Khalil et al., 1981), heart (Sordahl et al., 1974), smooth muscle (Vallières et al., 1975), uterus (Malmström and Carafoli, 1977), brain (Roman et al., 1981), and liver (Moreno-Sánchez, 1983). Among the different mechanisms suggested to explain such inhibition, it has been proposed that the matrix Ca^{2+} inhibits the ADP/ATP carrier (Malmström and Carafoli, 1977), induces loss of internal adenine nucleotides (Moreno-Sánchez, 1983), inhibits the release of the ATP inhibitor protein from the F_0F_1-ATPase complex (Gómez-Puyou et al., 1980), decreases the availability of matrix phosphate by causing calcium-phosphate precipitation (Abou-Khalil, 1981), or competes with Mg^{2+} for the formation of MgATP, the true substrate of the ATP synthase (Roman et al., 1981). Due to the potential significance of this process on the regulation of cell metabolism, we have studied the mechanism of this inhibition in intact liver mitochondria and submitochondrial particles (Fagian et al., 1986).

To avoid the deleterious effects on mitochondrial structure and function caused by Ca^{2+} accumulation in the presence of phosphate (Coelho and Vercesi, 1980), in the experiments with intact mitochondria we employed Sr^{2+} which is also accumulated by mitochondria and caused a similar inhibition of oxidative phosphorylation without inducing mitochondrial damage (Coelho and Vercesi, 1980). This inhibition was shown to be independent of significant alterations in the activity of the respiratory chain, the redox proton pump, the phosphate carrier, the F_0F_1-ATPase, and the ADP/ATP translocase (Fagian et al., 1986). Since Sr^{2+} or Ca^{2+} can bind to internal adenine nucleotides, these results suggested that the inhibition of ADP phosphorylation could be due to a decreased availability of adenine nucleotides to both the ADP/ATP carrier and the ATP synthase. This hypothesis was tested in experiments with inverted submitochondrial particles in order to eliminate the problems concerned with the measurements of the concentrations of free and bound forms of both adenine nucleotides and cations involved in the study. The ATPase activity of submitochondrial particles was measured in the presence of an uncoupler and different concentrations of Ca^{2+} or Sr^{2+}.

The hypothesis that Ca^{2+} or Sr^{2+} inhibits the ATPase activity by competing with Mg^{2+} for the binding to ATP was confirmed by the results of three types of experiments: (1) the concentrations of Ca^{2+} or Sr^{2+} that caused about 90% inhibition of ATPase activity (Fig. 2) decreased the concentration of MgATP from 0.91 mM to 0.16 mM, a value much lower than the Michaelis constant (K_m) (0.40 mM) of the reaction; (2) the inhibition of the ATPase activity by Sr^{2+} or Ca^{2+} could be overcome by increasing the concentration of Mg^{2+} in the reaction medium when the concentrations of Sr^{2+} or Ca^{2+} were maintained constant (Fig. 3); and (3) the double-reciprocal plots of experiments where the ATPase activity was studied in a series of constant Ca^{2+} concentrations at different concentrations of Mg^{2+} showed that Ca^{2+} inhibition gave straight lines with different slopes intercepting the ordinate axis at the same point.

The high Ca^{2+}/Mg^{2+} ratios of 2.5 or 25.0 that cause 50% and 90% inhibition of ATPase activity, respectively, in these studies strongly argue against a possible regulatory role of this process on cell metabolism under normal conditions. Rather, it may be a factor in the mechanism of cell toxicity associated to pathological increase in cytosolic Ca^{2+} concentration.

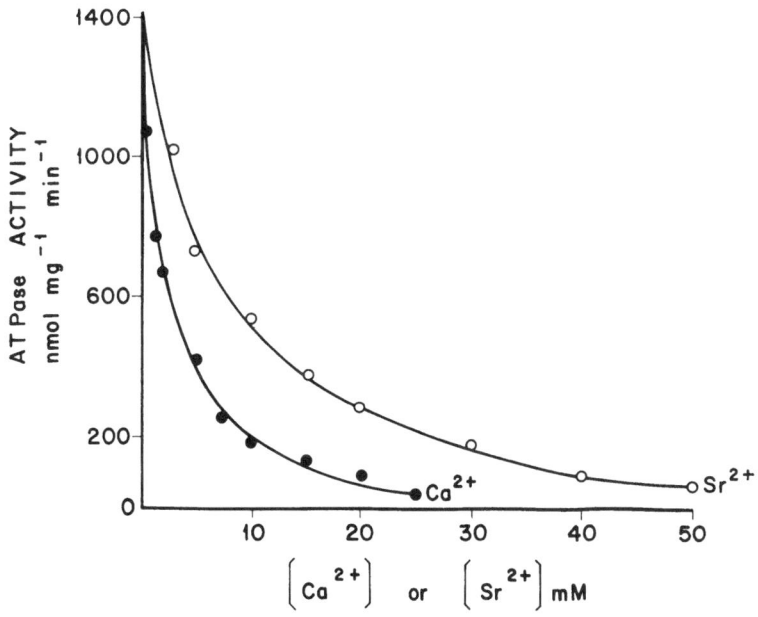

FIGURE 2. Effect of Ca^{2+} or Sr^{2+} concentration on the ATPase activity of rat liver submito-chondrial particles. The particles (0.13 mg) were incubated during 3 min in 1.0 ml of a medium containing 130 mM KCl, 3.0 mM Hepes buffer (pH 7.4), 2.0 mM ATP, 0.5 μM FCCP, 0.5 mM phosphoenolpyruvate, 200 nmol NADH, 3.7 U pyruvate kinase, 9.6 U lactic dehydrogenase, and different concentrations of Ca^{2+} or Sr^{2+}. The reactions were initiated by addition of 1.0 mM Mg^{2+} and changes in NADH fluorescence were followed using an Aminco-Bowman spectro-photofluorometer. (From Fagian et al. (1986), by permission of the copyright holder.)

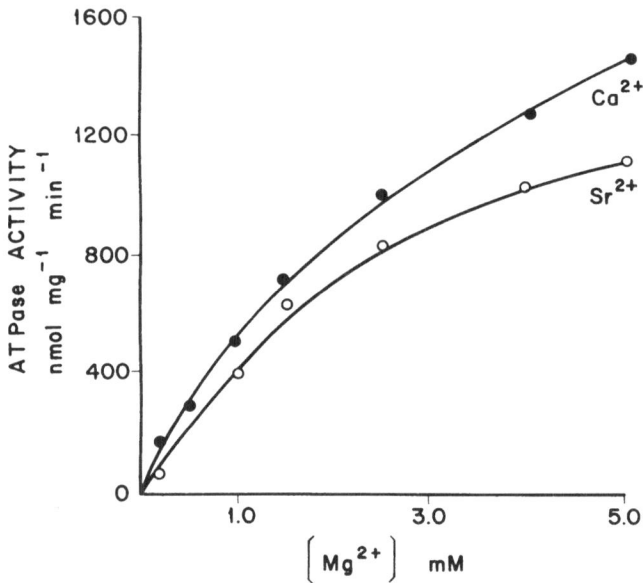

FIGURE 3. Effect of Mg^{2+} concentration on the ATPase activity of rat liver submitochondrial particles in the presence of Ca^{2+} or Sr^{2+}. The particles were preincubated during 3 min under the conditions of Fig. 2, in the presence of 2.5 mM Ca^{2+} or 10 mM Sr^{2+}. The reactions were initiated by the addition of different concentrations of Mg^{2+}. Control experiments showed that the activity of the ATP-regenerating system largely exceeded the ATPase activity in each Mg^{2+} concentration. (From Fagian et al. (1986), by permission of the copyright holder.)

4. Ca^{2+} TRANSPORT BY PLANT MITOCHONDRIA

It has been shown that Ca^{2+} transport by plant mitochondria, when it occurs, varies in many aspects between different plant species and operates differently from that of vertebrate tissue mitochondria (Douce, 1985). For instance, it has been claimed that Ca^{2+} uptake by plant mitochondria presents the following characteristics: (1) it is not associated to respiratory stimulation and depolarization of membrane potential; (2) it is substrate-dependent; (3) it has an absolute requirement for inorganic phosphate; and (4) it is insensitive to ruthenium red and Mg^{2+}. In addition, the high K_m and the low initial rate of Ca^{2+} uptake led to the conclusion that it is unlikely that Ca^{2+} transport by plant mitochondria represents a carrier-mediated transport (McCormack and Denton, 1986).

Recently, we developed a procedure that yields functionally intact preparations of plant mitochondria. With the exception of cabbage mitochondria, all other types of plant mitochondria studied (corn, soybean, bean, coffee) are much more active in respiration-coupled Ca^{2+} accumulation than those employed in most earlier studies (Martins and Vercesi, 1985; Martins et al., 1986; Carnieri et al., 1987).

These studies established that Ca^{2+} uptake by those plant mitochondria is accompanied by H^+ extrusion, small increase in the rate of respiration, and is inhibited by ruthenium red and Mg^{2+} (Fig. 4). The V_{max} for Ca^{2+} uptake by those mitochondria are in the range of

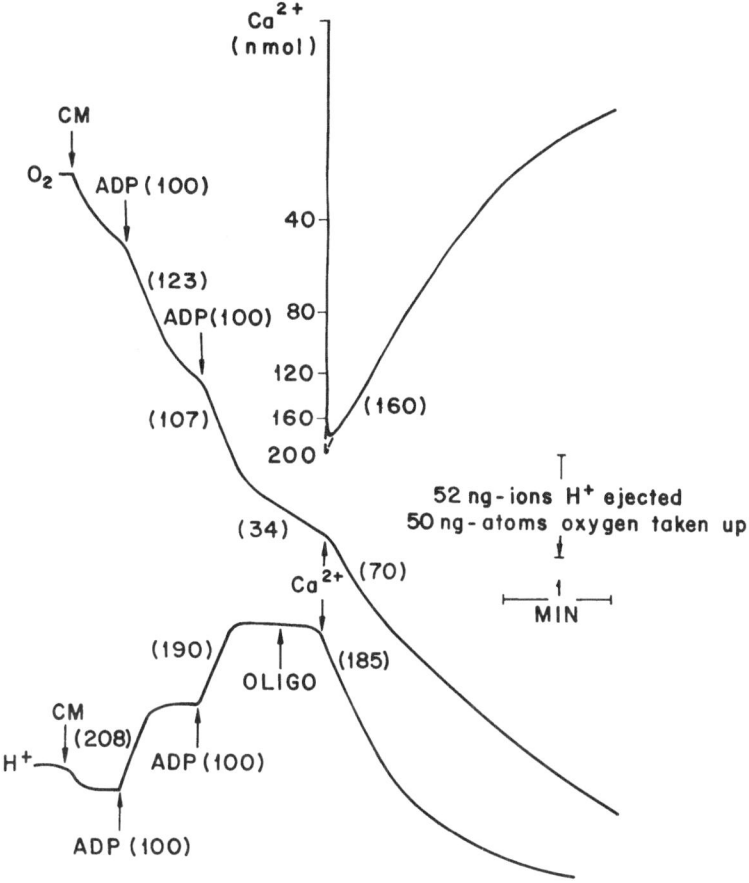

FIGURE 4. Alterations in the rates of O₂ consumption and pH of the reaction medium caused by the additions of ADP and Ca²⁺ to isolated corn mitochondria. Corn mitochondria (CM) (0.5 mg) were added to 1 ml of reaction medium containing 300 mM mannitol, 20 mM KCl, 0.1% bovine serum albumin, 2 mM Hepes buffer (pH 7.2), 5 mM succinate, 5 mM phosphate, and 5 μM rotenone. Oligomycin (Oligo) (2 μg/mg protein), ADP (100 nmol), and Ca²⁺ (200 nmol) were added where indicated. The numbers in parentheses refer to the velocities of O₂ consumption (ng atom O/min/mg), H⁺ uptake or extrusion (nmol H⁺/min/mg), or Ca²⁺ uptake (nmol Ca²⁺/min/mg). (From Martins et al. (1986), by permission of the copyright holder.)

140–180 nmol Ca²⁺/mg/min, which is much smaller than those observed for mitochondria from vertebrate tissues. The low activity of the Ca²⁺ influx transporter as compared to the activity of the redox proton pump in these mitochondria explains why Ca²⁺ accumulation is followed by only a small decrease in $\Delta\psi$ and small increase in the rate of respiration. In fact, a linear relationship was obtained between the changes in $\Delta\psi$ and the rates of succinate oxidation (ratios of H⁺ efflux) during Ca²⁺ uptake when succinate oxidation was gradually inhibited by increasing concentrations of malonate (Carnieri et al., 1987). All these mitochondria show the ability to buffer external free Ca^{2+} at concentrations in the range of 1.5–2.0 μM under experimental conditions in which rat liver mitochondria maintain external free Ca^{2+} in the range of 0.5–0.75 μM. Addition of Ca^{2+} or EGTA to these mitochondria under steady-state conditions was followed by Ca^{2+} uptake or release, respectively, tending to restore the original Ca^{2+} steady state indicating that, like mitochondria from vertebrate tissues, these plant mitochondria possess independent pathways for Ca^{2+} influx and efflux.

In contrast to mitochondria from vertebrate tissues, the three intramitochondrial dehydrogenases (PDH, NAD-ICDH, and OGDH) prepared from plant mitochondria do not exhibit Ca^{2+} sensitivity (McCormack and Denton, 1986; and also I. S. Martins, unpublished results) indicating that the Ca^{2+} transport system present in plant mitochondria is not linked to the regulation of these enzymes.

Although the concentrations at which these mitochondria buffer external free Ca^{2+} concentrations are relatively high when compared to mitochondria from vertebrate tissues, the data do not allow a definitive statement about the importance of these mitochondria in the regulation of Ca^{2+} distribution within plant cells since there are no data at present concerning the free Ca^{2+} concentration in the cytosol of these cells.

ACKNOWLEDGMENT. The results from this laboratory were supported by grants of the Brazilian agencies FAPESP and CNPq.

REFERENCES

Abou-Khalil, S., Abou-Khalil, W. H., and Yunis, A. A., 1981, Inhibition by Ca^{2+} of oxidative phosphorylation in myeloid tumor mitochondria, *Arch. Biochem. Biophys.* **209:**460–464.

Bellomo, G., Jewell, S. A., Thor, H., and Orrenius, S., 1982, Regulation of intracellular calcium compartmentation: Studies with isolated hepatocytes and t-butylhydroperoxide, *Proc. Natl. Acad. Sci. USA* **79:**6842–6846.

Bernardes, C. F., Pereira-da-Silva, L., and Vercesi, A. E., 1986, t-butylhydroperoxide-induced Ca^{2+} efflux from liver mitochondria in the presence of physiological concentrations of Mg^{2+} and ATP, *Biochim. Biophys. Acta* **850:**41–48.

Broekemeier, K. M., Schmid, P. C., Schmid, H. H. O., and Pfeiffer, D. R., 1985, Effects of phospholipase A_2 inhibitors on ruthenium red-induced Ca^{2+} release from mitochondria, *J. Biol. Chem.* **260:**105–113.

Carnieri, E. G. S., Martins, I. S., and Vercesi, A. E., 1987, Ca^{2+} transport by plant mitochondria: Aspects linked to the mechanism and biological role, *Brazilian J. Med. Biol. Res.* **20:**635–638.

Coelho, J. L. C., and Vercesi, A. E., 1980, Retention of Ca^{2+} by rat liver and heart mitochondria—Effect of phosphate, Mg^{2+} and NAD(P) redox state, *Arch. Biochem. Biophys.* **204:**141–147.

Douce, R., 1985, *Mitochondria in Higher Plants: Structure, Function, and Biogenesis*, Academic Press, Orlando.

Fagian, M. M., Pereira-da-Silva, L., and Vercesi, A. E., 1986, Inhibition of oxidative phosphorylation by Ca^{2+} or Sr^{2+}: A competition with Mg^{2+} for the formation of adenine nucleotide complexes, *Biochim. Biophys. Acta* **852:**262–268.

Fiskum, G., and Lehninger, A. L., 1979, Regulated release of Ca^{2+} from respiring mitochondria by $Ca^{2+}/2H^+$ antiport, *J. Biol. Chem.* **254:**6236–6239.

Fiskum, G., and Lehninger, A. L., 1981, *Calcium and Cell Functions*, Vol. 2 (W. Y. Cheung, ed.), Academic Press, New York, pp. 38–80.

Frei, B., Winterhalter, K. H., and Richter, C., 1985, Quantitative and mechanistic aspects of the hydroperoxide-induced release of Ca^{2+} from rat liver mitochondria, *Eur. J. Biochem.* **149:**633–639.

Gómez-Puyou, M. T., Gavilanes, M., Gómez-Puyou, A., and Ernster, L., 1980, Control of activity states of heart mitochondrial ATPase. Role of the proton-motive force and Ca^{2+}, *Biochim. Biophys. Acta* **592**:396–405.

Hansford, R. G., 1985, Relation between mitochondrial calcium transport and control of energy metabolism, *Rev. Physiol. Biochem. Pharmacol.* **102**:1–72.

Kosower, N. S., Zipser, Y., and Faltin, Z., 1982, Membrane thiol-disulfide status in glucose-6-phosphate dehydrogenase deficient red cells. Relationship to cellular glutathione, *Biochim. Biophys. Acta* **691**:345–352.

Lehninger, A. L., Vercesi, A. E., and Bababunmi, E. A., 1978, Regulation of Ca^{2+} release from mitochondria by the oxidation-reduction state of pyridine nucleotides, *Proc. Natl. Acad. Sci. USA* **75**:1690–1694.

Lê-Quôc, K., and Lê-Quôc, D., 1985, Crucial role of sulfhydryl groups in the mitochondrial inner membrane structure, *J. Biol. Chem.* **260**:7422–7428.

Malmström, R., and Carafoli, E., 1977, The interaction of Ca^{2+} with mitochondria from human myometrium, *Arch. Biochem. Biophys.* **182**:657–666.

Martins, I. S., and Vercesi, A. E., 1985, Some characteristics of Ca^{2+} transport in plant mitochondria, *Biochem. Biophys. Res. Commun.* **129**:943–948.

Martins, I. S., Carnieri, E. G. S., and Vercesi, A. E., 1986, Characteristics of Ca^{2+} transport by corn mitochondria, *Biochim. Biophys. Acta* **850**:49–56.

McCormack, J. G., and Denton, R. M., 1986, Ca^{2+} as a second messenger within mitochondria, *Trends Biochem. Sci.* **11**:258–262.

Moreno-Sánchez, R., 1983, Inhibition of oxidative phosphorylation by a Ca^{2+}-induced diminution of the adenine nucleotide translocation, *Biochim. Biophys. Acta* **724**:278–285.

Nicholls, D. G., and Brand, M. D., 1980, The nature of the calcium efflux induced in rat liver mitochondria by the oxidation of endogenous nicotinamide nucleotides, *Biochem. J.* **188**:113–118.

Panfili, E., Sottocasa, G. L., Sandri, G., and Liut, G., 1980, The Ca^{2+}-binding glycoprotein as the site of metabolic regulation of mitochondrial Ca^{2+} movements, *Eur. J. Biochem.* **105**:205–210.

Roman, I., Clark, A., and Swanson, P. D., 1981, The interaction of calcium transport and ADP phosphorylation in brain mitochondria, *Membrane Biochem.* **4**:1–9.

Sies, H., Graf, P., and Estrela, J. M., 1981, Hepatic calcium efflux during cytochrome P-450-dependent drug oxidations at the endoplasmic reticulum in intact liver, *Proc. Natl. Acad. Sci. USA* **78**:3358–3362.

Sordahl, L. A., 1974, Effects of magnesium, ruthenium red and antibiotic ionophore A-23187 on initial rates of calcium uptake and release by heart mitochondria, *Arch. Biochem. Biophys.* **167**:104–115.

Vallières, J., Scarpa, A., and Somlyo, A. P., 1975, Subcellular fractions of smooth muscle, *Arch. Biochem. Biophys.* **170**:659–669.

Vercesi, A. E., 1984a, Dissociation of NAD(P)$^+$-stimulated mitochondrial Ca^{2+} efflux from swelling and membrane damage, *Arch. Biochem. Biophys.* **232**:86–91.

Vercesi, A. E., 1984b, Possible participation of membrane thiol groups on the mechanism of NAD(P)$^+$-stimulated Ca^{2+} efflux from mitochondria, *Biochem. Biophys. Res. Commun.* **119**:305–310.

Villalobo, A., and Lehninger, A. L., 1980, Inhibition of oxidative phosphorylation in ascites tumor mitochondria and cells by intramitochondrial Ca^{2+}, *J. Biol. Chem.* **255**:2457–2464.

II

Phosphoinositide Metabolism

13

Inositol Lipids and Intracellular Communication

MICHAEL JOHN BERRIDGE

1. INTRODUCTION

Cells communicate with each other by means of chemical signals such as hormones and neurotransmitters. There has been a major interest in trying to unravel how cells detect these incoming signals and translate the information into the second messengers responsible for regulating many diverse physiological processes. An intracellular communication system based on the products of inositol lipid hydrolysis is central to many cellular control mechanisms. This signaling pathway is particularly responsible for altering the intracellular level of calcium, which is the major theme of this article. Initially, I shall examine how cells synthesize and hydrolyze the inositol lipid responsible for generating the second messengers inositol-1,4,5-trisphosphate (Ins1,4,5P$_3$) and diacylglycerol (DG). The way in which these two messengers function to modulate calcium homeostasis will be considered with particular emphasis on how they might interact with each other to produce a highly integrated signaling network. One consequence of such interactions is that the second messengers may oscillate at variable frequencies, which may be an important element in how they mediate their effects. Before considering these more dynamic aspects, it is necessary first to describe some basic biochemical features of this inositol lipid-signaling pathway.

2. SECOND MESSENGER FORMATION AND METABOLISM

2.1. Inositol Lipid Metabolism

The key feature of this signaling system is that the second messengers are generated by hydrolyzing the lipid phosphatidylinositol-4,5,-bisphosphate (PtdIns4,5P$_2$) located within the cytoplasmic leaflet of the plasma membrane. This substrate lipid is formed by the stepwise phosphorylation of phosphatidylinositol (PtdIns) first on the 4 position and then on the 5

MICHAEL JOHN BERRIDGE ● Unit of Insect Neurophysiology and Pharmacology, Department of Zoology, University of Cambridge, Cambridge CB2 3EJ, United Kingdom.

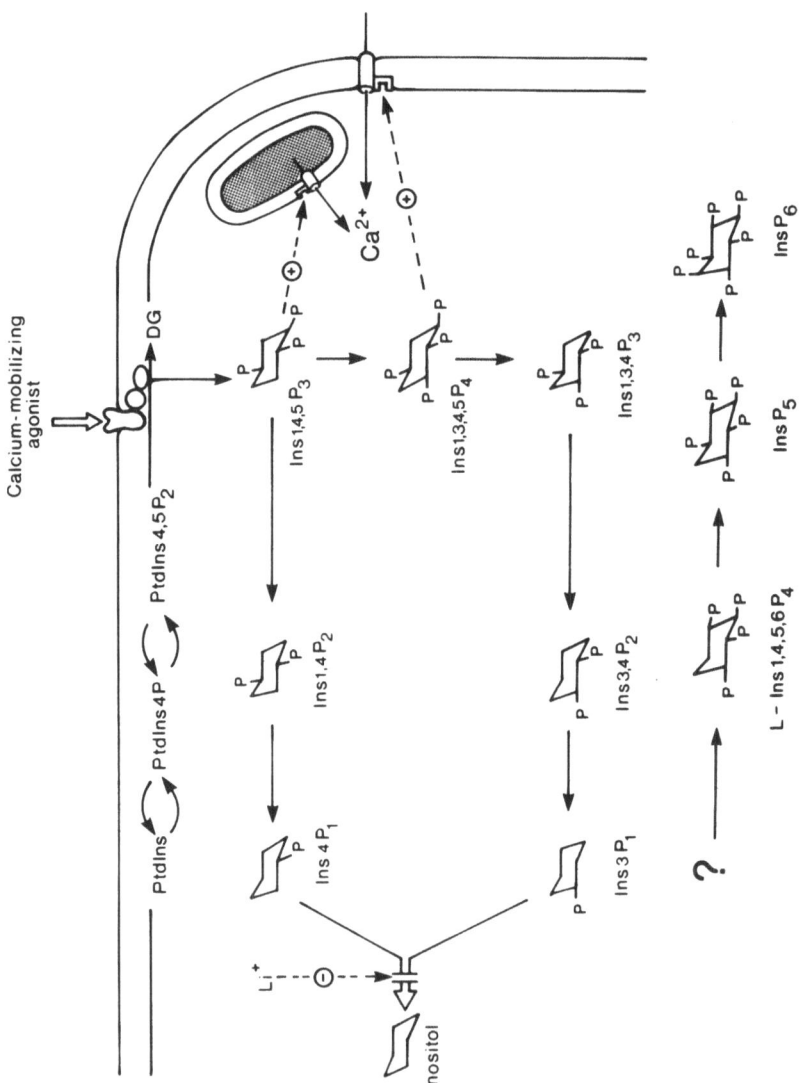

FIGURE 1. Agonist-dependent phosphoinositide metabolism.

position of the inositol headgroup. When agonists bind to the appropriate receptor, they stimulate a phosphoinositidase (Downes and Michell, 1985) that cleaves PtdIns4,5P$_2$ to release Ins1,4,5P$_3$ to the cytosol whereas the DG remains within the plane of the membrane. This DG is subsequently metabolized either by being hydrolyzed by a DG lipase to release arachidonic acid or by being phosphorylated to phosphatidic acid (PA) by a DG kinase. The PA is primed by interacting with CTP to produce the CMP. PA, which combines with free inositol to reform the parent molecule PtdIns. Originally, it was thought that this resynthesis of PtdIns took place within the endoplasmic reticulum, but now there is evidence that this lipid can be formed within the plasma membrane (Imai and Gershengorn, 1987).

2.2. Inositol Trisphosphate Metabolism

The metabolism of Ins1,4,5P$_3$ is complicated by the existence of two separate pathways (Fig. 1). In one pathway it is sequentially dephosphorylated to free inositol (Storey et al., 1984). The first step is carried out by an inositol trisphosphatase that specifically removes the phosphate from the 5 position which serves to terminate the calcium-mobilizing action of Ins1,4,5P$_3$ because the product Ins1,4P$_2$ is incapable of releasing calcium (Berridge and Irvine, 1984). The other pathway begins with a kinase that phosphorylates Ins1,4,5P$_3$ specifically on the 3 position to give inositol-1,3,4,5-tetrakisphosphate (Ins1,3,4,5P$_4$) (Irvine et al., 1986). The latter is then dephosphorylated to Ins1,3,4P$_3$ by specifically removing the phosphate from the 5 position (Batty et al., 1985), apparently by means of the same enzyme that hydrolyzes Ins1,4,5P$_3$ (Tennes et al., 1987). The Ins1,3,4P$_3$ is then dephosphorylated to Ins3,4P$_2$ and perhaps also to Ins1,3P$_2$ (Irvine et al., 1987). The significance of this inositol tris/tetrakis pathway is that it generates additional inositol polyphosphates that may have messenger functions as discussed in the following section.

3. SECOND MESSENGERS AND CALCIUM SIGNALING

3.1. Some General Properties of Cellular Calcium Homeostasis

At equilibrium, the net flow of calcium into the cytoplasmic compartment is exactly balanced by the net extrusion of calcium out of the cell or into the endoplasmic reticulum. There is thus a constant cycling of calcium across both the endoplasmic reticulum and the plasma membrane. There are two major mechanisms for raising the cytoplasmic level of calcium. The first mechanism is triggered by depolarization of the plasma membrane and depends on having a voltage sensor in the plasma membrane that is coupled to various effector systems. The best known are the voltage-operated calcium channels (VOC) located within the plasma membrane. There is less information concerning the phenomenon of excitation-contraction coupling where the voltage sensor within skeletal muscle membranes triggers the release of calcium from the sarcoplasmic reticulum (SR).

The problem is to explain how the signal derived from depolarizing the t tubule is transmitted across the 15- to 20-nm gap to trigger the release of calcium from the SR. Traditionally there have been two major hypotheses, one chemical and the other electrical. The chemical hypothesis suggests that some diffusible signal, such as calcium, carries information across the gap. Ins1,4,5P$_3$ has now been put forward as an alternative candidate for the chemical transmitter generated in the t tubule to diffuse across to the SR to release calcium (Vergara et al., 1985; Volpe et al., 1985).

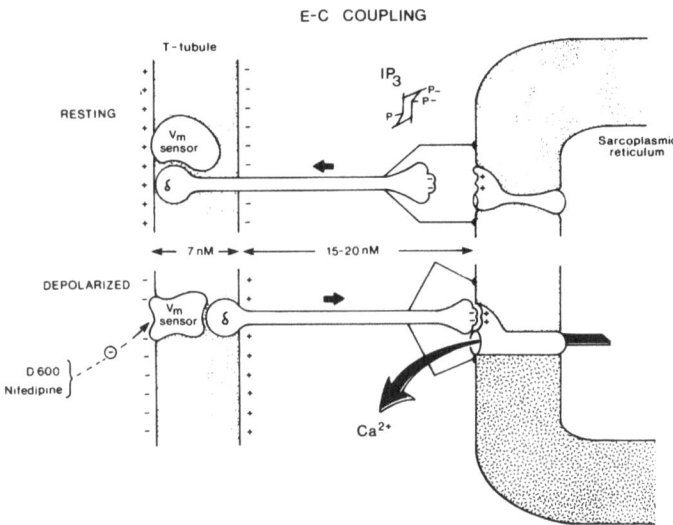

FIGURE 2. Proposed role of t-tubule feet in E-C coupling. Depolarization may induce a displacement of the t-tubule feet such that their ends, which may possess a negatively charged configuration similar to that of Ins1,4,5P₃, activate a receptor-operated calcium channel in the endoplasmic reticulum.

These chemical hypotheses do not take into account the t-tubule feet that span the gap and are incorporated into the electrical hypothesis, which considers that these feet may act as mechanotransducers that move in response to t-tubule depolarization and so distort the SR membrane sufficiently to release calcium. A possible mechanism whereby t-tubule movement might release SR calcium is shown in Fig. 2. It is proposed that the end of the t-tubule foot facing the SR membrane may have a molecular configuration resembling that of Ins1,4,5P₃ which is poised over an Ins1,4,5P₃ receptor similar to that already described in the endoplasmic reticulum. Depolarization of the t-tubule membrane will displace the t tubule sufficiently for its Ins1,4,5P₃-like headgroup to activate the receptor resulting in a rapid release of calcium (Fig. 2). Such a mechanism would have very rapid on-off times and would also explain the observation that Ins1,4,5P₃ can stimulate the contraction of skinned muscle cells (Vergara *et al.*, 1985; Volpe *et al.*, 1985). In this experimental situation, the Ins1,4,5P₃ may stimulate the receptors normally operated by the ends of the t-tubule feet and so does not necessarily imply that Ins1,4,5P₃ is the normal transmitter. The observation that Ins1,4,5P₃ can stimulate skeletal muscle to contract has attracted a lot of attention, but more evidence is required in order to establish its precise role in E-C coupling.

The other mechanism for generating calcium signals is mediated by receptors that are capable of releasing calcium from the endoplasmic reticulum and of stimulating influx across the plasma membrane. The nature of the calcium entry mechanism is still a matter of debate and may occur by at least two main mechanisms. Firstly, certain agonists such as glutamate may act to open a calcium channel directly and would thus be an example of a receptor-operated calcium channel (ROC). The second example is those calcium channels that are opened by receptors acting indirectly through second messengers. Such second messenger-operated channels (SMOC) are particularly important with regard to the action of calcium-mobilizing receptors in that the generation of inositol polyphosphates may play a role in regulating the influx of external calcium.

3.2. Mobilization of Internal Calcium by Ins1,4,5P$_3$

It is now generally accepted that Ins1,4,5P$_3$ is the second messenger responsible for the mobilization of calcium from internal reservoirs (Berridge and Irvine, 1984; Berridge, 1987; Putney, 1987). The endoplasmic reticulum has a specific receptor that binds to Ins1,4,5P$_3$ with high affinity (Guillemette *et al.*, 1987). This receptor appears to be connected in some way to a channel that allows calcium to escape from the endoplasmic reticulum. The properties of this Ins1,4,5P$_3$-induced release of calcium have been described in detail elsewhere (Berridge and Irvine, 1984; Berridge, 1987; Putney, 1987). The significance of this calcium release mechanism is apparent from the fact that the addition of Ins1,4,5P$_3$ to permeabilized cells can trigger a whole host of physiological processes including the contraction of smooth muscle (Bitar *et al.*, 1986), the release of cortical granules from sea urchin oocytes (Clapper and Lee, 1985), the formation of cyclic GMP and actin polymerization in slime mold (Europe-Finner and Newell, 1985, 1986), the activation of ornithine decarboxylase in T lymphocytes (Mustelin *et al.*, 1986), and the aggregation and formation of TXB$_2$ in blood platelets (Authi *et al.*, 1986). The activation of all these physiological responses can be accounted for by the release of calcium from the endoplasmic reticulum by Ins1,4,5P$_3$.

3.3. Entry of External Calcium

The other action of calcium-mobilizing receptors is to stimulate the entry of calcium across the plasma membrane. Just how the hydrolysis of inositol lipids results in the opening of a calcium channel is still not certain. Some of the proposed mechanisms are shown in Fig. 3, which illustrates that some of the mechanisms invoke a role for second messengers such as calcium or the inositol phosphates. In neutrophils, there is evidence for calcium-induced calcium influx, which might function to perpetuate the initial Ins1,4,5P$_3$-induced increase in intracellular calcium resulting from the mobilization of this ion from the endoplasmic reticulum (von Tscharner *et al.*, 1986). A more direct action of Ins1,4,5P$_3$ to open ion channels in the plasma membrane has been described in lymphocytes by Kuno and Gardner (1987). In these experiments, the addition of Ins1,4,5P$_3$ to isolated membrane patches was found to open ion channels identical to those opened in response to the addition of mitogens. Another complication is that the operation of SMOCs may depend in some way on the participation of the underlying endoplasmic reticulum.

There is morphological evidence from many cell types for a close association between the endoplasmic reticulum and the plasma membrane which may have a functional significance with regard to calcium entry. In his capacitative model, Putney (1986) suggested that Ins1,4,5P$_3$ has a primary site of action to release calcium from the endoplasmic reticulum that somehow then promotes the entry of calcium across the plasma membrane. The sudden release of calcium from the endoplasmic reticulum will stimulate the active accumulation of calcium, which may rapidly deplete the concentration of calcium within the narrow space between the two membranes. This in turn might destabilize the plasma membrane sufficiently to allow calcium to enter from outside. Alternatively, there may be an energy-coupled transfer of calcium, because the decrease in calcium concentration together with the formation of ADP and phosphate will create conditions that could lead to a reversal of the plasma membrane

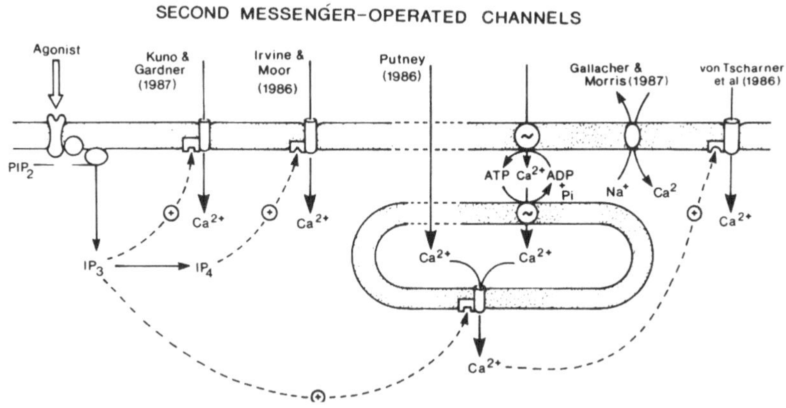

SECOND MESSENGER-OPERATED CHANNELS

FIGURE 3. Summary of some of the proposed mechanisms for controlling calcium entry through second messenger-operated calcium channels.

calcium pump, thus allowing calcium to enter the cell with the additional bonus of a net synthesis of ATP (Fig. 3). In this model, the entry of calcium would occur via a carrier mechanism rather than through a channel. The possibility that calcium might enter via a carrier rather than a channel has been raised by Gallacher and Morris (1987), who suggest that the Na^+-Ca^{2+} exchanger might play a role in regulating calcium entry in submandibular cells.

Another model for linking inositol lipid hydrolysis to the entry of external calcium envisages a role for Ins1,3,4,5P$_4$ to open calcium channels in the plasma membrane (Irvine and Moor, 1986). Evidence for such a mechanism has been obtained in sea urchin oocytes, where Ins1,3,4,5P$_4$ promotes the entry of external calcium through a process that seems to require prior calcium depletion of the endoplasmic reticulum (Irvine and Moor, 1986). Injection of Ins1,3,4,5P$_4$ alone has no effect but when combined with Ins2,4,5P$_3$, which causes a partial release of calcium from the endoplasmic reticulum, it induced a profound release of cortical granules. It would appear that the calcium that enters the cell in response to Ins1,3,4,5P$_4$ may be rapidly buffered by the underlying endoplasmic reticulum and cannot trigger a cellular response unless this calcium-sequestering system is short-circuited by an appropriate inositol trisphosphate. Once again, the endoplasmic reticulum is envisaged to play a crucial role in modulating calcium entry. It is clear that we require data from patch-clamp experiments to define more accurately the putative role of inositol polyphosphates in regulating calcium entry across the plasma membrane.

3.4. Diacylglycerol and Calcium Signaling

The DG that remains within the plasma membrane following the hydrolysis of PtdIns4,5P$_2$ also functions as a second messenger by activating protein kinase C (Nishizuka, 1984; Kikkawa and Nishizuka, 1986). One important function of this DG/C-kinase pathway is to modulate several aspects of calcium signaling and this could represent one of its primary functions (Berridge, 1987). It may influence a range of physiological processes by altering their sensitivity to calcium, which is particularly important in secretory cells. Another important action of protein kinase C is to reduce the level of intracellular calcium through actions at several steps along the calcium-signaling pathway such as the activation of calcium pumps, stimulation of Ins1,4,5P$_3$ hydrolysis, and inhibition of the transduction process in the plasma membrane that generates Ins1,4,5P$_3$ (Berridge, 1987).

The net result of all these reactions is that the DG/C-kinase pathway exerts a powerful negative feedback effect on the calcium-signaling pathway. This negative feedback component is counteracted by a positive feedback loop operating between the Ins1,4,5P$_3$/Ca^{2+} and DG/C-kinase pathway because calcium plays an important role in stimulating protein kinase C (Nishizuka, 1984). The existence of such positive and negative feedback components means that this is a highly integrated signaling system that may be somewhat unstable and could contribute to the oscillations in intracellular calcium that have been recorded in many different cell types.

4. OSCILLATIONS IN INTRACELLULAR CALCIUM

As techniques become available for monitoring intracellular calcium in single cells, it is becoming increasingly apparent that the level is seldom constant but oscillates, especially under conditions whereby cells are being stimulated. Some examples of such oscillations are summarized in Table I, where calcium was measured either indirectly through its ability to open endogenous ion channels or directly by using an appropriate detector such as aequorin. Of particular significance is the fact that not only is the cellular oscillator induced by natural stimuli but its frequency can vary with the concentration of the external stimulus (Rapp and Berridge, 1981; Woods et al., 1986). In the insect salivary gland, periodicity varied with the concentration of 5-hydroxytryptamine in the same range over which fluid secretion was being stimulated and led to the proposal that the rate of secretion might be a function of second messenger frequency rather than amplitude (Rapp and Berridge, 1981). Cells thus appear to have an *agonist-controlled oscillator* that sets up stable fluctuations in intracellular calcium which is decoded by effector systems to give various physiological responses such as secretion or contraction. This hypothesis of frequency-dependent control is amply supported by measurements of the calcium oscillations in hepatocytes, where increases in the intracellular level of vasopressin had no effect on the amplitude of the calcium fluctuations but markedly enhanced their frequency (Woods et al., 1986), thus supporting the view that cells have an agonist-controlled oscillator.

If cells are responding to oscillations in intracellular calcium rather than its amplitude,

TABLE I. Examples of Intracellular Calcium Oscillations in Different Cell Types Measured Either Indirectly by Monitoring Fluctuations in Membrane Potential Caused by the Opening of Ca^{2+}-Dependent Ion Channels[a] or Directly by Means of the Photoprotein Aequorin[b]

Tissue	Period (sec)	Reference
Pituitary	1–3[a]	Poulsen and Williams (1976)
Intestinal cells	10[a]	Yada et al. (1986)
L cells	12[a]	Okada et al. (1977)
Mouse oocytes	15–35[b]	Cuthbertson and Cobbold (1985)
Calliphora salivary gland	20–90[a]	Rapp and Berridge (1981)
Hepatocytes	40–60[b]	Woods et al. (1986)
Golden hamster eggs	60[a]	Miyazaki et al. (1986)
Sympathetic ganglion	180[a]	Morita et al. (1980)

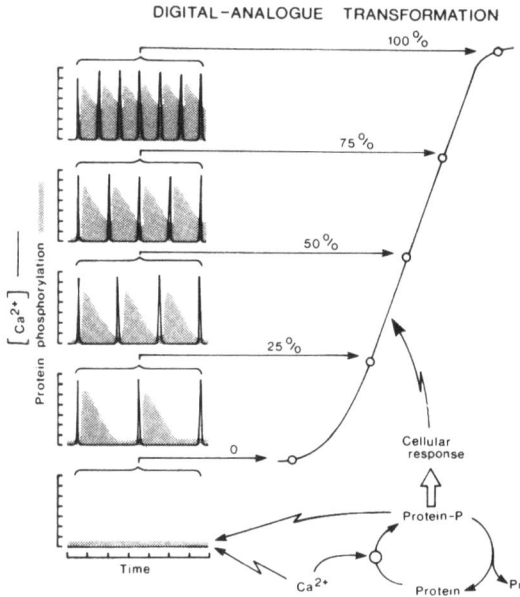

FIGURE 4. A digital-analog transformation model to explain how different intracellular calcium oscillation frequencies might be translated into a smooth change in some cellular response. Each calcium spike triggers a calcium-dependent phosphorylation of a protein (stippled area) that serves as a short-term "memory" that can be integrated over time to transform changes in frequency into variable cellular responses.

the next problem to consider is how variations in frequency are translated into a linear monotonic change in some physiological response. One obvious possibility is that the rise in calcium triggers some response which, for the sake of argument, might depend on the phosphorylation of a protein by the calcium/calmodulin-dependent protein kinase (Fig. 4). The calcium pulse is rapidly removed, leaving the protein phosphorylated, and the cell will remain active until the protein is dephosphorylated by a phosphatase. Each pulse of calcium is thus translated into a certain degree of cell activation. If the calcium pulses are widely spaced, the cellular response will be small, but as the frequency increases in response to an elevation in hormone concentration the proportion of the time that the protein remains phosphorylated will also rise, resulting in an increase in activity (Fig. 4). It is therefore proposed that the agonist-controlled oscillator encodes hormonal information into a digital form (the calcium spikes), which is then decoded by the effector system. The advantage of such frequency-dependent control is that it may provide for much greater accuracy, which will be less corrupted by noise than an amplitude-dependent system (Rapp and Berridge, 1981).

5. CONCLUSION

Many cell surface receptors are now known to act through a transduction mechanism that utilizes $PtdIns4,5P_2$ as a substrate to generate second messengers that have a major role in modulating the intracellular level of calcium. The primary $Ins1,4,5P_3$ released to the cytosol mobilizes calcium from the endoplasmic reticulum, which is a major source of signal calcium especially during early stimulation periods. When stimulation is sustained, calcium enters from outside, but little is known about the mechanism of agonist-induced calcium entry. In some cases, agonists may act on receptors coupled directly to ROCs, while there are other examples of SMOCs where the receptor acts indirectly through some internal messenger such as calcium or the inositol polyphosphates.

The DG that remains within the plane of the membrane activates protein kinase C, which can modulate calcium signaling either by altering the sensitivity of calcium effector systems or by reducing the level of signal calcium. Such interactions between these messenger pathways may contribute to the sustained oscillations in intracellular calcium that have been recorded in numerous cell types. It is proposed that cells have an agonist-controlled oscillator that encodes external signals into a digital form (calcium oscillations) which are then decoded by the final effector system. Such frequency-dependent signaling may provide a noise-free system capable of translating small variations in hormone concentration into stable changes in cellular activity.

REFERENCES

Authi, K. S., Evenden, B. J., and Crawford, N., 1986, Metabolic and functional consequences of introducing inositol 1,4,5-trisphosphate into saponin-permeabilized human platelets, *Biochem. J.* **233**:707–718.

Batty, I. R., Nahorski, S. R., and Irvine, R. F., 1985, Rapid formation of inositol (1,3,4,5) tetrakisphosphate following muscarinic stimulation of rat cerebral cortical slices, *Biochem. J.* **232**:211–215.

Berridge, M. J., 1987, Inositol trisphosphate and diacylglycerol: Two interacting second messengers, *Ann. Rev. Biochem.* **56**:159–193.

Berridge, M. J., and Irvine, R. F., 1984, Inositol trisphosphate, a novel second messenger in cellular signal transduction, *Nature* **312**:315–321.

Bitar, K. N., Bradford, P. G., Putney, J. W., and Makhlouf, G. M., 1986, Stoichiometry of contraction and Ca^{2+} mobilization by inositol 1,4,5-trisphosphate in isolated gastric smooth muscle cells, *J. Biol. Chem.* **216**:16591–16596.

Clapper, D. L., and Lee, H. C., 1985, Inositol, trisphosphate induces calcium release from nonmitochondrial stores in sea urchin egg homogenates. *J. Biol. Chem.* **260**:13947–13954.

Cuthbertson, K. S. R., and Cobbold, P. H., 1985, Phorbol ester and sperm activate mouse oocytes by inducing sustained oscillations in cell Ca^{2+}, *Nature* **316**:541–542.

Downes, C. P., and Michell, R. H., 1985, Inositol phospholipid breakdown as a receptor-controlled generator of second messengers, in: *Molecular Mechanisms of Transmembrane Signaling* (P. Cohen and M. D. Houslay, eds.), Elsevier, New York, pp. 3–56.

Europe-Finner, G. N., and Newell, P. C., 1985, Inositol 1,4,5-trisphosphate induces cyclic GMP formation in *Dictyostelium discoideum*, *Biochem. Biophys. Res. Commun.* **130**:1115–1122.

Europe-Finner, G. N., and Newell, P. C., 1986, Inositol 1,4,5-trisphosphate and calcium stimulate actin polymerization in *Dictyostelium discoideum*, *J. Cell. Sci.* **82**:41–51.

Gallacher, D. V., and Morris, A. P., 1987, The receptor-regulated calcium influx in mouse submandibular acinar cells is sodium dependent: A patch-clamp study, *J. Physiol.* **384**:119–130.

Guillemette, G., Balla, T., Baukal, A. J., Spat, A., and Catt, K. J., 1987, Intracellular receptors for inositol 1,4,5-trisphosphate in angiotensin II target tissues, *J. Biol. Chem.* **262**:1010–1015.

Imai, A., and Gershengorn, M. C., 1987, Independent phosphatidylinositol synthesis in pituitary plasma membrane and endoplasmic reticulum, *Nature* **325**:726–728.

Irvine, R. F., Letcher, A. J., Heslop, J. P., and Berridge, M. J., 1986, The inositol tris/tetrakisphosphate pathway: Demonstration of Ins(1,4,5)P_3-3-kinase activity in animal tissues, *Nature* **320**:631–634.

Irvine, R. F., Letcher, A. J., Lander, D. J., Heslop, J. P., and Berridge, M. J., 1987, Inositol (3,4) bisphosphate and inositol (1,3) bisphosphate in GH₄ cells: Evidence for complex breakdown of inositol (1,3,4) trisphosphate, *Biochem. Biophys. Res. Commun.* **143**:353–359.

Irvine, R. F., and Moor, R. M., 1986, Micro-injection of inositol (1,3,4,5) tetrakisphosphate activates sea urchin eggs by a mechanism dependent on external Ca^{2+}, *Biochem. J.* **240**:917–920.

Kikkawa, U., and Nishizuka, Y., 1986, The role of protein kinase C in transmembrane signalling, *Ann. Rev. Cell. Biol.* **2**:149–178.

Kuno, M., and Gardner, P., 1987, Ion channels activated by inositol 1,4,5-trisphosphate in plasma membrane of human T-lymphocytes, *Nature* **326**:301–304.

Miyazaki, S-I., Hashimoto, N., Yoshimoto, Y., Kishimoto, T., Igusa, Y., and Hiramoto, Y., 1986, Temporal and spatial dynamics of the periodic increase in intracellular free calcium at fertilization of golden hamster eggs, *Dev. Biol.* **118**:259–267.

Morita, K., and Kojetsu, K., 1980, Oscillation of $[Ca^{2+}]_i$-linked K^+ conductance in bullfrog sympathetic ganglion cell is sensitive to intracellular anions, *Nature* **283**:204–205.

Mustelin, T., Poso, H., and Andersson, L. C., 1986, Role of G-proteins in T cell activation: Non-hydrolyzable GTP analogues induce early ornithine decarboxylase activity in human T lymphocytes, *EMBO J.* **5**:3287–3290.

Nishizuka, Y., 1984, The role of protein kinase C in cell surface signal transduction and tumor promotion, *Nature* **308**:693–697.

Okada, Y., Doida, Y., Roy, G., Tsuchiya, W., Inouye, K., and Inouye, A., 1977, Oscillations of membrane potential in L cells, I. Basic characteristics. *J. Membrane Biol.* **35**:319–335.

Poulsen, J. H., and Williams, J. A., 1976, Spontaneous repetitive hyperpolarizations from cells in the rat adenohypophysis, *Nature* **263**:156–158.

Putney, J. W., 1986, A model for receptor-regulated calcium entry, *Cell* **7**:1–12.

Putney, J. W., 1987, Formation and actions of calcium-mobilizing messenger, inositol 1,4,5-trisphosphate, *Am. J. Physiol.* **252**:G149–G157.

Rapp, P. E., and Berridge, M. J., 1981, The control of transepithelial potential oscillations in the salivary gland of *Calliphora erythrocephala, J. Exp. Biol.* **93**:119–132.

Storey, D. J., Shears, S. B., Kirk, C. J., and Michell, R. H., 1984, Stepwise enzymatic dephosphorylation of inositol 1,4,5-trisphosphate to inositol in liver, *Nature* **312**:374–376.

Tennes, K. A., McKinney, J. S., and Putney, J. W. Jr., 1987, Metabolism of inositol 1,4,5-trisphosphate in guinea-pig hepatocytes, *Biochem. J.* **242**:797–802.

von Tscharner, W., Prod'hom, B., Baggiolini, M., and Reuter, H., 1986, Ion channels in human neutrophils activated by a rise in free cytosolic calcium concentration, *Nature* **324**:369–372.

Vergara, J., Tsien, R. Y., and Delay, M., 1985, Inositol 1,4,5-trisphosphate: A possible chemical link in excitation-contraction coupling in muscle, *Proc. Natl. Sci. U.S.A.* **82**:6352–6356.

Volpe, P., Salviati, G., Di Virgilio, F., and Pozzan, T., 1985, Inositol 1,4,5-trisphosphate induces calcium release from sarcoplasmic reticulum of skeletal muscle, *Nature* **316**:347–349.

Woods, N. M., Cuthbertson, K. S. R., and Cobbold, P. H., 1986, Repetitive transient rises in cytoplasmic free calcium in hormone-stimulated hepatocytes, *Nature* **319**:600–602.

Yada, T., Oiki, S., Veda, S., and Okada, Y., 1986, Synchronous oscillations of the cytoplasmic Ca^{2+} concentration and membrane potential in cultured epithelial cells (intestine 407), *Biochim. Biophys. Acta* **887**:105–112.

14

Receptor-Mediated Ca^{2+} Signaling
Role in Inositol-1,3,4,5-Tetrakisphosphate Generation

JOHN R. WILLIAMSON, ROY A. JOHANSON,
CARL A. HANSEN, and JONATHAN R. MONCK

1. INTRODUCTION

Over the past few years it has become apparent that Ca^{2+} signaling in many different cell types is initiated by an agonist-induced breakdown of inositol lipids in the plasma membrane (Berridge and Irvine, 1984; Williamson et al., 1985; Hokin, 1985). This is mediated by a receptor-coupled GTP-binding protein, which is thought to dissociate upon binding of suitable agonists to their receptors to a $\beta\gamma$-subunit complex and an α subunit containing bound GTP (Fitzgerald et al., 1986; Dickey et al., 1987). The GTP-bound α subunit then promotes the activity of an inositol lipid phosphodiesterase (phospholipase C) that splits phosphatidylinositol-4,5-bisphosphate to a lipophilic 1,2-diacylglycerol (the physiological activator of protein kinase C) and a water-soluble product, inositol-1,4,5-trisphosphate (Ins-1,4,5-P$_3$). This inositol polyphosphate isomer acts as a Ca^{2+}-mobilizing second messenger by opening a Ca^{2+} channel located in a specialized intracellular pool of calcium (Joseph and Williamson, 1986).

The size of the Ins-1,4,5-P$_3$-sensitive Ca^{2+} pool in most cells appears to be small relative to the total intracellular Ca^{2+} content but is sufficient to raise the cytosolic free Ca^{2+} to about 1 μM within a few seconds after maximal agonist stimulation (Thomas et al., 1984). Typically, the rise in cytosolic free Ca^{2+} is transient due to the removal of Ca^{2+} from the cell via the plasma membrane Ca^{2+}-ATPase and to its sequestration into other intracellular organelles, including mitochondria, where it activates enzymes associated with energy metabolism (Denton and McCormack, 1985). This phase of the agonist-induced Ca^{2+} transient is totally dependent on an increased concentration of Ins-1,4,5-P$_3$ but is essentially independent of extracellular Ca^{2+} provided that the time of exposure of cells to Ca^{2+}-free medium is sufficiently short to prevent depletion of intracellular Ca^{2+} pools. In the absence of extracellular Ca^{2+}, the agonist-induced cellular response is short-lived, in contrast to a longer

JOHN R. WILLIAMSON, ROY A. JOHANSON, CARL A. HANSEN, and JONATHAN R. MONCK
• Department of Biochemistry and Biophysics, University of Pennsylvania School of Medicine, Philadelphia, Pennsylvania 19104-6089.

acting effect observed with normal extracellular Ca^{2+}, which is associated with a small but sustained increase in the cytosolic free Ca^{2+} (Joseph et al., 1985). This latter response is caused by an increased rate of Ca^{2+} entry into the cell, which in the steady state is revealed by an increased rate of turnover of Ca^{2+} across the plasma membrane (Rasmussen and Barrett, 1984).

Agonist-induced perturbation of cellular Ca^{2+} homeostasis thus has three distinct components, namely intracellular Ca^{2+} mobilization, intracellular Ca^{2+} redistribution, and extracellular Ca^{2+} mobilization. The mechanism of receptor-mediated Ca^{2+} entry into cells by voltage-independent Ca^{2+} channels has not been ascertained but, as with intracellular Ca^{2+} mobilization, may be directly related to the stimulation of inositol lipid metabolism. One possibility that has attracted considerable attention is that a phosphorylated metabolite of Ins-1,4,5-P_3, namely Ins-1,3,4,5-P_4, may promote Ca^{2+} entry by activation of a ligand-gated Ca^{2+} channel in the plasma membrane (Irvine and Moor, 1986). Ins-1,3,4,5-P_4, which is formed from Ins-1,4,5-P_3 by the action of a soluble ATP-dependent inositol phosphate-3-kinase (Irvine et al., 1986; Hansen et al., 1986), has been shown to accumulate rapidly in cells treated with Ca^{2+} mobilizing agonists (see Williamson et al., 1988 for review). Both metabolites undergo hydrolysis by a 5-phosphomonoesterase (Connolly et al., 1987; Shears et al., 1987). Recent studies have shown that the 3-kinase activity in the cytosolic fraction of insulin-secreting RIN m5F cells was stimulated by Ca^{2+} in the range from 10^{-7} to 10^{-6} M (Biden and Wollheim, 1986), suggesting that the initial increase in cytosolic free Ca^{2+} induced by Ins-1,4,5-P_3 may have a role in redirecting its metabolism from degradative hydrolysis to formation of higher inositol phosphate products, potentially with different functional roles.

In this paper we present new data on the kinetics of agonist-stimulated increases in cytosolic free Ca^{2+} in single cells loaded with Fura 2, which show that in some instances intra- and extracellular Ca^{2+} mobilization can be kinetically separated. In addition, our studies on the purification and properties of enzymes involved in the metabolism of inositol polyphosphates are summarized, which help to define the detailed overall pathway for their metabolism and interconversion. Finally, data are presented to show that the Ca^{2+} sensitivity of the Ins-1,4,5-P_3 3-kinase can be accounted for by the ability of the purified catalytic subunit to bind calmodulin.

2. AGONIST-STIMULATED Ca^{2+} TRANSIENTS IN SINGLE CELLS

The development of several Ca^{2+}-sensitive fluorescent indicators (Tsien, 1980; Grynkiewicz et al., 1985) that can be loaded into cells without disruption of the plasma membrane has enabled measurement of changes of the cytosolic free Ca^{2+} in cells with minimal disturbance of cell function. Fura 2, which has a high quantum efficiency (Grynkiewicz et al., 1985), permitting use of low concentrations of the indicator, as well as rapid kinetics for Ca^{2+} association and dissociation, allowing it to respond to fast changes in intracellular Ca^{2+} distribution (Jackson et al., 1987), has been applied successfully to measurements of intracellular free Ca^{2+} in single cells from a variety of tissues (e.g., Williams et al., 1985; Tsien and Poenie, 1986; Connor et al., 1987). Measurement of Ca^{2+} changes in single cells by fluorescence microscopy has several advantages over those made with bulk cell suspensions. Thus measurements in individual cells allow determination of free Ca^{2+} changes in particular cells within mixed populations, as well as the detection of heterogeneous responses within apparently homogeneous cell populations and also the capability of detecting spatially restricted changes of Ca^{2+} within individual cells (Tsien and Poenie, 1986; Connor *et al.*, 1987). As illustrated below, studies with single cells have in fact produced unexpected findings that could not have been anticipated from measurements with bulk cell populations.

Figure 1 shows results obtained with single Fura 2-loaded vascular smooth-muscle A10 cells stimulated with a submaximally effective concentration (0.5 nM) of vasopressin. Mono-layers of cells grown on glass coverslips were loaded with Fura 2 by incubating with 5 μM Fura-2/AM in L15 tissue culture medium for 60 min at 37°C and transferred to a Dvorak flow cell for fluorescence imaging using a Nikon microscope equipped for epifluorescence. The cells were perfused with modified Hanks medium at a flow rate of 250 μl/min at 28°C, excited at a wavelength of 340 nM, and the fluorescence emission was collected over wavelengths of 470–600 nm. Video images were obtained using a SIT camera, digitized with averaging, and analyzed using a Joyce Loebl/Nikon Magiscan 2a image analysis system. For each experiment a series of images were collected at appropriate intervals and the total fluorescence was calculated by segmenting the stored images, integrating the intensity values over the whole-cell area, and subtracting the background signal. The fractional fluorescence change (F/F_0) obtained by dividing the fluorescence (340 nm excitation) at a specific time by the fluorescence value in the resting cell was used for calculation of changes of the cytosolic free Ca^{2+} (Monck *et al.*, 1988) after determining the resting free Ca^{2+} concentration by the 340 nm/380 nm ratio method of Grynkiewicz *et al.* (1985).

As shown in Fig. 1, when the A10 cells were incubated with 1.25 mM extracellular Ca^{2+} (filled circles), vasopressin induced an initial rapid rise in cytosolic free Ca^{2+} to 250 nM from a resting value of 68 ± 7 nM (mean of 10 determinations), which declined almost

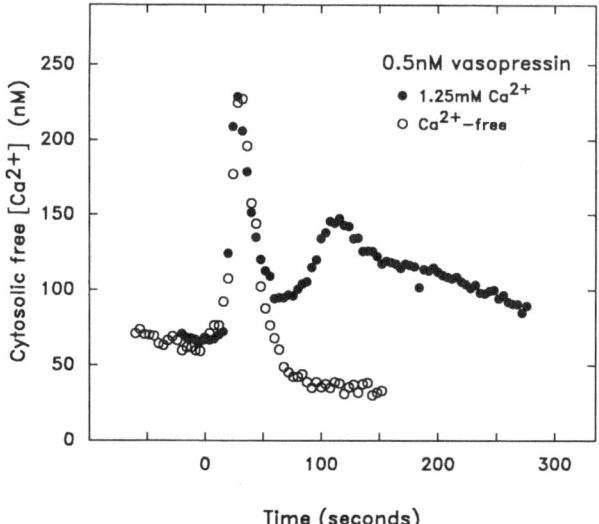

FIGURE 1. Changes in cytosolic free Ca^{2+} concentration in single A10 cells stimulated with 0.5 nM vasopressin. Results from two separate cells loaded with Fura 2 are shown for incubations in Ca^{2+}-free (5 μM) medium or with 1.25 mM Ca^{2+}. Successive images were taken 4 sec apart.

to resting values before increasing again after about 80 sec to give a second, smaller rise in free Ca^{2+}. This second Ca^{2+} transient was totally abolished when vasopressin was added to cells incubated in nominally Ca^{2+}-free medium (about 5 μM), but the initial Ca^{2+} transient was unaffected (Fig. 1, closed circles). The second Ca^{2+} transient was also attenuated when the cells were pretreated with 2 mM Co^{2+} as a nonspecific Ca^{2+} channel blocker but was not affected by diltiazem (10 μM), a specific antagonist of L-type voltage-sensitive Ca^{2+} channels (Fox et al., 1986). By analogy with other studies on smooth muscle (e.g., Nabika et al., 1985; Reynolds and Dubyak, 1986), the initial phase of the Ca^{2+} transient can be ascribed to mobilization of intracellular Ca^{2+} by agonist-dependent production of Ins-1,4,5-P_3, whereas the second peak reflects a stimulated influx of Ca^{2+} from the extracellular medium. The presence of two phases of Ca^{2+} mobilization has been demonstrated in many cell types (Rasmussen and Barrett, 1984; Williamson et al., 1985), but the temporal separation seen with single A10 cells, and also seen in single GH_3 cells (Kruskal et al., 1984), is unusual. Furthermore, a double-Ca^{2+} transient was not observed in single A10 cells stimulated with maximum vasopressin concentrations, with cell populations of A10 cells at submaximal vasopressin concentrations, or with single hepatocytes (Monck et al., 1988). At present the mechanism of receptor-mediated increased Ca^{2+} influx has not been ascertained, although activation of voltage-sensitive Ca^{2+} channels by G proteins has been described (Heschler et al., 1987). The temporal separation between intracellular and extracellular Ca^{2+} mobilization in a manner dependent on the agonist concentration is compatible with the suggestion that a sufficient concentration of a different second messenger (e.g., Ins-1,3,4,5-P_4) has to be generated in order to activate a ligand-gated Ca^{2+} channel in the plasma membrane. Alternatively, it is possible that the Ca^{2+} channel must first be primed by a prior phosphorylation step, perhaps mediated by Ca^{2+}/calmodulin-dependent protein kinase or protein kinase C, before ligand-stimulated Ca^{2+} influx can occur.

In contrast to the A10 cells, however, stimulation of single Fura 2-loaded hepatocytes at 28°C with submaximal concentrations of phenylephrine (Fig. 2) or vasopressin (Fig. 3) showed only an abrupt increase of cytosolic free Ca^{2+} followed by a plateau phase lasting

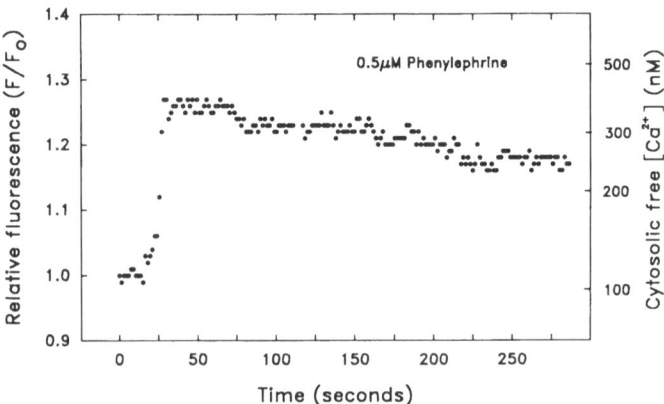

FIGURE 2. Changes in cytosolic free Ca^{2+} concentration in a single hepatocyte loaded with Fura 2 and stimulated with 0.5 μM phenylephrine. Successive images were taken at 1.5-sec intervals and time zero represents the estimated time of arrival of the agonist.

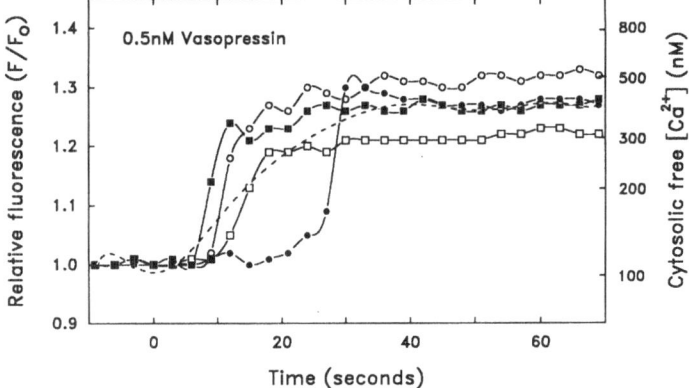

FIGURE 3. Changes in cytosolic free Ca^{2+} concentration in hepatocytes loaded with Fura 2 and stimulated with 0.5 nM vasopressin. Four hepatocytes were viewed simultaneously under the microscope and images were collected at 3-sec intervals. Only the early part of the transient is shown to emphasize the different latencies of response of the cells, but data were collected for a total of 4 min and showed that the Ca^{2+} remained elevated with no sign of periodic oscillations. The dashed line represents the mean change of the four individual cells. Time zero represents the estimated time of arrival of the agonist, which was determined by infusing a solution of Fura 2 free acid.

several minutes. As with the A10 cells, the sustained increase of cytosolic free Ca^{2+} was dependent on the presence of extracellular Ca^{2+}. Thus, it is evident that in hepatocytes, the triggering mechanism for activation of Ca^{2+} influx has a similar time of onset to that for intracellular Ca^{2+} mobilization. Studies with hepatocytes in bulk cell suspensions stimulated with phenylephrine (Williamson and Hansen, 1987) or vasopressin (Hansen et al., 1986) show that the accumulation of Ins-1,3,4,5-P$_4$ is almost as rapid as that of Ins-1,4,5-P$_3$. However, it must be stressed that to date no causal connection has been made between production of Ins-1,3,4,5-P$_4$ and the stimulation of Ca^{2+} flux across the plasma membrane.

Figure 3 illustrates a separate phenomenon that can only be observed by analysis of Ca^{2+} transients in single cells, namely a difference in latency of the response of different cells to agonist addition. In Fig. 3, the change in fluorescence of the Fura 2–Ca^{2+} complex was measured simultaneously in four individual hepatocytes stimulated with 0.5 nM vasopressin. Each of the cells responded with different delays of from 5 to 25 sec after the estimated time of exposure to the hormone (time zero in Fig. 3), although once initiated the increase in Ca^{2+} reached a peak rapidly after 9–12 sec. The mean change for these cells (dashed line) shows that the average of these four cells reproduces a response that closely resembles the population response seen in suspensions of hepatocytes (Thomas et al., 1984; Charest et al., 1985). Similar latencies in the response to individual cells were also observed with submaximal concentrations of phenylephrine, but with maximal agonist concentrations the latency was reduced to 1–3 sec and asynchrony of the responses was diminished. These data suggest that the variable latency preceding the rapid increase in cytosolic free Ca^{2+} in different cells may be explained by a threshold phenomenon (e.g., due to different receptor densities) in which it takes some cells longer than others to accumulate sufficient Ins-1,4,5-P$_3$ to trigger the Ca^{2+} response.

Finally, the data in Figs. 2 and 3 provide no evidence for repetitive spikes of increased Ca^{2+} with continuous agonist stimulation, as recently reported for single hepatocytes microinjected with aequorin (Woods *et al.*, 1986, 1987). In these latter experiments, 0.5 nM phenylephrine induced peaks of aequorin light emission, which returned to resting levels within 10 sec and were followed by repetitive increases 80- to 90-sec intervals. The reason for this discrepancy is presently not clear, but it may be related to differences between the mechanisms whereby the indicators report Ca^{2+} or their methods of introduction into the cell. Thus, microinjection of aequorin into cells may destabilize the plasma membrane to ion conductances, or Fura 2 may cause significant buffering of the Ca^{2+} transients such that all but the first are abolished. It is also possible that Fura 2 itself or its buffering effect on the Ca^{2+} rise interferes with the negative feedback mechanism that is responsible for generating the oscillatory behavior. Indirect evidence using two other techniques suggests that oscillations in Ca^{2+} concentration occur in liver. In livers perfused with low (10 μM) Ca^{2+} medium, agonist-induced oscillations of extracellular Ca^{2+} were seen after addition of vasopressin, phenylephrine, and angiotensin II, suggesting that the Ca^{2+} was derived from intracellular stores (Graf *et al.*, 1987). Moreover, in voltage-clamped guinea pig hepatocytes (Capiod *et al.*, 1987), norepinephrine stimulated an outward K^+ current, which exhibited cyclic fluctuations that could be blocked by inhibitors of Ca^{2+}-activated K^+ channels. Internal perfusion of Ins-1,4,5-P_3 gave similar oscillations whereas raising the intracellular free Ca^{2+} to 2 μM did not, suggesting that if oscillations in cytosolic free Ca^{2+} are occurring, the site of generation is distal to Ins-1,4,5-P_3 production. At present the mechanism and functional significance of hormonally induced oscillations of intracellular Ca^{2+} remain to be established.

3. PATHWAY AND ENZYMES OF INOSITOL PHOSPHATE METABOLISM

In order to understand the implications of the potential roles of inositol polyphosphates as intracellular second messengers or signaling modulators of cell function, it is essential to have knowledge of the properties and kinetics of the enzymes responsible for the formation and metabolism of the inositol phosphate products resulting from receptor-mediated activation of inositol lipid breakdown. Toward this end we have purified a number of enzymes that are involved in the metabolism of Ins-1,4,5-P_3 and Ins-1,3,4,5-P_4, and have characterized the enzymes and intermediates in their dephosphorylation pathway to myoinositol. Rat brain was used as a source of material for this purpose because of the relatively high enrichment of enzymes of the inositol lipid pathway in this tissue, but differences in the properties as well as the characteristics of the enzymes are possible between tissues and species.

Figure 4 schematizes current knowledge concerning the pathways of metabolism of Ins-1,4,5-P_3 based on our work and that of others. Ins-1,4,5-P_3 and in some tissues its 1,2-cyclic phosphate are formed by a phosphodiester cleavage of phosphatidylinositol-4,5-bisphosphate. The formation and metabolism of the 1,2-cyclic phosphate inositol phosphate derivatives have been reviewed recently by Majerus *et al.* (1986). They appear to be substrates of the same enzymes as those for noncyclic inositol phosphates, although with different kinetic constants. So far no distinct roles for the cyclic forms of the inositol phosphates have been described that differ significantly from those of their noncyclic counterparts, and their significance in receptor signaling is currently unknown.

As seen from Fig. 4, inactivation of Ins-1,4,5-P_3 as a Ca^{2+} mobilizing second messenger occurs by two routes. One is by a Mg^{2+}- and ATP-dependent 3-kinase to Ins-1,3,4,5-P_4 and

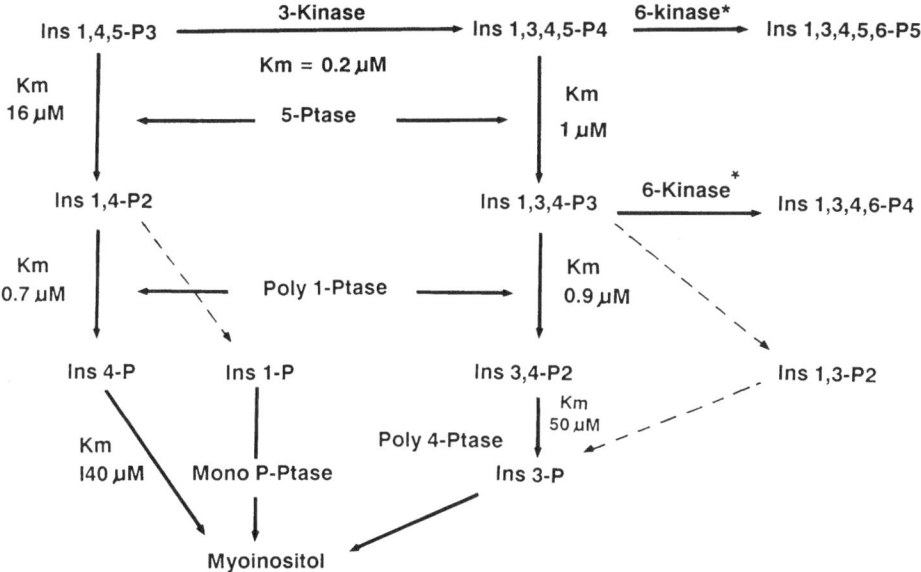

FIGURE 4. Pathways for metabolism of inositol phosphates. The putative inositol polyphosphate 6-kinase has not been characterized and its reaction with Ins-1,3,4,5-P$_4$ as substrate has not been demonstrated in mammalian cells.

the other is by phosphomonoesterase enzymes specific for hydrolysis of phosphates in the 5 position of inositol polyphosphates. Characteristically, the K_m for metabolism of Ins-1,4,5-P$_3$ to Ins-1,3,4,5-P$_4$ is much lower (up to 100-fold) than for its metabolism to Ins-1,4-P$_2$, although the total 5-phosphomonoesterase activity in the tissues that have been examined is greater than the 3-kinase activity. Brain tissue contains at least three enzymes with 5-phosphomonoesterase activity and, with Ins-1,3,4,5-P$_4$ as substrate, its K_m is generally lower than for Ins-1,4,5-P$_3$. The product of Ins-1,3,4,5-P$_4$ hydrolysis is Ins-1,3,4-P$_3$ (Batty et al., 1987), which is further degraded by a polyinositol phosphate-1-phosphomonoesterase to Ins-3,4-P$_2$ (Inhorn et al., 1987). The same enzyme also degrades Ins-1,4-P$_2$ to Ins-4-P, with the K_m for both substrates being approximately equal. Ins-3,4-P$_2$ is degraded to Ins-3-P by a poorly characterized 4-phosphomonoesterase that has a relatively high K_m, appears to have a broad substrate specificity, and probably accounts for the appearance of Ins-1,3-P$_2$ as well as Ins-3,4-P$_2$ in intact cells (Irvine et al., 1987). Finally, inositol monophosphates appear to be all hydrolyzed to myoinositol by Li$^+$-sensitive inositol monophosphate phosphatase with broad substrate specificity and relatively high K_m (Hallcher and Sherman, 1980). Figure 4 also illustrates possible routes for the formation of higher inositol phosphates by a putative inositol polyphosphate-6-kinase. This enzyme has not been characterized in mammalian tissues, but the major inositol polyphosphate in avian erythrocytes is known to be the Ins-1,3,4,5,6-P$_5$ isomer, which implies the presence of a kinase specific for the 6 position with Ins-1,3,4,5-P$_4$ as the presumed substrate (Chakrabarti and Biswas, 1981). A different IP$_4$, namely, Ins-1,4,5,6-P$_4$, has been identified by nuclear magnetic resonance in avian erythrocytes (Mayr and Dietrich, 1987), suggesting the presence of an inositol polyphosphate-

3-phosphomonoesterase in this tisssue. Yet another IP_4 isomer, Ins-1,3,4,6-P_4, has been shown to be produced by phosphorylation of Ins-1,3,4-P_3 by the putative 6-kinase (Shears et al., 1987).

3.1. Inositol-5-phosphomonoesterase Enzymes

Both soluble and particulate forms of inositol polyphosphate-5-phosphomonoesterase have been identified and the distribution of these forms is known to vary from tissue to tissue. In erythrocytes (Downes et al., 1982), liver (Storey et al., 1984; Seyfred et al., 1984; Joseph and Williams, 1985), macrophages (Kukita et al., 1986), and brain (Erneux et al., 1986), the major activity is in the particulate (plasma membrane) fraction, whereas in platelets (Connolly et al., 1985) and coronary artery smooth muscle (Sasagura et al., 1985), the principle activity is soluble. Two soluble forms of the enzyme with different molecular weights and kinetic characteristics have recently been purified from rat brain (Hansen et al., 1987). A 100,000g supernatant with a specific activity of 0.014 μmol/mg/min was purified by phosphocellulose column chromatography, which allowed separation of the 3-kinase from the 5-phosphomonoesterase activities with a 20-fold enrichment of the latter. A 0.3 M KCl gradient elution from a DEAE-Sepharose column separated the 5-phosphomonoesterase activity into two peaks (designated type 1 and type 2 enzymes), which were further purified by ATP affinity chromatography, hydroxylapatite, and Sephacryl size exclusion chromatography. The properties of these enzymes are shown in Table I and compared with those of the membrane-bound enzyme, which has not yet been purified sufficiently for an estimate to be made of its molecular weight. None of the enzyme activities were sensitive to inhibition by Li^+, and were little affected by Ca^{2+} between 10^{-7} and 10^{-5} M. The highly purified type 1 and type 2 enzymes had final specific activities of 1 and 15 μmol/mg/min, respectively. The membrane-bound and type 1 (M_r of 60,000) enzyme can use both Ins-1,4,5-P_3 and Ins-1,3,4,5-P_4 as substrates (which makes them mutually competitive inhibitors), while the type 2 enzyme (M_r of 160,000) will only utilize Ins-1,3,4,5-P_4 at high substrate concentrations. The type 1 enzyme has similar kinetic properties and M_r to the soluble 5-phosphomonoesterase purified from platelets (Connolly et al., 1985) and, as previously shown for the platelet enzyme (Connolly et al., 1986), is phosphorylated by protein kinase C with an increase of V_{max} but with no effect on the K_m for either substrate. At present it is not clear whether other tissues have multiple forms of 5-phosphomonoesterase. The

TABLE I. Characteristics of Rat Brain Inositol Phosphate-5-Phosphomonoesterase

Characteristic	Type 1	Type 2	Membrane bound (isolated synaptic membranes)
Distribution in crude homogenate	12%	15%	72%
SDS-PAGE M_r	60,000	160,000	?
Gel filtration M_r	60,000	160,000	?
K_m Ins-1,4,5-P_3	3 μM	18 μM	16 μM
K_m Ins-1,3,4,5-P_4	0.8 μM	>150 μM	3 μM
pH optima	7.5	6.5	7.0
Potentially regulated by protein kinase C	Yes, ↑ V_{max}	?	?

functional significance of the finding that some tissues have a predominantly membrane-bound form has also not been ascertained.

3.2. Inositol Polyphosphates 1- and 4-phosphomonoesterase

During purification of the 5-phosphomonoesterase, it was observed that phosphatase activity toward Ins-1,3,4-P$_3$ or Ins-1,4-P$_2$ was not retained by the phosphocellulose column. Consequently, this fraction was chromatographed on a DE-52 column, which upon elution with a 0.4 M KCl gradient allowed separation of the bulk of the heat stable (70°C) inositol monophosphate phosphomonoesterase activity. Upon further purification by hydroxylapatite chromatography and elution with a 0.2 M phosphate gradient, two different phosphomonoesterases could be separated. One, which retained activity toward both Ins-1,3,4-P$_3$ and Ins-1,4-P$_2$, produced only Ins-3,4-P$_2$ and Ins-4-P, respectively, as products and hence is termed inositol polyphosphate-1-phosphomonoesterase. The second enzyme activity degraded Ins-3,4-P$_2$ to Ins-3-P and also converted Ins-1,3,4-P$_3$ to Ins-1,3-P$_2$, and appears to be an inositol polyphosphate-4-phosphomonoesterase.

The identity of the reaction products was ascertained first by their characteristic elution profiles by high-performance liquid chromatography (HPLC) using a Partisil SAX column with different ammonium formate step gradients (Hansen et al., 1986) and by ^{1}H and ^{31}P NMR (Cerdan et al., 1986; Hansen et al., 1989). The degradation of [^{32}P]Ins-1,3,4-P$_3$ and [^{3}H]Ins-1,4-P$_2$ by the enzyme eluting from the DE-52 column is illustrated in Fig. 5, where it is seen that the major reaction products are Ins-3,4-P$_2$ and Ins-4-P, respectively. The K_m

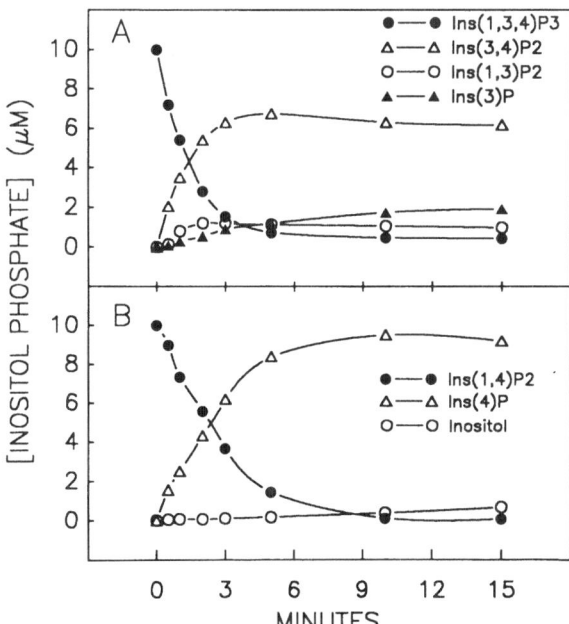

FIGURE 5. Degradation of [^{32}P]Ins-1,3,4-P$_3$ and [^{3}H]Ins-1,4-P$_2$ by partially purified inositol polyphosphate-1-phosphomonoesterase. The reaction products were analyzed by HPLC.

values of the inositol polyphosphate-1-phosphomonoesterase for Ins-1,3,4-P_3 and Ins-1,4-P_2 were 0.9 and 0.7 μM, respectively, while the cosubstrates behaved as competitive inhibitors, with K_i values of 0.5 and 0.9 μM for Ins-1,3,4-P_3 and Ins-1,4-P_2, respectively. Li$^+$ behaved as a noncompetitive inhibitor with K_i values of 230 and 320 μM with Ins-1,3,4-P_3 and Ins-1,4-P_2 as substrates, respectively.

3.3. Ins-1,4,5-P_3 3-kinase

Ins-1,4,5-P_3 3-kinase is located in the soluble fraction of the cell. It has been extensively purified from rat brain to a specific activity of 2.3 μmol/mg/min, with a M_r of 53,000 by SDS-PAGE and 45,000–50,000 by size exclusion chromatography (Johanson *et al.*, 1988). The purification scheme initially used involved chromatography with phosphocellulose, orange A dye ligand, and ATP affinity columns with a final hydroxylapatite concentration step, which yielded an enzyme with a specific activity of 1.3 μmol/mg/min. However, during the course of this work it was found that the 3-kinase would bind to a calmodulin column and could be eluted by decreasing the Ca^{2+} concentration of the buffer with excess EGTA. Consequently, this step replaced the ATP affinity column in the purification scheme finally adopted. The purified enzyme has a pH optimum at 8.0, a K_m for Ins-1,4,5-P_3 of 0.2 $\pm$ 0.07 μM (mean $\pm$ SE of four different enzyme preparations). It appears to have a narrow substrate specificity since it is inactive with Ins-1,4-P_2. A particularly interesting characteristic of the enzyme is its activation by Ca^{2+}/calmodulin as described below.

4. REGULATION OF THE INOSITOL TRIS/TETRAKISPHOSPHATE PATHWAY BY Ca^{2+}

Previous studies with crude preparations of Ins-1,4,5-P_3 3-kinase have shown that the effect of Ca^{2+} over a concentration range of 10^{-7} to 10^{-5} M was to cause either an inhibition (Irvine *et al.*, 1986) or an activation (Biden and Wollheim, 1986) of enzyme activity. Figure 6 shows that the 3-kinase in the 100,000g brain supernatant was activated threefold by an increase in free Ca^{2+} from 10^{-7} to 10^{-6} M, while a further increase in Ca^{2+} to 10^{-4} M showed some inhibition of enzyme activity.

These data suggested that the Ca^{2+} sensitivity of the enzyme might be conferred by calmodulin. In order to investigate this possibility, the 100,000g supernatant was first applied to a DEAE-Sepharose column equilibrated with buffer containing 10 mM Tris (pH 7.3), 1 mM $MgCl_2$, 1 mM dithiothreitol, and 1 mM EGTA. The enzyme eluted as a single peak with a linear KCl gradient and the pooled enzyme fraction containing approximately 0.1 M KCl was supplemented with Ca^{2+} to give a free Ca^{2+} concentration of about 0.1 mM and was applied to a calmodulin agarose affinity column (BioRad). As illustrated in Fig. 7, the enzyme was retained by the column and was eluted as a sharp peak with Ca^{2+}-free buffer. These two column steps produced an enzyme with a specific activity of 0.3 μmol/mg/min, thereby providing a convenient, alternative method of preparing a partially pure enzyme.

Figure 8 (open circles) shows that the enzyme eluting from the calmodulin column was not affected by Ca^{2+} over the concentration range from 10^{-7} to 10^{-6} M. However, when calmodulin (6 μM) was added to the enzyme (Fig. 8, closed circles), activation by Ca^{2+} in this range was restored, suggesting that calmodulin may be a subunit of the native enzyme. A similar effect has been observed with the 3-kinase from smooth muscle (Yamaguchi *et*

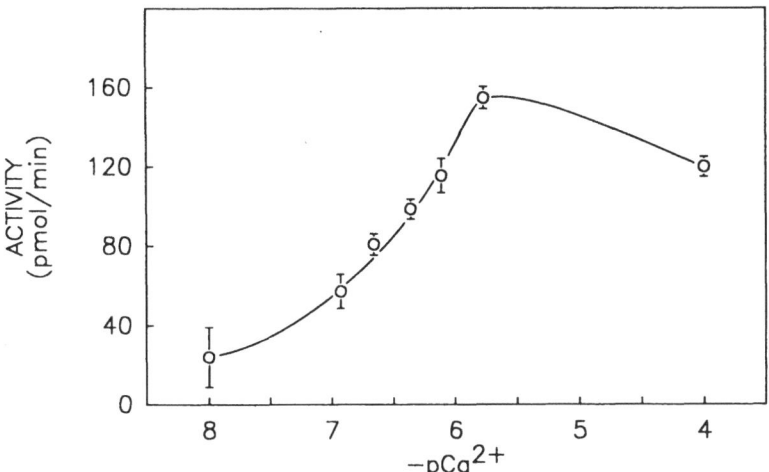

FIGURE 6. Effect of Ca^{2+} on Ins-1,4,5-P$_3$ 3-kinase activity in brain 100,000g supernatant. The supernatant was diluted 10-fold in 0.1 ml of buffer containing 20 mM Tris (pH 8.0), 100 mM KCl, 1 mM dithiothreitol, 2 mM MgATP^{2-}, 2 mM 2,3-bisphosphoglycerate, and 5 mM EGTA to which Ca^{2+} was added to give different free Ca^{2+} concentrations as measured with a Ca^{2+} electrode. The reaction was started by addition of 2 μM of [^{3}H]Ins-1,4,5-P$_3$ and after incubation for 2, 4, and 6 min at 30°C was quenched by addition of 20-μl aliquots to 40 μl of 12% (w/v) perchloric acid containing 3 mM EDTA. Ins-1,3,4,5-P$_4$ was separated using Dowex AG 1-X8 columns and ammonium formate/formic acid elution.

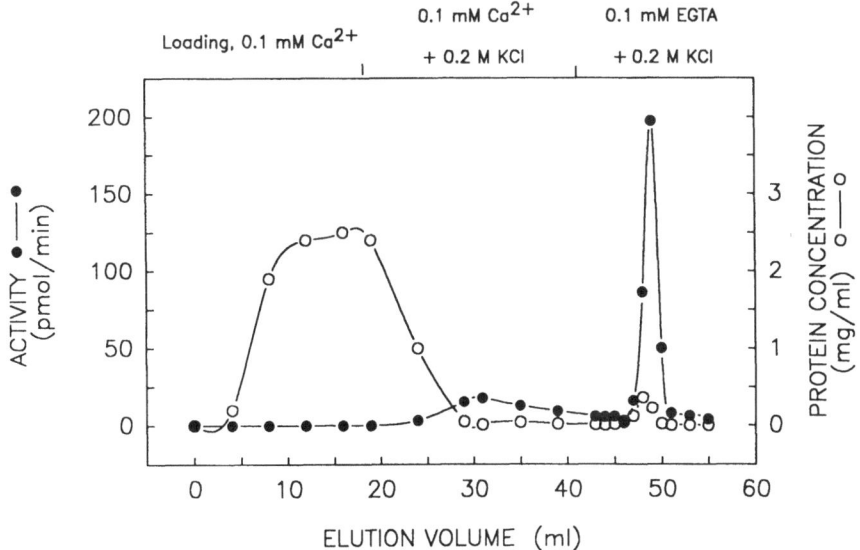

FIGURE 7. Calmodulin affinity chromatography of Ins-1,4,5-P$_3$ 3-kinase. The peak enzyme activity eluting from a DEAE-Sepharose column (20 ml) was loaded onto a 5-ml calmodulin agarose (BioRad) column in buffer containing 20 mM Tris (pH 7.3), 1 mM MgCl$_2$, 1 mM dithiothreitol, and 0.1 mM Ca^{2+}. After a Ca^{2+}-KCl buffer wash, the enzyme was eluted with buffer containing 0.1 mM EGTA in place of Ca^{2+}. Albumin (0.5 mg/ml) was added to fraction containing Ins-1,4,5-P$_3$ 3-kinase activity to stabilize the enzyme.

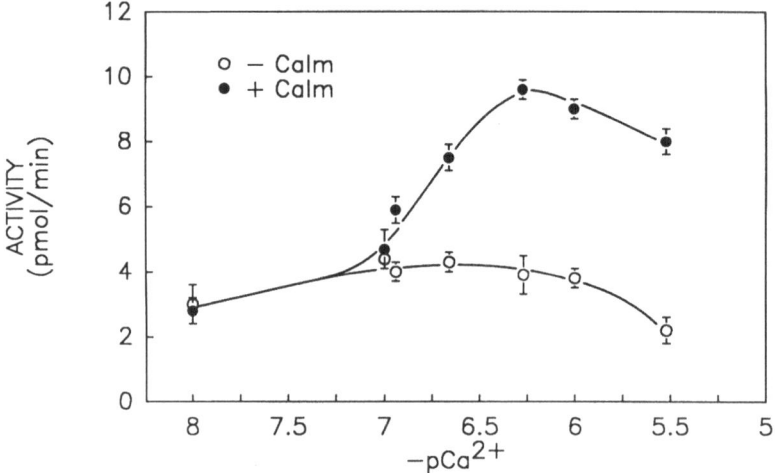

FIGURE 8. Effect of Ca^{2+} on purified Ins-1,4,5-P_3 3-kinase activity in the absence and presence of calmodulin. Enzyme eluting with EGTA buffer from the calmodulin affinity column (see Fig. 7) was incubated in 100 μl of buffer containing different free Ca^{2+} concentrations in the absence and presence of 6 μM calmodulin (Calm).

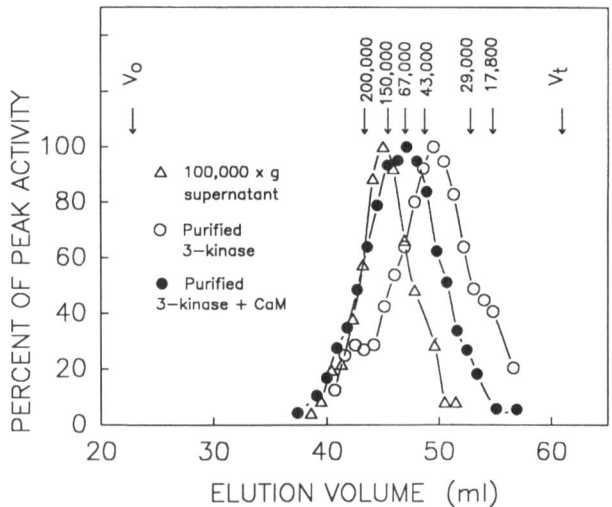

FIGURE 9. Sepharose 6B size exclusion chromatography. Samples containing Ins-1,4,5-P_3 3-kinase were chromatographed using a flow rate of 7.5 ml/hr on a 1.6 cm × 30.5 cm column equilibrated with 15 mM potassium phosphate (pH 7.4), 100 mM KCl, 1 mM $MgCl_2$, 1 mM dithiothreitol with sample collection at 7.5-min intervals. The column was standardized using blue dextran (V_0), potato β-amylase (200 kDa), alcohol dehydrogenase (150 kDa), bovine serum albumin (67 kDa), ovalbumin (43 kDa), carbonic anhydrase (29 kDa), sperm whale myoglobin (17.8 kDa), and acetone (V_t). The purified enzyme used was that eluting from the calmodulin affinity column (Fig. 7), incubated in the absence and presence of 8 μM calmodulin (CaM).

al., 1987). Preliminary studies suggest that the effect of calmodulin/Ca^{2+} activation of the 3-kinase is to increase the V_{max} of the enzyme without affecting its K_m for Ins-1,4,5-P$_3$.

Chromatography of the purified 3-kinase on a Sepharose 6B size exclusion column gave a peak of activity that corresponded to a M_r of about 45,000 (Fig. 9, open circles). When calmodulin (8 μM) was added to the 3-kinase prior to gel filtration, the enzyme activity eluted with a M_r of 90,000 (Fig. 9, closed circles). The enzyme/calmodulin complex formed in this manner, however, is of considerably lower M_r than the value of 140,000 obtained with the 100,000g brain supernatant (Fig. 9, open triangles), suggesting the formation of mixed monomeric and dimeric enzyme subunits with calmodulin.

As depicted in Fig. 10, the significance of Ca^{2+} activation of Ins-1,4,5-P$_3$ 3-kinase is that during the initial phase of agonist-induced Ca^{2+} mobilization, metabolism of Ins-1,4,5-P$_3$ is redirected from the dephosphorylation pathway toward generation of Ins-1,3,4,5-P$_4$ and its metabolic products. This effect may account for the striking overshoot of Ins-1,4,5-P$_3$ accumulation often observed within the first minute of agonist addition to cells, which correlates fairly closely with the kinetics of the Ca^{2+} transient (Williamson *et al.*, 1988). However, the regulation of Ins-1,4,5-P$_3$ levels appears to be more complex, since at least some forms of the 5-phosphomonoesterase (which utilize both Ins-1,4,5-P$_3$ and Ins-1,3,4,5-P$_4$ as substrates) are phosphorylated and activated by protein kinase C (see Fig. 10). At present the physiological significance of this effect in intact cells has not been evaluated, but the combined effects of Ca^{2+} activation of the 3-kinase and C-kinase activation of the 5-phosphomonoesterase would not only be to accelerate removal of Ins-1,4,5-P$_3$ but also to promote

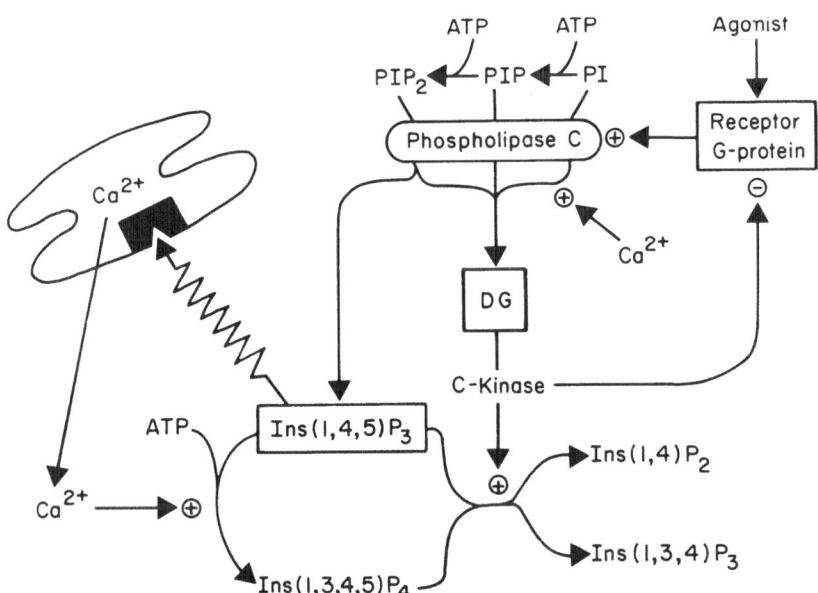

FIGURE 10. Scheme showing negative feedback effects via Ca^{2+} and C-kinase for possible regulation of Ins-1,4,5-P$_3$ and Ins-1,3,4,5-P$_4$ concentrations. The abbreviations used are: PIP$_2$, phosphatidylinositol-4,5,-bisphosphate; PIP, phosphatidylinositol-4-phosphate; PI, phosphatidylinositol; DG, 1,2-diacylglycerol; C-kinase, protein kinase C.

the formation of Ins-1,3,4-P$_3$ and its phosphorylated product Ins-1,3,4,6-P$_4$. Nevertheless, the steady-state level of Ins-1,4,5-P$_3$ after hormonal stimulation appears sufficient to produce a long-term depletion of the hormone-sensitive Ca^{2+} pool, as evidenced by the fact that addition of a second agonist without first removing or antagonizing the first agonist fails to produce a further mobilization of intracellular Ca^{2+} (Putney, 1987). Hence the real significance of the feedback effects on Ins-1,4,5-P$_3$ metabolism may reside in the as yet undiscovered possible signaling roles of other inositol polyphosphates. Likewise, further studies are required, particularly with single cells, to ascertain the significance of the oscillations of Ca^{2+} observed under some conditions after hormone stimulation and to what extent earlier studies, particularly with Quin2-loaded cells in bulk suspension, may have yielded misleading information. The complex feedback interactions via Ca^{2+} and C-kinase to multiple sites in the receptor–G protein–phospholipase C signaling pathway appear to provide several possibilities to explain the hormone-stimulated oscillations of Ca^{2+}. One possibility suggested by Woods et al. (1987) is negative feedback by diacylglycerol-activated protein kinase C on phospholipase C activity. An alternative mechanism is provided by increased Ins-1,4,5-P$_3$ removal by Ca^{2+} activation of the 3-kinase and/or protein kinase C activation of the 5-phosphomonoesterase.

ACKNOWLEDGMENT. This work was supported by grants DK-15120 and HL-14461 from the National Institutes of Health.

REFERENCES

Batty, I. R., Nahorski, S. R., and Irvine, R. F., 1987, Rapid formation of inositol 1,3,4,5-tetrakis-phosphate following muscarinic receptor stimulation of rat cerebral cortical slices, *Biochem. J.* **232:**211–215.

Berridge, M. J., and Irvine, R. F., 1984, Inositol trisphosphate, a novel second messenger in cellular signal transduction, *Nature* **312:**315–321.

Biden, T. J., and Wollheim, C. B., 1986, Ca^{2+} regulates the inositol tris/tetrakisphosphate pathway in intact and broken preparations of insulin secreting RIN m5F cells, *J. Biol. Chem.* **261:**11931–11934.

Capiod, T., Field, A. C., Ogden, D. C., and Sandford, C. A., 1987, Internal perfusion of guinea-pig hepatocytes with buffered Ca^{2+} or inositol 1,4,5-trisphosphate mimics noradrenaline activation of K$^+$ and Cl$^-$ conductances, *FEBS Lett.* **217:**247–252.

Cerdan, S., Hansen, C. A., Johanson, R., Inubushi, T., and Williamson, J. R., 1986, Nuclear magnetic resonance spectroscopic analysis of myoinositol phosphate including inositol 1,3,4,5-tetrakisphosphate, *J. Biol. Chem.* **261:**14676–14680.

Chakrabarti, S., and Biswas, B. B., 1981, Evidence for the existence of a phosphoinositol kinase in chicken erythrocytes, *Ind. J. Biochem. Biophys.* **18:**398–401.

Charest, R., Prpic, V., Exton, J. H., and Blackmore, P. F., 1985, Stimulation of inositol trisphosphate formation in hepatocytes by vasopressin, adrenaline and angiotensin II and its relationship to changes in cytosolic free Ca^{2+}, *Biochem. J.* **227:**79–90.

Connolly, T. M., Bansal, V. S., Bross, T. E., Irvine, R. F., and Majerus, P. W., 1987, The metabolism of tris- and tetraphosphates of inositol 5-phosphomonoesterase and 3-kinase enzymes, *J. Biol. Chem.* **262:**2146–2149.

Connolly, T. M., Bross, T. E., and Majerus, P. W., 1985, Isolation of a phosphomonoesterase from human platelets that specifically hydrolyzes the 5-phosphate of inositol 1,4,5-trisphosphate, *J. Biol. Chem.* **260:**7868–7874.

Connolly, T. M., Lawing, W. J. Jr., and Majerus, P. W., 1986, Protein kinase C phosphorylates human platelet inositol trisphosphate 5-phosphomonoesterase, increasing the phosphatase activity, *Cell* **46:**951–958.

Connor, J. A., Cornwall, M. C., and Williams, G. H., 1987, Spatially resolved cytosolic calcium response to angiotensin II and potassium in rat glomerulosa cells measured by digital imaging techniques, *J. Biol. Chem.* **262**:2919–2927.

Denton, R. M., and McCormack, J. G., 1985, Ca^{2+} transport by mammalian mitochondria and its role in hormone action, *Am. J. Physiol.* **249**:E543–E554.

Dickey, B. F., Fishman, J. B., Fines, R. E., and Navarro, J., 1987, Reconstitution of the rat liver vasopressin receptor coupled to guanine nucleotide binding proteins, *J. Biol. Chem.* **262**:8738–8742.

Downes, C. P., Mussat, M. C., and Michell, R. H., 1982, The inositol trisphosphate phospho-monoesterase of the human erythrocyte membrane, *Biochem. J.* **203**:169–177.

Erneux, C., Delvaux, A., Moreau, C., and Dumont, J. E., 1986, Characterization of D-myoinositol 1,4,5-trisphosphate phosphatase in rat brain, *Biochem. Biophys. Res. Commun.* **134**:351–358.

Fitzgerald, T. J., Uhing, R. J., and Exton, J. H., 1986, Solubilization of the vasopressin receptor from rat liver plasma membranes. Evidence for a receptor-GTP-binding protein complex, *J. Biol. Chem.* **261**:16871–16877.

Fox, A. P., Hess, P., Lansman, J. B., Nilius, B., Nowycky, M. C., and Tsien, R. W., 1986, Shifts between modes of calcium channel gating as a basis for pharmacological modulation of calcium influx in cardiac neuronal and smooth muscle derived cells, in: *New Insights into Cell and Membrane Transport Processes* (S. Poste and S. T. Crooke, eds.), Plenum Press, New York, pp. 99–124.

Graf, P., Dahl, S. V., and Sies, H., 1987, Sustained oscillations in extracellular calcium concentrations upon hormonal stimulation of perfused rat liver, *Biochem. J.* **241**:933–936.

Grynkiewicz, G., Poenie, M., and Tsien, R. Y., 1985, A new generation of Ca^{2+} indicators with greatly improved fluorescence properties, *J. Biol. Chem.* **260**:3440–3450.

Hallcher, L. M., and Sherman, W. R., 1980, The effects of lithium ion and other agents on the activity of myoinositol 1-phosphate from bovine brain, *J. Biol. Chem.* **255**:10896–10901.

Hansen, C. A., Inubushi, T., Williamson, M. T., and Williamson, J. R., 1989, *Biochem. Biophys. Acta*, in press.

Hansen, C. A., Johanson, R. A., Williamson, M., and Williamson, J. R., 1987, Purification and characterization of two types of soluble inositol phosphate 5-phosphomonoesterases from rat brain, *J. Biol. Chem.* **262**:17319–17326.

Hansen, C. A., Mah, S., and Williamson, J. R., 1986, Formation and metabolism of inositol 1,3,4,5-tetrakisphosphate in liver, *J. Biol. Chem.* **261**:8100–8103.

Hescheler, J., Rosenthal, W., Twautwein, W., and Schultz, G., 1987, The GTP-binding protein, Go, regulates neuronal calcium channels, *Nature* **325**:445–447.

Hokin, L. E., 1985, Receptors and phosphoinositide-generated second messengers, *Ann. Rev. Biochem.* **54**:205–235.

Inhorn, R. C., Bansal, V. S., and Majerus, P. W., 1987, Pathway for inositol 1,3,4-trisphosphate and 1,4-bisphosphate metabolism, *Proc. Natl. Acad. Sci. USA* **84**:2170–2174.

Irvine, R. F., and Moor, R. M., 1986, Micro-injection of inositol 1,3,4,5-tetrakisphosphate activates sea urchin eggs by a mechanism dependent on external Ca^{2+}, *Biochem. J.* **240**:917–920.

Irvine, R. F., Letcher, A. J., Heslop, J. P., and Berridge, M. J., 1986, The inositol tris/tetrakisphosphate pathway: Demonstrations of Ins-1,4,5-P$_3$ 3-kinase activity in animal tissue, *Nature* **320**:631–634.

Irvine, R. F., Letcher, A. J., Lander, D. J., Heslop, J. P., and Berridge, M. R., 1987, Inositol 3,4-bisphosphate and inositol 1,3-bisphosphate in GH$_4$ cells: Evidence for complex breakdown of inositol 1,3,4-trisphosphate, *Biochem. Biophys. Res. Commun.* **143**:353–359.

Jackson, A. P., Timmermann, M. P., Bagshaw, C. R., and Ashley, C. C., 1987, The kinetics of calcium binding to fura-2 and indo-1, *FEBS Lett.* **216**:35–39.

Johanson, R. A., Hansen, C. A., Coll, K. E., and Wiliamson, J. R., 1988, Purification and calmodulin sensitivity of inositol 1,4,5-trisphosphate 3-kinase from rat brain, *J. Biol. Chem.* **263**:7465–7471.

Joseph, S. K., and Williams, R. J., 1985, Subcellular localization and some properties of the enzymes hydrolyzing inositol polyphosphates in liver, *FEBS Lett.* **180**:150–154.

Joseph, S. K., and Williamson, J. R., 1986, Characteristics of inositol tris-phosphate-mediated Ca^{2+} release from permeabilized hepatocytes, *J. Biol. Chem.* **261**:14658–14664.

Joseph, S. K., Coll, K. E., Thomas, A. P., Rubin, R., and Williamson, J. R., 1985, The role of extracellular Ca^{2+} in the response of the hepatocyte to Ca^{2+}-dependent hormones, *J. Biol. Chem.* **260:**12508–12515.

Kruskal, B. A., Keith, C. H., and Maxfield, F. R., 1984, Thyrotropin-releasing hormone-induced changes in intracellular $[Ca^{2+}]$ measured by microspectrofluorometry on individual quin2-loaded cells, *J. Cell Biol.* **99:**1167–1172.

Kukita, M., Hirata, M. and Koga, T., 1986, Requirements of Ca^{2+} for the production and degradation of inositol 1,4,5-trisphosphate in macrophages, *Biochim. Biophys. Acta* **885:**121–128.

Majerus, P. W., Connolly, T. M., Deckmyn, H., Ross, T. S., Bross, T. E., Ishii, H., Bansal, V. S., and Wilson, D. B., 1986, The metabolism of phosphoinositide-derived messenger molecules, *Science* **234:**1519–1526.

Mayr, G. W. and Dietrich, W., 1987, The only inositol tetrakisphosphate detectable in avian erythrocytes is the isomer lacking phosphate at position 3: a NMR study, *FEBS Lett.* **213:**278–282.

Monck, J. R., Reynolds, E. E., Thomas, A. P., and Williamson, J. R., 1988, Novel kinetics of single cell Ca^{2+} transients in stimulated hepatocytes and A10 cells measured using fura 2 and fluorescent to videomicroscopy, *J. Biol. Chem.* **263:**4569–4575.

Nabika, T., Velletri, P. A., Lovenberg, W., and Beaven, M. A., 1985, Increase in cytosolic calcium and phosphoinositide metabolism induced by angirotensin II and [Arg]vasopressin in vascular smooth muscle cells, *J. Biol. Chem.* **260:**4661–4670.

Putney, J. W. Jr., 1987, Formation and actions of calcium-mobilizing messenger inositol 1,4,5-trisphosphate, *Am. J. Physiol.* **252:**G149–G157.

Rasmussen, H., and Barrett, P. Q., 1984, Calcium messenger system: An integrated view, *Physiol. Rev.* **64:**938–984.

Reynolds, E. E., and Dubyak, G. R., 1986, Agonist-induced calcium transients in cultured smooth muscle cells: Measurements with fura-2 loaded monolayers, *Biochem. Biophys. Res. Commun.* **136:**927–934.

Sasagura, T., Hirata, M., and Kuriyama, H., 1985, Dependence on Ca^{2+} of the activities of phosphatidylinositol 4,5-bisphosphate phosphodiesterase and inositol 1,4,5-trisphosphate phosphatase in smooth muscles of the porcine coronary artery, *Biochem. J.* **231:**497–503.

Seyfred, M. A., Farrell, L. E., and Wells, W. W., 1984, Characterization of D-myoinositol 1,4,5-trisphosphate in rat liver plasma membranes, *J. Biol. Chem.* **259:**13204–13208.

Shears, S. B., Parry, J. B., Tang, E. K. Y., Irvine, R. F., Michell, R. N., and Kirk, C. J., 1987, Metabolism of D-myo-inositol 1,3,4,5-tetrakisphosphate by rat liver, including the synthesis of a novel isomer of *myo*-inositol tetrakisphosphate, *Biochem. J.* **246:**139–147.

Shears, S. B., Storey, D. J., Morris, A. J., Cubitt, A. B., Parry, J. B., Michell, R. H., and Kirk, C. J., 1987, Dephosphorylation of myoinositol 1,4,5-trisphosphate and myoinositol 1,3,4-trisphosphate, *Biochem. J.* **242:**393–402.

Storey, D. J., Shears, S. B., Kirk, C. J., and Michell, R. H., 1984, Stepwise enzymatic dephosphorylation of inositol 1,4,5-trisphosphate to inositol in liver, *Nature* **312:**374–376.

Thomas, A. P., Alexander, J., and Williamson, J. R., 1984, Relationship between inositol polyphosphate production and the increase of cytosolic free Ca^{2+} induced by vasopressin in isolated hepatocytes, *J. Biol. Chem.* **259:**5574–5584.

Tsien, R. Y., 1980, New calcium indicators and buffers with high selectivity against magnesium and protons: Design, synthesis, and properties of prototype structures. *Biochemistry* **19:**2396–2404.

Tsien, R. Y., and Poenie, M., 1986, Fluorescence ratio imaging: A new window into intracellular ionic signalling, *Trends in Biochem. Sci.* **11:**450–455.

Williams, D. A., Fogarty, K. E., Tsien, R. Y., and Fay, F. S., 1985, Calcium gradients in single smooth muscle cells revealed by the digital imaging microscope using fura-2, *Nature* **318:**558–561.

Williamson, J. R., and Hansen, C. A., 1987, Signalling systems in stimulus-response coupling, in: *Biochemical Actions of Hormones,* Vol. 14 (G. Litwack, ed.), Academic Press, Orlando, pp. 29–80.

Williamson, J. R., Cooper, R. H., Joseph, S. K., and Thomas, A. P., 1985, Inositol trisphosphate and diacylglycerol as intracellular signal messengers, *Am. J. Physiol.* **248:**C203–C216.

Williamson, J. R., Hansen, C. A., Johanson, R. A., Coll, K. E., and Williamson, M., 1988, Formation and metabolism of inositol phosphates: The inositol tris/tetrakisphosphate pathway, in: *Regulation of Cellular Calcium Homeostasis* (D. R. Pfeiffer, ed.), Plenum Press, pp. 183–196.

Woods, N. M., Cuthbertson, K. S. R., and Cobbold, P. H., 1986, Repetitive transient rises in cytoplasmic free calcium in hormone-stimulated hepatocytes, *Nature* **319:**600–602.

Woods, N. M., Cuthbertson, K. S. R., and Cobbold, P. H., 1987, Agonist-induced oscillations in cytoplasmic free calcium concentration in single rat-hepatocytes, *Cell Calcium* **8:**79–100.

Yamaguchi, K., Hirata, M., and Kuriyama, H., 1987, Calmodulin activates inositol 1,4,5-trisphosphate 3-kinase activity in pig aortic smooth muscle, *Biochem. J.* **244:**787–791.

$$15$$

Metabolism of Ins(1,3,4)P$_3$ by Rat Liver Homogenates

C. J. KIRK, J. B. PARRY, R. F. IRVINE, R. H. MICHELL, and S. B. SHEARS

1. INTRODUCTION*

Receptor - mediated PtdIns(4,5)P$_2$ hydrolysis directly generates two intracellular messenger molecules: diacylglycerol, which activates protein kinase C, and Ins(1,4,5)P$_3$, which mobilizes Ca^{2+} from intracellular stores (see Berridge and Irvine, 1984; Downes and Michell, 1985 for review). Irvine *et al.* (1984) first recognized the existence of another inositol trisphosphate isomer, Ins(1,3,4)P$_3$. This has now been shown to be the major inositol trisphosphate present following prolonged receptor stimulation in several tissues (Irvine *et al.*, 1985; Burgess *et al.*, 1985; Hansen *et al.*, 1986). Ins(1,3,4)P$_3$ arises as a consequence of 5-phosphatase attack on Ins(1,3,4,5)P$_4$, which is itself the product of the phosphorylation of Ins(1,4,5)P$_3$ (Irvine *et al.*, 1986a; Hansen *et al.*, 1986). This pathway is probably a major route for the removal of the Ca^{2+}-mobilizing Ins(1,4,5)P$_3$ message from stimulated cells (Irvine *et al.*, 1986a). In addition, Ins(1,3,4,5)P$_4$ has been suggested to fulfill a second messenger role inside the cell, by stimulating Ca^{2+} entry across the plasma membrane (Irvine and Moor, 1986).

We previously reported on the metabolism of Ins(1,3,4,5)P$_4$ in liver homogenates (Shears *et al.*, 1987b). We now report studies on the further metabolism of Ins(1,3,4)P$_3$.

---

* Abbreviations used: InsP, InsP$_2$, Ins(1,4)P$_2$, Ins(1,3)P$_2$, Ins(3,4)P$_2$, InsP$_3$, Ins(1,3,4)P$_3$, Ins(1,4,5)P$_3$, InsP$_4$, Ins(1,3,4,5)P$_4$, and Ins(1,3,4,6)P$_4$ are myoinositol and its mono-, bis-, tris-, and tetrakisphosphate derivatives, locants designated where appropriate, and known or assumed to be D enantiomers, PtdIns(4,5)P$_2$, phosphatidylinositol-4,5-bisphosphate.

C. J. KIRK, J. B. PARRY, R. H. MICHELL, and S. B. SHEARS ● Department of Biochemistry, University of Birmingham, Birmingham B15 2TT, United Kingdom.   R. F. IRVINE ● Department of Biochemistry, AFRC Institute of Animal Physiology, Babraham, Cambridge CB2 4AT, United Kingdom.

2. THE HYDROLYSIS OF Ins(1,3,4)P₃ IN LIVER HOMOGENATES

[³H]Ins(1,3,4)P₃ was prepared from [³H]Ins(1,4,5)P₃ by the sequential action of liver cell Ins(1,4,5)P₃ kinase and erythrocyte inositol polyphosphate-5-phosphatase (Irvine *et al.*, 1986b; Shears *et al.*, 1987a). When this material was incubated with a diluted liver homogenate, the profile of the products separated on high-performance liquid chromatography (HPLC) (Shears *et al.*, 1987a) was as shown in Fig. 1. There were two distinct peaks in the InsP₂ region of the chromatogram (B and C). The smaller of these two peaks (B), which typically contained 2–6% of the total "InsP₂" radioactivity, coeluted with a [³²P]Ins(1,4)P₂ standard (Shears *et al.*, 1987a). In addition, there was a further peak of radioactivity that eluted at lower ionic strength (A) which was presumed to be an InsP.

In order to determine the isomeric configuration of the inositol phosphates corresponding to these individual peaks, we incubated a mixture of [³H]Ins(1,3,4)P₃ and [³²P-4]Ins(1,3,4)P₃ [prepared from [³²P-4,5]Ins(1,4,5)P₃ as described by Shears *et al.* (1987a) and Irvine *et al.* (1986b)] with diluted liver homogenate, and separated the products of this hydrolysis by HPLC as described above. Table I shows the proportion of the initial ³H and ³²P radioactivity recovered in the various HPLC fractions, which eluted in a manner similar to the experiment shown in Fig. 1. In incubations where 88% of the initial radioactivity in Ins(1,3,4)P₃ was metabolized to other compounds, the major InsP₂ product (corresponding to peak C in Fig. 1) had a ³²P/³H ratio similar to that of the Ins(1,3,4)P₃ substrate. This indicates that the major

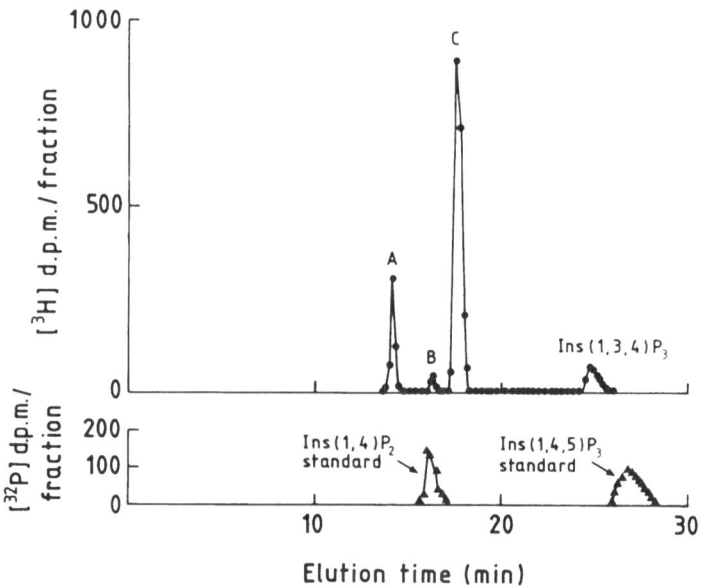

FIGURE 1. HPLC analysis of [³H]Ins(1,3,4)P₃ metabolism by liver homogenate. Homogenates were incubated with [³H]Ins(1,3,4)P₃ as described by Shears *et al.* (1987b). After 30 min, reactions were quenched with 0.2 ml 1.7 M HClO₄, neutralized, "spiked" with [³²P]Ins(1,4)P₂ and [³²P-4,5]Ins(1,4,5)P₃, and analyzed by HPLC. ● = ³H dpm/fraction, ▲ = ³²P dpm/fraction.

TABLE I. Metabolism of [³H]Ins(1,3,4)P$_3$ and [³²P-4]Ins(1,3,4)P$_3$ by
Liver Homogenates[a]

HPLC elution time (min)	Fraction (Fig. 1)	Percent initial substrate			Designation
		¹H	³²P	³²P/³H	
12	"A"	15	1.8	0.12	Ins3P, Ins4P
16	"B"	1.5	0.06	0.04	Ins(1,3)P$_2$
17.5	"C"	62	65	1.05	Ins(3,4)P$_2$
26	Ins(1,3,4)P$_3$	12	12	1	Ins(1,3,4)P$_3$

[a] A mixture of [³H]Ins(1,3,4)P$_3$ and [³²P-4]Ins(1,3,4)P$_3$ was incubated with diluted liver homogenate as described in the text. After 20 min the incubation was quenched (Shears *et al.*, 1987b), neutralized, and analyzed by HPLC. The ¹H and ³²P radioactivity recovered in each of the HPLC fractions shown in Fig. 1 is expressed as a percentage of that in the substrate at $t = 0$. Results are means of three determinations.

InsP$_2$ product of Ins(1,3,4)P$_3$ dephosphorylation retains the ³²P-labeled 4-phosphate. Hence this material may be either Ins(1,4)P$_2$ or Ins(3,4)P$_2$. Since it did not coelute with an Ins(1,4)P$_2$ standard, it must be Ins(3,4)P$_2$. The smaller InsP$_2$ peak, which coeluted with an Ins(1,4)P$_2$ standard, retained very little ³²P radioactivity. Hence we conclude that it must be Ins(1,3)P$_2$. The ³²P/³H ratio of the InsP fraction was 12% of that of the Ins(1,3,4)P$_3$ substrate. Assuming that this material was mainly derived from the major Ins(3,4)P$_2$ product of Ins(1,3,4)P$_3$ hydrolysis, we conclude that it comprised a mixture of Ins3P and Ins4P in a ratio of about 8:1.

3. INFLUENCE OF ATP ON Ins(1,3,4)P$_3$ HYDROLYSIS

We previously showed that ATP inhibits the hydrolysis of Ins(1,4,5)P$_3$ and Ins(1,3,4,5)P$_4$ by liver homogenates (Shears *et al.*, 1987b). When 5 mM ATP was added to incubations of dilute liver homogenates with [³H]Ins(1,3,4)P$_3$, the rate of metabolism of this substrate was inhibited by about 14% (Table II). However, the most interesting feature of the experiment shown in Table II was the appearance of a radioactive product in the "InsP$_4$" fraction

TABLE II. Effect of ATP on Ins(1,3,4)P$_3$
Metabolism[a]

Incubation time (min)	[ATP]	[³H] dpm/fraction	
		InsP$_3$	InsP$_4$
0	5 mM	952 ± 81	50 ± 4
10	0	330 ± 20	22 ± 7
10	5 mM	418 ± 36	227 ± 15

[a] [³H]Ins(1,3,4)P$_3$ was incubated with dilute liver homogenate containing 5 mM ATP (Shears *et al.*, 1987b). At the times indicated, incubations were quenched, neutralized, and applied to BioRad AG1-X8 (200–400 mesh) ion exchange columns. "InsP$_3$" and "InsP$_4$" fractions were separated as described by Shears *et al.* (1987b).

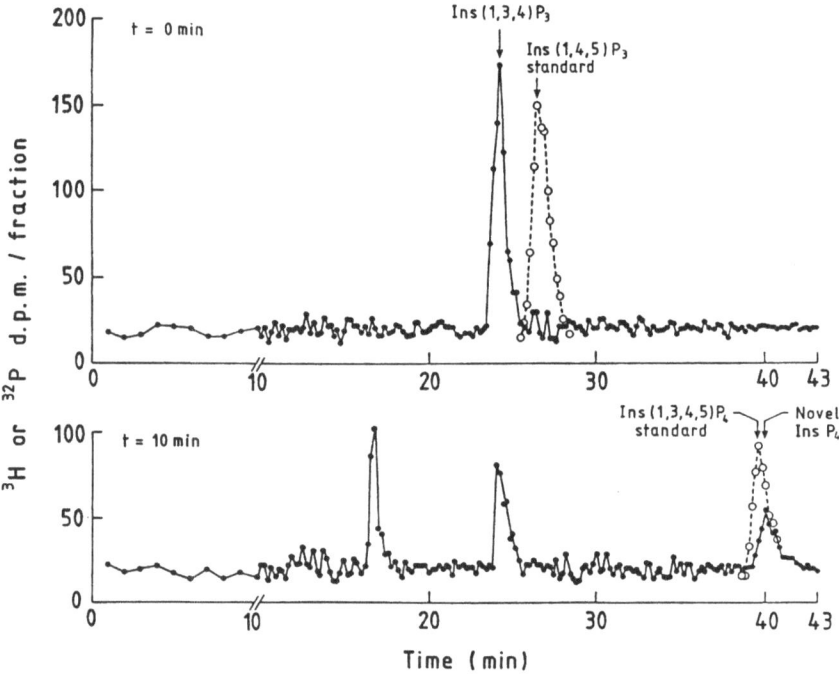

FIGURE 2. HPLC analysis of the products of Ins(1,3,4)P$_3$ metabolism in the presence of ATP. Liver homogenates (1.8%, w/v) were incubated for 0 and 10 min with [^{3}H]Ins(1,3,4)P$_3$ in the presence of 5 mM ATP. Immediately before HPLC samples were spiked with [^{32}P-4,5]Ins(1,4,5)P$_3$ and [^{32}P-4,5]Ins(1,3,4,5)P$_4$ as indicated. ● = ^{3}H dpm/fraction, ○ = ^{32}P dpm/fraction. Results are from a single experiment, typical of three.

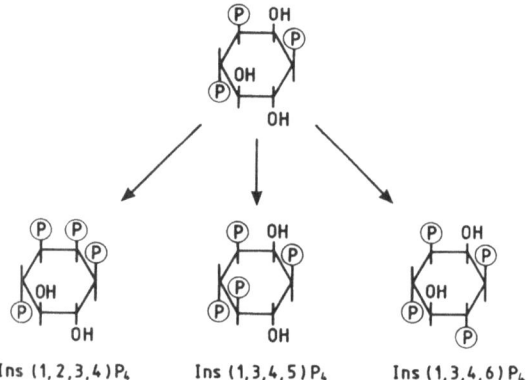

FIGURE 3. Possible products of Ins(1,3,4)P$_3$ kinase.

separated by ion exchange chromatography (Shears *et al.*, 1987b). This result indicates that liver homogenates contain an ATP-dependent Ins(1,3,4)P$_3$ kinase.

The HPLC profile of the products that accumulate when liver homogenates metabolize [^{3}H]Ins(1,3,4)P$_3$ in the presence of ATP is shown in Fig. 2. In addition to the substrate, and a major peak in the InsP$_2$ region of the chromatogram [Ins(3,4)P$_2$], there is a substantial InsP$_4$ peak. Assuming that phosphate migration does not occur during the kinase reaction, the possible products of Ins(1,3,4)P$_3$ phosphorylation are as shown in Fig. 3. Of these products, Ins(1,2,3,4)P$_4$ has a pair of vicinal *trans*-hydroxyl groups and would be expected to be susceptible to periodate oxidation (Tomlinson and Ballou, 1967; Lim and Tate, 1973). The InsP$_4$ generated in the experiment shown in Table II was incubated with 0.1 M periodate (pH 2) for 4 days in the dark (Tomlinson and Ballou, 1967). The products were reduced, dephosphorylated (Shears *et al.*, 1987b), and separated by ionophoresis (Frahn and Mills, 1959) or paper chromatography (Grado and Ballou, 1961). The only ^{3}H-labeled product found was inositol. Ins(1,2,3,4)P$_4$ would have been expected to yield altritol following this treatment, and we therefore conclude that the InsP$_4$ generated in the experiment described above must be Ins(1,3,4,5)P$_4$ or Ins(1,3,4,6)P$_4$. The former isomer is a substrate for the membrane-bound inositol polyphosphate-5-phosphatase of erythrocytes and other tissues (Shears *et al.*, 1987b). We incubated a sample of [^{32}P] Ins(1,3,4,5)P$_4$ (Shears *et al.*, 1987a) and our unknown InsP$_4$ with a sample of washed erythrocyte plasma membranes (Downes *et al.*, 1982). In conditions where 78% of the Ins(1,3,4,5)P$_4$ was hydrolyzed, our unknown InsP$_4$ remained almost completely resistant to periodate attack. We therefore conclude that liver homogenates contain an Ins(1,3,4)P$_3$ 6-kinase.

4. CONCLUDING REMARKS

Receptor-generated Ins(1,4,5)P$_3$ may undergo a variety of metabolic transformations in cells and cellular extracts. A major fate of this messenger in several tissues involves its phosphorylation to Ins(1,3,4,5)P$_4$ and the subsequent hydrolysis of this compound to Ins(1,3,4)P$_3$. The latter isomer, which may accumulate in significant quantities following prolonged stimulation of various tissues, may be further hydrolyzed to Ins(3,4)P$_2$ or re-phosphorylated to Ins(1,3,4,6)P$_4$. The possible intracellular functions of these novel inositol phosphates and their metabolic interrelationship to InsP$_5$ and InsP$_6$, which have been identified in several cells (Heslop *et al.*, 1985; Morgan *et al.*, 1987), remain to be elucidated.

REFERENCES

Berridge, M. J., and Irvine, R. F., 1984, Inositol trisphosphate, a novel second messenger in cellular signal transduction, *Nature* **312**:315–321.

Burgess, G. M., McKinney, J. S., Irvine, R. F., and Putney, J. W. Jr., 1985, Inositol 1,4,5-trisphosphate and inositol 1,3,4-trisphosphate formation in Ca^{2+}-mobilizing-hormone-activated cells, *Biochem. J.* **232**:237–243.

Downes, C. P., and Michell, R. H., 1985, In *Molecular Mechanisms of Transmembrane Signalling* (P. Cohen, and M. D. Houslay, eds.), pp. 3–56, Elsevier, Amsterdam.

Downes, C. P., Mussat, M. C., and Michell, R. H., 1982, The inositol trisphosphate phosphomonoesterase of the human erythrocyte membrane, *Biochem. J.* **203**:169–177.

Frahn, J. L., and Mills, A. J., 1959, *Aust. J. Chem.* **12**:65–89.

Grado, C., and Ballou, C. E., 1961, Myo-inositol phosphates obtained by alkaline hydrolysis of beef brain phosphoinositide, *J. Biol. Chem.* **236:**54–60.

Hansen, C. A., Mah, S., and Williamson, J. R., 1986, Formation and metabolism of inositol 1,3,4,5-tetrakisphosphate in liver, *J. Biol. Chem.* **261:**8100–8103.

Heslop, J. P., Irvine, R. F., Tashjian, A. M., and Berridge, M. J., 1985, Inositol tetrakis- and pentakisphosphates in GH4 cells, *J. Exp. Biol.* **119:**395–401.

Irvine, R. F., and Moor, R. M., 1986, Micro-injection of inositol 1,3,4,5-tetrakisphosphate activates sea urchin eggs by a mechanism dependent on external Ca^{2+}, *Biochem. J.* **240:**917–920.

Irvine, R. F., Letcher, A. J., Lander, D. J., and Downes, C. P., Inositol trisphosphates in carbachol-stimulated rat parotid glands, *Biochem. J.* **223:**237–243.

Irvine, R. F., Anggard, E. E., Letcher, A. J., and Downes, C. P., 1985, Metabolism of inositol 1,4,5-trisphosphate and inositol 1,3,4-trisphosphate in rat parotid glands *Biochem. J.* **229:**505–511.

Irvine, R. F., Letcher, A. J., Heslop, J. P., and Berridge, M. J., 1986a, The inositol tris/tetrakisphosphate pathway—demonstration of Ins(1,4,5)P₃-3 kinase activity in animal tissues, *Nature* **320:**631–634.

Irvine, R. F., Letcher, A. J., Lander, D. J., and Berridge, M. J., 1986b, Specificity of inositol phosphate-stimulated Ca^{2+} mobilization from swiss-mouse 3T3 cells, *Biochem. J.* **240:**301–304.

Lim, P. E., and Tate, M. E., 1973, The phytases II. Properties of phytase fractions F_1 and F_2 from wheat bran and the myoinositol phosphates produced by fraction F_2, *Biochem. Biophys. Acta* **302:**316–328.

Morgan, R. O., Chang, J. P., and Catt, K. J., 1987, Novel aspects of gondotrophin-releasing hormone action on inositol polyphosphate metabolism in cultured pituitary gonadotrophs, *J. Biol. Chem.* **262:**1166–1171.

Shears, S. B., Storey, D. J., Morris, A. J., Cubitt, A. B., Parry, J. B., Michell, R. H., and Kirk, C. J., 1987a, Dephosphorylation of myo-inositol 1,4,5-trisphosphate and myo-inositol 1,3,4-tris-phosphate, *Biochem. J.* **242:**393–402.

Shears, S. B., Parry, J. B., Tang, E. K. Y., Irvine, R. F., Michell, R. H., and Kirk, C. J., 1987b, Metabolism of D-myo-inositol 1,3,4,5-tetrakisphosphate by rat liver, including the synthesis of a novel isomer of myo-inositol tetrakisphosphate, *Biochem. J.* **246:**139–147.

Tomlinson, R. V., and Ballou, C. E., 1967, *Biochemistry*, **1:**66–70.

16

Evidence for the Existence of the Inositol Tris/Tetrakisphosphate Pathway in Rat Heart

D. RENARD and J. POGGIOLI

1. INTRODUCTION

In skeletal and cardiac muscles, polyphosphoinositide hydrolysis by phospholipase C is increased by electrical or hormonal stimulation, thus producing inositol-1,4,5-trisphosphate (Ins-1,4,5-P_3) and diacylglycerol (Vergara *et al.*, 1985; Poggioli *et al.*, 1986). Ins-1,4,5-P_3 was shown to release Ca^{2+} from sarcoplasmic reticulum and promote tension in skinned fibers (Hirata *et al.*, 1984; Nosek *et al.*, 1986; Movsesian *et al.*, 1985). Diglycerides activate protein kinase C, which in turn may influence certain ionic permeabilities (De Riemer *et al.*, 1985) and the sensitivity of myofilaments to Ca^{2+} (Endoh and Blinks, 1988).

Besides its rapid hydrolysis in inositol-1,4-bisphosphate (InsP_2) by a phosphomonoesterase, another pathway for Ins-1,4,5-P_3 metabolism was recently described in several tissues (Irvine *et al.*, 1986; Biden and Wollheim, 1986). It consists of the phosphorylation of Ins-1,4,5-P_3 by a kinase to inositol 1,3,4,5-tetrakisphosphate (InsP_4), which is then dephosphorylated to give inositol-1,3,4-trisphosphate. It has been shown that InsP_4 may play a key role in sea urchin egg activation by stimulating Ca^{2+} fluxes through the cell plasma membrane (Irvine and Moor, 1986). It is thus of interest to study further that inositol tris/tetrakisphosphate pathway.

In this work, we examine the formation of the two InsP_3 isomers following an α_1-adrenergic stimulation of rat heart. In addition, we demonstrate the presence of an InsP_3 kinase which is stimulated by Ca^{2+} in the soluble fraction of these cells.

D. RENARD and J. POGGIOLI ● Cellular Physiology and Pharmacology Research Unit, INSERM U-274, Université Paris-Sud (Bât. 443) F-91405 ORSAY Cedex, France.

2. MATERIALS AND METHODS

2.1. Materials

Materials included noradrenaline, propranolol, atropine, and 2,3-bisphosphoglycerate from Sigma, D-myoinositol-1,4,5-triphosphate from Amersham (France). Myo[2-^{3}H]inositol (619 GBq/mmol) was obtained from New England Nuclear, sodium [^{32}P]phosphate from CEA (France). D-myo[2-^{3}H]inositol-1,4,5-triphosphate and L-myo[U-^{14}C]inositol-L-phosphate were purchased from Amersham (France). All other chemicals were of reagent grade.

2.2. Tissue Incubation and Labeling

Isolated right ventricles from female Wistar rats (160–180g) were labeled with [^{3}H]myoinositol (1.11 MBq/ml) for 3 hr in a modified Krebs solution containing 116 mM NaCl, 5.4 mM KCl, 0.9 mM CaCl$_2$, 0.81 mM MgCl$_2$, 0.92 mM NaH$_2$PO$_4$, 25 mM NaHCO$_3$ supplemented with glucose (1 g/liter) and mannitol (0.4 g/liter) under an atmosphere of O$_2$/CO$_2$ (19:1) as previously described (Poggioli *et al.*, 1986). They were rinsed, preincubated in Krebs solution containing 10 mM LiCl, 10 μM propranolol, and 10 μM atropine, and then stimulated or not with 50 μM noradrenaline for 10 sec, 30 sec, or 2 min. The incubation was stopped by freeze clamping and [^{3}H]inositol phosphates were extracted using HClO$_4$ as in (Poggioli *et al.*, 1986).

2.3. Separation of [^{3}H]Inositol Phosphates

Neutralized samples of perchloric acid-soluble material were analyzed by high-pressure liquid chromatography (HPLC) on Partisyl SAX 10 column (Irvine *et al.*, 1985). After sample injection, the column was washed for 3 min with H$_2$O and the [^{3}H]inositol polyphosphates were subsequently eluted by three successive convex gradients (Waters 4, 2, and 1, respectively) of increasing ammonium formate buffer (0–1.7 M) adjusted to pH 3.7 with orthophosphoric acid. The flow rate was 1.2 ml/min and the eluent was collected in 1-min fractions over the first 20 min and 0.2-min fractions over the next 15 min. [^{32}P]Ins-1,4-P$_2$ and [^{32}P]Ins-1,4,5-P$_3$ were prepared from red blood cell ghosts activated by Ca^{2+} (Downes *et al.*, 1982). [2-^{3}H]Ins-1,3,4,5-P$_4$ was made from [2-^{3}H]Ins-1,4,5-P$_3$ using a soluble fraction of rat liver (Irvine *et al.*, 1986). Both labeled standards were purified by anion exchange chromatography on a Dowex 1 × 8 column (formate form) as in Downes and Michell (1981), diluted 5 times with water, and desalted by lyophilization.

2.4. Enzyme Assay

Both whole ventricles from one rat heart were homogenized in 2 ml of an ice-cold buffer containing 250 mM sucrose, 5 mM Hepes pH 7.5, and centrifuged for 90 min at 4°C (100,000g). The resultant supernatant was designed as the soluble fraction. The buffer used to test the kinase activity contained 250 mM sucrose, 10 mM ATP, 20 mM MgCl$_2$, 50 mM Hepes (pH 7.3), 5 mM Na pyrophosphate, 10 mM Na 2,3-bisphosphoglycerate, 2 mM EGTA, 0.2 mg/ml saponin. The buffer used to test the phosphomonoesterase activity contained 250 mM sucrose, 50 mM Hepes (pH 7.3), 1 mM MgCl$_2$, 10 mM glucose, 50 U/ml hexokinase, 0.2 mg/ml saponin, and 0.5 mM EGTA. To study the phosphomonoesterase activity in the particulate fraction, the 100,000g, pellet was resuspended in 500 μl of 83 mM saccharose containing 50 mM Hepes (pH 7.3). The incubation buffer was the same as that used for the soluble phosphomonoesterase. For both soluble and particulate phosphomonoester-ases, 10 mM Na 2,3-bisphosphoglycerate were added in some experiments. For assays of both enzymes $CaCl_2$ was added to give free Ca^{2+} concentrations in the range of 1 nM to 1 μM (Bartfai, 1979). They were performed at 35°C with 50 μl of particulate or cytosolic fraction in a total volume of 500 μl. Incubations were started by addition of Ins-1,4,5-P_3 (0.5–10 μM) and [2-^{3}H] Ins-1,4,5-P_3 (0.2 μCi/ml) or [^{32}P]Ins-1,4,5-P_3 (10,000 dpm/ml). They were stopped with 33 μl of ice-cold 50% (v/v) C1O_4H. Labeled products were analyzed as explained above. The kinase activity was calculated as follows: cpm of InsP$_4$ formed/specific activity of [^{3}H]- or [^{32}P]Ins-1,4,5-P_3 × time of incubation. The phosphomonoesterase activity was calculated as follows: cpm of hydrolyzed Ins-1,4,5-P_3/specific activity of [^{32}P]Ins-1,4,5-P_3 × time of incubation.

3. RESULTS AND DISCUSSION

3.1. Presence of More Than One InsP$_3$ Isomer in Heart Cells

Experiments were performed to determine whether or not the InsP$_3$ produced by hormone stimulation in rat heart contained the two different isomeric forms (Ins-1,4,5-P_3 and Ins-1,3,4-P_3) as already observed in other tissues (Burgess *et al.*, 1985; Batty *et al.*, 1985; Merrit *et al.*, 1986). [^{3}H]labeled isolated right ventricles were stimulated for periods of either 10 sec, 30 sec, or 2 min with a maximal dose of noradrenaline (50 μM).

Figure 1 shows the ^{3}H products eluted from heart extracts. In the absence of stimulation

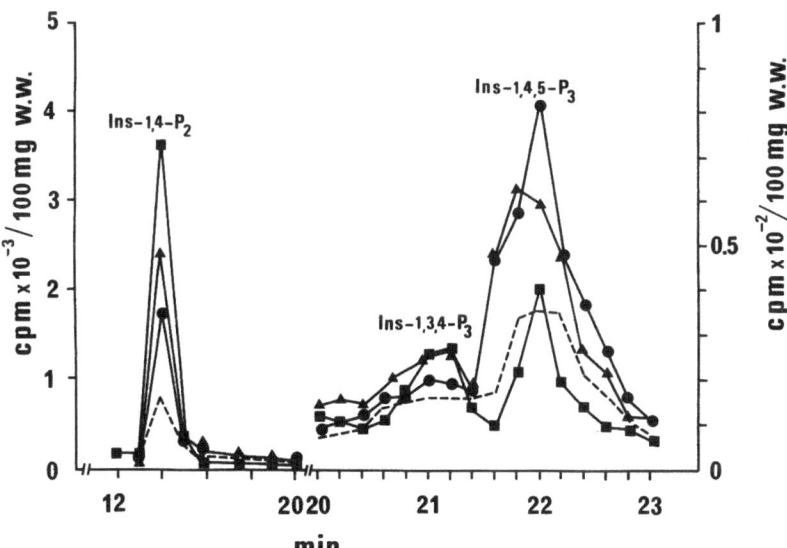

FIGURE 1. Elution profile of inositol phosphates by HPLC. For experimental details see Sec. 2. (A) The samples analyzed were neutralized perchloric acid extracts of right ventricles (100 mg wet weight) prelabeled with [^{3}H]myoinositol and stimulated or not (---) with 50 μM noradrenaline for 10 sec (●), 30 sec (▲), or 2 min (■). Note the change in scale for InsP$_3$. Identification of inositol phosphates is based on comparison with the elution profiles of [^{32}P]Ins-1,4-P_2, [^{32}P]Ins-1,4,5-P_3, and [^{3}H]Ins-1,4,5-P_3.

four peaks could be observed; the figure shows only the last three. They were characterized by coelution with appropriate standards: inositol monophosphate (Ins-1-P, not shown), Ins-1,4-P$_2$, and Ins-1,4,5-P$_3$. The two InsP$_3$ isomers were indeed identified accordingly to a previous report (Irvine *et al.*, 1985), Ins-1,3,4-P$_3$ was eluted with or very close to ATP, and Ins-1,4,5-P$_3$ was eluted shortly afterward. In stimulated cells, the ^{3}H peak corresponding to Ins-1,4,5-P$_3$ (22 min) predominated at the shorter time of stimulation (10 sec). The other isomer (Ins-1,3,4-P$_3$, 21 min) was much increased following a longer stimulation period (30 sec). This figure also shows that in heart InsP$_3$ isomers did not accumulate much; correlatively, an increase in InsP$_2$ appeared very soon suggesting the presence of an efficient phosphomonoesterase in heart.

Although InsP$_4$ could not be detected in heart cells *in vivo*, the presence of Ins-1,3,4-P$_3$, which has been repeatedly shown to result from InsP$_4$ dephosphorylation in other tissues (Downes *et al.*, 1986; Hansen *et al.*, 1986; Rossier *et al.*, 1986), led us to investigate the presence of an Ins-1,4,5-P$_3$ kinase in the soluble fraction of heart.

3.2. Detection of Inositol Tetrakisphosphate in the Soluble Fraction of Heart Cells

To further examine the possibility that InsP$_4$ was being produced in rat heart, we used subcellular fractions. Figure 2 shows a time course analysis of the metabolism of [^{3}H]Ins-1,4,5-P$_3$ (16 μM) in the presence of ATP in the soluble fraction. The data show that, 1 min after the addition of [^{3}H]Ins-1,4,5-P$_3$ to the 100,000g supernatant, a new, more polar inositol-containing peak appeared, which was identified as InsP$_4$ by coelution with [^{3}H]InsP$_4$ prepared from hepatocytes. InsP$_4$ was not formed when ATP was absent. Simultaneously with the formation of InsP$_4$, a shoulder arose on the Ins-1,4,5-P$_3$ peak. This coeluted with ATP and represented Ins-1,3,4-P$_3$ (see 3.1). InsP$_4$ accumulated and the ratio of the two InsP$_3$ isomers inversed with the length of incubation. Figure 2 shows that InsP$_2$ and InsP were formed too. Only one peak of InsP$_2$ was observed which coeluted with a standard of [^{32}P]Ins-1,4-P$_2$ prepared from ^{32}P-labeled red cell ghosts activated by Ca^{2+} (see Sec. 2). This suggests that InsP$_2$ formation results from the attack of Ins-1,4,5-P$_3$ by a phosphomonoesterase removing PO$_4^{2-}$ from position 5. As shown in Table I, both particulate and soluble phosphomonoesterases were detected in heart. They were activated by Mg^{2+} and inhibited by the 2,3-bisphosphoglycerate. In agreement with the above results, InsP$_3$ phosphomonoesterase has been reported to be soluble in platelets partly soluble in hepatocytes and in the particulate fraction in other tissues (Connolly *et al.*, 1985; Storey *et al.*, 1984; Raval and Allen, 1985). This phosphomonoesterase may hydrolyze Ins-1,4,5-P$_3$ and Ins-1,3,4,5-P$_4$ (Connolly *et al.*, 1987). Its activity might explain why we have been unable to detect InsP$_4$ *in vivo* since, as shown above, as soon as it is formed InsP$_4$ is dephosphorylated in Ins-1,3,4-P$_3$. The addition of phosphomonoesterase inhibitors under *in vitro* conditions reduced InsP$_4$ turnover and allowed it to accumulate.

3.3. Modulation of InsP$_3$ Kinase Activity by Ca^{2+}

Cytosolic free Ca^{2+} varies from 200 to 300 times/min in rat heart cells. [Ca^{2+}]$_i$ is also dependent on hormonal stimulation (Powell *et al.*, 1984; Thomas *et al.*, 1986). In that prospect, the possible dependence of Ins-1,4,5-P$_3$ kinase activity on Ca^{2+} is of physiological relevance. Further experiments were performed to test this possibility.

As shown before, the experiments were complicated by the presence of soluble en-

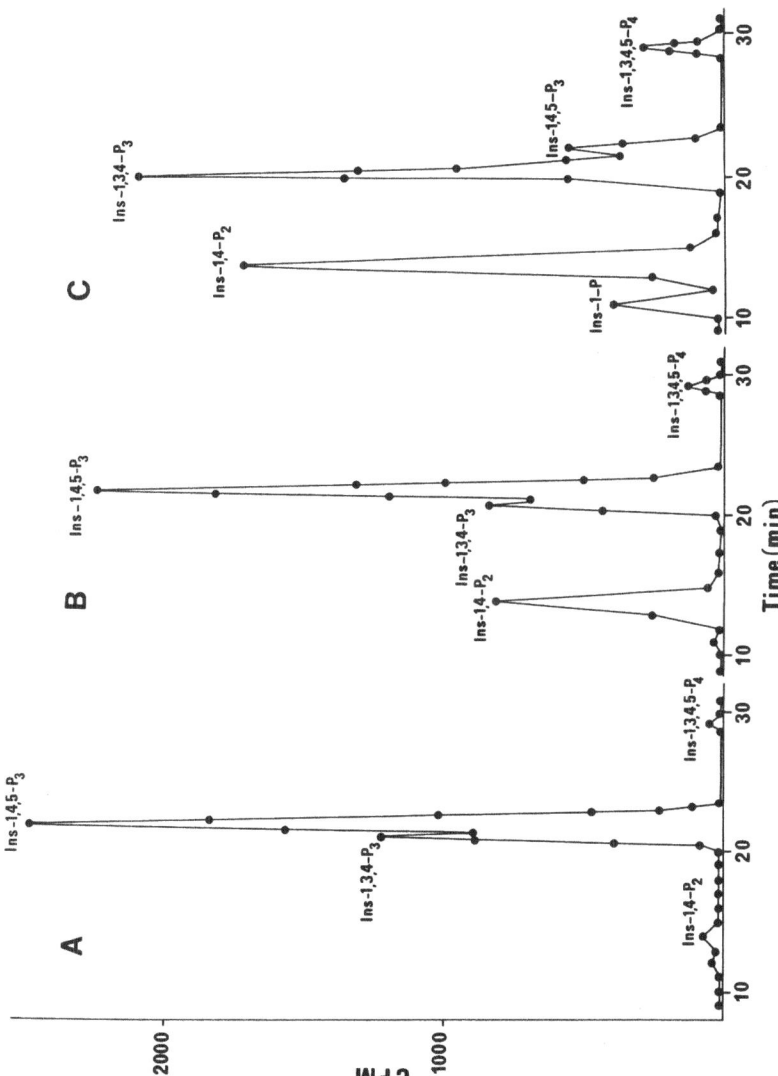

FIGURE 2. Time course of [³H]Ins-1,4,5-P₃ conversion to [³H]InsP₄. An aliquot of the soluble fraction (450 μg of protein in 50 μl) was added to 500 μl of an incubation buffer containing 250 mM sucrose, 10 mM ATP, 20 mM MgCl₂, 50 mM Hepes (pH 7.3), 5 mM Na pyrophosphate, 2.5 mM Na 2,3-bisphosphoglycerate, 0.5 mM EDTA, 0.2 mg/ml saponin, 16 μM Ins-1,4,5-P₃ and [³H]Ins-1,4,5-P₃ (8.14 kBq/ml). CaCl₂ was added to give a free Ca²⁺ concentration of 10 μM. Incubations were performed at 35°C and were stopped after 1 min (A), 3 min (B), or 5 min (C) by addition of 50 μl of 50% HClO₄. Similar results were obtained in three separate experiments.

TABLE I. Activator and Inhibitor of Phosphomonoesterase Activity

MgCl₂ (mM)	0	1	1
EDTA (mM)	1	0	0
Na 2,3-bisphosphoglycerate (mM)	0	0	10
Soluble activity	60.6 ± 5.76	608.3 ± 48.8	62.2 ± 4.05
n	8	8	7
Particulate activity	47.0 ± 3.4	691.3 ± 118.7	51.7 ± 7.4
n	5	5	5

[4.5 – ³²P] Ins(1,4,5)P₃ was prepared and incubated with soluble or particulate fraction as described in Section 2. Incubations were terminated after 2 min and inositol phosphates were separated by anion exchange chromatography. Results are the means ± SEM of n experiments and expressed in pmol of Ins-P₃ hydrolyzed/min per mg of protein.

zyme(s) that dephosphorylates Ins-1,4,5-P₃ and Ins-1,3,4,5-P₄. We tentatively eliminated the phosphomonoesterase effects by including 10 mM bisphosphoglycerate in incubating solutions. Under the conditions used (see Sec. 2), the reaction rate was linear at least over a 10-min period (Fig. 3, inset). The kinase activity was doubled by raising the free Ca²⁺ concentration from 1 nM to 1 μM (EC₅₀ for Ca²⁺ approximately 120 nM, Fig. 3). This effect was due to an increase in the apparent V_{max} of the enzyme from 135.1 ± 25.4 to 253.5 ± 26.5 pmol of InsP₄ formed/min/mg of protein ($n = 3$, $P < 0.05$) in the presence

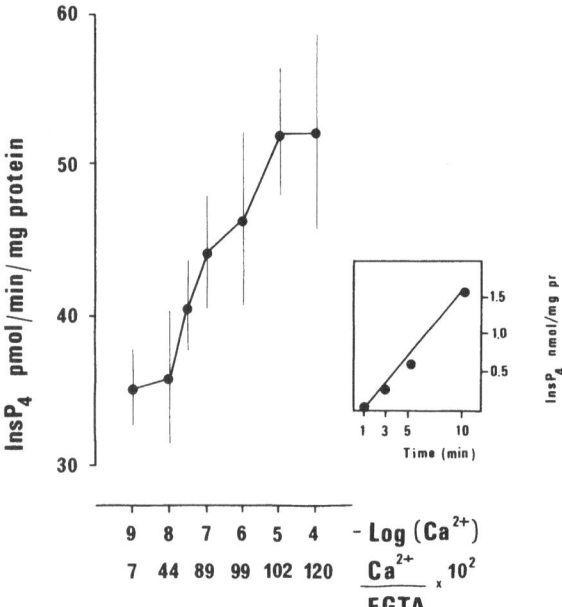

FIGURE 3. Ca²⁺ dependence of Ins-1,4,5-P₃ kinase activity in a soluble fraction from rat heart. The protocol was identical to that described in legend of Fig. 2 except for Ins-1,4,5-P₃ and 2,3-bisphosphoglycerate, which were 10 μM and 10 mM, respectively. Final free Ca²⁺ concentrations are those plotted on the abscissa. The incubations lasted 3 min. Results are the mean ± SEM of four to six experiments.

of 1 nM or 1 μM Ca^{2+}, respectively. It has been reported that, in RIN m5F cells, Ins-1,4,5-P_3 kinase was stimulated by Ca^{2+} over the physiological concentration range (Biden and Wollheim, 1986) while in brain no relevant changes in kinase activity have been observed over physiological range (Irvine et al., 1986). The Ca^{2+} concentration required for half-maximal kinase activity is in agreement with the internal Ca^{2+} concentration reported in vivo (Powell et al., 1984; Thomas et al., 1986). The apparent half-maximal substrate concentration 5.61 ± 1.59 μM (n = 3, 1 nM Ca^{2+}) and 5.58 ± 0.19 μM (n = 3, 1 μM Ca^{2+}) was unchanged by Ca^{2+}. The apparent K_m of the kinase for Ins-1,4,5-P_3 reported here is in keeping with Ins-1,4,5-P_3 concentrations, determined chemically or by biological assay, in activated cells (Rittenhouse and Sasson, 1985; Bradford and Rubin, 1986). Effective Ins-1,4,5-P_3 concentrations required for Ca^{2+} release from sarcoplasmic reticulum are also in the 10-μM range (Hirata et al., 1984; Nosek et al., 1986).

The present study provides the first evidence that Ins-1,3,4-P_3 is produced together with Ins-1,4,5-P_3 in hormone-stimulated heart as previously reported for other cell types (Burgess et al., 1985; Batty et al., 1985; Merrit et al., 1986). It has been shown that the interconversion of the two InsP₃ isomers depends on the formation of a higher phosphorylated intermediate InsP₄. InsP₄ is synthesized from Ins-1,4,5-P_3 in the presence of ATP by a kinase as reported in brain, liver, adrenal glomerulosa cells, and RIN m5F cells (Biden and Wollheim, 1986; Hansen et al., 1986; Irvine et al., 1986; Rossier et al., 1986). The kinase appears to be Ca^{2+}-sensitive. This alternative pathway may lead to inactivation of the well-characterized messenger Ins-1,4,5-P_3 and the formation of a putative other messenger InsP₄ (Houslay, 1987; Mitchell, 1986).

REFERENCES

Bartfai, T., 1979, Preparation of metal-chelate complexes and the design of steady-state kinetic experiments involving metal nucleotide complexes, Adv. Cyclic Nucleotide Res. 10:219–242.

Batty, I. R., Nahorski, S. R., and Irvine, R. F., 1985, Rapid formation of inositol 1,3,4,5-tetrakisphosphate following muscarinic receptor stimulation of rat cerebral cortical slices, Biochem. J. 232:211–215.

Biden, T. J., and Wollheim, C. B., 1986, Ca^{2+} regulates the inositol tris/tetrakisphosphate pathway in intact and broken preparations of insulin-secreting RINm5F cells, J. Biol. Chem. 261:11931–11934.

Bradford, P. G., and Rubin, R. P., 1986, Quantitative changes in inositol 1,4,5-trisphosphate in chemoattractant stimulated neutrophils. J. Biol. Chem. 261:15644–15647.

Burgess, G. M., McKinney, J. S., Irvine, R. F., and Putney, J. W. Jr., 1985, Inositol 1,4,5-trisphosphate and inositol 1,3,4-trisphosphate formation in Ca^{2+}-mobilizing-hormone-activated cell, Biochem. J. 232:237–243.

Connolly, T. M., Bross, T. E., and Majerus, P. W., 1985, Isolation of phosphomonoesterase from human platelets that specifically hydrolyzes the 5-phosphate of inositol 1,4,5-trisphosphate, J. Biol. Chem. 260:7868–7874.

Connolly, T. M., Bansal, V. S., Bross, T. E., Irvine, R. F., and Majerus, P. W., 1987, The metabolism of tris- and tetraphosphates of inositols by 5-phosphomonoesterase and 3-kinase enzymes, J. Biol. Chem. 262:2146–2149.

De Riemer, S. A., Strong, J. A., Albert, K. A., Greengard, P., and Kaczmarek, L. K., 1985, Enhancement of calcium current in Aplysia neurones by phorbol ester and protein kinase C. Nature 313:313–316.

Downes, C. P., Hawkins, P. T., and Irvine, R. F., 1986, Inositol 1,3,4,5-tetrakisphosphate and not phosphatidylinositol 3,4-bisphosphate is the probable precursor of inositol 1,3,4-trisphosphate in agonist-stimulated parotid gland, Biochem. J. 238:501–506.

Downes, C. P., and Michell, R. H., 1981, The polyphosphoinositide phosphodiesterase of erythrocyte membranes, *Biochem. J.* **198**:133–140.

Downes, C. P., Mussat, M. C., and Michell, R. H., 1982, The inositol trisphosphate phosphomonoesterase of the human erythrocyte membrane, *Biochem. J.* **203**:169–177.

Endoh, M., and Blinks, J. R., 1988, Actions of sympathomimetic amines, on the Ca^{2+} transients and contractions of rabbit myocardium: Reciprocal changes in myofibrillar responsiveness to Ca^{2+} mediated through α- and β-adrenoceptors, *Circ. Res.* **62**:247–265.

Hansen, C. A., Mah, S., and Williamson, J. R., 1986, Formation and metabolism of inositol 1,3,4,5-tetrakisphosphate in liver, *J. Biol. Chem.* **261**:8100–8103.

Hirata, M., Suematsu, E., Hashimoto, T., Hamachi, T., and Koga, T., 1984, Release of Ca^{2+} from a non-mitochondrial store site in peritoneal macrophages tested with saponin by inositol 1,4,5-trisphosphate, *Biochem. J.* **223**:229–236.

Houslay, M. D., 1987, Egg activation unscrambles a potential role for IP4, *Trends Biochem. Sci.* **12**:1–2.

Irvine, R. F., Anggard, E. E., Letcher, A. J., and Downes C. P., 1985, Metabolism of inositol 1,4,5-trisphosphate and inositol 1,3,4-trisphosphate in rat parotid glands, *Biochem. J.* **229**:505–511.

Irvine, R. F., and Moor, R. M., 1986, Micro-injection of inositol 1,3,4,5-tetrakisphosphate activates sea urchin eggs by a mechanism dependent on external Ca^{2+}, *Biochem. J.* **240**:917–920.

Irvine, R. F., Letcher, A. J., Heslop, J. P., and Berridge, M. J., 1986, The inositol tris/tetrakisphosphate pathway-demonstration of Ins(1,4,5)P3 3-kinase activity in animal tissues, *Nature* **320**:631–634.

Merrit, J. E., Taylor, C. W., Rubin, R. P., and Putney, J. W. Jr., 1986, Isomers of inositol trisphosphate in exocrine pancreas, *Biochem. J.* **238**:825–829.

Michell, B., 1986, A second messenger function for inositol tetrakisphosphate, *Nature* **324**:613.

Movsesian, M. A., Thomas, A. P., Selak, M., and Williamson, J. R., 1985, Inositol trisphosphate does not release Ca^{2+} from permeabilized cardiac myocytes and sarcoplasmic reticulum, *FEBS Lett.* **185**:328–332.

Nosek, T. M., Williams, M. F., Zeigler, S. T., and Godt, R. E., 1986, Inositol trisphosphate enhances calcium release in skinned cardiac and skeletal muscle, *Am. J. Physiol.* **250**:C807–C811.

Poggioli, J., Sulpice, J. C., and Vassort, G., 1986, Inositol phosphate production following α_1-adrenergic, muscarinic or electrical stimulation in isolated rat heart, *FEBS Lett.* **206**:292–297.

Powell, T., Tatham, P. E. R., and Twist, V. W., 1984, Cytoplasmic free calcium measured by quin2 fluorescence in isolated ventricular myocytes at rest and during potassium depolarization, *Biochim. Biophys. Res. Commun.* **122**:1012–1020.

Raval, P. J., and Allan, D., 1985, Ca^{2+}-induced polyphosphoinositide breakdown due to phosphomonoesterase activity in chicken erythrocytes, *Biochem. J.* **231**:179–182.

Rittenhouse, S. E., and Sasson, J. P., 1985, Mass changes in myoinositol trisphosphate in human platelets stimulated by thrombin, *J. Biol. Chem.* **260**:8657–8660.

Rossier, M. F., Dentand, I. A., Lew, P. D., Capponi, A. M., and Vallonton, M. B., 1986, Interconversion of inositol (1,4,5)-trisphosphate to inositol (1,3,4,5)-tetrakisphosphate and (1,3,4)-trisphosphate in permeabilized adrenal glomerulosa cells is calcium-sensitive and ATP-dependent. *Biochim. Biophys. Res. Commun.* **139**:259–265.

Storey, D. J., Shears, S. B., Kirk, C. J., and Michell, R. H., 1984, Stepwise enzymatic dephosphorylation of inositol 1,4,5-trisphosphate to inositol in liver. *Nature* **312**:374–376.

Thomas, A. P., Selak, M., and Williamson, J. R., 1986, Measurement of electrically-induced Ca^{2+} transients in quin2-loaded cardiac myocytes. *J. Molec. Cell. Cardiol.* **18**:541–545.

Vergara, J., Tsien, R. Y., and Delay, M., 1985, Inositol 1,4,5-trisphosphate: A possible chemical link in excitation-contraction coupling in muscle. *Proc. Natl. Acad. Sci.* **82**:6352–6356.

17

Mechanisms of Intracellular Calcium Movement Activated by Guanine Nucleotides and Inositol-1,4,5-Trisphosphate

DONALD L. GILL, JULIENNE M. MULLANEY, TARUN K. GHOSH, and SHEAU-HUEI CHUEH

1. INTRODUCTION

It is now well established that the intracellular second messenger inositol-1,4,5-trisphosphate (IP_3) is involved in the release of Ca^{2+} from a Ca^{2+}-sequestering organelle, widely considered to be the endoplasmic reticulum (ER) (Berridge and Irvine, 1984; Gill, 1985; Majerus *et al.*, 1986). In a series of recent studies, we observed that a highly sensitive and specific guanine nucleotide regulatory process induces a release of Ca^{2+} in cells that appears very similar to that mediated by IP_3 (Gill *et al.*, 1986; Ueda *et al.*, 1986; Chueh and Gill, 1986). Our initial studies were conducted using either permeabilized cells or isolated microsomal membrane vesicles derived from the N1E-115 neuronal cell line; GTP-dependent Ca^{2+} release was observed to be very similar in the two preparations (Gill *et al.*, 1986; Ueda *et al.*, 1986). Recent studies (Henne and Söling, 1986; Jean and Klee, 1986; Chueh *et al.*, 1987) have extended the number of diverse cell types in which the same GTP-activated Ca^{2+} release process is observed. In each cell type, submicromolar GTP concentrations rapidly effect a substantial release of Ca^{2+} sequestered via internal Ca^{2+}-pumping activity within a non-mitochondrial organelle, believed to be the ER. The Ca^{2+}-accumulating properties of this intracellular organelle have been described in detail in earlier studies with permeabilized cells (Gill and Chueh, 1985).

In this chapter, the characteristics of the GTP-activated Ca^{2+} release process will be described first. In view of the earlier studies by Dawson and his colleagues (Dawson, 1985; Dawson *et al.*, 1986), in which it was observed that GTP alters the effectiveness of IP_3 in

DONALD L. GILL, JULIENNE M. MULLANEY, TARUN K. GHOSH, and SHEAU-HUEI CHUEH
● Department of Biological Chemistry, University of Maryland School of Medicine, Baltimore, Maryland 21202.

liver microsomes, subsequent sections of this chapter will address the relationship between GTP-activated Ca^{2+} release and that activated by IP_3. Lastly, we will consider the possible mechanisms by which GTP induces the translocation of Ca^{2+} ions within cells, and our recently formulated model (Mullaney *et al.*, 1987), which may explain the actions of GTP and its possible modifying effects of Ca^{2+} release induced by IP_3.

2. GTP-ACTIVATED CALCIUM RELEASE

2.1. Sensitivity and Nucleotide Specificity

The guanine nucleotide-activated release of Ca^{2+} that has been observed using either permeabilized cells (Gill *et al.*, 1986) or microsomes derived from cells (Ueda *et al.*, 1986) has remarkably high sensitivity to GTP, with a K_m for GTP of 0.75 μM. Release is not observed with GMP, cyclic GMP, GTPγS, or GppNHp. Also, other nucleoside triphosphates including ITP, UTP, and CTP have no effect on Ca^{2+} movements. Submicromolar GTP concentrations function to release Ca^{2+} in the presence of millimolar ATP concentrations, indicating the considerable specificity of the GTP-activated release process. GDP induces release, but only after conversion to GTP via nucleoside diphosphokinase activity. GDP competitively blocks the action of GTP with a K_i of approximately 3 μM; GTPγS also blocks the effect of GTP. These latter effects indicate that GTP hydrolysis is required for the activation of Ca^{2+} release. Very slow release activated by GTPγS may be consistent with slow cleavage of the phosphorothioate residue.

2.2. Cellular and Subcellular Specificity

It was important to establish whether the effectiveness of GTP in directly inducing Ca^{2+} release was an anomaly, perhaps restricted to the N1E-115 neuronal cell line used in early studies. Experiments undertaken on a quite unrelated cell type, the DDT_1MF-2 smooth-muscle cell line derived from hamster vas deferens (Norris *et al.*, 1974), suggest that this is not the case. Thus, a sensitive, specific, and substantial GTP-dependent release of Ca^{2+} is observed using permeabilized DDT_1MF-2 cells loaded with Ca^{2+}, with pronounced effectiveness of as low as 0.1 μM GTP in the presence of 1 mM ATP. In addition to the DDT_1MF-2 cell line, we have measured almost identical effects of GTP on Ca^{2+} release using permeabilized cells from the rat BC_3H-1 smooth-muscle cell line and from the human WI-38 normal embryonic lung fibroblast cell line. Using microsomal membrane vesicle fractions prepared from DDT_1MF-2 cells by methods similar to those described for N1E-115 cell-derived microsomes (Ueda *et al.*, 1986), we have observed GTP effects on Ca^{2+} release almost identical to those seen with permeabilized cells. Furthermore, using microsomes derived from guinea pig parotid gland, Henne and Söling (1986) observed very similar effects on release of accumulated Ca^{2+} induced by GTP; in this study, GTP-activated Ca^{2+} movements were followed using Ca^{2+} electrodes. The observations of Jean and Klee (1986) on GTP- and IP_3-mediated Ca^{2+} release from microsomes derived from NG108-15 neuroblastoma X glioma hybrid cells are also very consistent with our studies.

The process of Ca^{2+} release is specific to a nonmitochondrial Ca^{2+}-sequestering organelle (believed to be endoplasmic reticulum); no effects of guanine nucleotides or IP_3 can be observed on Ca^{2+} fluxes across the mitochondrial membrane or isolated plasma membranes. The observation that less than 100% of Ca^{2+} release from ER is effected by GTP or IP_3 suggests that only a subcompartment of ER contains the activatable efflux mechanisms. Although we have no direct proof that ER is a source of GTP-releasable Ca^{2+}, interpretation of the effects of oxalate, a known permeator of the ER membrane, may indicate that ER is indeed a site of action of GTP, as described in the next section.

2.3. Reversibility of Action of GTP

An important question concerns the nature of the Ca^{2+} translocation process activated by GTP. Either of two distinct possibilities appears likely: first, GTP could activate a channel process to permit the flow of Ca^{2+} out of the organelle(s) into which Ca^{2+} is sequestered; alternatively, GTP could activate a fusion between organelle membranes and result in the release or transfer of Ca^{2+}. In the latter case, it is very unlikely that such a process would be reversible. Recently, we reported that GDP reverses the prior effectiveness of GTP at least partially (Gill *et al.*, 1986). Since then, a more definitive indication of the reversibility of the effect of GTP has come from a simpler study involving washing of cells after GTP activation. Thus, it has been observed that after activation of the GTP-dependent Ca^{2+} release process (with up to 100 μM GTP), the effectiveness of GTP can be substantially (more than 70%) reversed by simple washing of permeabilized cells with GTP-free medium. Thus, after such washing, the ability of ER to accumulate Ca^{2+} within cells is largely restored; furthermore, release of Ca^{2+} can be reactivated by a further application of GTP. It would be extremely difficult to reconcile this observed reversibility with a simple membrane fusion process activated by GTP; in other words, the effects of a direct membrane fusion event would not be reversed by washing and result in the restoration of almost normal Ca^{2+} retention, as observed.

2.4. Membrane Interactions

Recent electron microscopic analysis of membrane vesicles treated with GTP has suggested that the action of GTP, while unlikely to involve membrane fusion, may be promoted by close association between membranes. Thus, it is now well established that the effects of GTP on Ca^{2+} release are promoted by 1–3% polyethylene glycol (PEG) (Gill *et al.*, 1986; Ueda *et al.*, 1986; Chueh and Gill, 1986). Although PEG is a known fusogen above 25% (Hui *et al.*, 1985), the effect of PEG in enhancing Ca^{2+} release does not appear to involve membrane fusion. Thus, our recent studies analyzed by electron microscopy the appearance of isolated microsomal membrane vesicles derived from N1E-115 cells after treatment with GTP with or without PEG. GTP was without any effect on vesicle appearance, whereas 3% PEG induced a very clear coalescence of vesicles into tightly associated conglomerates with very few free or unattached vesicles. The effect of PEG was not visibly altered by GTP. It may therefore be concluded that GTP itself does not induce any observable alteration in vesicle structure or association. However, the striking effectiveness of PEG is good evidence to suggest that the effect of GTP in inducing Ca^{2+} movements is promoted by a condition that increases the close association between membranes. This may be an important clue to the action of GTP, as discussed in Section 4.

3. RELATIONSHIP BETWEEN IP₃- AND GTP-ACTIVATED CALCIUM RELEASE

3.1. Distinctions in Mechanism of Action

A number of distinctions exist between the actions of IP_3 and GTP on Ca^{2+} release, as described in our recent report (Chueh and Gill, 1986). First, IP_3-mediated release is unaffected by either GDP or GTPγS, both of which block the action of GTP on Ca^{2+} release. Second, polyethylene glycol, which considerably promotes GTP-activated release, does not alter the action of IP_3. Third, IP_3-induced Ca^{2+} release is modified by the free-Ca^{2+} concentration (normally 0.1 μM in experiments); thus, the effect of IP_3 is reduced by 50% at 1 μM free Ca^{2+} and completely abolished with 10 μM Ca^{2+}. In contrast, GTP induces identical fractional Ca^{2+} release over this entire range of free Ca^{2+}. Lastly, the actions of IP_3 and GTP are distinct with regard to temperature dependency. Thus, the rate of IP_3-induced Ca^{2+} release is decreased by only 20% when the temperature is decreased from 37 to 4°C; this contrasts with the complete abolition of the effectiveness of GTP at the lower temperature. From these results we would conclude that the activation of Ca^{2+} release by GTP or IP_3 occurs via distinct mechanisms. Several of these distinctions between the actions of GTP and IP_3 have also been reported by Henne and Söling (1986) using either liver- or parotid-derived microsomes, and by Jean and Klee (1986) using microsomes derived from NG108-15 neuroblastoma X glioma hybrid cells. It is concluded that the relative temperature insensitivity and rapidity of IP_3-induced Ca^{2+} release are consistent with its probable direct activation of a Ca^{2+} channel in ER, a conclusion in agreement with the observations of others (Smith *et al.*, 1985; Muallem *et al.*, 1985). In contrast, GTP appears to effect release by a temperature-sensitive process that probably involves the enzymic hydrolysis of the terminal phosphate from GTP.

3.2. IP₃- and GTP-Releasable Calcium Pools

Both the IP_3- and GTP-induced Ca^{2+} release processes function on a similar type of intracellular Ca^{2+}-sequestering compartment. Yet the size of the releasable pools of Ca^{2+} are distinct, that induced by GTP being approximately twice the size of the IP_3-releasable pool. Thus, as shown in Fig. 1 using permeabilized N1E-115 cells, following maximal Ca^{2+} release by GTP, IP_3 is not effective in releasing further Ca^{2+}; however, following maximal release by IP_3 (approximately 30% of accumulated Ca^{2+}), GTP effects a further release of Ca^{2+} down to the level GTP could induce when added alone (approximately 60% of accumulated Ca^{2+}). These results suggest that three compartments exist: one sensitive to both GTP and IP_3, another releasing only in response to GTP, and a third releasing in response to neither agent. Thus it is apparent that, although the GTP-releasable pool differs from the IP_3-releasable pool in being larger, at least a significant proportion of accumulated Ca^{2+} lies within a pool that can be released by both agents. In other words, it appears that all of the Ca^{2+} within the IP_3-sensitive Ca^{2+} pool is also releasable by the GTP-activated process, even if additional GTP-releasable Ca^{2+} also exists. This implies a probable proximal relationship between the IP_3- and GTP-activated Ca^{2+} release processes, and permits us to consider the existence of possible coupling events linking their modes of action.

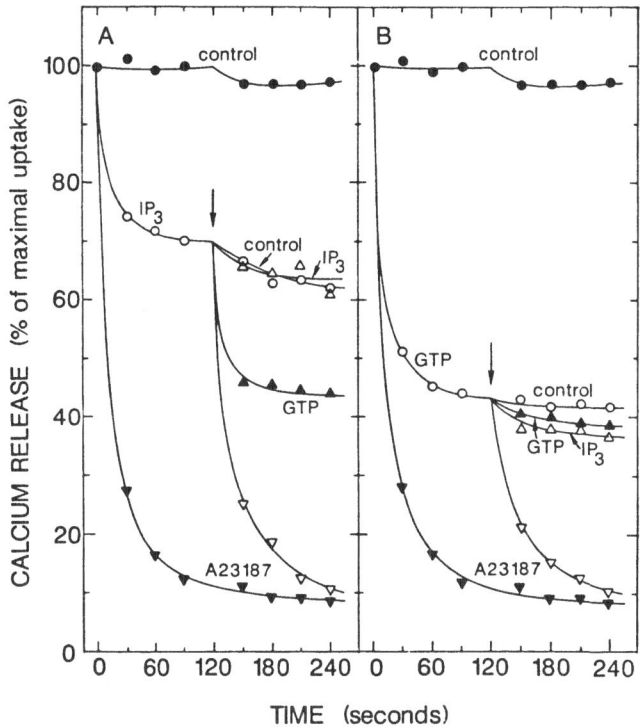

FIGURE 1. Effects of sequential addition of IP$_3$ and GTP on Ca^{2+} release from permeabilized N1E-115 neuroblastoma cells. Ca^{2+} release was measured after loading for 5 min in the presence of 0.1 μM free Ca^{2+}, under the standard conditions described by Chueh et al. (1987). (A) Immediately following uptake, release was observed after addition of either 10 μM IP$_3$ (○), 5 μM A23187 (▼), or buffer control (●); after 120 sec of release in the presence of IP$_3$ measurement of release was continued after further additions of either 10 μM IP$_3$ (△), 10 μM GTP (▲), 5 μM A23187 (▽), or buffer control (○). (B) Immediately following uptake, release was observed after addition of either 10 μM GTP (○), 5 μM A23187 (▼), or buffer control (●); after 120 sec of release in the presence of GTP, release was continued after further addition of either 10 μM IP$_3$ (△), 10 μM GTP (▲), 5 μM A23187 (△), or buffer control (○). In each case, samples of the Ca^{2+}-loaded permeabilized cell suspension were removed followed by rapid filtration and washing as described by Chueh and Gill (1986).

4. GTP-ACTIVATED CALCIUM UPTAKE

4.1. Oxalate Effects

The evidence provided above suggests that simple membrane fusion is not likely to account for the observed release of Ca^{2+} induced by GTP. A recent and dramatic observation gives perhaps a better clue to the mechanism of action of GTP. As shown in our previous studies (Gill and Chueh, 1985), and well known for many cells (Henkart et al., 1978; McGraw et al., 1980; Wakasugi et al., 1982; Burton and Laveri, 1985), the ER is permeable to anions such as oxalate, which can diffuse into the ER lumen and hence promote a large increment in Ca^{2+} uptake due to formation of the insoluble Ca^{2+}-oxalate complex. In order to further investigate how GTP activates Ca^{2+} release, we tested to see if oxalate-precipitated Ca^{2+} within ER could be released by GTP; thus a negative result would again militate against a simple membrane fusion event accounting for release and would instead argue in favor of a more selective channel mechanism through which passage of precipitated Ca^{2+} would not be expected. However, as shown in Fig. 2, an unexpected profound increase in Ca^{2+} uptake was observed in the presence of oxalate, a remarkable and entirely opposite effect to that observed in the absence of oxalate. The effect is observed with concentrations of oxalate (2 mM) that show very little effect on uptake of Ca^{2+} in the absence of GTP (Fig. 2C). When oxalate is present at a concentration that induces linear uptake of Ca^{2+} (Fig. 2E), GTP still activates an additional increase in the rate of uptake. An identical effect of GTP

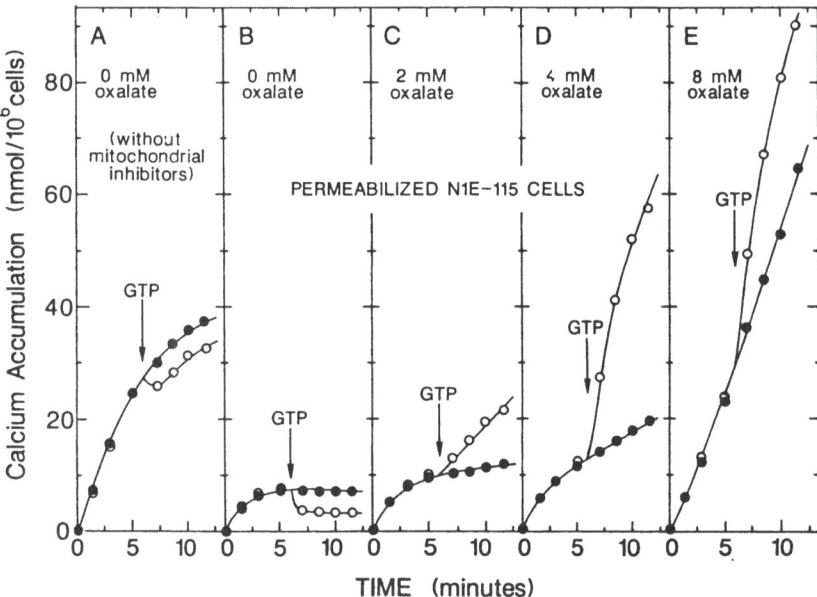

FIGURE 2. GTP-activated movements of Ca^{2+} in permeabilized N1E-115 cells with increasing oxalate concentrations. Details of the permeabilization of N1E-115 cells, loading with labeled Ca^{2+}, and conditions for Ca^{2+} release were described by Mullaney et al. (1987). During uptake, cells suspended at approximately 2×10^5 cells/ml were gently stirred at 37°C in uptake medium [140 mM KCl, 10 mM NaCl, 2.5 mM $MgCl_2$, 10 mM Hepes-KOH pH (7.0)] with 1 mM ATP, 3% PEG, and 30 μM $CaCl_2$ (containing 80 Ci/mol $^{45}Ca^{2+}$) without EGTA. Experiments were undertaken either in the absence of mitochondrial inhibitors (A), or in the presence of 5 μM ruthenium red and 10 μM oligomycin. K-oxalate was either absent (A and B) or present from the beginning of uptake at a final concentration of 2 mM (C), 4 mM (D), or 8 mM (E). After 6 min of uptake, 10 μM GTP (○) or control buffer (●) was added to the permeabilized cell suspensions. 200-μl aliquots from the incubation vials were taken at the times shown, and Ca^{2+} remaining within cells was determined by rapid La^{3+} quenching and filtration as described by Chueh et al. (1987).

on Ca^{2+} uptake in the presence of oxalate has been observed using both N1E-115 and DDT_2MF-2 cells.

4.2. Relationship to GTP-Activated Calcium Release

In view of this paradoxically opposite effect of GTP in the presence as opposed to the absence of oxalate, it was important to establish whether both GTP-mediated events resulted from a common mechanism activated by GTP. It now very much appears that this is the case. Thus, as shown in Fig. 3, the GTP dependence of Ca^{2+} uptake induced in the presence of oxalate is almost identical to the release induced without oxalate. In this experiment, only the GTP-activated uptake component is shown. From linearization of these data, a K_m for GTP of 0.9 μM is obtained, very close to the value of 0.75 μM derived from Ca^{2+} release data reported in earlier studies (Gill et al., 1986). Further studies revealed that the uptake of Ca^{2+} induced by GTP in the presence of oxalate is promoted by PEG in just the same way as GTP-activated Ca^{2+} release is augmented by PEG without oxalate. Moreover, GTP-activated Ca^{2+} uptake in the presence of oxalate is not activated by $GTP\gamma S$ or GppNHp; however, $GTP\gamma S$ (but not GppNHp) completely blocks the action of GTP. GDP gives a delayed uptake response that is blocked by ADP, indicating that its action arises from conversion to GTP. In the presence of ADP, GDP blocks the action of GTP; $GDP\beta S$, which does not activate uptake, also blocks the action of GTP. These data reveal almost exact correlation between parameters affecting GTP-activated uptake and release. A summary of these effects is given in Table I. Such data provide very strong evidence to suggest that the same GTP-activated process mediates uptake and release of Ca^{2+} in the presence and absence of oxalate, respectively. The only divergence between the two processes is the effectiveness of vanadate, which only blocks GTP-induced uptake. However, as discussed below, the proposed model for the actions of GTP accounts for this difference.

5. CONCLUSIONS AND MODEL FOR THE ACTIONS OF GTP AND IP₃

Although it is likely that IP_3 operates by directly activating a channel to release Ca^{2+} from ER, the effect of GTP is clearly distinct. The enhancement of GTP-activated Ca^{2+} release by PEG suggests that a possible interaction between membranes may be involved in the process; indeed, our electron microscopic analyses directly reveal that PEG treatment causes significant vesicle coalescence. Although the induction of close contact between membranes promotes the actions of GTP, such studies do not reveal any visible fusion of membranes, with or without GTP. Moreover, as stated above, the reversibility experiments and oxalate effects are inconsistent with a membrane fusion event being involved in the action of GTP. Instead, we believe that the effects of GTP are consistent with its promoting a "conveyance" of Ca^{2+} from one organelle to another. Such organelles may be distinct with regard to their permeability to oxalate. Thus, in many cells, oxalate freely traverses the ER membrane probably via a mechanism analogous to the nonselective anion transporter that operates to transport oxalate across the sarcoplasmic reticulum membrane of muscle (see Martonosi, 1982). It is quite clear that other membranes, e.g., the plasma membrane, do not permit oxalate passage, as we reported in earlier studies (Gill et al., 1984). We hypothesize that GTP may promote a rather selective conveyance of Ca^{2+} between organelles with

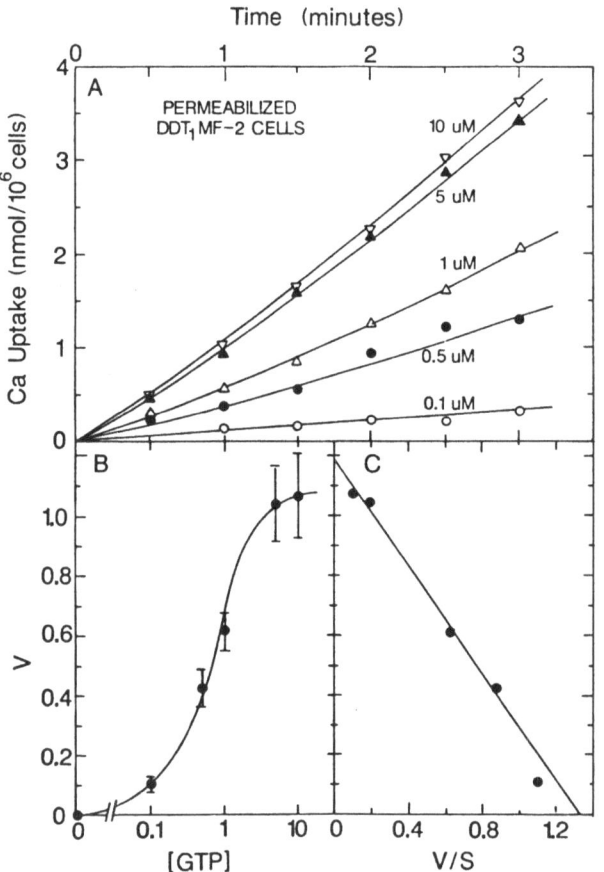

FIGURE 3. GTP dependence of GTP-activated Ca^{2+} uptake in permeabilized DDT_1MF-2 cells. Cells were incubated in uptake medium under the conditions described in Fig. 2, with 10 μM oligomycin, 4 mM K-oxalate, and 3% PEG present in all incubations. Ca^{2+} uptake into cells proceeded without any GTP addition for 6 min, at which time GTP was added to give final GTP concentrations of 0.1 μM (○), 0.5 μM (●), 1 μM (△), 5 μM (▲), or 10 μM (▽). GTP-dependent uptake of Ca^{2+} (i.e., uptake above that observed over the same time period without GTP addition) is plotted against time after GTP addition in panel A. Mean values ± standard deviation calculated by linear regression of the rate of GTP-activated Ca^{2+} uptake (in units of nmol $Ca^{2+}/10^6$ cells/min) over the 3-min uptake period are plotted against GTP concentration in panel B. Eadie-Hofstee analysis of the same data is shown in panel C from which a K_m value for GTP of 0.9 μM is obtained.

different oxalate permeabilities, as shown in Fig. 4. In this scheme, sequestered Ca^{2+} within an organelle impermeable to oxalate might be conveyed into a compartment where oxalate can gain access (perhaps the ER), resulting in the observed stimulation of uptake. How does GTP activate Ca^{2+} release in the absence of oxalate? We would asssume that a similar conveyance of sequestered Ca^{2+} is instead promoted between enclosed Ca^{2+}-pumping or-

TABLE I. Summary of the Effects of GTP-Activated Calcium Release and Uptake in Permeabilized Cells in the Absence and Presence of Oxalate, Respectively

Parameter or condition[a]	Calcium release	Calcium uptake
K_m for GTP	0.75 μM	0.9 μM
GDP	Delayed full effect	Delayed full effect
GDP (+ ADP)	No effect	No effect
GDPβS	No effect	No effect
GTP + high GDP (+ ADP)	GTP effect blocked	GTP effect blocked
GTP + high GDPβS	GTP effect blocked	GTP effect blocked
GTPγS	Slight effect	Slight effect
GTP + high GTPγS	GTP effect blocked	GTP effect blocked
GppNHp	No effect	No effect
GTP + high GppNHp	GTP effect not blocked	GTP effect not blocked
PEG	Stimulated	Stimulated
Vanadate	No effect	Blocked

[a] Each of the parameters for GTP-activated Ca^{2+} uptake observed in the presence of oxalate refers to data presented in the current report. The observations relating to Ca^{2+} release (in the absence of oxalate) were published in previous reports (Gill et al., 1986; Ueda et al., 1986; Chueh and Gill, 1986). Explanations and details of the conditions described are given in the text.

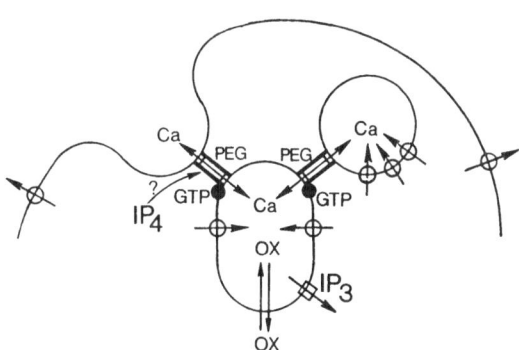

FIGURE 4. Hypothetical scheme for GTP-activated transmembrane "conveyance" of Ca^{2+}. The diagram depicts a scheme that proposes that GTP activates the movement of Ca^{2+} across membranes and between intracellular organelles mediated by hypothetical junctional processes perhaps activated in a GTP-dependent manner. In this scheme, PEG may promote GTP-dependent activation of such Ca^{2+} conveyance by inducing close appositions between membranes. IP$_3$ is depicted as inducing release by operating on a channel that releases Ca^{2+} from ER, presumed to be the organelle also permeable to oxalate. The nature of the other hypothetical Ca^{2+} pumping organelle that may be impermeable to oxalate is unknown. Also yet to be determined is the involvement of the plasma membrane in GTP-activated Ca^{2+} release and any connection with the possible action of IP$_4$. ATP-dependent Ca^{2+} pumping activity in the scheme is represented by circles with arrows. See text for details and implications of such a scheme.

ganelles and nonclosed membranes; in this case, the latter could easily be the plasma membrane. We would also assume that such release would not occur when most of the Ca^{2+} within the ER is precipitated with oxalate; thus, accumulation in the presence of oxalate would not be limited by release.

It is possible that such a scheme is linked to the movements of Ca^{2+} that might occur within cells to replenish the Ca^{2+} lost from ER during stimulation. Thus, Putney (1986) suggested that the frequently observed prolonged responses to external signals that are dependent on external Ca^{2+} are likely to reflect external entry of Ca^{2+} across the plasma membrane, but directed into the ER from where it can be released by IP_3. It is possible that this same process that conveys external Ca^{2+} into the ER could account for the GTP-activated release of Ca^{2+} observed in our experiments with nonintact cells. Close apposition of membranes, perhaps induced by PEG, could enhance this effect, whereas actual conveyance of Ca^{2+} ions may occur only when some junctional process is effected, presumably in a GTP-dependent manner. This type of transfer of Ca^{2+} between pools and across membranes would easily account for the enhanced IP_3-mediated release observed by Dawson and his colleagues (Dawson, 1985; Dawson et al., 1986), where we would assume that an increased pool size was being effected by GTP. It is also intriguing to consider that the effects of IP_4 recently reported by Irvine and Moor (1986), which may indicate IP_4's enhancement of external Ca^{2+} conveyance to an IP_3-sensitive pool, could reflect a process similar to, or even connected with, that which we believe may account for the effects of GTP. Further study of this possibility is clearly important.

ACKNOWLEDGMENTS. This work was supported by grant NS19304 from the National Institutes of Health and grant DCB-8510225 from the National Science Foundation. We thank Dr. Robin Irvine for generously supplying high-purity IP_3.

REFERENCES

Berridge, M. J., and Irvine, R. F., 1984, Inositol trisphosphate, a novel second messenger in cellular signal transduction, *Nature* **312**:315–321.

Burton, P. R., and Laveri, L. A., 1985, The distribution, relationships to other organelles, and calcium-sequestering ability of smooth endoplasmic reticulum in frog olfactory axons, *J. Neurosci.* **5**: 3047–3060.

Chueh, S. H., and Gill, D. L., 1986, Inositol 1,4,5-trisphosphate and guanine nucleotides activate calcium release from endoplasmic reticulum via distinct mechanisms, *J. Biol. Chem.* **261**:13883–13886.

Chueh, S. H., Mullaney, J. M., Ghosh, T. K., Zachary, A. L., and Gill, D. L., 1987, GTP and inositol 1,4,5-trisphosphate-activated intracellular calcium movements in neuronal and smooth muscle cell lines, *J. Biol. Chem.* **262**:13857–13864.

Dawson, A. P., 1985, GTP enhances inositol trisphosphate-stimulated Ca^{2+} release from rat liver microsomes, *FEBS Lett.* **184**:147–150.

Dawson, A. P., Comerford, J. G., and Fulton, D. V., 1986, The effect of GTP on inositol 1,4,5-trisphosphate-stimulated Ca^{2+} efflux from a rat liver microsomal fraction, *Biochem. J.* **234**:311–315.

Gill, D. L., 1985, Receptors coupled to calcium mobilization, *Adv. Cyclic Nucleotide Protein Phos. Res.* **19**:195–212.

Gill, D. L., and Chueh, S. H., 1985, An intracellular (ATP + Mg^{2+})-dependent calcium pump within the N1E-115 neuronal cell line, *J. Biol. Chem.* **260**:9289–9297.

Gill, D. L., Chueh, S. H., and Whitlow, C. L., 1984, Functional importance of the synaptic plasma membrane calcium pump and sodium-calcium exchanger, *J. Biol. Chem.* **259**:10807–10813.

Gill, D. L., Ueda, T., Chueh, S. H., and Noel, M. W., 1986, Ca^{2+} release from endoplasmic reticulum is mediated by a guanine nucleotide regulatory mechanism, *Nature* **320**:461–464.

Henkart, M. P., Reese, T. S., and Brinley, F. J., 1978, Endoplasmic reticulum sequesters calcium in the squid giant axon, *Science* **202:**1300–1303.

Henne, V., and Söling, H-D., 1986, Guanosine 5'-triphosphate releases calcium from rat liver and guinea pig parotid gland endoplasmic reticulum independently of inositol 1,4,5-trisphosphate, *FEBS Lett.* **202:**267–273.

Hui, S. W., Isac, T., Boni, L. T., and Sen, A., 1985, Action of polyethylene glycol on the fusion of human erythrocyte membranes, *J. Membrane Biol.* **84:**137–146.

Irvine, R. F., and Moor, R. M., 1986, Micro-injection of inositol 1,3,4,5-tetrakisphosphate activates sea urchin eggs by a mechanism dependent on external Ca^{2+}, *Biochem. J.* **240:**917–920.

Jean, B., & Klee, C. B., 1986, Calcium modulation of inositol 1,4,5-trisphosphate-induced calcium release from neuroblastoma X glioma hybrid (NG108-15) microsomes, *J. Biol. Chem.* **261:** 16414–16420.

Majerus, P. W., Connolly, T. M., Deckmyn, H., Ross, T. S., Bross, T. E., Ishii, H., Bansal, V. S., and Wilson, D. B., 1986, The metabolism of phosphoinositide-derived messenger molecules, *Science* **234:**1519–1526.

Martonosi, A. N., 1982, Transport of calcium by sarcoplasmic reticulum, in: *Calcium in Cell Function,* Vol. 3 (W. Y. Cheung, ed.), Academic Press, New York, pp. 37–102.

McGraw, C. F., Somlyo, A. V., and Blaustein, M. P., 1980, Localization of calcium in presynaptic nerve terminals. *J. Cell Biol.* **85:**228–241.

Muallem, S., Schoeffield, M., Pandol, S., and Sachs, G., 1985, Inositol trisphosphate modification of ion transport in rough endoplasmic reticulum, *Proc. Natl. Acad. Sci. USA* **82:**4433–4437.

Mullaney, J. M., Chueh, S. H., Ghosh, T. K., and Gill, D. L., 1987, Intracellular calcium uptake activated by GTP: Evidence for a possible guanine nucleotide-induced transmembrane conveyance of intracellular calcium, *J. Biol. Chem.* **262:**13865–13872.

Norris, J. S., Gorski, J., and Kohler, P. O., 1974, Androgen receptors in a Syrian hamster ductus deferens tumour cell line, *Nature* **248:**422–424.

Putney, J. W., 1986, A model for receptor-regulated calcium entry, *Cell Calcium* **7:**1–12.

Smith, J. B., Smith, L., and Higgins, B. L., 1985, Temperature and nucleotide dependence of calcium release by myo-inositol 1,4,5-trisphosphate in cultured vascular smooth muscle cells, *J. Biol. Chem.* **260:**14413–14416.

Ueda, T., Chueh, S. H., Noel, M. W., and Gill, D. L., 1986, Influence of inositol 1,4,5-trisphosphate and guanine nucleotides on intracellular calcium release within the N1E-115 neuronal cell line, *J. Biol. Chem.* **261:**3184–3192.

Wakasugi, H., Kimura, T., Haase, W., Kribben, A., Kaufmann, R., and Schulz, I., 1982, Calcium uptake into acini from rat pancreas: evidence for intracellular ATP-dependent calcium sequestration, *J. Membrane Biol.* **65:**205–220.

18

Activation of the Inositol-1,4,5-Trisphosphate Signaling System by Acute Ethanol Treatment of Rat Hepatocytes

ANDREW P. THOMAS, JAN B. HOEK, RAPHAEL RUBIN, and EMANUEL RUBIN

1. INTRODUCTION

A large number of hormones and other agonists bring about their intracellular effects through an elevation of cytosolic free Ca^{2+} concentration. The mechanism by which this elevation of cytosolic Ca^{2+} is achieved recently became much clearer as a result of the elucidation of the role of inositol lipids in transmitting hormonal signals to the interior of the cell (see Berridge, 1984; Berridge and Irvine, 1984; Williamson *et al.*, 1985 for reviews). It is now widely accepted that the primary event following receptor activation is the stimulation of an inositol lipid-specific phospholipase C that cleaves phosphatidylinositol-4,5-bisphosphate [PtdIns(4,5)P$_2$] to yield inositol-1,4,5-trisphosphate [Ins(1,4,5)P$_3$] and diacylglycerol. Evidence obtained using subcellular systems has indicated that a GTP-binding protein (G protein) is probably involved in coupling the occupied receptor to the phospholipase C (Cockcroft, 1987). Once formed in this primary reaction, both Ins(1,4,5)P$_3$ and diacylglycerol have distinct second messenger functions. Diacylglycerol is a potent activator of protein kinase C (Nishizuka, 1984) and Ins(1,4,5)P$_3$ is able to trigger Ca^{2+} release from an intracellular ATP-dependent Ca^{2+} storage pool (Streb *et al.*, 1983; Joseph *et al.*, 1984).

Ethanol has multiple actions in biological systems but relatively little is known about the mechanisms responsible for these effects. One common facet of many of the different actions of ethanol is that they appear to involve interactions between ethanol and biological membranes (Taraschi and Rubin, 1985). It is known that alcohols perturb the physical and chemical properties of both natural membranes and artificial membranes prepared from pure lipids, and it has been suggested that this may represent a general mode of action for the

---

ANDREW P. THOMAS, JAN B. HOEK, RAPHAEL RUBIN, and EMANUEL RUBIN ● Department of Pathology and Cell Biology, Thomas Jefferson University, Philadelphia, Pennsylvania 19107.

pharmacological effects of ethanol (Taraschi and Rubin, 1985). In this paper we describe an effect of ethanol that apparently results from alterations in the interaction between proteins within the cellular plasma membrane and that may have important consequences for the function of these cells. We have shown that ethanol is able to directly activate the inositol lipid signaling system with the generation of Ins(1,4,5)P_3 and diacylglycerol in both rat hepatocytes and human platelets. This response to ethanol is dependent on the activation of a GTP-binding protein and may also require the presence of a fully coupled (but unoccupied) hormone receptor/G-protein complex. As a result of the ethanol-induced production of these second messengers, cytosolic free Ca^{2+} is elevated and protein kinase C is activated. Among the consequences of these changes is an activation of phosphorylase kinase in the hepatocyte and induction of shape change and 40-kDa protein phosphorylation in the platelet.

2. ELEVATION OF CYTOSOLIC Ca^{2+} BY ETHANOL

The first indication that ethanol might mimic some of the effects of certain hormones in liver cells came from the finding that acute ethanol treatment of hepatocytes led to an activation of glycogen phosphorylase (Hoek *et al.*, 1987b). This was manifested as an increase in the proportion of phosphorylase *a*, presumably reflecting a stimulation of phosphorylase kinase. It is known that ethanol can interact with the adenyl cyclase system at very high concentrations but the ethanol-induced activation of glycogen phosphorylase was found to occur in the absence of changes of cellular cAMP. An alternative method by which phosphorylase kinase can be activated is through an increase of cytosolic Ca^{2+} concentration. We therefore examined the effects of ethanol on cytosolic free Ca^{2+} using the fluorescent Ca^{2+} indicator Quin2.

The measurement of cytosolic free Ca^{2+} using Quin2 can be interfered with by reduced pyridine nucleotides due to the similar excitation and emission wavelengths of these compounds. Measurements of cytosolic Ca^{2+} changes in response to ethanol in liver cells are especially sensitive to this problem because of the presence of alcohol dehydrogenase. This problem could be overcome by pretreating the cells with 4-methylpyrazole to inhibit alcohol dehydrogenase. An alternative solution to this problem was to carry out the measurements at two excitation wavelengths: 339 nm for measurement of Quin2-Ca, and 357 nm, which is isosbestic for Quin2 and Quin2-Ca so that only pyridine nucleotide changes were observed. It was then possible to correct the 339-nm fluorescence signals for the contribution from pyridine nucleotides. Using both methods it was found that ethanol caused a rapid elevation of the cytosolic free Ca^{2+} concentration. The effective concentration range for ethanol-induced Ca^{2+} mobilization and phosphorylase activation was 25–500 mM. Figure 1 shows the Ca^{2+} changes measured in Quin2-loaded hepatocytes treated with a range of ethanol concentrations in the presence of 4-methylpyrazole. The Ca^{2+} increase obtained with 260 mM ethanol was similar in magnitude to an approximately half-maximal dose of a classic Ca^{2+}-mobilizing hormone such as vasopressin in these hepatocytes. It should be noted that alcohol dehydrogenase was saturated at less than 10 mM ethanol, a concentration that was without any detectable effect on cytosolic Ca^{2+}.

The increase in cytosolic Ca^{2+} concentration brought about by ethanol was unaffected by the removal of extracellular Ca^{2+} with EGTA shortly before ethanol addition. This indicates that ethanol acts by mobilizing Ca^{2+} from an intracellular storage pool. Similar results have been obtained for vasopressin and α_1-adrenergic agents in liver cells (Blackmore *et al.*, 1982; Joseph *et al.*, 1985). Experiments in which ethanol and other agonists were

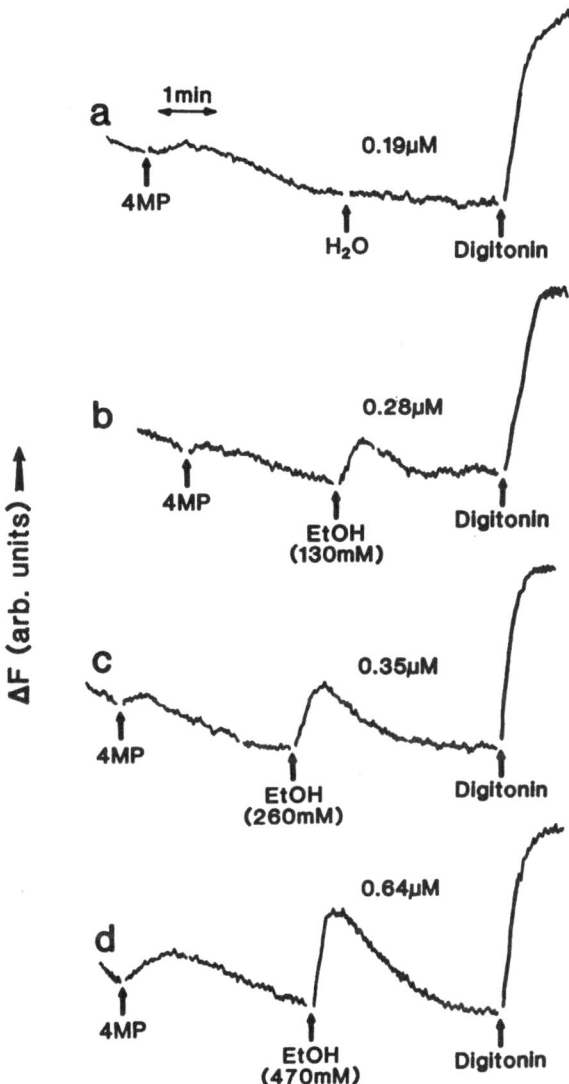

FIGURE 1. Ethanol-induced increases of cytosolic Ca^{2+} measured using Quin2. Isolated rat hepatocytes were loaded with Quin2 and fluorescence measurements of cytosolic Ca^{2+} changes were carried out as described previously (Hoek et al., 1987b). The cells (2.5 mg cell protein/ml) were incubated at 37°C with stirring, and the excitation and emission monochromators were set to 339 and 495 nm, respectively. Additions were made as indicated: 4MP, 15 mM 4-methyl-pyrazole; digitonin, 5 nmol/ml digitonin; EtOH, ethanol at the concentrations indicated. The numbers on each trace following the addition of ethanol or the H_2O blank represent the peak cytosolic free-Ca^{2+} levels in micromolar calculated from the Quin2 fluorescence.

added in different sequences showed that ethanol releases Ca^{2+} from the same storage pool as the hormones (Hoek *et al.*, 1987b). However, the ethanol-induced Ca^{2+} release differed from the responses to hormones in that the cytosolic Ca^{2+} increase was transient and decayed back to basal levels over 2–3 min. This could be because ethanol does not stimulate a secondary influx of Ca^{2+} from the extracellular medium as occurs with the hormones (Joseph *et al.*, 1985), but it seems more likely that ethanol may feedback-inhibit its own effects through the activation of protein kinase C (see Section 4).

3. ACTIVATION OF THE INOSITOL PHOSPHATE SIGNAL TRANSDUCTION SYSTEM BY ETHANOL

The simplest manner in which ethanol might release Ca^{2+} from the hormone-sensitive intracellular Ca^{2+} storage pool would be by a direct effect at the level of the Ca^{2+} storage organelle involved. To examine this possibility, experiments were carried out using digitonin-permeabilized hepatocytes. This system was previously used to study intracellular Ca^{2+} pools and in particular to demonstrate the existence of an $Ins(1,4,5)P_3$-sensitive Ca^{2+} release system (Joseph *et al.*, 1984). While ethanol did cause a small inhibition of ATP-dependent Ca^{2+} uptake in the permeabilized hepatocyte, these effects were found to be too small and too slow to account for the cytosolic Ca^{2+} changes observed in the intact cell. We therefore decided to determine whether ethanol might bring about its effects on intracellular Ca^{2+} homeostasis by activating the production of $Ins(1,4,5)P_3$. This question was addressed by examining the effects of ethanol on inositol lipid breakdown and the release of inositol phosphate headgroups.

In experiments using ^{32}Pi-labeled hepatocytes, it was found that ethanol (200 mM) caused a small but significant decrease in the level of $PtdIns(4,5)P_2$. This was accompanied by an elevation of phosphatidic acid, indicative of enhanced diacylglycerol formation. There was no detectable change in the level of $[^{32}P]PtdIns$ during these short incubations and ethanol did not influence the labeling of other major phospholipids such as phosphatidyl-choline. $PtdIns(4)P$ was actually found to increase as a result of ethanol treatment, an effect that may reflect activation of PtdIns kinase as part of a mechanism to stimulate the resynthesis of $PtdIns(4,5)P_2$. The inositol lipid changes showed a similar dose response for ethanol to that observed for ethanol-induced cytosolic Ca^{2+} increases.

In order to measure inositol phosphates, changes in response to ethanol hepatocytes were prelabeled for 90 min with $[^{3}H]myo$inositol. The ^{3}H-labeled inositol phosphates were separated from perchloric acid extracts of the cells using high-performance liquid chroma-tography (HPLC). The HPLC system was based on that described by Irvine *et al.* (1985) and allowed the separation of $Ins(1,4,5)P_3$ from the inactive isomer, $Ins(1,3,4)P_3$. It is difficult to calculate $Ins(1,4,5)P_3$ concentration changes directly from $[^{3}H]Ins(1,4,5)P_3$ levels because the specific activity of the parent lipid is unknown. For this reason the effects of ethanol on $Ins(1,4,5)P_3$ levels were compared to the changes obtained with a vasopressin concentration that gave a similar rate and magnitude of cytosolic Ca^{2+} increase. In Fig. 2 the effects of 300 mM ethanol and 1 nM vasopressin are compared with respect to the formation of $Ins(1,4,5)P_3$ and the elevation of cytosolic free Ca^{2+}. At these concentrations the two agents gave similar increases of cytosolic free Ca^{2+}, although the effect of vasopressin was more sustained. As was shown previously (Thomas *et al.*, 1984), the vasopressin-induced increase of cytosolic Ca^{2+} was accompanied by a rapid rise in the level of $Ins(1,4,5)P_3$ (Fig. 2B).

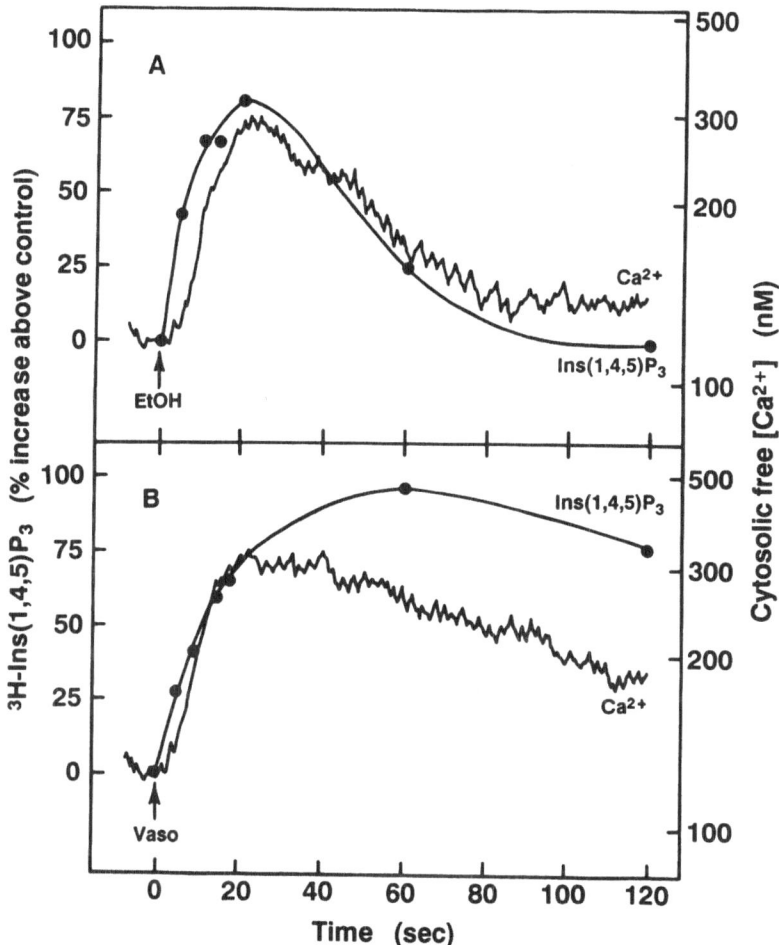

FIGURE 2. Time course of Ins(1,4,5)P_3 formation and the elevation of cytosolic Ca^{2+} in response to ethanol and vasopressin. Cytosolic free-Ca^{2+} levels were measured out as described for Fig. 1. For Ins(1,4,5)P_3 measurements hepatocytes were labeled with [^{3}H]inositol for 90 min and then briefly washed prior to the addition of 300 mM ethanol (panel A) or 1 nM vasopressin (panel B). At the indicated times samples were quenched with perchloric acid. After neutralization, inositol phosphates were separated from the deproteinated samples by HPLC and the ^{3}H content of these fractions was determined by liquid scintillation counting. Full details of these methods have been given elsewhere (Hoek et al., 1987b). The [^{3}H]Ins(1,4,5)P_3 levels (circles) are the mean of values from at least four separate experiments. The Quin2-Ca^{2+} fluorescence traces are from a representative experiment.

Ethanol caused a similar increase of Ins(1,4,5)P$_3$, clearly preceding the increase of cytosolic Ca^{2+} (Fig. 2A). Both ethanol and vasopressin also increased the levels of Ins(1,3,4)P$_3$ and InsP$_2$, although these changes lagged somewhat behind the accumulation of Ins(1,4,5)P$_3$ (Hoek *et al.*, 1987b). Interestingly, and in contrast to the effect of vasopressin, the ethanol-induced increase of Ins(1,4,5)P$_3$ was completely transient, returning to basal within 2 min (Fig. 2A). This is consistent with the transient nature of the cytosolic Ca^{2+} response to ethanol.

The ability of ethanol to activate the inositol lipid signaling system is not limited to hepatocytes. It is known that ethanol inhibits the activation of platelets by various stimuli (Fenn and Littleton, 1982; Mikhailidis *et al.*, 1983). However, we recently found that ethanol can also cause a direct activation of human platelets in the absence of other agonists. Acute addition of ethanol (50–300 mM) led to a rapid elevation of the cytosolic Ca^{2+} concentration that was associated with a shape change response (measured as an increase in light scattering by the platelet suspension) and a phosphorylation of myosin light chain (20 kDa). Ethanol did not cause aggregation or secretion in human platelets. The fact that ethanol did not cause the full spectrum of platelet responses could be due to the inhibitory effects of ethanol noted above. In agreement with the data obtained using isolated hepatocytes, the Ca^{2+} increase brought about by ethanol addition to platelets was associated with the breakdown of PtdIns(4,5,)P$_2$ to yield Ins(1,4,5)P$_3$. In addition to the indirect evidence for diacylglycerol formation obtained from the ethanol-induced increase of phosphatidic acid, it was also possible to show that ethanol increased the phosphorylation of the well-known 40-kDa protein substrate for protein kinase C in the platelet system.

4. INHIBITION OF THE ACTIONS OF ETHANOL BY PROTEIN KINASE C

The Ca^{2+} mobilization in response to several hormones can be blocked by pretreating cells with phorbol esters that activate protein kinase C (Lynch *et al.*, 1985; Cooper *et al.*, 1985). In hepatocytes the actions of α_1-adrenergic agonists are particularly sensitive to inhibition by phorbol ester and it has been proposed that this is due to a direct phosphorylation of the receptor (Lundberg-Leeb *et al.*, 1985). The data of Fig. 3 show that the ethanol-induced elevation of cytosolic Ca^{2+} is almost totally inhibited in the presence of 12-*O*-tetradecanoatephorbol-13-acetate (TPA). The effect of TPA was dose-dependent, with half-maximal inhibition of the ethanol response occurring at 5–10 nM. Further studies demonstrated that the locus of TPA action was close to the primary events associated with phospholipase C stimulation. Thus it was found that TPA pretreatment of hepatocytes prevented both the PtdIns(4,5)P$_2$ breakdown and Ins(1,4,5)P$_3$ production responses to ethanol (Hoek *et al.*, 1987a). Similar results were obtained when TPA was used to inhibit the actions of the α_1-adrenergic agonist phenylephrine. These data provide further evidence for common mechanisms involved in the actions of ethanol and Ca^{2+}-mobilizing hormones.

It has recently been shown that protein kinase C can be inhibited by the isoquinoline-sulfonamide derivative H7 (Kawamoto and Hidaka, 1984). This compound was able to prevent the inhibitory effect of TPA on ethanol-induced Ca^{2+} mobilization in isolated hepatocytes. Most interestingly, however, it was found that when H7 was added to hepatocytes incubated in the absence of TPA, the cytosolic Ca^{2+} increase in response to ethanol was less transient. While this effect clearly requires further study, it is interesting to speculate

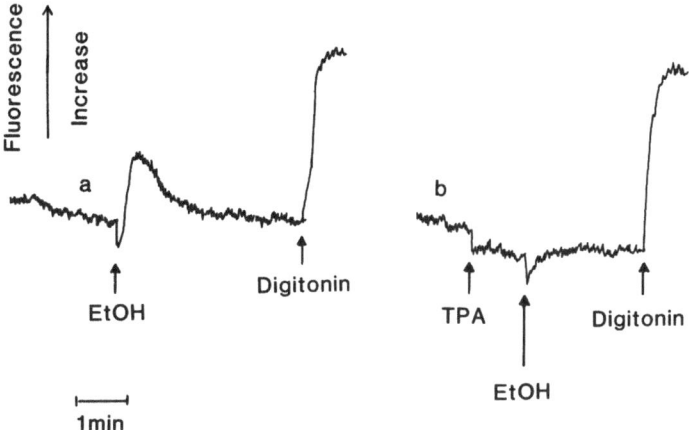

FIGURE 3. Inhibition of the ethanol-induced Ca^{2+} mobilization by phorbol ester. Cytosolic Ca^{2+} measurements were carried out using Quin2 as described in the legend to Fig. 1. Additions were made as follows: EtOH, 250 mM ethanol; digitonin, 5 μg/ml digitonin; TPA, 0.1 μM 12-O-tetradecanoatephorbol-13-acetate.

that the transient nature of the responses to ethanol may represent negative feedback resulting from the formation of diacylglycerol and consequent activation of protein kinase C.

5. ROLE OF A G PROTEIN IN THE STIMULATION OF PHOSPHOLIPASE C BY ETHANOL

We have attempted to investigate the effects of ethanol on inositol lipid breakdown by phospholipase C in a variety of subcellular systems. In digitonin-permeabilized hepatocytes it was possible to activate the phospholipase C with Ca^{2+} or GTPγS, but this preparation was not sensitive to ethanol (Rubin *et al.*, 1987). However, it was also found that the ability of hormones to stimulate phospholipase C was lost after hepatocyte permeabilization. Similar results were obtained with liver plasma membranes isolated from [^{3}H]inositol-labeled livers, where the formation of inositol polyphosphates could be stimulated by GTPγS but not by vasopressin or ethanol. In contrast to the results obtained using the liver systems, ethanol was found to activate the breakdown of PtdIns(4,5)P$_2$ to Ins(1,4,5)P$_3$ in a saponin-permeabilized platelet system based on that described by Brass *et al.* (1986) and Lapetina *et al.* (1985). In this preparation thrombin activated the inositol-lipid-specific phospholipase C in a GTPγS-dependent manner. The effect of ethanol was also completely dependent on the presence of GTPγS, with a similar half-maximal sensitivity at 1 μM GTPγS. It is interesting to note that the effects of both ethanol and thrombin were lost as the saponin concentration was increased, while the activation by GTPγS in the absence of other agonists was much more stable.

From the data outlined above it is apparent that ethanol does not directly activate phospholipase C. The dependence on GTPγS (or GTP) clearly indicates that a G protein is required for the effects of ethanol to be expressed. Furthermore, the ability of ethanol to

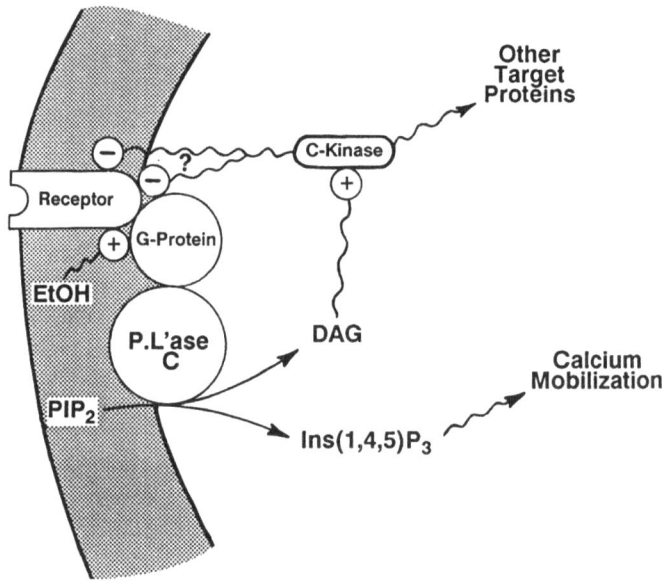

FIGURE 4. Scheme depicting the postulated site at which ethanol may activate the inositol lipid signaling system.

stimulate phospholipase C appears to be correlated with the existence of competent coupling between receptors, the relevant G protein, and the phospholipase. Based on these observations we have developed a working hypothesis for the mechanism of ethanol action that is depicted in Fig. 4. This scheme postulates that the presence of ethanol in the membrane causes an anomalous interaction between unoccupied receptors and a G protein, leading to activation of phospholipase C in the absence of agonist binding. At this stage we have no evidence that any specific receptor is involved. Taking into account the different patterns of receptors in hepatocytes and platelets, it would seem most likely that the effect of ethanol is relatively nonspecific with respect to receptor type. It is also unlikely that there is a specific receptor site for ethanol, since other alcohols and even unrelated hydrophobic solvents can also cause Ca^{2+} mobilization and activation of phospholipase C (Hoek et al., 1987b). The rank order of potency for a series of aliphatic alcohols (1-butanol > 1-propanol > ethanol > methanol) suggests that alcohols bring about their effects by interfering with hydrophobic interactions within the membrane. Elucidation of the precise site of ethanol action will require purification and reconstitution of an ethanol-sensitive phospholipase C.

REFERENCES

Berridge, M. J., and Irvine, R. F., 1984, Inositol trisphosphate, a novel second messenger in cellular signal transduction, Nature 312:315–321.

Berridge, M. J., 1984, Inositol trisphosphate and diacylglycerol as second messengers, Biochem. J. 220:345–360.

Blackmore, P. F., Hughes, B. P., Shuman, E. A., and Exton, J. H., 1982, α-Adrenergic activation of phosphorylase in liver cells involves mobilization of intracellular calcium without influx of extracellular calcium, *J. Biol. Chem.* **257**:190–197.

Brass, L. F., Laposata, M., Banga, H. S., and Rittenhouse, S. E., 1986, Regulation of the phosphoinositide hydrolysis pathway in thrombin-stimulated platelets by a pertussis toxin-sensitive guanine nucleotide-binding protein, *J. Biol. Chem.* **261**:16838–16847.

Cockcroft, S., 1987, Polyphosphoinositide phosphodiesterase: Regulation by a novel guanine nucleotide binding protein, G_p, *Trends. Biochem. Sci.* **12**:75–78.

Cooper, R. H., Coll, K. E., and Williamson, J. R., 1985, Differential effects of phorbol ester on phenylephrine and vasopressin-induced Ca^{2+} mobilization in isolated hepatocytes, *J. Biol. Chem.* **260**:3281–3288.

Fenn, C. G., and Littleton, J. M., 1982, Inhibition of platelet aggregation by ethanol in vitro shows specificity for aggregating agent used and is influenced by platelet lipid composition, *Thromb, Haemostas,* **48**:49–53.

Hoek, J. B., Rubin, R., and Thomas, A. P., 1987a, Phorbol esters inhibit ethanol-induced calcium mobilization and phospholipid turnover in isolated hepatocytes. *Ann. N.Y. Acad. Sci.* **492**:245–247.

Hoek, J. B., Thomas, A. P., Rubin, R., and Rubin, E., 1987b, Ethanol-induced mobilization of calcium by activation of phosphoinositide-specific phospholipase C in intact hepatocytes, *J. Biol. Chem.* **262**:682–691.

Irvine, R. F., Anggard, E. E., Letcher, A. J., and Downes, C. P., 1985, Metabolism of inositol 1,4,5-trisphosphate and inositol 1,3,4-trisphosphate in rat parotid glands, *Biochem. J.* **229**:505–511.

Joseph, S. K., Coll, K. E., Thomas, A. P., Rubin, R., and Williamson, J. R., 1985, The role of extracellular Ca^{2+} in the response of the hepatocytes to Ca^{2+}-dependent hormones, *J. Biol. Chem.* **260**:12508–12515.

Joseph, S. K., Thomas, A. P., Williams, R. J., Irvine, R. F., and Williamson, J. R., 1984, Myoinositol 1,4,5-trisphosphate, *J. Biol. Chem.* **259**:3077–3081.

Kawamoto, S., and Hidaka, H., 1984, 1-(5-isoquinolinesulfonyl)-2-methylpiperazine (H-7) is a selective inhibitor of protein kinase C in rabbit platelets, *Biochem. Biophys. Res. Commun.* **125**:258.

Lapetina, E. G., Silios, J., and Ruggiero, M., 1985, Thrombin induces serotonin secretion and aggregation independently of inositol phospholipids hydrolysis and protein phosphorylation in human platelets permeabilized with saponin, *J. Biol. Chem.* **260**:7078–7083.

Lundberg-Leeb, L. M. F., Cotecchia, S., Lomasney, J. W., DeBernardis, J. F., Lefkowitz, R. J., and Caron, M. G., 1985, Phorbol esters promote α_1-adrenergic receptor phosphorylation and receptor uncoupling from inositol phospholipid metabolism, *Proc. Natl. Acad. Sci. USA* **82**:5651–5655.

Lynch, C. J., Charest, R., Bocckino, S. B., Exton, J. H., and Blackmore, P. F., 1985, Inhibition of hepatic α_1-adrenergic effects and binding by phorbol myristate acetate, *J. Biol. Chem.* **260**:2844–2851.

Mikhailidis, D. P., Jeremy, J. Y., Barradas, M. A., Green, N., and Dandona, P., 1983, Effect of ethanol on vascular prostacyclin (prostaglandin I_2) synthesis, platelet aggregation, and platelet thromboxane release, *Br. Med. J.* **287**:1495–1498.

Nishizuka, Y., 1984, The role of protein kinase C in cell surface signal transduction and tumor promotion, *Nature* **308**:693–698.

Rubin, R., Thomas A. P., and Hoek, J. B., 1987, Ethanol does not stimulate guanine nucleotide-induced activation of phospholipase C in permeabilized hepatocytes, *Arch. Biochem. Biophys.* **256**: 29–38.

Streb, H., Irvine, R. F., Berridge, M. J., and Schulz, I., 1983, Release of Ca^{2+} from a nonmitochondrial intracellular store in pancreatic acinar cells by inositol-1,4,5-trisphosphate, *Nature* **306**:67–69.

Taraschi, T. F., and Rubin, E., 1985, Biology of disease. Effects of ethanol on the chemical and structural properties of biologic membranes, *Lab. Invest.* **52**:120–131.

Thomas, A. P., Alexander, J., and Williamson, J. R., 1984, Relationship between inositol polyphosphate production and the increase of cytosolic free Ca^{2+} induced by vasopressin in isolated hepatocytes, *J. Biol. Chem.* **259**:5574–5584.

Williamson, J. R., Cooper, R. H., Joseph, S. K., and Thomas, A. P., 1985, Inositol trisphosphate and diacylglycerol as intracellular second messengers in liver, *Am. J. Physiol.* **248**:C203–C216.

19

Regulation of Hepatic Glycogenolysis by Calcium-Mobilizing Hormones

PETER F. BLACKMORE, CHRISTOPHER J. LYNCH, STEPHEN B. BOCCKINO, and JOHN H. EXTON

1. INTRODUCTION

The hormonal regulation of glycogen metabolism in the liver has been a subject of investigation for more than 30 years. The intracellular mediators responsible for this regulation have been identified as cAMP and Ca^{2+}. In the liver, several hormones such as vasopressin, angiotensin II, epidermal growth factor (Bosch et al., 1986), glucagon (Blackmore and Exton, 1986), α_1-adrenergic agonists and P_2 purinergic agonists (Charest et al., 1985a,b) increase free cytosolic Ca^{2+} ($[Ca^{2+}]_i$). Each of these hormones binds to specific cell surface receptors; this interaction then leads to the activation of a guanine nucleotide-binding protein (G_p) (e.g., Blackmore et al., 1985; Uhing et al., 1986). In the case of glucagon and epidermal growth factor, the mechanism of activation of G_p is not known but probably involves phosphorylation of G_p (Bosch et al., 1986; Johnson et al., 1986; Blackmore and Exton, 1986). This coupling protein then activates a specific phospholipase C which catalyzes the breakdown of phosphatidylinositol-4,5-bisphosphate (PI-4,5-P_2) (Creba et al., 1983; Rhodes et al., 1983; Thomas et al., 1983; Litosch et al., 1983). The hydrolysis of PI-4,5-P_2 yields myoinositol-1,4,5-trisphosphate (Ins-1,4,5-P_3) (Thomas et al., 1984) and 1,2-diacylglycerol (DAG) (Bocckino et al., 1985). The Ins-1,4,5-P_3 releases Ca^{2+} from the endoplasmic reticulum into the cytoplasm (Joseph et al., 1984), while DAG activates a Ca^{2+}- and phospholipid-dependent protein kinase (protein kinase C) in the plasma membrane (Nishizuka, 1984; Berridge, 1984).

In liver the resulting increase in $[Ca^{2+}]_i$ causes many cellular changes, most of which are mediated by the Ca^{2+}-binding protein calmodulin, which modifies the activity of a variety of enzymes resulting in physiological responses such as glycogen breakdown. Targets of Ca^{2+}-calmodulin-dependent phosphorylation in liver include phosphorylase, glycogen syn-

PETER F. BLACKMORE ● Department of Pharmacology, Eastern Virginia Medical School, Norfolk, Virginia 23501.   CHRISTOPHER J. LYNCH, STEPHEN B. BOCCKINO, and JOHN H. EXTON ● Howard Hughes Medical Institute Laboratories, and Department of Molecular Physiology and Biophysics, Vanderbilt University School of Medicine, Nashville, Tennessee 37232.

thase (see Exton, 1987 for references), pyruvate kinase (Schworer et al., 1985), and acetyl CoA carboxylase (Woodgett et al., 1983).

The activation of protein kinase C by DAG also elicits some of the actions of Ca^{2+}-mediated agonists; however, the exact intracellular substrates of this kinase are not known. Some evidence suggests that the α_1-adrenergic receptor (Leeb-Lundberg et al., 1985), glycogen synthase (Roach and Goldman, 1983; Blackmore et al., 1986), acetyl CoA carboxylase (Vaartjes et al., 1987), and the Na^+-K^+-ATPase (Lynch et al., 1986) are targets for this kinase, either directly or indirectly.

In this chapter the role of G_p in controlling $[Ca^{2+}]_i$ will be explored using AlF_4^- as an activator of guanine nucleotide-binding proteins (Blackmore et al., 1985; Sternweis and Gilman, 1982).

2. ROLE OF GUANINE NUCLEOTIDE-BINDING PROTEINS IN Ca^{2+}-MOBILIZING HORMONE ACTIONS

Several types of evidence indicate that a guanine nucleotide-binding protein is involved in coupling the receptors for Ca^{2+}-mobilizing hormones to PI-4,5-P_2-specific phospholipase C. Binding of agonists to their receptors is regulated by GTP such that GTP converts the receptors from a high affinity to a low affinity for agonist (e.g., El-Refai et al., 1979; Lynch et al., 1985).

Ca^{2+}-mobilizing agonists stimulate a low K_m GTPase activity in plasma membranes (Fitzgerald et al., 1986). In intact cells NaF stimulates the breakdown of PI-4,5-P_2 to Ins-1,4,5-P_3 and DAG with a resultant increase in $[Ca^{2+}]_i$ (Blackmore et al., 1985). The effect of F^- is potentiated by Al^{3+} and it is probable that AlF_4^- is the true activator (Sternweis and Gilman, 1982).

Finally, GTP or poorly hydrolyzable GTP analogs cause the hydrolysis of endogenous PI-4,5-P_2 when added to plasma membranes (e.g., Uhing et al., 1986). When Ca^{2+}-mobilizing agonists are added to plasma membranes in the presence of GTP and low concentrations of Ca^{2+}, there is a synergistic breakdown of PI-4,5-P_2.

3. ROLE OF G_p IN CONTROLLING $[Ca^{2+}]_i$: STUDIES USING AlF_4^-

Previous studies show that AlF_4^- is able to increase hepatocyte $[Ca^{2+}]_i$ and activate phosphorylase (Blackmore et al., 1985). AlF_4^- by activating G_p is able to mimic the effect of the Ca^{2+}-mobilizing hormones without hormone receptor occupation. The relative contribution of extracellular versus intracellular Ca^{2+} mobilization was explored in hepatocytes by using the Ca^{2+} chelator EGTA. When hepatocytes were loaded with the fluorescent Ca^{2+} indicator Fura 2 (Grynkiewicz et al., 1985), the addition of NaF produced an increase in fluorescence after a delay of approximately 40 sec that remained elevated for several minutes (Fig. 1). When Ca^{2+} influx was inhibited by the addition of 3.0 mM EGTA just prior to the addition of NaF (the concentration of Ca^{2+} in the medium was 2.4 mM), the initial increase in $[Ca^{2+}]_i$ was not modified or was slightly potentiated; however, $[Ca^{2+}]_i$ declined to basal values within several minutes (Fig. 1). This result suggests that AlF_4^- is able to elicit internal Ca^{2+} mobilization, consistent with its increasing Ins-1,4,5-P_3, as well as

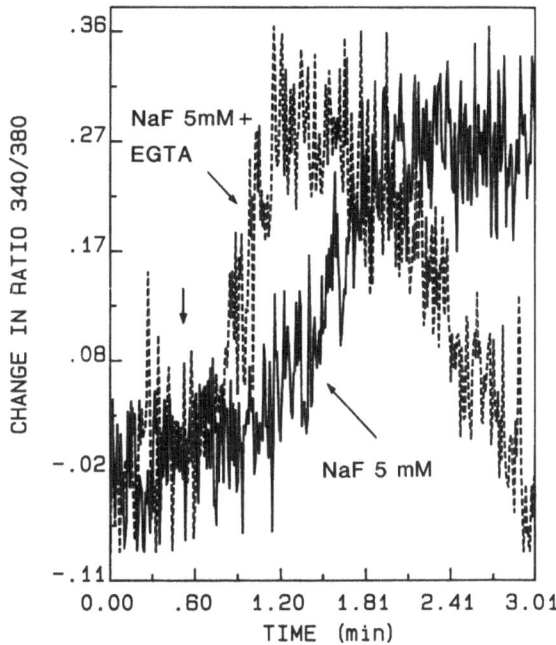

FIGURE 1. Effect of NaF (5 mM) on $[Ca^{2+}]_i$ in the presence and absence of extracellular Ca^{2+} measured using Fura 2. The medium contained 2.4 mM Ca^{2+}. In the low-Ca^{2+} condition, 3.0 mM EGTA was added at zero time. NaF was able to increase $[Ca^{2+}]_i$ even when free extracellular Ca^{2+} was eliminated by chelation with EGTA. However, the increase in $[Ca^{2+}]_i$ was transient because Ca^{2+} influx was unable to occur and since intracellular hormone-sensitive Ca^{2+} pools are of limited magnitude. The changes in $[Ca^{2+}]_i$ are expressed as the ratio of emission obtained at 500 nm when the cells are excited at 340 and 380 nm, respectively.

producing Ca^{2+} influx. Similarly, hormones elicit both Ca^{2+} mobilization and influx (Charest *et al.*, 1985a). The stimulation of G_p by AlF_4^- can lead to an influx of Ca^{2+} either by stimulating the Ca^{2+} channel directly or by producing a ligand (e.g., Ins-1,4,5-P_3, Ins-1,3,4,5-P_4) that may activate the channel (Kuno and Gardner, 1987; Irvine and Moor, 1986).

Further evidence to support the finding that stimulation of a guanine nucleotide-binding protein by AlF_4 can stimulate Ca^{2+} influx comes from experiments in which AlF_4 was added together with cAMP. This is based on our finding showing that when a Ca^{2+}-mobilizing hormone (e.g., vasopressin or angiotensin II) was added to hepatocytes together with a hormone that increases cAMP (e.g., glucagon), there was a net uptake of Ca^{2+} (Morgan *et al.*, 1983). The addition of 5 mM F^- alone to hepatocytes caused a time-dependent decrease in cell Ca^{2+} content. We believe this decrease is a result of the extrusion by the plasmalemmal Ca^{2+} pump of Ca^{2+} mobilized by Ins-1,4,5-P_3 (Prpic *et al.*, 1984). When the cAMP analog 8-(p-chlorophenylthio)cAMP (8-CPTcAMP, 10 μM) was added alone to hepatocytes there was a net loss of cell Ca^{2+}, consistent with internal mobilization (Blackmore and Exton, 1986). When 8-CPTcAMP was added together with NaF there was a net uptake of Ca^{2+} (approximately 20% increase at 5 min and a 110% increase at 10 min). Maximum uptake of cell Ca^{2+} was observed with 10 mM NaF.

Ca^{2+}-mobilizing hormones also promote net uptake of Ca^{2+} in hepatocytes when extracellular pH is raised to 8.0 or above (Blackmore *et al.*, 1984). The addition of 5 mM NaF to hepatocytes incubated at pH 8.0 produced a net uptake of Ca^{2+} (100% increase in total cell Ca^{2+} at 15 min with 5 mM NaF) similar to that observed with vasopressin, angiotensin II, and epinephrine (Blackmore *et al.*, 1984). These data suggest that the plasma membrane contains a Ca^{2+}/H$^+$ antiport (Blackmore *et al.*, 1984).

These results show that AlF$_4^-$ is able to stimulate Ca^{2+} influx in hepatocytes. This is observed when either the extracellular pH is raised to 8.0 or above or when 8-CPTcAMP was added together with AlF$_4^-$. Thus AlF$_4^-$ is able to mimic hormone effects on Ca^{2+} influx, either directly by G$_p$ or by a phospholipid metabolite activating the Ca^{2+} channel.

4. POSSIBLE REGULATORS OF Ca^{2+} CHANNEL ACTIVITY

Regarding the possible involvement of a metabolite-regulated Ca^{2+} channel, some experiments with sea urchin eggs suggest that Ins-1,3,4,5-P$_4$ stimulates Ca^{2+} influx (Irvine and Moor, 1986). The calcium-mobilizing hormones and NaF both produce increases in Ins-1,3,4,5-P$_4$ in hepatocytes. For example, 100 nM vasopressin produced an 11-fold increase in Ins-1,3,4,5-P$_4$ at 5 min whereas 10 mM NaF plus 10 μM AlCl$_3$ increased Ins-1,3,4,5-P$_4$ by seven-fold. If Ins-1,3,4,5-P$_4$ is responsible for activating Ca^{2+} channels, then the observed increases induced by hormones and NaF in hepatocytes would be consistent with this hypothesis.

Another phospholipid metabolite that has been proposed to be responsible for Ca^{2+} influx is phosphatidic acid (PA), although there is much evidence against it (e.g., Holmes and Yoss, 1983). Treatment of hepatocytes with 10 nM vasopressin produced a maximum three-fold increase in PA at 2 min; little change in DAG was seen at this time although large two- to threefold increases were observed at later times (Bocckino *et al.*, 1985). Increases in PA could be seen as early as 15 sec (Bocckino *et al.*, 1987). All of the Ca^{2+}-mobilizing hormones tested (vasopressin, angiotensin II, epinephrine, epidermal growth factor, ATP, and ADP) produced rapid increases in PA (Bocckino *et al.*, 1987). In addition, treatment of hepatocytes with phospholipase D (from *Streptomyces chromofuscus*) increased PA threefold and increased [Ca^{2+}]$_i$ twofold. The rapid increase in PA that preceded any increase in DAG suggests that the hormones activate a phospholipase D activity. This was confirmed by showing that incubation of plasma membranes with GTPγS produced an increase in PA that was not dependent on ATP (Bocckino *et al.*, 1987). At submaximal concentrations of GTPγS, the various Ca^{2+}-mobilizing hormones further stimulated PA formation. Only phosphatidylcholine decreased in the membranes in response to GTPγS. These data suggest that the Ca^{2+}-mobilizing hormones mainly increase PA levels in hepatocytes by activating a guanine nucleotide-binding protein coupled to phospholipase D and not be phosphorylation of diacylglycerol.

5. SUMMARY

Studies utilizing AlF$_4^-$ in hepatocytes suggest that activation of G$_p$ in hepatocytes is responsible for increasing [Ca^{2+}]$_i$. Some of this Ca^{2+} comes from internal stores since AlF$_4^-$ increases Ins-1,4,5-P$_3$ levels. Influx of Ca^{2+} from the external medium is also stim-

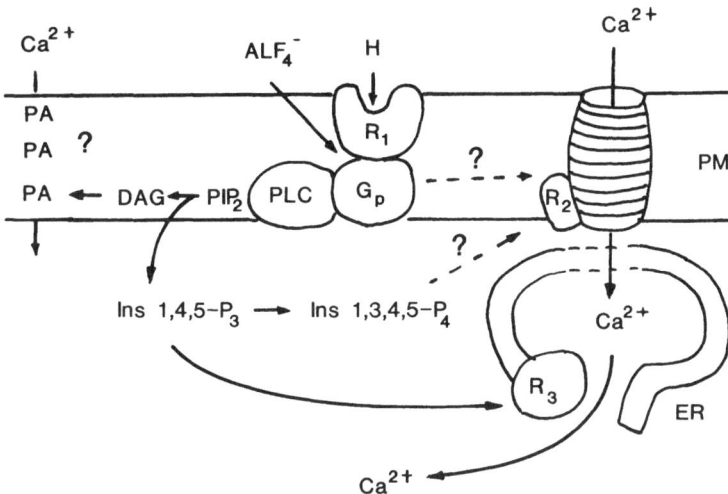

FIGURE 2. Scheme depicting the proposed mechanism by which Ca^{2+}-mobilizing hormones (H) interact with their receptors (R_1) in the plasma membrane (PM) and increase cytosolic free Ca^{2+}. The receptors R_1 interact with and activate a guanine nucleotide-binding protein (G_p) which activates a PI-4,5-P_2-specific phospholipase C (PLC) to produce DAG and Ins-1,4,5-P_3. Ins-1,4,5-P_3 interacts with a receptor (R_3) on the endoplasmic reticulum (ER) to release Ca^{2+}. Ins-1,4,5-P_3 can also be phosphorylated to Ins-1,3,4,5-P_4; this ester has been proposed to be involved with Ca^{2+} influx, presumably by binding to a receptor (R_2) on the Ca^{2+} influx channel or antiport. Alternatively, G_p or one of its subunits may activate the channel directly. G_p can be activated by AlF_4^- directly. DAG can be converted to PA, which may be responsible for Ca^{2+} entry, either directly by an ionophoretic activity, or indirectly by activating the Ca^{2+} channel. Although not depicted in the scheme, PA can be derived directly from phosphatidylcholine by the action of a phospholipase D activity that is stimulated by hormones acting through a guanine nucleotide-binding protein (Bocckino et al., 1987).

ulated. This may be due to a guanine nucleotide-binding protein, such as G_p, activating a Ca^{2+} channel directly (Fig. 2). Alternatively, a phospholipid metabolite such as Ins-1,3,4,5-P_4, Ins-1,4,5-P_3, or PA may alter Ca^{2+} channel activity (Fig. 2). If this is the case then the Ca^{2+} channel is not a receptor operated Ca^{2+} channel (ROC) but rather a second message operated channel (SMOC) or a guanine nucleotide-binding regulatory protein operated channel (GNOC). The nature of the Ca^{2+} influx process in unknown; however, it could be a Na^+/Ca^{2+} antiport, a H^+/Ca^{2+} antiport, or simply a gated Ca^{2+} channel. Putney (1986) suggested that the Ca^{2+} that enters the cell goes directly to the Ins-1,4,5-P_3-sensitive Ca^{2+} pool, the endoplasmic reticulum (Fig. 2).

The Ca^{2+}-mobilizing hormones in hepatocytes regulate glycogenolysis by increaseing $[Ca^{2+}]_i$; some of this Ca^{2+} comes from internal stores while the remainder enters the cell through a plasma membrane Ca^{2+} channel that is regulated either directly or indirectly by a guanine nucleotide-binding protein. The increase in $[Ca^{2+}]_i$ allosterically activates the calmodulin-containing enzyme phosphorylase kinase which then phosphorylates phosphorylase b to convert it to the active phosphorylase a form, thereby promoting the breakdown of glycogen (Exton, 1987).

REFERENCES

Berridge, M. J., 1984, Inositol trisphosphate and diacylglycerol as second messengers, *Biochem. J.* **220**:345–360.

Blackmore, P. F., and Exton, J. H., 1986, Studies on the hepatic calcium-mobilizing activity of aluminum fluoride and glucagon. Modulation by cAMP and phorbol myristate acetate, *J. Biol. Chem.* **261**: 11056–11063.

Blackmore, P. F., Strickland, W. G., Bocckino, S. B., and Exton, J. H., 1986, Mechanism of hepatic glycogen synthase inactivation induced by Ca^{2+} mobilizing hormones, *Biochem. J.* **237**:235–242.

Blackmore, P. F., Waynick, L., Blackman, G. E., Graham, C. W., and Sherry, R. S., 1984, α- and β-adrenergic stimulation of parenchymal cell Ca^{2+} influx: influence of extracellular pH, *J. Biol. Chem.* **259**:12322–12325.

Blackmore, P. F., Bocckino, S. B., Waynick, L. E., and Exton, J. H., 1985, Role of a guanine nucleotide-binding regulatory protein in the hydrolysis of hepatocyte phosphatidylinositol 4,5-bisphosphate by calcium-mobilizing hormones and the control of cell calcium. Studies utilizing aluminum fluoride, *J. Biol. Chem.* **260**:14477–14483.

Bocckino, S. B., Blackmore, P. F., Wilson, P. and Exton, J. H., 1987, Phosphatidate accumulation in hormone-treated hepatocytes via a phospholipase D mechanism, *J. Biol. Chem.* **262**:15309–15315.

Bocckino, S. B., Blackmore, P. F., and Exton, J. H., 1985, Stimulation of 1,2-diacylglycerol accumulation in hepatocytes by vasopressin, epinephrine and angiotensin II, *J. Biol. Chem.* **260**: 14201–14207.

Bosch, F., Bouscarel, B., Slaton, J., Blackmore, P. F., and Exton, J. H., 1986, Epidermal growth factor mimics insulin effects in rat hepatocytes, *Biochem. J.* **239**:523–530.

Charest, R., Prpic, V., Exton, J. H., and Blackmore, P. F., 1985a, Stimulation of inositol trisphosphate formation in hepatocytes by vasopressin, epinephrine and angiotensin II and its relationship to changes in cytosolic free Ca^{2+}, *Biochem. J.* **227**:79–90.

Charest, R., Blackmore, P. F., and Exton, J. H., 1985b, Characterization and responses of isolated rat hepatocytes to ATP and ADP, *J. Biol. Chem.* **260**:15789–15794.

Creba, J. A., Downes, C. P. K., Hawkins, P. T., Brewster, G., Michell, R. H., and Kirk, C. J., 1983, Rapid breakdown of phosphatidylinositol 4-phosphate and phosphatidylinositol 4,5-bisphosphate in rat hepatocytes stimulated by vasopressin and other Ca^{2+}-mobilizing hormones, *Biochem. J.* **212**:733–747.

El-Refai, M. F., Blackmore, P. F., and Exton, J. H., 1979, Evidence for two α-adrenergic binding sites in liver plasma membranes. Studies with [³H]epinephrine and [³H]dihydroergocryptine, *J. Biol. Chem.* **254**:4375–4386.

Exton, J. H., 1987, Mechanisms of α_1-adrenergic and related responses: Roles of calcium, phosphoinositides, guanine nucleotides, diacylglycerol, calmodulin and changes in protein phosphorylation, in: *Cell Membranes: Methods and Reviews* (E. L. Elson, W. A. Frazier, and L. Glaser, eds.). Plenum Press, New York **3**:113–182.

Fitzgerald, T. J., Uhing, R. J., and Exton, J. H., 1986, Solubilization of the vasopressin receptor from liver plasma membranes. Evidence for a receptor GTP binding protein complex, *J. Biol. Chem.* **261**:16871–16877.

Grynkiewicz, G., Poenie, M., and Tsien, R. Y., 1985, A new generation of Ca^{2+} indicators with greatly improved fluorescence properties, *J. Biol. Chem.* **260**:3440–3450.

Holmes, R. P., and Yoss, N. L., 1983, Failure of phosphatidic acid to translocate Ca^{2+} across phosphatidylcholine membranes, *Nature* **305**:637–638.

Irvine, R. F., and Moor, R. M., 1986, Micro-injection of inositol 1,3,4,5-tetrakisphosphate activates sea urchin eggs by a mechanism dependent of external Ca^{2+}, *Biochem. J.* **240**:917–920.

Johnson, R. M., Connelly, P. A., Sisk, R. B., Pobiner, B.F., Hewlett, E. L., and Garrison, J. C., 1986, Pertussis toxin or phorbol 12-myristate 13-acetate can distinguish between growth factor- and angiotensin-stimulated signals in hepatocytes, *Proc. Natl. Acad. Sci. USA* **83**:2032–2036.

Joseph, S. K., Thomas, A. P., Williams, R. J., Irvine, R. F., and Williamson, J. R., 1984, Myoinositol 1,4,5-trisphosphate: A second messenger for the hormonal mobilization of intracellular Ca^{2+} in liver, *J. Biol. Chem.* **259**:3077–3081.

Kuno, M., and Gardner, P., 1987, Ion channels activated by inositol 1,4,5-trisphosphate in plasma membrane of human T-lymphocytes, *Nature* **326**:301–304.

Leeb-Lundberg, L. M. F., Cotecchia, S., Lomasney, J. W., Debernadis, J. F., Lefkowitz, R. J., and Caron, M. G., 1985, Phorbol esters promote α_1-adrenergic receptor phosphorylation and receptor uncoupling from inositol phospholipid metabolism, *Proc. Natl. Acad. Sci. USA* **82**:5651–5655.

Litosch, I., Lin, S. H., and Fain, J. N., 1983, Rapid changes in hepatocyte phosphoinositides induced by vasopressin, *J. Biol. Chem.* **258**:13827–13732.

Lynch, C. J., Charest, R., Blackmore, P. F., and Exton, J. H., 1985, Studies on the hepatic α_1-adrenergic receptor. Modulation of guanine nucleotide effects by calcium temperature and age. *J. Biol. Chem.* **260**:1593–1600.

Lynch, C. J., Wilson, P. B., Blackmore, P. F., and Exton, J. H., 1986, The hormone-sensitive hepatic Na$^+$-pump. Evidence for regulation by diacylglycerol and tumor promoters. *J. Biol. Chem.* **261**:14551–14556.

Morgan, N. G., Blackmore, P. F., and Exton, J. H., 1983, Modulation of the α_1-adrenergic control of hepatocyte calcium redistribution by increases in cAMP, *J. Biol. Chem.* **258**:5110–5116.

Nishizuka, Y., 1984, The role of protein kinase C in cell surface signal transduction and tumour promotion, *Nature* **308**:693–698.

Prpic, V., Green, K. C., Blackmore, P. F., and Exton, J. H., 1984, Vasopressin-, angiotensin II-, and α_1-adrenergic-induced inhibition of Ca^{2+} transport by rat liver plasma membrane vesicles, *J. Biol. Chem.* **259**:1382–1385.

Putney, J. W., 1986, A model for receptor-regulated calcium entry, *Cell Calcium* **7**:1–12.

Rhodes, D., Prpic, V., Exton, J. H., and Blackmore, P. F., 1983, Stimulation of phosphatidylinositol 4,5-bisphosphate hydrolysis in hepatocytes by vasopressin, *J. Biol. Chem.* **258**:2770–2773.

Roach, P. J., and Goldman, M., 1983, Modification of glycogen synthase activity in isolated rat hepatocytes by tumor-promoting phorbol esters: Evidence for differential regulation of glycogen synthase and phosphorylase, *Proc. Natl. Acad. Sci. USA* **80**:7170–7172.

Schworer, C. M., El-Maghrabi, M. R., Pilkis, S. J., and Soderling, T. R., 1985, Phosphorylation of L-type pyruvate kinase by a Ca^{2+}/calmodulin-dependent protein kinase, *J. Biol. Chem.* **260**:13018–13022

Sternweis, P. C., and Gilman, A. G., 1982, Aluminum: A requirement for activation of the regulatory component of adenylate cyclase by fluoride, *Proc. Natl. Acad. Sci. USA* **79**:4888–4891.

Thomas, A. P., Alexander, J., and Williamson, J. R., 1984, Relationship between inositol polyphosphate production and the increase of cytosolic free Ca^{2+} induced by vasopressin in isolated hepatocytes, *J. Biol. Chem.* **259**:5574–5584.

Thomas, A. P., Marks, J. S., Coll, K. E., and Williamson, J. R., 1983, Quantitation and early kinetics of inositol lipid changes induced by vasopressin in isolated and cultured hepatocytes, *J. Biol. Chem.* **258**:5716–5725.

Uhing, R. J., Prpic, V., Jiang, H., and Exton, J. H., 1986, Hormone stimulated polyphosphoinositide breakdown in rat liver plasma membranes: roles of guanine nucleotides and calcium, *J. Biol. Chem.* **261**:2140–2146.

Vaartjes, W. J., deHaas, G. G. M., Geelen, M. J. H., and Bijleveld, C., 1987, Stimulation by a tumor-promoting phorbol ester of acetyl-CoA carboxylase activity in isolated rat hepatocytes, *Biochem. Biophys. Res. Commun.* **142**:135–140.

Woodgett, J. R., Davison, M. T., and Cohen, P., 1983, The calmodulin-dependent glycogen synthase kinase from rabbit skeletal muscle: purification, subunit structure and substrate specificity, *Eur. J. Biochem.* **136**:481–487.

20

Receptor Desensitization Is a Major Regulator of Hormone-Induced Ca^{2+} Mobilization in the Human Platelet

MICHAEL F. CROUCH and EDUARDO G. LAPETINA

1. INTRODUCTION

In many mammalian cell types, stimulation of specific receptors by agonists is accompanied by a nonvoltage-regulated mobilization of intracellular Ca^{2+} stores. This results in the elevation of the cytosolic Ca^{2+} concentration, which can evoke cellular responses such as, in the platelet, secretion, shape change, and aggregation. The Ca^{2+} appears to originate from specific, nonmitochondrial sites in the cell, probably endoplasmic reticulum or the platelet-dense tubular system.

The mechanism by which agonists cause a receptor-mediated Ca^{2+} mobilization has been extensively studied over recent years. In large part, it appears that the transduction mechanism of occupied receptors is the phospholipase C–catalyzed hydrolysis of inositol phospholipid, specifically phosphatidylinositol-4,5-biphosphate (PIP2). The immediate products of this reaction are diacylglycerol (DAG), which can activate protein kinase C (Nishizuka, 1984), and inositol trisphosphate (IP3), which will mobilize the hormone-sensitive Ca^{2+} stores when applied to permeabilized cells (Berridge, 1984).

Therefore, the connection between receptor and Ca^{2+} store mobilization has been quite convincingly established. The questions we wanted to address were as follows: (1) What are the kinetics of Ca^{2+} mobilization in the platelet? (2) Did these correlate with changes in the IP3 levels? (3) At what points in the chain of events following receptor occupation by an agonist is the release of Ca^{2+} modulated or controlled? We have discussed these points elsewhere (Crouch and Lapetina, 1988).

MICHAEL F. CROUCH ● Department of Pharmacology, Australian National University, Canberra, A.C.T. 2601, Australia   EDUARDO G. LAPETINA ● Division of Cell Biology, Burroughs Wellcome Co., Research Triangle Park, North Carolina 27709.

2. Ca^{2+} MOBILIZATION IN THE PLATELET

We chose three different agonists—α-thrombin, γ-thrombin, and platelet-activating factor (PAF)—to study Ca^{2+} mobilization in the platelet. Each of these agents was able to mobilize Ca^{2+} from intracellular Ca^{2+} stores, but to ranging degrees and rates (Fig. 1). α-Thrombin was most effective, whereas γ-thrombin produced a smaller and slower response (Fig. 1). PAF produced a rapid Ca^{2+} signal, but of smaller magnitude than that of γ-thrombin (Fig. 1).

Each of these responses was quite transient despite the continued presence of agonist, and after the return to baseline levels, the cytosolic Ca^{2+} level could not be reevaluated by a second addition of the same agonist at the same concentration (Fig. 1). Thus, the loss of response was not due to degradation of the agonist and could not be explained by the agonists fully depleting the Ca^{2+} stores since PAF was a much less effective agonist than α- or γ-thrombin, but the Ca^{2+} response to PAF returned most quickly to basal levels and could not be restimulated with PAF.

Rather, this appeared to be a desensitization of the receptors or of a postreceptor mechanism. To determine which of these two possibilities was more likely, the ability of a second agonist to elicit release of Ca^{2+} stores was examined. After the α- or γ-thrombin-

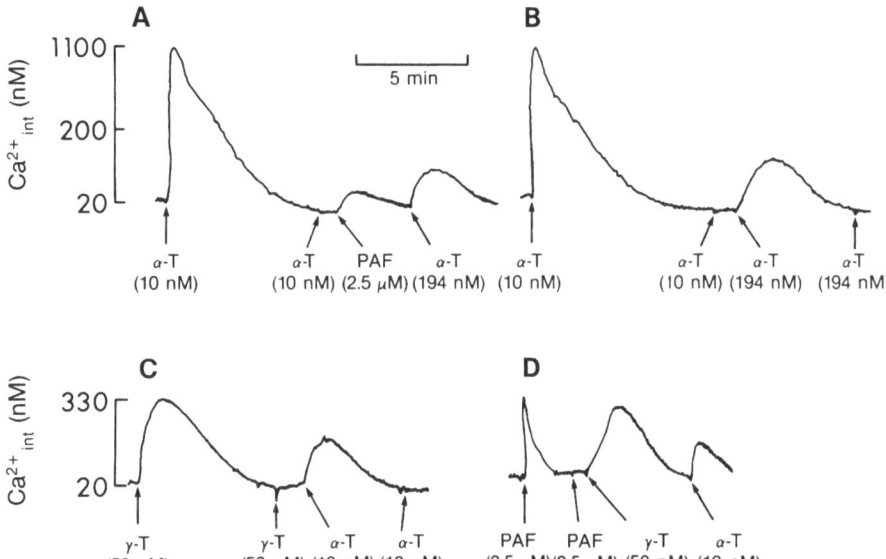

FIGURE 1. The effect of α-thrombin, γ-thrombin, and platelet-activating factor on intracellular Ca^{2+} mobilization in washed human platelets: homologous desensitization of each agonist by prolonged activation. Washed human platelets were treated with aspirin and then incubated for 45 min with 5 μM Indo-1 AM. After washing away the free Indo-1, platelets were resuspended in buffer, aliquots of platelets (2 ml) placed into plastic cuvettes, and test agents added. Changes in fluorescence were monitored with a fluorimeter. The responses shown were highly reproducible with different donors. α-Thrombin, α-T; γ-thrombin, γ-T; platelet-activating factor, PAF; epinephrine, EPIN. For details, see Crouch and Lapetina (1988).

induced Ca^{2+} release was complete, the cytosolic Ca^{2+} concentration had returned to basal levels, and addition of the same agonist at the same concentration was shown to have no effect, PAF was still able to induce a small but significant release of Ca^{2+} stores (Fig. 1). The smaller response to PAF after thrombin than to PAF alone probably occurs because thrombin has already depleted much of the Ca^{2+} store and PAF can act to release only a much reduced pool of Ca^{2+}.

3. DESENSITIZATION OF THROMBIN ACTION IS HOMOLOGOUS

Since thrombin and PAF are both thought to couple through the same second messenger system (i.e., phospholipase C), it seems likely that the desensitization observed is homologous and at the receptor-effector rather than a postreceptor site. It was shown previously (Molina y Vedia and Lapetina, 1986) that activation of protein kinase C, as occurs during stimulation of platelets with thrombin or PAF, is able to increase degradation of cytosolic IP3 by elevating IP3 phosphatase activity. This could decrease receptor-activated Ca^{2+} release. However, this does not appear to be the mechanism of the desensitization observed here since this

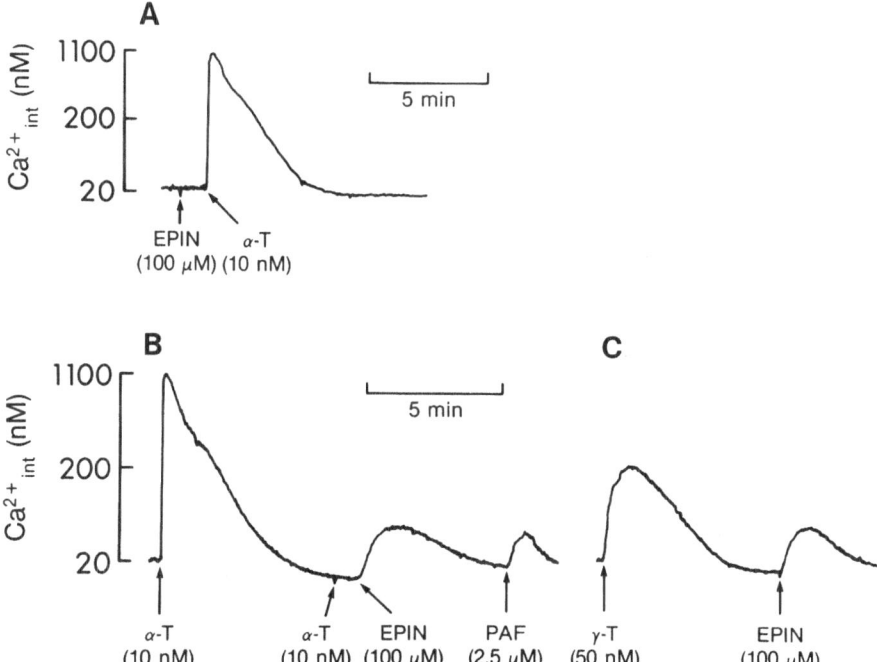

FIGURE 2. Epinephrine-induced resensitization of α-thrombin-stimulated Ca^{2+} mobilization is mediated by α-adrenergic receptors. Receptor-induced release of intracellular Ca^{2+} stores was measured using Indo-1 AM-loaded human platelets, as described in Fig. 1. For details, see Crouch and Lapetina (1988).

would cause a heterologous desensitization. The desensitization is better explained by being related to the generation of the Ca^{2+}-mobilizing signal (i.e., IP3) rather than to the fate of IP3 once formed.

4. DESENSITIZATION OF THE THROMBIN-INDUCED Ca^{2+} RELEASE IS RESENSITIZED BY EPINEPHRINE

To examine this effect further we used epinephrine, which is able to potentiate the action of many platelet agonists without itself having a direct effect on platelet responses. Epinephrine acts via α_2-adrenergic receptors, since yohimbine was found to totally inhibit its actions. We found that epinephrine alone had no effect on platelet Ca^{2+} levels. However, when added to platelets previously desensitized to thrombin, epinephrine could elicit a relatively large mobilization of Ca^{2+} stores (Fig. 2). Thus, epinephrine was able to resensitize the thrombin receptor to the generation of a Ca^{2+} signal.

5. FORMATION OF INOSITOL TRISPHOSPHATE

In support of the role of IP3 in mobilizing intracellular Ca^{2+} stores, the desensitization of the thrombin-induced Ca^{2+} release and resensitization by epinephrine were paralleled by desensitization and resensitization of the α-thrombin-induced IP3 formation (Fig. 3). This showed us that the thrombin receptor did desensitize at the level of receptor activation of phospholipase C and hydrolysis of PIP2, but that the α_2-adrenergic receptor could resensitize thrombin receptor coupling. Those effects were able to fully explain our Ca^{2+} data.

6. EFFECT OF PROTEIN KINASE C ON RECEPTOR DESENSITIZATION

The next part of our study was to try to assess what caused the homologous desensitizations. Since each desensitization was coupled to specific receptors but showed similar

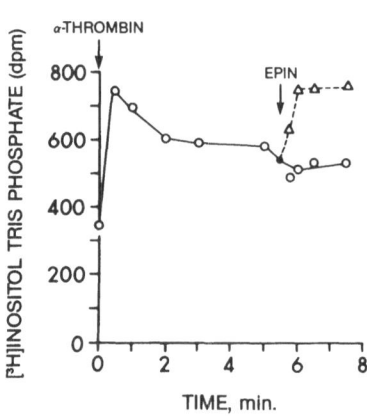

FIGURE 3. α-Thrombin-induced inositol trisphosphate formation is desensitized by prolonged stimulation and is reversed by epinephrine. Inositol phospholipids of aspirin-treated washed human platelets were labeled by incubation of cells with myo[2-^{3}H]inositol. Stimulation of these cells with α-thrombin (α-T, 10 nM, $t = 0$) resulted in formation of inositol trisphosphate (IP3). After 5.5 min, epinephrine (EPIN, 100 μM) or buffer was added and further samples were assayed for formation of inositol trisphosphates. For details, see Crouch and Lapetina (1988).

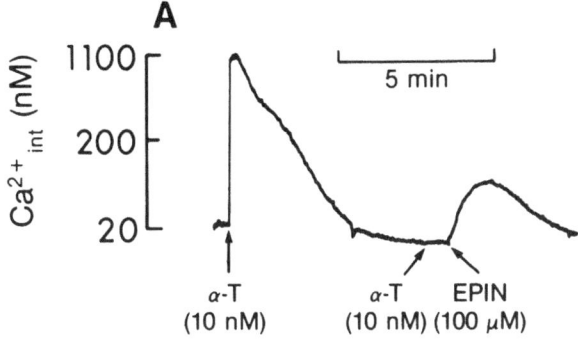

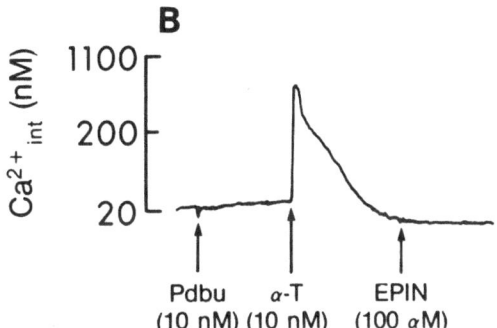

FIGURE 4. Phorbol esters reduce α-thrombin-induced mobilization of intracellular Ca^{2+} stores and abolish resensitization of the desensitized thrombin receptor by epinephrine. The release of intracellular Ca^{2+} stores by agonists was measured as detailed in Fig. 1. Phorbol-12,13-dibutyrate (Pdbu) was added either 3 min before α-thrombin (as shown) or after α-thrombin and 60 sec before epinephrine (not presented). For details, see Crouch and Lapetina (1988).

characteristics, it was likely that the agonists induced a local activation of a second messenger that was itself responsible for the receptor inactivation. It is known from studies of other cell types that protein kinase C activation can be inhibitory to hormone-induced phospholipase C. In the platelet, diacylglycerol, the endogenous protein kinase C activator, is rapidly converted to phosphatidic acid. Thus diacylglycerol may well be produced as a local event and locally activate protein kinase C near the occupied receptors.

When we treated platelets with the protein kinase C activating phorbol ester, phorbol-12,13-dibutyrate (Pdbu), we found a reduction in the thrombin-induced release of Ca^{2+} stores and an abolition of resensitization of thrombin action by epinephrine (Fig. 4). This effect of Pdbu was half-maximal at 1 nM and occurred within 30 sec, both parameters being consistent with a specific activation of protein kinase C. From these results, we suggest that a local elevation in the DAG near the occupied receptor causes a local activation and translocation of protein kinase C so that only the occupied receptor is densensitized.

7. CONCLUSION

We believe that, in the human platelet, receptor desensitization represents an important negative feedback system in the control of platelet responses, including the mobilization of intracellular Ca^{2+} stores and activation of phospholipase C. This desensitization appears to be homologous and mediated by hormone-induced activation of protein kinase C. We are currently investigating the pathway through which α_2-adrenergic receptors are able to re-sensitize Ca^{2+}-mobilizing receptor function.

REFERENCES

Berridge, M. J., and Irvine, R. F., 1984, Inositol trisphosphate, a novel second messenger in cellular transduction, *Nature* **312**:315–321.

Crouch, M. F., and Lapetina, E. G., 1988, A role for G$_i$ in control of thrombin receptor-phospholipase C coupling in human platelets, *J. Biol. Chem.* **263**:3363–3371.

Molina y Vedia, L., and Lapetina, E. G., 1986, Phorbol 12,13-dibutyrate and 1-oleyl 1-2-acetyldi-acylglycerol stimulate inositol trisphosphate dephosphorylation in human platelets, *J. Biol. Chem.* **261**:10493–10495.

Nishizuka, Y., 1984, The role of protein kinase C in cell surface signal transduction and tumor pro-duction, *Nature* **308**:693–698.

21

Specific Receptors for Inositol-1,4,5-Trisphosphate in Endocrine Target Tissues

GAETAN GUILLEMETTE, TAMAS BALLA, ALBERT J. BAUKAL, and KEVIN J. CATT

1. INTRODUCTION

The hydrolysis of inositol lipids by phospholipase C is believed to be the primary mechanism by which many hormones elicit calcium-mediated metabolic and secretory responses in their respective target tissues. The initiating event in this signaling system is the phosphodiesteratic cleavage of phosphatidylinositol-4,5-bisphosphate to generate the inositol-1,4,5-trisphosphate (IP_3), which can mobilize intracellular Ca^{2+} (Berridge and Irvine, 1984), and 1,2-diacylglycerol, which stimulates protein kinase C (Nishizuka, 1984). IP_3 has been shown to release Ca^{2+} from nonmitochondrial stores in a wide variety of cells (Streb et al., 1983; Joseph et al., 1984; for review see Abdel-Latif, 1986). The mechanism of IP_3-induced Ca^{2+} release is not yet known but it has been postulated that IP_3 interacts with a specific receptor on the endoplasmic reticulum. The purpose of this chapter is to describe the properties of specific binding sites for IP_3 in subcellular preparations of the adrenal cortex, the anterior pituitary gland, and the liver. The physiological relevance of these putative receptors is assessed by the study of Ca^{2+}-releasing activity of IP_3 in the same tissue preparations.

2. CHOICE OF THE CONDITIONS FOR THE BINDING OF IP_3

An essential requirement for valid binding studies is the availability of an appropriate radioactive ligand. This ligand should possess the structural and binding properties of the homologous ligand and be of sufficiently high specific radioactivity to detect high-affinity, low-capacity receptor sites. For this purpose, $[^{32}P]IP_3$ was prepared according to Spat et al.

GAETAN GUILLEMETTE, TAMAS BALLA, ALBERT J. BAUKAL, and KEVIN J. CATT • Endocrinology and Reproduction Research Branch, National Institute of Child Health and Human Development, National Institutes of Health, Bethesda, Maryland 20892.

(1986a) from human erythrocyte ghosts by a method based on their endogenous Mg^{2+}-dependent phosphatidylinositol kinase and Ca^{2+}-dependent phospholipase C activities. The specific radioactivity of the tracer (measured by self-displacement in the binding system) varied between 50 and 200 Ci/mmol.

In developing optimal incubation conditions for the analysis of IP_3 binding, a major consideration was the extremely rapid and extensive degradation of the ^{32}P-labeled ligand. For this reason, the incubation buffer included 1 mM EDTA (and omission of Mg^{2+}, an essential cofactor for the enzyme inositol trisphosphate phosphatase) and high concentrations of inorganic phosphate (for nonspecific inhibition of phosphatases).

Under these conditions, no significant degradation of the ligand was observed after incubation with subcellular particles for 30 min at 0°C (Guillemette et al., 1987a). Binding studies were performed for short periods of time (5–30 min) at 0°C in a final volume of 500 μl. Incubations were terminated by filtration through presoaked glass fiber filters (Whatman GF/B) and rapid washing with 2.5 ml of cold incubation buffer (described in the legend to Fig. 1). The particle-bound radioactivity was analyzed by liquid scintillation spectrometry.

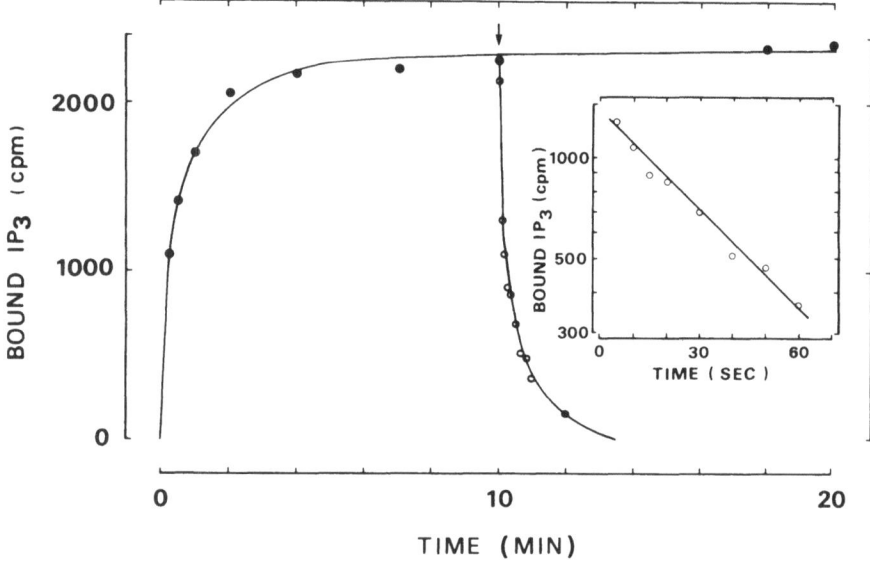

FIGURE 1. Association and dissociation of [^{32}P]IP_3 with adrenocortical binding sites as a function of time. Bovine adrenal cortices were dissected free of medullary tissue and homogenized with 10 strokes of a Dounce homogenizer in 20 mM $NaHCO_3$. After stirring for 5 min and centrifuging at 500g for 15 min, the supernatant was centrifuged at 25,000g for 20 min. The pellet was washed and used for binding studies. Binding was initiated at 0°C by the addition of the particulate suspension (500 μg protein) to assay buffer (500 μl final volume) containing sodium phosphate, 25 mM (pH 7.4); KCl, 100 mM; NaCl, 20 mM; EDTA, 1 mM; bovine serum albumin, 1.0 mg/ml; [^{32}P]IP_3, ~0.2 nM. Nonspecific binding was determined in the presence of 1 μM IP_3. Dissociation was initiated by adding 1 μM unlabeled IP_3 (arrow). At the indicated times, aliquots were removed and filtered through presoaked Whatman GF/B filters and washed with 2.5 ml of cold incubation buffer. The bound radioactivity was analyzed by liquid scintillation spectrometry. The dissociation data expressed on a logarithmic scale are shown in the inset. The experiment was performed in duplicate.

3. CHARACTERIZATION OF THE BINDING SITES FOR IP₃

Specific binding of $[^{32}P]IP_3$ to the particulate fractions was proportional to the amount of membrane added and was abolished by preexposure of the membranes to trypsin (100 μg/ml) for 30 min or by heating at 95°C for 15 min (Baukal *et al.*, 1985), indicating the protein nature of IP_3 binding sites.

Kinetic studies as shown in Fig. 1 revealed that specific binding of IP_3 to bovine adrenal cortex particles was half-maximal within about 10 sec at 0°C and reached a plateau after 5 min, then remained constant for up to 20 min. As expected of a dynamic ligand-receptor interaction, the binding was reversible and the addition of 10^{-6} M unlabeled IP_3 was followed by rapid dissociation of the bound tracer with a half-time of approximately 30 sec, according to a single exponential function (inset of Fig. 1). The association rate data were fitted to a second-order equation and the association rate constant $(k + 1)$ calculated from this equation was 0.4 $pmol^{-1}$ min^{-1}. The dissociation rate constant $(k - 1)$ calculated according to a first-order rate equation was 0.39 min^{-1}. The K_d calculated from the ratio of the rate constants for dissociation and association was 1.9 nM, in agreement with the value obtained from steady-state experiments (see Fig. 2, upper panel). Specific binding of IP_3 showed similar kinetic properties in bovine anterior pituitary gland (Guillemette *et al.*, 1987b) and rat liver (Guillemette *et al.*, 1988) subcellular preparations. These rapid kinetics of IP_3 binding and dissociation from microsomal sites are consistent with its ability to release calcium within seconds.

In saturation studies (Fig. 2, upper panel), tracer binding to bovine adrenal cortex particles was reduced by about 10% in the presence of 0.1 nM unlabeled IP_3 and was proportionally reduced by increasing concentrations up to 100 mM IP_3, where it was completely inhibited. Scatchard analysis of the binding data was consistent with a single set of high-affinity sites (Fig. 2, upper panel, inset) with K_d of 1.26 $\pm$ 0.55 nM and binding capacity of 104 $\pm$ 48 fmol/mg protein ($n = 14$). In the anterior pituitary gland, binding sites for IP_3 showed a K_d of 1.1 $\pm$ 0.4 nM and B_{max} of 28 $\pm$ 15 fmol/mg protein ($n = 6$) whereas in liver plasma membranes IP_3 was bound with a K_d of 1.7 $\pm$ 1.0 nM and B_{max} of 239 $\pm$ 91 fmol/mg protein ($n = 12$). While the affinity of IP_3 for its receptor is the same in these different tissues, the variations in the total amount of binding sites is probably a reflection of the purity (or enrichment) of the IP_3 receptor-bearing organelle in the different preparations. Saturation studies were also performed by incubating subcellular particles with increasing amounts of $[^3H]IP_3$ of low specific radioactivity (2–4 Ci/mmol). This method provides a more accurate determination of the receptor-binding profile at high ligand concentration, where low-affinity sites would be detected. The results of these experiments were similar to those obtained with $[^{32}P]IP_3$ and confirmed that IP_3 binds to a single class of high-affinity, low-capacity receptor sites (see Guillemette *et al.*, 1988).

The criterion for specificity is fulfilled by the fact that increasing concentrations of unlabeled IP_3 completely inhibit the binding of $[^{32}P]IP_3$. The specificity of IP_3 binding sites was also analyzed in competitive binding experiments performed with structural analogs of IP_3. As shown in Fig. 2 (lower panel), myoinositol and inositol-2-phosphate were unable to displace the tracer at any concentration tested up to 1 mM, showing that the affinity of IP_3 for its receptor is not due to a feature of the polyalcohol ring. Fructose-1,6-bisphosphate (FP2) and phytic acid (not shown, $K_d \sim 10^{-4}$ M) were unable to inhibit the binding of IP_3, but at potencies 10,000–100,000 times lower than the homologous ligand. The effect of these compounds emphasizes the importance of phosphate groups positioned on vicinal carbons for recognition of IP_3 by its binding site. In the bovine adrenal cortex, inositol-1,4-bisphosphate (IP_2) showed a 100-fold lower affinity than IP_3. This observation indicates that

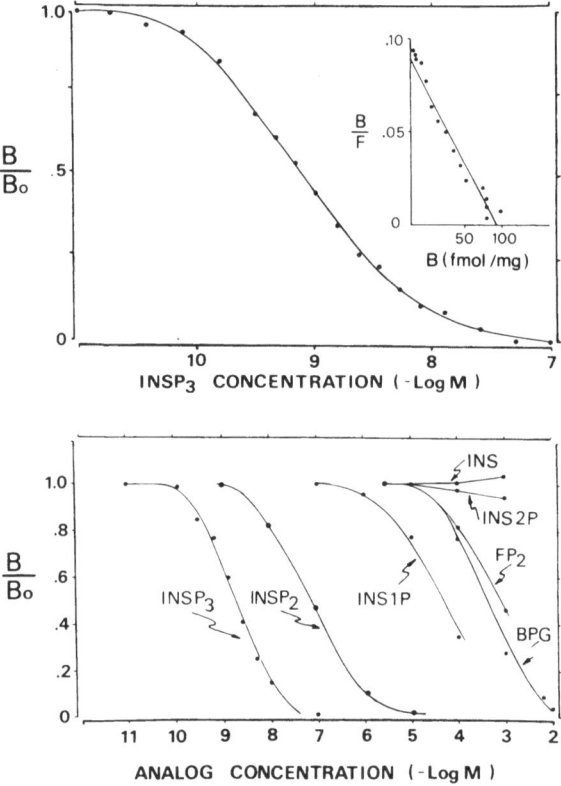

FIGURE 2. Competitive binding of IP₃ and analogs in bovine adrenal cortex. Upper panel, adrenal particles (500 μg protein) were incubated at 0°C in medium containing ~0.1 nM [^{32}P]IP₃ and increasing concentrations of unlabeled ligand. After 5 min the incubations were stopped as indicated in the legend of Fig. 1. The data are expressed as values relative to the total binding observed without unlabeled ligand (603 cpm) and corrected for nonspecific binding (86 cpm). The Scatchard plot of the same binding data is shown in the inset. Lower panel, similar conditions as in the above experiment. Nonradioactive IP₃ and several structural analogs were used as competitive ligands: myoinositol (INS); inositol-2-phosphate (INS2P); fructose-1,6-bisphosphate (FP2); 2,3-bisphosphoglycerate (BPG); inositol-1-phosphate (INS1P); inositol-1,4-bisphosphate (INSP2); and inositol-1,4,5-trisphosphate (INSP3). Each binding-inhibition curve was performed in duplicate.

removal of the phosphoryl group from the 5′ position of IP₃ decreases receptor affinity by 100-fold. However, the apparent displacement by IP₂ might also be attributable to contamination with 1% IP₃. A recent report from Spat *et al.* (1986a) indicated that binding of [^{32}P]IP₃ to liver microsomes was not inhibited by 10 μM IP₂ obtained from a different source. On the other hand, Fig. 2 shows that 100 μM inositol-1-phosphate (which is not likely to be contaminated with IP₃) substantially decreases the binding of [^{32}P]IP₃. It is noteworthy that phytic acid and inositol-1-phosphate displayed similar low affinities for the receptor, indicating that removal or addition of phosphoryl groups to IP₃ substantially impairs its binding affinity. The shapes of the curves shown in the lower panel of Fig. 2 and their parallelism with the IP$_3$ curve are good indications of a competitive interaction of these compounds with the IP$_3$ binding sites. Taken together, these data emphasize the high degree of selectivity of the IP$_3$ binding sites in the bovine adrenal cortex. Identical results were obtained from similar experiments using subcellular particles of the anterior pituitary gland and liver plasma membranes.

The enzymes inositol-1,4,5-trisphosphate-5-phosphatase (Seyfred et al., 1984; Storey et al., 1984) and inositol-1,4,5-trisphosphate-3-kinase (Hawkins et al., 1986; Irvine et al., 1986; Hansen et al., 1986) are involved in the metabolism of IP$_3$. These activities were studied in the adrenal cortex in order to evaluate the possibility that IP$_3$-binding activity might reflect interaction with these enzymes. Since the phosphatase showed a K_m of 17 μM (Guillemette et al., 1987a) and the kinase showed a K_m of 0.4 μM (Guillemette et al., 1987c; Balla et al., 1988), these activities cannot account for the high affinity of the IP$_3$ binding sites. Furthermore, the IP$_3$-kinase is a cytosolic enzyme, whereas the binding sites are located in the particulate fraction of the tissue.

4. BIOLOGICAL FUNCTION OF THE BINDING SITES FOR IP$_3$

If the binding sites described in this study are of physiological relevance, IP$_3$-induced Ca^{2+} release should be present in the particulate fractions in proportion to their IP$_3$-binding activities. Figure 3 (upper panel) shows that adrenal cortex particles can decrease the ambient Ca^{2+} concentration from a level above 1 μM to a value around 100 nM (upward deflection of the trace). This Ca^{2+} uptake process was dependent on the addition of ATP (1 mM) and was completely inhibited by 100 μM vanadate. To exclude a mitochondrial component of Ca^{2+} uptake, all experiments were performed in the presence of oligomycin, a blocker of mitochondrial ATPase activity. Addition of IP$_3$ caused immediate release of Ca^{2+} (downward deflection of the trace) with a peak at about 5 sec, followed by reuptake from the medium. The amount of Ca^{2+} released by IP$_3$ was calibrated by comparison with the deflection evoked upon addition of known amounts of Ca^{2+} (CaCO$_3$) to the mixture. Addition of the Ca^{2+} ionophore, ionomycin, immediately released all of the sequestered Ca^{2+}. The requirement of ATP for Ca^{2+} uptake and the ability of ionomycin to release all the sequestered Ca^{2+} are indications of the vesicular nature of the IP$_3$-sensitive Ca^{2+}-releasing structure. The dose–response relationship between IP$_3$ and Ca^{2+} mobilization in the adrenal cortex is shown in Fig. 3 (lower panel). The threshold response was elicited by 50 nM IP$_3$ and the ED$_{50}$ for Ca^{2+} mobilization was around 800 nM. Similar results were obtained with microsomal preparations from bovine anterior pituitary gland and rat liver. The discrepancy between the nanomolar binding affinity of IP$_3$ and its submicromolar Ca^{2+}-mobilizing potency may be largely attributable to the different experimental conditions necessary for the two assays. The degradation of IP$_3$ by inositol trisphosphate phosphatase is extremely rapid in the Ca^{2+}-mobilizing assay due to incubation at 37°C and in the presence of Mg^{2+}, both of which are necessary for Ca^{2+} uptake. For this reason it is likely that the actual concentration of IP$_3$ decreases rapidly in the Ca^{2+}-mobilizing system and is far below the calculated value. Another important factor is the requirement for ATP in the Ca^{2+}-mobilizing assay to energize Ca^{2+} uptake at a concentration (1 mM) that markedly impairs IP$_3$ binding (Guillemette et al., 1987b). Finally, the difference might also reflect decreased efficiency of the Ca^{2+}-gating mechanism due to loss or impairment of a putative regulatory component during the preparation of the subcellular fractions. Such a component could be a guanine nucleotide regulatory protein associated with the calcium release mechanism. In liver microsomes, Dawson (1985) showed that GTP enhances IP$_3$-stimulated Ca^{2+} release. However, in our hands, the effect of GTP on Ca^{2+} release in the adrenal cortex and the anterior pituitary as well as in the liver was independent of IP$_3$. As observed in other studies (Chueh and Gill, 1986; Jean and Klee, 1986; Hamachi *et al.*, 1987), GTP by itself was able to release Ca^{2+} but only when the viscosity of the incubation medium was increased by an agent such as polyethyleneglycol (3% w/v). The effect of GTP on microsomal release of Ca^{2+} was also dependent on its hydrolysis, since the stable analog GTPγS did not release Ca^{2+}. In fact, high doses of GTPγS were found to block the effect of a subsequent addition of GTP (Guillemette *et al.*, 1987b). More work is necessary to clarify the mechanism of action of GTP on microsomal Ca^{2+} release and the importance of this effect in the cellular response to Ca^{2+}-mobilizing stimuli. Nevertheless, it seems that this mechanism is distinct from the mechanism of action of IP$_3$.

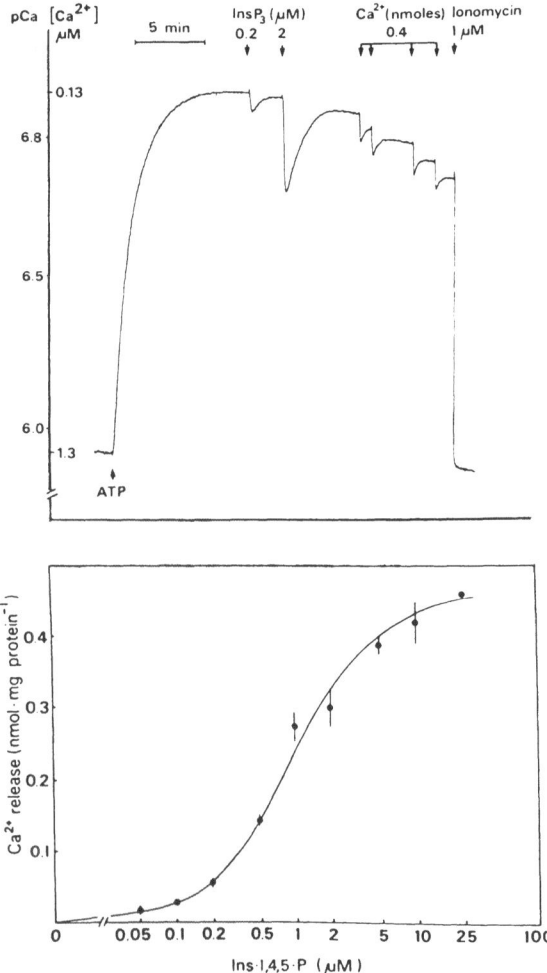

5. SUBCELLULAR DISTRIBUTION OF IP$_3$ RECEPTORS

An interesting question related to the mechanism of action of IP$_3$ concerns the nature and location of the pool from which Ca^{2+} is mobilized. It has been widely assumed that the endoplasmic reticulum is the major source for release of Ca^{2+} from intracellular stores. However, IP$_3$ may also promote Ca^{2+} mobilization and/or Ca^{2+} influx at the level of the plasma membrane during its action on the redistribution of intracellular Ca^{2+}. Since liver can be readily fractionated into enriched organelles, it was employed for a more detailed analysis of the subcellular distribution of IP$_3$ binding sites. Three fractions designated as plasma membrane, endoplasmic reticulum, and mitochondria were prepared according to a procedure combining the methods of Neville (1968) and Dawson and Irvine (1984). As shown in Fig. 4, the plasma membrane fraction was the most enriched in 5'-nucleotidase activity (a marker for plasma membranes) and the least enriched in glucose-6-phosphatase activity (a marker for endoplasmic reticulum). On the other hand, as expected, the endoplasmic reticulum showed an opposite pattern of enrichment. The mitochondrial fraction, although the most enriched in succinate reductase activity (a marker for mitochondria), showed substantial contamination with both glucose-6-phosphatase and 5'-nucleotidase activities. Angiotensin II-binding capacity, which provided a further marker for the plasma membrane, was consistent with the data obtained by enzyme assays. The three subcellular fractions were also analyzed for their IP$_3$-binding properties. As shown in Fig. 4, the plasma membrane fraction showed a much higher binding capacity for IP$_3$ than the endoplasmic reticulum and mitochondria. Scatchard analysis of the binding data indicates that the fractions showed similar binding affinity for IP$_3$ but differed markedly in their concentrations of binding sites for IP$_3$. The plasma membrane fraction contained a single class of high-affinity binding sites, with K_d of 1.7 ± 1.0 nM and concentration of 239 ± 91 fmol/mg protein ($n = 12$).

←——

FIGURE 3. IP$_3$-induced Ca^{2+} release from a subcellular preparation of the bovine adrenal cortex. Bovine adrenal cortices were homogenized with 10 strokes of a Dounce homogenizer in a medium containing KCl, 110 mM; NaCl, 10 mM; MgCl$_2$, 2 mM; Hepes, 20 mM; KH$_2$PO$_4$, 5 mM; dithiothreitol, 1 mM; EGTA, 2 mM (pH 7.2). After centrifugation at 500g for 15 min, the supernatant was centrifuged at 35,000g for 20 min. The pellet was washed with the same medium with EGTA and centrifuged at 35,000g for 20 min. The pellet was taken up in the incubation medium containing KCl, 110 mM; NaCl, 10 mM; MgCl$_2$, 2 mM; KH$_2$PO$_4$, 5 mM; Hepes, 20 mM (pH 7.2); creatine-phosphate, 10 mM; creatine phosphokinase, 20 IU/ml. Oligomycin (2.5 μg/ml) was added to block mitochondrial ATPase. Adrenal cortical particles (3 mg protein) were incubated at 37°C and their Ca^{2+} uptake and release activities were monitored by using Fura 2 free acid (3 μM) in a Perkin Elmer LS-5 fluorescence spectrophotometer; the excitation wavelength was 340 nm, and the emission was recorded at 500 nm. ATP, IP$_3$, and Ca^{2+} were added in small volume in a final incubation volume of 2 ml. The actual Ca^{2+} concentration of the medium was calculated from the F_{max} and F_{min} values obtained by adding excess Ca^{2+} and Mn^{2+}, respectively, after treatment with 1 μM ionomycin. The upper panel shows a typical trace in which the Ca^{2+} taken up by an ATP-dependent process is partially released (in a dose-dependent fashion) upon IP$_3$ addition. The amounts of Ca^{2+} added (nmol) for calibration are indicated in the figure. The Ca^{2+} concentration is expressed in pCa ($-\log$ M) on the left and in μM on the right side of the ordinate. The lower panel shows the relationship between the amount of Ca^{2+} released and the concentration of IP$_3$ added. The results are the mean $\pm$ SD of at least three observations.

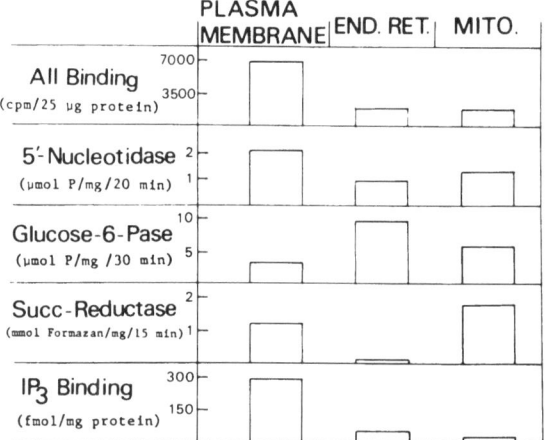

FIGURE 4. Characterization of subcellular fractions prepared from rat liver. Livers from 250- to 400-g male albino rats were minced and homogenized with 10 strokes of a Dounce homogenizer in a buffer containing Hepes/KOH, 20 mM (pH 7.2); KCl, 110 mM; NaCl, 10 mM; MgCl₂, 2 mM; KH₂PO₄, 5 mM; EGTA, 2 mM; dithiothreitol, 1 mM. After stirring for 5 min, filtration through cheese cloth, and centrifugation at 1500g for 20 min, the pellet was resuspended in homogenization buffer and adjusted to 44% in sucrose. From that point, the plasma membrane fraction was purified according to the method of Neville (1968) up to step 11. The mitochondrial fraction was prepared by centrifugation of the 1500g supernatant at 8000g for 10 min. The pellet was then resuspended in the homogenization buffer without EGTA (buffer A), sedimented at 8000g for 10 min, and taken up at a high-protein concentration (~30 mg/ml) in buffer A. The microsomal fraction was prepared by centrifugation of the 8000g supernatant at 35,000g for 20 min. This pellet was resuspended in buffer A, recentrifuged at 35,000g for 20 min, and taken up at a high protein concentration in buffer A. Small aliquots of each fraction were analyzed for their enzymatic and binding activities. Angiotensin II binding was estimated by incubating 25 μg of each fraction with [¹²⁵I]angiotensin II (0.5 nM). Of each fraction, 50 μg was used for the determination of 5′-nucleotidase, glucose-6-phosphatase, and succinate-tetrazolium reductase activities (see Guillemette et al., 1987c for details of the enzymatic characterization). IP₃ binding was evaluated by Scatchard analysis of competitive dose displacement experiments using 500 μg, 1 mg, and 2 mg of protein for the plasma membrane, endoplasmic reticulum, and mitochondria, respectively.

In contrast, the endoplasmic reticulum and mitochondrial fractions were, respectively, 7- and 20-fold less abundant in IP₃ binding sites (Guillemette et al., 1988).

   If the binding sites for IP₃ characterized in the liver are related to mobilization of intracellular Ca^{2+}, then we should observe IP₃-induced Ca^{2+} release in the plasma membrane preparation. As expected, the plasma membrane fraction was able to take up Ca^{2+} in the presence of ATP and to release it upon addition of IP₃. The endoplasmic reticulum and mitochondria behaved similarly but showed smaller responses to IP₃ stimulation. The average values for maximal IP₃-induced Ca^{2+} release were 174 ± 67 pmol Ca^{2+}/mg protein ($n = 8$), 45 ± 10 pmol Ca^{2+}/mg protein ($n = 3$), and 48 ± 7 pmol Ca^{2+}/mg protein ($n = 3$) in plasma membranes, endoplasmic reticulum, and mitochondria, respectively (Guillemette et al., 1988). The large IP₃-induced Ca^{2+} release observed in the plasma membrane is in agreement with the high concentration of IP$_3$ binding sites in this fraction and further suggests that these binding sites are the receptors through which IP$_3$ triggers the release of Ca^{2+} from an intracellular store.

An interesting question is whether the IP$_3$ receptors present in the plasma membrane fraction serve to control Ca^{2+} influx directly across the plasma membrane in addition to release from a Ca^{2+}-containing vesicular structure associated with the plasma membrane. A recent report on the activation of Ca^{2+} channels by IP$_3$ in the plasma membrane of T lymphocytes (Kuno and Gardner, 1987) provides evidence for such an influx. Although our data cannot rule out this possibility, they show that Ca^{2+} is released by IP$_3$ from vesicular structures after uptake by an ATP-dependent process (which is inhibited by 100 μM vanadate and can be completely reversed upon addition of a Ca^{2+} ionophore). Although such structures could represent inside-out vesicles originating from the plasma membranes during homogenization, this would not account for the Ca^{2+}-releasing activity of IP$_3$ in permeabilized hepatocytes (Joseph and Williamson, 1986), which do not contain inside-out vesicles. The most likely explanation is that IP$_3$ binds to and triggers the release of Ca^{2+} from a vesicular system that is closely associated with the plasma membrane, which purifies with the membrane when subcellular fractions are prepared. A model that integrates the two processes of intracellular Ca^{2+} mobilization and Ca^{2+} influx during cell stimulation was recently proposed by Putney (1986). Such a mechanism is supported by the present study, with direct evidence that the site of action of IP$_3$ is associated with the plasma membrane.

In summary, we have characterized binding sites for IP$_3$ in three endocrine target tissues: the adrenal cortex, the pituitary gland, and the liver. Such binding sites have properties that are consistent with their role of intracellular receptors through which the calcium-mobilizing effects of IP$_3$ are expressed during target cell activation by hormonal ligands. It is noteworthy that binding sites for IP$_3$ have also been described in other tissues such as macrophages (Hirata *et al.*, 1985), neutrophils (Spat *et al.*, 1986b), and brain (Worley *et al.*, 1987). It is most likely that such binding sites are ubiquitous or at least present in a wide variety of tissues, since inositol lipid turnover is a mechanism of action common to many cells in response to hormones and other extracellular stimuli.

6. SUMMARY

Receptors for inositol-1,4,5-trisphosphate (IP$_3$) have been characterized by bovine adrenal cortex, bovine anterior pituitary gland, and rat liver by binding studies with IP$_3$ labeled with ^{32}P to high specific radioactivity. In these three tissues, specific binding of [^{32}P]IP$_3$ was linearly proportional to the amount of microsomes added. Kinetic analysis revealed that specific IP$_3$ binding reached equilibrium in a few minutes at 0°C. Rapid dissociation of the bound ligand, according to a single exponential function, occurred upon addition of 1 μM unlabeled IP$_3$. In saturation studies, Scatchard analyses of the binding data were consistent with a single set of high-affinity sites with K_d around 1 nM and maximal binding capacity of 104 fmol/mg protein, 28 fmol/mg protein, and 239 fmol/mg protein in the adrenal cortex, the anterior pituitary gland, and the liver, respectively. The specificity of these binding sites was demonstrated by a markedly lower affinity for structural analogs such as inositol-1-phosphate, phytic acid, fructose-1,6-bisphosphate, 2,3-bisphosphoglycerate and inositol-1,4-bisphosphate. These binding sites are distinct from the enzymes IP$_3$-5-phosphatase and IP$_3$-kinase. The functional relevance of these binding sites was assessed by studying the Ca^{2+}-releasing activity of IP$_3$ in microsomal fractions from the individual target tissues. In the presence of ATP and oligomycin, Ca^{2+} movements were monitored with the fluorescent indicator Fura 2. In all three tissues, IP_3 released Ca^{2+} from microsomes with an ED_{50} around 0.5 μM. To study the intracellular distribution of binding sites for IP_3, subcellular fractions of the rat liver were analyzed. The plasma membrane fraction was much richer in binding sites for IP_3 than either the endoplasmic reticulum or the mitochondrial fraction. This observation was consistent with a higher IP_3-induced Ca^{2+} release in the liver plasma membrane fraction. These results suggest that IP_3 interacts with specific binding sites located on a vesicular system that is closely associated with the plasma membrane. These binding sites appear to be the receptors through which IP_3 triggers the mobilization of Ca^{2+} during cell stimulation by Ca^{2+}-mobilizing hormones and other stimuli.

REFERENCES

Abdel-Latif, A. A. 1986, Calcium-mobilizing receptors, polyphosphoinositides, and the generation of second messengers, *Pharmacol. Revs.* **38**:227–272.

Balla, T., Guillemette, G., Baukal, A. J., and Catt, K. J., 1988, Multiple pathways of inositol polyphosphate metabolism in angiotensin-stimulated adrenal glomerulosa cells, *J. Biol. Chem.* **263**:4083–4091.

Baukal, A. J., Guillemette, G., Rubin, R., Spat, A., and Catt, K. J., 1985, Binding sites for inositol trisphosphate in the bovine adrenal cortex, *Biochem. Biophys. Res. Commun.* **133**:532–538.

Berridge, M. J., and Irvine, R. F., 1984, Inositol trisphosphate, a novel second messenger in cellular signal transduction, *Nature* **312**:315–321.

Chueh, S.-H., and Gill, D. L., 1986, Inositol 1,4,5-trisphosphate and guanine nucleotides activate calcium release from endoplasmic reticulum via distinct mechanisms, *J. Biol. Chem.* **261**:13883–13886.

Dawson, A. P., and Irvine, R. F., 1984, Inositol trisphosphate-promoted Ca^{2+} release from microsomal fractions of rat liver, *Biochem. Biophys. Res. Commun.* **120**:858–864.

Dawson, A. P., 1985, GTP enhances inositol trisphosphate-stimulated Ca^{2+} release from rat liver microsomes, *FEBS Lett.* **185**:147–150.

Guillemette, G., Balla, T., Baukal, A. J., Spat, A., and Catt, K. J., 1987a, Intracellular receptors for inositol 1,4,5-trisphosphate in angiotension II target tissues, *J. Biol. Chem.* **262**:1010–1015.

Guillemette, G., Balla, T., Baukal, A. J., and Catt, K. J., 1987b, Inositol-1,4,5-trisphosphate binds to a specific receptor and releases microsomal calcium in the anterior pituitary gland, *Proc. Natl. Acad. Sci. USA* **84**:8195–8199.

Guillemette, G., Balla, T., Baukal, A. J., and Catt, K. J., 1988, Characterization of inositol 1,4,5-trisphosphate receptors and calcium mobilization in a hepatic plasma membrane fraction, *J. Biol. Chem.* **263**:4541–4548.

Guillemette, G., Baukal, A. J., Balla, T., and Catt, K. J., 1987c, Angiotension-induced formation and metabolism of inositol polyphosphates in bovine adrenal glomerulosa cells, *Biochem. Biophys. Res. Commun.* **142**:15–22.

Hamachi, T., Hirata, M., Kimura, Y., Ikebe, T., Ishimatsu, T., Yamaguchi, K., and Koga, T., 1987, Effect of guanosine triphosphate on the release and uptake of Ca^{2+} in the saponin-permeabilized macrophages and the skeletal-muscle sarcoplasmic reticulum, *Biochem. J.* **242**:253–260.

Hansen, C. A., Mag, S., and Williamson, J. R., 1986, Formation and metabolism of inositol 1,3,4,5-tetrakisphosphate in liver, *J. Biol. Chem.* **261**:8100–8103.

Hawkins, P. T., Stephens, L., and Downes, C. P., 1986, Rapid formation of inositol 1,3,4,5-tetrakisphosphate and inositol 1,3,4-trisphosphate in rat parotid glands may both result indirectly from receptor-stimulated release of inositol 1,4,5-trisphosphate from phospatidylinositol 4,5-bisphosphate, *Biochem J.* **238**:507–516.

Hirata, M., Sasaguri, T., Hamashi, T., Hashimoto, T., Kukita, M., and Koga, T., 1985, Irreversible inhibition of Ca^{2+} release in saponin-treated macrophages by the photoaffinity derivative of inositol-1,4,5-trisphosphate, *Nature* **317**:723–725.

Irvine, R. F., Letcher, A. J., Heslop, J. P., and Berridge, M. J., 1986, The inositol tris/tetrakisphosphate pathway-demonstration of ins (1,4,5)P₃ 3-kinase activity in animal tissues, *Nature* **320**:631–634.

Jean, T., and Klee, C. B., 1986, Calcium modulation of inositol 1,4,5-trisphosphate-induced calcium release from neuroblastoma X glioma hybrid (NG108-15) microsomes, *J. Biol. Chem.* **261**: 16414–16420.

Joseph, S. K., Thomas, A. P., Williams, R. J., Irvine, R. F., and Williamson, J. R., 1984, Myo-inositol 1,4,5-trisphosphate: A second messenger for the hormonal mobilization of intracellular Ca^{2+} in liver, *J. Biol. Chem.* **259**:3077–3081.

Joseph, S. K., and Williamson, J. R., 1986, Characteristics of inositol trisphosphate-mediated release from permeabilized hepatocytes, *J. Biol. Chem.* **261**:14658–14664.

Kuno, M., and Gardner, P., 1987, Ion channels activated by inositol 1,4,5-trisphosphate in plasma membrane of human T-lymphocytes, *Nature* **326**:301–304.

Neville, D. M., Jr., 1968, Isolation of an organ specific protein antigen from cell-surface membrane of rat liver, *Biochem. Biophys. Acta* **154**:540–552.

Nishizuka, Y., 1984, The role of protein kinase C in cell surface signal transduction and tumor promotion, *Nature* **308**:693–698.

Putney, J. W., Jr., 1986, A model for receptor-regulated calcium entry, *Cell Calcium* **7**:1–12.

Seyfred, M. A., Farrell, L. E., and Wells, W. W., 1984, Characterization of D-myo-inositol 1,4,5-trisphosphate phosphatose in rat liver plasma membranes, *J. Biol. Chem.* **259**:13204–13208.

Spat, A., Fabiato, A., and Rubin, R. P., 1986a, Binding of inositol trisphosphate by a liver microsomal fraction, *Biochem. J.* **233**:929–932.

Spat, A., Bradford, P. G., McKinney, V. S., Rubin, R. P., and Putney, J. W. Jr., 1986b, A saturable receptor for ³²Pinositol-1,4,5-trisphosphate in hepatocytes and neutrophyls, *Nature* **319**:514–516.

Storey, D. J., Shears, S. B., Kirk, C. J., and Michell, R. H.,1984, Stepwise enzymatic dephosphorylation of inositol 1,4,5-trisphosphate to inositol in liver, *Nature* **312**:374–376.

Streb, H., Irvine, R. F., Berridge, M. J., and Schulz, I., 1983, Release of Ca^{2+} from a nonmitochondrial intracellular store in pancreatic acinar cells by inositol 1,4,5-trisphosphate, *Nature* **306**:67–69.

Worley, P. F., Baraban, J. M., Colvin, J. S., and Snyder, S. H.,1987, Inositol trisphosphate receptor localization in brain: variable stoichiometry with protein kinase C, *Nature* **325**:159–161.

22

Intracellular Calcium Transients Associated with Endothelium-Derived Relaxing Factor Release May Be Mediated by Inositol-1,4,5-Trisphosphate

ALEX L. LOEB, NICHOLAS J. IZZO, Jr.,
RANDOLPH M. JOHNSON, JAMES C. GARRISON, and
MICHAEL J. PEACH

A large number of vasoactive agents induce relaxation by stimulating endothelial cells to release an endothelium-derived vascular relaxing factor (EDRF), an unstable and as yet unidentified compound. EDRF was first described by Furchgott and Zawadzki (1980) to relax preconstricted vascular smooth muscle. Since then, this compound has been demonstrated to stimulate the soluble form of guanylate cyclase, but not adenylate cyclase, in vascular smooth muscle (Rapoport and Murad, 1983). Endothelium-derived vascular relaxing factor is not a product of cyclooxygenase, since indomethacin pretreatment does not affect EDRF-induced smooth-muscle relaxation (Furchgott and Zawadzki, 1980) or cyclic GMP accumulation (Peach et al., 1985). The compound has a short half-life (< 1 min; Griffith et al., 1984; Luckhoff et al., 1987) and its actions are blocked by a wide variety of substances without a common mechanism of action (Furchgott, 1983; Griffith et al., 1984; Peach et al., 1985).

Early observations indicated that extracellular calcium was necessary for endothelium-dependent relaxation, since the actions of muscarinic agonists were inhibited by pretreatment with calcium channel blockers (Singer and Peach, 1982) or by incubation of vascular tissues in buffer without added calcium (Singer and Peach, 1982; Long and Stone, 1985). However, the phospholipase activator melittin will induce EDRF release (Loeb et al., 1988) and raise

ALEX L. LOEB ● Department of Anesthesia, University of Pennsylvania School of Medicine, Philadelphia, Pennsylvania 19104. NICHOLAS J. IZZO, Jr. ● Cardiovascular Division, Brigham and Women's Hospital, Boston, Massachusetts 02115. RANDOLPH M. JOHNSON ● Department of Biomolecular Pharmacology, Genentech, Inc., South San Francisco, California 94080. JAMES C. GARRISON and MICHAEL J. PEACH ● Department of Pharmacology, University of Virginia School of Medicine, Charlottesville, Virginia 22908.

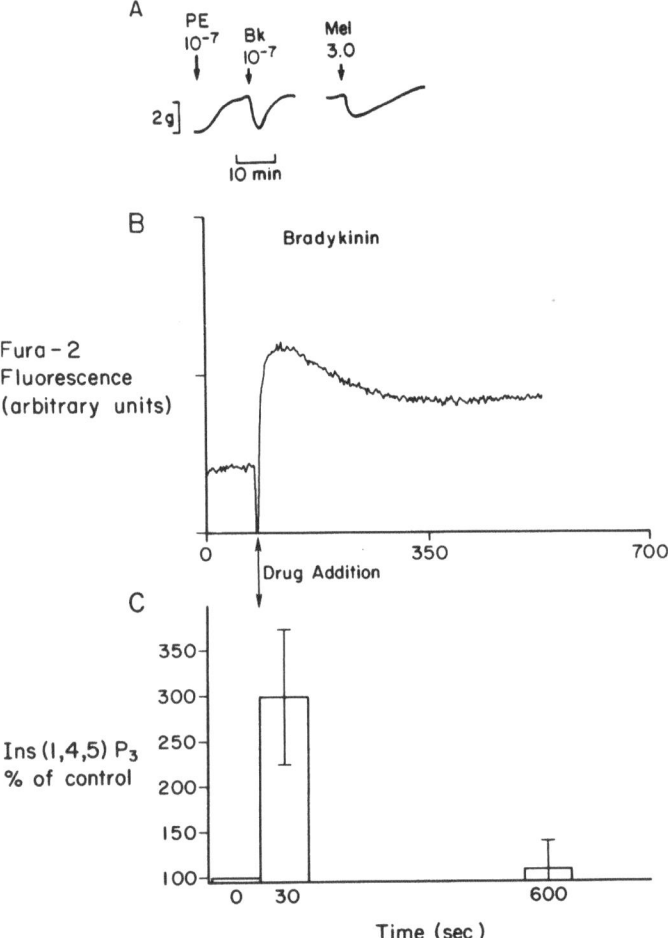

FIGURE 1. (A) EDRF-induced relaxation of a phenylephrine (PE)-preconstricted rabbit aortic ring. Addition of BK or Mel to the endothelial cell column induced relaxations of the bioassay ring. Note the longer time course of relaxation induced by MEL in this typical tracing. (B) Typical examples of agonist-induced Fura 2 fluorescence in bovine aortic endothelial cells. BK (30 nM) and Mel (3 μg/ml) stimulated increased cytosolic free-calcium concentrations as detected by

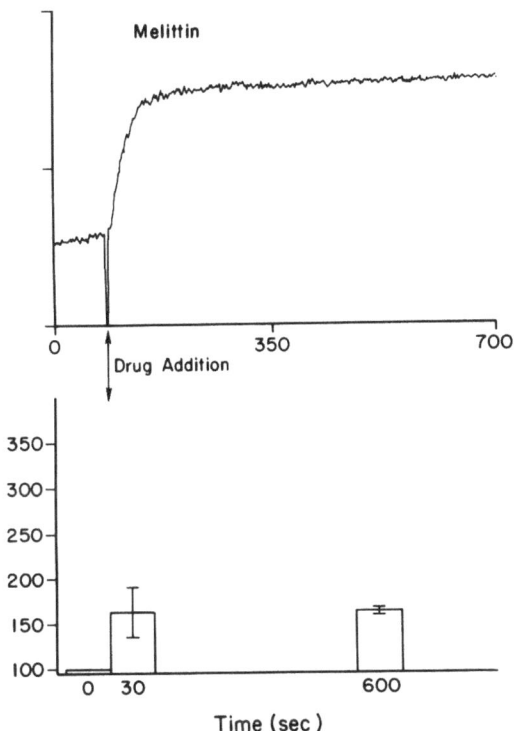

increased calcium-dependent fluorescence in suspensions of Fura 2/AM (1 μM)-loaded endo-thelial cells. Cell concentration was 10^6/ml. (C) Stimulation of Ins(1,4,5)P$_3$ accumulation by BK (1 μM) and Mel (3 μg/ml) in confluent endothelial cell monolayers. Data are expressed as percent increase (mean $\pm$ SEM) in [^{3}H]Ins(1,4,5)P$_3$ in stimulated cells compared to vehicle-treated cells (100%).

intracellular calcium levels in cultured endothelial cells (Peach et al., 1987) in the absence of extracellular calcium. These findings suggested that an increase in intracellular calcium within the endothelial cell could be associated with EDRF release. The purpose of these studies was to investigate whether the well-known mediator of intracellular calcium release, inositol trisphosphate (IP_3), could be involved in the increase in intracellular calcium and EDRF release from cultured bovine aortic endothelial cells.

We have developed methodologies for detection of EDRF release, measurement of changes in intracellular calcium levels, and phosphatidylinositol turnover in cultured bovine aortic endothelial cells (BAEC). To assess whether these cells were able to release EDRF, endothelial cells grown on microcarrier beads were placed in a column and superfused with oxygenated physiological buffer (Loeb et al., 1987; Peach et al., 1987). The column effluent was cascaded onto a preconstricted rabbit aortic ring denuded of endothelium, or onto cultured rat aortic smooth-muscle cells. Endothelium-derived vascular relaxing factor production was assayed by relaxation of the vascular ring or by stimulation of cyclic GMP in the cultured smooth muscle. As shown in Fig. 1A, infusion of bradykinin (BK) or melittin (Mel) over the cell column produced a dose-dependent relaxation. Pretreatment of the cells with indomethacin (28 μM) had no effect on the relaxations, suggesting that the relaxing factor was not a product of cyclooxygenase. When BK or Mel were superfused over the bioassay ring without BAEC in the column, no relaxations were detected. Bradykinin and melittin-induced release of EDRF from the endothelial cells was detected as an increase in cyclic GMP in the smooth-muscle cells (Fig. 2). The increase in cyclic GMP was temporally correlated with the relaxations shown in Figure 1A.

To determine whether changes in intracellular calcium might be involved in EDRF release, we loaded dispersed endothelial cells with the fluorescent calcium indicator dye Fura 2 as previously described (Peach et al., 1987). Thus, changes in endothelial cell intracellular free-calcium concentrations in response to EDRF agonists could be detected by monitoring the Fura 2 fluorescence signal from the cell suspensions. Typical calcium fluorescence responses to BK and Mel are shown in Fig. 1B. Note that the BK-induced calcium signal was biphasic with a rapid transient peak falling off to a sustained plateau. In contrast to the BK response, Mel induced a rapid and sustained rise in calcium that was maintained for at least 10 min. Pretreatment with 1.5 mM EGTA for 2 min prior to addition of agonist to the cell suspension had no qualitative effect on either the melittin-induced increase in intracellular free calcium or the initial phase of the bradykinin-induced transient, although the plateau phase of the bradykinin response was abolished (Peach et al., 1987). Higher concentrations of EGTA (up to 10 mM) had no additional effect (not shown). The ability of bradykinin and melittin to increase intracellular free calcium in the presence of calcium chelation suggested that these agents could release intracellular stores of calcium.

Inositol-1,4,5-trisphosphate [Ins(1,4,5)P_3] acts as an intracellular second messenger to release calcium from intracellular stores (Berridge and Irvine, 1984). Therefore, we wanted to see whether changes in Ins(1,4,5)P_3 could be associated with the increased intracellular calcium transients induced by agonists in cells known to release EDRF. For these studies, confluent monolayers of endothelium were incubated with [^{3}H]myoinositol (10 μCi/ml, 61 Ci/mmol) and unlabeled inositol (1 μM) for 24 hr in normal growth medium [Waymouth's medium (Gibco) + 10% fetal calf serum + antibiotics]. Before assay, the cells were washed twice with physiological salt solution and incubated for 20 min in LiCl (10 mM). Agonists were then added for 30 sec or 10 min. The reaction was stopped by addition of ice-cold perchloric acid (PCA, 0.5 M), and Ins(1,4,5)P_3 quantified by high-performance liquid chromatography (Johnson and Garrison, 1987).

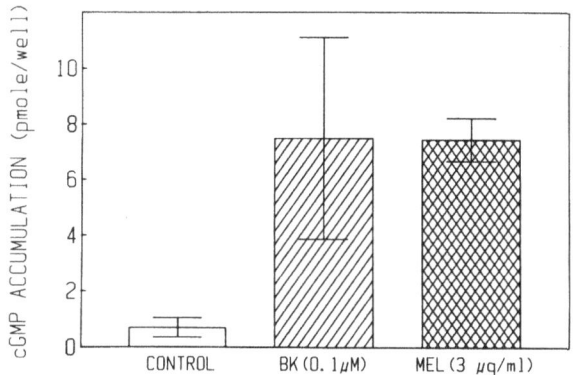

FIGURE 2. Biochemical assay for EDRF. Confluent rat aortic smooth muscle cells grown in 12 well culture plates (Owens *et al.*, 1986) were suspended beneath the endothelial cell column (Loeb *et al.*, 1987). Cyclic GMP (cGMP) accumulation by the smooth-muscle cells was used as evidence for EDRF production by the endothelial cells. Column effluent was collected for 30 sec into wells of smooth cells before (control) and 2 min after stimulation of the cell column with BK or MEL. Column effluent was incubated with smooth muscle for 90 sec, aspirated from the well, and 1 ml 0.1 N HCl added to extract cyclic GMP from the smooth-muscle cells. Cyclic GMP was determined by radioimmunoassay. There were approximately 8.0 × 10^5 cells per well. Data are expressed as mean ± SEM from three experiments.

Bradykinin and melittin both induced rapid increases in Ins(1,4,5)P$_3$ by 30 sec, as shown in Fig. 1C. These findings suggest that the rise in cytosolic free calcium stimulated by these agents may be mediated by an Ins(1,4,5)P$_3$-dependent release of calcium from intracellular stores. Interestingly, the Mel-stimulated increase in Ins(1,4,5)P$_3$ was sustained for at least 10 min. This sustained Mel-induced rise in Ins(1,4,5)P$_3$ was unexpected but correlated well with the sustained increase in intracellular calcium also induced by this compound.

Several investigators have shown that EDRF production by vascular endothelium may be dependent on extracellular calcium (Singer and Peach, 1982; Long and Stone, 1985; Griffith *et al.*, 1986). We demonstrated that cultured endothelial cells can be shown to release EDRF, which can be detected by functional (relaxation) and biochemical (cyclic GMP) assays. The relaxing factor produced by these cells is unstable, and it is either destroyed or its action is blocked by the EDRF inhibitor hydroquinone (Loeb *et al.*, 1987b). When the endothelial cells are stimulated to release EDRF by BK and Mel, there are rapid increases in Ins(1,4,5)P$_3$ and cytosolic free calcium within the cells.

Lambert *et al.* (1987b) showed that BK increased inositol trisphosphate in endothelial cells derived from porcine aorta but could not quantify the specific isomers made. Our studies demonstrated a link between BK stimulation, increased Ins(1,4,5)P$_3$, increased cytosolic free calcium, and EDRF release, all within the same cell type.

We postulate that agonist stimulation of endothelial cells causes release of inositol trisphosphate from the plasma membrane, which then triggers release of calcium from intracellular stores, as has been demonstrated in other systems (for review, see Berridge and Irvine, 1984). Cytosolic free calcium is then elevated, initially by release of calcium from intracellular stores through Ins(1,4,5)P$_3$, and then by a calcium influx from the extracellular space. The increase in cytosolic calcium triggers the synthesis and/or release of the relaxing factor. This conclusion is consistent with our findings that Ins(1,4,5)P$_3$ and intracellular calcium concentrations rise and fall in parallel. We have no explanation for the Mel-induced, sustained rise in Ins(1,4,5)P$_3$. The sustained rise in Ins(1,4,5)P$_3$ is very unusual and we speculate that it may be due to a melittin-induced inhibition of Ins(1,4,5)P$_3$ breakdown or to prolonged stimulation of phospholipase C. It is interesting to note that melittin can induce EDRF release in intact tissues as well as from cultured endothelium in the absence of extracellular calcium (Loeb et al., 1988), and that its relaxant effect is more prolonged (see Fig. 1A) than that of BK. The more prolonged relaxation induced by melittin is probably due to a longer lasting release of EDRF, as suggested by the sustained increase in inositol trisphosphate and intracellular calcium within the endothelial cells.

REFERENCES

Berridge, M. J., and Irvine, R. F., 1984, Inositol trisphosphate, a novel second messenger in cellular signal transduction, Nature 312:315–321.

Furchgott, R. F., 1983, Role of endothelium in responses of vascular smooth muscle, Circ. Res. 53: 557–573.

Furchgott, R. F., and Zawadzki, J. V., 1980, The obligatory role of the endothelial cells in the relaxation of arterial smooth muscle by acetylcholine, Nature 288:373–376.

Griffith, T. M., Edwards, D. H., Lewis, M. J., Newby, A. C., and Henderson, A. H., 1984, The nature of the endothelium-derived vascular relaxant factor, Nature 308:645–647.

Griffith, T. M., Edwards, D. H., Newby, A. C., Lewis, M. J., and Henderson, A. H., 1986, Production of endothelium-derived relaxation factor is dependent on oxidative phosphorylation and extracellular calcium, Cardiovasc. Res. 20:7–12.

Johnson, R. M., and Garrison, J. C., 1987, Epidermal growth factor and angiotensin II stimulate formation of inositol-(1,4,5) and inositol-(1,3,4)-trisphosphate in hepatocytes: Differential inhibition by pertussis toxin and phorbol-12-myristate-13-acetate, J. Biol. Chem. 262:17285–17293.

Lambert, T. L., Kent, R. S., and Whorton, A. R., 1986, Bradykinin stimulation of inositol polyphosphate production in porcine aortic endothelial cells, J. Biol. Chem. 261:15288–15293.

Loeb, A. L., Johns, R. A., and Peach, M. J., 1988, Extracellular calcium is not required for melittin-induced endothelium-dependent relaxation, in Vasodilation: Vascular smooth muscle, peptides, autonomic nerves, and endothelium (P. M. Vanhoutte, ed.), Raven Press, New York 1988:453–462.

Loeb, A. L., Johns, R. A., Milner, P. M., and Peach, M. J., 1987, Endothelium-derived relaxing factor in cultured cells, Hyptertension, 9(Suppl. III):188–192.

Long, C. J., and Stone, T. W., 1985, The release of endothelium-derived vascular relaxant factor is calcium dependent, Blood Vessels 22:205–208.

Luckhoff, A. Busse, R. Winter, I., and Bassenge, E., 1987, Characterization of vascular relaxant factor released from cultured endothelial cells, Hypertension 9:295–303.

Owens, G. K., Loeb, A. L., Gordon, D., and Thompson, M. M., 1986, Expression of smooth muscle-specific alpha-isoactin in cultured vascular smooth muscle cells: Relationship between growth and cytodifferentiation, J. Cell Biol. 102:343–352.

Peach, M. J., Loeb, A. L., Singer, H. A., and Saye, J. A., 1985, Endothelium-derived vascular relaxing factor, Hypertension 7(Suppl. I):94–100.

Peach, M. J., Singer, H. A., Izzo, N. J., and Loeb, A. L., 1987, Role of calcium in endothelium-dependent relaxation of arterial smooth muscle, Am. J. Cardiol. 59:35A–43A.

Rapoport, R. M., and Murad, F., 1983, Agonist-induced endothelium-dependent relaxation in rat thoracic aorta may be mediated through cGMP, Circ. Res. 52:352–357.

Singer, H. A., and Peach, M. J., 1982, Calcium- and endothelial-mediated vascular smooth muscle relaxation in rabbit aorta, Hypertension 4(Suppl II):19–25.

23

Protein Kinase C in Rat Skeletal Muscle

PERRY J. F. CLELAND, ERIK A. RICHTER,
STEPHEN RATTIGAN, ERIC Q. COLQUHOUN, and
MICHAEL G. CLARK

1. INTRODUCTION

Activation of skeletal muscle contraction under physiological conditions involves depolarization of transverse tubular regions of the sarcolemma (Huxley and Taylor, 1958; Hodgkin and Horowicz, 1960; Costantin and Taylor, 1973; Costantin, 1975) leading to Ca^{2+} release from the terminal cisternae of the sarcoplasmic reticulum (Stephenson, 1981). The transverse tubular-terminal cisternae transmembrane signaling mechanism apparently is non-electrical (Donaldson, 1985; Donaldson et al., 1987 and references therein) and there have been numerous studies aimed at examining the chemical triggering of sarcoplasmic reticulum Ca^{2+} release. In this regard some researchers have examined the role of inositol phosphates in the mobilization of intracellular skeletal muscle Ca^{2+} (Volpe et al., 1985; Vergara et al., 1985; Thieleczek and Heilmeyer, 1986; Nosek et al., 1986; Donaldson et al., 1987). Such studies were based on findings that inositol trisphosphate (IP_3) elicited Ca^{2+} release from the endoplasmic reticulum of a wide variety of cells (Berridge, 1981; Suematsu et al., 1984; Hirata et al., 1984; Hokin, 1985). A recent study showed that locally applied microinjected 1 μM IP_3 stimulated Ca^{2+} release from peeled skeletal muscle fibers and, although it did not directly activate the contractile apparatus (Donaldson et al., 1987), its role in excitation-contraction coupling was concluded to involve propagation of Ca^{2+} release acting beyond the step of transverse tubule depolarization (Donaldson et al., 1987) at the sarcoplasmic reticulum (Volpe et al., 1985).

An effect of IP_3 to stimulate Ca^{2+} release from, and associated force generation of, skeletal muscle fibers implies that excitation-contraction coupling involves an initial activation of phospholipase C and the production of diacylglycerol. If this is the case, protein kinase C may be activated by translocation from the cytosol to the plasma membrane or particulate fraction (Nishizuka, 1984).

PERRY J. F. CLELAND, STEPHEN RATTIGAN, ERIC Q. COLQUHOUN, and MICHAEL G. CLARK • Department of Biochemistry, University of Tasmania, Hobart, Australia.   ERIK A. RICHTER • August Krogh Institute, University of Copenhagen, Denmark.

A subsequent role for protein kinase C in the phosphorylation and regulation of enzymes and processes that adjust in response to increased muscle contraction is possible. The glucose transporter protein from human red blood cells is a substrate for brain kinase C (Witters *et al.*, 1985). In addition, activation of protein kinase C in 3T3L1 adipocytes by phorbol esters leads to increased glucose transport as well as phosphorylation of the glucose transporter (Gibbs *et al.*, 1986). Other substrates that relate to increased glucose transport and are potential candidates for phosphorylation are phosphofructokinase (Hofer *et al.*, 1985) and glycogen synthase (Bouscarel and Exton, 1986).

Thus, in the present study, we have developed an assay for protein kinase C in skeletal muscle preparations and examined the distribution in relation to fiber types and the effect of contraction on the activation of protein kinase C reflected by translocation of this enzyme from cytosol to the plasma membrane or particulate fraction.

2. METHODS

Fed, male, hooded Wistar rats (250 g) were anaesthetized with pentobarbitol (12.5 mg). The skin was removed from the left hindlimb and the sciatic nerve exposed and cut to allow positioning of the distal end in a suction electrode. The knee was secured by the tibiopatellar ligament and the Achilles tendon attached via a steel wire to a Harvard Apparatus isometric transducer. Tension development was recorded during electrical stimulation (200-msec trains of 100 Hz applied every 2 sec) adjusted (10–20 V) to attain full-fiber recruitment. Initial tension was 1065 g which decreased to 600 g after 1 min and then remained constant for the remaining 4 min. After 0, 1, 2, or 5-min stimulation the gastrocnemius-plantaris-soleus muscle group (approximately 1.3 g), representing mainly fast-twitch red and white muscle (Ariano *et al.*, 1973), was rapidly removed and immediately homogenized in 4 ml ice-cold 20 mM Tris-HCl buffer, pH 7.5, containing 250 mM sucrose, 1 mM dithiothreitol, and 20 μg/ml trypsin inhibitor (soya bean) using an Ultra-Turrax homogenizer (2 × 30 sec). The homogenate was centrifuged for 1 hr at 100,000g. The pellet (particulate fraction) was extracted with 30 ml 20 mM Tris-HCl buffer, pH 7.5, containing 2 mM EDTA, 0.5 mM EGTA, 1 mM dithiothreitol, 0.2% (v/v) Triton X-100, or with 2 × 6 ml 20 mM Tris-HCl buffer, pH 7.5, containing 2 mM EDTA, 1 mM dithiothreitol, 2.0% (v/v) Triton X-100, and 10 mM EGTA. To the supernatant was added $\frac{1}{10}$ volume of 20 mM Tris-HCl buffer, pH 7.5, containing 250 mM sucrose, 20 mM EDTA, 5 mM EGTA, and 1 mM dithiothreitol. Finely powdered ammonium sulfate was added to give a concentration of 21% (w/v) and the protein precipitate was removed by centrifugation at 8000g for 20 min. A further addition of ammonium sulfate was made to the supernatant [final concentration 45% (w/v)] and the protein pellet, recovered after centrifugation, was dissolved in 0.2 ml 20 mM Tris-HCl buffer, pH 7.5, containing 2 mM EDTA, 0.5 mM EGTA, and 1 mM dithiothreitol. After desalting on a 7-ml column of Bio-Gel P6DG (Bio-Rad, USA) all fractions containing protein were applied to a 3.5 ml column of Whatman DE-52 and eluted using a linear gradient of 36 ml of 0–150 mM NaCl (Averdunk and Gunther, 1986). The Triton X-100 extracts of the particulate fraction were applied directly to DE-52 columns and eluted in a similar manner. Fractions were assayed for protein kinase C (Kikkawa *et al.*, 1982) using histone IIIs as the substrate and Whatman phosphocellulose P81 paper to collect the acid-precipitable material.

3. RESULTS

3.1. Fiber-Type Distribution of Protein Kinase C

In the course of experiments aimed at recovering protein kinase C activity from skeletal muscle, it became apparent that routine procedures using 0.1–0.3% (v/v) Triton X-100 containing buffers (Kuo, 1980; Kikkawa *et al.*, 1982) were insufficient to fully extract activity associated with the particulate fraction. Thus modification of the extraction procedure was required and buffers containing a higher concentration of Triton X-100 (e.g., 2%) as well as 10 mM EGTA were found to fully extract particulate activity. Table I shows protein kinase C activities in soleus and gastrocnemius white muscles. Soleus muscle contained significantly ($P < 0.01$) more enzyme activity than gastrocnemius white muscle both as total and particulate activity. The percentage of the total contained by the particulate fraction also tended to be greater for the soleus muscle, 73.3 ± 6.1 ($n = 6$) versus 57.6 ± 10.4 ($n = 6$) but was not significant. Furthermore, values for the total content and particulate activity of protein kinase C for the gastrocnemius red and plantaris muscles were less than soleus and more than gastrocnemius white. The percentage of the total contained by the particulate fraction for gastrocnemius red and plantaris muscles were similar to that of the gastrocnemius white (approximately 60%; data not shown). Table I also shows the published values for fiber-type composition (Ariano *et al.*, 1973) and glucose metabolic index values (Rg') for nonexercising and exercising states (James *et al.*, 1985). The latter values were determined for rats *in vivo*. Of the two muscles shown, soleus has the highest content of red slow oxidative fibers and the highest value for Rg'. In addition, the soleus muscle increases its value for Rg' the most during exercise.

3.2. Contraction-Induced Activation of Protein Kinase C

Figure 1 shows the effect of electrically induced contraction of the gastrocnemius-plantaris-soleus muscle group on the distribution of protein kinase C between the cytosolic and particulate fractions. Approximately 60% of the total protein kinase C activity was located in the particulate fraction prior to contraction. This value increased in a time-dependent manner with contraction to reach a maximum of 83% at 2 min with a corresponding decrease in cytosolic activity. After 2 min there was no further loss of activity from the cytosol but the particulate showed some decline. Approximately 13% of the total activity remained unaccounted for at 5 min.

Contraction is associated with increased blood flow to the gastrocnemius-plantaris-soleus muscle group and the weight of this group was found to increase from 1.3 to 1.5, 1.6, and 1.6 g at 1, 2, and 5 min, respectively. To rule out the possibility that the increased blood content of the muscle group contributed to the observed translocation of protein kinase C, 0.3 g blood from stimulated animals (cardiac puncture) was added to noncontracted muscle during homogenization. This had no effect on either the distribution of protein kinase C between the cytosolic and particulate fractions or on the total activity (data not shown). Similarly, homogenization of noncontracted muscle in buffer containing 1 mM $CaCl_2$ also had no effect on these values. It is noteworthy that EGTA could not be included in the initial homogenizing buffers as this displaced the particulate enzyme and diminished the translocation; however, when used in conjunction with Triton X-100, EGTA improved the extraction of protein kinase C from the 100,000g particulate material.

TABLE I. Comparison of Protein Kinase C Activities, Slow Oxidative Red Fiber Content, and Glucose Metabolic Indices of Soleus and Gastrocnemius White Muscles of Rat

Muscle	Slow oxidative[a] fiber content (%)	Index of glucose[b] metabolism		PKc activity pmol/min per g muscle[c]		
		Rg' rest	Rg' exercise	Cytosol	Particulate	Total
Soleus	84	2.8 ± 0.6	90.4 ± 5.7	804 ± 186	2482 ± 410	3287 ± 426
Gastrocnemius white	5	1.3 ± 0.1	7.0 ± 0.8	602 ± 155	1039 ± 265	1642 ± 220
				n.s.	$P < 0.01$	$P < 0.01$

[a] Data from Ariano *et al.* (1973).
[b] Data from James *et al.* (1985).
[c] Values are means ± SE for six animals. Significance of difference between soleus and gastrocnemius white muscle is indicated by *P* values using paired t test.

Figure 1 also shows the results of a separate series of experiments where the particulate fraction was extracted using a lower, more conventional concentration of 0.2% Triton X-100. The particulate activity extracted under these conditions increased approximately twofold from 147 ± 28 pmol/min per g muscle to reach a maximum at 2 min and coincided with a corresponding decrease in cytosolic activity of 630 pmol/min per g muscle. The effect of contraction on the percentage content of the particulate fraction was an increase from 11.6 ± 1.2 to $28.0 \pm 4.4\%$ $(P < 0.05)$ at 2 min (Fig. 1C). However, while this change reflected a significant translocation of protein kinase C from cytosol to the particulate (membrane) fraction, 76% of the decrease in cytosolic activity remained unaccounted for. As indicated above, this was due to incomplete extraction of the particulate activity. Multiple extractions of the particulate fraction with 0.2% Triton X-100 were unsuccessful (data not shown).

3.3. Characteristics of Skeletal Muscle Protein Kinase C

Since protein kinase C activities noted in the present study were considerably greater than the published value of 373 pmol/min per g muscle (Kuo et al., 1980), it was considered relevant to establish the identity of the enzyme as protein kinase C. Anion exchange chromatography using DEAE-cellulose (Whatman DE-52) indicated that both the cytosolic and solubilized particulate activity eluted at 40–50 mM NaCl (data not shown). In addition, ligand dependency of cytosolic as well as particulate activity was assessed. Table II shows that the properties of the two activities were similar, although the cytosolic activity appeared to be more dependent on diolein. Full activity for histone phosphorylation was dependent on the presence of diolein, phosphatidyl serine, and Ca^{2+}. Omission of these substances as well as histone completely inhibited protein phosphorylation. However, when histone alone was omitted, endogenous protein phosphorylation by endogenous kinase(s) was evident. The data of Table II appear to be characteristic of protein kinase C (Inoue et al., 1977; Takai et al., 1979).

4. DISCUSSION

Three findings emerge from the present study. First, comparison of the activities of protein kinase C in soleus and gastrocnemius white muscle indicates that the former has both a greater total and a greater particulate activity. Also for soleus muscle approximately 73% of the total activity was located in the particulate fraction compared with 58% for the gastrocnemius white muscle.

A greater total content and a greater apparent content of active (particulate) protein kinase C may relate to the metabolic and physiological differences between these two muscle types. Soleus muscle, which is postural, is rich in slow oxidative red fibers (Ariano et al., 1973) and rich in mitochondrial enzymes. The soleus muscle also has a high resting capacity for glucose uptake and metabolism, and also shows a marked increase in glucose uptake during exercise (James et al., 1985). Recent reports that the glucose transporter is phosphorylated by protein kinase C (Witters et al., 1985) and that increased glucose transport coincided with phosphorylation of the transporter in other systems (Gibbs et al., 1986) raise the interesting possibility that control of glucose transport in muscle may be mediated by protein kinase C. The data of Table I lend further support to this view. However, it is not yet known whether phosphorylation of this protein by protein kinase C in fact alters its transport properties.

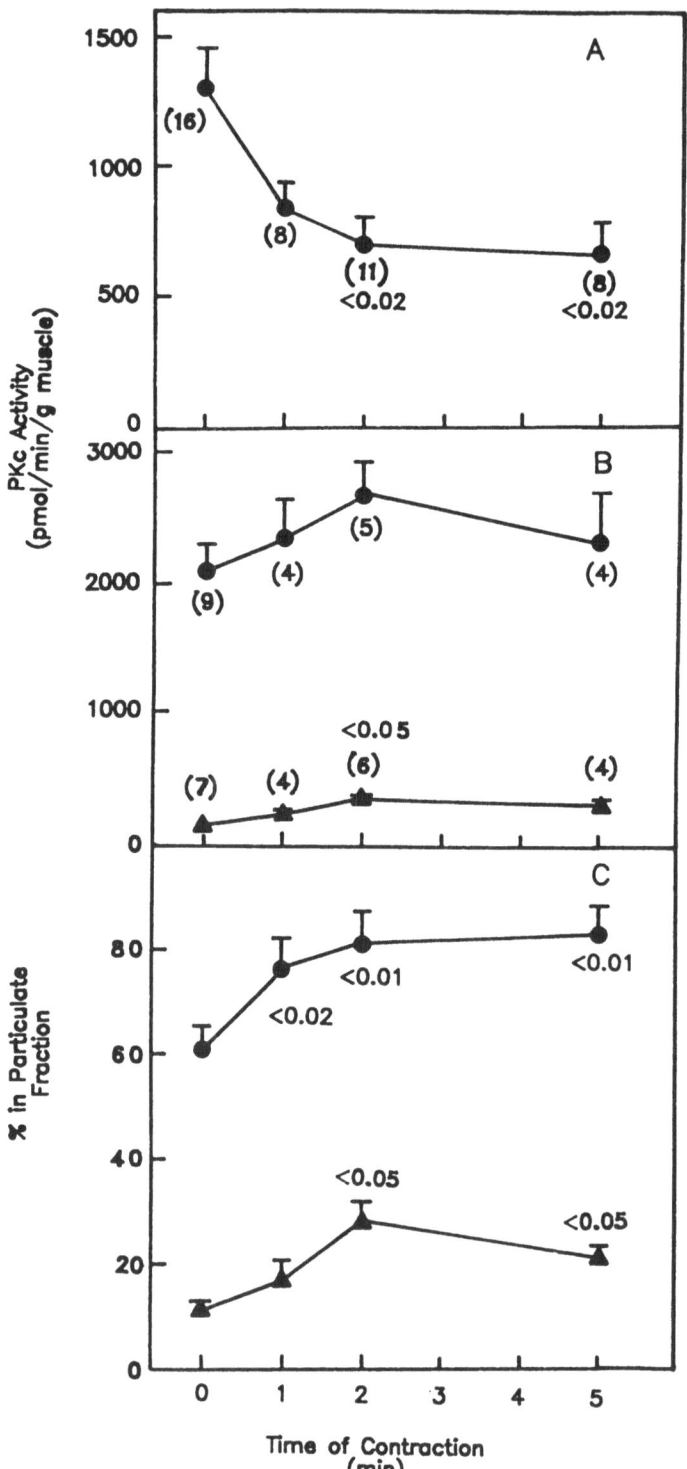

TABLE II. Ligand Dependency of Protein Kinase C from the Cytosolic and
Particulate Fractions of Skeletal Muscle

Conditions	PKc activity[a] (pmol/min per g muscle)	
	Cytosolic	Particulate
Complete assay	202 ± 9	530 ± 30
Histone	13 ± 10	27 ± 4
Phosphatidyl serine	15 ± 9	23 ± 7
Diolein	30 ± 4	288 ± 10
Phosphatidylserine, -diolein	10 ± 3	23 ± 3
Ca^{2+}, +7.3 mM EGTA	8 ± 1	68 ± 9
Ca^{2+}, -phosphatidyl serine, -diolein, +7.3 mM EGTA	5 ± 6	38 ± 6
Ca^{2+}, -phosphatidyl serine, -diolein, -histone, +7.3 mM EGTA	1 ± 1	1 ± 1

[a] Activity was determined on fractions from DE-52 chromatography. Values shown are means ± SE for three preparations.

The second finding from this study represents the first report of a contraction-associated translocation of skeletal muscle protein kinase C from the cytosolic to the particulate fraction. These findings support the "muscle IP$_3$ hypothesis" (Volpe et al., 1987) that links phosphatidylinositol turnover with Ca^{2+} release from the sarcoplasmic reticulum. If this hypothesis is correct, phosphorylation of membrane-bound substrates or substrates occasionally associated with particulate material by protein kinase C could then account for some of the changes in metabolism which occur as a result of increased muscle contraction. These include the glucose transporter (Witters et al., 1985; Gibbs et al., 1986), phosphofructokinase (Hofer et al., 1985), and glycogen synthase (Bouscarel and Exton, 1986).

The third finding from this study is that muscle contains a high proportion of its total protein kinase C activity in the particulate fraction. Complete extraction of this activity required relatively high concentrations of Triton X-100 (2%) and EGTA (10 mM). The presence of such a high proportion of the total activity present in what is assumed to be the membrane component implies a high state of activation of this enzyme even under resting conditions. There was no evidence to indicate that the exercise-induced translocated activity could be selectively extracted with lower concentrations of Triton X-100.

Experiments are under way to explain this apparent high state of activation and to determine whether the activity is in fact located in the plasma membrane fraction.

←

FIGURE 1. Effect of electrical stimulation-induced contraction of muscle on the distribution of protein kinase C activity in cytosolic (A) and particulate (B) fractions. Particulate activity was extracted with buffer containing 0.2% Triton X-100 (▲) or 2.0% Triton X-100 and 10 mM EGTA (●). In part C the distribution of activity in the particulate fraction is expressed as a percentage of the total activity for the corresponding Triton extract. Values are means ± SE with the number of experiments given in parentheses. P values compared to 0 min using Student's t test are also shown.

5. SUMMARY

Protein kinase C activities in red and white muscles were compared and the effect of electrical stimulation on the distribution of protein kinase C between cytosolic and particulate fractions of the gastrocnemius-plantaris-soleus muscle group was studied. Full extraction of the particulate activity required the combination of high concentrations of Triton X-100 and EGTA. Greater than 60% of the total activity was found to be located in the particulate fraction of resting skeletal muscle suggesting a high state of activation of this enzyme under basal conditions. Soleus muscle, which is rich in slow oxidative red fibers and possesses a high rate of glucose disposal under noncontracting and contracting states, was found to have a twofold greater total activity and a twofold greater particulate activity of protein kinase C than gastrocnemius white muscle. Contraction of the gastrocnemius-plantaris-soleus muscle group by electrical stimulation of the sciatic nerve *in vivo* led to a time-dependent translocation of protein kinase C from the muscle cytosol to the particulate fraction. Maximum translocation of 20% of the total activity occurred after 2 min of intermittent, short, tetanic contractions. It is concluded that protein kinase C of skeletal muscle may play a role in the maintenance of membrane processes in both noncontracting and contracting states.

ACKNOWLEDGMENTS. We thank Geoffrey Appleby for expert technical assistance. This work was supported in part by grants from the National Health and Medical Research Council of Australia, the Ramaciotti Foundations, and the Danish Medical Research Council.

REFERENCES

Ariano, M. A., Armstrong, R. B., and Edgerton, V. R., 1973, Hindlimb muscle fibre populations of five mammals, *J. Hist. Cytochem.* **21**:51–55.

Averdunk, R., and Gunther, T., 1986, Protein kinase C in cytosol and cell membranes of concanavalin A-stimulated rat thymocytes, *FEBS Lett.* **195**:357–361.

Berridge, M. J., 1981, Phosphatidylinositol hydrolysis: A multifunctional transducing mechanism, in *Molec. Cell. Endocrinol.* **24**:115–140.

Bouscarel, B., and Exton, J. H., 1986, Regulation of hepatic glycogen phosphorylase and glycogen synthase by calcium and diacylglycerol, *Biochem. Biophys. Acta* **888**:126–134.

Costantin, L. L., 1975, Contractile activation in skeletal muscle, *Prog. Biophys. and Mol. Biol.* **29**:197–224.

Costantin, L. L., and Taylor, S. R., 1973, Graded activation in frog muscle fibres, *J. Gen. Physiol.* **61**:424–443.

Donaldson, S. K. B., 1985, Peeled mammalian skeletal muscle fibres. Possible stimulation of Ca^{2+} release via a transverse tubule sarcoplasmic reticulum mechanism, *J. Gen. Physiol.* **86**:501–525.

Donaldson, S. K., Goldberg, N. D., Walseth, T. F., and Huetteman, D. A., 1987, Inositol trisphosphate stimulates calcium release from peeled skeletal muscle fibres, *Biochim. Biophys. Acta* **927**:92–99.

Gibbs, E. M., Allard, W. J., and Lienhard, G. E., 1986, The glucose transporter in 3T3-L1 adipocytes is phosphorylated in response to phorbol ester but not in response to insulin, *J. Biol. Chem.* **261**:16597–16603.

Hirata, M., Suematsu, E., Hashimoto, T., Hamachi, T., and Koga, T., 1984, Release of Ca^{2+} from a non-mitochondrial store site in peritoneal macrophages treated with saponin by inositol 1,4,5-trisphosphate, *Biochem. J.* **223**:229–236.

Hodgkin, A. L., and Horowicz, P., 1960, Potassium contractures in single muscle fibres, *J. Physiol. (Lond.)* **153**:386–403.

Hofer, H. W., Schlatter, S., and Graefe, M., 1985, Phosphorylation of phosphofructokinase by protein kinase C changes the allosteric properties of the enzyme, *Biochem. Biophys. Res. Commun.* **129:** 892–897.

Hokin, L. E., 1985, Receptors and phosphoinositide-generated second messengers, *Ann. Rev. Biochem.* **54:**205–235.

Huxley, A. F., and Taylor, R. E., 1958, Local activation of striated muscle fibres, *J. Physiol. (Lond.)* **144:**426–441.

Inoue, M., Kishimoto, A., Takai, Y., and Nishizuka, Y., 1977, Studies on a cyclic nucleotide-independent protein kinase and its proenzyme in mammalian tissues, *J. Biol. Chem.* **252:**7610–7616.

James, D. E., Kraegen, E. W., and Chisholm, D. J., 1985, Muscle glucose metabolism in exercising rats: comparison with insulin stimulation, *Am. J. Physiol.* **248:**E575–E580.

Kikkawa, U., Takai, Y., Minakuchi, R., Inohara, S., and Nishizuka, Y., 1982, Calcium-activated, phospholipid-dependent protein kinase from rat brain, in *J. Biol. Chem.* **257:**13341–13348.

Kuo, J. F., Andersson, R. G. G., Wise, B. C., Mackerlova, L., Salomonsson, L., Brackett, N. L., Katoh, N., Shoji, M., and Wrenn, R. W., 1980, Calcium-dependent protein kinase: Widespread occurrence in various tissues and phyla of the animal kingdom and comparison of effects of phospholipid, calmodulin, and trifluoperazine, *Proc. Natl. Acad. Sci. USA* **77:**7039–7043.

Nishizuka, Y., 1984, The role of protein kinase C in cell surface signal transduction and tumour promotion, *Nature* **308:**693–698.

Nosek, T. M., Williams, M. F., Zeigler, S. T., and Godt, R. E.,1986, Inositol trisphosphate enhances calcium release in skinned cardiac and skeletal muscle, *Am. J. Physiol.* **250:**C807–C811.

Stephenson, E. W., 1981, Activation of fast skeletal muscle: Contributions of studies on skinned fibres, *Am. J. Physiol.* **240:**C1–C19.

Suematsu, E., Hirata, M., Hashimoto, T., and Kuriyama, H., 1984, Inositol 1,4,5-trisphosphate releases Ca^{2+} from intracellular store sites in skinned single cells of porcine coronary artery, *Biochem. Biophys. Res. Commun.* **120:**481–485.

Takai, Y., Kishimoto, A., Kikkawa, U., Mori, T., and Nishizuka, Y., 1979, Unsaturated diacylglycerol as a possible messenger for the activation of calcium-activated, phospholipid-dependent protein kinase system, *Biochem. Biophys. Res. Commun.* **91:**1218–1224.

Thieleczek, R., and Heilmeyer, L. M. G. Jr., 1986, Inositol 1,4,5-trisphosphate enhances Ca^{2+} sensitivity of the contractile mechanism of chemically skinned rabbit skeletal muscle fibres, *Biochem. Biophys. Res. Commun.* **135:**662–669.

Vergara, J., Tsien, R. Y., and Delay, M., 1985, Inositol 1,4,5-trisphosphate: a possible chemical link in excitation-contraction coupling in muscle, *Proc. Natl. Acad. Sci. USA* **82:**6352–6356.

Volpe, P., Di Virgilio, F., and Pozzan, T., 1987, Inositol trisphosphate and muscle: caution is a must, *Trends. Biochem. Sci.* **12:**139–140.

Volpe, P., Salviati, G., Di Virgilio, F., and Pozzan, T., 1985, Inositol 1,4,5-trisphosphate induces calcium release from sarcoplasmic reticulum of skeletal muscle, *Nature* **316:**347–349.

Witters, L. A., Vater, C. A., and Lienhard, G. E., 1985, Phosphorylation of the glucose transporter in vitro and in vivo by protein kinase C, *Nature* **315:**777–778.

24

Activation of Phosphoinositide-Specific Phospholipase C by Ligands in the Presence of Guanine Nucleotides

JOHN N. FAIN, MICHAEL A. WALLACE, RICHARD J. H. WOJCIKIEWICZ, and DEJAN BOJANIC

1. THE CALCIUM ROAD TO PHOSPHOINOSITIDE BREAKDOWN IN HEPATOCYTES

Our interest in this area arose from studies on the mechanisms by which vasopressin and α_1-catecholamine agonists activate rat hepatocyte glycogen phosphorylase and gluconeogenesis. In 1973 we postulated that α-adrenergic stimulation of gluconeogenesis in rat hepatocytes was not secondary to elevations of cyclic AMP (Tolbert *et al.*, 1973). Since we found no evidence for an involvement of cyclic GMP in hepatic glycogenolysis (Pointer *et al.*, 1976), we next focused on calcium as a possible second messenger. Michell (1975) in a seminal review suggested that phosphoinositide turnover was linked to elevations in intracellular calcium. Kirk *et al.* (1977) soon found that vasopressin specifically stimulated the uptake of ^{32}P into phosphoinositides of rat liver. Fain (1978) postulated that vasopressin increased the breakdown of phosphatidylinositol resulting in release of ''trigger'' calcium that mediated the activation of glycogen phosphorylase. Subsequently we found that vasopressin increased the breakdown of phosphatidylinositol in the plasma membrane (Lin and Fain, 1981). Wallace *et al.* (1982) were able to show a direct activation of phosphatidylinositol breakdown after addition of vasopressin to isolated rat liver plasma membranes incubated in buffer containing deoxycholate. However, Michell *et al.* (1981) suggested that vasopressin preferentially stimulated the breakdown of phosphatidylinositol-4,5-bisphosphate (PIP$_2$) and this was linked in some way to elevations in intracellular Ca^{2+}. This hypothesis turned out to be correct. There was a period when some investigators thought that all the increases in phosphoinositide breakdown in the liver, including even that of PIP$_2$, were secondary to elevations of intracellular calcium (Exton *et al.*, 1983).

The link between polyphosphoinositide breakdown and intracellular calcium was established by the discovery that inositol-1,4,5-trisphosphate (IP$_3$), one of the two products

JOHN N. FAIN, MICHAEL A. WALLACE, RICHARD J. H. WOJCIKIEWICZ, and DEJAN BO-JANIC • Department of Biochemistry, College of Medicine, University of Tennessee, Memphis, Tennessee 38163.

derived from phospholipase C action on PIP_2, released calcium from intracellular nonmitochondrial stores in many cells (Berridge, 1984). This effect was readily reproduced in a variety of cells and is now well accepted.

Calcium is not the only intracellular regulator derived as a result of phosphoinositide breakdown. The diacylglycerol resulting from degradation of any phosphoinositide activates protein kinase C (Nishizuka, 1986; Bell, 1986). Furthermore the tumor-promoting phorbol esters act as nonhydrolyzable analogs of diacylglycerol to directly activate protein kinase C.

2. EVIDENCE FOR INVOLVEMENT OF A GUANINE NUCLEOTIDE-BINDING PROTEIN IN THE ACTION OF HORMONES THAT ACTIVATE PHOSPHOINOSITIDE BREAKDOWN

Our recent research has focused on the possible involvement of a unique guanine nucleotide-binding protein (G_x or G_p) in phospholipase C activation (Fig. 1) that plays a role similar to that of N_s (G_s) or N_i (G_i) in adenylate cyclase regulation (see reviews by Litosch and Fain, 1986 and Cockcroft, 1987).

2.1. Pertussis Toxin

In 1983 we found an effect of pertussis toxin on phosphoinositide turnover in adipocytes (Moreno et al., 1983). In adipocytes taken from rats injected 3 days previously with pertussis

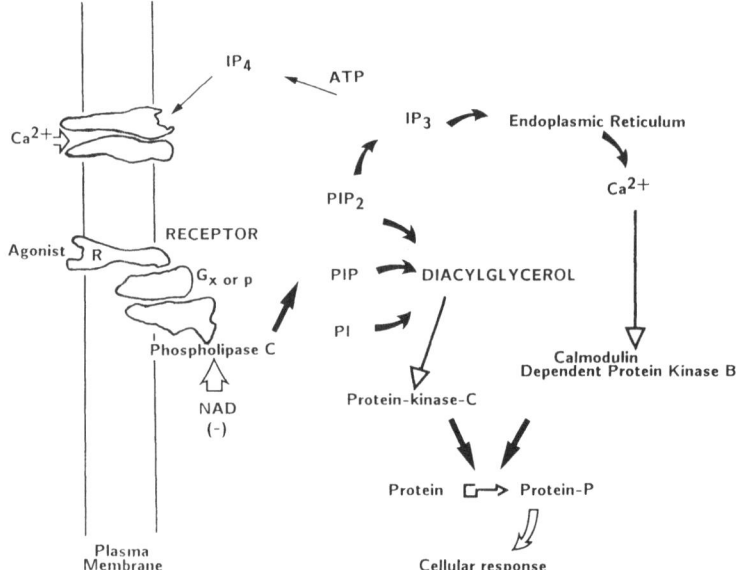

FIGURE 1. The 1987 model for the relationship between phosphoinositide turnover and the accumulation of calcium and diacylglycerol in cells.

toxin, the incorporation of labeled inorganic phosphate into phosphoinositides due to α-adrenergic stimulation, but not that due to insulin, was markedly decreased. However, Masters *et al.* (1985) found that in cultured chick heart cells or 1321N1 human astrocytoma cells the effects of muscarinic cholinergic stimulation were unaffected by pertussis toxin. A similar insensitivity to pertussis toxin was noted with respect to thrombin, angiotensin II, and platelet-activating factor in 3T3 cells (Murayama and Ui, 1985). There was also no effect of pertussis toxin on phosphoinositide breakdown due to thyrotropin-releasing hormone (TRH) in rat pituitary tumor cells (Martin *et al.*, 1986a,b; Hinkle *et al.*, 1986; Aub *et al.*, 1986; and Wojciekiewicz *et al.*, 1986). Clearly, the stimulation of phosphoinositide breakdown due to ligands is unaffected by pertussis toxin in many cells.

However, in human leukemic cells (Brandt *et al.*, 1985), rat renal mesangial cells (Pfeilschifter and Bauer, 1986), mast cells (Nakamura and Ui, 1985), and neutrophils (Smith *et al.*, 1985; Bradford and Rubin, 1985), it has been possible to see pertussis toxin-sensitive phosphoinositide breakdown. These data suggest that G_i, or a protein like G_i that is linked to phospholipase C, may be ADP-ribosylated by pertussis toxin in some cells.

2.2. Guanine Nucleotide Effects on Ligand Binding

The first evidence of a role for GTP in the action of ligands that stimulate phosphoinositide breakdown came from vasopressin-binding studies using rat liver membranes (Cantau *et al.*, 1980). Guanine nucleotides modulate the binding of agonists linked to stimulation of phosphoinositide breakdown including muscarinic cholinergic agonists (Evans *et al.*, 1985), vasopressin (Bojanic and Fain, 1986), α_1-adrenergic amines (Goodhardt *et al.*, 1982), and fMet-Leu-Phe (Snyderman *et al.*, 1984).

Delean *et al.* (1980) postulated that occupation of receptors by agonists whose mode of action involves activation of adenylate cyclase results in the formation of a high-affinity complex consisting of the agonist, receptor, and G_s protein. If GTP is added, the complex should dissociate to a GTP-liganded G_s protein and a low-affinity ligand-receptor complex. Vasopressin binds to only a single set of high-affinity binding sites in rat liver membranes and there are no known effects of vasopressin on processes involving the G_s or G_i proteins in these membranes. Bojanic and Fain (1986) found that the nonhydrolyzable guanine nucleotide analog GppNHp enhanced the release of tritiated vasopressin from its receptor if soluble extracts were prepared from rat liver plasma membranes previously labeled with vasopressin in the absence of guanine nucleotide. The molecular weight of the photoaffinity-labeled hepatic vasopressin receptor is around 70,000 (Boer and Faahrenholz, 1985). In the absence but not in the presence of guanine nucleotides, tritiated vasopressin bound to receptors eluted from molecular sizing columns as a single peak with an apparent molecular weight of approximately 260,000. These data are compatible with the hypothesis that the vasopressin-receptor complex also contains a guanine nucleotide-binding protein.

2.3. Activation of GTPase Activity

If a guanine nucleotide-binding protein is involved in phosphoinositide breakdown, it should have high-affinity GTPase activity that can be activated by ligands. Hinkle and Phillips (1984) found that TRH activated a high-affinity GTPase activity in membranes from rat pituitary tumor cells. Fain *et al.* (1985) subsequently observed that direct addition of vasopressin to isolated membranes prepared for cultured rat hepatocytes stimulated high-affinity GTPase activity. The stimulation by TRH of GTPase activity in membranes from (GH_3)

pituitary tumor cells is unaffected by either cholera or pertusiss toxins (Hinkle et al., 1986; Wojcikiewicz et al., 1986). However, it remains to be proven that the specific GTPase activity stimulated by the agonists reflects GTPase activity of the putative guanine nucleotide-binding protein G_p, thought to be involved in coupling ligand-receptor complexes to phospholipase C.

2.4. Fluoride Activation of Phospholipase C

There is evidence that aluminum fluoride interacts with the GDP bound to transducin and mimics the role of the γ-phosphate of GTP (Bigay et al., 1985). Aluminum fluoride increased polyphosphoinositide breakdown in hepatocytes (Blackmore et al., 1985). Taylor et al. (1986) found an increased accumulation of inositol trisphosphate in permeabilized rat parotid cells exposed to fluoride. Martin et al. (1986b) also observed activation by fluoride of the breakdown of phosphoinositides in membranes derived from cells previously labeled with inositol. Whether fluoride directly activates a phospholipase C by binding to the putative G_p protein in cells is unclear at this time.

3. EFFECTS OF GUANINE NUCLEOTIDES ON PHOSPHOLIPASE C

Gomperts (1983) found effects of guanine nucleotides on calcium-induced exocytotic secretion in permeabilized mast cells. Haslam and Davidson (1984a,b) subsequently found that guanine nucleotides enhanced the increase in diacylglycerol formation due to thrombin as well as secretion in permeabilized platelets.

3.1. Permeabilized Cells and Membranes from Cells Labeled by Prior Incubation with Tritiated Inositol

Cockcroft and Gomperts (1985) reported that in plasma membrane preparations from human neutrophils the breakdown of prelabeled polyphosphoinositides was enhanced by GTPγS but they reported no effects of ligands on polyphosphoinositide breakdown. In 1985 we reported that GTP was required for hormone-induced phospholipase C activation in cell-free extracts of fly salivary glands (Litosch et al. 1985). Smith et al. (1985) found an effect of fMet-Leu-Phe on breakdown of prelabeled polyphosphoinositides in the presence of guanine nucleotides using membranes from human polymorphonuclear leukocytes. Wallace and Fain (1985), Melin et al. (1986), and Uhing et al. (1985) reported that guanine nucleotides enhanced polyphosphoinositide breakdown in rat liver plasma membranes. Uhing et al. (1986) found that vasopressin activation of the breakdown of prelabeled polyphosphoinositides in rat liver plasma membranes was dependent on guanine nucleotides and insensitive to either cholera or pertussis toxin. Similarly, Guillon et al. (1986) prelabeled WRK1 cells with inositol and then prepared partially purified plasma membranes. The addition of vasopressin stimulated the accumulation of inositol trisphosphate in the presence of guanine nucleotides. Guanine nucleotides also potentiated the activation by TRH of polyphosphoinositide breakdown in isolated membranes prepared from rat pituitary tumor cells previously incubated with inositol (Martin et al., 1986b; Aub et al., 1986; Straub and Gershengorn, 1986; and Wojcikiewicz et al., 1987, unpublished).

Haslam and Davidson (1984a,b) utilized electrically permeabilized cells for studies on regulation of diacylglycerol formation by guanine nucleotides. It has been difficult to obtain ligand-responsive formation of inositol trisphosphate in cells prelabeled with inositol after permeabilization of the cells. However, Martin *et al.* (1986a) saw a TRH stimulation of inositol trisphosphate formation in electrically permeabilized (GH$_3$) rat pituitary tumor cells in the presence of added ATP. In their studies, calcium alone activated phosphoinositide breakdown at concentrations above 1 μM but TRH and guanine nucleotide effects were seen at low calcium concentrations (less than 0.1 μM). Similar results were seen by Merritt *et al.* (1986) with electrically permeabilized rat pancreatic acinar cells and by Taylor *et al.* (1986) with rat parotid cells. Martin *et al.* (1986a) also saw no effect of AppNHp on TRH action in permeabilized GH$_3$ cells indicating that nonhydrolyzable analogs of ATP cannot substitute for ATP. The effect of TRH in the presence of Mg ATP on phosphoinositide breakdown was markedly potentiated by GTPγS.

3.2. Assay of Phospholipase C Using Exogenous Labeled PIP$_2$

Studies in permeabilized cells, homogenates or plasma membrane-rich fractions prepared from cells in which the polyphosphoinositides are labeled with ^{32}P or inositol do not indicate whether the ligands are activating the substrate or a phospholipase C. However, Litosch and Fain (1985) found that serotonin (5-HT) increased the breakdown of exogenous tritium-labeled phosphatidylinositol-4,5-bisphosphate (PIP$_2$) via a process with an absolute dependence on added guanine nucleotides. These data suggested that 5-HT activates a phospholipase C enzyme. An effect of 5-HT was also seen on breakdown of exogenous phosphatidyl-inositol, but of lesser magnitude, indicating that the enzyme or enzymes activated by ligands are not absolutely specific for PIP$_2$. Similar results were seen by Sasaguri *et al.* (1985), who found that carbachol increased breakdown of exogenous polyphosphoinositides in the presence of guanine nucleotides by homogenates of porcine coronary arteries.

Litosch *et al.* (1986) found that the polyphosphoinositides of crude blowfly salivary gland membranes could be labeled by incubation with γ-labeled ATP. The subsequent addition of ligand in the presence of guanine nucleotides resulted in a 50% loss of label from PIP$_2$ and PIP. The effect of ligand in the presence of GTPγS was seen as a decrease in net accumulation of label in PIP$_2$ and PIP if labeled ATP was added at the same time as the ligand (Litosch *et al.*, 1986). Half-maximal effects of ligand required only 0.1 μM GTPγS and there was no effect on the accumulation of labeled phosphatidic acid. These results demonstrated that in blowfly salivary gland membranes any stimulation by hormone of PI or PIP kinases was more than compensated for by the activation of phospholipasec.

It is easy to label PA, PIP$_2$, and PIP of isolated membranes in the presence of [^{32}P-γ]-labeled ATP. However, it is often difficult to see subsequent degradation of PIP$_2$ or PIP unless high concentrations of calcium and bile salts are present. Plantavid *et al.* (1986) used human platelet plasma membranes incubated with γ-labeled ATP and found that even 1 mM calcium did not stimulate the accumulation of labeled inositol bis- or trisphosphate. However, in the presence of 5 mM sodium deoxycholate 60–75% of the PIP and PIP$_2$ were degraded and inositol phosphate accumulation was seen with as little as 10^{-7} M calcium. There was little effect of added cytosol indicating that the membranes as isolated contain sufficient amounts of phospholipase C to sustain a marked degradation of PIP$_2$ and PIP if they are rendered accessible by deoxycholate.

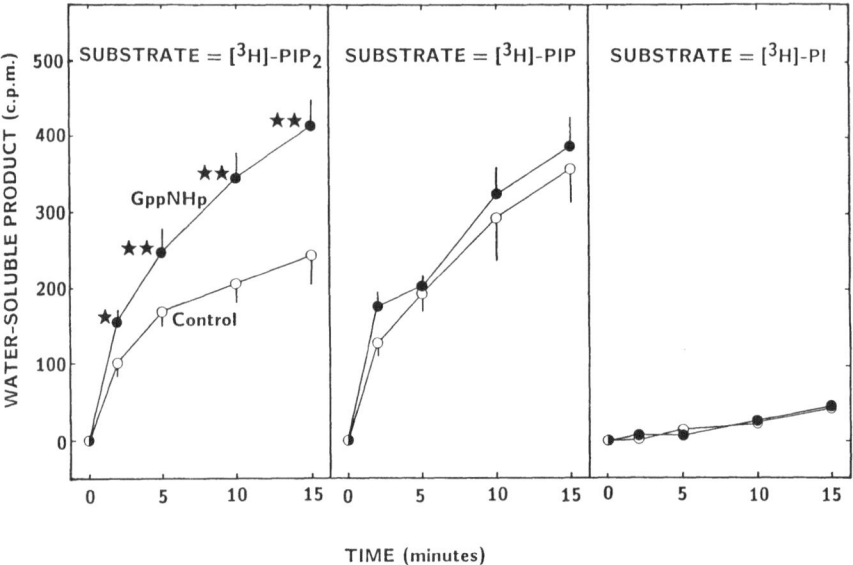

TIME (minutes)

FIGURE 2. Stimulation by GppNHp of PIP₂ degradation using GH₃ cell membranes.
Membranes (10 μg) prepared according to Wojcikiewicz et al. (1986) were incubated for various times with approximately 9000 cpm of [³H]PIP₂ (25 nM), [³H]PIP (38 nM), or [3H]PI (9 nM) in the presence (●) or absence (○) of 100 μM GppNHp. The buffer contained 60 mM Tris (pH 7.0) and 2 mM sodium cholate. Water-soluble radioactivity was measured. Mean ± SEM of results from at least four independent experiments are shown. **($p < 0.005$) and *($p < 0.05$) denote significance of stimulation calculated by Students paired t test (Wojcikiewicz et al., 1987, unpublished).

It is now accepted that the polyphosphoinositides are preferentially degraded in contrast to phosphatidylinositol in intact cells or isolated membranes after the addition of ligands in the presence of guanine nucleotides. A similar preferential degradation of PIP₂ and PIP was seen using inositol prelabeled membranes from human platelets or rat corneas in the presence of deoxycholate (Chung et al., 1985). In rat corneas the pH optimum in the presence of 7.5 mM deoxycholate was around 7 and little breakdown was seen in the presence of 2 mM EGTA. Maximal activation of breakdown was seen with 5 mM deoxycholate.

Deoxycholate also stimulates breakdown of phosphatidylinositol in brain synaptosomes and rat liver membranes. Manning and Sun (1983) found a specific breakdown of phosphatidylinositol (phosphatidylcholine and phosphatidylethanolamine content was unaffected) by rat brain synaptosomes in the presence of deoxycholate or taurocholate. Calcium was required for the effects of deoxycholate (maximal activation was seen at 3.6 mM) but other divalent cations such as copper and zinc were inhibitory at concentrations as low as 25 μM (Manning and Sun, 1983).

Wallace et al. (1982) found a similar specific activation of phosphatidylinositol degradation by the addition of 1.2 mM deoxycholate to rat liver plasma membranes. This effect could be seen in buffer containing 0.5 mM EGTA and no added Ca^{2+}. A stimulation of phosphatidylinositol breakdown due to vasopressin was seen in the presence but not in the absence of deoxycholate (Wallace *et al.*, 1982).

When we attempted to utilize exogenous PIP_2 for studies of guanine nucleotide effects on hormonal activation of phospholipase C using mammalian membranes, we encountered difficulties. It was necessary to either use deoxycholate or prepare liposomes containing PIP_2 in order to see breakdown. As yet we have been unable to see effects of ligands on breakdown of exogenous PIP_2 by mammalian membranes from a variety of tissues. We did see effects of guanine nucleotides (Bojanic *et al.*, 1987) as reported by others (Jackowski *et al.*, 1986; Deckmyn *et al.*, 1986; Rock and Jackowski, 1987).

GppNHp specifically activated the breakdown of PIP_2 as shown in Fig. 2 using membranes from rat pituitary tumor cells. However, a nonhydrolyzable analog of ATP (AppNHp) was equipotent and subsequently we observed that pyrophosphate was able to mimic the effects of nucleotides (Bojanic *et al.*, 1987). The effects of pyrophosphate or nucleotides containing a pyrophosphate group cannot be taken as evidence for involvement of a guanine nucleotide-binding protein in phospholipase C assays using deoxycholate. We have seen a similar effect of pyrophosphate on the breakdown of labeled PIP_2 (50 nM) in liposomes containing 120 μM PE, 60 μM PS, and 12 μM PI (Wallace *et al.*, 1987, unpublished). The effect of pyrophosphate, while specific for PIP_2 breakdown, was apparently on substrate access to the enzyme as all effects were reversed by increasing PIP_2 concentration in the assays from 0.05 to 30 μM. Furthermore we have confirmed that there are no effects of AppNHp or of pyrophosphate on ligand-induced breakdown of PIP_2 in permeabilized rat pituitary tumor cells if the PIP_2 was labeled by prior incubation with inositol.

The available evidence suggests the involvement of a guanine nucleotide-binding protein in phospholipase C activation. However, caution should be exercised in studies using exogenous PIP_2 as there are effects of guanine nucleotides that are nonspecific.

REFERENCES

Aub, D. L., Frey, E. A., Sekura, R. D., and Cote, T. E., 1986, Coupling of the thyrotropin-releasing hormone receptor to phospholipase C by a GTP-binding protein distinct from the inhibitory or stimulatory GTP-binding protein, *J. Biol. Chem.* **261**:9333–9340.

Bell, R. M., 1986, Protein kinase C activation by diacylglycerol second messengers, *Cell* **45**:631–632.

Berridge, M. J., 1984, Inositol trisphosphate and diacylglycerol as second messengers, *Biochem J.* **220**: 345–360.

Bigay, J., Deterre, P., Pfister, C., and Chabre, M., 1985, Fluoroaluminates activate transducin-GDP by mimicking the gamma-phosphate of GTP in its binding site, *FEBS Lett.* **191**:181–185.

Blackmore, P. F., Bocckino, S. B., Waynick, L. E., and Exton, J. H., 1985, Role of a guanine nucleotide-binding regulatory protein in the hydrolysis of hepatocyte phosphatidylinositol 4,5-bisphosphate by calcium-mobilizing hormones and the control of cell calcium, *J. Biol. Chem.* **260**: 14477–14483.

Boer, R., and Fahrenholz, F., 1985, Photoaffinity labeling of the V_1 vasopressin receptor in plasma membranes from rat liver, *J. Biol. Chem.* **260**:15051–15054.

Bojanic, D., and Fain, J. N., 1986, Guanine nucleotide regulation of [^{3}H]vasopressin binding to liver membranes and solubilized receptors: Evidence for the involvement of a guanine nucleotide regulatory protein, *Biochem. J.* **240**:361–365.

Bojanic, D., Wallace, M. A., Wojcikiewicz, R. J. H., and Fain, J. N., 1987, Guanine nucleotide and pyrophosphate activate exogenous phosphatidylinositol 4,5-bisphosphate hydrolysis in rat liver plasma membranes, *Biochem. Biophys. Res. Comm.* **147:**1088–1094.

Bradford, P. G., and Rubin, R. P., 1985, Pertussis toxin inhibits chemotactic factor-induced phospholipase C stimulation and lysosomal enzyme secretion in rabbit neutrophils, *FEBS Lett.* **183:**317–320.

Brandt, S. J., Dougherty, R. W., Lapetina, E. G., and Niedel, J. E., 1985, Pertussis toxin inhibits chemotactic peptide-stimulated generation of inositol phosphates and lysosomal enzyme secretion in human leukemic (HL-60) cells, *Proc. Natl. Acad. Sci. USA* **82:**3277–3280.

Cantau, B., Keppens, S., deWulf, H., and Jard, S., 1980, [^{3}H]Vasopressin binding to isolated rat hepatocytes and liver membranes: Regulation by GTP and relation to glycogen phosphorylase activation, *J. Receptor Res.* **1:**137–168.

Chung, S. M., Proia, A. D., Klintworth, G. K., Watson, S. P., and Lapetina, E. G., 1985, Deoxycholate induces the preferential hydrolysis of polyphosphoinositides by human platelet and rat corneal phospholipase C, *Biochem. Biophys. Res. Commun.* **129:**411–416.

Cockcroft, S., 1987, Polyphosphoinositide phosphodiesterase: Regulation by a novel guanine nucleotide binding protein, G_p, *Trends Biochem. Sci.* **12:**75–78.

Cockcroft, S., and Gomperts, B. D., 1985, Role of guanine nucleotide binding protein in the activation of polyphosphoinositide phosphodiesterase, *Nature* **314:**534–536.

Deckmyn, H., Tu, S-M, and Majerus, P. W., 1986, Guanine nucleotides stimulate soluble phosphoinositide-specific phospholipase C in the absence of membranes, *J. Biol. Chem.* **261:**16553–16558.

DeLean, A., Stadel, J. M., and Lefkowitz, R. J., 1980, A ternary complex model explains the agonist-specific binding properties of the adenylate cyclase-coupled beta-adrenergic receptor, *J. Biol. Chem.* **255:**7108–7117.

Evans, T., Martin, M. W., Hughes, A. R., and Harden, T. K., 1985, Guanine nucleotide-sensitive, high affinity binding of carbachol to muscarinic cholinergic receptors of 1321N1 astrocytoma cells is insensitive to pertussis toxin, *Molec. Pharmacol.* **27:**32–37.

Exton, J. H., Chrisman, T. D., Strickland, W. G., Prpic, V., and Blackmore, P. F., 1983, Mechanisms involved in the actions of calcium-dependent hormones in liver, in *Isolation, Characterization, and Use of Hepatocytes* (R. A. Harris and N. W. Cornell, eds.), Elsevier, Amsterdam, pp. 401–410.

Fain, J. N., 1978, Hormones, membranes and cyclic nucleotides, in *Receptors and Recognition Series 6A* (P. Cuatrecasas and M. F. Greaves, eds.), Chapman & Hall, London, pp. 1–62.

Fain, J. N., Brindley, D. N., Pittner, R. A., and Hawthorne, J. N., 1985, Stimulation of specific GTPase activity by vasopressin in isolated membranes from cultured rat hepatocytes, *FEBS Lett.* **192:**215–254.

Gomperts, B. D., 1983, Involvement of guanine nucleotide-binding protein in the gating of Ca^{2+} by receptors, *Nature* **306:**64–66.

Goodhardt, M., Ferry, N., Geynet, P., and Hanoune, J., 1982, Hepatic alpha$_1$-adrenergic receptors show agonist-specific regulation by guanine nucleotides, *J. Biol. Chem.* **257:**11577–11583.

Guillon, G., Balestre, M. N., Mouillac, B., and Devilliers, G.,1986, Activation of membrane phospholipase C by vasopressin: A requirement for guanyl nucleotides, *FEBS Lett.* **196:**155–159.

Haslam, R. J., and Davidson, M. M. L., 1984a, Guanine nucleotides decrease the free [Ca^{2+}] required for secretion of serotonin from permeabilized blood platelets. Evidence of a role for a GTP-binding protein in platelet activation, *FEBS Lett.* **174:**90–95.

Haslam, R. J., and Davidson, M. M. L., 1984b, Receptor-induced diacylglycerol formation in permeabilized platelets; Possible role for a GTP-binding protein, *J. Receptor Res.* **4:**605–629.

Hinkle, P. M., Hewlett, C. L., and Gershengorn, M. C., 1986, Thyroliberin action in pituitary cells is not inhibited by pertussis toxin, *Biochem. J.* **237:**181–186.

Hinkle, P. M., and Phillips, W. J., 1984, Thyrotropin-releasing hormone stimulates GTP hydrolysis by membranes from GH$_4$C$_1$ rat pituitary tumor cells, *Proc. Natl. Acad. Sci. USA* **81:**6183–6187.

Jackowski, S., Rettenmier, C. W., Sherr, C. J., and Rock, C. O.,1986, A guanine nucleotide-dependent phosphatidylinositol 4,5-diphosphate phospholipase C in cells transformed by the v-fms and v-fes oncogenes. *J. Biol. Chem.* **261:**4978–4985.

Kirk, C. J., Verrinder, T. R., and Hems, D. A., 1977, Rapid stimulation, by vasopressin and adrenaline, of inorganic phosphate incorporation into phosphatidyl inositol in isolated hepatocytes, *FEBS Lett.* **83**:267–271.

Lin, S. H., and Fain, J. N., 1981, Vasopressin and epinephrine stimulation of phosphatidylinositol breakdown in the plasma membrane of rat hepatocytes, *Life Sci.* **18**:1905–1912.

Litosch, I., Calista, C., Wallis, C., and Fain, J. N., 1986, 5-Methyltryptamine decreases net accumulation of ^{32}P into the polyphosphoinositides from [γ-^{32}P]ATP in a cell-free system from blowfly salivary glands, *J. Biol. Chem.* **261**:638–643.

Litosch, I., and Fain, J. N., 1985, 5-Methyltryptamine stimulates phospholipase C-mediated breakdown of exogenous phosphoinositides by blowfly salivary gland membranes, *J. Biol. Chem.* **260**: 16052–16055.

Litosch, I., Fain, J. N., 1986, Regulation of phosphoinositide breakdown by guanine nucleotides, *Life Sci.* **39**:187–194.

Litosch, I., Wallis, C., and Fain, J. N., 1985, 5-Hydroxytryptamine stimulates inositol phosphate production in cell-free system from blowfly glands: Evidence for a role of GTP in coupling receptor activation to phosphoinositide breakdown, *J. Biol. Chem.* **260**:5464–5471.

Manning, R., and Sun, G. Y., 1983, Detergent effects on the phosphatidylinositol-specific phospholipase C in rat brain synaptosomes, *J. Neurochem.* **41**:1735–1743.

Martin, T. F. J., Lucas, D. O., Bajjalieh, S. M., and Kowalchyk, J. A., 1986a, Thyrotropin-releasing hormone activates a Ca^{2+}-dependent polyphosphoinositide phosphodiesterase in permeable GH$_3$ cells, *J. Biol. Chem.* **261**:2918–2927.

Martin, T. F. J., Bajjalieh, S. M., Lucas, D. O., and Kowalchyk, J. A., 1986b, Thyrotropin-releasing hormone stimulation of polyphosphoinositide hydrolysis in GH$_3$ cell membranes is GTP dependent but insensitive to cholera or pertussis toxin, *J. Biol. Chem.* **261**:10041–10049.

Masters, S. B., Martin, M. W., Harden, T. K., and Brown, J. H., 1985, Pertussis toxin does not inhibit muscarinic-receptor-mediated phosphoinositide hydrolysis or calcium mobilization, *Biochem. J.* **227**:933–937.

Melin, P.-M., Sundler, R., and Jergil, B., 1986, Phospholipase C in rat liver plasma membranes: Phosphoinositide specificity and regulation by guanine nucleotides and calcium, *FEBS Lett.* **198**: 85–88.

Merritt, J. E., Taylor, C. W., Rubin, R. P., and Putney, J. W. Jr., 1986, Evidence suggesting that a novel guanine nucleotide regulatory protein couples receptors to phospholipase C in exocrine pancreas, *Biochem. J.* **236**:337–343.

Michell, R. H., 1975, Inositol phospholipids and cell surface receptor function. *Biochim. Biophys. Acta* **415**:81–147.

Michell, R. H., Kirk, C. J., Jones, L. M., Downes, C. P., and Creba, J. A., 1981, The stimulation of inositol lipid metabolism that accompanies calcium mobilization in stimulated cells: Defined characteristics and unanswered questions, *Phil. Trans. R. Soc. Lond. B* **296**:123–137.

Moreno, F. J., Mills, I., Garcia-Sainz, J. A., and Fain, J. N., 1983, Effects of pertussis toxin treatment on the metabolism of rat adipocytes, *J. Biol. Chem.* **258**:10938–10943.

Murayama, T., and Ui, M., 1985, Receptor-mediated inhibition of adenylate cyclase and stimulation of arachidonic acid release in 3T3 fibroblasts, *J. Biol. Chem.* **260**:7226–7233.

Nakamura, T., and Ui, M., 1985, Simultaneous inhibitions of inositol phospholipid breakdown, arachidonic acid release, and histamine secretion in mast cells by islet-activating protein, pertussis toxin, *J. Biol. Chem.* **260**:3584–3593.

Nishizuka, Y., 1986, Studies and perspectives of protein kinase C, *Science* **233**:305–312.

Pfeilschifter, J., and Bauer, C., 1986, Pertussis toxin abolishes angiotensin II-induced phosphoinositide hydrolysis and prostaglandin synthesis in rat renal mesangial cells, *Biochem. J.* **236**:289–294.

Plantavid, M., Rossignol, L., Chap, H., and Douste-Blazy, L., 1986, Studies of endogenous polyphosphoinositide hydrolysis in human platelet membranes. Evidence that polyphosphoinositides remain inaccessible to phosphodiesterase in the native membranes, *Biochim. Biophys. Acta* **875**: 147–156.

Pointer, R. H., Butcher, F. R., and Fain, J. N., 1976, Studies on the role of cyclic GMP and extracellular Ca^{2+} in the regulation of glycogenolysis in rat liver cells, *J. Biol. Chem.* **251**:2987–2992.

Rock, C. O., Jackowski, S., 1987, Thrombin- and nucleotide-activated phosphatidylinositol 4,5-bisphosphate phospholipase C in human platelet membranes, *J. Biol. Chem.* **262**:5492–5498.

Sasaguri, T., Hirata, M., and Kuriyama, H., 1985, Dependence on Ca^{2+} of the activities of phosphatidylinositol 4,5-bisphosphate phosphodiesterase and inositol 1,4,5-trisphosphate phosphatase in smooth muscles of the porcine coronary artery, *Biochem. J.* **231**:497–503.

Smith, C. D., Lane, B. C., Kusaka, I., Verghese, and Snyderman, R., 1985, Chemoattractant receptor-induced hydrolysis of phosphatidylinositol 4,5-bisphosphate in human polymorphonuclear leukocyte membranes, *J. Biol. Chem.* **260**:5875–5878.

Snyderman, R., Pike, M. C., Edge, S., and Lane, B., 1984, A chemoattractant receptor on macrophages exists in two affinity states regulated by guanine nucleotides, *J. Cell. Biol.* **98**:444–448.

Straub, R. E., Gershengorn, M. C., 1986, Thyrotropin-releasing hormone and GTP activate inositol trisphosphate formation in membranes isolated from rat pituitary cells, *J. Biol. Chem.* **261**:2712–2717.

Taylor, C. W., Merritt, J. E., Putney, J. W. Jr., and Rubin, R. P., 1986, A guanine nucleotide-dependent regulatory protein couples substance P receptors to phospholipase C in rat parotid gland, *Biochem. Biophys. Res. Commun.* **136**:362–368.

Tolbert, M. E. M., Butcher, F. R., and Fain, J. N., 1973, Lack of correlation between catecholamine effects on cyclic adenosine 3',5'-monophosphate and gluconeogenesis in isolated rat liver cells, *J. Biol. Chem.* **248**:5686–5692.

Uhing, R. J., Jiang, H., Prpic, V., and Exton, J. H., 1985, Regulation of a liver plasma membrane phosphoinositide phosphodiesterase by guanine nucleotides and calcium, *FEBS Lett.* **188**:317–320.

Uhing, R. J., Prpic, V., Jiang, H., and Exton, J. H., 1986, Hormone-stimulated polyphosphoinositide breakdown in rat liver plasma membranes: Roles of guanine nucleotides and calcium, *J. Biol. Chem.* **261**:2140–2146.

Wallace, M. A., and Fain, J. N., 1985, Guanosine-5'-0-thiotriphosphate (GTP gamma S) stimulates phospholipase C activity in plasma membranes of rat hepatocytes, *J. Biol. Chem.* **260**:9527–9530.

Wallace, M. A., Randazzo, P., Li, S. Y., and Fain, J. N., 1982, Direct stimulation of phosphatidylinositol degradation by addition of vasopressin to purified rat liver plasma membranes. *Endocrinology* **111**:341–343.

Wojcikiewicz, R. J. H., Kent, P. A., and Fain, J. N., 1986, Evidence that thyrotropin-releasing hormone-induced increases in GTPase activity and phosphoinositide metabolism in GH_3 cells are mediated by a guanine nucleotide-binding protein other than G_s or G_i, *Biochem. Biophys. Res. Commun.* **138**:1383–1389.

25

Guanine Nucleotide-Dependent Release of Arachidonic Acid in Permeabilized Inflammatory Cells

YUKIO OKANO, SHIGERU NAKASHIMA, TOYOHIKO TOHMATSU, KOUJI YAMADA, KOH-ICHI NAGATA, and YOSHINORI NOZAWA

1. INTRODUCTION

Phosphoinositide metabolism has been known to play a crucial role in signal transduction systems of various cells, producing two second messengers (Nishizuka 1984; Berridge and Irvine, 1984). One is 1,2-diacylglycerol (1,2-DG), which directly activates protein kinase C, and the other is inositol-1,4,5-trisphosphate, which releases Ca^{2+} from internal storage sites. Recently, several lines of evidence have been revealing the involvement of guanine nucleotide-binding regulatory protein (G protein) in agonist-induced activation of phospholipase C, analogous to the adenylate cyclase system (Cockcroft, 1987; Haslam and Davidson, 1984; Cockcroft and Gomperts, 1985; Bradford and Rubin, 1986; Brass et al., 1986; Kikuchi et al., 1986; Lapetina, 1986). Upon stimulation of various types of inflammatory cells, arachidonic acid is released and converted to physiologically active substances, prostaglandins and leukotrienes. Although the coupling of phospholipase C with G protein has been intensively studied, only limited information is available regarding the involvement of G protein in phospholipase A_2 activation (Burch et al., 1986; Jelsema, 1987; Benjamin et al., 1987; Nakashima et al., 1987). Possible involvement of G protein in phospholipase A_2 activation was investigated by examining the effects of guanine nucleotides (GTP or its nonhydrolyzable analog, GTPγS) on generation of free arachidonic acid in permeabilized inflammatory cells, including saponin-permeabilized human platelets and ATP-permeabilized rat peritoneal mast cells.

YUKIO OKANO, SHIGERU NAKASHIMA, TOYOHIKO TOHMATSU, KOUJI YAMADA, KOH-ICHI NAGATA, and YOSHINORI NOZAWA ● Department of Biochemistry, Gifu University School of Medicine, Tsukasamachi 40, Gifu, 500 Japan.

2. RAT PERITONEAL MAST CELLS

Mast cells resuspended in divalent cation-free buffer were permeabilized with ATP (150 μM) (Cockcroft and Gomperts, 1979) either in the presence or in the absence of GTPγS (100 μM). At 6 min, 1 μM MgCl$_2$ was added to reseal the cells. Ten minutes after addition of MgCl$_2$, 3 μM CaCl$_2$ was added to the cells to initiate exocytotic responses. The permeabilized resealed cells were further incubated for 10 min before terminating the reaction to examine histamine secretion or arachidonic acid liberation.

The exocytotic histamine secretion from permeabilized resealed cells was approximately 25% at 10 min after CaCl$_2$ addition. The presence of GTPγS dramatically enhanced the released histamine content up to 60%, indicating involvement of G protein in exocytotic responses (Nakamura and Ui, 1984a,b; Cockcroft and Gomperts, 1985; Okano *et al.*, 1987).

[^{3}H]Arachidonic acid liberation from permeabilized and nonpermeabilized cells was examined in the presence of BW755C (100 μM), an inhibitor of both cyclooxygenase and lipoxygenase (Smith *et al.*, 1985), to gain high recovery of released arachidonate (Fig. 1, Okano *et al.*, 1987). Compared with control intact cells, permeabilized cells in the presence of Ca^{2+} induced a significant increase in ^{3}H radioactivity of the free arachidonate fraction. Entrapment of GTPγS further enhanced the level of free [^{3}H]arachidonate radioactivity. Intact cells with CaCl$_2$ or permeabilized cells without Ca^{2+} caused little or small arachidonate release, whereas permeabilized cells with GTPγS elicited a significant arachidonate release in the absence of Ca^{2+}. Although this appeared to be the Ca^{2+}-sparing effect of GTPγS observed in various cells, the effect may not have been solely due to the reduction of the Ca^{2+} requirement. The temporal changes in [^{3}H]arachidonate radioactivity in various lipid

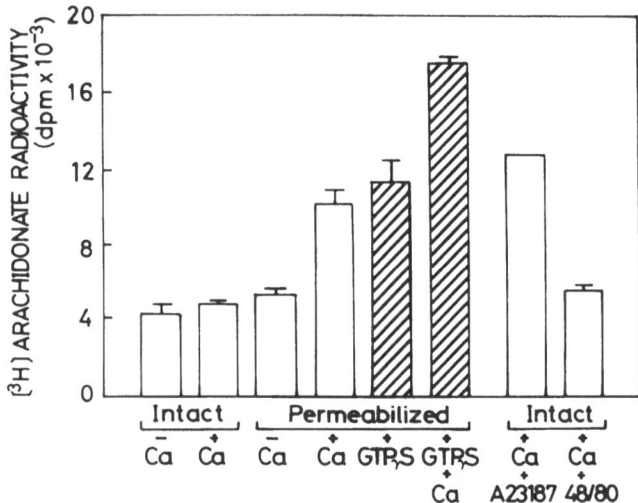

FIGURE 1. Arachidonic acid liberation in permeabilized and nonpermeabilized mast cells. [^{3}H]Arachidonate-labeled cells were incubated for 6 min at 37°C either in the presence or absence of ATP (150 μM) with or without GTPγS (100 μM). Then, after incubation with MgCl$_2$ for resealing, CaCl$_2$ (3 mM) was added to the resealed cells. Incubations were terminated 10 min after addition of CaCl$_2$. The intact [^{3}H]arachidonate-labeled cells were stimulated with either A23187 (1 μM) or compound 48/80 (5 μg/ml). (Adapted from Okano *et al.*, 1987 with permission.)

fractions showed a progressive increase in free fatty acid and a corresponding decrease in the phosphatidylcholine (PC) fraction. Phosphatidic acid and 1,2-DG exhibited small but significant increases in both GTPγS-entrapped and nontrapped cells. Breakdown of phosphatidylinositol was not clearly shown, which may be due to its rapid resynthesis. Alternative substrates for phospholipase C may be phosphatidylinositol-4,5-bisphosphate or PC. A novel pathway of PC-specific phospholipase C was recently reported by several investigators (Besterman et al., 1986; Daniel et al., 1986; Martin et al., 1987).

In order to obtain more information regarding the molecular mechanism of arachidonic acid liberation, experiments were undertaken with a DG lipase inhibitor, RHC 80267 (Sutherland and Amin, 1982). A dose-dependent inhibition of arachidonic acid liberation was observed in the inhibitor-entrapped cells in both the presence and absence of GTPγS, suggesting contribution of the DG lipase pathway to arachidonic acid liberation from activated mast cells, in addition to the phospholipase A$_2$ pathway.

Nakamura and Ui (1984a, b) previously reported the inhibitory actions of pertussis toxin on compound 48/80-stimulated and guanine nucleotide-dependent histamine secretion, phosphoinositide metabolism, and arachidonate release. Taken together with these results, our present observation obtained with permeabilized rat peritoneal mast cells provides further evidence for the involvement of G protein in arachidonate liberation (Okano et al., 1987).

3. HUMAN PLATELETS

Human platelets prelabeled with [^{3}H]arachidonate were permeabilized with saponin (20 μg/ml) to introduce membrane-impermeable guanine nucleotides. The saponin-permeabilized cells, having the equivalent responsiveness to various agonists as intact platelets, were activated by the addition of thrombin, releasing arachidonic acid and producing 1,2-DG. The results indicated the thrombin-induced activation of phospholipase C hydrolyzing inositol phospholipids. The thrombin-stimulated cells in the presence of GTP or GTPγS exhibited dose-dependent enhancement of the level of free [^{3}H]arachidonate. GTPγS showed a dose-dependent increase in 1,2-DG formation, while GTP at a concentration of 1 mM failed to enhance thrombin-induced 1,2-DG generation. Furthermore, the addition of GTP or GTPγS alone to saponin-permeabilized platelets in the absence of thrombin also induced a small but significant release of [^{3}H]arachidonate (Nakashima et al., 1987).

It is generally accepted that liberated arachidonic acid is derived via two major pathways: phospholipase A$_2$ (Bills et al., 1978) or sequential action of phospholipase C and DG lipase (Bell et al., 1979). In order to determine which pathway plays a major role in arachidonic acid liberation, permeabilized platelets were treated with a phospholipase C inhibitor, neomycin (Schacht, 1976) or with a DG lipase inhibitor, RHC 80267 (Sutherland and Amin, 1982). RHC 80267 caused no significant changes in the levels of liberated arachidonate and generated 1,2-DG in thrombin-activated platelets in both the presence and absence of GTPγS (Fig. 2; Nakashima et al., 1987). Neomycin had little effect on [^{3}H]arachidonate release in thrombin-stimulated permeabilized platelets, whereas higher concentrations (more than 100 μM) enhanced arachidonate liberation. These results suggest that the DG lipase pathway may not play a major role in thrombin-stimulated and guanine nucleotide-dependent arachidonic acid liberation. Although it is generally considered that millimolar concentrations of Ca^{2+} are required for the activation of phospholipase A$_2$, the finding that GTPγS-stimulated arachidonic acid liberation was most effectively observed in the absence of added Ca^{2+} suggested the Ca^{2+}-sparing effect of guanine nucleotide.

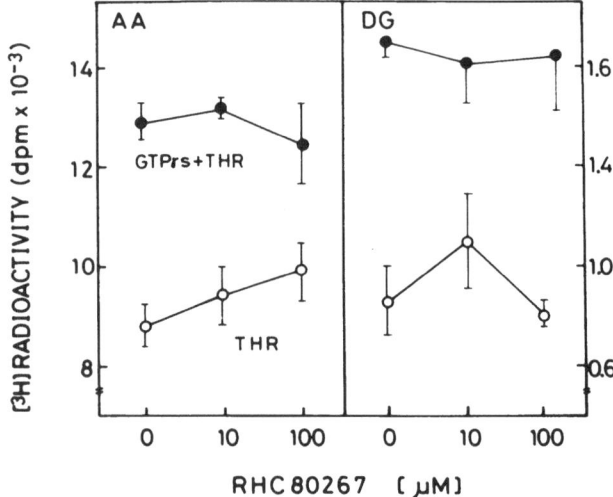

FIGURE 2. Effects of RHC 80267 on [³H]arachidonate liberation and [³H]arachidonoyl DG formation in saponin-permeabilized platelets. The [³H]arachidonate-labeled human platelets permeabilized with saponin (20 µg/ml) were incubated in the presence of indicated concentrations of RHC 80267 for 5 min with thrombin (2 U/ml, ○) or GTPγS (100 µM) plus thrombin (●). (Adapted from Nakashima *et al.*, 1987 with permission.)

In order to obtain further evidence for the involvement of G protein in arachidonic acid liberation, permeabilized platelets were pretreated with GDPβS (1 mM) and/or GTPγS (100 µM) and then stimulated with thrombin (Fig. 3). GTPγS-dependent thrombin-induced arachidonic acid liberation was significantly repressed by the pretreatment with GDPβS. It is thus reasonable to consider that a G protein is involved in the arachidonic acid release, probably via phospholipase A₂, in permeabilized human platelets (Nakashima *et al.*, 1987).

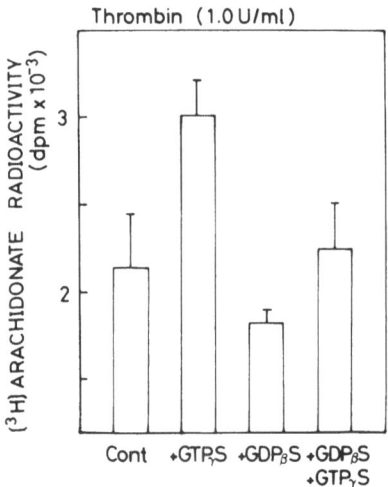

FIGURE 3. Effects of GTPγS and GDPβS on thrombin-induced arachidonate liberation. The [³H]arachidonate-labeled human platelets permeabilized with saponin (20 µg/ml) were stimulated with thrombin (1.0 U/ml) in the presence or absence of GTPγS (100 µM) and/or GDPβS (1 mM). The reactions were terminated after 10 min incubation at 37°C.

4. CONCLUSION

Guanine nucleotide-dependent arachidonic acid liberation was observed in various types of permeabilized inflammatory cells, such as platelets and mast cells. The results obtained with inhibitors for either phospholipase A_2 or phospholipase C/DG lipase pathway strongly suggested a major contribution of phospholipase A_2 to arachidonic acid liberation in these cells. The inhibitory effects of pertussis toxin on arachidonic acid liberation from mast cells (Nakamura and Ui, 1984a,b) and neutrophils (Bokoch and Gilman, 1984) support the view of involvement of G protein in arachidonic acid liberation, though the effects of the toxin on platelets are controversial (Brass *et al.*, 1986; Lapetina, 1986). Therefore, it is conceivable that the phospholipase A_2 responsible for arachidonic acid liberation may be coupled with one of the G-protein family. However, further investigations are necessary to define whether distinct proteins are coupled with phospholipases A_2 and C.

ACKNOWLEDGMENT. This study was supported in part by a research grant from the Ministry of Education, Culture and Science of Japan.

REFERENCES

Bell, R. L., Kennerly, D. A., Stanford, N., and Majerus, P. W., 1979, Diglyceride lipase: A pathway for arachidonate release from human platelets, *Proc. Natl. Acad. Sci. USA* **76:**3238–3241.

Benjamin, C. W., Tarpley, W. G., and Gorman, R. R., 1987, Loss of platelet-derived growth factor-stimulated phospholipase activity in NIH-3T3 cells expressing the EJ-ras oncogene, *Proc. Natl. Acad. Sci. USA* **84:**546–550.

Berridge, M. J., and Irvine, R. F., 1984, Inositol trisphosphate, a novel second messenger in cellular signal transduction, *Nature* **312:**315–321.

Besterman, J. M., Duronio, V., and Cuatrecasas, P., 1986, Rapid formation of diacylglycerol from phosphatidylcholine: A pathway for generation of a second messenger, *Proc. Natl. Acad. Sci. USA* **83:**6785–6789.

Bills, T. K., Smith, J. B., and Silver, M. J., 1977, Selective release of arachidonic acid from the phospholipids of human platelets in response to thrombin, *J. Clin. Invest.* **60:**1–6.

Bokoch, G. M., and Gilman, A. G., 1984, Inhibition of receptor-mediated release of arachidonic acid by pertussis toxin, *Cell* **39:**301–308.

Bradford, P. G., and Rubin, R. P., 1986, Guanine nucleotide regulation of phospholipase C activity in permeabilized rabbit neutrophils, *Biochem. J.* **239:**97–102.

Brass, L. F., Laposata, M., Banga, H. S., and Rittenhouse, S. E.,1986, Regulation of the phosphoinositide hydrolysis pathway in thrombin-stimulated platelets by a pertussis toxin-sensitive guanine nucleotide-binding protein: Evaluation of its contribution to platelet activation and comparisons with the adenylate cyclase inhibitory protein, Gi, *J. Biol. Chem.* **261:**16838–16847.

Burch, R. M., Luini, A., and Axelrod, J., 1986, Phospholipase A_2 and phospholipase C are activated by distinct GTP-binding proteins in response to α_1-adrenergic stimulation in FRTL5 thyroid cells, *Proc. Natl. Acad. Sci. USA* **83:**7201–7205.

Cockcroft, S., 1987, Polyphosphoinositide phosphodiesterase: Regulation by a novel guanine nucleotide binding protein, Gp, *Trends Biochem. Sci.* **12:**75–78.

Cockcroft, S., and Gomperts, B. D., 1979, Activation and inhibition of calcium-dependent histamine secretion by ATP ions applied to rat mast cells, *J. Physiol. (Lond.)* **296:**229–243.

Cockcroft, S., and Gomperts, B. D., 1985, Role of nucleotide binding protein in the activation of polyphosphoinositide phosphodiesterase, *Nature* **314:**534–536.

Daniel, L. W., Waite, M., and Wykle, R. L., 1986, A novel mechanism of diglyceride formation:

12-O-Tetradecanoylphorbol-13-acetate stimulates the cyclic breakdown and resynthesis of phosphatidylcholine, *J. Biol. Chem.* **261**:9128–9132.

Haslam, R. J., and Davidson, M. M. L., 1984, Receptor-induced diacylglycerol formation in permeabilized platelets: Possible role for a GTP-binding protein, *J. Receptor Res.* **4**:605–629.

Jelsema, C. L., 1987, Light activation of phospholipase A_2 in rod outer segments of bovine retina and its modulation by GTP-binding proteins, *J. Biol. Chem.* **262**:163–168.

Kikuchi, A., Kozawa, O., Kaibuchi, K., Katada, T., Ui, M., and Takai, Y., 1986, Direct evidence for involvement of a guanine nucleotide-binding protein in chemitactic peptide-stimulated formation of inositol bisphosphate and trisphosphate in differentiated human leukemic (HL-60) cells: Reconstitution with Gi or Go of the plasma membrane ADP-ribosylated by pertussis toxin, *J. Biol. Chem.* **261**:11558–11562.

Lapetina, E. G., 1986, Effect of pertussis toxin on the phosphodiesteratic cleavage of the polyphosphoinositides by guanosine 5'-O-thiotriphosphate and thrombin in permeabilized human platelets, *Biochem. Biophys. Acta* **884**:219–224.

Litosch, I., Wallis, C., and Fain, J. N., 1985, 5-Hydroxytryptamine stimulates inositol phosphate production in a cell-free system from blowfly salivary glands: Evidence for a role of GTP in coupling receptor activation to phosphoinositide breakdown, *J. Biol. Chem.* **260**:5464–5471.

Martin, T. W., Wysolmerski, R. B., and Lagunoff, D., 1987, Phosphatidylcholine metabolism in endothelial cells: evidence for phospholipase A and a novel Ca^{2+}-independent phospholipase C, *Biochim. Biophys. Acta* **917**:296–307.

Nakamura, T., and Ui, M., 1984a, Simultaneous inhibitions of inositol phospholipid breakdown, arachidonic acid release, and histamine secretion in mast cells by islet-activating protein, pertussis toxin, *J. Biol. Chem.* **260**:3584–3593.

Nakamura, T., and Ui, M., 1984b, Islet-activating protein, pertussis toxin, inhibits Ca^{2+}-induced and guanine nucleotide-dependent releases of histamine and arachidonic acid from rat mast cells, *FEBS Lett.* **173**:414–418.

Nakashima, S., Tohmatsu, T., Hattori, H., Suganuma, A., and Nozawa, Y., 1987, Guanine nucleotides stimulate arachidonic acid release by phospholipase A_2 in saponin-permeabilized human platelets, *J. Biochem.* **101**:1055–1058.

Nishizuka, Y., 1984, The role of protein kinase C in cell surface signal transduction and tumour promotion, *Nature* **308**:693–698.

Okano, Y., Yamada, K., Yano, K., and Nozawa, Y., 1987, Guanosine 5'-(γ-thio)triphosphate stimulates arachidonic acid liberation in permeabilized rat peritoneal mast cells, *Biochem. Biophys. Res. Commun.* **145**:1267–1275.

Ragab-Thomas, J. M. F., Hullin, F., Chap, H., and Douste-Blazy, L., 1987, Pathways of arachidonic acid liberation in thrombin and calcium ionophore A23187-stimulated human endothelial cells: Respective roles of phospholipids and triacylglycerol and evidence for diacylglycerol generation from phosphatidylcholine, *Biochim. Biophys. Acta* **917**:388–397.

Schacht, J., 1976, Inhibition of neomycin of polyphosphoinositide turnover in subcellular fractions of guinea-pig cerebral cortex in vitro, *J. Neurochem.* **27**:1119–1124.

Smith, J. B., Dangelmaier, C., and Mauco, G., 1985, Measurement of arachidonic acid liberation in thrombin-stimulated human platelets. Use of agents that inhibit both the cyclooxygenase and lipoxygenase enzymes, *Biochim. Biophys. Acta* **835**:344–351.

Sutherland, C. A., and Amin, D., 1982, Relative activities of rat and dog platelet phospholipase A_2 and diglyceride lipase, *J. Biol. Chem.* **257**:14006–14010.

III

Cell Proliferation and Differentiation

26

Measurement of Free Intracellular Calcium (Ca$_i$) in Fibroblasts
Digital Image Analysis of Fura 2 Fluorescence

R. W. TUCKER, K. MEADE-COBUN, and H. LOATS

1. INTRODUCTION

Measurement of intracellular calcium has long been a goal in many areas of biology. Calcium has been heralded as an intracellular messenger because of its unique structure and binding properties (Kretsinger *et al.*, 1980), which allow Ca^{2+} ion to bind and activate specific regulatory molecules, such as calmodulin (Cheung, 1980). Ca^{2+}-activated processes have been implicated in signal transduction in metabolism, neural transmission, excitation-contraction coupling, and mitosis, and, more recently, in the stimulation of early mitogenic events in cells as diverse as fibroblasts (Boynton *et al.*, 1980; Chafouleas *et al.*, 1982; Morris *et al.*, 1984; McNeil *et al.*, 1985; Tucker *et al.*, 1986) and lymphocytes (Hesketh *et al.*, 1983; Gelfand *et al.*, 1986).

Recently, the study of these Ca^{2+}-activated processes has been given new impetus by technological advances. First, calcium-sensitive fluorescent probes were developed that could easily be introduced into the intracellular compartments of living cells. The first probe was Quin2 (Pozzan *et al.*, 1982), which proved to be very useful for spectrophotometric studies of cell populations (Pozzan *et al.*, 1982). Unfortunately, the low quantum efficiency and high bleaching rate of Quin2 made single-cell measurements impractical (Fay and Tucker, 1985). Another probe, Indol 1, could be used for flow cytometric measurement of Ca$_i$ in single cells (Rabinovitch *et al.*, 1986), but changes in Ca$_i$ could not be followed in the same cell as a function of time. The newest generation of Ca^{2+} probes includes Fura 2, an indicator that has a high quantum efficiency and low bleaching rate (Grynkiewicz *et al.*, 1985). These properties allow investigators to use lower concentrations of Fura 2 in order to provide a relatively stable intracellular fluorescent signal that is sensitive to Ca^{2+} changes. The low concentration of indicator also means that Ca^{2+} transients will not be buffered as much as they are with Quin2 loading. In fact, Quin2 has become more useful as an intracellular chelator (Pershadsingh *et al.*, 1987; Nemeth and Scarpa, 1986), while Fura 2 has provided an improved method for detecting changes in Ca$_i$ (Ratan *et al.*, 1986; Poenie *et al.*, 1986;

R. W. TUCKER and K. MEADE-COBUN • Johns Hopkins Oncology Center, Baltimore, Maryland 21205. H. LOATS • Loats Associates, Inc., Westminster, Maryland 21157.

Williams et al., 1985; Connor, 1986; Lemasters et al., 1987; Millard et al., 1986; Wier et al., 1987).

A second technical advance that has facilitated the study of calcium transients in single cells is the recent development of digital image analysis (Williams et al., 1985; Agard and Sedat, 1983; Bright et al., 1987; Tsien et al., 1985). As applied to the measurement of Ca_i in individual cells, digital image analysis allows the investigator to digitize images of intracellular Fura 2 fluorescence and to use the computer to solve explicitly for Ca_i values at each part (pixel) of the cellular image. At any instant of time an image of Ca_i can be obtained for any cell that can be visualized in a fluorescent microscope. The Ca_i changes can then be followed simultaneously in different cellular regions as a function of time. In the course of using this approach to study Ca_i transients in fibroblasts stimulated with platelet-derived growth factor (PDGF) (Tucker and Loats, 1987), we encountered several complications that subsequently form the focus for this discussion. We found that accurate interpretation of Ca_i changes using digital image analysis of Fura 2 fluorescence requires (1) accounting for incomplete loading (deesterification) of Fura 2 and (2) careful interpretation of spatial gradients of Ca_i within living cells.

2. MATERIALS AND METHODS

2.1. Cell Culture and Materials

FeSin cells, embryonic human fibroblasts (American Type Culture Collection, Rockville, MD), and an original stock of BALB/c 3T3 cells (clone A-31, obtained by G. Todaro) were grown as previously described (Tucker et al., 1986; Tucker and Loats, 1987). After FeSin cells ($2 \times 10^4/2$ ml/35-mm dish) were allowed to attach for 2 hr, the cells were washed twice and placed in Dulbecco's modified eagles medium (DME) + 0.1% platelet-poor plasma (PPP) for 48 hr. BALB/c 3T3 cells (8×10^4/dish) were plated and allowed to become quiescent over 6 days of growth in DME + 0.5% CS.

Purified PDGF was obtained from outdated human platelets (Dr. W. J. Pledger, Vanderbilt University) and used as previously described (Tucker et al., 1986).

2.2. Fura 2 Loading

Cells were loaded with Fura 2 either by incubation with Fura 2/AM (Molecular Probes, Junction City, OR) or by hypoosmolar (HOSTing) loading using Fura 2 acid. Fura 2/AM loading of FeSin cells was carried out as previously described (Tucker and Loats, 1987). Similar loading procedures were not successful with BALB/c 3T3 cells, and a dispersing agent (Pluronic F127, BASF Wyandotte Corp.) was used as recently described (Poenie et al., 1986). BALB/c 3T3 cells were incubated for 2 hr at room temperature with 2.5 μM Fura 2/AM in 0.03% Pluronic, 0.01% dimethylsulfoxides and 0.75% PPP. In both cell lines, Fura 2 loading under these conditions did not enhance or inhibit DNA synthesis stimulated by 20% serum as assessed by autoradiography of cells labeled continuously with [^{3}H]thymidine (data not shown). Intracellular Fura 2 concentration (130 μM in FeSin; 750 μM in BALB/c 3T3) was estimated from the fluorescence of material released from cells permeabilized with 0.001% digitonin in Krebs/Hepes buffer for 20 min. The second method utilized hypoosmolar treatment to introduce Fura 2 acid directly into FeSin cells as previously described (Tucker et al., 1986).

2.3. Digital Image Analysis

Ca$_i$ distribution was calculated from the fluorescence of intracellular Fura 2 in cells on the microscope stage as previously described (Tucker and Loats, 1987). Cell cultures in Krebs/Hepes solutions were observed for 5–10 min on the microscope stage, and then the fluid was changed to Krebs/Hepes containing PDGF (4–8 U/ml). Paired excitation images (340 and 380 nm) were taken every 15–30 sec, and Ca$_i$ was calculated as described (Tucker and Loats, 1987).

2.4. Calibration Curves

In order to determine an *in vitro* calibration curve for defining K_d, Fura 2 solutions containing different free-calcium concentrations were measured on the microscopic stage. Aliquots of Fura 2 were diluted with MOPS buffer (in mM: KCl 100, KMOPS 10, with varying proportions of K_2H_2 EGTA 10 or K_2Ca EGTA 10). The free calcium was determined by ion-selective electrode (no. F2210CA, Radiometer, Copenhagen) in collaboration with Drs. William B. Guggino and M. Cornejo-Palez, Johns Hopkins Medical School (Cornejo *et al.*, 1987). Standard Ca^{2+} solutions were made by dissolving $CaCl_2$ (Aldrich) in 17.8-megaohm distilled water (hydroorganic ultrapure water, model C2-18, Hydro Services and Supplies Inc., Research Triangle Park, NC) with 135 mM KCl (Gold Label, Aldrich) specified to contain Ca^{2+} impurities less than 10^{-9} M. Electrode was linear to a Ca^{2+} concentration of 10^{-7} M, with a slope of 28.4 mV/pCa. The pH and free calcium of each solution was measured simultaneously and pH was adjusted to 7.2 with KOH.

The adequacy of Fura 2 loading was checked by studying fluorescent material released from loaded cells by digitonin permeabilization (0.001%). Excitation spectra were measured using a Perkin-Elmer fluorescence spectrophotometer (4-nm slits) with a xenon lamp in collaboration with Dr. Reynafarje, Johns Hopkins Medical School.

3. RESULTS

3.1. Fura 2 Loading

Fura 2-loaded FeSin cells illuminated at 380 nm had a brightly fluorescent nucleus and occasional hyperfluorescent particles in the cytoplasm, as well as vesicular areas that were not fluorescent and thus contained little or no dye (Fig. 1). These FeSin cells loaded by incubation with Fura 2/AM resembled BALB/c 3T3 cells microinjected with Fura 2 (Fay and Tucker, 1985). However, the average Ca$_i$ calculated for FeSin cells (36 ± 3 nM; n = 23) was lower than that determined for BALB/c 3T3 cells loaded using Fura 2/AM with Pluronic (55 ± 3 nM, n = 61). To distinguish between the possibilities that the two cell types have truly different basal Ca$_i$, or that the calculated Ca$_i$ values were sensitive to the loading conditions, we loaded FeSin cells directly with Fura 2 acid using hypoosmolar (HOSTing) treatment (Fig. 1) (Tucker *et al.*, 1986). HOSTed FeSin cells had a higher calculated Ca$_i$ (62 ± 13 nM, n = 15) than Fura 2/AM-loaded FeSin cells (36 ± 3 nM), but a level similar to that of BALB/c 3T3 cells incubated with Fura 2/AM plus Pluronic (55 ± 3 nM). The Ca$_i$ of HOSTed FeSin cells and that of microinjected BALB/c 3T3 cells (48 ± 5 nM, n = 29) (Fay and Tucker, 1985) were both lower than that in growing PtK$_2$ (Ratan *et al.*, 1986) and PtK$_1$ (Poenie *et al.*, 1986) cells, perhaps reflecting physiological differences in Ca$_i$ between growing and quiescent (or unstimulated) cells.

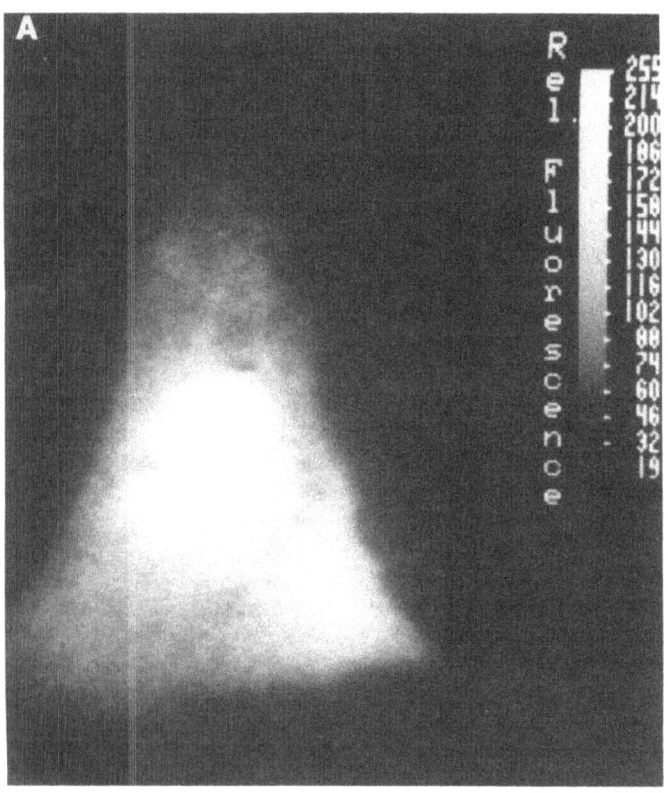

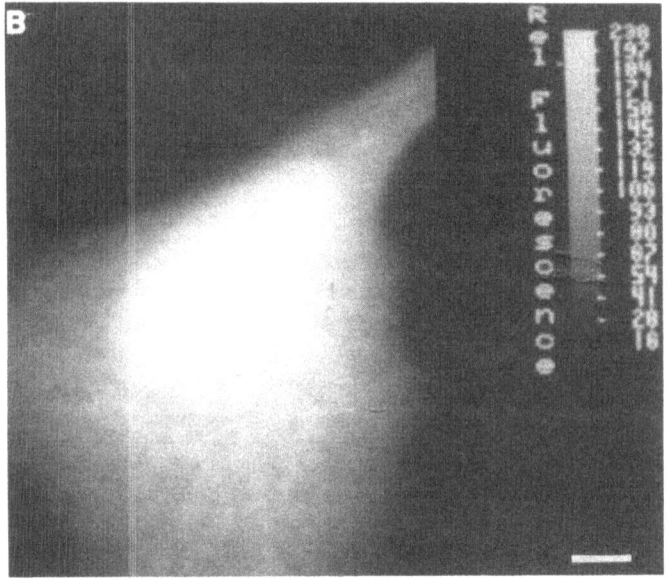

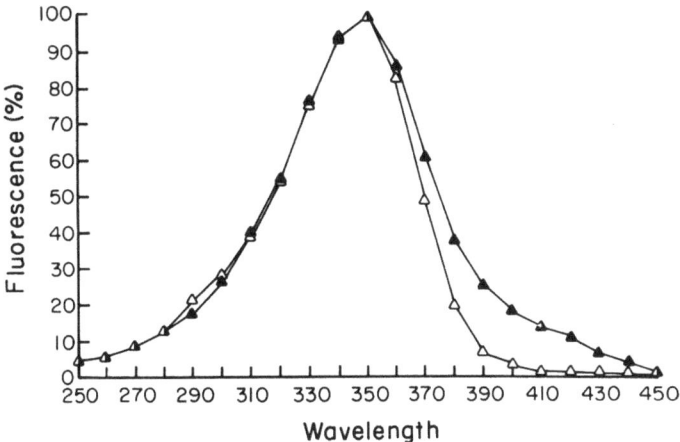

FIGURE 2. Excitation spectra of Fura 2 and material from permeabilized cells. Emission was measured at 500 nm, and excitation spectra obtained for both Fura 2 ($\triangle$) and cell material ($\blacktriangle$).

Since incubating some cells (FeSin) with Fura 2/AM results in a lower calculated Ca$_i$ than that for the same cells loaded directly with Fura 2 acid, incubation with Fura 2/AM must not load Fura 2 directly and reliably into all cells. In order to investigate this possibility further, we studied the fluorescent material obtained from permeabilized FeSin cells loaded by incubation with Fura 2/AM. As Fig. 2 shows, the excitation spectrum of material from Fura 2-loaded cells was not identical to that of Fura 2 acid: the cell material had increased relative fluorescence at 380 nm excitation. The extra fluorescence was probably due to a partially deesterified Fura 2/AM component, since further incubation of the cell material with esterase (Sigma) produced material that had an excitation spectrum identical to that for Fura 2 acid. Short incubations of Fura 2/AM with esterase also produced components whose excitation spectrum was identical to that of the cellular material, while a longer incubation with esterase produced fluorescent material with an excitation spectrum identical to that of Fura 2 acid (data not known). Experimental results obtained in mouse polymorphonuclear cells (Scanlon *et al.*, 1987) also suggested that Fura 2/AM-loaded cells contained a Ca^{2+}-insensitive fluorescent component. We conclude that the incomplete deesterification of Fura 2/AM in FeSin cells produced a fluorescent Ca^{2+}-insensitive component that was more fluorescent at 380 nm excitation than was Fura 2 acid. This component thus results in a lower calculated cell ratio (340:380) and lower calculated Ca$_i$.

3.2. Calibration

In order to quantitate Ca$_i$, we determined a calibration curve for Fura 2 using the digital imaging system and the fluorescent microscope. We measured fluorescent ratios of Fura 2

FIGURE 1. Images of Fura 2-loaded cells (380 nm excitation). (A) FeSin cells loaded with hypoosmolar treatment and Fura 2 acid. (B) FeSin cells incubated with Fura 2/AM. Bar equals 10 μm.

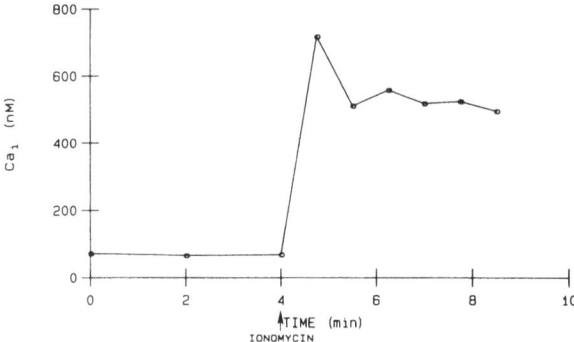

FIGURE 3. Ca_i changes during ionomycin treatment of Fura 2-loaded FeSin cells. Cells loaded by incubation with Fura 2/AM as described. Ionomycin (10 μM) was added, and average cellular Ca_i was followed as a function of time. Ca_i increased rapidly, then decreased to a Ca_i above original basal values and remained elevated for the duration of stimulus (>5 min) or until EGTA was added to chelate intracellular Ca^{2+}.

solutions ($R = 340:380$) visualized in the microscope and compared them with [Ca^{2+}] measured in the same Fura 2 solutions using an ion-selective electrode (Cornejo et al., 1987). A plot of Ca_i versus R closely approximates the line given by Eq. (1), in which $K_d = 220$, a value similar to that used in other studies (Williams et al., 1985; Grynkiewicz et al., 1985).

In vivo calibrations using a Ca^{2+} ionophore, ionomycin, to equilibrate extracellular Ca^{2+} and Ca_i has been successful in other cell types (Williams et al., 1985) but has not proved useful in the fibroblasts we used because of the complicated Ca_i transients induced by ionophores (Fig. 3). However, ionomycin was useful in increasing Ca_i to maximal levels (>2 μM) and in saturating intracellular Fura 2 in order to measure maximum cell ratios (340:380). There was good agreement between the maximum fluorescence ratio (340:380) in ionomycin-treated cells measured in the microscope and the fluorescence ratio (340:380) of permeabilized cell material determined in the spectrophotometer.

The calibration curve for the material from permeabilized FeSin cells indicates that a calculated basal Ca_i of 40 nM would actually be 65 nM, a finding in good agreement with experimental results, while a calculated 220 nM would actually be 1 μM. Thus, a combination of excitation spectra and calibration curves for Fura 2 acid and permeabilized cell material can be used to qualitatively estimate the errors of calculated Ca_i in different cells. In this report Ca_i values are expressed as fold changes (stimulated Ca_i/basal Ca_i) that are independent of K_d.

3.3. Ca_i Image Heterogeneity

The spatial heterogeneity of Ca_i distribution within individual unstimulated fibroblasts was determined by analysis of average Ca_i in different regions (100 pixel^2) of the same cell. About 70% of both HOSTed and Fura 2/AM-loaded FeSin cells had spatial gradients of Ca_i equal to 30 nM or a 1.6-fold change. For a total of 10 cells examined in this way, the range of Ca_i gradients ranged from 1.0 to 3.0, with a mean of 1.6. The fact that HOSTed and Fura 2/AM-loaded cells had similar Ca_i gradients indicates that Ca_i gradients are not an artifact caused solely by regional differences in the partially deesterified Fura 2/AM com-

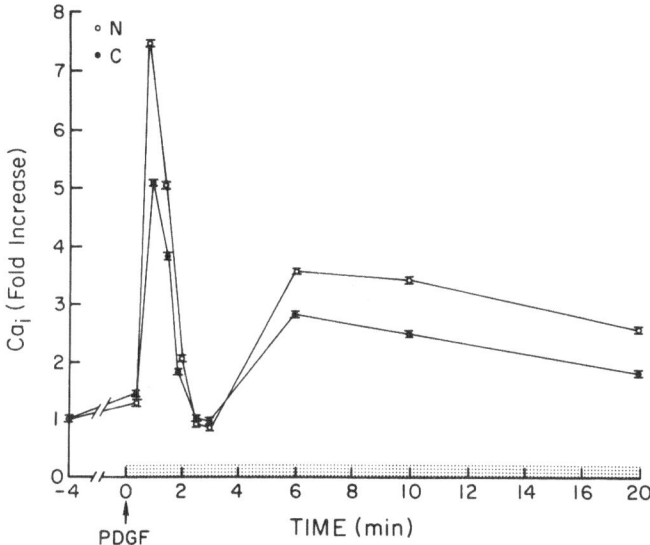

FIGURE 4. Effect of PDGF on average Ca$_i$ in Fura 2-loaded FeSin cells. Increase (*n*-fold) of Ca$_i$ was biphasic in both the nucleus and cytoplasm of this particular cell. Basal Ca$_i$ was particularly low (20 nM) in this cell because of low signal noise in the 380-nm image.

ponent. However, sequestration of Fura 2 in organelles with higher [Ca^{2+}] could contribute to Ca$_i$ gradients, as could regional differences in out-of-focus fluorescence from Fura 2. Indeed, digitonin treatment of cells sometimes reveals residual granular fluorescence even after >80% of cellular Fura 2 fluorescence has been released (data not shown).

3.4. Platelet-Derived Growth Factor Stimulation

When Fura 2-loaded fibroblasts were stimulated with platelet-derived growth factor (PDGF) (8 U/ml), Ca$_i$ increased (up to eightfold) in both cytoplasm and nucleus. Regions of the cell with high basal Ca$_i$ were also relatively high during PDGF stimulation. The time course of average Ca$_i$ increased in parallel in all areas of the cytoplasm and nucleus, and was biphasic in some cells (Fig. 4). An initial transient early rise was complete within 2 min and was then followed by a second gradual increase over the ensuing 18 min. Since the Ca$_i$ gradient was preserved during the Ca$_i$ increase, a partially deesterified component was unlikely to contribute to the Ca$_i$ gradient. Calculated Ca$_i$ would have increased more slowly in the region containing the deesterified compound, and this was not observed.

4. DISCUSSION

We have investigated the utility of measuring Ca$_i$ in individual Fura 2-loaded fibroblasts using digital image analysis. Two problems were encountered: (1) in some cells Ca$_i$ measurements were underestimates because of incomplete deesterification of Fura 2/AM, and

(2) interpretation of spatial Ca_i gradients was complicated by possible organellar entrapment of Fura 2 components.

Quantitation of Ca_i was a problem only in some cell lines that did not completely deesterify Fura 2/AM (FeSin; human fibroblasts). The resulting partially deesterified compound was Ca^{2+}-insensitive and contributed to an artificially lowered calculated Ca_i. Approximate corrections in the magnitude of the calculated Ca_i could be obtained from the cell material released from permeabilized Fura 2-loaded cells. In the case of BALB/c 3T3 cells, most of the intracellular fluorescent material had the excitation spectrum of Fura 2, and corrections in calculated Ca_i were only necessary above 350 nM. On the other hand, FeSin cells had substantially more partially deesterified component, and corrections to Ca_i were probably necessary even at basal Ca_i (40 nM). By necessity, the calibration curve of cell material represents an average over the entire cell population. Any individual cell may have different proportions of Ca^{2+}-sensitive and Ca^{2+}-insensitive components, so that a precise adjustment of Ca_i in each individual cell is currently not possible.

A spatial gradient of Ca_i (1.6- to 2.0-fold) is one aspect of the Ca_i distribution that is not affected by the loading procedure. Ca_i gradients have been found in other cells (Ratan *et al.*, 1986; Connor, 1986; Connor *et al.*, 1987), but such gradients need to be interpreted cautiously since there are multiple reasons for apparent spatial gradients of Ca_i in Fura 2/AM-loaded cells. Localized binding of Fura 2 or deesterified Fura 2/AM products and concentration of Fura 2 in organelles containing a high Ca^{2+} concentration may all contribute to apparent localized Ca_i changes.

Perhaps one of the most significant findings was the increase in Ca_i in both the cytoplasm and in the nucleus. Thus, Ca_i increases in either the cytoplasm or nucleus could contribute to stimulation of mitogenesis. Indeed, there are several Ca^{2+}-sensitive processes in the nucleus that could be directly stimulated by Ca^{2+}. Ca^{2+}-activated contraction of actin and myosin could contribute to nuclear shape alterations associated with functional changes (Simons *et al.*, 1986), and Ca^{2+} activation of calmodulin could influence phosphorylation of nuclear proteins. Moreover, calcium- or metal ion-sensitive regions of DNA have been described in the regulation of gene expression (Sequin *et al.*, 1987; Resendez *et al.*, 1985; White, 1985). Ca^{2+} may also activate nuclear proteases that cause DNA damage, thereby directly affecting genome stability and the potential expression of new genes (Cerutti, 1988).

The most exciting applications for digital image analysis of intracellular fluorescent probes lie ahead. One can now measure Ca_i and pH_i in the same cells in which effects of physiological stimuli can be defined. The temporal and spatial heterogeneity of these ionic changes can be explicitly considered in interpreting the role of intracellular messengers in various functions. Use of fluorescent probes for lipid components, mitochondria, and cytoskeleton will allow detailed correlations between structure and ionic changes in living cells, thereby helping us understand the generation and effects of ionic second messengers. Digital image analysis of intracellular fluorescent compounds will thus facilitate experimental approaches to many important research questions.

5. SUMMARY

The development of fluorescent calcium probes and digital image analysis has made it possible to measure free intracellular calcium (Ca_i) in individual living cells. Here we discuss some of the practical problems that occur in measuring Ca_i in human and mouse fibroblasts. Foremost among the problems is the incomplete deesterification of the membrane-permeant ester Fura 2/AM. A related problem concerns dye sequestration and out-of-focus fluorescence, which complicates the interpretation of spatial patterns of Ca_i in single cells. Despite these complications, digital image analysis of Fura 2-loaded cells can provide new and important information about the Ca_i changes in different regions of an individual cell. In this paper, our work with this approach is discussed in reference to fibroblasts stimulated with platelet-derived growth factor (PDGF).

ACKNOWLEDGMENT. This work was supported by NIH grant CA34472. The dedicated technical assistance of David Chang and the professional secretarial services of Donna Kinsella are greatly appreciated.

REFERENCES

Agard, D. A., and Sedat, J. W., 1983, Three-dimensional architecture of a polytene nucleus, *Nature* **302**:676–681.

Boynton, A. L., Whitfield, J. R., and MacManus, J. P., 1980, Calmodulin stimulates DNA synthesis in rat liver cells, *Biochem. Biophys. Res. Commun.* **95**:745–749.

Bright, G. R., Fisher, G. W., Rogowska, J., and Taylor, D. L., 1987, Fluorescence ratio imaging microscopy: temporal and spatial measurement of cytoplasmic pH, *J. Cell Biol.* **104**:1019–1033.

Cerutti, P., 1988, Oxidant tumor promotors, *Toxicol. Pathol.* **16**:73–74.

Chafouleas, J. G., Bolton, W. E., Hidaka, H., Boyd, A. E., and Means, A. R., 1982, Calmodulin and the cell cycle: Involvement in regulation of cell-cycle progression, *Cell* **28**:41–50.

Cheung, W. Y., 1980, Calmodulin—An introduction, in: *Calcium and Cell Function*, Vol. 1 (W. Y. Cheung, ed.), Academic Press, New York, pp. 1–9.

Connor, J. A., 1986, Digital imaging of free calcium changes and of spatial gradients in growing processes in single, mammalian central nervous system cells, *Proc. Natl. Acad. Sci. USA* **83**: 6179–6183.

Connor, J. A., Cornwall, M. C., and Williams, G. H., 1987, Spatially resolved cytosolic calcium response to angiotensin II and potassium in rat glomerulosa cells measured by digital imaging techniques, *J. Biol. Chem.* **262**:2919–2927.

Cornejo, M., Guggino, S., and Guggino, W. B., 1987, Modification of calcium-activated potassium channel in cultured medullary thick ascending limb cells by N-bromoacetamide, *J. Memb. Biol.* **99**:147–155.

Fay, F. S., and Tucker, R. W., 1985, Studies of free cytosolic calcium (Ca_i) in quiescent BALB/c 3T3 cells using Quin2 and Fura 2, *J. Cell Biol.* **101**:476a.

Gelfand, E. W., Cheung, R. K., Grinstein, S., and Mills, G. B., 1986, Characterization of the role for calcium influx in mitogen-induced triggering of human T cells. Identification of calcium-dependent and calcium-independent signals, *Eur. J. Immunol.* **16**:907–912.

Grynkiewicz, G., Poenie, M., and Tsien, R. Y., 1985, A new generation of Ca^{2+} indicators with greatly improved fluorescence properties, *J. Biol. Chem.* **260**:3440–3450.

Hesketh, T. R., Smith, G. A., Moore, J. P., Taylor, M. V., and Metcalfe, J. C., 1983, Free cytoplasmic calcium concentration and the mitogenic stimulation of lymphocytes, *J. Biol. Chem.* **258**:4876–4882.

Kretsinger, R. H., 1980, Structure and evolution of calcium-modulated proteins, *CRC Crit. Rev. Biochem.* **8**:114–174.

Lemasters, J. J., DiGuiseppi, J., Nieminen, A.-L., and Herman, B., 1987, Blebbing, free Ca^{2+} and mitochondrial membrane potential preceding cell death in hepatocytes, *Nature* **325**:78–81.

McNeil, P. L., McKenna, M. P., and Taylor, D. L., 1985, A transient rise in cytosolic calcium follows stimulation of quiescent cells with growth factors and is inhibitable with phorbol myristate acetate, *J. Cell. Biol.* **101**:372–379.

Millard, P., Gross, D., Webb, W., and Fewtrell, C., 1988, Imaging asynchronous changes in intracellular Ca^{2+} in individual stimulated tumor mast cells, *Proc. Natl. Acad. Sci.* **85**:1854–1858.

Morris, J. D. H., Metcalfe, J. C., Smith, G. A., Hesketh, T. R., and Taylor, M. V., 1984, Some mitogens cause rapid increases in free calcium in fibroblasts, *FEBS Lett.* **169**:189–193.

Nemeth, E. F. and Scarpa, A., 1986, Cytosolic Ca^{2+} and the regulation of secretion in parathyroid cells, *FEBS Lett.* **203**:15–19.

Pershadsingh, H. A., Shade, D. L., Delfert, D. M., and McDonald, J. M., 1987, Chelation of intracellular calcium blocks insulin action in the adipocyte, *Proc. Natl. Acad. Sci. U.S.A.* **84**:1025–1029.

Poenie, N., Alderton, J., Steinhardt, R., and Tsien, R. Y., 1986, Calcium rises abruptly and briefly throughout the cell at the onset of amaphase, *Science* **233**:886–889.

Pozzan, T. Arslan, P., Tsien, R. Y., and Rink, T. J., 1982, Anti-immunoglobulin, cytoplasmic free calcium, and capping in B lymphocytes, *J. Cell Biol.* **94**:335–340.

Rabinovitch, P. S., June, C. H., Grossman, A., and Ledbetter, J. A., 1986, Heterogeneity among T cells in intracellular free calcium responses after mitogen stimulation with PHA or Anti-CD3. Simultaneous use of Indo-1 and immunofluorescence with flow cytometry. *J. Immol.* **137**:952–961.

Ratan, R. R., Shelanski, M. L., and Maxfield, F. R., 1986, Transition from metaphase to anaphase is accompanied by local changes in cytoplasmic free calcium in Pt K2 kidney epithelial cells, *Proc. Natl. Acad. Sci. USA* **83**:5136–5140.

Resendez, E. Jr., Attenello, J. W., Grafsky, A., Chang, C. S.,and Lee, A. S., 1985, Calcium ionophore A23187 induces expression of glucose-regulated genes and their heterologous fusion genes, *Molec. Cell. Biol.* **5**:1212–1219.

Sequin, C. and Hamer, D. H., 1987, Regulation in vitro of metallothionein gene binding factors, *Science* **235**:1383–1387.

Scanlon, M., Williams, D. A., and Fay, F. S., 1987, A Ca^{2+} insensitive form of Fura-2 associated with polymorphonuclear leukocytes. Assessment and accurate Ca^{2+} measurement, *J. Biol. Chem.* **262**:6308–6312.

Simons, J. W., Noga, S. J., Colombani, P. M., Beschorner, W. E., Coffey, D. S., and Hess, A. D., 1986, Cyclosporine A, an *in vitro* calmodulin antagonist, induces nuclear lobulations in human T cell lymphocytes and monocytes, *J. Cell Biol.* **102**:145–150.

Tucker, R. W., and Loats, H., 1987, Analysis of intracellular free cytosolic calcium in human fibroblasts using Fura-2 and digital image analysis with a microcomputer, *Ann. N.Y. Acad. Sci.* **494**:283–286.

Tucker, R. W., Snowdowne, K. W., and Borle, A. B., 1986, Cytosolic free calcium and DNA synthesis in BALB/c 3T3 cells: aequorin luminescence studies, *Eur. J. Cell Biol.* **41**:347–351.

Tsien, R. Y., Rink, T. J., and Poenie, M., 1985, Measurement of cytosolic free Ca^{2+} in individual small cells using fluorescent microscopy with dual excitation wavelengths, *Cell Calcium* **6**:145–157.

Wier, W. G., Cannell, M. B., Berlin, J. R., Marban, E., and Leaderer, W. J., 1987, Cellular and subcellular heterogeneity of $[Ca^{2+}]_i$ in single heart cells revealed by Fura-2, *Science* **235**:325–328.

White, B. A., 1985, Evidence for a role of calmodulin with regulation of prolactin gene expression, *J. Biol. Chem.* **260**:1213–1217.

Williams, D. A., Fogarty, K. E., Tsien, R. Y., and Fay, F. S., 1985, Calcium gradients in single smooth muscle cells revealed by the digital imaging microscope using Fura-2, *Nature* **318**:558–561.

27

Growth Factors, Oncogenes, and Protein Kinase C

IAN G. MACARA, GEORGE GRAY, JAMES GAUT,
ANNA COCO, THERESA WINGROVE, DONNA FALETTO,
and ALAN WOLFMAN

1. PHOSPHATIDYLINOSITOL METABOLISM IN TRANSFORMED CELLS

1.1. Diacylglycerol Levels

Subconfluent cultures of 3T3 fibroblasts were labeled to isotopic equilibrium with [³H]glycerol and analyzed by thin-layer chromatography after extraction of lipids to determine the levels of [³H]diacylglycerol (DAG) in comparison to the total amount of ³H incorporated into cellular lipid. The transformed cells used in this study were always compared to their true parental clone and were of recent origin to decrease the possibility that observed changes are merely reflecting long-term cell divergence.

In every case—using cells transformed by Ki-*ras*, Ha-*ras*, v-*src*, and v-*fms*—the level of DAG was found to be significantly elevated in the transformed cells as compared to the controls, under identical conditions of exponential growth (Table I). Moreover, when the levels of [³H]DAG were examined in a 3T3 cell line transfected with a temperature-sensitive v-*src* oncogene, a significant increase in DAG was detected within 30 min of switching the cells from the restrictive (40°C) to the permissive temperature (36°C) (Fig. 1). No significant increase was detected in [³H]DAG using cells labeled with [³H]myristate instead of [³H]glycerol (not shown).

By contrast, the induction of terminal differentiation of transformed erythroid cells (Friend murine erythroleukemia cells) is accompanied by a rapid fall in DAG. Addition of exogenous synthetic DAG, or phorbol ester, blocks differentiation, implicating a high DAG level in the maintenance of the transformed phenotype in these cells (Faletto *et al.*, 1985).

IAN G. MACARA, GEORGE GRAY, JAMES GAUT, ANNA COCO, THERESA WINGROVE, DONNA FALETTO, and ALAN WOLFMAN ● Environmental Health Sciences Center, Department of Biophysics, University of Rochester Medical Center, Rochester, New York 14642.

TABLE I. Changes in Diacylglycerol and Protein Kinase C Activity in
Oncogene-Transformed 3T3 Fibroblasts

Cell line	Percent DAG	Percent 80-kDa-^{32}P		Amount of 80-kDa protein	[^{3}H]PDBu binding	^{86}Rb$^+$ Uptake +PMA (% stimulation)
		− PMA	+ PMA			
NIH 3T3	100	<2	100	100	100	100
Hal 3T3	125	10	39	40	75	26
KSV 3T3	210	53	90	ND	ND	38
Src 3T3	130	17	77	64	26	17
Fms 3T3	195	<2	7	57	51	0

Values are means of several independent determinations of diacylglycerol content (DAG), basal 80-kDa phosphoryl-ation, and phorbol ester-stimulated 80-kDa phosphorylation (±PMA), cellular level of 80-kDa protein by immuno-blotting, kinase C quantitation by [^{3}H]phorbol ester binding ([^{3}H]PDBu), and phorbol ester stimulation of Na$^+$/H$^+$ exchange as assayed by ouabain-sensitive ^{86}Rb$^+$ influx. Values are expressed as percentages of the control cell values. Data are taken from Wolfman and Macara (1987) and Wolfman *et al.* (1987).

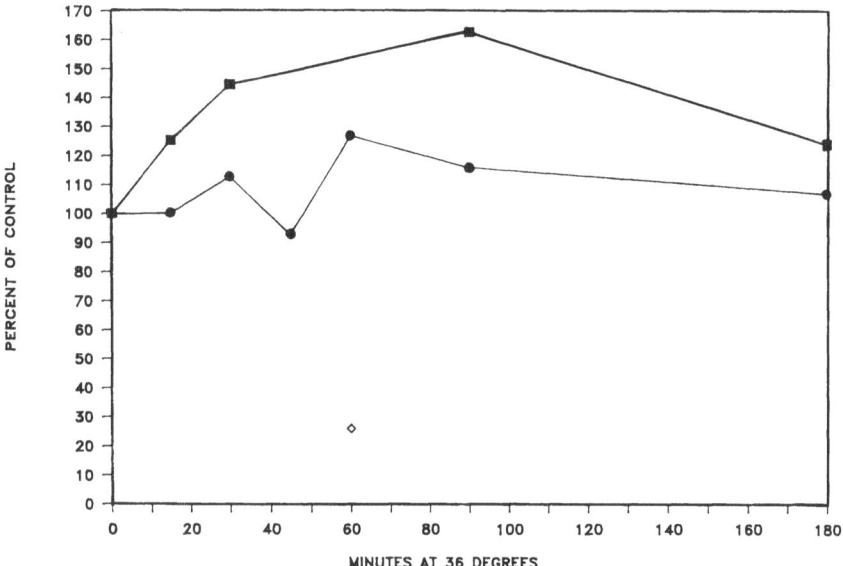

FIGURE 1. [^{3}H]Glycerol-labeled diacylglycerol (■) and [^{3}H]inositol phosphate levels (●, ◇) in NIH 3T3 cells transfected with a v-*src* oncogene encoding a temperature-sensitive p60^{v-src} gene product. Cells were grown and labeled to isotopic equilibrium at the restrictive temperature (40°C), then switched to the permissive temperature (36°C) at the start of the experiment. The [^{3}H]inositol experiment was performed in the presence (●) or absence (◇) of 10 mM Li$^+$ to block inositol monophosphate hydrolysis. [^{3}H]Diacylglycerol and [^{3}H]inositol phosphates were assayed as described elsewhere (Faletto *et al.*, 1985; Wolfman and Macara, 1987).

1.2. Inositol Phosphates

To determine whether the observed changes in DAG originated from increased phosphatidylinositol (PI) breakdown, we measured steady-state levels of [^{3}H]inositol-labeled inositol phosphates in *ras*-transformed 3T3 fibroblasts and in the temperature-sensitive v-*src* transformed 3T3 cells. The *ras*-transformed cells displayed a small increase in inositol bisphosphate (IP$_2$) but no significant differences in the levels of the other inositol phosphates, nor of the phosphoinositides (Wolfman and Macara, 1987; Wolfman *et al.*, 1987). Nor was any increase in total, unseparated [^{3}H]inositol phosphates detectable within 2 hr of switching the temperature-sensitive v-*src* cells to 36°C, even in the presence of 10 mM LiCl to block dephosphorylation of the inositol monophosphate (Fig. 1). Finally, although a small (10–20%) decrease in IP$_3$ accompanies the induction of Friend erythroleukemia cell differentiation, it occurs more gradually than does the larger decrease in DAG, perhaps implying that the two events are not closely coupled. These observations suggest that the elevated DAG in transformed cells does not arise *in toto* from increased phosphatidylinositol turnover. A suggestion that the DAG is generated by elevated phosphatidylcholine hydrolysis in *ras*-transformed cells (Lacal *et al.*, 1987) has not been substantiated (Macara, 1989). Other sources have not yet been identified, but an attractive possibility is *de novo* synthesis from glycerol-3-phosphate, a pathway that is substantially stimulated in insulin-treated BC3H-1 cells (Farese *et al.*, 1987).

2. CHANGES IN PROTEIN KINASE C ACTIVITY

2.1. 80-kDa Phosphorylation

To determine if the observed changes in DAG correlated with changes in the activity of protein kinase C (PKC), we examined the phosphorylation of an 80-kDa protein substrate of the enzyme in ^{32}P-labeled cells (Rozengurt *et al.*, 1983). Cell extracts were analyzed by sodium dodecyl sulfate (SDS)-gel electrophoresis followed by autoradiography and densitometry. As shown in Table I, the parental lines had almost undetectable basal levels of 80-kDa phosphorylation, but a dramatic increase in phosphorylation occurred in response to phorbol ester addition. By contrast, the *ras*- and *src*-transformed cells exhibited a high basal level of phosphorylation of an 80-kDa band, the peptide map of which was identical to that of the band phosphorylated in response to phorbol ester (Wolfman and Macara, 1987). These results suggest that PKC is partially activated in these transformed cells. Surprisingly, however, the phosphorylation response of all the transformed lines to phorbol ester addition was profoundly attenuated, as compared to the parental lines (Table I). This attenuation could be mimicked by prolonged pretreatment of the parental cells with phorbol ester, suggesting that it reflects a partial downregulation of PKC activity.

The elevation of basal 80-kDa phosphorylation was not caused by the secretion of autocrine factors, since incubation of ^{32}P-labeled parental cells with conditioned medium from Ha-*ras*-transformed cells had no detectable effect on 80-kDa phosphorylation.

2.2. Stimulation of ^{86}Rb$^+$ Influx

To determine whether the downregulation of PKC was of functional significance in the transformed cells, we examined the response of ouabain-sensitive ^{86}Rb$^+$ uptake to phorbol esters. A large increase in uptake has been demonstrated in many cell types in response to serum, growth factors, and phorbol esters (Rozengurt, 1981), reflecting a stimulation of the Na,K-ATPase. The stimulation is probably a consequence of increased Na^+-H^+ exchange, which elevates cytosolic Na^+ (Moolenar et al., 1982). This interpretation is supported by our observation that the stimulation of ouabain-sensitive $^{86}Rb^+$ uptake by phorbol esters in NIH 3T3 cells is completely abolished by 100 μM dimethylamiloride, a potent and specific inhibitor of Na^+-H^+ exchange (Wolfman et al., 1987). Interestingly, while phorbol ester induced about a twofold increase in $^{86}Rb^+$ influx into the control cells, no stimulation was detected in any of the oncogene-transformed cell lines (Table I; Wolfman et al., 1987).

2.3. Downregulation of PKC and Its 80-kDa Substrate

The attenuation of PKC functions in the transformed cells indicates that the enzyme has been partially downregulated. To test this possibility, we quantitated the binding of the phorbol ester, phorbol-12,13-dibutyrate ([^{3}H]PDBu), to the various cell lines and found that in each case the transformed lines possessed lower PDBu-binding capacity than the corresponding parental lines (Table I).

We also quantitated the level of the 80-kDa protein substrate using a polyclonal antibody provided by Perry Blackshear (Blackshear et al., 1986) to perform immunoblots on gels of cellular extracts after transfer to nitrocellulose paper. Again, in every transformed line examined, the cellular level of the 80-kDa protein was significantly lower than in the corresponding parental lines (Table I; Wolfman et al., 1987). This decrease could be mimicked by prolonged pretreatment of the parental cells with phorbol ester, suggesting that the activity of PKC controls the level of 80-kDa expression.

3. DISCUSSION

These results indicate that the elevation of DAG, and consequent changes in PKC activity, are common features of transformation by a number of different oncogenes. The changes in activity are of functional significance, as exemplified by the loss of responsiveness of $^{86}Rb^+$ uptake and of [^{3}H]thymidine incorporation to stimulation by phorbol esters, and the downregulation of the 80-kDa substrate of PKC. Why does this attenuation occur? One exciting possibility is that it represents an unsuccessful attempt by the transformed cell to block the constitutive proliferative signals generated by the oncogene products. This hypothesis is presented in Fig. 2A. It predicts that in revertant cells the attenuation is large enough to be successful (Fig. 2B), and evidence in support of this mechanism was obtained, although in another context, by Kamata et al. (1987), who found a complete loss of 80 kDa phosphorylation in a revertant of a Kirsten sarcoma virus infected 3T3 cell line, and a larger downregulation of PKC in the revertant than in the transformed line. We have recently determined that the molecular basis for the loss of responsiveness in the revertant cell-line lies in a combination of decreased abundance of the 80-kDa protein and the constitutive basal phosphorylation of the protein (Wojtaszek and Macara, unpublished observations). The important question is whether these alterations are causally related to the revertant phenotype. It will also be important to determine which of the many parameters of the transformed phenotype can be ascribed to changes in PKC activity, and which are independent of this system, and to determine by what mechanism the level of diacylglycerol is regulated in oncogene-transformed cells. Finally, using antibodies specific to the three known isoforms

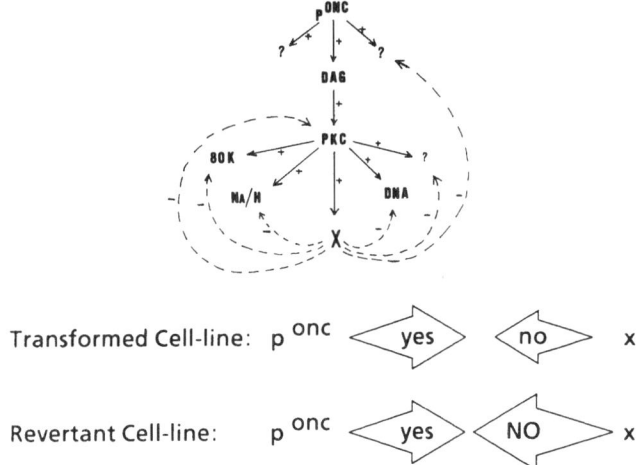

Transformed Cell-line: p ^{onc} ⟨ yes ⟩ ⟨ no ⟩ x

Revertant Cell-line: p ^{onc} ⟨ yes ⟩ ⟨ NO ⟩ x

FIGURE 2. (A) Hypothetical scheme showing negative feedback modulation of proliferative signals generated by an oncogene protein. The model suggests that this modulator (X) is activated or generated by kinase C, although other mechanisms are conceivable. (B) Predicted mechanism of oncogenic reversion. In the revertant cells the negative modulation described in A is amplified to a point where it can successfully overcome the oncogene-generated proliferative signals.

of PKC, we hope to determine whether specific isoforms of this enzyme are responsive to oncogenic transformation.

4. SUMMARY

One mechanism by which oncogene proteins might transform cells is a deregulation of the phosphatidylinositol cycle or of other metabolic pathways that generate diacylglycerol (DAG), which is the endogenous activator of protein kinase C (PKC). This enzyme is a high-affinity receptor for the tumor-promoting phorbol esters, which can mimic many of the phenotypic changes of oncogenic transformation in cultured cells. The unregulated generation of DAG might be expected to constitutively activate PKC and produce responses similar to those of the tumor promoters (Macara, 1985). In support of this hypothesis, we have shown that NIH 3T3 cells transformed by Ki-*ras*, Ha-*ras*, v-*src*, and v-*fms* all possess elevated levels of DAG as compared to control cells, even under conditions of exponential growth. In contrast, there is a decrease in the DAG level of erythroleukemia cells induced to differentiate, and exogenous DAG or phorbol esters block differentiation, implicating PKC in the maintenance of the transformed phenotype (Faletto *et al.*, 1985). In addition to the elevation in DAG, the oncogene-transformed 3T3 cells also exhibit changes in PKC activity. The basal level of phosphorylation of an 80-kDa endogenous substrate of PKC is significantly elevated in the Ki-*ras*, Ha-*ras*, and v-*src* cells. Surprisingly, however, all of the oncogene-transformed cells exhibited a profound attenuation of 80-kDa phosphorylation in response to added phorbol esters. This effect is in part a consequence of a downregulation both of

PKC and of its 80-kDa substrate. Transformation by at least two classes of oncogene is therefore associated with increased DAG levels, which appear to partially activate and partially downregulate protein kinase C (Wolfman and Macara, 1987; Wolfman et al., 1987). The downregulation may reflect an attempt by the transformed cells to suppress the constitutive proliferative signals generated by the oncogene products, and in revertant cell lines this attempt is successful.

ACKNOWLEDGMENTS. The work described above was supported by PHS grant CA-38888 from the National Cancer Institute, DHHS (to I.G.M), NIEHS Center grant ES 01247, research fellowship CA-07791 (to A.W.), and postdoctoral training fellowship DK-07092 (to T.G.W.).

REFERENCES

Blackshear, P. J., Wen, L., Glynn, B. P., and Witters, L. A., 1986. Protein kinase C-stimulated phosphorylation in vitro of an M_r 80,000 protein phosphorylated in response to phorbol esters and growth factors in intact fibroblasts, *J. Biol. Chem.* **261**:1459–1469.

Faletto, D. L., Arrow, A. S., and Macara, I. G., 1985, An early decrease in phosphatidylinositol turnover occurs on induction of Friend cell differentiation and precedes the decrease in c-myc expression, *Cell* **43**:315–325.

Farese, R. V., Konda, T. S., Davis, J. S., Standaert, M. L., Pollet, R. J., and Cooper, D. R., 1987, Insulin rapidly increases diacylglycerol by activating de novo phosphatidic acid synthesis, *Science* **236**:586–589.

Kamata, T., Sullivan, N. F., and Wooten, M. W., 1987, Reduced protein kinase C activity in a *ras*-resistant cell line derived from KiMSV transformed cells, *Oncogene* **1**:37–46.

Lacal, J., Moscat, J., and Aaronson, S. A., 1987, Novel Source of 1,2-diacylglycerol elevated in cells, transformed by the Ha-*ras* oncogene, *Nature* **330**:269–271.

Macara, I. G., 1985, Oncogenes, ions and phospholipids, *Am. J. Physiol.* **248**:C3–C11.

Macara, I. G., 1989, Elevated phosphocholine in *ras*-transformed 3T3 cells from increased choline kinase activity, not from phosphatidylcholine breakdown, *Mol. Cell. Biol.* (in press).

Moolenar, W. H., Yarden, Y., deLaat, S. W., and Schlessinger, J., 1982, Epidermal growth factor induces electrically silent Na^+ influx in human fibroblasts, *J. Biol. Chem.* **257**:8502–8506.

Rozengurt, E., 1981, Stimulation of Na^+ influx, Na-K pump activity and DNA synthesis in quiescent cultured cells, *Adv. Enz. Regul.* **19**:61–85.

Rozengurt, E., Rodriquez-Pena, M., and Smith, K. A., 1983, Phorbol esters, phospholipase C and growth factors rapidly stimulate the phosphorylation of a M_r 80,000 protein in intact quiescent 3T3 cells, *Proc. Natl. Acad. Sci. USA* **80**:7244–7248.

Wolfman, A., and Macara, I. G., 1987, Elevated levels of diacylglycerol and decreased phorbol ester sensitivity in *ras*-transformed fibroblasts, *Nature* **325**:359–361.

Wolfman, A., Wingrove, T., Blackshear, P. J., and Macara, I. G., 1987, Down-regulation of protein kinase C and of an endogenous 80-kDa substrate in transformed fibroblasts, *J. Biol. Chem.* **262**: 16546–16552.

28

Insulin-like Growth Factor II Stimulates Calcium Influx in Competent Balb/c 3T3 Cells Primed with Epidermal Growth Factor

IKUO NISHIMOTO, ETSURO OGATA, and ITARU KOJIMA

1. INTRODUCTION

Insulin-like growth factors (IGFs) are potent mitogens in mammalian cells (Zapf *et al.*, 1981). The mode of action of IGFs has been extensively studied in Balb/c 3T3 cells. Thus, IGFs promote cell cycle progression specifically in competent Balb/c 3T3 cells while IGFs have essentially no effect in Go-arrested cells (Pledger *et al.*, 1977; Stiles *et al.*, 1979). To promote cell cycle progression, IGFs should be exposed to competent cells continuously throughout the G_1 phase (Pledger *et al.*, 1977). Hence, it is presumed that IGFs may generate a continuous mitogenic signal. This observation together with the fact that extracellular calcium is required for cell cycle progression have led to the consideration that calcium influx may be a mitogenic signal of IGFs.

We recently showed that IGF-II increases cytoplasmic free-calcium concentration, $[Ca^{2+}]_c$, in competent Balb/c 3T3 cells pretreated with epidermal growth factor (Nishimoto *et al.*, 1987). The action of IGF-II on $[Ca^{2+}]_c$ is dependent on the cell cycle and is presumably due to IGF-II-induced calcium influx. In the present study, we measured the unidirectional calcium influx rate and characterized IGF-II effects on cell calcium. We also examined whether IGF-II stimulates DNA synthesis by increasing calcium influx.

IKUO NISHIMOTO, ETSURO OGATA, and ITARU KOJIMA ● Cell Biology Research Unit, Fourth Department of Internal Medicine, University of Tokyo School of Medicine, 3-28-6 Mejirodai, Bunkyo-ku, Tokyo 112, Japan.

2. MATERIALS AND METHODS

2.1. Cell Culture

Balb/c 3T3 cells (clone A31) were grown in Dulbecco's modified Eagle's solution containing 25 mM Hepes/NaOH (pH 7.4) and 2.5 mM NaHCO$_3$ (DME) supplemented with 10% calf serum (CS), 100 U/ml penicillin G, and 100 μg/ml streptomycin in a humidified atmosphere of 95% O$_2$ and 5% CO$_2$ at 37°C. Quiescent cells were obtained as described by Tominaga and Lengyel (1985). In short, cells were detached and seeded at a 1:5 split ratio and cultured in DME containing 10% CS for 5 days without renewing the medium. Density arrested monolayers were further incubated for 24 hr in DME containing 5% platelet-poor plasma (PPP) to confirm quiescence. All experiments were performed within 4 weeks after thawing each lot of cells frozen in liquid nitrogen.

2.2. Measurement of Calcium Influx Rate

Calcium influx rate was determined by measuring an initial uptake of ^{45}Ca (Mauger et al., 1984; Kojima et al., 1985). Cells were grown in a 24-well plate. Calcium uptake was initiated by adding 1 ml of modified Hanks solution containing 5 μCi ^{45}CaCl$_2$ in the presence or absence of IGF-II. Cells were incubated for 15, 45, 75, or 105 sec and the reaction was terminated by aspirating the medium. Cells were immediately washed three times with 1.5 ml of washing solution containing 144 mM NaCl, 5 mM CaCl$_2$, and 5 mM Tris/HCl (pH 7.4). After washing, cells were lysed by adding 1 M NaOH. The radioactivity and protein content were measured. Calcium uptake was linear within these time points indicating that redistribution of calcium was negligible (Kojima et al., 1985). The calcium influx rate was calculated by using a slope of the linear regression line of calcium uptake. The influx rate obtained as indicated above is designated as calcium influx rate at time 0. In some experiments, cells were incubated in modified Hanks solution containing IGF-II for indicated times and calcium uptake was then initiated by changing the medium to modified Hanks solution containing ^{45}Ca and IGF-II. When the extracellular calcium concentration was reduced to 10 μM, calcium-EGTA buffer was employed (Waisman et al., 1981). When lanthanum was included, phosphate was removed from the Hanks solution to avoid precipitation. Protein was measured by the method of Bradford (1976) using bovine γ-globulin as standard.

2.3. Measurement of DNA Synthesis

DNA synthesis was assessed by measuring [^{3}H]thymidine incorporation into TCA-precipitable materials. Quiescent cells were obtained in a 24-well plate as described above. Cells were incubated with DME containing 20 U/ml platelet derived-growth factor (PDGF) for 3 hr. After washing with DME, cells were incubated with 10 nM EGF for 20 min. Cells were then washed once with DME containing 28 mM β-mercaptoethanol and twice with DME. These primed competent cells were then cultured for 24 hr in DME containing 1 μCi/ml [^{3}H]thymidine in the presence or absence of IGF-II. [^{3}H]Thymidine incorporation into TCA-precipitable materials was measured by the method of McNiel et al. (1985). When BAY K 8644 was employed, all procedures were done under a dimmed light.

2.4. Measurement of Binding of [^{125}I]IGF-II

IGF-II was iodinated by using the chloramine T method to a specific activity of 800 μCi/mol (Rechler *et al.*, 1977).

Binding assays were done as described by Lee *et al.* (1986). Monolayer cells were detached with 1 mM EDTA in saline. Cells were pelleted by centrifuging at 100 *g* for 10 min, washed, resuspended in cold buffer containing 50 mM Tris/HCl (pH 7.4), 1 mM EDTA, 0.25 M sucrose, 4 mM iodoacetic acid, homogenized with Dounce homogenizer, and centrifuged at 30,000 *g* for 40 min. The resultant membrane pellet was resuspended in binding buffer containing 100 mM Hepes/NaOH (pH 7.5), 120 mM NaCl, 5 mM KCl, 1.2 mM MgSO$_4$, 15 mM CH$_3$COONa, 1 mM EDTA, 10 mM dextrose, and 0.5% bovine serum albumin and stored frozen at -20°C. Aliquots containing 20 μg membrane protein were incubated with 10 pM iodinated IGF-II, in the presence and absence of unlabeled IGF-II at 4°C for 16 hr. The membranes were washed in several volumes of cold buffer and bound radioactivity was measured by a γ counter. Nonspecific binding was determined in the presence of 1 μM IGF-II. When the effect of calcium was examined, varying concentrations of CaCl$_2$ were added instead of EDTA.

2.5. Materials

Partially purified PDGF was prepared from outdated human platelet-rich plasma by employing CM-Sephadex chromatography and Sephadex Blue-Sepharose chromatography (Hasegawa-Sasaki, 1985). Rat IGF-II (multiplication-stimulating activity) was prepared from conditioned medium of BRL cells (Moses *et al.*, 1980). Synthetic IGF-I was supplied by Fujisawa Pharmaceuticals (Osaka). Platelet-poor plasma was prepared from human PPP as described by Pledger *et al.* (1978). EGF and anti-EGF antibodies were obtained from Collaborative Research (Lexington, MA). [^{3}H]Thymidine, Na^{125}I, and ^{45}CaCl$_2$ were obtained from New England Nuclear (Boston, MA).

3. RESULTS

3.1. Effect of IGF-II on Calcium Influx in Primed Competent Balb/c 3T3 Cells

In the present study, we directly measured the action of IGF-II on calcium influx. Resting calcium influx in quiescent cells was 0.52 $\pm$ 0.15 nmol/min/mg protein (mean $\pm$ SE, $n = 8$). In primed competent cells, resting calcium influx was 0.58 $\pm$ 0.10 nmol/min/mg protein ($n = 12$), which was identical to that observed in quiescent cells. IGF-II did not stimulate calcium influx in either quiescent cells or competent cells not primed with EGF (data not shown). In contrast, an addition of 1 nM IGF-II resulted in a rapid approximately two-fold increase in calcium influx in primed competent cells (Fig. 1). The action of IGF-II continued for at least 3 hr. After removal of IGF-II, calcium influx rapidly returned to the resting value. Thus, IGF-II caused sustained stimulation of calcium influx specifically in primed competent cells.

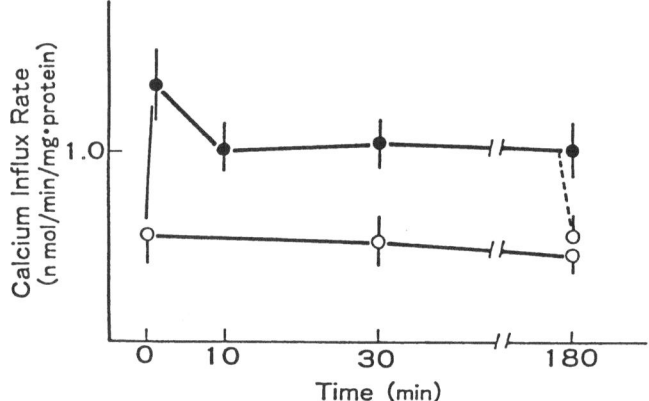

FIGURE 1. Time course of IGF-II-induced calcium influx in primed competent cells. Primed competent cells were incubated for indicated times with (●) or without (○) 1 nM IGF-II. In some experiments (---○) IGF-II was removed at 179 min. Calcium influx was measured as described in Section 2. Values are the mean ± SE for four experiments.

3.2. Effect of PDGF and EGF Treatment in Subsequent IGF-II Action in Competent Balb/c 3T3 Cells

In our previous study [4], we showed that priming with EGF is indispensable for IGF-II to increase $[Ca^{2+}]_c$ in competent cells. In addition, a few minutes is required for EGF to exert its priming action. In the present study, we examined the time course of priming action of EGF. Platelet-derived growth factor-treated competent cells were treated with 10 nM EGF for various periods and then IGF-II was added. Without EGF treatment, IGF-II did not affect calcium influx. When competent cells were treated with EGF for 3 min, IGF-II did not stimulate calcium influx. In contrast, IGF-II increased calcium influx when cells were treated with EGF for 5 min or more. When competent cells treated with EGF for 10 min were washed with DME three times, IGF-II increased calcium influx even in the presence of anti-EGF antibody. Thus, once primed, competence was acquired and no EGF was required. Furthermore, when primed competent cells were incubated in DME alone for 2 hr, subsequent addition of IGF-II increased calcium influx even after a 2-hr interval.

3.3. Effect of Changes in Extracellular Calcium in IGF-II-Mediated Calcium Influx

Figure 2 demonstrates the dependency of IGF-II-mediated calcium influx on extracellular calcium in primed competent cells. At 10 μM extracellular calcium, basal calcium influx was 0.10 ± 0.06 nmol/min/mg protein and IGF-II did not increase calcium influx in this condition. When extracellular calcium was 100 μM, IGF-II had a small stimulatory effect on calcium influx; however, the rate was no greater than that of resting influx at 1 mM extracellular calcium. In the presence of IGF-II, calcium influx increased as a function of the extracellular calcium concentration.

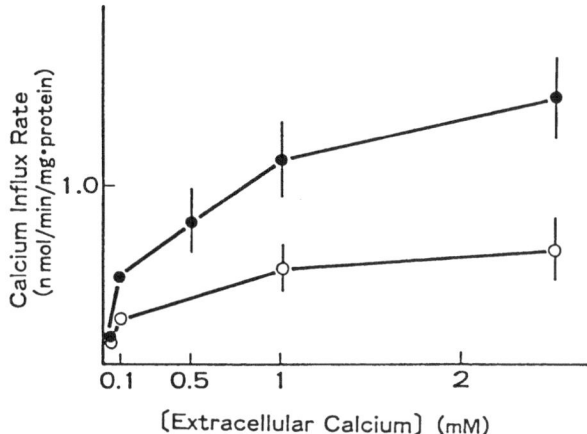

FIGURE 2. Dependency of IGF-II-induced calcium influx on extracellular calcium. Primed competent cells were incubated with (●) or without (○) 1 nM IGF-II in the presence of various concentrations of extracellular calcium. Values are the mean ± SE for four experiments.

3.4. Effect of Lanthanum and Cobalt on IGF-II-Mediated Calcium Influx in Primed Competent Balb/c 3T3 Cells

We examined the effects of lanthanum and cobalt on IGF-II-mediated calcium influx in primed competent cells. As shown in Fig. 3A, the action of IGF-II on calcium influx was completely blocked by 10 μM lanthanum. The ID_{50} for lanthanum-induced inhibition was 3 μM. Likewise, IGF-II-mediated calcium influx was inhibited by cobalt in a concentration-dependent manner (Fig. 3B). The ID_{50} for cobalt-induced inhibition was approximately 220 μM.

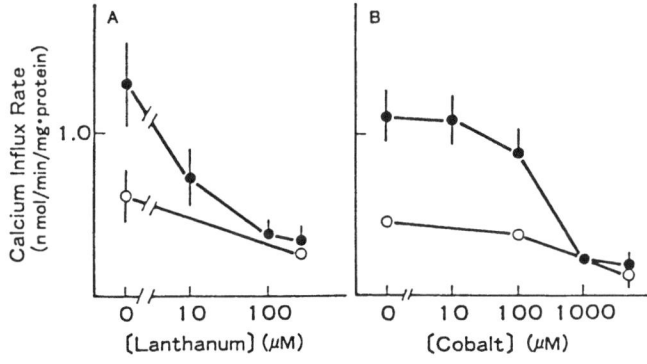

FIGURE 3. Effect of lanthanum and cobalt on IGF-II-induced calcium influx. Primed competent cells were incubated with (●) or without (○) 1 nM IGF-II in the presence of varying concentrations of lanthanum (A) or cobalt (B). Values are the mean ± SE for three experiments.

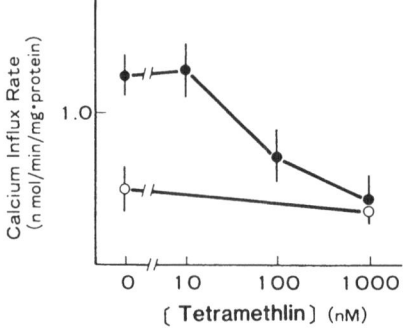

FIGURE 4. Effect of tetramethrin on IGF-II-induced calcium influx. Primed competent cells were incubated with (●) or without (○) 1 nM IGF-II in the presence of varying concentrations of tetramethlin. Values are the mean ± SE for four experiments.

3.5. Effect of Nitrendipine and Tetramethrin on IGF-II-Mediated Calcium Influx in Primed Competent Cells

Nitrendipine, a dihydropyridine calcium channel blocker, did not inhibit IGF-II-mediated calcium influx in primed competent cells (data not shown). In contrast, IGF-II-mediated calcium influx is sensitive to tetramethrin, a compound which inhibits IGF-II-mediated increase in $[Ca^{2+}]_c$ in Balb/c 3T3 cells (Nishimoto and Kojima, unpublished observation). As shown in Fig. 4, tetramethrin inhibited the IGF-II action on calcium influx dose-dependently. At 500 nM, tetramethrin almost completely blocked IGF-II-stimulated calcium influx without affecting basal calcium influx.

An agonist of long-lasting-type, voltage-dependent calcium channels, BAY K 8644, increased calcium influx in Balb/c 3T3 cells (Table I). Although the concentration of BAY K 8644 which stimulated calcium influx was considerably higher than that observed in other tissues, BAY K 8644 may act on long-lasting type-voltage-dependent calcium channels since the action of BAY K 8644 was blocked by 1 nM nitrendipine. In contrast, tetramethrin does not inhibit calcium entry via long-lasting-type calcium channels (data not shown).

TABLE I. Effect of Nitrendipine and Tetramethrin on BAY K 8644-Induced Calcium Influx and [3H]Thymidine Incorporation[a]

Addition	Calcium influx (nmol/min/mg protein)	[3H]Thymidine incorporation (cpm/well)
None	0.06 ± 0.10	24,300 ± 2311
BAY K 8644	1.05 ± 0.18	58,721 ± 3410
+ nitrendipine	0.65 ± 0.21	26,211 ± 2031
+ tetramethrin	1.12 ± 0.22	57,362 ± 4811

[a] Primed competent cells were stimulated with 10 μM BAY K 8644 in the presence and absence of 1 μM nitrendipine and 100 nM tetramethrin. Values are the mean ± SE for four determinations.

3.6. Effect of Changes in Extracellular Calcium on IGF-II-Stimulated [³H]Thymidine Incorporation in Primed Competent Cells

We showed previously that IGF-II stimulates [³H]thymidine incorporation specifically in primed competent cells (Nishimoto *et al.*, 1987). In the present study, dependency of IGF-II-stimulated [³H]thymidine incorporation on extracellular calcium was determined. As shown in Fig. 5, reduction of the extracellular calcium concentration resulted in a decrease in IGF-II-mediated [³H]thymidine incorporation. When the extracellular calcium concentration was 10 μM, the effect of IGF-II on [³H]thymidine incorporation was abolished.

3.7. Effects of Cobalt and Tetramethrin on IGF-II-Stimulated [³H]Thymidine Incorporation in Primed Competent Cells

As shown in Fig. 6A, cobalt inhibited IGF-II-stimulated [³H]thymidine incorporation in primed competent cells in a dose-dependent manner. The ID_{50} was 260 μM. At concentrations higher than 500 μM, [³H]thymidine incorporation was less than that in cobalt-free medium incubated without IGF-II. Although cobalt greatly reduced [³H]thymidine incorporation, the cells started growing again when cobalt was removed by changing the medium (data not shown).

Tetramethrin also inhibited [³H]thymidine incorporation in a dose-dependent manner (Fig. 6B). At 500 nM, tetramethrin almost completely blocked IGF-II-stimulated [³H]thymidine incorporation without reducing the basal value. The ID_{50} for tetramethrin-mediated inhibition was 60 nM.

3.8. Effect of BAY K 8644 on [³H]Thymidine Incorporation in Primed Competent Cells

If calcium influx is a mitogenic signal of IGF-II, it would be expected that stimulation of calcium influx would result in an increase in [³H]thymidine incorporation. By employing BAY K 8644, an agonist of voltage-dependent calcium channels, we examined whether

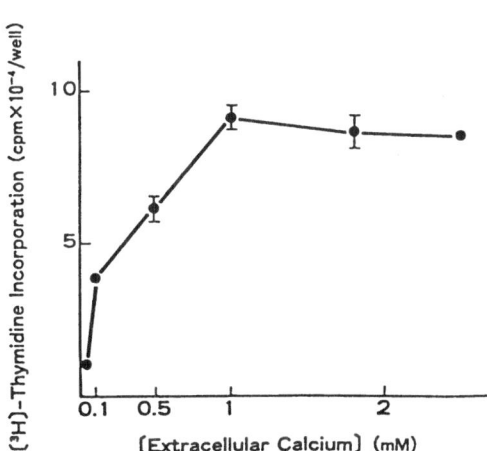

FIGURE 5. Dependency of IGF-II-induced [³H]thymidine incorporation on extracellular calcium. Primed competent cells were incubated with 1 nM IGF-II and 1 μCi/ml [³H]thymidine for 24 hr in the presence of varying concentrations of extracellular calcium. Values are the mean ± SE for four determinations.

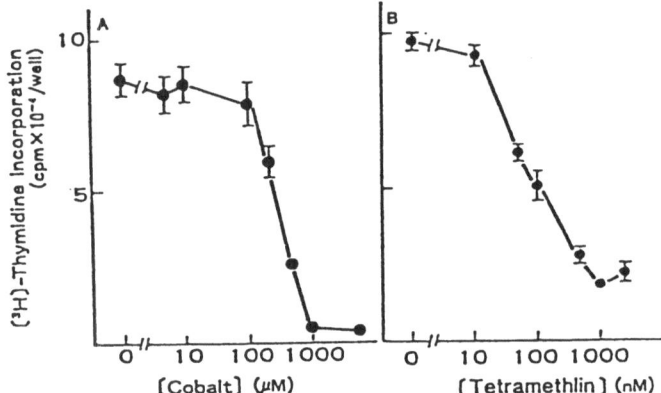

FIGURE 6. Effect of cobalt and tetramethlin on IGF-II-induced [³H]thymidine Incorporation. Primed competent cells were incubated with 1 nM IGF-II, 1 μCi/ml [³H]thymidine, and varying concentrations of cobalt (A) or tetramethlin (B) for 24 hr. Values are the mean ± SE for four determinations.

stimulation of calcium influx resulted in an increase in [³H]thymidine incorporation in primed competent cells. As shown in Table I, BAY K 8644 stimulated [³H]thymidine incorporation. When primed competent cells were incubated with BAY K 8644 for 40 hr, the cell number increased by 147%. It should be noted that BAY K 8644 did not stimulate [³H]thymidine incorporation in either quiescent or competent cells, although BAY K 8644 stimulated calcium influx to the same extent (data not shown).

3.9. Binding of [¹²⁵I]IGF-II in Balb/c 3T3 Cells

The binding of [¹²⁵I]IGF-II was studied in the membrane fraction of Balb/c 3T3 cells. The binding of [¹²⁵I]IGF-II was replaced by unlabeled IGF-II. IGF-I was less effective in replacing [¹²⁵I]IGF-II binding. In contrast, [¹²⁵I]IGF-II binding was barely affected by insulin. It is reported that binding of IGF-I is enhanced in the absence of calcium while that of insulin is reduced. Calcium did not affect the binding of [¹²⁵I]IGF-II (data not shown). Furthermore, [¹²⁵I]IGF-II binding was not affected by lanthanum, cobalt, or tetramethlin (data not shown).

4. DISCUSSION

The present results demonstrate that IGF-II stimulates calcium influx in primed competent Balb/c 3T3 cells. The IGF-II-induced calcium influx may be brought about via calcium channels since both lanthanum and cobalt, ionic inhibitors of calcium channels, effectively inhibit IGF-II-induced calcium influx. IGF-II-sensitive calcium channels are, however, resistant to nitrendipine, an inhibitor of long-lasting-type, voltage-dependent calcium channels. By contrast, IGF-II-sensitive calcium channels are sensitive to tetramethrin. Although the specificity of the inhibitory action of tetramethrin is not totally clear, a recent study done in our laboratory using the patch-clamp technique shows that tetramethrin blocks IGF-II-sensitive cation channels (Matsunaga *et al.*, 1988). Taken together, these results indicate that IGF-II stimulates calcium influx by acting on tetramethrin-sensitive calcium channels.

IGF-II stimulates calcium influx in a cell cycle-dependent manner: IGF-II does not stimulate calcium influx in either G_0-arrested or competent cells, whereas IGF-II augments calcium influx in competent cells primed with EGF. Such dependency totally agrees with the dependency of IGF-II-mediated DNA synthesis on the cell cycle. Furthermore, the following observations support the idea that calcium influx is causally related to DNA synthesis. First, both IGF-II-induced calcium influx and DNA synthesis are dependent on extracellular calcium. Second, inhibition of calcium influx by either cobalt or tetramethlin results in a reduction of DNA synthesis with identical ID_{50} values. Finally, pharmacological stimulation of calcium influx by BAY K 8644 leads to an increase in DNA synthesis in primed competent cells. Thus, stimulation of calcium influx, through either IGF-II-sensitive calcium channels or BAY K 8644-sensitive channels, leads to the promotion of DNA synthesis. It should be emphasized that BAY K 8644 does not stimulate DNA synthesis in G_0-arrested or competent cells despite the finding that BAY K 8644 increases calcium influx independent of the cell cycle. Indeed, calcium sensitivity of the machinery for DNA synthesis is also regulated in cell cycle-dependent manner. Therefore, IGF-II action is regulated in a cell cycle-dependent manner in two senses: first, generation of calcium signal is regulated in a cell cycle-dependent manner and, second, the calcium-sensing system is operated in a cell cycle-dependent manner. Further studies are needed to define the mechanism for such cell cycle dependencies.

REFERENCES

Baxter, R. C., and Williams, P. F., 1983, *Biochem. Biophys. Res. Commun.* **116:**62–67.

Bradford, M., 1976, A rapid and sensitive method for the quantitation of microgram quantities of protein utilizing the principle of protein-dye binding, *Anal. Biochem.* **72:**248–253.

Hasegawa-Sasaki, H., 1985, Early changes in inositol lipids and their metabolites induced by platelet-derived growth factor in quiescent Swiss 3T3 cells, *Biochem. J.* **232:**99–109.

Kojima, I., Kojima, K., and Rasmussen, H., 1985, role of calcium fluxes in the sustained phase of angiotensin II-induced aldosterone secretion from adrenal glomerulosa cells, *J. Biol. Chem.* **262:** 9177–9183.

Lee, P. D. K., Hodges, D., Hintz, R. L., Wyche, J. H., and Rosefeld, R., 1986, Identification of receptors for insulin-like growth factor-II in two insulin-like growth factor-II producing cell lines, *Biochem. Biophys. Res. Commun.* **134:**595–600.

Mauger, J. P., Poggioli, J., Guesdon, F., and Claret, M., 1984, Noradrenalin, vasopressin and angiotensin increase Ca^{2+} influx opening a common pool of Ca^{2+} channels in isolated rat liver cells, *Biochem. J.* **221:**121–127.

Matsunaga, H., Nishimoto, I., Kojima, I., Yamashita, N., Kurokawa, K. and Ogata, E. (1988) Activation of a calcium-permeable cation channel by insulin-like growth factor-II in Balb/c 3T3 cells. *Am. J. Physiol.*

McNiel, P. L., McKenna, M. P., and Taylor, D. L., 1985, A transient rise in cytosolic calcium follows stimulation of quiescent cells with growth factors and is inhibitable with phorbol myristate acetate, *J. Cell. Biol.* **101:**372–379.

Moses, A. C., Nissley, P., Short, P. A., Rechler, M. M., and Podskalny, J. M., 1980, Purification and characterization of multiplication-stimulating activity. Insulin-like growth factors purified from cat-liver-cell-conditioned medium, *Eur. J. Biochem.* **103:**387–392.

Nishimoto, I., Ohkuni, Y., Ogata, E., and Kojima, I., 1987, Insulin-like growth factor-II increases cytoplasmic free calcium in competent Balb/c 3T3 cells treated with epidermal growth factor, *Biochem. Biophys. Res. Commun.* **142:**275–286.

Pledger, W. J., Stiles, C. D., Antoniades, H. N., and Scher, C. D., 1977, Introduction of DNA synthesis in Balb/c 3T3 cells by serum components: Reevaluation of the commitment process, *Proc. Natl. Acad. Sci. USA* **74**:4481–4485.

Pledger, W. J., Stiles, C. D., Antoniades, H. N., and Scher, C. D., 1978, An ordered sequence of events is required before Balb/c 3T3 cells become committed to DNA synthesis, *Proc. Natl. Acad. Sci. USA* **75**:2839–2843.

Rechler, M. M., Podskalny, J. M., and Nissley, S. P., 1977, Characterization of the binding of multiplication stimulation activity to a receptor for growth in chick embryo fibroblasts, *J. Biol. Chem.* **252**:3989–3909.

Stiles, C. D., Capone, G. T., Sche, C. D., Antoniades, H. N., Van Wyk, J. J., and Pledger, W. J., 1979, Dual control of cell growth by somatomedins and platelet-derived growth factor, *Proc. Natl. Acad. Sci. USA* **76**:1279–1283.

Tominaga, S., and Lengyel, P., 1985, Beta-interferon alters the pattern of proteins secreted from quiescent and platelet-derived growth factor-treated BALB/C 3T3 cells, *J. Biol. Chem.* **260**:1975–1978.

Waisman, D. M., Gimble, J. M., Goodman, D. B. P., and Rosmusse, H., 1981, Studies on the Ca^{2+} transport mechanism of human erythrocyte inside-out plasma membrane vesicle. I. Regulation of the Ca^{2+} pump by calmodulin, *J. Biol. Chem.* **256**:248–253.

Zapf, J., Froesch, E. R., and Humble, R. E., 1981, The insulin-like growth factors of human serum. Chemical and biological characterization and aspects of their possible physiological role. *Curr. Top. Cell. Reg.* **19**:257–309.

29

Small Cell Lung Cancer Bombesin Receptors Utilize Calcium as a Second Messenger

TERRY W. MOODY, ANNE N. MURPHY, SAMIRA MAHMOUD, and GARY FISKUM

1. INTRODUCTION

Neuropeptide receptors mediate their action through the use of second messenger systems. Vasoactive intestinal polypeptide (VIP) and secretin, which are 28 and 27 amino acid peptides, respectively, utilize cAMP as a second messenger (Jensen and Gardner, 1984). Each of these peptides binds to distinct polypeptide receptors which are coupled to a stimulatory guanine nucleotide regulatory subunit (Ns) which interacts positively with the membrane-bound enzyme adenylate cyclase. In contrast, somatostatin (SRIF) receptors interact with an inhibitory guanine nucleotide regulatory subunit (Ni) which interacts negatively with adenylate cyclase.

Other neuropeptide receptors utilize Ca^{2+} as a second messenger. In guinea pig pancreative acini, distinct receptors exist for substance P, cholecystokinin, and bombesin (BN), which are 11, 33, and 14 amino acid peptides, respectively. Each of these receptors when activated stimulates ^{45}Ca efflux, elevates cGMP levels, and causes secretion of the enzyme amylase from exocrine pancreas cells (Jensen and Gardner, 1986).

Recently, BN receptors were extensively characterized for second-messenger production using a clone of Swiss 3T3 fibroblasts, which have 100,000 receptors per cell (Zachary and Rozengurt, 1985). Bombesin stimulates phosphatidylinositol turnover resulting in the production of diacylglycerol and inositol-1,4,5-trisphosphate (Heslop et al., 1986). The former agent, which activates protein kinase C, stimulates phosphorylation of an 80,000-dalton protein (Zachary et al., 1986). The latter agent triggers the release of Ca^{2+} from intracellular stores such as the endoplasmic reticulum. Fluorescent Quin2 measurements of intracellular

TERRY W. MOODY, ANNE N. MURPHY, and SAMIRA MAHMOUD • Department of Biochemistry, George Washington University School of Medicine and Health Sciences, Washington, D.C. 20037. GARY FISKUM • Departments of Biochemistry and Emergency Medicine, George Washington University School of Medicine and Health Sciences, Washington, D.C. 20037.

free Ca^{2+} have demonstrated that nanomolar concentrations of BN rapidly elevate cytosolic Ca^{2+} in 3T3 fibroblasts (Mendoza *et al.*, 1986). Because receptors for BN-like peptides are also present on human small-cell lung cancer (SCLC) cells (Moody *et al.*, 1985), we investigated whether BN-like peptides elevated cytosolic Ca^{2+} levels in these pathologically important cells.

2. RECEPTOR-BINDING STUDIES

Radiolabeled (Tyr[4])BN, a potent BN agonist (Rivier and Brown, 1978), was employed as a SCLC BN receptor probe. (^{125}I-Tyr[4])BN bound with high affinity (K_d = 0.5 nM) to a single class of sites (1500/cell) using SCLC cell line NCI-H345. Binding was specific, saturable, and reversible. The specificity of the binding was investigated. Table I shows that (Tyr[4])BN was very potent and inhibited 50% of the specific (^{125}I-Tyr[4])BN-binding activity (IC_{50}) at a 0.7 nM concentration. Gastrin-releasing peptide (GRP), a 27 amino acid peptide that is structurally similar to BN and has 9 of the 10 same C-terminal amino acid residues, was slightly less potent than (Tyr[4])BN and had an IC_{50} value of 5 nM. Ac-GRP[20-27] and GRP[14-27] were slightly more potent than GRP with IC_{50} values of 1 and 2 nM, respectively, whereas GRP[1-16] was inactive. These data indicate that the C terminal of GRP is essential for high-affinity binding activity. GRP[21-27], GRP[22-27], and GRP[23-27] were dramatically less potent with IC_{50} values of 300, 10,000, and >20,000 nM, respectively. Thus the His at position 21 of GRP and the Trp at position 22 of GRP are important for high-affinity binding. Other weak compounds, similar to GRP[21-27] and GRP[22-27], were (des-His[12])BN, (D-Leu[13])BN, (D-Val[10])BN, and (Pro[11])BN, which had IC_{50} values of 300, 500, 2000, and 5000 nM,

TABLE I. Biological Activity of BN-Like Peptides on SCLC Cell Line NCI-H345[c]

Peptides	IC_{50} (μm)	Ca^{2+} response
(Tyr[4])BN	0.7	+ +
Ac-GRP[20-27]	1.0	n.d.
BN	1.5	+ +
GRP[14-27]	2.0	+ +
GRP	5.0	+ +
(des-His[12])BN	300	n.d.
GRP[21-27]	300	+ +
(D-Arg[1], D-Pro[2], D-Trp[7], Leu[11])SP	1,000	− −
(D-Leu[13])BN	500	n.d.
(D-Val[10])BN	2,000	n.d.
(Pro[11])BN	5,000	n.d.
(D-Pro[4], D-Trp[7,9,11])SP[4-11]	7,000	− −
GRP[22-27]	10,000	+
GRP[23-27]	>10,000	−
(D-Trp[8])BN	>10,000	−
(Des-Leu[13], Met[14])BN	>10,000	n.d.
GRP[1-16]	>10,000	−

[a] The ability of various peptides to inhibit specific (^{125}I-Tyr[4])BN binding was determined and the IC_{50} value calculated. The ability of various peptides to elevate cytosolic Ca^{2+} levels was determined using the fluorescent Ca^{2+} indicator Fura 2. Strong agonist (+ +), weak agonist (+), inactive (−), antagonist (− −), not determined (n.d.).

TABLE II. Structures of BN and GRP[a]

BN	pQQRLGNQWAVGHLM*
GRP	APVSVGGGTVLAKMYPRGNHWAVGHLM*

[a] The amino acid sequence is indicated by single letter abbreviations; pQ, pyro-glutamate; *, NH_2.

respectively. Thus deletion of the His at position 12 of BN, substitution of D-Leu or D-Val for the natural L-amino acids at the 13 or 10 positions of BN, or substitution of a Pro for a Gly at the 11 position of BN reduces the binding affinity by over two orders of magnitude. These data indicate that the Val, His, and Leu at the 10, 12, and 13 positions of BN are important for high-affinity binding activity. Because Pro disrupts the α-helix secondary structure of BN, the C terminal of BN may be in an α helix when it binds to BN receptors. Other inactive compounds, similar to GRP[23-27], include (D-Trp[8])BN and (des-Leu[13], Met[14])BN, which have IC_{50} values >20,000 nM. Therefore substitution of D-Trp for the natural L-Trp at the 8 position or deletion of the Leu and Met at the 13 and 14 positions of BN results in loss of high-affinity binding activity. In summary, the C-terminal octapeptide of GRP or BN is essential for high-affinity receptor interactions. The structures of BN and GRP are shown in Table II.

SCLC is a neuroendocrine tumor which produces high concentrations of BN-like peptides (Moody et al., 1981; Wood et al., 1981; Erisman et al., 1982; Yamaguchi et al., 1983). These peptides are secreted from SCLC cells (Sorenson et al., 1982; Moody et al., 1983; Korman et al., 1986) and may bind to receptors present on the cell surface (Moody et al., 1985). Also, in a clonagenic assay which utilizes a serum-free medium, addition of exogenous BN (50 nM) stimulates the growth of SCLC cell lines (Carney et al., 1987). The observation that the growth of SCLC is inhibited by an anti-BN monoclonal antibody in vitro and in the nude mouse in vivo indicates that BN-like peptides may function as autocrine growth factors (Cuttitta et al., 1985). Thus one possible assay for the biological activity of BN-like peptides in SCLC cells is their ability to stimulate growth. Using the clonagenic assay in the absence of serum, BN stimulates the growth of SCLC cells in a concentration-dependent manner. Using 10 nM BN, the growth rate is increased threefold using celline NCI-N592. In contrast, the inactive (des-Leu[13], Met[14])BN (10 nM) had no effect on the growth rate of SCLC cells (Carney et al., 1987). These observations strongly suggest that BN receptors may regulate the growth of SCLC cells. The putative BN receptor antagonist (D-Arg[1], D-Pro[2], D-Trp[7,9], Leu[11])substance P reduced the growth rate of cell line NCI-N592, which has approximately 1500 receptors/cell, by approximately 50% in the absence or presence of 10 nM BN (Moody et al., 1987a). (D-Arg[1], D-Pro[2], D-Trp[7,9], Leu[11])substance P inhibited specific (^{125}I-Tyr[4])BN-binding activity with an IC_{50} value of 1 μM (Table I). Therefore (D-Arg[1], D-Pro[2], D-Trp[7,9], Leu[11])substance P may function as a SCLC BN receptor antagonist.

3. CYTOSOLIC CALCIUM DETERMINATIONS

While it is important to ascertain the growth characteristics of BN-like peptides, the clonagenic assay, which takes approximately 2 weeks to perform, is too time consuming to scan a large number of BN-like peptides. As an alternative approach, we tested the ability of BN-like peptides to rapidly elevate the cytosolic Ca^{2+} levels of SCLC cells. SCLC cell line NCI-H345 was cultured in a serum-supplemented medium and then harvested in HITES medium (RPMI-1640 containing 10^{-8} M hydrocortisone, 5 μg/ml bovine insulin, 10 μg/ml human transferrin, 10^{-8} M β-estradiol, and 3 $\times$ 10^{-8} M Na_2SeO_3) which contained 20 mM HEPES/NaOH (pH 7.4). Cell suspensions (2.5 $\times$ 10^6 cells/ml) were loaded with the fluorescent Ca^{2+} indicator Fura 2/AM (5 μM) at 37°C for 30 min (Grynkiewicz *et al.*, 1985). Unloaded Fura 2/AM was removed by centrifugation and the cells were resuspended (2.5 $\times$ 10^6 cells/ml) and transferred to a spectrofluorometer. The excitation and emission wavelengths were 340 and 510 nm respectively, and the cytosolic Ca^{2+} concentrations were calculated using methods similar to those used for Ca^{2+} indicator Quin2 (Tsien *et al.*, 1982).

Figure 1 shows that GRP caused a transient increase in the cytosolic free-Ca^{2+} levels in the SCLC cells in a dose-dependent manner. The basal cytosolic Ca^{2+} level in these cells was 150 $\pm$ 40 nM. Gastrin-releasing peptide (0.1 nM) caused a moderate increase in the cytosolic Ca^{2+} after 1 min (Fig. 1B). This slow time course in the fluorescence was indicative of a weak Ca^{2+} response. Using 1 nM or 10 nM GRP, the fluorescence increased rapidly after 15 sec to maximal values and then slowly decreased (Fig. 1C, D). This rapid increase in the fluorescence followed by a slow decrease was indicative of a strong Ca^{2+} response. These data are very similar both qualitatively and quantitatively to results we obtained with these cells in response to the addition of similar concentrations of BN (Moody *et al.*, 1987b). Surprisingly, the IC_{50} for GRP is 5 nM (Table I). Thus at 0.1 nM GRP, which induces a weak Ca^{2+} response, theoretically only 2% of the receptors should be occupied by GRP. At 1 and 10 nM GRP, which induces a strong Ca^{2+} response, 17 and 63% of the receptors are theoretically occupied by GRP. These data suggest that there may be spare receptors for BN-like peptides. Similarly, using pancreatic acini, nanomolar concentrations of BN stimulated cellular ^{45}Ca efflux, elevated cGMP levels, and induced release of amylase from the cells (Jensen *et al.*, 1978). Because the ED_{50} to induce these biological responses was only 0.2 nM whereas the IC_{50} to inhibit specific binding was substantially greater, it was hy-

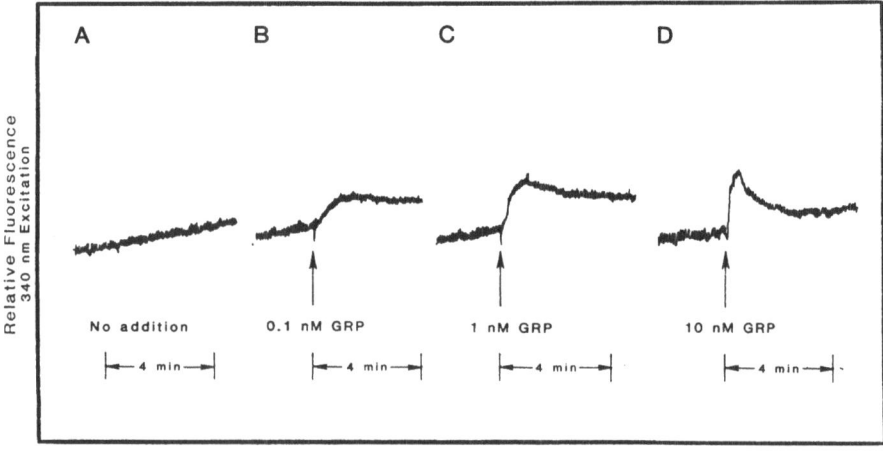

FIGURE 1. Dose dependence of GRP on the cytosolic Ca^{2+} concentration of small-cell lung cancer cells. Fura 2 measurements were made of cytosolic Ca^{2+} in SCLC cells in the absence (A) and the presence of 0.1 nM GRP (B), 1.0 nM GRP (C), and 10 nM GRP (D).

pothesized that there were spare receptors for BN and that occupation of only 20% of these receptors was sufficient to induce a biological response.

Using Swiss 3T3 cells, Mendoza *et al.* (1986) observed that BN produced a dose-dependent increase in cytosolic Ca^{2+} levels. The ED_{50} was 0.3 nM for BN and 2.5 nM for GRP whereas GRP^{1-16} was inactive. The threefold increase in the cytosolic Ca^{2+} levels caused by BN (3.1 nM) occurred within 15 sec. Also, in quiescent cultures of Swiss 3T3 cells that were preloaded with ^{45}Ca for 12–24 hr, BN (6.2 nM) dramatically increased the rate of ^{45}Ca efflux. Because EGTA (3 mM) had no effect on the BN-stimulated release of ^{45}Ca from the Swiss 3T3 cells, it appears likely that BN stimulates release of Ca^{2+} from intracellular pools. Similarly, it is hypothesized that the inositol trisphosphate produced in Swiss 3T3 cells by BN may trigger release of Ca^{2+} from intracellular stores such as the endoplasmic reticulum (Heslop *et al.*, 1986). In this regard we investigated if manipulation of extracellular Ca^{2+} alters the response of the SCLC cells to GRP. The addition of the Ca^{2+} chelator EGTA (5 mM) to the medium which contained 0.4 mM Ca^{2+} caused an immediate decrease in the fluorescent signal due to the conversion of Ca^{2+}-bound to Ca^{2+}-unbound extracellular Fura 2 (Fig. 2B). The subsequent addition of 100 nM GRP was followed by a rapid rise in the cytosolic Ca^{2+} that was similar in rate and magnitude to that observed in the presence of extracellular Ca^{2+}. These results suggest that the main source of the GRP-induced elevated cytosolic Ca^{2+} is from intracellular stores, possibly the endoplasmic reticulum, rather than plasmalemmal Ca^{2+} influx.

The pharmacology of the increase in cytosolic Ca^{2+} in SCLC cell line NCI-H345 was also investigated. $(Tyr^4)BN$, GRP, and GRP^{14-27}, which are potent BN receptor agonists,

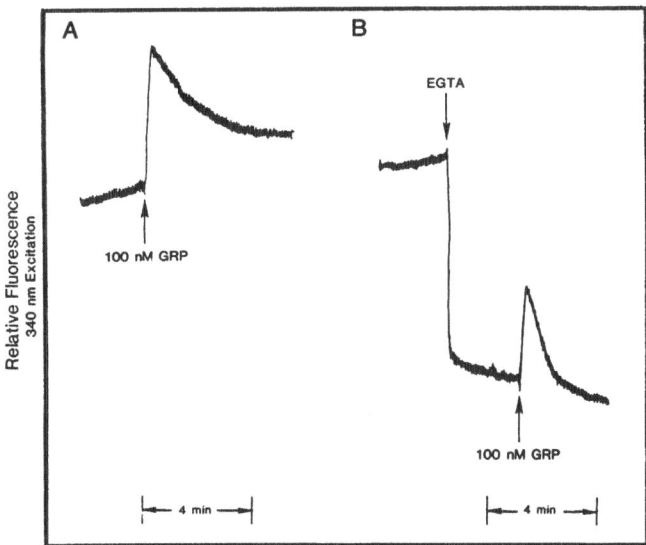

FIGURE 2. Response of cytosolic Ca^{2+} to GRP in the presence and absence of extracellular free Ca^{2+}. Fura 2 measurements were made of cytosolic Ca^{2+} in SCLC cells in the presence of 0.4 mM $CaCl_2$ in the absence (A) and presence (B) of 5 mM EGTA. Gastrin-releasing peptide was added at the concentration of 100 nM. A sufficient quantity of NaOH was added together with EGTA to counteract the acidification caused by the binding of Ca^{2+} to EGTA.

rapidly increased the cytosolic Ca^{2+} levels when each was added at a 1 μM concentration (Table I). At this concentration of BN, GRP, or GRP[14-27], over 99% of the receptors should be occupied. Similarly, GRP[21-27] and GRP[22-27], which are weak BN receptor agonists, increased the fluorescence signal rapidly and slowly, respectively. At a 1 μM concentration, 67 and 10% of the receptors should theoretically be occupied by GRP[21-27] and GRP[23-27], respectively. In contrast, a 1 μM concentration of GRP[1-16], GRP[23-27], or (D-Trp[8])BN, which are inactive, did not affect the cytosolic Ca^{2+} levels. Previously, we found that VIP and SRIF receptors are present on SCLC cell line NCI-H345. Each of these peptides had no effect on the cytosolic Ca^{2+} levels; however, they do alter cAMP levels (Kee et al., 1987), Korman et al., 1986). Thus numerous BN receptor agonists but not other peptides alter cytosolic Ca^{2+} levels in SCLC cells.

The effect of antagonists on the cytosolic Ca^{2+} levels of SCLC cells was also investigated. (D-Pro[4], D-Trp[7,9,11])substance P[4-11] had no effect on cytosolic Ca^{2+} at a peptide concentration of 30 or 100 μM (Fig. 3). If 10 nM BN was subsequently added, there was a slight increase in the cytosolic Ca^{2+} levels indicative of a weak Ca^{2+} response. In contrast, addition of 100 nM BN rapidly elevated the Ca^{2+} levels followed by a slow decrease, which is indicative of a strong Ca^{2+} response. These data indicate that (D-Pro[4], D-Trp[7,9,11])substance P[4-11] has no BN receptor agonist activity but does partially antagonize the response to BN. (D-Pro[4], D-Trp[7,9,11])substance P[4-11] inhibited specific ([125I-Tyr[4])BN-binding activity with an IC_{50} value of 10 μM. In contrast, (D-Arg[1], D-Pro[2], D-Trp[7,9], Leu[11])substance P was more potent and inhibited specific ([125I-Tyr[4])BN-binding activity with an IC_{50} value of 1 μM. (D-Arg[1], D-Pro[2], D-Trp[7,9], Leu[11])substance P (30 μM) had no effect on the Fura 2 fluorescence by itself but totally antagonized the Ca^{2+} response to 10 or 100 nM BN (data not shown). Similar data were obtained if SCLC cells were first treated with inactive peptides, such as GRP[1-16] followed by addition of BN. These data indicate that by using this assay one can readily distinguish between BN agonists, BN antagonists, and inactive compounds.

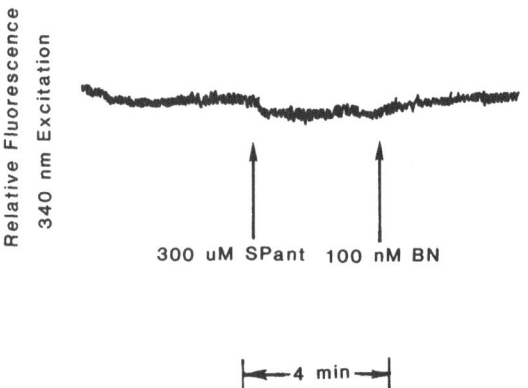

FIGURE 3. Inhibition of BN-induced increase in cytosolic Ca^{2+} by a BN receptor antagonist. Fura 2 measurements were made of the effect of 100 nM bombesin on the cytosolic Ca^{2+} concentration in the presence of 300 μM of the substance P antagonist (SPant), (D-Pro[4], D-Trp[7,9,11])substance P.

4. SUMMARY

Bombesin-like peptides function as autocrine growth factors in SCLC cells. High levels of endogenous GN-like peptides are present in and secreted from SCLC cells. Also, receptors for BN-like peptides are present on the cell surface and addition of exogenous BN stimulates the growth of SCLC *in vitro*. Here BN or the structurally related GRP elevated cytosolic Ca^{2+} levels as determined using Fura 2. Nanomolar concentrations of GRP elevated cytosolic Ca^{2+} levels in the presence or absence of extracellular Ca^{2+}. Other potent BN receptor agonists such as GRP^{14-27} elevated cytosolic Ca^{2+} levels whereas inactive peptides such as GRP^{1-16} did not. Also BN receptor antagonists such as (D-Pro[4], D-Trp[7,9,11]) substance P^{4-11} had no effect on the Ca^{2+} levels but antagonized the increase in Ca^{2+} caused by 10 nM BN. These data suggest that BN receptors may utilize Ca^{2+} as a second messenger in SCLC cells.

ACKNOWLEDGMENTS. This research is supported by NCI grants CA-33767 (T.W.M.) and CA-32946 (G.F.).

REFERENCES

Carney, D. N., Cuttitta, F., Moody, T. W., and Minna, J. D., 1987, Selective stimulation of small cell lung cancer clonal growth by bombesin and gastrin releasing peptides, *Cancer Res.* **46:** 1214–1218.

Cuttitta, F., Carney, D. N., Mulshine, J., Moody, T. W., Fedorko, J., Fischler, A., and Minna, J. D., 1985, Bombesin-like peptides can function as autocrine growth factors in human small cell lung cancer, *Nature* **316:**823–825.

Erisman, M. R., Linnoila, O., Hernandez, R., DiAugistine, R., and Lazarus, L., 1982, Human lung small cell carcinoma contains bombesin, *Proc. Natl. Acad. Sci. USA* **79:**2379–2383.

Grynkiewicz, G., Poenic, M., and Tsien, R. Y., 1985, A new generation of Ca^{2+} indicators with greatly improved fluorescence properties. *J. Biol. Chem.* **260:**3440–3450.

Heslop, J. P., Blakely, D. M., Brown, K. D., Irvine, R. F., and Berridge, M. J., 1986, Effects of bombesin and insulin on inositol (1,4,5) trisphosphate and inositol (1,3,4) trisphosphate formation in Swiss 3T3 cells, *Cell* **47:**703.

Jensen, R. T., Moody, T., Pert, C., Rivier, J. E., and Gardner, J. D., 1978, Interaction of bombesin and litorin with specific membrane receptors on pancreatic acinar cells, *Proc. Natl. Acad. Sci. USA* **75:**6139–6143.

Jensen, R. T., and Gardner, J. D., 1984, Identification and characterization of receptors for secretagogues on pancreatic acinar cells, *Fed. Proc.* **40:**2486–2496.

Jensen, R. T., and Gardner, J. D., 1986, Structure-function studies of agonists and antagonists for gastrointestinal peptide receptors on pancreatic acinar cells, in *Neural and Endocrine Peptides and Receptors* (Moody, T. W., ed.), Plenum Press, New York, pp. 485–500.

Kee, K. A., Finan, T. M., Korman, L. Y., and Moody, T. W., 1987, Somatostatin inhibits the secretion of bombesin-like peptides from small cell lung cancer cells, *Peptides* **9**(Suppl. 1):257–261.

Korman, L. Y., Carney, D. N., Citron, M. L., and Moody, T. W., 1986, Secretin/VIP stimulated secretion of bombesin-like peptides from human small cell lung cancer, *Cancer Res.* **47:**821–825.

Mendoza, S. A., Schneider, J. A., Lopez-Rivas, A., Sinnet-Smith, J. W., and Rozengurt, E., 1986, Early events elicited by bombesin and structurally related peptides in quiescent swiss 3T3 cells. II. Changes in Na^+ and Ca^{2+} fluxes, Na^+K^+ pump activity and intracellular pH, *J. Cell Biol.* **102:**2223–2233.

Moody, T. W., Pert, C., Gazdar, A., Carney, D., and Minna, J., 1981, High levels of intracellular bombesin characterize human small cell lung cancer, *Science* **214:**1246–1248.

Moody, T. W., Russel, E. K., O'Donohue, T. L., Linden, C. D., and Gazdar, A., 1983, Bombesin-like peptides in small cell lung cancer: Biochemical characterization and secretion from a cell line, *Life Sci.* **32:**487–493.

Moody, T. W., Carney, D. N., Cuttitta, R., Quattrocchi, K., and Minna, J. D., 1985, High affinity receptors for bombesin/GRP-like peptides on human small cell lung cancer cells, *Life Sci.* **36:** 105–113.

Moody, T. W., Mahmoud, S., Koros, A., Cuttitta, F., Willey, J., Rotsch, M., Zeymer, U., and Bepler, G., 1987a, Substance P analogues antagonize small cell lung cancer bombesin receptors, *Fed. Proc.* **46:**2201.

Moody, T. W., Murphy, A. N., Mahmoud, S., and Fiskum, G., 1987b, Bombesin-like peptides elevate cytosolic calcium in small lung cancer cells, *Biochem. Biophys. Res. Commun.* **147:**189–195.

Rivier, J. and Brown, M. R., 1978, Bombesin, bombesin analogs, and related peptides: Effects on thermoregulation, *Biochemistry* **17:**1766–1771.

Sorenson, G. D., Pettengill, O. S., Cate, C. C., Ghatei, M. A., Molyneus, K. E., Gosselin, E. J., and Bloom, S. R., 1982, Bombesin and calcitonin secretion by pulmonary carcinoma is modulated by cholinergic receptors, *Life Sci.* **33:**1939.

Tsien, R. Y., Pozzan, T., and Rink, R. J., 1982, Calcium homeostasis in intact lymphocytes: Cytoplasmic free Ca^{2+} monitored with new, intracellularly trapped fluorescent indicator, *J. Cell. Biol.* **94:**325–334.

Wood, S., Wood, J., Ghatei, M., Lee, U., Shaughnessy, D., and Bloom, S., 1981, Bombesin, somatostatin and neurotensin-like immunoreactivity in bronchial carcinoma, *J. Clin. Endocrin.* **53:** 1310–1312.

Yamaguchi, K., Abe, K., Kameya, T., Adachi, I., Taguchi, S., Otsubo, K., and Tanaihara, K., 1983, Production and molecular size heterogeneity of immunoreactive gastrin releasing peptide in fetal and adult lungs and primary tumors, *Cancer Res.* **43:**3932–3939.

Zachary, I., and Rozengurt, E., 1985, High affinity receptors for peptides of the bombesin family in swiss 3T3 cells, *Proc. Natl. Acad. Sci. USA* **82:**7616–7620.

Zachary, I., Sinnett-Smith, J. W., and Rozengurt, E., 1986, Early events solicited by bombesin and structurally related peptides in quiescent swiss 3T3 cells. I. Activation of protein kinase C and inhibition of epidermal growth factor binding, *J. Cell. Biol.* **102:**2211–2222.

30

Extracellular Ca²⁺ and Cell Cycle Transitions
Effect of Protein Kinase C and InsP₃

ALTON L. BOYNTON, JEAN ZWILLER, TIMOTHY D. HILL,
THOMAS NILSSON, PER ARKHAMMER, and
PER-OLOF BERGGREN

1. INTRODUCTION

Extracellular Ca^{2+} has been demonstrated to be required for proliferation of a wide variety of nonneoplastic cells of mesenchymal and epithelial origin both *in vivo* and *in vitro* (Boynton *et al.*, 1974,1981; Boynton and Whitfield, 1976; Hennings *et al.*, 1980; Swierenga *et al.*, 1980; Lechner and Kaighn, 1981; Swierenga, 1984). Specific stages of the growth–division cycle require extracellular Ca^{2+} as well as intracellular Ca^{2+} transients (Boynton *et al.*, 1985a; Hazelton *et al.*, 1979; Hesketh *et al.*, 1987). Neoplastic cells, on the other hand, require 10- to 50-fold less extracellular Ca^{2+} than their neoplastic counterparts for their proliferative activity regardless of the means of neoplastic transformation (Swierenga *et al.*, 1980; Boynton *et al.*, 1981; Whitfield *et al.*, 1987). However, it is not known if neoplastic cells require intracellular Ca^{2+} transients for cell cycle progression. It was first demonstrated in 1965 that adenovirus-infected cells could be selected from uninfected cells by culturing them in Ca^{2+}-deficient medium (Freeman *et al.*, 1965, 1966). Balk *et al.* (1973) extended this observation in 1973 to Rous sarcoma virus-infected chicken fibroblasts, and Boynton and Whitfield (1976) found that neoplastic cells transformed by DNA or RNA viruses, by chemicals, or by spontaneous means all required 10- to 50-fold less extracellular Ca^{2+} for proliferative activity.

This chapter will focus on a model system built around the extracellular Ca^{2+} requirement for proliferative activity of nonneoplastic T51B rat liver epithelial cells.

ALTON L. BOYNTON, JEAN ZWILLER, and TIMOTHY D. HILL ● Cancer Research Center of Hawaii, University of Hawaii, Honolulu, Hawaii 96813.   THOMAS NILSSON, PER ARKHAMMER, and PER-OLOF BERGGREN ● Department of Medical Cell Biology, Biomedicum, University of Uppsala, S751 23 Uppsala, Sweden.

2. EXTRACELLULAR Ca^{2+} AND IN VITRO CELL PROLIFERATION OF RAT LIVER EPITHELIAL CELLS

T51 rat liver cells were originally derived from an adult Fisher rat in the laboratory of Dr. M. Farber, University of Toronto. T51B is a cell line cloned from T51 (Swierenga et al., 1978) and is probably of oval cell origin. These epithelial cells are classified as non-neoplastic because they do not express any of the common phenotypic markers of neoplastic cells including growth in soft agar, growth in Ca^{2+}-deficient medium, or tumor formation in athymic nude mice (Boynton et al., 1984).

The extracellular Ca^{2+} requirement for proliferation can be demonstrated by several methods. First, simply reducing the extracellular Ca^{2+} concentration in the medium-serum mixture from the usual 1.8 mM to only 0.025 mM prevents the formation of colonies (Boynton et al., 1984). These Ca^{2+}-deprived cells are still viable after 2 weeks because the readdition of Ca^{2+} (final concentration of 1.25 mM) to the medium-serum enabled them to proliferate and form colonies.

Second, reducing the extracellular Ca^{2+} concentration in exponentially growing cultures from the usual 1.8 mM to only 0.025 mM blocked cells in the G$_1$ phase of their cell cycle (Boynton et al., 1985a). This is illustrated by determining their DNA contents using propidium iodide staining of the cellular DNA and quantitation by flow cytometry. To illustrate where in the G$_1$ phase these cells were blocked, Ca^{2+} was added back to the medium-serum mixture to a final concentration of 1.25 mM and the flow of cells into S phase was determined at hourly intervals using autoradiographic analysis of DNA synthesis (Boynton et al., 1985a). There was an almost immediate increase in S-phase cells that peaked by 4 hr, which was followed by a second wave of DNA synthetic activity at 12–14 hr. These results suggested that Ca^{2+} deprivation of T51B rat liver cells blocked them in the G$_1$ phase of their cell cycle both at the G$_1$ → S boundary and also at the G$_0$ → G$_1$ boundary.

The third method of illustrating extracellular Ca^{2+} dependence of G$_1$ progression uses confluent T51B cells, which are proliferatively quiescent with a labeling index of about 2–5% (Boynton et al., 1985b). A fresh 80% BME/20% BCS (BME, Eagles Basal Medium; BCS, Bovine Calf Serum) fluid change stimulates cells to synchronously progress through the prereplicative G$_1$ phase of their growth–division cycle and begin replicating their chromosomes 12–14 hr later. As with regeneration of rat liver hepatocytes in vivo (Whitfield et al., 1985), T51B rat liver cells in vitro require extracellular Ca^{2+} for successful G$_1$ progression (Boynton et al., 1985a). T51B cells have two distinct and transient extracellular Ca^{2+}-dependent states in their G$_1$ phase (Fig. 1; Boynton et al., 1985b). The initial G$_0$ → G$_1$ transition is known to require extracellular Ca^{2+} since fresh 80% BME/20% BCS containing between 0.02 and 0.1 mM Ca^{2+} will not stimulate this transition. Further studies revealed that the cells grown in low-Ca^{2+} medium-serum do not replicate their chromosomes and that delaying the raising of the medium-serum Ca^{2+} concentration (to 1.25 mM) until either 4 or 12 hr after adding the fresh low-Ca^{2+} medium-serum delayed the initiation of DNA synthesis by the same amount of time. This was interpreted to mean that these epithelial cells required serum- and Ca^{2+}-dependent events for successful G$_0$ → G$_1$ transition and that either factor alone was insufficient to stimulate this transition. The extracellular Ca^{2+}-initiated events were brief because a 30-min pulse of fresh 80% BME/20% BCS containing 1.8 mM Ca^{2+} followed by fresh 80% BME/20% BCS containing only 0.02–0.075 mM Ca^{2+} was sufficient to stimulate G$_0$ → G$_1$ transition. Addition of Ca^{2+} (to a final concentration of 1.25 mM) 8 hr later enabled the cells to replicate their chromosomes at the normal time (i.e., 12–14 hr from the original medium-serum change). This addition of Ca^{2+} 8 hr after

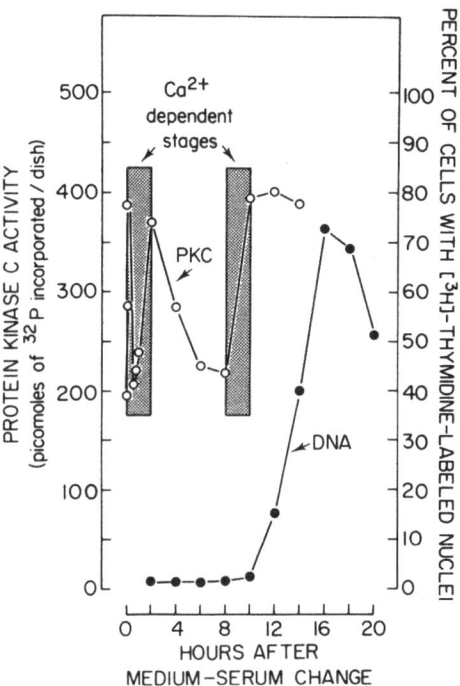

FIGURE 1. The association of protein kinase C surges with the two Ca²⁺-dependent stages of G₁ phase of serum-stimulated confluent cultures of T51B rat liver epithelial cells. Confluent and proliferatively quiescent T51B cells were stimulated to transit $G_0 \rightarrow G_1$ and $G_1 \rightarrow S$ and start replicating their DNA by replacing the old BME/BCS medium with fresh 80% BME 20% BCS. At various intervals, protein kinase C and DNA synthetic activity were determined as previously described (Boynton et al., 1985b).

the initial stimulation also indicated a late G_1 requirement for extracellular Ca²⁺ because if the Ca²⁺ concentration was not raised to 1.25 mM at this time the cells did not progress into their S phase. Furthermore, the late G_1 extracellular Ca²⁺ requirement was found to be transitory because if the medium-serum Ca²⁺ concentration was raised to 1.25 mM at 10 hr or later, rather than at 8 hr, the fraction of cells able to replicate their chromosomes was considerably reduced (from about 70% to only 20–30%). Thus, there exists within the G_1 prereplication period of nonneoplastic T51B rat liver cells two extracellular Ca²⁺-dependent stages. The first corresponds to a brief 30-min $G_0 \rightarrow G_1$ transition and the second to a transient 2-hr window immediately preceding $G_1 \rightarrow S$ transition. The intervening events between $G_0 \rightarrow G_1$ and $G_1 \rightarrow S$ are independent of extracellular Ca²⁺ (Boynton et al., 1985a). It should be noted that WI-38 human fibroblasts also have two extracellular Ca²⁺-dependent stages in the G_1 phase of their growth–division cycle (Hazelton et al., 1979).

3. MECHANISMS OF Ca²⁺ ACTION

Several lines of evidence suggest that calmodulin, cyclic AMP, and cyclic AMP-dependent protein kinases are mechanistically involved in the progression of cells through the G_1 phase of their cell cycle (see Boynton et al., 1985a; Whitfield et al., 1985 for review). Here we want to concentrate on two relatively new second messengers: diacylglycerol (DAG) and inositol-1,4,5-trisphosphate (InsP₃). DAG and InsP₃ result from the hydrolysis of phosphatidylinositol-4,5-bisphosphate (PIP₂) by the action of a GTP requiring phospholipase C

(Gomperts, 1984). Diacylglycerol remains in the membrane domain and activates the Ca^{2+}/phospholipid-dependent protein kinase or protein kinase C (Takai et al., 1982). $InsP_3$, on the other hand, is responsible for the rapid and transient increase in intracellular free Ca^{2+} (Streb et al., 1983). Thus, the dual action of protein kinase C and Ca^{2+} may be necessary for specific cell cycle transitions.

As already discussed, the G_1 phase of serum-stimulated T51B rat liver epithelial cells has two transient extracellular Ca^{2+}-dependent phases (Boynton et al., 1985b). Protein kinase C is a Ca^{2+}/phospholipid-dependent enzyme, and because of the rapid appearance of the phosphatidylinositol hydrolytic products DAG and $InsP_3$ during G_1, it would not be unexpected to see alterations in this protein kinase during the G_1 phase of the cell cycle. Two brief surges of ethylenediaminetetraacetate (EDTA)-extractable protein kinase C activity occur during the Ca^{2+}-dependent $G_0 \rightarrow G_1$ (i.e., the initial 30 min following stimulation) transition, the first of which was within 2.5 min of the fresh medium-serum change (Fig. 1). A third and much more prolonged surge of EDTA-extractable protein kinase C activity begins 6–8 hr later and peaks during the second Ca^{2+}-dependent period in the late G_1 phase (Fig. 1; Boynton et al., 1985b). These observations suggest that protein kinase C is somehow involved in these two critical Ca^{2+}-dependent stages of the cell cycle and, if so, any compound that stimulates the enzyme by sensitizing it to Ca^{2+} should be able to stimulate G_1 transit in Ca^{2+}-deficient medium. Two compounds were used to test this hypothesis: first, the potent tumor promoter and protein kinase C activator 12-o-tetradecanoylphorbol-13-acetate (TPA) (50–100 ng/ml), and second, a synthetic diacylglycerol, 1-oleoyl-2-acetylglycerol (OAG; 200 μg/ml). Neither TPA nor OAG alone induced quiescent confluent T51B cells to undergo the $G_0 \rightarrow G_1$ transition and replicate their chromosomes. However, they did enable these cells to transit $G_0 \rightarrow G_1$ and $G_1 \rightarrow S$ and replicate their chromosomes when added immediately after fresh 80% BME/20% BCS containing only 0.07 mM Ca^{2+} (a Ca^{2+} concentration which is usually ineffective in stimulating G_1 transit). 4α-phorbol didecanoate, a TPA analog and nonactive, non-tumor-promoting phorbol ester, was unable to stimulate G_1 transit and was furthermore unable to activate protein kinase C (Boynton et al., 1985b).

Events other than protein kinase C activation such as extracellular or intracellular Ca^{2+}-dependent events must also be required for TPA or OAG action. Thus, reducing the Ca^{2+} concentration to 0.02 mM in the stimulating mixture (80% BME/20% BCS) was sufficient to prevent TPA or OAG from stimulating G_1 transit (Boynton et al., 1985b). If the extracellular Ca^{2+} deprivation was affecting the intracellular Ca^{2+} levels and preventing G_1 transit, it should be possible to raise the intracellular Ca^{2+} level by adding either the divalent cation ionophore A23187 or perhaps by adding the intracellular Ca^{2+}-mobilizing agent $InsP_3$. A23187 or $InsP_3$ alone did not stimulate G_1 development or enable 80% BME/20% BCS containing 0.02 mM Ca^{2+} to stimulate G_1 transit and chromosome replication of T51B cells (Fig. 2). However, adding the protein kinase C activator OAG or TPA along with an intracellular Ca^{2+}-elevating agent such as A23187 (10^{-7} M) or $InsP_3$ (10^{-7} M) immediately after 80% BME 20% BCS containing 0.02 mM Ca^{2+} enabled a significant fraction of the cells to transit G_1 and start replicating their DNA even in this severely Ca^{2+}-deficient medium (Fig. 2; Zwiller and Boynton, unpublished observations).

In order to determine if TPA, A23187, or $InsP_3$ increased intracellular Ca^{2+} of these T51B cells, we loaded them with the fluorescent Ca^{2+} probe Fura 2 AM. The acetoxymethyl ester derivative is necessary for entry into the cells and free Fura 2 is rapidly formed by nonspecific esterase hydrolysis of the acetoxymethyl ester derivative. Fura 2, when not complexed with Ca^{2+}, excites at 380 nM and has an emission wavelength of 510 nM. Binding of Ca^{2+} to the fluorescent probe shifts the excitation wavelength to 340 nM, while the

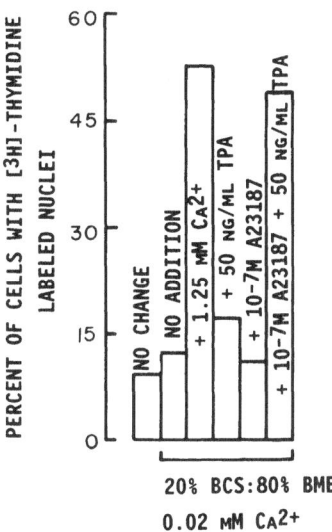

FIGURE 2. The synergistic requirement of protein kinase C and Ca^{2+} in stimulation of $G_1 \rightarrow S$ transition of T51B cells. Confluent and proliferatively quiescent T51B cells were stimulated to transit their $G_0 \rightarrow G_1$ by fresh 80% BME/20% BCS for a period of 1 hr. At this time, the medium-serum was replaced with 80% BME/20% BCS containing only 0.02 mM Ca^{2+} and incubated an additional 7 hr at which time either no addition was made; Ca^{2+} was added to a final concentration of 1.25 mM; TPA was added at 50 ng/ml; A23187 was added at 10^{-7} M; or TPA (50 ng/ml) plus A23187 (10^{-7} M) were both added. DNA synthesis was estimated by incubating the cells with [³H]thymidine (5 μCi) for 12 to 20 hr and the percentage of labeled nuclei determined by autoradiography.

emission wavelength remains the same at 500 nM. Thus, by calculating the ratio of emission fluorescence at both excitation wavelengths, it is possible to estimate the intracellular free Ca^{2+} (Grynkiewicz *et al.*, 1985).

The addition of 100 nM A23187 to these Ca^{2+}-deprived T51B cells in the late G_1 phase of their growth–division cycle increased the intracellular level of Ca^{2+} by about twofold (Fig. 3). TPA alone (50 ng/ml) had no effect on the intracellular level of Ca^{2+} and also did not affect the ability of A23187 to increase the levels of intracellular Ca^{2+} (Fig. 4). This effect of TPA is consistent with its role in activating protein kinase C and functioning separately from $InsP_3$ mobilization of Ca^{2+}. It should be noted that we were recently able to demonstrate that the exogenous addition of $InsP_3$ (1,4,5 isomer; 10^{-7}–10^{-6} M; as well as inositol-1,3,4,5-tetrakisphosphate) to these T51B cells increases their intracellular level of Ca^{2+}. The effect is totally dependent on the presence of extracellular Ca^{2+} because the removal of extracellular Ca^{2+} abolishes the effect, as does the Ca^{2+} channel blocker La^{3+} (Hill, Kindmark, and Boynton, unpublished observations). This suggests the intriguing possibility that several inositol phosphates are able to open plasma membrane Ca^{2+} channels.

It is now feasible to conclude that protein kinase C and Ca^{2+}-dependent events combine to mediate the mitogenic actions of various growth factors by activating the components of a plasma membrane cell surface signaling system. This signaling system would appear to begin with the binding of a growth factor with its cell surface receptor thus coupling phos-

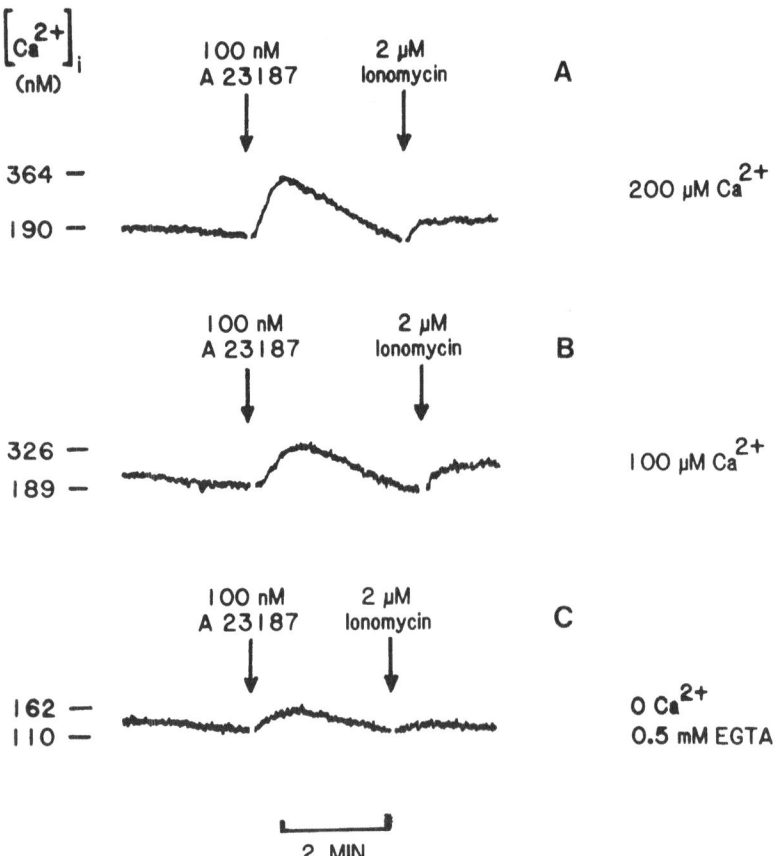

FIGURE 3. The ability of A23187 to increase the intracellular free Ca^{2+} concentration of Fura 2-loaded T51B cells. T51B cells were brought to their Ca^{2+}-dependent $G_1 \rightarrow S$ transition as described in the legend to Fig. 2. Cells were then loaded with 5 μM Fura 2 AM for 30 min and the effect of A23187 (10^{-7} M) on intracellular Ca^{2+} determined in various amounts of extracellular Ca^{2+} using a Perkin Elmer 650-10S spectrofluorometer. The result is a typical tracing from a total of 10 separate experiments.

pholipase C with its substrate, PIP_2, in a GTP-dependent process. The result of this hydrolysis is the production of two potentially important second messengers: diacylglycerol which stimulates protein kinase C activity and $InsP_3$ which mobilizes intracellular Ca^{2+}. In turn, these compounds could activate other protein kinases, proteases, phosphatases, or function by modulating calmodulin activity. The increased protein kinase C activity may be responsible for the down modulation of growth factor receptors through phosphorylation. In addition, protein kinase C stimulates adenylate cyclase activity (Naghshineh et al., 1986) and increases cyclic AMP (Boynton et al., 1985a), which in turn increases cyclic AMP-dependent protein kinase activity (Campos-Ganzalez and Boynton, unpublished observations), which is known to activate several genes (Bonney et al., 1983; Whitfield et al., 1987).

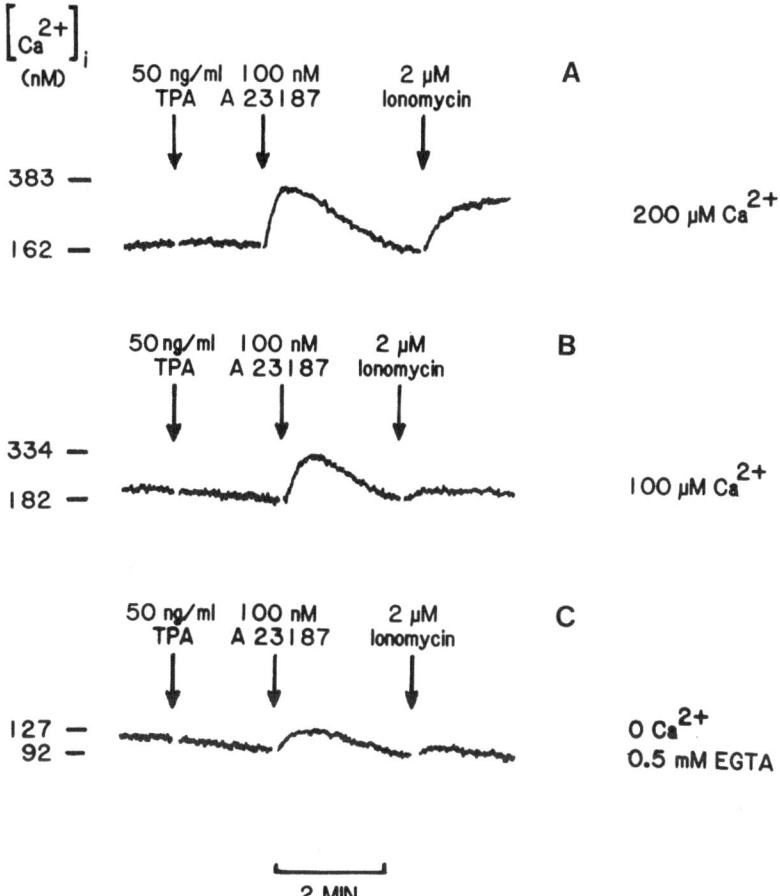

FIGURE 4. The inability of TPA to increase intracellular Ca^{2+} and its inability to affect the A23187-induced Ca^{2+} increase. Culture conditions were exactly as described in the legend to Fig. 3. The result is a typical tracing from a total of five separate experiments.

4. NEOPLASTIC TRANSFORMATION

Neoplastic transformation induced by DNA or RNA viruses, by chemicals, by physical carcinogens, or occurring spontaneously results in the aberrant proliferative activity of cells grown in an environment that usually does not support cell proliferation. Therefore, regardless of the means of neoplastic transformation, these cells are able to proliferate in Ca^{2+}-deficient media (Swierenga *et al.*, 1980; Boynton *et al.*, 1981). We have neoplastically transformed T51B rat liver epithelial cells with aflatoxin B$_1$ and selected several clones which are still sensitive to density-dependent inhibition of proliferation. When stimulated to proliferate by fresh 80% BME 20% BCS containing only 0.025 mM Ca^{2+}, unlike their nonneoplastic counterparts, they progress through the G$_0$ → G$_1$ and G$_1$ → S transitions. It follows that if we knew the mechanism by which nonneoplastic cells regulate or control their passage through these two Ca^{2+}-dependent cell cycle transitions, we would have a basis on which to rationally investigate the mechanism(s) of neoplastic transformation.

Perhaps a clue to this phenomenon of the ability to proliferate in Ca^{2+}-deficient media by tumor cells comes from an investigation of protein kinase C. Neoplastic cells, unlike their nonneoplastic counterparts, are able to maintain a high level of EDTA-extractable protein kinase C activity in Ca^{2+}-deficient medium, where they also are able to maintain a high rate of proliferative activity (Boynton et al., 1983; Bossi et al., 1985). The inference is that in nonneoplastic cells the extracellular Ca^{2+} is somehow linked to the turnover of phosphatidylinositol and the production of the two second messengers DAG and $InsP_3$. On the other hand, extracellular Ca^{2+} may be linked to the transcription of mRNA species and/or the translation of critical cellular proteins. In fact, in Ras-transformed cells there is an indication of increased levels of phosphatidylinositol-4,5-bisphosphate and its catabolites (Fleischman et al., 1986).

Some oncogenic viruses produce products which function as growth factors, while others make proteins which function as growth factor receptors or protein kinases (see Whitfield et al., 1987 for review). These oncogenic products are able to affect the normal cell surface signaling system and therefore maintain the proliferative activity of the cell. For example, the p21 product of cellular ras oncogene are GTP-binding GTPases related to the β subunits of G_s proteins and may function to open membrane Ca^{2+} channels, activate adenylate cyclase, and conceivably stimulate the hydrolysis of phosphoinositides (Whitfield et al., 1987).

5. CONCLUSIONS

The transit of nonneoplastic cells through the G_1 phase of their growth–division cycle is dependent on extracellular Ca^{2+} at two brief and distinct points. The first Ca^{2+}-dependent state occurs immediately after stimulation of quiescent cells with serum and is referred to as the $G_0 \rightarrow G_1$ transition, while the second Ca^{2+}-dependent state is late in the prereplicative period immediately before DNA synthesis, and is referred to as the $G_1 \rightarrow S$ transition. Evidence has been presented which strongly implicates PI turnover, protein kinase C activation and/or translocation, and intracellular Ca^{2+} release as the events responsible for these two transitions. Additional evidence also suggests that cyclic AMP and cyclic AMP-dependent protein kinases are involved in these transitions and that protein kinase C may be linked to adenylate cyclase activity and the increase in cyclic AMP-dependent protein kinase activity. Thus, a model is built which utilizes as a core the plasma membrane components adenylate cyclase, phospholipase C, the phosphatidylinositol hydrolytic products DAG and $InsP_3$ and growth factor receptors. The system is activated by the binding of a growth factor to its receptor that in turn activates a lipid kinase and the Ca^{2+}/GTP-dependent phospholipase C. The lipid kinase and phospholipase C catalyze the phosphorylation of phosphoinositides and the hydrolysis of phosphatidylinositol-4,5-bisphosphate (PIP_2), respectively. The diacylglycerol product of PIP_2 breakdown remains in the membrane domain and activates protein kinase C (Takai et al., 1982), while $InsP_3$ is thought to increase intracellular free Ca^{2+} by mobilizing it from the endoplasmic reticulum (Streb et al., 1983). The mechanism by which protein kinase C and Ca^{2+} turn on genes and induce specific cell cycle transitions is not yet clear.

Neoplastic transformation reduces or eliminates the need for extracellular Ca^{2+} and for growth factors by altering various parts of this signal transduction mechanism. Thus, some oncogenic viruses produce products which function as growth factors, while others make proteins which function as growth factor receptors, protein kinases, or GTP-binding proteins that are necessary for coupling enzymes with substrates.

Through the understanding of the normal signal transduction processes we are finally obtaining useful information about the mechanisms involved in neoplastic transformation. It is increasingly clear that neoplasia can result via alteration of several independent pathways, each, however, with the same result–aberrant proliferative behavior. It is only by understanding these mechanisms that we will find a better rationale for modulating and treating the disease.

ACKNOWLEDGMENT. Supported in part by NIH/NCI grant CA39745 and CA42942.

REFERENCES

Balk, S. D., Whitfield, J. F., Youdale, T., and Braun, A. C., 1973, Roles of calcium, serum, plasma, and folic acid in the control of proliferation of normal and Rous sarcoma virus-infected chicken fibroblasts, *Proc. Natl. Acad. Sci.* **70**:675–679.

Bonney, C., Fink, D., Schlichter, D., Carr, K., and Wicks, W., 1983, Direct evidence that the protein kinase catalytic subunit mediates the effects of cAMP on tyrosine aminotransferase synthesis, *J. Biol. Chem.* **258**:4911–4918.

Bossi, D., Whitfield, J. F., and Boynton, A. L., 1985, The influence of extracellular calcium on the distribution of protein kinase C in non-neoplastic and neoplastic rat liver cells. *Cancer Letts.* **26**:303–310.

Boynton, A. L., and Whitfield, J. F., 1976, Different calcium requirements for proliferation of conditionally and unconditionally tumorigenic mouse cells, *Proc. Natl. Acad. Sci.* **73**:1651–1654.

Boynton, A. L., Kleine, L. F., and Whitfield, J. F., 1983, Ca^{2+}/phospholipid-dependent protein kinase activity correlates to the ability of transformed liver cells to proliferate in Ca^{2+}-deficient medium, *Biochem. Biophys. Res. Commun.* **115**:383–390.

Boynton, A. L., Kleine, L. P., and Whitfield, J. F., 1984, Relation between colony formation in calcium-dependent medium, colony formation in soft agar, and tumor formation by T51B rat liver cells, *Cancer Lett.* **21**:293–302.

Boynton, A. L., Kleine, L. P., and Whitfield, J. F., 1985a, Cyclic AMP elevators stimulate the initiation of DNA synthesis by calcium-deprived rat liver cells, in: *Control of Animal Cell Proliferation*, Vol. I (A. L. Boynton and H. L. Leffert, eds.), Academic Press, Orlando, pp. 122–150.

Boynton, A. L., Kleine, L. P., Whitfield, J. F., and Bossi, D., 1985b, Involvement of the Ca^{2+}/phospholipid-dependent protein kinase in the G1 transit of T51B rat liver epithelial cells, *Exp. Cell Res.* **160**:197–205.

Boynton, A. L., Swierenga, S. H. H., and Whitfield, J. F., 1981, The calcium independence of neoplastic cell proliferation: A promising tool for carcinogen detection, in: *Short-term Tests for Chemical Carcinogens* (H. F. Stich and R. H. San, eds.), Springer-Verlag, New York, pp. 362–371.

Boynton, A. L., Whitfield, J. F., Isaacs, R. J., and Morton, H. J., 1974, Control of 3T3 cell proliferation by calcium, *In Vitro* **10**:12–17.

Fleischman, L. F., Chahwala, S. B., and Cantley, L., 1986, Ras-transformed cells: altered levels of phosphatidylinositol-4,5-bisphosphate and catabolites, *Science* **231**:309–312.

Freeman, A. E., Calisher, C. H., Price, P. J., Turner, H. C., and Huebner, R. J., 1966, Calcium sensitivity of cell cultures derived from adenovirus-induced tumors, *Proc. Soc. Exp. Biol. Med.* **122**:835–840.

Freeman, A. E., Hollings, S., Price, P. J., and Calisher, C. H., 1965, The effect of calcium on cell lines derived from adenovirus type 12-induced hamster tumors, *Exp. Cell Res.* **39**:259–264.

Gomperts, B. D., 1984, Involvement of guanine nucleotide-binding protein in the gating of Ca^{2+} receptors, *Nature,* **306**:64–66.

Grynkiewicz, G., Poenie, M., and Tsien, R. Y., 1985, A new generation of Ca^{2+} indicators with greatly improved fluorescence properties, *J. Biol. Chem.* **260:**3440–3450.

Hazelton, B., Mitchell, B., and Tupper, J., 1979, Calcium, magnesium, and growth control in the WI-38 human fibroblast cell, *J. Cell. Biol.* **83:**487–498.

Hennings, H., Michael, D., Chen, C., Steinert, P., Holbrook, K., and Yuspa, S., 1980, Calcium regulation of growth and differentiation of mouse epidermal cells in culture, *Cell* **19:**245–254.

Hesketh, R., Smith, G. A., and Metcalfe, J. C., 1987, Intracellular calcium and normal eukaryotic cell growth, in: *Control of Animal Cell Proliferation* (A. L. Boynton and H. L. Leffert, eds.), Academic Press, Orlando, pp. 395–434.

Lechner, J. F., and Kaighn, M. E., 1981, Reduction of the calcium requirement of normal human epithelial cells by EGF, *Exp. Cell. Res.* **121:**432–435.

Naghshineh, S., Noguchi, M., Huang, K. P., and Londos, C., 1986, Activation of adipocyte adenylate cyclase by protein kinase C, *J. Biol. Chem.* **261:**14534–14538.

Streb, H., Irvine, R. F., Berridge, M. J., and Schulz, I., 1983, Release of Ca^{2+} from a non-mitochondrial intracellular store in pancreatic acinar cells by inositol-1,4,5-trisphosphate, *Nature* **306:**67–69.

Swierenga, S. H. H., 1984, Use of low calcium medium in carcinogenicity testing: studies with rat liver cells, in: *In Vitro Models for Cancer Research* Vol. II (M. M. Weber and L. T. Skely, eds.), CRC Press, Boca Raton, pp. 61–89.

Swierenga, S. H. H., Whitfield, J. F., Boynton, A. L., MacManus, J. P., Rixon, R. H., Sikorska, M., Tsang, B. K., and Walker, P. R., 1980, Regulation of proliferation of normal and neoplastic rat liver cells by calcium and cyclic AMP, *Ann. N.Y. Acad. Sci.* **349:**294–311.

Swierenga, S. H. H., Whitfield, J. F., and Karasaki, S., 1978, Loss of proliferative calcium dependence: Simple *in vitro* indicator of tumorigenicity, *Proc. Natl. Acad. Sci.* **75:**6069–6072.

Takai, Y., Kishimoto, A., and Nishizuka, Y., 1982, Calcium and phospholipid turnover as transmembrane signalling for protein phosphorylation, in: *Calcium and Cell Function* (W. Y. Cheung, ed.) Academic Press, Orlando, pp. 385–412.

Whitfield, J. F., Boynton, A. L., Rixon, R. H., and Youdale, T., 1985, The control of cell proliferation by calcium, Ca^{2+}-calmodulin, and cyclic AMP, in: *Control of Animal Cell Proliferation*, Vol. I, (A. L. Boynton and H. L. Leffert, eds.), Academic Press, Orlando, pp. 332–366.

Whitfield, J. F., Durkin, J. P., Franks, D. J., Kleine, L. P., Raptis, L., Rixon, R. H., Sikorska M., and Walker, P. R., 1987, Calcium, cyclic AMP and protein kinase C-partners in mitogenesis, Cancer and Metastasis Reviews, **5:**205–250.

31

Role and Properties of Ligand-Induced Calcium Fluxes in Lymphocytes

SERGIO GRINSTEIN, STEPHEN MACDOUGALL, ROY K. CHEUNG, and ERWIN W. GELFAND

Binding of foreign antigens to the surface of resting B lymphocytes activates and induces them to proliferate, producing an expanded and functionally responsive population. *In vitro*, the initial steps of this activation process can be mimicked by crosslinking surface immunoglobulins (Ig), which function as antigen receptors on the plasma membrane of B lymphocytes. Crosslinking can be readily achieved using antibodies directed against Ig (anti-Ig). In most of the experiments discussed below, human tonsillar B cells were activated with antibodies against IgM, which is present on the surface of over 60% of the cells.

The molecular events that underlie activation of B lymphocytes by antigens or by anti-Ig are not completely understood. It is clear, however, that crosslinking of Ig leads to an increase in cytoplasmic free Ca^{2+} ($[Ca^{2+}]_i$). This was originally inferred from measurements of isotopic Ca^{2+} distribution (Braun *et al.*, 1979) and later confirmed by direct determinations of $[Ca^{2+}]_i$ by means of fluorescent probes such as Quin2 and indo-1 (Pozzan *et al.*, 1982; Bijsterbosch *et al.*, 1986). The increased $[Ca^{2+}]_i$ originates at least in part from release of Ca^{2+} from intracellular stores, inasmuch as anti-Ig produces an elevation of $[Ca^{2+}]_i$ in cells suspended in Ca^{2+}-free media containing divalent cation chelators (Pozzan *et al.*, 1982; Bijsterbosch *et al.*, 1986). The source of intracellular Ca^{2+} is believed to be the endoplasmic reticulum, which has been shown to release the cation when challenged with inositol-1,4,5-trisphosphate. The latter compound is known to be liberated in the cytoplasm of B cells following treatment with anti-Ig (Ransom *et al.*, 1986).

A second but sustained component of the anti-Ig-induced increase in $[Ca^{2+}]_i$ is dependent on the presence of extracellular Ca^{2+}. Though the precise mode of entry has not been elucidated, it is generally assumed that uptake occurs through ligand-activated Ca^{2+} channels. The purpose of the studies outlined below was to test this hypothesis experimentally. Two criteria must be fulfilled if the sustained increase in $[Ca^{2+}]_i$ is mediated by ligand-induced Ca^{2+} channels: first, the $[Ca^{2+}]_i$ changes must be a function of the electrochemical Ca^{2+}

SERGIO GRINSTEIN, STEPHEN MACDOUGALL, ROY K. CHEUNG, and ERWIN W. GELFAND ● Divisions of Cell Biology and Immunology and Rheumatology, Research Institute, Hospital for Sick Children, Department of Biochemistry, University of Toronto, Ontario M5G IX8, Canada.

gradient, the force driving the flux of this cation through a conductive pathway; second, a Ca^{2+}-selective conductance increase must be detectable following crosslinking of Ig.

To test the first prediction, we measured the changes in $[Ca^{2+}]_i$ induced by anti-Ig in normal and depolarized cells. Unlike fluxes through electroneutral pathways, changing the membrane potential and thereby the electrochemical Ca^{2+} gradient is likely to affect the magnitude of the flux through Ca^{2+} channels. As shown in Fig. 1, depolarization by elevation of the extracellular K^+ concentration significantly depressed the magnitude of the $[Ca^{2+}]_i$ change (cf. Fig. 1A and C). These experiments were performed in cells loaded with 1,2-bis(2-aminophenoxy)ethane-N,N,N',N'-tetraacetate (BAPTA), a Ca^{2+} chelating agent that obliterates the $[Ca^{2+}]_i$ change in response to Ca^{2+} release from stores, thereby facilitating the analysis of the sustained, external Ca^{2+}-dependent uptake mechanism (see below). The decrease in Ca^{2+} uptake was not due to the omission of Na^+, since replacement with impermeant organic cations, which preserve or even slightly hyperpolarize the membrane potential, did not inhibit Ca^{2+} uptake (Fig. 1B). It is noteworthy that the baseline $[Ca^{2+}]_i$ was not significantly affected by depolarization, suggesting that the Ca^{2+} uptake pathway differs from the well-documented voltage-gated channels of excitable tissues. Similar results have been reported in T lymphocytes, where lectin-induced $[Ca^{2+}]_i$ changes are decreased not only by increasing $[K^+]$, but also by depolarization with ionophores such as gramicidin and nystatin (Gelfand et al., 1984). More recently, depolarization was also shown to affect ligand-induced Ca^{2+} uptake in rat basophilic leukemia cells (Mohr and Fewtrell, 1987) and in neutrophils (DiVirgilio et al., 1987). These observations are consistent with opening of conductive Ca^{2+}-selective channels upon interaction of the ligands with their receptors.

The opening of Ca^{2+} channels by crosslinking of surface receptors should be associated with an increase in membrane conductance. In T lymphocytes, a macroscopic increase in conductance was indeed detected in activated cells (DeCoursey et al., 1984; Matteson and Deutsch, 1984). The change was, however, attributed to an elevated K^+ channel activity.

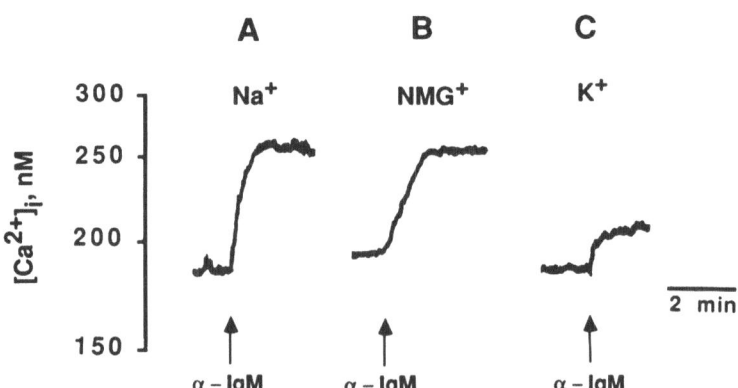

FIGURE 1. Effect of depolarization of the membrane potential on the $[Ca^{2+}]_i$ change induced by anti-Ig. Human tonsillar B lymphocytes were loaded with indo-1 and BAPTA by incubation in the presence of 1 and 10 μM, respectively, of the acetoxymethylester precursors of these agents. The cells were then suspended in media containing 140 mM of either Na^+ (A), NMG^+ (B), or K^+ (C). Where indicated by the arrows, the cells were treated with goat anti-human IgM (α-IgM). $[Ca^{2+}]_i$ was measured monitoring indo-1 fluorescence as described (Gelfand et al., 1986).

Because both mitogen-induced K^+ channel activity and proliferation are blocked by 4-aminopyridine, tetraethylammonium, and some Ca^{2+} channel blockers, it was suggested that Ca^{2+} enters the cells via these K^+ channels. However, direct measurements of $[Ca^{2+}]_i$ later demonstrated that the inhibitors did not significantly block mitogen-induced Ca^{2+} uptake, and that their inhibitory effect on proliferation was not specific, i.e., it was independent of changes in $[Ca^{2+}]_i$ (Gelfand et al., 1986). More recently, patch-clamping experiments in T lymphocytes revealed the opening of Ca^{2+}-selective channels upon treatment with mitogenic lectins (Kuno et al., 1986), and it has been suggested that inositol-1,4,5-trisphosphate may mediate this process (Kuno and Gardner, 1987).

To our knowledge, direct electrophysiological recordings have not been performed in B cells. However, indirect measurements using fluorescent dyes have been reported for murine B lymphocytes (Monroe and Cambier, 1983). In these studies, a sizable depolarization was found when the cells were treated with anti-Ig, a finding consistent with opening of Ca^{2+} channels, considering the positive reversal potential of this ion. Unfortunately, the ionic basis of the reported depolarization was not investigated. Moreover, because flow cytometry was used, the early stages of the time course of the membrane potential changes was not explored in detail. In these studies, we decided to reanalyze the changes in membrane potential that accompany the activation of B cells. For this purpose, we made continuous recordings of the fluorescence of bis-oxonol, an anionic membrane potential probe that is not accumulated by mitochondria. Contrary to our expectations, based on earlier reports, we found that treatment of human tonsillar B cells with anti-Ig resulted in a moderate (<10 mV) *hyperpolarization*, not depolarization. This potential change was still present when N-methyl-D-glucammonium$^+$ (NMG^+) was used to substitute extracellular Na^+, indicating that the latter cation is not involved in the hyperpolarization process (MacDougall et al., 1987). Several lines of evidence suggested that the hyperpolarization was due to a Ca^{2+}-sensitive K^+ conductance: First, a similar potential change was obtained when $[Ca^{2+}]_i$ was elevated to comparable levels by means of ionomycin, a nonconductive ionophore. Second, the shape of the hyperpolarization mirrored the changes in $[Ca^{2+}]_i$. Under normal conditions, the biphasic increase in $[Ca^{2+}]_i$ was paralleled by a biphasic hyperpolarization. Delaying the development of the $[Ca^{2+}]_i$ response by increasing the Ca^{2+}-buffering power with BAPTA delayed the hyperpolarization. Conversely, eliminating the sustained phase of the $[Ca^{2+}]_i$ response by omission of external Ca^{2+} abolished the prolonged hyperpolarization. As predicted, the hyperpolarization was completely prevented when cells loaded with BAPTA were treated with anti-Ig in Ca^{2+}-free medium (MacDougall et al., 1987). Taken together, these results suggest that Ca^{2+}-activated K^+ channels are present in B cells, but provide no evidence for the existence of a conductive Ca^{2+} pathway.

When B cells suspended in Ca^{2+}-free medium were activated with anti-Ig, not only was the sustained hyperpolarization absent, but a marked *depolarization* was unmasked. This depolarization was found to be strictly dependent on extracellular Na^+, as it was eliminated by substitution by NMG^+. By analogy with voltage-gated Ca^{2+} channels (Hess and Tsien, 1984), it is conceivable that omission of extracellular Ca^{2+} altered the selectivity of the anti-Ig-induced Ca^{2+} channels, increasing their conductance to Na^+. This possibility was reinforced by experiments using verapamil or dihydropyridine channel blockers. Though much less sensitive than the voltage-sensitive Ca^{2+} channels, ligand-activated Ca^{2+} uptake in B cells was also inhibited by verapamil and by nifedipine and nitrendipine (at micromolar concentrations). Similarly, the Na^+-dependent depolarization induced by anti-Ig in cells suspended in Ca^{2+}-free solutions was also blocked by micromolar concentrations of these drugs. This further suggests that Na^+ permeates through ligand-gated Ca^{2+} channels when the cells are activated in the absence of Ca^{2+}.

Though indirect, these data are consistent with the activation of Ca^{2+} channels by anti-Ig. Failure to detect a depolarization associated with rheogenic Ca^{2+} influx can be readily explained by the existence of a large resting K^+ conductance, which is further increased when $[Ca^{2+}]_i$ is elevated (see above). We reasoned that the induction of Ca^{2+} channels could be unmasked by either reducing the conductance to other ions (e.g., by means of blockers) or by increasing the magnitude of the Ca^{2+} current. The experiments discussed below used the latter approach. The rationale of the experimental design will be described initially, followed by the results obtained.

Figure 2A shows that, as mentioned above, the $[Ca^{2+}]_i$ change induced by anti-Ig is biphasic, being composed of an early, transient phase that can be largely attributed to release from internal stores and a sustained phase that is external Ca^{2+}-dependent. The latter reflects transmembrane Ca^{2+} uptake, likely through conductive channels. As the cells are loaded with progressively higher concentrations of BAPTA (traces *b–d*), the early phase is gradually obliterated. In contrast, the sustained plateau phase, though attained later, remains otherwise unaltered (Figs. 2A and B). Because the Ca^{2+}-buffering power of the cells increases markedly along with the intracellular concentration of BAPTA, the constancy of the plateau $[Ca^{2+}]_i$ level implies that increasing amounts of Ca^{2+} must have entered the cell. In other words, considerably more Ca^{2+} must have entered the cells to attain the same final $[Ca^{2+}]_i$ while overcoming a much larger buffering power. If the uptake mechanism were conductive, this maneuver would be expected to increase the associated inward current, favoring detectability.

Based on the above premise, we tried to detect an inward Ca^{2+} current in BAPTA-loaded cells by monitoring bisoxonol fluorescence. Typical results are shown in Fig. 3. Pretreatment with BAPTA did not significantly affect the resting potential. Upon treatment with anti-Ig, however, a small but reproducible depolarization was observed in BAPTA-loaded cells, followed by a moderate hyperpolarization. The latter likely reflects Ca^{2+}-induced K^+ efflux and is equivalent to the hyperpolarization recorded in untreated (BAPTA-free) cells (see p. 285). The initial depolarization could represent Ca^{2+} uptake. This possibility was analyzed by ion substitution and by inhibition studies with verapamil and dihydropyridines. A depolarization could be caused by increased permeability to ions with equilibrium potentials more positive than the resting potential, such as Na^+, Cl^-, or Ca^{2+}. As shown in Fig. 3, the transient initial depolarization induced by anti-Ig was unaffected when Na^+ was replaced by NMG^+ (Fig. 3B) or when Cl^- was replaced by gluconate$^-$ (Fig. 3C). Similar results were obtained when the cells were preincubated in gluconate$^-$, resulting in depletion of intracellular Cl^-. These results are not compatible with a role for Na^+ or Cl^- in the depolarization but are consistent with a Ca^{2+} conductance.

As mentioned before, removal of extracellular Ca^{2+} prior to stimulation of the cells leads to a marked depolarization upon addition of anti-Ig. This response was also observed in BAPTA-loaded cells (Fig. 3D). However, in this instance, reintroduction of Ca^{2+} to the medium induced a further, transient depolarization, which was then followed by a repolarization that probably reflects opening of the Ca^{2+}-sensitive K^+ channels. The initial depolarization is dependent on extracellular Na^+, as it is not observed in cells suspended in NMG^+ (Fig. 3E). However, the depolarization noted upon addition of Ca^{2+} was present in both cases, suggesting entry of Ca^{2+} through anti-Ig-induced Ca^{2+} channels. This depolarization could be inhibited by high (micromolar) doses of verapamil and by dihydropyridine blockers (not illustrated). Taken together, these data are consistent with the opening of Ca^{2+} conductive channels in B cells upon crosslinking of surface immunoglobulins.

Two hypotheses could explain the observation that, after addition of anti-Ig, $[Ca^{2+}]_i$ attains a constant level regardless of the amount of BAPTA trapped in the cells (Fig. 2).

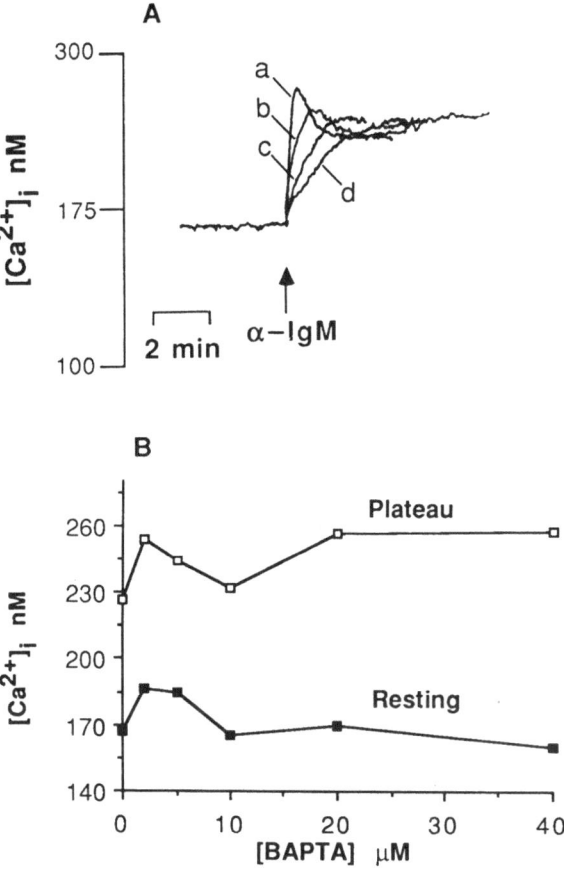

FIGURE 2. Effect of increasing BAPTA concentrations on the $[Ca^{2+}]_i$ changes induced by anti-Ig. (A) Cells were loaded in media containing 1 μM indo-1 AM in the absence (curve a) or presence (curves b–d) of increasing concentrations of BAPTA-AM (2.5–4.0 μM). The cells were then suspended in Ca^{2+}-containing Na^+-rich medium and, where indicated by the arrow, stimulated by addition of goat anti-human IgM (α-IgM). $[Ca^{2+}]_i$ was measured as in Fig. 1. (B) Summary of $[Ca^{2+}]_i$ measurements in cells loaded with the concentration of BAPTA-AM indicated in the abscissa. The resting values (solid symbols) refer to measurements made prior to addition of anti-Ig. Plateau values (open symbols) refer to the steady levels attained following stimulation with anti-Ig.

The constancy of $[Ca^{2+}]_i$ could indicate a fortuitous equilibrium between influx through the resting plus the activated Ca^{2+} pathways and efflux through the Ca^{2+} pump (no evidence exists in lymphoid cells for Na^+–Ca^{2+} exchange). However, it is also possible that the rate of entry of Ca^{2+} through the channels is regulated by $[Ca^{2+}]_i$, such that channel closure occurs at a threshold $[Ca^{2+}]_i$ level. The latter model predicts that treatment with anti-Ig will have little effect on the channels if $[Ca^{2+}]_i$ is increased, by alternative means, to or above the threshold prior to addition of the antibody. To test this hypothesis, cells were pulsed

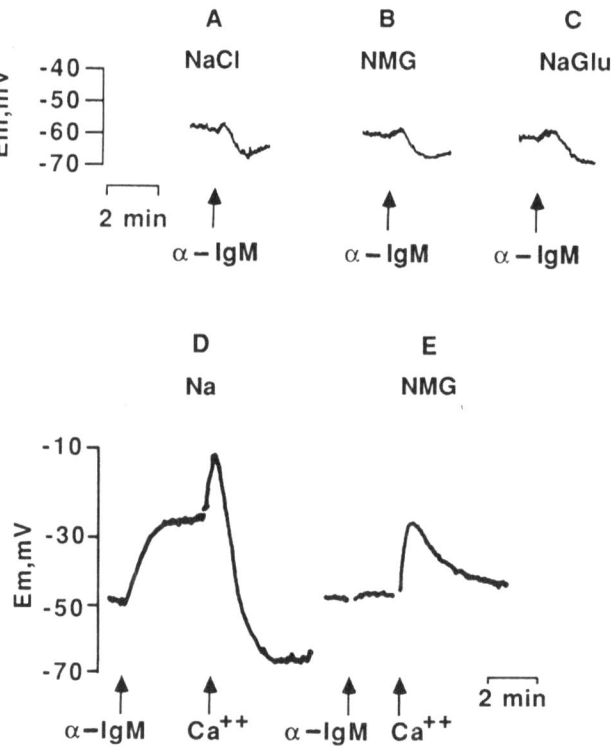

FIGURE 3. Changes in membrane potential induced by anti-Ig in BAPTA-loaded B lymphocytes. Cells loaded by incubation with 10 μM BAPTA-AM were equilibrated with bis-oxonol in media containing 140 mM of either NaCl (A and D), NMG-Cl (B and E), or Na gluconate (C) in the presence (A–C) or absence (D–E) of 1 mM CaCl$_2$. Ca^{2+}-free media were supplemented with 1 mM EGTA. Where indicated, the cells were stimulated with anti-Ig. In D and E, this was followed by the addition of 2 mM CaCl$_2$. Membrane potential was estimated from the fluorescence emission of bis-oxonol, and calibrated using gramicidin and mixtures of Na$^+$ and NMG$^+$ as in Gelfand et al. (1986).

with low concentrations of ionomycin before addition of anti-Ig, while continuously monitoring [Ca^{2+}]$_i$. A representative experiment is illustrated in Fig. 4. When the cells were prepulsed with ionomycin, the antibody still produced an increase in [Ca^{2+}]$_i$. However, the two components of the [Ca^{2+}]$_i$ response were affected differently: the transient increase persisted even at the higher concentration of ionomycin, while the sustained increase was gradually reduced as the baseline [Ca^{2+}]$_i$ was elevated. Data from several determinations are summarized in the bottom panel of Fig. 4, which shows that the plateau response was virtually eliminated when the baseline [Ca^{2+}]$_i$ was increased by $\approx$150 nM, a value comparable to the sustained increase normally produced by anti-Ig. A slight decrease in the magnitude of the transient response was also noted, which could be due to partial depletion

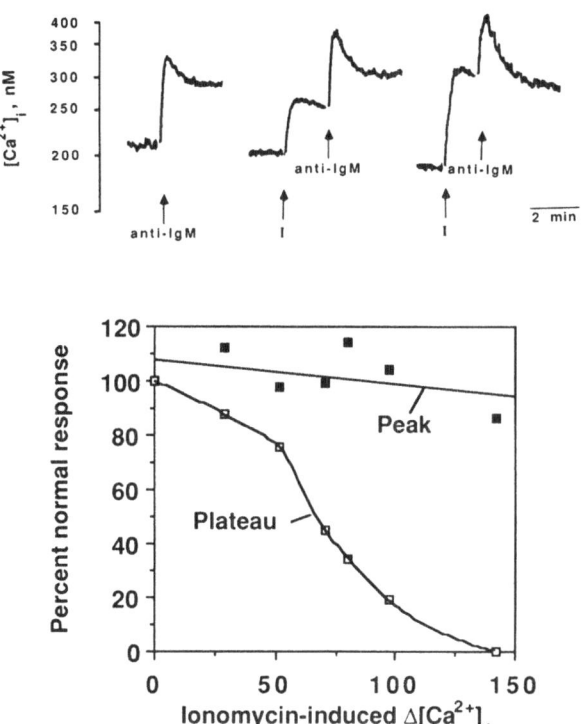

FIGURE 4. Effects of anti-Ig on $[Ca^{2+}]_i$ in cells pretreated with varying concentrations of io-nomycin. $[Ca^{2+}]_i$ was monitored in cells loaded with indo-1 and suspended in Ca^{2+}-containing Na^+ medium. Where indicated, the cells were treated with 5 and 10 nM ionomycin (I; middle and right trace, respectively). The cells were then activated with goat anti-human anti-IgM. Lower panel: the fractional responses induced by anti-IgM are plotted as a function of the increase in $[Ca^{2+}]_i$ produced by the ionomycin pretreatment. The peak of the transient response (solid symbols) and the level of the sustained response (plateau; open symbols) are plotted independently.

of the intracellular stores by the ionophore. The progressive inhibition of the sustained response to anti-Ig by previous elevations in $[Ca^{2+}]_i$ is consistent with the model in which the ligand-activated channels are regulated by $[Ca^{2+}]_i$.

What is the relevance of increased Ca^{2+} uptake to the biological responses induced by ligand binding? In T lymphocytes, it is well established that omission of extracellular Ca^{2+} impairs the proliferative response induced by mitogenic lectins. However, it is not clear whether this results from the elimination of Ca^{2+} influx from the medium, or from the depletion of intracellular stores, which need to be mobilized during activation. We have tried to establish whether the increase in $[Ca^{2+}]_i$ produced by release of internal stores is required for proliferation, by measuring thymidine incorporation into cells loaded with BAPTA. At concentrations where the chelator completely prevented the $[Ca^{2+}]_i$ transient associated with release of intracellular stores, cellular proliferation was unaffected (Gelfand *et al.*,

1988). We therefore conclude that influx of extracellular Ca^{2+} is the essential component of the Ca^{2+} response elicited by mitogenic lectins.

In summary, we have presented preliminary evidence supporting the existence of conductive Ca^{2+} channels activated in B lymphocytes by crosslinking of surface receptors. These ligand-activated channels resemble the voltage-gated Ca^{2+} channels of excitable cells in that they are selective yet can become permeable to Na^+ in the absence of extracellular Ca^{2+}. Moreover, like the voltage-sensitive channels, the ligand-activated pathway also appears to be regulated by $[Ca^{2+}]_i$. However, the latter channel differs from the voltage-gated channels in that it cannot be activated by depolarization and is susceptible to inhibition by verapamil and dihydropyridines only at micromolar concentrations. Much lower concentrations are usually required to block the L-type voltage-sensitive channels.

It is not clear whether the channels that are activated by binding of ligands to their receptors are functional in cells in the resting state (i.e., in the absence of ligand). However, it is tempting to speculate that the same channels do operate in resting cells, contributing to basal $[Ca^{2+}]_i$ homeostasis, and that ligand binding simply shifts their $[Ca^{2+}]_i$ dependence such that a higher $[Ca^{2+}]_i$ threshold must be reached to induce channel closure.

ACKNOWLEDGMENTS. The original work by the authors summarized in this chapter was supported by the National Cancer Institute (Canada) and the Medical Research Council of Canada. S. G. is the recipient of a Medical Research Council Scientist Award.

REFERENCES

Bijsterbosch, M. K., Rigley, K. P., and Klaus, G. G. B., 1986, Crosslinking of surface immunoglobulin on B lymphocytes induces both intracellular Ca^{2+} release and Ca^{2+} influx: analysis with indo-1, *Biochem. Biophys. Res. Commun.* 137:500–506.

Braun, J., Sha'afi, R. I., and Unanue, E. R., 1979, Crosslinking by ligands to surface immunoglobulins triggers mobilization of intracellular $^{45}Ca^{2+}$ in B lymphocytes, *J. Cell. Biol.* 82:755–782.

DeCoursey, T. E., Chandy, K. G., Gupta, S., and Cahalan, M. D.,1984, Voltage-gated K^+ channels in human T lymphocytes: A role in mitogenesis?, *Nature* 307:465–467.

DiVirgilio, F., Lew, D. P., Andersson, T., and Pozzan, T., 1987, Plasma membrane potential modulates chemotactic peptide-stimulated cytosolic free Ca^{2+} changes in human neutrophils, *J. Biol. Chem.* 262:4574–4579.

Gelfand, E. W., Cheung, R. K., and Grinstein, S, 1984, Role of membrane potential in the regulation of lectin-induced calcium uptake, *J. Cell. Physiol.* 121:533–539.

Gelfand, E. W., Cheung, R. K., and Grinstein, S., 1986, Mitogen-induced changes in Ca^{2+} permeability are not mediated by voltage-gated K^+ channels, *J. Biol. Chem.* 261:11520–11523.

Gelfand, E. W., Cheung, R. K., and Grinstein, S., 1988, Uptake of extracellular calcium, and not release from internal stores, is essential for T-lymphocyte proliferation, *Eur. J. Immun.* 18:917–922.

Hess, P., and Tsien, R. W., 1984, Mechanism of ion permeation through calcium channels, *Nature* 309:453–456.

Kuno, M., and Gardner, P., 1987, Ion channels activated by inositol 1,4,5-trisphosphate in plasma membrane of human T lymphocytes, *Nature* 326:301–304.

Kuno, M., Goronzy, J., Weyand, C. M., and Gardner, P., 1986, Single-channel and whole cell recordings of mitogen-regulated inward currents in human cloned helper T lymphocytes, *Nature* 323:269–273.

MacDougall, S. L., Grinstein, S., and Gelfand, E. W., 1987, Activation of Ca^{2+}-dependent K^+ channels in human B lymphocytes by anti-immunoglobulin, *J. Clin. Invest.* 81:449–454.

Matteson, D. R., and Deutsch, C., 1984, K⁺ channels in T lymphocytes: A patch-clamp study using monoclonal antibody adhesion, *Nature* **307**:468–470.

Mohr, F. C., and Fewtrell, C., 1987, Depolarization of rat basophilic leukemia cells inhibits calcium uptake and exocytosis, *J. Cell. Biol.* **104**:783–792.

Monroe, J. C., and Cambier, J. C., 1983, B cell activation. I. Receptor cross-linking by anti-immunoglobulin antibodies induces a rapid decrease in B cell plasma membrane potential, *J. Exp. Med.* **157**:2073–2080.

Pozzan, T., Arslan, P., Tsien, R. Y., and Rink, T. J., 1982, Anti-immunoglobulin, cytoplasmic free calcium and capping in B lymphocytes, *J. Cell Biol.* **94**:335–353.

Ransom, J. T., Harris, L. K., and Cambier, J. C., 1986, Anti-Ig induces release of inositol 1,4,5 trisphosphate which mediates mobilization of intracellular Ca^{2+} stores in B lymphocytes, *J. Immunol.* **137**:708–713.

32

Modulation of c-*fos* and c-*myc* Expression by Effectors of Ion Movements

ANTHONY F. CUTRY, ALAN J. KINNIBURGH,
KIRK J. LEISTER, and C. E. WENNER

1. INTRODUCTION

The role of ion movements in signal transduction and macromolecular biosynthesis has been well substantiated. For example, Ca^{2+} mobilization is an essential component of growth factor signaling where there is an involvement of protein kinase C. K^+ movements, mediated by the Na^+, K^+-ATPase membrane pump, are necessary not only for maintaining the membrane potential gradient but also for macromolecular synthesis, i.e., protein synthesis and subsequent DNA replication.

In our studies we have attempted to further clarify the roles that these two particular cations play in cell cycle activation and associated early genomic events. Apparently, the degree of cellular responsiveness to these ions varies greatly from cell type to cell type. For example, consider when extracellular calcium influx is involved. "Excitable" cells, such as muscle or nerve cells, almost always possess voltage-sensitive calcium channels, while "nonexcitable" cells, such as fibroblasts, may not possess these gates. The absence or presence of such calcium gates can have a major impact on how a cell responds to a given stimulus. This is only one example of how different cell types may display differential responses to effectors of ion movements, in this instance, calcium movements.

Calcium influx has previously been demonstrated to activate transcription of the c-*fos* protooncogene in PC-12 cells (Morgan and Curran, 1986). More recently, c-*fos* activation by epidermal growth factor (EGF) has been postulated to occur via EGF stimulation of the Na^+–H^+ antiport system (Leffert et al., 1987), an event closely followed by activation of the Na^+, K^+-ATPase. Hence, monovalent as well as divalent cation movements may play an integral role in initiation of early genetic events associated with cell cycle activation.

Here, we report the effects of various modulators of ion movements on the aforemen-

ANTHONY F. CUTRY, KIRK J. LEISTER, and C. E. WENNER • Department of Biochemistry, Roswell Park Memorial Institute, Buffalo, New York 14263.   ALAN J. KINNIBURGH • Department of Human Genetics, Roswell Park Memorial Instiute, Buffalo, New York 14263.

tioned genetic events (c-*fos* and c-*myc* activation) in C3H 10T1/2 mouse fibroblasts. We have tried to delineate the effects of such ion movements on macromolecular synthetic events (DNA and protein synthesis). We have attempted to correlate these findings with those observed at the level of gene expression. Our results suggest that ion movements can play an integral role in cell cycle progression and activation of early events associated with cell cycle progression.

2. ROLE OF CALCIUM

2.1. Calcium and Activation of c-*fos* and c-*myc* Protooncogenes

Following stimulation with a variety of growth factors, it has been observed by several groups that there is a rapid and transient increase in the mRNA levels of both the c-*fos* and c-*myc* oncogenes (Kelly *et al.*, 1983; Campisi *et al.*, 1984; Greenberg and Ziff, 1984; Verma, 1986). Both genes encode proteins that localize in the nucleus. Both c-*myc* and c-*fos* have counterparts in acutely transforming retroviruses, hence their involvement in neoplasia. Several lines of evidence indicate that c-*myc* plays an integral role in proliferation. Antibodies to c-*myc* inhibit DNA replication. *Myc* may also activate a DNA polymerase, if indeed the antibody used in these studies is specific for c-*myc* and not a blocker of DNA polymerase activity in general (Studzinski *et al.*, 1986). Finally, the observation that constitutive *myc* expression favors a proliferative modality for the cell over differentiation supports the idea that continued c-*myc* expression produces an "immortalized" phenotype (Langdon *et al.*, 1986). The function of the *fos* gene is enigmatic at this time. It is activated rapidly by growth factors that result in cell proliferation (Greenberg and Ziff, 1984) as well as differentiation agents (Kruijer *et al.*, 1985). The actual function of the *fos* gene product may be to prime the genome for forthcoming genetic events associated with proliferation and/or differentiation.

Clearly, it is of interest to attempt to identify pathways which are responsible for the activation of these two genes given their intimate ties to cellular functions in proliferation and differentiation. Furthermore, since the way in which mitogens trigger the activation of these genes and other "early genes" such as those identified by Lau and Nathans (1987) remains unclear, we must identify the components of the growth factor signaling pathways that can affect these genetic elements.

We set out to investigate the role of ion movements in the stimulation of these proto-oncogenes, since changes in ion movements occur very rapidly following stimulation of quiescent cells (Hesketh *et al.*, 1985). While this idea is not novel, since other previously mentioned groups have demonstrated induction of c-*fos* with calcium, we have obtained results which are unique to our cell type. These may help to illustrate the differential responses of individual cell types to the same effectors.

We used the C3H 10T1/2 mouse fibroblast cell line and their methylcholanthrene (MCA)-transformed counterparts in our experiments. The cells are plated and allowed to reach confluency, after which we deprive them of growth factors by allowing them to exhaust the supply of these elements in the serum. Thus, we have a synchronous population of cells arrested in G_1/G_0. All experiments were performed with cell growth arrested in this manner.

Figure 1 shows the results of a set of representative experiments on c-*myc* (1a, top) and c-*fos* (1b, bottom) expression. For this part of the chapter, we will concern ourselves only with the effect of calcium on these two genes. At 5.8 mM $[K^+]_o$, which represents the unstimulated control, there is a very low level of c-*myc* mRNA and an even lower level of c-*fos* mRNA detectable. Stimulating the cells with the phorbol ester tumor promoter 12-O-tetradecanoyl-13-phorbol acetate (TPA), an activator of protein kinase C, gives a seven-fold increase in c-*myc* mRNA levels and a 9.5-fold increase in c-*fos* mRNA levels. If we treat the cells instead with the calcium ionophore A23187, a threefold increase and a 12-fold increase in c-*myc* and c-*fos* mRNA levels, respectively, is observed. Thus, there appears to be a calcium-mediated pathway through which stimulation of these protooncogenes can be achieved. After obtaining these results, we decided to try to stimulate the mRNA levels of these genes using calcium channel agonists and depolarizing concentrations of potassium, since this should cause the opening of voltage-sensitive calcium gates as demonstrated by Morgan and Curran (1986). We were not able to observe any stimulation of c-*fos* or c-*myc* with the calcium channel agonist BAY K 8644 or with 50 mM K$^+$. The results shown are for cells incubated in 50 mM K$^+$ for 2 hr. Experiments were also performed with 1-hr incubations and no increase in protooncogene mRNA levels was obtained. This result is consistent with C3H 10T1/2 cells being a nonexcitable cell type which probably does not possess voltage-sensitive calcium channels. It had been demonstrated previously that BAY K 8644 exerts its calcium influx activity through these voltage-sensitive calcium channels, otherwise referred to as "L" channels (Armstrong and Erxleben, 1986). Table I summarizes the fold inductions of c-*myc* and c-*fos* by the agents used, as determined by densitometric analysis of the autoradiographs.

A very interesting result obtained in these studies came from work done with the synthetic diacylglycerol, dicaproin. When the cells were treated with dicaproin, we observed a 10-fold increase in c-*fos* mRNA levels, but only a 1.5-fold increase in c-*myc* mRNA levels. This synthetic diacylglycerol is presumed to be exerting its effects through protein kinase C. We might expect to observe a greater stimulation of c-*myc*, but because of the constraints imposed on the experiment due to the time at which we isolated the RNA following addition of dicaproin (60 min, as indicated in the legend for Fig. 1), we only observed this marginal stimulation of c-*myc*. It is possible that dicaproin is being rapidly broken down by cellular enzymes in such a manner that we are able to observe significant stimulation of c-*fos* but not c-*myc* by the synthetic diacylglycerol. However, the more interesting result is obtained when dicaproin and A23187 are used simultaneously. In this case, a synergistic effect of the two agents on c-*fos* mRNA levels is observed. There is a 50-fold increase in c-*fos* mRNA levels over control values. Surprisingly, there is no such synergy with respect to *myc* expression; indeed, there appears to be an antagonistic effect of the two agents on c-*myc* expression. Control levels of c-*myc* mRNA (1.3-fold increase over control) were observed when the synthetic diacylglycerol and calcium ionophore are given together. We expected to find the induction we observed with A23187 alone, since the concentration and exposure time was the same when we used this agent alone or in conjunction with dicaproin (see the figure legend for concentrations used and exposure times), but that is clearly not the case. This may indicate that the pathway activated by A23187, presumably a calcium-calmodulin pathway, and the pathway activated by dicaproin, presumably protein kinase C, may interact with one another to play a regulatory role in the activation and deactivation of these genes.

3. ROLE OF POTASSIUM

3.1. Potassium and Modulation of c-*myc* and c-*fos* Expression

We have also investigated the role of potassium in the control of c-*myc* and c-*fos* expression in C3H 10T1/2 fibroblasts and their MCA 10T1/2 transformed counterparts. The

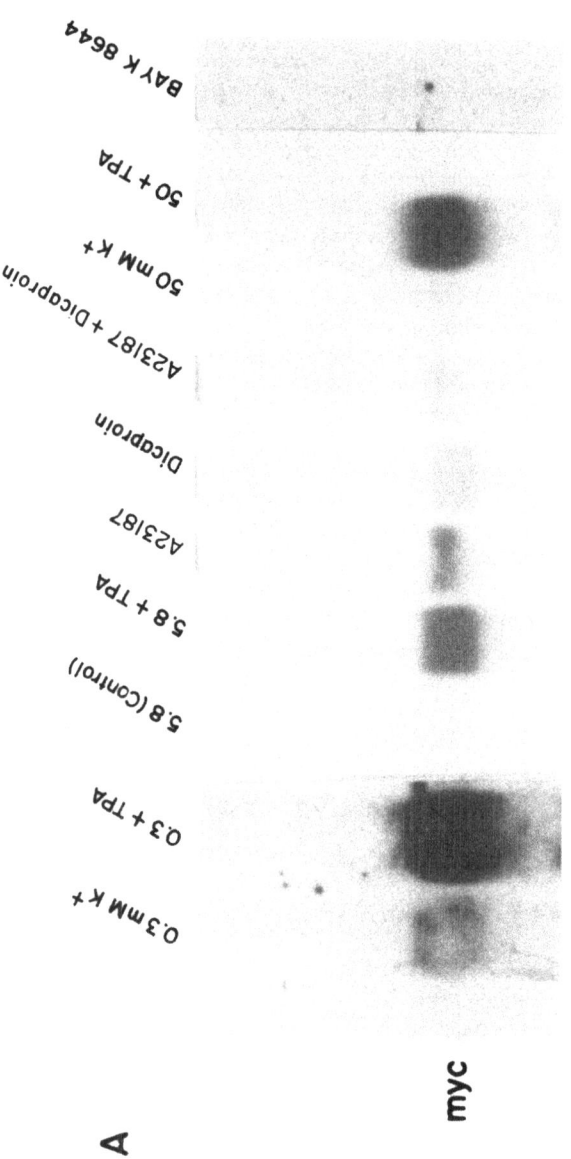

A

myc

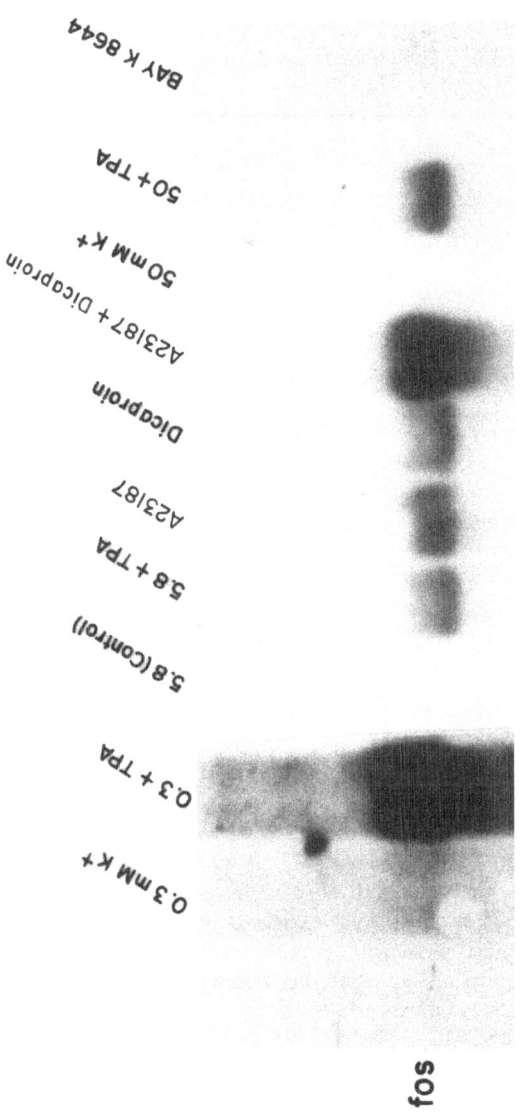

FIGURE 1. Modulation of c-*myc* (A) and c-*fos* (B) by effectors of ion movements. Postconfluent G_1-arrested C3H 10T1/2 fibroblasts were treated with the following reagents for the indicated times: A23187 (3 µg/ml, 60 min), dicaproin (100 µg/ml, 60 min), A23187 + dicaproin (same concentrations added simultaneously, 60 min), BAY K 8644 (600 nM, 60 min). All TPA treatments (10^{-7} M) were for 60 min. For 0.3 mM K^+, the cells were incubated in this media for 3 hr prior to isolating the RNA or adding TPA. For 50 mM K^+, samples were allowed to equilibrate in this media for 2 hr prior to harvesting. TPA was added to these samples for an additional 60 min following the 2-hr equilibration period. RNA was isolated by the method of Chirgwin et al. (1979). 10 µg of denatured (glyoxal) whole-cell RNA was electrophoresed on a 1.1% agarose gel, transferred to a nylon membrane (Hybond-N), and hybridized to random primer labeled [^{32}P]c-*myc* (A) or c-*fos* cDNA probe (B) at 68°C for 18 hr. The hybridized membrane was washed and then exposed to autoradiographic film for 3 days. To control for RNA amounts loaded, the bbt was also hybridized to a [^{32}P] triose phosphate isomerase cDNA probe (not shown), whose mRNA levels are relatively invariant throughout the cell cycle. Results showed no quantitative variation in the amount of RNA on each lane.

TABLE I. Densitometric Evaluation of c-*fos* and c-*myc* Stimulation by Effectors
of Ion Movements[a]

Sample	TPA	Relative c-*myc* mRNA	Relative c-*fos* mRNA
Experiment 1			
5.8 mM $[K^+]_o$ (control)	−	1.0	1.0
5.8 mM $[K^+]_o$	+	7.0	9.5
+ A23187	−	3.0	12.0
+ Dicaproin	−	1.5	10.7
+ BAY K 8644	−	1.0	1.0
+ A23187 + dicaproin	−	1.3	51.0
50 mM $[K^+]_o$	−	0.9	1.0
50 mM $[K^+]_o$	+	11.3	15.0
Experiment 2			
5.8 mM $[K^+]_o$ (control)	−	1.0	1.0
5.8 mM $[K^+]_o$	+	7.0	23.5
0.3 mM $[K^+]_o$	−	20.9	102.8
0.3 mM $[K^+]_o$	+	193.0	353.0

[a] Each band was scanned and the control (5.8 mM K^+) was designated as "1." The area under the curve generated for each band was quantified on an integration scale, from which fold-increases in the intensity of each band over the control level were calculated.

hypothesis we entertained is that activation of the Na^+, K^+-ATPase pump, which causes a concomitant change in intracellular K^+ concentration, may be a key component of the signal-transducing mechanisms since it is one of the earliest events observed upon mitogenic stimulation (Leister *et al.*, 1985). Upon lowering the $[K^+]_o$ to 0.3 mM, which effectively shuts down the pump, we are able to observe an increase in the c-*fos* and c-*myc* mRNA levels (Fig. 1). Quantitatively, these increases are approximately 20-fold for c-*myc* and 100-fold for c-*fos*. When cells are treated with media containing TPA and low concentrations of potassium, superinduction of c-*myc* and c-*fos* mRNA levels is observed; there is nearly a 200-fold increase in c-*myc* mRNA and a 350-fold increase in c-*fos* mRNA. We believe that both of these effects are due in large part to the inhibition of protein synthesis at this low $[K^+]_o$ level (Table II). Studies on the stability or "half-life" of the mRNA species help to bear out this conclusion. At the low potassium concentration, if we inhibit transcription with actinomycin D 30–60 min following stimulation with TPA, c-*myc* and c-*fos* mRNAs are still detectable long after they can be detected when the same experiment is performed in the presence of physiological extracellular potassium, 5.8 mM (data not shown). This indicates that at low extracellular potassium, we are interfering with the normal degradation of these mRNAs, and the resultant increase in half-life of the messages allows the super-induction effect with TPA. We have obtained similar results with ouabain (an inhibitor of the Na^+, K^+-ATPase pump which competes for the potassium-binding site) and cyclo-heximide (a widely used inhibitor of protein synthesis). These findings are also in accord with those of Makino *et al.* (1984).

It was noted previously that depolarizing concentrations of potassium (50 mM K^+) in cell culture media had no effect on c-*fos* and c-*myc* expression in our system. However, we noted a most interesting effect when we subsequently treated the cells with TPA. As shown in Fig. 1, when we treat cells incubated in media containing 50 mM K^+ with TPA we see

TABLE II. Effect of Varied $[K^+]_o$ on DNA Synthesis $\pm$ TPA[a]

Additions	Cumulative [³H]dThd incorporation (cpm) with $T = 24$			
	Experiment 1			
	0.3	0.7	1.2	5.8
Control	759 ± 112 (0.33)	1,342 ± 22 (0.59)	2,112 ± 77 (0.92)	2,290 ± 120 (1.0)
+ TPA(10^{-7} M)	816 ± 85 (0.36)	49,499 ± 2,937 (21.6)	44,362 ± 4,331 (19.4)	40,808 ± 1,840 (17.8)
	Experiment 2			
	5.8	15	25	50
Control	2,433 ± 80 (1.0)	18,402 ± 11,352 (7.6)	48,036 ± 1,614 (19.7)	34,657 ± 181 (15.1)
+ TPA(10^{-7} M)	52,467 ± 3,219 (21.5)	86,804 ± 6,093 (35.7)	124,505 ± 1,630 (54.4)	100,104 ± 4,936 (41.1)

[a] Postconfluent C3H 10T1/2 fibroblasts were incubated in BME containing varied $[K^+]_o$ for 3 hr prior to [³H]dThd addition. TPA was added simultaneously with the label where indicated. Incubation was carried out at 37°C for 24 hr following addition of the radiolabel. The reaction was stopped with two rinses with cold phosphate-buffered saline (PBS). Cells were then treated with 5% trichloroacetic acid (TCA) for 1 hr, then solubilized in 0.5 N NaOH for 2 hr. The acid-insoluble fraction was then collected and counted in a liquid scintillation counter. Each group was performed in quadruplicate or triplicate. Values in parentheses are fold changes from untreated control with 5.8 mM K^+ in the respective experiment.

a small superinduction of c-*fos* and c-*myc* mRNAs. There is an 11-fold increase in c-*myc* mRNA and a 15-fold increase in c-*fos* mRNA under these conditions. This compares to a 7-fold increase in c-*myc* and a 9.5-fold increase in c-*fos* mRNA levels when we treat the cells with TPA at physiological $[K^+]_o$ (5.8 mM). This superinduction apparently is not due to an inhibition of protein synthesis at this potassium concentration, as substantiated by the [³H]leucine incorporation data (Table III). Although we cannot rule out the involvement of calcium in the production of this effect, we do not believe this to be the case, since we see no effect on either c-*myc* or c-*fos* with 50 mM K^+ alone. Furthermore, if calcium was involved in the observed superinduction, one would expect to see results similar to those discussed earlier, where usage of a calcium ionophore in conjunction with a synthetic diacylglycerol had an antagonistic effect on c-*myc* expression. That is clearly not the case here.

3.2. Effect of Varied $[K^+]_o$ on DNA Synthesis in the Presence and Absence of TPA

The effect of $[K^+]_o \pm$ TPA on [³H]thymidine incorporation into the acid-insoluble fraction of C3H 10T1/2 cells was examined as described in Table II. Postconfluent fibroblasts reached a state of quiescence by the fourth day following a media change as determined by the low level of endogenous DNA synthesis. With the establishment of this condition, media was changed to BME with the designated K^+ concentration. Quadruplicate plates were used for each test at the designated $[K^+]_o \pm$ TPA (10^{-7} M). One microcurie of [³H]thymidine was added at $T = 0$ and following incubation at 37°C in 5% CO_2–95% air, the reaction was stopped with two rinses of cold, phosphate-buffered saline and then treatment with 5%

trichloroacetic acid (TCA) for 1 hr. Following removal of TCA, the cells were solubilized for 2 hr in 0.5 N NaOH and counted in a Beckman liquid scintillation counter.

At low $[K^+]_o$, DNA synthesis is minor compared with physiological concentrations (5.8 mM) even when cells are treated with TPA. In experiment 1, little or no synthesis was observed with TPA addition when the potassium concentrations were below 0.7 mM. This is in accord with previous data which indicated a ouabain-sensitive block in DNA synthesis up until 2 hr prior to S-phase entry (Leister et al., 1985).

In experiment 2, increased $[K^+]_o$ by itself was capable of increasing [^{3}H]thymidine incorporation into the acid-insoluble fraction. Optimal radioactivity incorporation was observed with 25 mM $[K^+]_o$. This was also seen when TPA was used to enhance DNA synthesis. With TPA at 25 mM $[K^+]_o$, [^{3}H]thymidine incorporation was increased 50-fold over control values (i.e., in the absence of TPA with 5.8 mM $[K^+]_o$).

3.3. Effect of Varied $[K^+]_o$ on [^{3}H]Leucine Incorporation into the Acid Insoluble Fraction

As a measure of the effect of $[K^+]_o$ on protein synthesis, postconfluent C3H 10T1/2 cells were incubated for 1–2 hr at varied times after media change with designated $[K^+]_o$ ± TPA (10^{-7} M) as described in Table III. In experiment 1, cells were incubated for 2 hr in media containing varied $[K^+]$ in the presence or absence of TPA; cells were then pulsed with [^{3}H]leucine for an additional hour and then worked up for liquid scintillation counting as described in Table II. As reported for DNA synthesis, protein synthesis was strongly inhibited at 0.3 mM $[K^+]_o$ whether TPA was present or not.

TABLE III. Effect on Varied $[K^+]_o$ ± TPA on [^{3}H]Leucine Incorporation
Into Acid-Soluble Fraction at Varied $[K^+]^a$

Additions	Experiment 1 Pulse incorporation of [^{3}H]leu at $T = 24$ (1 hr) cpm, $[K^+]_o$ (mM)			
	0.3	0.7	1.2	5.8
None	619 ± 133 (0.36)	1,952 ± 53 (1.13)	1,709 ± 38 (0.99)	1,730 ± 5 (1.0)
TPA(10^{-7} M)	870 ± 6 (0.50)	4,458 ± 169 (2.58)	3,455 ± 246 (2.00)	3,233 ± 135 (1.87)

Additions	Experiment 2 Pulse incorporation of [^{3}H]leu at $T = 0$ (2 hr), cpm, $[K^+]_o$ (mM)			
	5.8	15	25	50
None	2,422 ± 124 (1.00)	2,592 ± 340 (1.07)	2,735 ± 214 (1.13)	3,020 ± 427 (1.25)
TPA(10^{-7} M)	2,766 ± 181 (1.14)	2,898 ± 91 (1.20)		

^{a} Experiment 1: Postconfluent 10T1/2 cells were incubated for 24 hr in media containing varying $[K^+]$ alone or with TPA where indicated. Cells were then pulsed with [^{3}H]leu for an additional 1 hr, after which the reaction was terminated and the acid-insoluble fraction was worked up for liquid scintillation counting as described in Table II. Experiment 2: Same as above, except that the cells were pulsed with labeled leucine for 2 hr beginning at the time of media change and TPA addition where indicated. Values in parentheses are fold changes from untreated controls with 5.8 mM $[K^+]_o$.

In experiment 2, increasing $[K^+]_o$ raised [^{3}H]leucine incorporation into the acid-insoluble fraction as did TPA by itself. The most significant isotope incorporation was observed at the higher potassium concentrations, notably 50 mM $[K^+]_o$.

3. CONCLUSIONS

From the data presented, it is apparent that effectors of ion movements are potent modulators of c-*myc* and c-*fos* expression in C3H 10T1/2 fibroblasts and their MCA-transformed counterparts. Calcium appears to play a more integral role in the activation of these genes than do monovalent cations such as potassium and sodium. Some reports, however, have indicated that activation of the Na^+, H^+ antiport system is essential for stimulation of c-*fos* (Leffert *et al.*, 1987), while others claim that the alkalinization of the cytosol which results from the activation of the antiport has no role in activation of these genes (Moore *et al.*, 1985). The c-*fos* mRNA appears to be more calcium-labile than c-*myc* mRNA, as evidenced by experiments done with the calcium ionophore A23187. It should be noted, however, that the time frame in which the mRNA levels of these genes were assayed may not have allowed us to view the maximal stimulation of c-*myc* by A23187. We are proposing that the stimulation of these two genes by A23187 is mediated by a calcium-calmodulin kinase. This is supported by the observations that there is a synergistic effect of A23187 and dicaproin on stimulation of c-*fos*, while the same two compounds are antagonistic with regard to c-*myc* expression. Such observations are consistent with the hypothesis that the two agents are stimulating two independent branches of the growth factor signaling pathway, a calcium-calmodulin branch by A23187, and the protein kinase C component by dicaproin. However, further experiments are needed to completely rule out that the actions of A23187 are due to a stimulation of protein kinase C via interaction of endogenous low levels of diacylglycerol with the increase in cytosolic calcium.

Both protein synthesis and DNA synthesis exhibit an exquisite sensitivity to external potassium concentrations. At low $[K^+]_o$, both protein synthesis and DNA synthesis are strongly inhibited, even when the cells are treated with TPA. This large decrease in protein synthesis is responsible for the superinduction effect of TPA on c-*myc* and c-*fos* at low $[K^+]_o$, presumably by one of the two following mechanisms: (1) the synthesis of a K^+-labile RNase or "repressor" protein is inhibited, thus allowing for a stabilization of c-*myc* and c-*fos* mRNAs which are then susceptible to superinduction by the phorbol ester, or (2) stabilization of the mRNAs by "freezing" them on polysome complexes. Peptidyltransferase A has a K^+ dependency, so that when $[K^+]_i$ drops, the ribosomes stall on the mRNAs, protecting them from degradation by ribonucleases. Again, this stabilization renders c-*myc* and c-*fos* susceptible to superinduction by TPA. This also offers further documentation that activation of c-*fos* and c-*myc* does not unconditionally commit the cell to progress through S phase, since the data in Table III show that DNA synthesis is completely inhibited at low external potassium concentrations.

C3H 10T1/2 fibroblasts apparently lack calcium L channels, otherwise referred to as voltage-sensitive calcium channels. This is evidenced by the fact that BAY K 8644 (which has been shown to act on these L channels) as well as 50 mM $[K^+]_o$ (which should solicit the opening of these voltage-sensitive calcium channels via membrane depolarization) fail to cause any detectable increase in c-*myc* or c-*fos* transcripts in these fibroblasts. The ability of 50 mM $[K^+]_o$ to stimulate DNA synthesis (Table III) may be due to its activation of the Na^+, K^+-ATPase and subsequent macromolecular synthetic events. By direct activation of this enzyme, the primary event activated by growth factors, amiloride-sensitive Na^+ movements, may be circumvented, thus allowing for progression through the cell cycle. The superinduction effect of c-*myc* and c-*fos* by TPA at this external potassium concentration may be explained in part by the increase in protein synthesis observed at 50 mM $[K^+]_o$. This increase could lead to a larger pool of "transcriptional activator proteins" as proposed by Morgan and Curran (1986), which when phosphorylated could give rise to a larger than normal induction of c-*myc* and c-*fos* mRNAs by TPA. While the involvement of transcriptional activator proteins in c-*myc* stimulation is still being investigated, involvement of such proteins in c-*fos* activation has been demonstrated (Treisman, 1986). While we believe this to be an attractive model, we cannot as yet rule out changes in posttranscriptional mechanisms due to 50 mM $[K^+]_o$, which could in turn be responsible for the superinduction of c-*myc* and c-*fos* mRNAs at this potassium concentration. We do not believe that the superinduction is a calcium-mediated event for reasons outlined earlier. It is obvious from our results that differential responses to the same agents observed in different cell types can be due to the absence (as in our cells) or the presence (as in PC12 cells) of voltage-sensitive calcium gates.

An apparent synergy between protein kinase C and a calcium-calmodulin kinase is involved in the very large stimulation of c-*fos* by A23187 used in conjunction with the synthetic diacylglycerol dicaproin. However, this is not observed with c-*myc*; in fact, c-*myc* stimulation seen with the ionophore alone is decreased to nearly control levels when the ionophore is used along with the synthetic diacylglycerol. The most attractive explanation for these differential effects of c-*myc* and c-*fos* is an antagonistic effect of the aforementioned signalling kinases with regard to c-*myc*. Such cross-talk between signalling pathways has been proposed for the protein kinase A and protein kinase C systems (Yoshimasa *et al.*, 1987), so it may not be unlikely that the same kind of cross-talk is involved between kinase C and a calcium-calmodulin kinase. Further studies are needed to fully dissect this paradox, which at the very least presents evidence that the idea of "coordinate regulation" of c-*myc* and c-*fos* may not be entirely correct.

Although we have not been able to demonstrate any effect of A23187 on DNA synthesis at this time, others have shown stimulation of DNA synthesis by short treatments with A23187 in NIH 3T3 fibroblasts (Andersson and Norrby, 1977). Moore *et al.* (1985) showed increases in DNA synthesis with the plant lectin concanavalin A, which presumably acts via calcium influx, in thymocytes as well. Our failure to observe a stimulation of DNA synthesis in C3H 10T1/2 fibroblasts with the ionophore may be due to the induction of a futile cycle by A23187 (i.e., the activation of Ca^{2+} ATPases in the membrane to maintain calcium homeostasis or by mitochondrial pumping), thus inhibiting DNA synthesis by energy depletion.

We believe that the data presented here provide a strong case for the role of ion movements, particularly calcium, in the stimulation of early events associated with cell cycle progression. This is not to say that calcium is unequivocally responsible for the early events associated with cell cycle progression, but it appears that it does play a role. The degree to which calcium is involved in such events may indeed be cell type-specific, particularly when considering excitable versus nonexcitable cells. The final verdict on the role of calcium, as well as other ions, in cellular proliferation and differentiation awaits a more complete characterization of ion channels and how these respond and operate when subjected to a variety of stimuli.

REFERENCES

Andersson, R. G. G., and Norrby, K., 1977, Induction of proliferation in dense non-starved 3T3 cells by Ca^{++} ionophore A23187, *Virchows Arch.* **23:**185–194.

Armstrong, D., and Erxleben, C., 1986, BAY K 8644 only modifies the gating of phosphorylated calcium channels, in *Programs and Abstracts, VIth International Conference on Cyclic Nucleotides, Calcium and Protein Phosphorylation: Signal Transduction in Biological Systems,* May 1986, Bethesda, Maryland.

Campisi, J., Gray, H. E., Pardee, A. B., Dean, M., and Sonenshein, G. E., 1984, Cell-cycle control of c-myc but not c-ras expression is lost following chemical transformation, *Cell.* **36:**241–247.

Greenberg, M. E., and Ziff, E. B., 1984, Stimulation of 3T3 cells induces transcription of the c-fos proto-oncogene, *Nature* **311:**433–438.

Hesketh, T. R., Moore, J. P., Morris, J. D. H., Taylor, M. V., Rogers, J., Smith, G. A., and Metcalfe, J. C., 1985, A common sequence of calcium and pH signals in the mitogenic stimulation of eukaryotic cells, *Nature* **313:**482–484.

Kelly, K., Cochran, B. H., Stiles, C. D., and Leder, P., 1983, Cell-specific regulation of the c-myc gene by lymphocyte mitogens and platelet-derived growth factor, *Cell* **35:**603–610.

Kruijer, W., Schubert, D., and Verma, I. M., 1985, Induction of the proto-oncogene fos by nerve growth factor, *Proc. Natl. Acad. Sci. USA* **82:**7330–7334.

Langdon, W. Y., Harris, A. W., Cory, S., and Adams, J. M., 1986, The c-myc oncogene perturbs B lymphocyte development in E_{u}-myc transgenic mice, *Cell* **47:**11–18.

Lau, L. F., and Nathans, D., 1987, Expression of a set of growth related immediate early genes in BALB/c 3T3 cells: coordinate regulation with c-fos or c-myc, *Proc. Natl. Acad. Sci. USA* **84:** 1182–1186.

Leffert, H. L., Koch, K. S., Shapiro, I. P., Skelly, H., Wolff, J., Yee, J.-K., and Friedmann, T., 1987, Growth control of rat hepatocytes in primary culture, in abstracts of *Regulation of Liver Gene Expression,* Cold Spring Harbor Laboratories, New York.

Leister, K. J., Wenner, C. E., and Tomei, L. D., 1985, Correlation of ouabain-sensitive ion movements with cell-cycle activation, *Proc. Natl. Acad. Sci. USA* **82:**1599–1603.

Makino, R., Hayashi, K., and Sugimura, T., 1984, c-myc transcript is induced in rat liver at a very early stage of regeneration or by cycloheximide treatment, *Nature* **310:**697–698.

Moore, J. P., Todd, J. A., Hesketh, T. R., and Metcalfe, J. C., 1985, c-fos and c-myc gene activation, ionic signals, and DNA synthesis in thymocytes, *J. Biol. Chem.* **261:**8158–8162.

Morgan, J. I., and Curran, T., 1986, Role of ion flux in the control of c-fos expression, *Nature* **322:** 552–555.

Studzinski, G. P., Brelvi, Z. S., Feldman, S. C., and Watt, R. A., 1986, Participation of c-myc protein in DNA synthesis in human cells, *Science* **234:**467–470.

Treisman, R., 1986, Identification of a protein binding site that mediates transcriptional responses of the c-fos gene to serum factors, *Cell* **46:**567–574.

Verma, I. M., 1986, *Trends Genet.* **2:**93–96.

Yoshimasa, T., Sibley, D. R., Bouvier, M., Lefkowitz, R. J., and Caron, M. G., 1987, Cross-talk between cellular signalling pathways suggested by phorbol-ester-induced adenylate cyclase phosphorylation, *Nature* **327:**67–70.

33

Regulation of c-*fos* Expression by Voltage-Dependent Calcium Channels

JAMES I. MORGAN and TOM CURRAN

1. INTRODUCTION

The determination of cell function is mediated to a significant degree by dynamic alterations in the intracellular concentration of free calcium. To date, the study of the role of calcium as a second messenger has been largely confined to the analysis of rapid events triggered by this cation that mostly involve posttranslational modification. However, recently it has been recognized that agents that provoke an influx of calcium ions into PC12 cells elicit a rapid, transient, transcriptional activation of the *fos* protooncogene (Morgan and Curran, 1986; Greenberg *et al.*, 1986). This has led to the proposition (Fig. 1) that c-*fos* is but one member of a family of cellular immediate-early genes that are induced following stimulation and that act to modulate the long-term responses of a cell. As c-*fos* encodes a nuclear protein (Curran *et al.*, 1984), it is assumed that these inducible genes are themselves involved in the activation and/or repression of further sets of genes that are responsible for such phenomena as plasticity, adaptation, long-term potentiation, etc. This study will elaborate the evidence for the involvement of calcium in the regulation of both c-*fos* expression and the posttranslational modification of its protein product (Fos).

1.1. The *fos* Protooncogene

The *fos* protooncogene (c-*fos*) is the normal cellular homolog of the transforming gene (v-*fos*) of the FBJ and FBR murine osteogenic sarcoma viruses (Curran and Teich, 1982; Curran *et al.*, 1982; Curran and Verma, 1984; Van Beveren *et al.*, 1984). The protein products of both v-*fos* and c-*fos* are found in the cell nucleus (Curran *et al.*, 1984) where they are associated with chromatin (Sambucetti and Curran, 1986). The precise function of Fos is unknown at present. While c-*fos* is expressed constitutively in a small number of tissues and cell types (e.g., amnion and macrophages), it is rapidly induced by a wide range

JAMES I. MORGAN • Department of Neurosciences, Roche Institute of Molecular Biology, Roche Research Center, Nutley, New Jersey 07110. TOM CURRAN • Department of Molecular Oncology, Roche Institute of Molecular Biology, Roche Research Center, Nutley, New Jersey 07110.

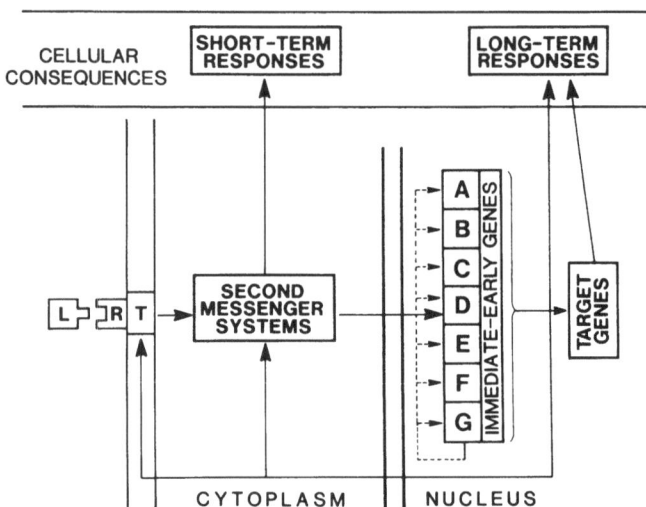

FIGURE 1. Hypothetical scheme linking extracellular stimuli to short- and long-term cellular responses. Extracellular ligands (L) interact with receptors (R) and activate second-messenger systems via membrane-transducing components (T). Second messengers elicit rapid alterations that constitute a short-term response. In addition, the same second messengers, directly or indirectly, act to induce transcription of cellular immediate-early genes. We have used c-*fos* as a marker for the immediate-early cellular response; however, we propose that a set of such genes exists (nominally designated A–G here). This transcriptional activation could have a number of consequences. First, immediate-early gene products may be components or regulators of the signal transduction cascade (e.g., receptors, G proteins, kinases, phosphatases, etc.). This would permit the cell to modify its response to a subsequent stimulation by the same agent. Second, immediate-early genes may promote the transcriptional activation of further target genes that cause changes in long-term responses. For example, in the nervous system these target genes might encode ion channels, neuropeptide precursors, synaptic proteins, or extracellular matrix proteins. It seems likely that the immediate-early genes like c-*fos* are under stringent regulation by some form of negative feedback, as indicated by the dotted line.

of agents in many cell types *in vitro* (for review, see Curran, 1988). Typically c-*fos* mRNA is detectable within 5 min of exposure to an inducing stimulus; levels then rise to a maximum at 30 min and subsequently decline to resting values over the next 30–60 min (Muller *et al.*, 1984). The Fos protein is somewhat longer lived, being present in stimulated cells for at least 3–4 hr postinduction.

1.2. The PC12 Pheochromocytoma

The PC12 cell line was derived from a transplantable rat pheochromocytoma (Greene and Tischler, 1976) and retains many properties characteristic of neurons (Greene and Tischler, 1982). Critical among these is an ability to respond to the neurotrophic polypeptide nerve growth factor (NGF), as well as neurotransmitter substances such as nicotine and depolarizing agents that mimic neurotransmitters (Greene and Tischler, 1976, 1982; Traynor and Schubert, 1984). Thus, the PC12 line has been used extensively as a model neuronal system to study the coupling mechanisms linking extracellular stimuli to intracellular responses.

2. ROLE OF CALCIUM IN THE REGULATION OF c-*fos* EXPRESSION

Treatment of PC12 cells with NGF leads to the rapid induction of c-*fos* mRNA and protein (Fig. 2). This response is independent of extracellular calcium and is insensitive to the blockage of voltage-dependent calcium channels with dihydropyridine (DHP) antagonists such as nisoldipine (Morgan and Curran, 1986). In contrast, induction of c-*fos* expression by elevated extracellular potassium, which depolarizes PC12 cells, does require calcium and is blocked by calcium channel antagonists (Morgan and Curran, 1986). This situation led us to suppose that the influx of calcium induced the c-*fos* gene. To solidify this notion, PC12 cells were treated with veratridine, an alkaloid that depolarizes neurons by holding sodium channels in their open state. Veratridine also produced a calcium-dependent, DHP-blockable induction of c-*fos* (Morgan and Curran, 1986). Furthermore, a classical antagonist of veratridine, tetrodotoxin, abolished veratridine-induced c-*fos* expression (Morgan and Curran, 1986). Thus, two independent methods of depolarization elicit a calcium-dependent stimulation of c-*fos*.

To determine the role of calcium in the induction process, PC12 cells were treated with an agonist of the voltage-dependent calcium channel, BAY K 8644. Like elevated potassium and veratridine, BAY K 8644 stimulated c-*fos* expression in a calcium-dependent manner and was blocked by an excess of the DHP antagonist nisoldipine (Morgan and Curran, 1986). Thus, it was concluded that the influx of calcium triggered c-*fos* transcription and the role of depolarization was to elicit this ingress of calcium by opening voltage-sensitive calcium channels. The overall scheme of activation is depicted in Fig. 3.

Once intracellular calcium levels are elevated a number of pathways could account for transcriptional activation. In the first, calcium would act directly on c-*fos*. Second, calcium could activate the calcium/calmodulin kinase pathway with a resulting phosphorylation/dephosphorylation of transcriptional regulatory proteins. Third, elevated intracellular calcium might influence the activity of voltage- and calcium-dependent potassium channels. Finally, calcium might impinge on the calpain and protein kinase C pathways again, altering the phosphorylation of endogenous transcriptional activators. To address these questions, PC12 cells were exposed to depolarizing concentrations of potassium in the presence of pharmacological agents that modulate elements of the above transduction pathways. One universal finding was that calmodulin inhibitors such as trifluoperazine always blocked depolarization-induced c-*fos* expression, whereas stimulation by NGF or phorbol esters was unaffected (Morgan and Curran, 1986). This finding would tend to suggest that (1) the calmodulin pathway was involved in potassium-evoked c-*fos* expression and (2) the C-kinase pathway was not involved since at the doses of calmodulin inhibitors used phorbol ester induction was unaltered. However, a number of studies have pointed out that calmodulin inhibitors also affect voltage-dependent calcium channels and may even displace DHP binding. Thus, the conclusion that the calmodulin pathway is involved must await further confirmation but is used here as a working hypothesis. These data seem to exclude the central involvement of protein kinase C in c-*fos* expression in PC12 cells under these circumstances.

It was reasoned that if intracellular calcium activated a potassium channel, then inhibitors

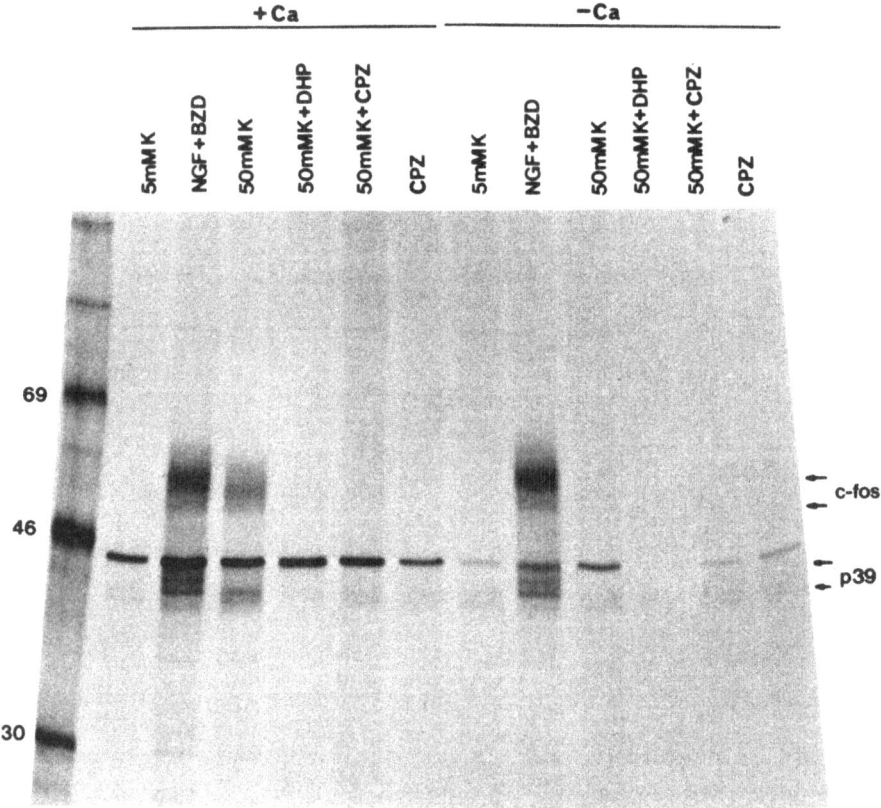

FIGURE 2. Induction of c-*fos* protein by 50 mM potassium is calcium- and calmodulin-dependent. PC12 cells were incubated in basal medium with the following additions: 5 mM potassium
chloride (5 mM K), 200 ng/ml NGF plus 100 μM benzodiazepine (Ro7-3351) (NGF + BZD),
50 mM potassium chloride (50 mM K), 50 mM potassium chloride plus 15 mM dihydropyridine
(nisoldipine, 1,4-dihydro-2,6-dimethyl-4(2-nitrophenyl)-3,5-pyridenedicarboxylate) (50 mM K +
DHP), 50 mM potassium chloride plus 30 μM chlorpromazine (50 mM K + CPZ) or 30 μM
chlorpromazine (CPZ), in the presence of 1.1 mM calcium chloride (+Ca). The drugs chlorpromazine and nisoldipine were added to cultures 5 min before the other reagents. In the
experiment shown, nisoldipine was used at a saturating concentration (15 μM); however, 50 nM
nisoldipine also attenuated potassium-induced c-*fos* expression (data not shown). Incubation
was continued for 30 min at 37°C, then 300 μCi/ml [^{35}S] methionine (approximately 800 μCi/mmol;
Amersham) was added for a further 15 min. Extracts were prepared and immunoprecipitated
with *fos*-specific antibodies. The immunoprecipitation products from 10^7 cpm of TCA-insoluble
proteins were analyzed by sodium dolecyl sulfate polyacrylamide gel electrophoresis (SDS-
PAGE). Arrows indicate the position of the c-*fos* protein. The numbers on the left indicate the
relative molecular masses (M_r) of the ^{14}C-methylated marker proteins (Amersham). (Reprinted
with permission from *Nature*, *322*, 1986, 552.)

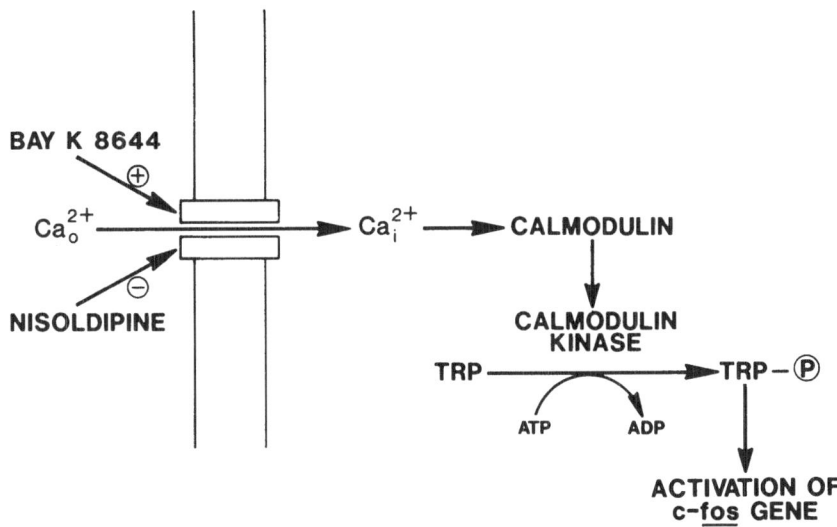

FIGURE 3. Coupling of voltage-dependent calcium channels to c-fos expression. Extracellular calcium (Ca_o^{2+}) can be gated into PC12 cells via a voltage-dependent calcium channel. This channel may be opened by depolarizing agents such as veratridine or elevated potassium or specific agonists such as BAY K 8644. The influx of calcium provoked by the above can be blocked by the calcium channel antagonist nisoldipine. The elevation of intracellular calcium ion (Ca_i^{2+}) concentration apparently activates the calmodulin/calmodulin kinase system. We propose that this results in phosphorylation of a hypothetical substrate TRP (transcription regulatory protein) and that this protein (TRP-P) directly or indirectly induces c-fos transcription.

of this type of conductance should block induction of c-fos. However, this was not the case. In fact, one inhibitor of potassium channels, barium, elicited a powerful induction of c-fos (Curran and Morgan, 1986). This induction of c-fos by barium was blocked both by calcium channel antagonists and calmodulin inhibitors (Curran and Morgan, 1986). This led to the conclusion that barium induced c-fos by entering the cell through the voltage-dependent calcium channel. Further evidence for this conclusion was furnished by the finding that extracellular calcium attenuated barium induction of c-fos, presumably by competing for transport through the channel (Curran and Morgan, 1986). Since barium is not known to regulate calmodulin, it is postulated that once in the cell, barium causes an elevation of free calcium, perhaps from intracellular stores, which in turn induces the c-fos gene. It should also be pointed out that barium probably depolarizes the PC12 cell under these conditions; however, this may be irrelevant (at least with regard to calcium influx) since extracellular calcium is not required for the induction of c-fos by barium (Curran and Morgan, 1986).

During the studies of c-fos induction by barium, it was noted that the molecular weight of the Fos protein obtained following barium administration was lower than when induction was achieved with NGF or phorbol ester (Curran and Morgan, 1986). The upward shift in molecular weight is a consequence of posttranslational modification, including phosphorylation. In vitro studies suggest that some of these phosphorylations are catalyzed by the cyclic AMP (cAMP)-dependent protein kinase. It is proposed, therefore, that NGF elevates endogenous cAMP which in turn leads to an activation of the cAMP-dependent kinase, the catalytic subunit of which phosphorylates Fos. Since Fos is a nuclear protein, this may mean that the catalytic subunit migrates to the nucleus during activation to phosphorylate Fos. Calcium-dependent inducers of c-*fos* are deficient in this activation of cAMP-kinase, but posttranslational modification may be rescued by combining barium with NGF. Thus, if these calcium-dependent agents do act via calmodulin kinase, it must be concluded that this enzyme cannot phosphorylate Fos at appropriate sites. The significance of this phosphorylation of Fos is not known but could conceivably be related to the regulation of its function. The data do show, however, that transcriptional regulation may be separated from posttranslational regulation.

   While NGF does not appear to require extracellular calcium to induce c-*fos*, preliminary evidence does suggest that the simultaneous modulation of intracellular calcium levels by other drugs may influence NGF-stimulated Fos expression. It had been found previously that cotreatment of PC12 cells with NGF and so-called peripheral-type benzodiazepines (pBZDs) led to a superinduction of c-*fos* (Curran and Morgan, 1985). Indeed pBZDs do not induce c-*fos* in the absence of NGF (Curran and Morgan, 1985). While this was a marked and structurally stereospecific action of the pBZDs (Curran and Morgan, 1985), the mechanism by which they brought about the superinduction of c-*fos* was unknown. Recently, it was found that superinduction with pBZDs was, like the induction by NGF, independent of extracellular calcium. Furthermore, the action of the pBZDs was resistant to levels of calmodulin inhibitors that completely blocked potassium- and barium-induced c-*fos* expression. These results appeared to indicate that calcium was irrelevant to the action of pBZDs. However, it was next discovered that the superinduction of c-*fos* by pBZDs but not the induction with NGF alone was blocked by both the dihydropyridine, nisoldipine (at micromolar concentrations), and quinacrine hydrochloride. These data suggested that the pBZDs might indeed influence a calcium channel or calcium transport system. It is our working hypothesis that the targets of the pBZDs are either involved in eliminating calcium from the cytoplasm via the plasma membrane (i.e., the calcium ATPase or calcium exchanger) or are components of an intracellular calcium sequestration mechanism.

3. REGULATION OF c-*fos* EXPRESSION IN THE CENTRAL NERVOUS SYSTEM

   The obvious significance of the foregoing findings is that the membrane depolarization and concomitant calcium influx which in cultured PC12 cells results in an induction of c-*fos* could have the same outcome in the nervous system *in vivo*. To test this hypothesis we treated mice with the convulsant drug Metrazole® (pentylenetetrazol) and monitored subsequent c-*fos* expression. Metrazole® produces a rapid and transient induction of c-*fos* mRNA exclusively in brain. Immunocytochemically, all the Fos immunoreactivity can be localized to the nuclei of neurons in distinct brain regions. Furthermore, c-*fos* induction by Metrazole® is completely abolished if the mice are pretreated with the anticonvulsant agent diazepam. Thus, seizure-like activation of neurons, which involves an influx of calcium, does elicit a c-*fos* induction. It should be noted that such a response is a normal part of the repertoire of a neuron *in vivo*, since we have observed a significant number of Fos-positive neurons in the brains of untreated mice, most notably in regions related to olfaction. Clearly, the expression of c-*fos* will be of some utility both in assessing the precise cellular targets of neuropharmacological agents and as a molecular end point for drug interactions *in vivo*.

Indeed, present studies are focused on the role of voltage-dependent calcium channels in the control of c-*fos* expression in the rodent central nervous system.

Our studies on c-*fos* expression in mouse brain suggest a hitherto unsuspected level of regulation of stimulus-transcription coupling in the central nervous system. After a Metrazole®-induced convulsion, there ensues a transient period of elevated c-*fos* mRNA levels followed by a protracted phase of subbasal levels. During this time, a subsequent Metrazole®-induced convulsion elicits a muted c-*fos* mRNA induction, particularly in the period of subbasal expression. However, throughout this entire time course Fos is present at high levels in neurons. Thus, the appearance of c-*fos* mRNA is subject to a negative feedback regulatory mechanism. The molecular basis of this feedback modulation could be a direct or indirect consequence of Fos or the products of other immediate-early genes. Alternatively, the mechanism could be transcription-independent inasmuch as regulation may be effected by posttranslational modification of constitutive cellular regulatory components. In any event, the appearance and levels of c-*fos* mRNA and protein are exquisitely regulated by both the intra- and extracellular milieux.

Recently, attention has been focused on the involvement of intracellular calcium in the pathological actions of excitatory amino acid neurotransmitters such as N-methyl-D-aspartate (NMDA). Of particular relevance to our investigation are the roles of NMDA and calcium in the dentate gyrus and hippocampus in seizure and cerebral ischemia. In these latter situations the release of excitatory transmitters such as NMDA provokes the voltage-dependent gating of calcium ions that may, in extreme cases, lead to neuronal death. In less severe situations, these agents cause adaptive changes in, for instance, the dentate gyrus. Indeed, it is well established that repeated seizures, as occurs for example in the kindling model of epilepsy, lead to alterations in receptor numbers (Valdes *et al.*, 1982; Shin *et al.*, 1985). Since c-*fos* is induced by both calcium influx and seizure activity, we propose that Fos and its related immediate-early inducible genes are involved in the modulation of these adaptive phenomena. Further, we suggest that an alteration in the intraneuronal free-calcium concentration is a key element in stimulus-transcription coupling in neurons.

REFERENCES

Curran, T., 1988, The *fos* oncogene, in: *The Oncogene Handbook* (E. P. Reddy, A. M. Skalka, and T. Curran, eds.), pp. 307–325, Elsevier, Amsterdam.

Curran, T., and Morgan, J. I., 1985, Superinduction of c-*fos* by nerve growth factor in the presence of peripherally active benzodiazepines, *Science* **229**:1265–1268.

Curran, T., and Morgan, J. I., 1986, Barium modulates c-*fos* expression and post-translational modification, *Proc. Natl. Acad. Sci. USA* **83**:8521–8524.

Curran, T., and Teich, N. M., 1982, Candidate product of the FBJ murine osteosarcoma virus oncogene: Characterization of 55,000 dalton phosphoprotein, *J. Virol.* **42**:114–122.

Curran, T., and Verma, I. M., 1984, The FBR murine osteosarcoma virus. I. Molecular analysis and characterization of a 75,000 Da gag-fos fusion product, *Virology* **135**:218–228.

Curran, T., Peters, G., Van Beveren, C., Teich, N. M., and Verma, I. M. 1982, FBJ murine osteosarcoma virus: Identification and molecular cloning of biologically active proviral DNA, *J. Virol.* **44**:674–682.

Curran, T., Miller, A. D., Zokas, L., and Verma, I. M., 1984, Viral and cellular fos proteins: A comparative analysis, *Cell* **36**:259–268.

Greenberg, M. E., Ziff, E. B., and Greene, L. A., 1986, Stimulation of neuronal acetylcholine receptors induces rapid gene transcription, *Science* **234**:80–83.

Greene, L. A., and Tischler, A. S., 1976, Establishment of a noradrenergic clonal cell line of rat adrenal pheochromocytoma cells which respond to nerve growth factor, *Proc. Natl. Acad. Sci. USA* **73:** 2424–2428.

Greene, L. A., and Tischler, A. S., 1982, PC12 pheochromocytoma cultures in neurobiological research, *Advanc. Cell. Neurobiol.* **3:**373–414.

Morgan, J. I., and Curran, T., 1986, Role of ion flux in the control of c-fos expression, *Nature* **322:** 552–555.

Muller, R., Bravo, R., Burckhardt, J., and Curran, T., 1984, Induction of c-fos gene by protein growth factors precedes activation of c-myc, *Nature* **312:**716–720.

Sambucetti, L. C., and Curran, T., 1986, The Fos protein complex is associated with DNA in isolated nuclei and binds to DNA cellulose, *Science* **234:**1417–1419.

Shin, C., Pedersen, H. B., and McNamara, J. O., 1985, γ-aminobutyric acid and benzodiazepine receptors in the kindling model of epilepsy: A quantitative radiohistochemical study, *J. Neurosci.* **5:**2696–2701.

Traynor, A., and Schubert, D., 1984, Phospholipases elevate cyclicAMP levels and promote neurite extension in a clonal nerve cell line, *Dev. Brain Res.* **14:**197–204.

Valdes, F., Dashieff, R. M., Birmingham, F., Crutcher, K. A., and McNamara, J. O., 1982. Benzodiazepine receptor increases after repeated seizures: Evidence for localization to dentate granule cells, *Proc. Natl. Acad. Sci. USA* **79:**193–197.

Van Beveren, C., Enami, S., Curran, T., and Verma, I. M., 1984, The FBR murine osteosarcoma virus. II. Nucleotide sequence of the provirus reveals that the genome contains sequences derived from two cellular genes, *Virology* **135:**229–243.

34

Calcium Pulses, Waves, and Gradients in Early Development

LIONEL F. JAFFE

Free Ca^{2+} ions are among the half dozen or so small ions or molecules whose changing patterns within cells is in immediate control of their growth, development, behavior, and death. This chapter focuses on the role of free Ca^{2+} in activating eggs and in the development of pattern.

1. EGG ACTIVATION WAVES

Parthenogenesis tells us that fertilization is not needed to continue development. Nevertheless sexual forms generally arrest development to let new genes into the egg. The nature of the lock which arrests egg development remains unknown; but it is now clear that a giant pulse of free calcium ions serves to blast the lock off. Moreover, in the vertebrate line, this pulse takes the form of a traveling explosion which crosses the egg at about 10 μm/sec via a chain reaction mediated by calcium-stimulated calcium release (Fig. 1a).

The best evidence for this conclusion began with the work of Toki-o Yamamoto on medaka fish eggs during the dark days of the Second World War. Yamamoto observed a wave of exocytosis which began at the point of sperm entry (at the animal pole) and then crossed this millimeter-diameter egg at about 10 μm/sec so as to reach the vegetal pole a few minutes later. A series of ingenious experiments showed that beneath the visible exocytotic wave lay an invisible "fertilization wave." For example, he observed that exocytosis could be largely blocked by gently centrifuging the (normally exocytosed) cortical vesicles off the egg's surface and toward one pole; yet such eggs would develop into fish when fertilized. Indeed, he even guessed that the fertilization wave was in fact a free-calcium wave. This guess was then confirmed a decade ago by observing medaka eggs injected with the chemiluminescent, calcium-specific, photoprotein aequorin (Gilkey *et al.*, 1978). Such eggs exhibit a remarkably bright wave of luminescence which travels from pole to pole at 10 μm/sec and indicates a wave of high free Ca^{2+}. At its peak this wave reaches the remarkable level of 30 μM Ca^{2+} for about half a minute. That this wave is needed to blast

---

LIONEL F. JAFFE ● Marine Biological Laboratory, Woods Hole, Massachusetts 02543.

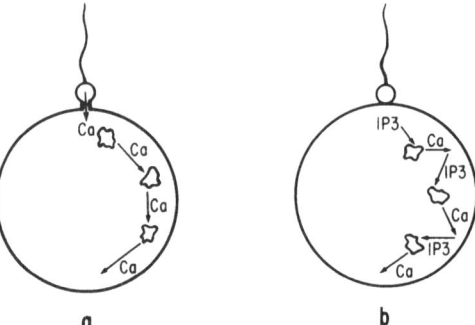

FIGURE 1. Proposed mechanisms to start and propagate calcium waves through activating eggs on the vertebrate line. (a) Started by sperm calcium; propagated by direct calcium-stimulated calcium release from the endoplasmic reticulum. (b) Started by IP3 in turn released by sperm action; propagated by a chain reaction in which IP3 releases calcium from the endoplasmic reticulum and calcium in turn releases IP3 from the plasma membrane.

loose the block to development was confirmed by injections of calcium buffers. A buffer set at 1 μM or less of Ca^{2+} blocks egg activation altogether if introduced near the point of sperm entry and blocks it locally if introduced elsewhere. That the wave is actively propagated by calcium-stimulated calcium release was shown by injecting buffers above the trigger level of about 5 μM Ca^{2+}. Since the wave was scarcely affected by removing external calcium, it was obviously supported by internally released calcium. Since comparable exocytotic waves have been reported among eggs in most groups on the vertebrate line, it was inferred that similar calcium waves serve to activate all or most eggs on the vertebrate line (Jaffe, 1983, 1985).

More recently, 10 μm/sec calcium waves were directly observed during the activation of tunicate (Speksnijder et al., 1986; Fig. 3), sea urchin (Swann and Whitaker, 1986; Hafner et al., 1988), sand dollar (Yoshimoto et al., 1986), frog (Busa and Nuccitelli, 1985; Kubota et al., 1987), and hamster eggs (Miyazaki et al., 1986) as well as being seen again in medaka eggs (Yoshimoto et al., 1986). The calcium waves seen in tunicate and hamster eggs are of particular interest since these eggs, unlike most eggs on the vertebrate line, do not show an exocytotic wave during activation. Nevertheless they show calcium waves comparable to those of the medaka fish; so it is increasingly certain that such waves blast off the lock to egg development throughout the vertebrate line of organisms.

1.1. Role of Inositol-1,4,5-trisphosphate

However, the mechanisms which serve to detonate (as well as propagate) these calcium waves may include inositol-1,4,5-trisphosphate (IP3) as well as Ca^{2+}. It is clear that gamete activation is a process which usually starts in the sperm (as indicated by the exocytotic acrosome reaction) and then spreads to the egg during fertilization. Since free calcium seems to rise greatly in activated sea urchin sperm, it was first proposed that the sperm is a "calcium bomb" which detonates the calcium wave simply by injecting calcium at the trigger level (Jaffe, 1983). Evidence that IP3 may also be involved in detonation of the calcium wave includes observations that injection of IP3 into sea urchin eggs at *pipette* concentrations of about 0.1–1 μM will fully activate half of them (Whitaker and Irvine, 1984; Turner et al., 1986).* This is comparable to the IP3 concentration needed for half-maximal release of

* Pipette, rather than final concentrations, are used here in agreement with Whitaker and Irvine (as opposed to Turner et al.) because of evidence of local effects of injected IP3 as well as considerations of its diffusion constant, its intracellular instability, and the radical dilution which even a stable injectate would have undergone in these experiments.

calcium from isolated sea urchin cortices (Oberdorf *et al.*, 1986) as well as a wide variety of permeabilized cells (Berridge, 1987). It therefore suggests that the sperm detonates a calcium wave via calcium released by IP3 (in the manner of other agonists) as well as introducing calcium directly.

Moreover, Swann and Whitaker (1986) provided evidence suggesting that in the sea urchin egg IP3 acts as an intermediate in the process which propagates the calcium wave through this egg. In particular, calcium releases IP3 from isolated egg plasma membranes in concentrations comparable to those reached during the calcium wave; moreover, 10 mM neomycin blocks wave propagation. Together with the evidence that IP3 may naturally act to release calcium from the endoplasmic reticulum in these eggs, this suggests that IP3 may be a natural intermediate in the process (of calcium-stimulated calcium release) which propagates the calcium wave. These possibilities are illustrated in Fig. 1b.

1.2. Postactivation Pulses

Aequorin-loaded *Ciona* and *Phallusia* eggs both show huge pulses of luminescence during normal activation by sperm. These indicate a rise in free cytosolic calcium to peak

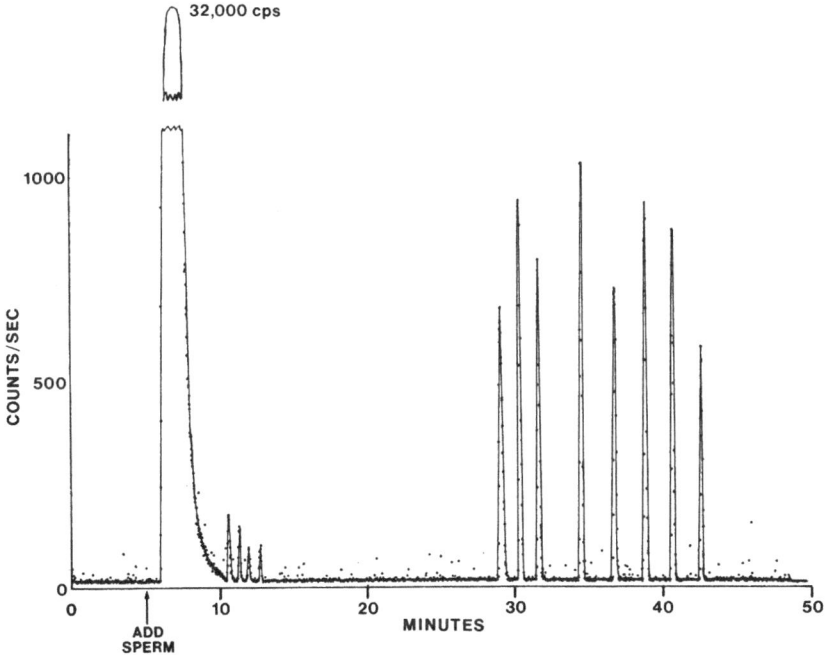

FIGURE 2. Typical fertilization and postfertilization calcium pulses in an aequorin-loaded *Ciona* egg (experiment 14-1 on 12/19/85). The enormous fertilization pulse of luminescence indicates a 300-fold rise in free cytosolic calcium; the four subsequent ones—during first polar body formation?—rises of 3- to 10-fold; the eight later ones—during second polar body formation?—rises of 30- to 100-fold. Note that the calcium pulse heights during each of the two postfertilization groups must have been much more constant than the data show because luminescence varies with the 2.5 power of free calcium and because of the small number of counts per data point. (From work in progress by J. E. Speksnijder, D. W. Corson, T. H. Qiu, and L. F. Jaffe.)

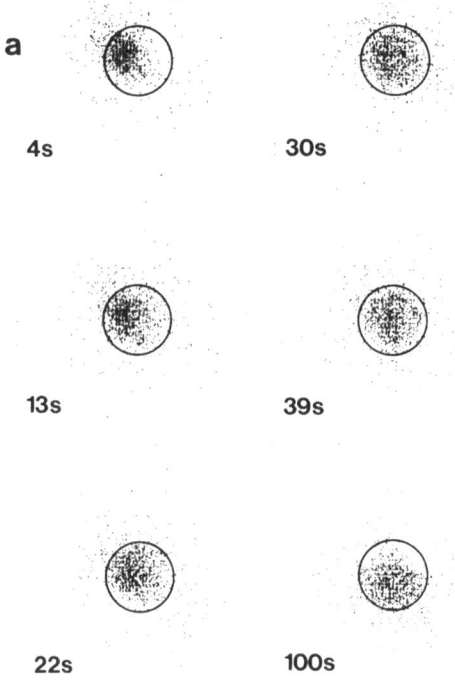

FIGURE 3(a). Fertilization and postfertilization calcium waves in an aequorin-loaded *Phallusia* egg. The six panels above show successive stages of the fertilization pulse at various times after it began. The egg may have rotated 45° clockwise during this sequence. (b) These panels show the beginnings of postfertilization waves number 4, 5, 6, 13, and 14 at the indicated times after fertilization, as well as a later stage of wave number 13 (at 17.5 m). Note that successive waves 4, 5, and 6, as well as 13 and 14, start at opposite poles. (From work in progress by J. E. Speksnijder, C. Sardet, and L. F. Jaffe.)

levels of about 10 and 3 μM, respectively. These huge rises last a few minutes and cross these eggs as a wave moving at about 7 μm/sec. As discussed above, these findings add the tunicates to the phyla on the vertebrate line whose eggs are known to be activated by large free-calcium waves.

More surprising was the discovery of a remarkable series of 12–16 smaller calcium pulses which occur between the end of the activation pulse and first cleavage in these ascidian eggs. These postactivation pulses occur in two groups which seem to accompany first and second polar body formation. They have a characteristic shape, are semiperiodic, have relatively constant peak levels of the order of 1–3 μM, start locally and then often travel across the egg. Successive pulses either start in the same region or in an antipodal one— like an echo (Figs. 2 and 3).

These postactivation pulses in the ascidian egg are quite reminiscent of the postinsemination calcium pulses recently reported in zona-free hamster eggs by Miyazaki *et al.* (1986).

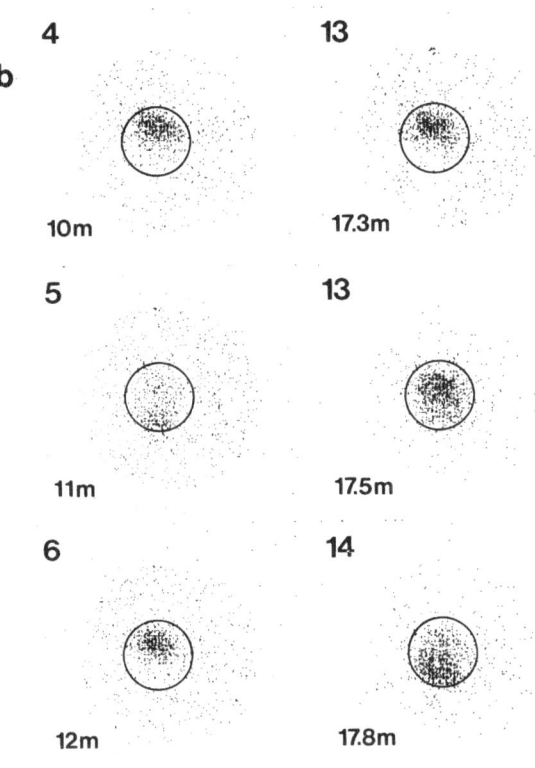

FIGURE 3. (*Continued*)

However, these ascidian pulses are more likely to be natural, developmental events than the hamster ones, since they were observed in eggs which generally underwent several cleavages and sometimes developed into swimming tadpoles, whereas the hamster pulses were observed in eggs which never cleaved, let alone developed.

It is too early to discuss the developmental significance of these postactivation pulses, but they do remind one of a report of agonist-induced calcium pulses in rat hepatocytes by Woods *et al.* (1987). These authors suggest that the frequency rather than the amplitude of these calcium pulses "is the principal determinant of the amplitude of the cellular response to calcium-mobilizing agonists." Stated another way, the emerging idea is that calcium pulses in various nonneural cells may play a role comparable to that of action potentials in nerve axons.

2. PATTERNING

The development of pattern in many eggs (and other developing systems) involves "gradients" of well-defined developmental consequences but of rather obscure physical basis.

Moreover, many eggs are known to undergo more or less dramatic episodes of "ooplasmic segregation" which act to localize developmental determinants. My overall aim is to explore the roles of free cytosolic calcium gradients in representative cases of these poorly understood phenomena.

Five such systems are on my current agenda: in the fucoid egg, I would aim to explore the role of calcium gradients in localizing the rhizoidal pole; in the medaka egg, the role of such gradients in segregating the bulk of the cytoplasm (and thus the embryo proper) at the animal pole and the oil droplets at the vegetal one; in the echinoderm egg, their role in localizing the differentation of mesodermal cells at the extreme vegetal pole, ectodermal in the animal half, and endodermal in between; in the ascidian eggs, their role in the well-known segregation of the main larval determinants right after fertilization; in the slime mold, *Dictyostelium*, their role in localizing prestalk cells at the slug's front and prespore ones at its rear.

Rather little is known of the role of calcium gradients in the development of pattern. Here I would include the developmental phenomena of determination, of morphogenesis, and of differentiation but not those of simple growth and cell division.

However, as discussed above, it *is* known that the fertilization of many and perhaps all eggs on the vertebrate line involves an intense free-calcium wave which starts at the point of sperm entry and then crosses the egg at about 10 μm/sec. This then raises the interesting and quite unanswered question of whether the direction of the fertilization waves is involved in establishing pattern in various eggs.

Moreover, one might consider Child's so-called metabolic gradients (Child, 1941). Operationally, the "high" ends of these gradients were actually sites of early cytolysis and/or mitochondrial dye reduction in response to anoxia or poisons like cyanide. Modern cell physiology suggests that this occurs where free cytosolic calcium first reaches levels high enough to activate autolytic proteases and to collapse mitochondrial membrane potentials; this in turn suggests that calcium first reaches these extreme levels where it was highest under normal conditions (Jaffe, 1982). So the extensive old literature on "metabolic gradients" may yet offer some clues about natural free-calcium gradients. Beyond that, little is known except in the fucoid egg (but see Jaffe, 1986 regarding medaka eggs).

2.1. Fucoid Eggs

Relatively direct evidence for polarization by natural intracellular calcium gradients is only available for fucoid eggs (reviewed in Jaffe, 1982). The most important components of this evidence are as follows: (1) Fucoid eggs drive large, steady calcium currents through themselves with calcium ions entering the future growth, or rhizoid pole, hours before this is either visible for irreversibly established. (2) If still unpolarized eggs are exposed to external gradients of calcium or of a calcium ionophore, they show a strong tendency to form their growth pole toward the end into which calcium is driven. (3) The earliest structural changes in the polarizing fucoid egg—ones seen many hours before the cell exhibits localized growth and is irreversibly polarized—are indicators of localized exocytosis and secretion at the future rhizoid pole. This includes both freeze-fracture evidence as well as the visualization of the localized thickening of a remarkably tenuous extracellular jelly via decoration with cationic beads. (4) Polarization of the fucoid eggs by various external vectors can be rationalized by assuring that they act via internal free-calcium gradients. For example, when eggs are exposed to external gradients of dinitrophenol, they grow toward the high dinitrophenol (DNP) end. This can be rationalized by assuming that more calcium is internally released from the high-DNP end.

Put together, these older observations strongly suggested that a natural cytosolic calcium gradient is part of the positive feedback loop which establishes growth and then a rhizoid cell at a particular region of the initially unpolarized egg. More recently, Brawley and Robinson (1985) obtained evidence that localized actin filament formation somehow feeds back to favor calcium entry through the growing rhizoid tip; Brownlee and Wood (1986) used intracellular electrodes to directly measure steady free-calcium levels of a few micromolar (!) near this tip; while our own work in progress with injected calcium buffers indicates that this steady high-calcium zone is essential for local growth initiation and may attain levels even higher than those measured by Brownlee and Wood (1986).

Central to my thinking has been the idea that the initiation of tip growth at one pole of these eggs (as well as the subsequent differentiation of a rhizoid cell there) requires a steady localized rise of cytosolic free calcium in this region. If this is true, it should be possible to block local growth and differentiation simply by injecting enough of the right calcium buffer into these cells; so we are well along in a study of the effects of injecting various concentrations of Roger Tsien's so-called BAPTA [bis(o-Aminopheroxy)ethane-N,N,N',N'-tetraacetic acid] buffers into these cells (see Tsien, 1980; Haugland, 1985).

It turns out that over a narrow and reproducible final intracellular concentration range, all BAPTA buffers tried can completely block local growth (and, indeed, any visible differentiation) for periods of days or even weeks. Such critically "baptized" cells do not die; they may divide, but they do not differentiate. The critical inhibitory concentration varies with the calcium dissociation constant or K_D of the particular BAPTA buffer used. So far, the weaker the buffer, the less it takes to suspend development: 5,5'-dibromobapta with a K_D of about 4 μM, is the most effective one tried so far. It has a critical final intracellular buffer concentration of 1 mM. In contrast, 5,5'-dimethylbapta, with a K_D of 0.4 μM, is ineffective below about 7 mM.

These remarkable effects cannot be due to the immediate effects of injecting the buffer since they show little or no dependence on the pCa or free-calcium concentration of the injected buffer. Nor are they at all likely to be mediated by damping out natural calcium pulses: They are injected at 6 hr after fertilization and seem to act during a stage when extracellular recording argues against the occurrence of any natural calcium pulses in fucoid eggs (Nuccitelli and Jaffe, 1974). Rather, I suspect that these buffer injections work by damping out natural steady calcium gradient by *shuttling* calcium from regions of high free calcium to ones of relatively low free calcium. If this is true—and we are vigorously pursuing this project—such "baptism" should provide an important tool for studying the nature and consequences of intracellular free-calcium gradients. For example, a buffer which acts by such a shuttle mechanism should be more effective when its dissociation constant lies somewhere between those of the natural high- and low-calcium regions. So the high effectiveness of dibromobapta suggests a natural standing high-calcium region (or regions) of the order of 10 μM.

One other point: Cell *death* is eventually produced by buffer concentrations above the critical inhibitory level. I would tentatively attribute this second interesting phenomenon to the existence of a standing natural radial gradient of free calcium, which the whole peripheral, subplasmalemmal level being well above the internal, subcortical one. High buffer should damp out such gradients, lower subplasmalemmal calcium, reactivate calcium channels (see Eckert and Chad, 1984), and thus raise calcium influx to the point where calcium toxicity kills the cell.

REFERENCES

Berridge, M. J., 1987, Inositol trisphosphate and diacylglycerol: Two interacting second messengers, *Ann. Rev. Biochem.* **56:**159–193.

Brawley, S. J., and Robinson, K. R., 1985, Cytochalasin treatment disrupts the endogenous currents associated with cell polarization in fucoid zygotes, *J. Cell Biol.* **100:**1173–1184.

Brownlee, C., and Wood, J. W., 1986, A gradient of cytoplasmic free calcium in growing rhizoid cells of *Fucus serratus, Nature* **320:**624–626.

Busa, W. B., and Nuccitelli, R., 1985, An elevated free cytosolic Ca^{2+} wave follows fertilization in eggs of the frog, *Xenopus laevis, J. Cell Biol.* **100:**1325–1329.

Child, C. M., 1941, *Patterns and Problems of Development*, University of Chicago Press, Chicago.

Eckert, R., and Chad, J. E., 1984, Inactivation of Ca channels, *Prog. Biophys. Molec. Biol.* **44:** 215–267.

Gilkey, J. C., Jaffe, L. J., Ridgway, E. B., and Reynolds, G. T., 1978, A free calcium wave traverses the activating medaka egg, *J. Cell Biol.* **76:**448–466.

Hafner, M., Petzelt, C., Nobiling, R., Pawley, J. B., Kramp, D., and Schatten, G., 1988, Wave of free calcium at fertilization in the sea urchin egg visualized with Fura-2, *Cell Motility and the Cytoskeleton*, **9:**271–277.

Haugland, R. P., 1985, *Molecular Probes Handbook of Fluorescent Probes*, Molecular Probes, Junction City, Oregon.

Jaffe, L. F., 1982, Developmental current, voltages, and gradients, *Symp. Soc. Dev. Biol.* **40:**183–218.

Jaffe, L. F., 1983, Sources of calcium in egg activation. A review and hypothesis, *Dev. Biol.* **99:** 267–276.

Jaffe, L. F., 1985, The role of calcium explosions, waves, and pulses, in activating eggs, in: *Biology of Fertilization*, Vol. 3 (C. B. Metz and A. Monroy, eds.), Academic Press, New York, pp. 127–165.

Jaffe, L. F., 1986, Calcium and morphogenetic fields, *CIBA Symp.* **122:**217–288.

Kubota, H. Y., Yoshimoto, Y., Yoneda, M., and Hiramoto, Y., 1987, Free calcium wave upon activation in *Xenopus* eggs, *Dev. Biol.* **119:**129–136.

Miyazaki, S–I., Hashimoto, N., Yoshimoto, Y., Kishimoto, T., Igusa, Y., and Hiramoto, Y., 1986, Temporal and spatial dynamics of the periodic increase in intracellular free calcium at fertilization of golden hamster eggs, *Dev. Biol.* **118:**259–267.

Nuccitelli, R., and Jaffe, L. F., 1974, Spontaneous current pulses through developing fucoid eggs, *Proc. Natl. Acad. Sci. USA* **71:**4855–4859.

Oberdorf, J. A., Head, J. A., and Kaminer, B., 1986, Calcium uptake and release by isolated cortices and microsomes from the unfertilized egg of the sea urchin *Strongylocentrotus droebachiensis, J. Cell Biol.* **102:**2205–2210.

Speksnijder, J. E., Corson, D. W., Jaffe, L. F., and Sardet, C., 1986, Calcium pulses and waves through ascidian eggs, *Biol. Bull.* **171:**488.

Swann, K., and Whitaker, M., 1986, The part played by inositol trisphosphate and calcium in the propagation of the fertilization wave in sea urchin eggs, *J. Cell Biol.* **103:**2333–2342.

Tsien, R. Y., 1980, New calcium indicators and buffers with high selectivity against magnesium and protons, *Biochemistry* **19:**2396–2404.

Turner, P. R., Jaffe, L. A., and Fein, A., 1986, Regulation of cortical vesicle exocytosis in sea urchin eggs by inositol 1,4,5-trisphosphate and GTP-binding protein, *J. Cell Biol.* **102:**70–76.

Whitaker, M. J., and Irvine, R. F., 1984, Inositol (1,4,5) trisphosphate microinjection activates sea urchin eggs, *Nature* **312**:636–639.

Woods, N. M., Cuthbertson, K. S. R., and Cobbold, P. H., 1987, Agonist-induced oscillations in cytoplasmic free calcium concentration in single rat hepatocytes, *Cell Calcium* **8**:79–100.

Yoshimoto, Y., Iwamatsu, T., Hirano, K.-I., and Hiromoto, Y., 1986, The wave pattern of free calcium release upon fertilization in medaka and sand dollar eggs, *Develop. Growth and Differ.* **28**:583–596.

IV

Regulation of Metabolism

35

Calcium Ions, Hormones, and Mammalian Oxidative Metabolism

JAMES G. McCORMACK and RICHARD M. DENTON

1. INTRODUCTION

Many hormones and other external stimuli are known to act on mammalian cells by causing increases in the cytoplasmic concentration of Ca^{2+}. In most if not all cases, the effects brought about, e.g., contraction, secretion, and so on, require an increased supply of ATP. One possible mechanism whereby this increased energy demand may be met in some tissues is by a concomitant enhancement by Ca^{2+} of glycogen breakdown and hence glycolysis through the activation of phosphorylase kinase. In this paper we review the evidence for another potential mechanism whereby increases in Ca^{2+} may affect rates of metabolism and hence ATP synthesis, and which may be of more general importance. This mechanism involves the relay of the increases in cytoplasmic Ca^{2+} into the mitochondrial matrix resulting in the activation of three oxidative dehydrogenases which play key roles in the supply of NADH to the respiratory chain for ATP production.

2. THE Ca^{2+}-SENSITIVE INTRAMITOCHONDRIAL DEHYDROGENASES

There are three exclusively intramitochondrial dehydrogenases in mammalian cells which can be activated by increases in Ca^{2+} in the range 0.1–10 μM (see Denton and McCormack, 1980, 1985 for reviews) with $K_{0.5}$ values (half-maximally effective concentrations) for Ca^{2+} of around 1 μM in each case. They are the pyruvate (PDH), NAD$^+$-isocitrate (NAD-ICDH), and 2-oxoglutarate (OGDH) dehydrogeneases. Ca^{2+} increases the amount of active, nonphosphorylated PDH (PDH$_a$) by activating PDH phosphate phosphatase (PDHP-Pase) and perhaps also by inhibiting PDH$_a$ kinase. The effects on the phosphatase may involve both an increase in V_{max} and a reduction in the enzyme's K_m for Mg^{2+} (Midgley et

JAMES G. McCORMACK ● Department of Biochemistry, University of Leeds, Leeds LS2 9JT, United Kingdom.   RICHARD M. DENTON ● Department of Biochemistry, University of Bristol Medical School, University Walk, Bristol BS8 1TD, United Kingdom.

al., 1987). Ca^{2+} activates NAD-ICDH and OGDH more directly by causing marked decreases in their respective K_m values for *threo-D_s*-isocitrate and 2-oxoglutarate. The effects of Ca^{2+} can be mimicked by Sr^{2+}, but at approximately 10-fold higher concentrations in each case.

These three dehydrogenases can all also be activated by increases in the $NAD^+/NADH$ and ADP/ATP concentration ratios, and NAD-ICDH and OGDH can also be activated by decreases in pH in the physiological range (see Denton and McCormack, 1980; McCormack and Denton, 1986). However, their regulation by Ca^{2+} is largely distinct and independent of that of the other, more local effectors, suggesting that Ca^{2+} could override the effects of these metabolites.

The three dehydrogenases exhibit Ca^{2+} sensitivity in extracts of all vertebrate tissues so far examined, but not in extracts of nonvertebrate tissues such as plants or insects (see McCormack and Denton, 1986), which do, however, still exhibit nucleotide sensitivity. Interestingly, there appears to be a functional evolutionary link between the occurrence of the Ca^{2+} sensitivity of the dehydrogenases and the ability of mitochondria to take up Ca^{2+} by a specific pathway (McCormack and Denton, 1986).

3. DEHYDROGENASE REGULATION BY Ca^{2+} WITHIN INTACT MITOCHONDRIA

Studies on intact, fully coupled mitochondria from rat heart, liver, skeletal muscle, brain, and adipose tissue, and within permeabilized pig lymphocytes, have now fully established that these intramitochondrial Ca^{2+}-sensitive enzymes can be activated as a result of increases in extramitochondrial Ca^{2+} within the expected physiological range (i.e., approximately 0.05–5 μM) (see Denton and McCormack, 1985; Hansford, 1985). This is accomplished by incubation with physiological concentrations of Mg^{2+} and Na^+ which inhibit mitochondrial Ca^{2+} uptake and promote mitochondrial Ca^{2+} egress, respectively (see Nicholls and Akerman, 1982) (Table I). The absence of either or both of these effectors of mitochondrial Ca^{2+} transport decreases the effective extramitochondrial Ca^{2+} range (Table I). In uncoupled mitochondria, where the intra- and extramitochondrial Ca^{2+} pools are in equilibrium, the enzymes exhibit similar Ca^{2+} sensitivity as they do in extracts (Table I). This suggests that the gradient of Ca^{2+} ions (in/out) across the mitochondrial inner membrane is only around 2–3, which fits well with measurements using X-ray probe microanalysis

TABLE I. $k_{0.5}$ Values for Extramitochondrial Ca^{2+} (nM) in the Activation of PDH and OGDH within Rat Heart and Liver Mitochondria Incubated under Various Conditions[a]

	Heart		Liver	
Incubation condition	PDH	OGDH	PDH	OGDH
Control (coupled)	39	21	109	121
Na^+ (10–15 mM)	189	82	184	202
Mg^{2+} (0.5–1 mM)	175	96	356	420
Na^+ plus Mg^{2+}	464	328	484	560
Uncoupled (1 μM FCCP[b])	980	940	1059	—

[a] Data are taken from Denton *et al.* (1980) and McCormack (1985a).
[b] FCCP, carbonylcyanide *p*-trifluoromethoxyphenyl hydrazone.

showing that the total mitochondrial Ca content of normal cells *in situ* is very similar to that in the cytoplasm (e.g., Somlyo *et al.*, 1985; Wendt-Gallitelli, 1986).

4. EVIDENCE FOR THE HORMONAL REGULATION OF MITOCHONDRIAL OXIDATIVE METABOLISM THROUGH CHANGES IN INTRAMITOCHONDRIAL Ca²⁺

The most extensively studied systems to date have been the effects of adrenaline and other positive inotropic agents acting on rat heart, and those of α-adrenergic agonists, vasopressin, and glucagon acting on rat liver. In each of these cases, the hormones increase not only cytoplasmic $[Ca^{2+}]$, but also oxygen uptake, citrate cycle flux, and the amount of tissue PDH$_a$ (see Denton and McCormack, 1985). Moreover, there is little evidence for sustained decreases in ATP/ADP and NADH/NAD$^+$ ratios, which indeed under some circumstances actually increase, and whole-tissue or cell contents of oxoglutarate, glutamate, and citrate have often been observed to decline (see Denton and McCormack, 1985).

Table II summarizes some observations which support the view that the activation of PDH under these conditions involves increases in intramitochondrial Ca^{2+}. For example, in the rat heart, the effects of adrenaline on PDH$_a$ can be blocked by perfusing with ruthenium red, a potent inhibitor of mitochondrial Ca^{2+} uptake (McCormack and England, 1983); in contrast, the stimulation by adrenaline of the cytoplasmic Ca^{2+}-dependent processes, contraction and phosphorylase activity, are unaffected by ruthenium red. Hansford (1987) working with Quin2-loaded myocytes also demonstrated that ruthenium red can block PDH activation without affecting the rise in cytoplasmic Ca^{2+} under various conditions. In both rat heart and liver, evidence has been obtained using cell fractionation that the hormones increase the total amount of Ca associated with mitochondrial fractions (Table II), provided that precautions are taken to avoid the loss and uptake of Ca during mitochondrial preparation and that contamination by other intracellular organelles is minimized (Crompton *et al.*, 1983; Assimacopoulos-Jeannet *et al.*, 1986). It is worth noting that similar basal values have been determined for both liver and heart mitochondrial Ca content *in situ* using X-ray probe microanalysis (Somlyo *et al.*, 1985; Wendt-Gallitelli, 1986), and that in the latter instance

TABLE II. Effects of Hormones and Other Treatments on the Amounts of PDH$_a$ in Rat Heart and Liver Tissue, and Amounts of Total Calcium Associated with Subsequently Prepared Mitochondrial Fractions[a]

Perfused rat tissue	Treatment	Amount of PDH$_a$ (as % total PDH)	Ca content of mitochondrial fraction (nmol/mg protein)
Heart	Control	10	1.8
	Adrenaline	41	4.2
	Adrenaline + ruthenium red	11	—
Liver	Control	5	1.2
	Vasopressin	27	2.1
	Glucagon + vasopressin	45	4.9

[a] Data are taken from McCormack and England (1983), Crompton *et al.* (1983) (heart), and Assimacopoulos-Jeannet *et al.* (1986) (liver).

there is some evidence for increased content as the result of treatments designed to increase cytoplasmic Ca^{2+}. Moreover, studies with isolated rat heart and liver mitochondria indicate that changes in the activities of PDH and OGDH occur as total Ca is increased over the 0–5 nmol/mg protein range and that above this range their Ca^{2+}-dependent activations are saturated (see Hansford, 1985; Assimacopoulos-Jeannet et al., 1986); this matches well with the in vivo values (Table II).

However, by far the most convincing and direct evidence for the role of intramitochondrial Ca^{2+} has followed the realization that the hormone-induced increase in PDH_a in both heart and liver can persist through the preparation of mitochondria and also their subsequent incubation for several minutes at 30°C in KCl-based media containing respiratory substrates and EGTA (Table III) (McCormack and Denton, 1984; McCormack, 1985b; Assimacopoulos-Jeannet et al., 1986). This is evident provided that both the uptake and release of Ca^{2+} by the mitochondria is blocked by the presence of EGTA and absence of Na^+ ions, respectively. If the mitochondria are incubated with Na^+ to allow Ca^{2+} egress, or else with sufficient Ca^{2+} to result in the saturation of Ca^{2+}-dependent activation of the enzyme, then the persistent effects of the hormones are rapidly lost (Table III). In addition, these effects of Na^+ are blocked by diltiazem, an inhibitor of mitochondrial $Na^+–Ca^{2+}$ exchange (Vághy et al., 1982). Moreover, at the level of isolated mitochondria, the activity of OGDH (at subsaturating 2-oxoglutarate concentrations) can also be assayed and thus also be used as an intramitochondrial Ca^{2+} probe; this activity was found to parallel that of PDH within the mitochondria incubated under various conditions (Table III). However, it should be noted that the application of similar approaches to those described above to try and determine the role of intramitochondrial Ca^{2+} in the activation of PDH by insulin in adipose tissue yielded negative results, i.e., there was no evidence for a role of Ca^{2+} in this case (Marshall et al., 1984).

TABLE III. Persistence of the Activations of PDH and OGDH within Incubated Mitochondria Isolated from Perfused Rat Heart and Liver (in vivo) Previously Exposed to Adrenaline[a]

Additions to mitochondrial incubation media	PDH_a (as % total PDH) after:		OGDH activity (as % V_{max}) after:	
	No hormone	Adrenaline	No hormone	Adrenaline
Perfused rat heart				
None	8	20	23	35
Na^+ (10 mM)	8	7	25	24
Na^+ + diltiazem (300 μM)	8	20	24	32
Saturating Ca^{2+} (150 nM)	45	47	—	—
Liver				
None	12	20	8	13
Na^+	13	14	8	8
Na^+ + diltiazem	13	23	9	15
Saturating Ca^{2+} (400 nM)	49	51	33	35

[a] Data are taken from McCormack and Denton (1984) and McCormack (1985b).

5. CONCLUSIONS AND OTHER COMMENTS

The observations summarized above strongly suggest that hormones which use increases in cytoplasmic Ca^{2+} to stimulate energy-requiring processes in mammalian tissues, as a result also bring about parallel or secondary increases in intramitochondrial Ca^{2+} concentration, and thus meet the enhanced energy demand, at least in part, by activating the dehydrogenases and hence oxidative metabolism and ATP production. The major advantage of such a role for Ca^{2+} as a second messenger in mitochondria would be to the homeostasis of the cell's energy metabolism, as it would allow increased formation of NADH for oxidative phosphorylation without the important NADH/NAD$^+$ and ATP/ADP ratios being diminished at times when it would be desirable to maintain or even increase them.

The main role of the Ca^{2+}-transport system of the mitochondria of mammalian cells, therefore, is probably to ensure that changes in cytoplasmic Ca^{2+} concentration are relayed into the mitochondrial matrix. Also of interest in this context are the observations that α-adrenergic agonists appear to increase the Ca^{2+} uptake pathway of both heart (Crompton et al., 1983) and liver (Taylor et al., 1980) mitochondria, and that the Na$^+$-dependent egress pathway for mitochondrial Ca^{2+} may be inhibited by physiological concentrations of extramitochondrial Ca^{2+} (Hayat and Crompton, 1982). This may allow augmentation of the increases in intramitochondrial Ca^{2+} in comparison to those in the cytoplasm. Goldstone et al. (1983) also reported that the activity of the Na$^+$-dependent egress pathway may be stimulated by glucagon.

However, the above viewpoint on the role of mitochondrial Ca^{2+}-transport is not compatible with the earlier suggestions that mitochondria may buffer or set cytoplasmic Ca^{2+} concentration (see, e.g., Nicholls and Akerman, 1982). It now appears that buffering behavior will only be exhibited at concentrations of Ca^{2+} where Ca^{2+}-sensitive processes in both the mitochondria and the cytoplasm would be saturated (see Denton and McCormack, 1985). Under such circumstances, mitochondria may play a protective role to restrict increases in cytoplasmic Ca^{2+} (e.g., after ischemic episodes); however, derangement of mitochondrial function can also result if they too are overloaded.

ACKNOWLEDGMENTS. Work in the authors' laboratories was supported by grants from the Medical Research Council, the British Diabetic Association, and the British Heart Foundation.

REFERENCES

Assimacopoulos-Jeannet, F. D., McCormack, J. G., and Jeanrenaud, B., 1986, Vasopressin and/or glucagon rapidly increases mitochondrial calcium and oxidative enzyme activities in the perfused rat liver, J. Biol. Chem. 261:8799–8804.

Crompton, M., Kessar, P., and Al-Nasser, I., 1983, The α-adrenergic mediated activation of the mitochondrial Ca^{2+} uniporter and its role in the control of intramitochondrial Ca^{2+} in vivo, Biochem. J. 216:333–342.

Denton, R. M., and McCormack, J. G., 1980, On the role of the calcium transport cycle in heart and other mammalian mitochondria, FEBS Lett. 119:1–8.

Denton, R. M., and McCormack, J. G., 1985, Ca^{2+}-transport by mammalian mitochondria and its role in hormone action, Am. J. Physiol. 249:E543–E554.

Denton, R. M., McCormack, J. G., and Edgell, N. J., 1980, Role of calcium ions in the regulation of intramitochondrial metabolism: Effects of Na$^+$, Mg^{2+} and ruthenium red on the Ca^{2+}-stimulated oxidation of oxoglutarate and on pyruvate dehydrogenase activity in intact rat heart mitochondria, *Biochem. J.* **190**:107–117.

Goldstone, T. P., Duddridge, R. J., and Crompton, M., 1983, The activation of the Na^+-dependent efflux pathway of Ca^{2+} from liver mitochondria by glucagon and β-adrenergic agonists, *Biochem. J.* **210**:463–472.

Hansford, R. G., 1985, Relation between mitochondrial calcium transport and control of energy metabolism, *Rev. Physiol. Biochem. Pharmacol.* **102**:1–72.

Hansford, R. G., 1987, Relation between cytosolic free Ca^{2+} concentration and the control of pyruvate dehydrogenase in isolated cardiac myocytes, *Biochem. J.* **241**:145–151.

Hayat, L. H., and Crompton, M., 1982, Evidence for the existence of regulatory sites for Ca^{2+} on the Na^+/Ca^{2+} carrier of cardiac mitochondria, *Biochem. J.* **176**:627–629.

Marshall, S. E., McCormack, J. G., and Denton, R. M., 1984, Role of Ca^{2+} ions in the regulation of intramitochondrial metabolism in rat epididymal adipose tissue: Evidence against a role for Ca^{2+} in the activation of pyruvate dehydrogenase by insulin, *Biochem. J.* **218**:249–260.

McCormack, J. G., 1985a, Characterisation of the effects of Ca^{2+} on the intramitochondrial Ca^{2+}-sensitive enzymes from rat liver and within rat liver mitochondria, *Biochem. J.* **231**:581–595.

McCormack, J. G., 1985b, Studies on the activation of rat liver pyruvate dehydrogenase and 2-oxoglutarate dehydrogenase by adrenaline and glucagon: Role of increases in intramitochondrial Ca^{2+} concentration, *Biochem. J.* **231**:597–608.

McCormack, J. G., and Denton, R. M., 1984, Role of Ca^{2+} in the regulation of intramitochondrial metabolism in rat heart: Evidence from studies with isolated mitochondria that adrenaline activates the pyruvate dehydrogenase and 2-oxoglutarate dehydrogenase complexes by increasing the intramitochondrial concentration of Ca^{2+}, *Biochem. J.* **218**:235–247.

McCormack, J. G., and Denton, R. M., 1986, Ca^{2+} as a second messenger within mitochondria, *Trends Biochem. Sci.* **11**:258–262.

McCormack, J. G., and England, P. J., 1983, Ruthenium red inhibits the activation of pyruvate dehydrogenase caused by positive inotropic agents in the perfused rat heart, *Biochem. J.* **214**:581–585.

Midgley, P. J. W., Rutter, G. A., Thomas, A. P. and Denton, R. M., 1987, Effects of Ca^{2+} and Mg^{2+} on the activity of pyruvate dehydrogenase phosphate phosphatase within toluene-permeabilized mitochondria, *Biochem. J.* **241**:371–377.

Nicholls, D. G., and Akerman, K. E. O., 1982, Mitochondrial calcium transport, *Biochim. Biophys. Acta* **683**:57–88.

Somlyo, A. P., Bond, M., and Somlyo, A. V., 1985, Calcium content of mitochondria and endoplasmic reticulum in liver frozen rapidly in vivo, *Nature* **314**:622–625.

Taylor, W. M., Prpic, V., Exton, J. H., and Bygrave, F. L., 1980, Stable changes to calcium fluxes in mitochondria isolated from rat livers perfused with adrenergic agonists and with glucagon, *Biochem. J.* **188**:443–450.

Vághy, P. L., Johnson, T. D., Matlib, M. A., Wang, T., and Schwarz, A., 1982, Selective inhibition of Na^+-induced Ca^{2+} release from heart mitochondria by diltiazem and certain other Ca^{2+} antagonist drugs, *J. Biol. Chem.* **257**:6000–6002.

Wendt-Gallitelli, M. F., 1986, Ca-pools involved in the regulation of cardiac contraction under positive inotropy. X-ray microanalysis on rapidly-frozen ventricular muscles of guinea-pig, *Basic Res. Cardiol.* **81**:25–32.

36

Regulation of Pyruvate Dehydrogenase in Isolated Cardiac Myocytes and Hepatocytes by Cytosolic Calcium

RICHARD G. HANSFORD, RAFAEL MORENO-SÁNCHEZ, and JAMES M. STADDON

1. INTRODUCTION

It is well established that the Ca^{2+} ion plays a central role as a mediator of excitation-contraction and stimulus-secretion coupling (see, e.g., Katz, 1970; Rasmussen, 1981). Perhaps less well known is the role that this ion also plays in activating mitochondrial oxidation and thereby making available ATP at an increased rate to match the increased energy demands associated with contraction and secretion. This activation has been most clearly established for glycerol-3-phosphate dehydrogenase (Hansford and Chappell, 1967), pyruvate dehydrogenase phosphatase (Denton *et al.*, 1972; Pettit *et al.*, 1972), NAD-isocitrate dehydrogenase (Denton *et al.*, 1978), and 2-oxoglutarate dehydrogenase (McCormack and Denton, 1979). In the case of pyruvate dehydrogenase phosphatase, activation by micromolar concentrations of Ca^{2+} ions leads to the generation of an increased amount of the active, dephospho, form of the pyruvate dehydrogenase complex (PDH_A) (for reviews, see Hansford, 1980; Wieland, 1983). In the case of the other three dehydrogenases, activation is of an allosteric nature, and results in a decreased K_m for the substrate.

After the original work with the purified pyruvate dehydrogenase phosphatase (Denton *et al.*, 1972; Pettit *et al.*, 1972), it was subsequently shown that pyruvate dehydrogenase interconversion is also responsive to Ca^{2+} when the enzyme complex is in its intramitochondrial milieu. Indeed, the content of active PDH_A of respiring, coupled suspensions of rat heart mitochondria responds to changes in extramitochondrial free Ca^{2+} concentration which are plausible for the cytosol of muscle cells, namely, the range 10^{-7}–10^{-6} M (Hansford and Cohen, 1978; Denton *et al.*, 1980; Hansford, 1981). The question arises as to whether this mechanism operates in the intact cell, with the mitochondria exposed to cytosolic

RICHARD G. HANSFORD, RAFAEL MORENO-SÁNCHEZ, and JAMES M. STADDON • Energy Metabolism and Bioenergetics Section, Laboratory of Cardiovascular Science, Gerontology Research Center, National Institute on Aging, National Institutes of Health, Baltimore, Maryland 21224.

factors, known and unknown, that may affect the activities of the mitochondrial Ca^{2+} transport cycle.

This paper presents recent work from our laboratory in which we follow the response of PDH_A content to the elevation of $[Ca^{2+}]_c$, to determine whether this is consistent with a cause–effect relationship. The cell types employed are freshly isolated Ca^{2+}-tolerant cardiac myocytes from adult animals, in which we raise $[Ca^{2+}]_c$ by plasma membrane depolarization, and freshly isolated hepatocytes, in which case we use the Ca^{2+}-mobilizing hormones phenyl-ephrine, vasopressin, and glucagon.

2. METHODS

The procedures for the isolation of Ca^{2+}-tolerant cardiac myocytes from adult rats, the loading of these cells with Quin2, the studies of Quin2 fluorescence and of the measurement of pyruvate dehydrogenase have all been described by Hansford (1987).

The procedures for the isolation of hepatocytes, the loading of these cells with Quin2 and the measurement of the fluorescence of intracellular Quin2 have been described by Staddon and Hansford (1986). Procedures for the measurements of the fluorescence of intracellular indo-1 and of NAD(P)H and for the extraction and measurement of hepatocyte pyruvate dehydrogenase have been described by Staddon and Hansford (1987).

3. RESULTS AND DISCUSSION

3.1. Effect of Plasma Membrane Depolarization on $[Ca^{2+}]_c$ and PDH_A Content of Cardiac Myocytes

It is shown in Table I that the elevation of the extracellular K^+ concentration from 5 to 80 mM, a procedure which leads to a graded depolarization and increase in $[Ca^{2+}]_c$ (Powell *et al.*, 1984; Sheu *et al.*, 1986), also gives rise to a progressive increase in PDH_A content. Equally, the treatment of the myocytes with veratridine plus ouabain, a procedure which might be expected to lead to depolarization and Na^+ overload (Escueta and Appel, 1969; Ohta *et al.*, 1973), gives rise to a large increase in PDH_A.

The response of $[Ca^{2+}]_c$ to these interventions is shown in Fig. 1. Raising $[K^+]$ to 40 mM gives a larger increase (a) than does raising $[K^+]$ to 20 mM (b). Ouabain alone, added to inhibit the Na^+,K^+-ATPase (Escueta and Appel, 1969), gives a very modest increase in $[Ca^{2+}]_c$, consistent with the known insensitivity of the rat to cardiac glycosides (c). However, ouabain does potentiate the effect of a nonsaturating concentration of veratridine (e), which acts through holding open Na-channels (Ohta *et al.*, 1973). Veratridine alone gives a large, rapid increase in $[Ca^{2+}]_c$ (d) when added at the high concentration (25 μM) used for the pyruvate dehydrogenase studies.

It is clear that there is a relation between the degree of increase in $[Ca^{2+}]_c$ shown in Fig. 2 and the degree of increase in PDH_A content shown in Table I. However, in our opinion it is not possible to define the changes in $[Ca^{2+}]_c$ more quantitatively, as the calibration of the fluorescence signal from Quin2 is marred by the apparent presence of more than one compartment for the dye within the cell (see Hansford, 1987 and Staddon and Hansford, 1986 for more discussion).

Furthermore, it would be premature to conclude from these data that Ca^{2+} is directly

TABLE I. Effect of Ruthenium Red and Ryanodine on the Increase in PDH_A Content of Cardiac Myocytes Induced by KCl and Veratridine Plus Ouabain[a]

Condition	PDH_A (% of total)
5 mM KCl	31.7 ± 1.4 (32)
20 mM KCl	39.5 ± 0.6 (4)
40 mM KCl	49.5 ± 4.6 (6)
80 mM KCl	60.6 ± 3.9 (11)
5 mM KCl + ruthenium red	31.5 ± 2.7 (3)
40 mM KCl + ruthenium red	38.7 ± 2.3 (3)
80 mM KCl + ruthenium red	37.8 ± 3.3 (7)[b]
20 mM KCl + ryanodine	39.8 ± 4.1 (3)
40 mM KCl + ryanodine	56.0 ± 6.9 (8)
80 mM KCl + ryanodine	72.7 ± 9.2 (8)
5 μM veratridine	37.6 ± 4.5 (6)
25 μM veratridine	51.0 ± 1.9 (10)
25 μM veratridine + 0.2 mM ouabain	57.6 ± 3.9 (8)
0.2 mM ouabain	45.6 ± 5.0 (5)
25 μM veratridine + 0.2 mM ouabain + ruthenium red	35.8 ± 2.0 (10)[c]
25 μM veratridine + 0.2 mM ouabain + ryanodine	52.6 ± 4.1 (3)

[a] From Hansford (1987).
    Values of PDH_A are significantly lower in the presence of ruthenium red than in otherwise identical incubations omitting ruthenium red: [b] $p < 0.005$; [c] $p < 0.001$ (from Hansford, 1987).
Note: Suspensions of rat cardiac myocytes were incubated and sampled for PDH_A content as described in Hansford (1987). The basal medium contained 5 mM KCl: higher concentrations of K^+ were achieved by mixing volumes of an isotonic medium in which K^+ replaced Na^+ ions. Sampling was 10 min after the addition to the suspension of K^+-containing medium or of veratridine plus ouabain, as appropriate. Where indicated, ruthenium red and ryanodine were added 3 min before the K^+-containing medium or veratridine, to give concentrations of 12 and 1 μM, respectively. Data are presented as the mean ± SEM, with the number of preparations in parentheses.

responsible for the activation of pyruvate dehydrogenase. Thus, whenever $[Ca^{2+}]_c$ is sufficiently elevated, a muscle will undergo contraction and the availability of ADP to the mitochondria will increase. Pyruvate dehydrogenase interconversion is sensitive to changes in the mitochondrial ATP/ADP ratio, as ADP inhibits the pyruvate dehydrogenase kinase (Hucho *et al.*, 1972; see also Hansford, 1976) and so the question arises as to how important are changes in $[Ca^{2+}]_c$ per se in signaling metabolic demand to this enzyme system, compared to changes in the ATP/ADP ratio which occur consequent to muscle contraction. The following arguments bear on this point.

(1) The presence of ruthenium red, an inhibitor of the uptake of Ca^{2+} by the mitochondria (Moore, 1971), largely prevents any increase in PDH_A content due to high concentrations of K^+ or to treatment with veratridine + ouabain (Table I). Yet ruthenium red does not attenuate the response of $[Ca^{2+}]_c$ to these interventions (result not shown). Nor does ruthenium red alter the frequency of spontaneous localized contractile waves, "waving," which occur in a significant fraction of these cell populations and are thought to reflect values of $[Ca^{2+}]_c$.

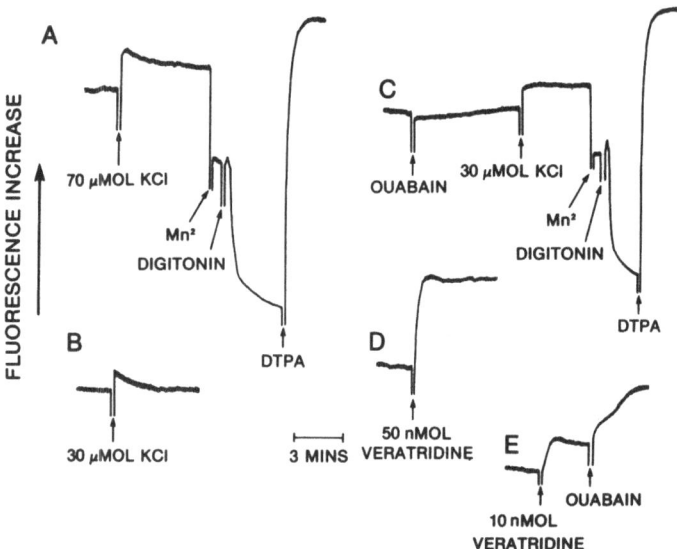

FIGURE 1. Effect of plasma membrane depolarization on the fluorescence of suspensions of cardiac myocytes loaded witth Quin2. Cells were incubated at 3.5 mg of protein/ml in medium containing 7.5 μM Quin2/AM for 30 min at 37°C. This generated cells containing 0.73 nmol of Quin2/mg protein. After this loading step, the cells were washed by centrifugation and used for the fluorescence studies shown. Where indicated, MnCl$_2$ was added to 0.1 mM to quench the fluorescence of extracellular Quin2, digitonin was added to 5 μM to allow the Mn^{2+} to react with intracellular Quin2, and diethylenetriaminepentaacetic acid (DTPA) was added to 1 mM to remove Mn^{2+} from the Quin2/Mn^{2+} chelate and therefore generate maximal fluorescence. Full details are given in the original paper. From Hansford (1987).

(2) Treatment of the cell suspensions with ryanodine, an inhibitor of sarcoplasmic reticulum Ca^{2+} efflux (Sutko and Kenyon, 1983), renders the cells totally quiescent when viewed under the microscope but has no effect on the elevation of PDH$_A$ content in response to the interventions shown in Table I.

(3) Loading of the cells with the Ca^{2+}-chelating agent Quin2, to an intracellular content of 0.5–0.7 nmol/mg of protein, makes the cells totally quiescent but does not diminish the response of the steady-state content of PDH$_A$ to procedures which elevate [Ca^{2+}]$_c$ (result not shown). There is, however, a slowing of the rate of increase of PDH$_A$ content upon depolarization when the cells are very heavily loaded with Quin2 (Hansford, 1987), consistent with chelation of Ca^{2+} by the Quin2.

From these results, it seems quite clear that increased mechanical work is not a prerequisite for the activation of pyruvate dehydrogenase and that the enzyme system is capable of responding to changes in [Ca^{2+}]$_c$ alone. However, this finding does not rule out a supplementary effect of a decreased mitochondrial ATP/ADP ratio in actively contracting heart muscle and it is noted that the degree of mechanical work performed by the waving cells in

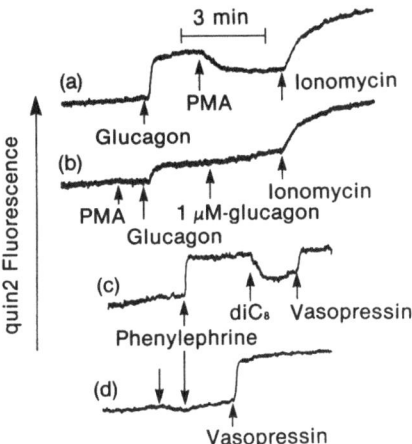

FIGURE 2. Response of hepatocyte $[Ca^{2+}]_c$ to treatment with glucagon, phenylephrine, and vasopressin. Hepatocytes were loaded with Quin2 by incubation of the cells (2.5 mg protein/ml) with 100 μM Quin2/AM for 15 min at 37°C. This procedure gave cells containing 1 nmol/mg of cell protein, or approximately 1 mM intracellular, Quin2. Fluorescence measurements were made as described by Staddon and Hansford (1986). In (a), glucagon was added to 10 nM, 4β-phorbol-12-myristate-13-acetate (PMA) to 500 nM, and ionomycin to 10 μM. In (b) PMA was added prior to glucagon: the second addition of glucagon gave 1 μM. In (c) phenylephrine was added to 25 μM, and L-1-α-1,2-dioctanoylglycerol (di-C$_8$) was added to 10 μM; this unnatural diacylglycerol is used to activate protein kinase C, in analogy with the use of PMA. A subsequent addition of vasopressin, to 25 nM, was effective in elevating $[Ca^{2+}]_c$. In (d), treatment of the hepatocytes with PMA (500 nM) abolished the response to phenylephrine (25 μM) but allowed an undiminished response to vasopressin (25 nM). The medium contained 2.5 mM Ca^{2+} in these studies.

this study is likely to be very small compared to their work load in the animal. Nevertheless, the efficacy of the Ca^{2+} signal is established.

3.2. Effect of the Ca^{2+}-Mobilizing Hormones Glucagon, Vasopressin, and Phenylephrine on $[Ca^{2+}]_c$ and PDH_A Content of Isolated Hepatocytes

It is known that α_1-adrenergic stimulation and the hormone vasopressin raise $[Ca^{2+}]_c$ in hepatocytes (Charest *et al.*, 1983) as well as stimulating the pathway of gluconeogenesis. Recently, it was found that the hormone glucagon, which also stimulates gluconeogenesis but which had formerly been considered to act solely through the elevation of cyclic AMP, also gives rise to an increase in $[Ca^{2+}]_c$ (Charest *et al.*, 1983; Sistare *et al.*, 1985; Staddon and Hansford, 1986). For this reason, we chose to investigate whether or not the PDH_A content of hepatocytes responded to these agents and, if so, whether there was a quantitative relationship between changes in $[Ca^{2+}]_c$ and changes in PDH_A content. Early in this investigation we discovered that the activation of protein kinase C by phorbol esters results in an attenuation of the increase in $[Ca^{2+}]_c$ which normally occurs on exposure to glucagon (Staddon and Hansford, 1986).

TABLE II. Activation of Pyruvate Dehydrogenase by Phenylephrine,
Vasopressin, and Glucagon in Hepatocytes and the Influence of
4β-Phorbol-12-myristate-13-acetate[a]

Hormone	PDH activity, nmol/min per mg cell protein	
	None	Plus PMA (500 nM)
—	1.01 ± 0.05 (8)	0.98 ± 0.06 (8)
Phenylephrine (25 μM)	1.20 ± 0.09 (6)[c]	1.01 ± 0.08 (6)
Vasopressin (25nM)	1.69 ± 0.13 (8)[d]	1.58 ± 0.11 (8)[d]
Glucagon (10 nM)	2.27 ± 0.22 (8)[d]	1.25 ± 0.13 (8)[b]

[a] From Staddon and Hansford (1987).
Note: Pyruvate dehydrogenase activity was measured as described in the original paper. Values
shown are mean ± SEM with the number of cell preparations indicated in parentheses. Statistical
significance for the difference between values in each vertical column was determined by a paired
t test: [b] $p < 0.05$; [c] $p < 0.01$; [d] $p < 0.001$. In comparison between columns: 4β-phorbol-12-myristate-
13-acetate (PMA) was without effect on unstimulated PDH$_A$ content; PMA was without effect on
the vasopressin-induced and phenylephrine-induced increase in PDH$_A$ content; PMA significantly
($p < 0.001$) diminished the activation due to glucagon.

Thus, glucagon, vasopressin, and phenylephrine provided tools allowing the elevation
of $[Ca^{2+}]_c$, and the phorbol ester 4β-phorbol-12-myristate-13-acetate (PMA) provided a tool
for reversing this effect, when stimulation was by glucagon or phenylephrine. Table II shows
that phenylephrine, vasopressin, and glucagon each gave an increase in PDH$_A$ content, with
an increasing effectiveness in that order. The response to phenylephrine was abolished by
pretreatment with PMA, whereas the response to glucagon was severely attenuated and that
to vasopressin was unaffected (Table II).

When $[Ca^{2+}]_c$ was monitored in Quin2-loaded hepatocytes (Fig. 2), it was found that
glucagon, vasopressin, and phenylephrine each gave a large increase in $[Ca^{2+}]_c$, which was
sustained provided that the cells were incubated in a medium containing physiologically
appropriate levels of Ca^{2+}. Prior treatment with PMA abolished the effect of phenylephrine,
as had previously been shown by Cooper et al. (1985) and Lynch et al. (1985), and severely
attenuated the response to glucagon. The response to vasopressin was unaffected.

The general correspondence between changes in $[Ca^{2+}]_c$ and changes in PDH$_A$ content
is consistent with the thesis that the former is a determinant of the latter, in the response of
hepatocytes to these hormones. An apparent anomaly, which may be informative, is that
vasopressin gives as large an increase in $[Ca^{2+}]_c$ as glucagon but less of an activation of
pyruvate dehydrogenase (Fig. 2, Table II). Conceivably, glucagon has the additional effect
of resetting the mitochondrial Ca^{2+} transport cycle, such that uptake is favored; alternatively,
vasopressin might reset the cycle in the favor of efflux (see Nicholls and Åkerman, 1982;
Hansford, 1985, for reviews on the mitochondrial Ca^{2+} transport cycle).

3.3. Balance between Dehydrogenase and Respiratory Chain Activity

Activation of pyruvate dehydrogenase, NAD-linked isocitrate dehydrogenase, and
2-oxoglutarate dehydrogenase by Ca^{2+} would be expected to increase the rate of provision
of NADH to the respiratory chain and, in the absence of an activation of the chain itself or
of phosphorylation reactions (which is, however, a possibility; see Moreno-Sánchez, 1985a,b),

to lead to an increased steady-state content of mitochondrial NADH. By contrast, activation of energy-utilizing processes in the cytosol might be expected to lower mitochondrial NADH content by increasing the availability of ADP to the mitochondrion and enhancing respiratory chain activity, according to the classic theory of Chance and Williams (1956). We were curious to see what the response of mitochondrial NADH content would be, therefore, to the perturbations described above.

Addition of KCl (30–80 mM) to suspensions of cardiac myocytes resulted in either negligible change in the redox state of mitochondrial nicotinamide nucleotides, or in a slight oxidation (less than 8% of the total span defined by nucleotide susceptible to oxidation by the mitochondrial uncoupling agent carbonyl cyanide p-trifluoromethoxyphenylhydrazone, FCCP, and reducible by the respiratory chain inhibitor rotenone). Addition of veratridine to give 25 μM, a concentration which is maximally effective in stimulating O_2 uptake (unpublished observations), gave rise to a substantial oxidation of the nicotinamide nucleotide associated with the mitochondria by the criteria described above. This is shown in Fig. 3. This indicates a predominance of activation of the respiratory chain over activation of the dehydrogenases under these conditions. That the stimulated O_2 uptake seen in response to 25 μM veratridine reflects oxidative phosphorylation and increased availability of ADP to the mitochondria is evident from the near-total sensitivity to oligomycin (unpublished observations). A plausible interpretation of this result is that the activation of the actomyosin ATPase by the rise in cytosolic Ca^{2+} shown in Fig. 1 results in a decrease in the ATP/ADP ratio of the adenine nucleotides available for translocation into the mitochondria. The decreased intramitochondrial ATP/ADP ratio in turn enhances the activity of ATP-synthase and allows more rapid transfer of reducing equivalents to O_2, as in the classic formulation by Chance and Williams (1956). This is not to deny that increased intramitochondrial ADP will activate the pyruvate dehydrogenase complex, NAD-isocitrate dehydrogenase, and 2-oxoglutarate dehydrogenase: this has been amply documented in experiments with rat heart mitochondria, as reviewed by Hansford (1980). It does, however, show that the action of ADP at the respiratory chain level is predominant under these conditions.

The experiments of Fig. 3 do, however, provide a fascinating glimpse of dehydrogenase activation by the Ca^{2+} ion. When the cardiac myocytes are relatively substrate-poor, as in Fig. 3 where the substrate is mainly endogenous plus only 1 mM glucose, the addition of submaximal concentrations of veratridine (5 μM) results in a much larger oxidation of nicotinamide nucleotide in the presence of ruthenium red than in its absence. The inference is that higher total mitochondrial dehydrogenase activity can be maintained if the mitochondria are allowed to preserve their calcium content: in the steady state, the latter will be depleted by exposure to ruthenium red. When the energy demand is sufficiently high, as when maximally effective concentrations of veratridine are used (25 μM, Fig. 3), ruthenium red is without discernible effect on the mitochondrial redox state. These conditions generate maximal O_2 uptake by the myocytes, in analogy to state 3 respiration by isolated mitochondria. It is not uncommon for metabolic control processes to be balanced quite differently at 100% of the state 3 respiratory rate from the balance seen in the range 50–85% of state 3 (Hansford and Cohen, 1978; Hansford, 1980), a region which may correspond more to physiology.

In summary, increased work load in cardiac myocytes in general gives rise to an oxidation of nicotinamide nucleotide, as has been seen when the work load on the perfused isolated heart is raised (Illingworth et al., 1975). Dehydrogenase activation occurs but respiratory chain activation is predominant under the experimental conditions investigated.

By contrast, the elevation of $[Ca^{2+}]_c$ in hepatocytes leads to a marked increase in nicotinamide nucleotide fluorescence (Fig. 4). Glucagon, phenylephrine, and vasopressin

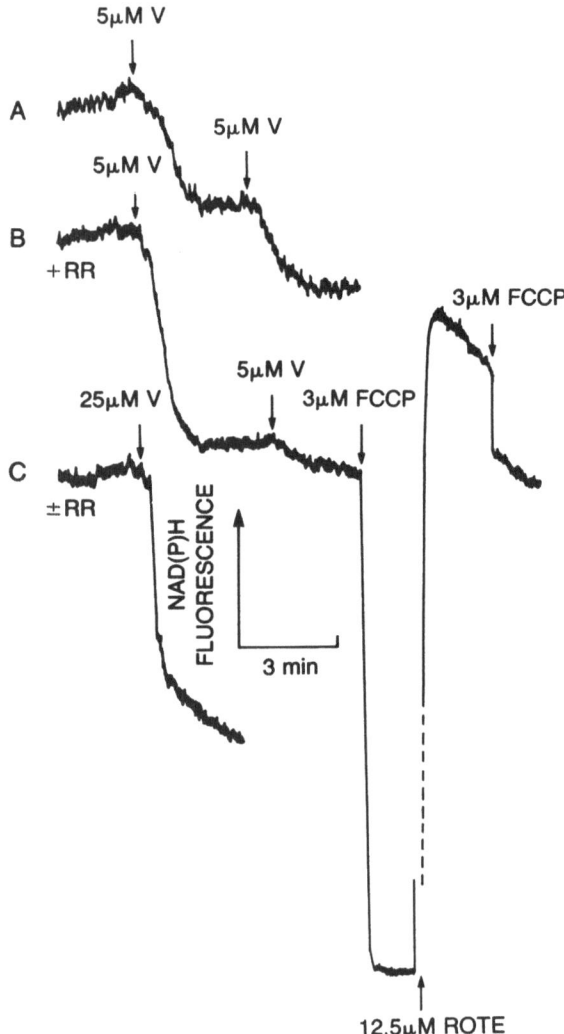

FIGURE 3. Effect of veratridine on nicotinamide nucleotide fluorescence of suspensions of cardiac myocytes. Myocytes (1.9 mg of protein/ml) were suspended in a Hepes-buffered medium, pH 7.4, 37°C with 1 mM glucose present as the only exogenous substrate. Cellular fluorescence was excited at 333 nm and collected at wavelengths greater than 460 nm. Where indicated, veratridine (V), FCCP, and rotenone (ROTE) were added at the concentrations shown. In addition, ruthenium red was added to 25 μM, 25 min prior to veratridine in trace B. The response to 25 μM vertridine (C) was identical in the presence and absence of ruthenium red.

(separately) lead to a rapid increase in fluorescence, which is not fully sustained. This is attributed to an increase in mitochondrial NADH content on the basis of control experiments (not shown) in which cytosolic NAD is reduced with ethanol and the transfer of reducing agents into the mitochondria blockaded with aminooxyacetate. Under these conditions results similar to those shown in Fig. 4 are obtained. An increase in NAD(P)H fluorescence in

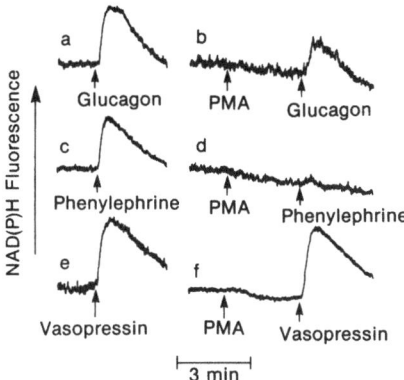

FIGURE 4. Response of the NAD(P)H fluorescence of a suspension of hepatocytes to Ca^{2+}-mobilizing hormones. The fluorescence of a suspension of hepatocytes was monitored as described for Fig. 3. The final concentrations of added compounds were glucagon, 10 nM; PMA, 500 nM; phenylephrine, 25 μM; vasopressin, 25 nM. The Ca^{2+} concentration of the medium was 2.5 mM. The maximum increases in fluorescence caused by the agonists, expressed as a percentage of the increase in fluorescence obtained by adding 5 μM rotenone plus 20 mM DL-3-hydroxybutyrate, were (a) glucagon 37%; (b) glucagon 17%; (c) phenylephrine 51%; (e) vasopressin 44%.

response to glucagon has also been reported by Sugano *et al.* (1980), Balaban and Blum (1982), and Sistare *et al.* (1985) in work with perfused liver or isolated hepatocytes. 4β-phorbol-12-myristate-13-acetate prevents the rise in NADH due to phenylephrine and attenuates that due to glucagon, in striking analogy to the results for $[Ca^{2+}]_c$ (Fig. 2) and for PDH_A content (Table II).

An increase in $[Ca^{2+}]_c$ is a sufficient mechanism to cause the increased mitochondrial $NADH/NAD^+$ ratio seen in response to these hormones, as shown in experiments in which $[Ca^{2+}]_c$ was artificially raised by use of the divalent cation ionophore ionomycin (Staddon and Hansford, 1987).

We suggest that the clear difference in response to an elevation of $[Ca^{2+}]_c$ in myocytes and hepatocytes reflects the great potential activity of the Ca^{2+}-activated actomyosin ATPase in the former. Whereas in muscle cells the activated state is associated with increased availability of both Ca^{2+} and ADP to the mitochondria, in hepatocytes hormonal activation may result in an increase in $[Ca^{2+}]_c$ without any decrease in ATP/ADP ratio. The advantage to the animal of dehydrogenase level activation is that the tendency to maintain (muscle) or even elevate (liver) mitochondrial $NADH/NAD^+$ ratios allows more rapid rates of oxidative phosphorylation without the necessity of allowing ATP/ADP $\times$ Pi ratios to fall to unacceptable levels. For a fuller discussion, the reader is referred to Hansford (1980) or to Hansford and Staddon (1987).

REFERENCES

Balaban, R. S., and Blum, J. J., 1982, Hormone-induced changes in NADH fluorescence and O₂ consumption of rat hepatocytes, *Am. J. Physiol.* **242**:C172–C177.

Chance, B., and Williams, G. R., 1956, The respiratory chain and oxidative phosphorylation, *Adv. Enzymol.* **17**:65–134.

Charest, R., Blackmore, P. F., Berthon, B., and Exton, J. H., 1983, Changes in free cytosolic Ca^{2+} in hepatocytes following α_1-adrenergic stimulation. Studies on Quin-2-loaded hepatocytes, *J. Biol. Chem.* **258**:8769–8773.

Cooper, R. H., Coll, K. E., and Williamson, J. R., 1985, Differential effects of phorbol ester on phenylephrine and vasopressin-induced Ca^{2+} mobilization in isolated hepatocytes, *J. Biol. Chem.* **260**:3281–3288.

Denton, R. M., McCormack, J. G., and Edgell, N. J., 1980, Role of calcium ions in the regulation of intramitochondrial metabolism. Effects of Na^+, Mg^{2+} and ruthenium red on the Ca^{2+}-stimulated oxidation of oxoglutarate and on pyruvate dehydrogenase activity in intact rat heart mitochondria, *Biochem. J.* **190**:107–117.

Denton, R. M., Randle, P. J., and Martin, B. R., 1972, Stimulation by Ca^{2+} of pyruvate dehydrogenase phosphate phosphatase, *Biochem. J.* **128**:161–163.

Denton, R. M., Richards, D. A., and Chin, J. G., 1978, Calcium ions and the regulation of NAD^+-linked isocitrate dehydrogenase from the mitochondria of rat heart and other tissues, *Biochem. J.* **176**:899–906.

Escueta, A. V., and Appel, S. H., 1969, Biochemical studies of synapses in vitro. II. Potassium transport, *Biochemistry* **8**:725–733.

Hansford, R. G., 1976, Studies on the effects of coenzyme A-SH: Acetyl coenzyme A, nicotinamide adenine dinucleotide: Reduced nicotinamide adenine dinucleotide and adenosine diphosphate: Adenosine triphosphate ratios on the interconversion of active and inactive pyruvate dehydrogenase in isolated rat heart mitochondria, *J. Biol. Chem.* **251**:5483–5489.

Hansford, R. G., 1980, Control of mitochondrial substrate oxidation, *Cur. Top. Bioenerg.* **10**:217–278.

Hansford, R. G., 1981, Effect of micromolar concentrations of free Ca^{2+} ions on pyruvate dehydrogenase interconversion in intact rat heart mitochondria, *Biochem. J.* **194**:721–732.

Hansford, R. G., 1985, Relation between mitochondrial calcium transport and energy metabolism, *Rev. Physiol. Biochem. Pharmacol.* **102**:1–72.

Hansford, R. G., 1987, Relation between cytosolic free Ca^{2+} and the control of pyruvate dehydrogenase in isolated cardiac myocytes, *Biochem. J.* **241**:145–151.

Hansford, R. G., and Chappell, J. B., 1967, The effect of Ca^{2+} on the oxidation of glycerol phosphate by blowfly flight-muscle mitochondria, *Biochem. Biophys. Res. Commun.* **27**:686–692.

Hansford, R. G., and Cohen, L., 1978, Relative importance of pyruvate dehydrogenase interconversion and feed-back inhibition in the effect of fatty acids on pyruvate oxidation by rat heart mitochondria, *Arch. Biochem. Biophys.* **191**:65–81.

Hucho, F., Randall, D. D., Roche, T. E., Burgett, M. W., Pelley, J. W., and Reed, L. J., 1972, α-Keto acid dehydrogenase complexes. XVII. Kinetic and regulatory properties of pyruvate dehydrogenase kinase and pyruvate dehydrogenase phosphatase from bovine kidney and heart, *Arch. Biochem. Biophys.* **151**:328–340.

Illingworth, J. A., Ford, W. C. L., Kobayashi, K., and Williamson, J. R., 1975, Regulation of myocardial energy metabolism, in: *Recent Advances in Studies on Cardiac Structure and Metabolism*, Vol. 8 (P.-E. Roy and P. Harris, eds.), University Park Press, Baltimore, pp. 271–290.

Katz, A. M., 1970, Contractile proteins of the heart, *Physiol. Rev.* **50**:63–158.

Lynch, C. J., Charest, R., Bocckino, S. B., Exton, J. H., and Blackmore, P. F., 1985, Inhibition of hepatic α_1-adrenergic effects and binding by phorbol myristate acetate, *J. Biol. Chem.* **260**:2844–2851.

McCormack, J. G., and Denton, R. M., 1979, The effects of calcium ions and adenine nucleotides on the activity of pig heart 2-oxoglutarate dehydrogenase complex, *Biochem. J.* **180**:533–544.

Moore, C. L., 1971, Specific inhibition of mitochondrial Ca^{2+} transport by ruthenium red, *Biochem. Biophys. Res. Commun.* **42**:298–305.

Moreno-Sánchez, R., 1985a, Regulation of oxidative phosphorylation in mitochondria by external free Ca^{2+} concentrations, *J. Biol. Chem.* **260**:4028–4034.

Moreno-Sánchez, R., 1985b, Contribution of the translocator of adenine nucleotides and the ATP synthase to the control of oxidative phosphorylation and arsenylation in liver mitochondria, *J. Biol. Chem.* **260**:12554–12560.

Nicholls, D., and Åkerman, K., 1982, Mitochondrial calcium transport, *Biochim. Biophys. Acta* **683**: 57–88.

Ohta, M., Narahashi, T., and Keeler, R. F., 1973, Effect of veratrum alkaloids on membrane potential and conductance of squid and crayfish giant axons, *J. Pharmacol. Exp. Ther.* **184**:143–154.

Pettit, F. H., Roche, T. E., and Reed, L. J., 1972, Function of calcium ions in pyruvate dehydrogenase phosphatase activity, *Biochem. Biophys. Res. Commun.* **49**:563–571.

Powell, T., Tatham, P. E. R., and Twist, V. W., 1984, Cytoplasmic free calcium measured by Quin 2 fluorescence in isolated ventricular myocytes at rest and during potassium-depolarization, *Biochem. Biophys. Res. Commun.* **122**:1012–1020.

Rasmussen, H., 1981, *Calcium and cAMP as Synarchic Messengers*. John Wiley and Sons, New York.

Sheu, S.-S., Sharma, V. K., and Uglesity, A., 1986, Na^+-Ca^{2+} exchange contributes to increase of cytosolic Ca^{2+} concentration during depolarization in heart muscle, *Am. J. Physiol.* **250**:C651–C656.

Sistare, F. D., Picking, R. A., and Haynes, R. C., Jr., 1985, Sensitivity of the response of cytosolic calcium in Quin-2-loaded rat hepatocytes to glucagon, adenine nucleosides and adenine nucleotides, *J. Biol. Chem.* **260**:12744–12747.

Staddon, J. M., and Hansford, R. G., 1986, 4-β-Phorbol 12-myristate 13-acetate attenuates the glucagon-induced increase in cytoplasmic free Ca^{2+} concentration in isolated rat hepatocytes, *Biochem. J.* **238**:737–743.

Staddon, J. M., and Hansford, R. G., 1987, The glucagon-induced activation of pyruvate dehydrogenase in hepatocytes is diminished by 4β-phorbol 12-myristate 13-acetate: A role for cytoplasmic Ca^{2+} in dehydrogenase regulation, *Biochem. J.* **241**:729–735.

Sugano, T., Shiota, M., Khono, H., Shimada, M., and Oshino, N., 1980, Effects of calcium ions on the activation of gluconeogenesis by norepinephrine in perfused rat liver, *J. Biochem.* **87**:465–472.

Sutko, J. L., and Kenyon, J. L., 1983, Ryanodine modification of cardiac muscle responses to potassium-free solutions. Evidence for inhibition of sarcoplasmic reticulum calcium release, *J. Gen. Physiol.* **82**:385–404.

Wieland, O. H., 1983, The mammalian pyruvate dehydrogenase complex: Structure and regulation, *Rev. Physiol. Biochem. Pharmacol.* **96**:124–170.

37

Calcium as a Hormonal Messenger for Control of Mitochondrial Functions

ROBERT C. HAYNES, Jr.

During the past decade it has become generally accepted that hormones such as norepinephrine, angiotensin II, and vasopressin affect metabolism of the rat liver by increasing free calcium $[Ca^{2+}]$ of the cytosol. In contrast, a role for calcium as a mediator of the action of glucagon has been controversial. Reports that glucagon alters fluxes of calcium in and out of the liver cell and that it lowers mitochondrial calcium were nullified to some degree by other reports indicating that these effects were observed only when pharmacological doses of glucagon were administered and that cytosolic $[Ca^{2+}]$ is not altered by glucagon. The unsettled nature of the field was described in the comprehensive review of Williamson et al. (1981). Things began to fall into place when the fluorescent probe for $[Ca^{2+}]$, Quin2, became available, and Charest et al. (1983) demonstrated that glucagon added at the high-dose level of 10 nM raised cytosolic $[Ca^{2+}]$ of hepatocytes from a basal concentration of about 0.2 μM to 0.6 μM.

This finding of Charest et al. (1983) overturned the concept that glucagon does not alter cytosolic $[Ca^{2+}]$. We were stimulated by this iconoclastic observation to wonder whether, if one concept in the field could fall so easily, another might also. We therefore decided to reexamine dose-response relationships between glucagon and $[Ca^{2+}]$ using Quin2 to monitor cytosolic $[Ca^{2+}]$. We quickly found that the cytosolic $[Ca^{2+}]$ of hepatocytes responds to glucagon at half-maximal concentrations of about 0.2 to 0.3 nM (Fig. 1), the same sensitivity reported by numerous investigators for activation of glycogenolysis and gluconeogenesis and inhibition of pyruvate kinase (Sistare et al., 1985). Our estimates of cytosolic $[Ca^{2+}]$ for resting cells was 136 $\pm$ 2 nM and 581 $\pm$ 157 nM after maximal stimulation by glucagon, values quite close to those of Charest et al. (1983). It was found that glucagon was able to elevate cytosolic $[Ca^{2+}]$ even when hepatocytes were incubated in a medium with no added calcium and in the presence of EGTA, although the duration of the effect was shortened under these conditions. This decreased duration of response suggested that, in addition to increasing release of calcium from an internal pool, glucagon also increases entry of extracellular calcium into the hepatocyte. Combettes et al. (1986) confirmed this observation and

---

ROBERT C. HAYNES, JR. • Department of Pharmacology, University of Virginia, Charlottesville, Virginia 22908.

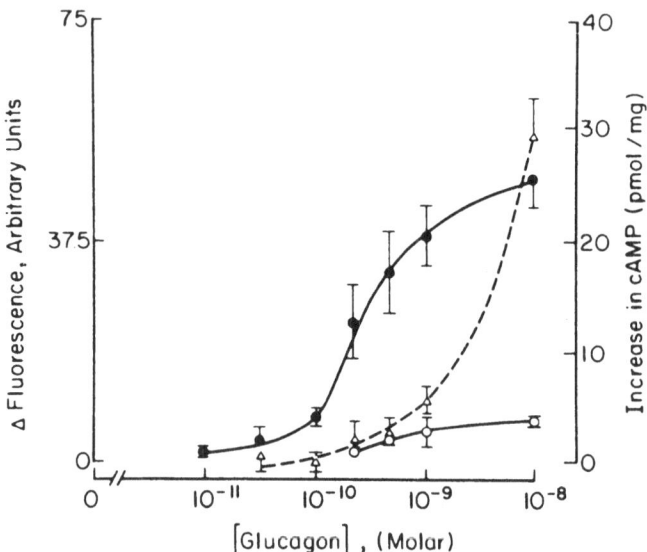

FIGURE 1. The response of [Ca^{2+}] and cAMP to glucagon. Hepatocytes were loaded with Quin2 and incubated as described by Sistare *et al.* (1985). Increments in fluorescence were determined at 90 sec after adding glucagon, and at the same time samples were taken for cAMP determination. Closed circles represent the mean increase in fluorescence produced by glucagon. Open circles indicate the increase in fluorescence in cells incubated without Quin2. Open triangles represent mean values of cAMP. Each mean is presented ± SE for three or four experiments.

reached a similar conclusion. In considering the effect of glucagon on cytosolic calcium, we tested forskolin, which imitated glucagon as did cAMP. To investigate the specificity of cAMP in evoking this response we tested a number of related compounds and found that while cAMP was active with a K_{act} between 30 and 100 μM, 5-AMP, ATP, and ADP were considerably more active with K_{act} values of approximately 10, 0.3, and 0.3 μM, respectively. In a study of phosphorylase activation in hepatocytes via purinergic receptors, Keppens and De Wulf (1985) reported similar sensitivities to adenine derivatives and suggested that the effect of agonists active at purinergic receptors is probably mediated by calcium.

The question of relative roles and possible interactions of the cAMP and [Ca^{2+}] responses to glucagon remain uncertain. A recent paper of Connelly *et al.* (1987) concludes that activation of calcium-dependent protein kinases is a minor component of the hepatocyte response to glucagon.

We have studied one mitochondrial phenomenon that results from glucagon treatment, and the evidence is strong that this is mediated by the rise in cytosolic [Ca^{2+}]. This effect of glucagon is an acute increase in mitochondrial adenine nucleotides that was noted by several laboratories, including ours in the late 1970s. The one-to-one exchange of nucleotides facilitated by the adenine nucleotide translocase is the best known nucleotide transport system of mitochondria, but there is considerable evidence that net movement of the adenine nucleotides across the mitochondrial membrane also occurs. This was first discovered by

Carafoli *et al.* (1965), who noted during investigation of the uptake of large amounts of calcium and phosphate by rat liver mitochondria that there was also an uptake of adenine nucleotides. These workers hypothesized that the nucleotide accumulation was essential to maintain ionic balance in mitochondria containing large amounts of calcium and phosphate in a molar ratio of 1.67. In 1968 Meisner and Klingenberg showed that rat liver mitochondria experience a net loss of adenine nucleotides if incubated in the presence of phosphate, magnesium ions, and ethylene diamine tetraacetic acid (EDTA). These demonstrations of net uptake and net loss of adenine nucleotides by mitochondria were apparently not explored further for several years, and their biological significance was obscure. Interest was renewed in these phenomena when it was reported that hepatic mitochondria of newborn rats have a low respiratory control index and a deficiency of adenine nucleotides (Nakazawa *et al.*, 1973). Within an hour after birth the respiratory index is increased and the mitochondrial adenine nucleotides are doubled while there is no significant change in total hepatic adenine nucleotides (Sutton and Pollak, 1978). This, of course, indicates that a net translocation of the nucleotides into the mitochondria occurs in the neonatal period.

Aprille and her colleagues studied the net transport of adenine nucleotides into mitochondria using a simple *in vitro* system that permits net uptake of nucleotides and have made significant contributions to its understanding. Their work indicated that ATP in most circumstances is preferentially transported relative to ADP because of a lower K_m for ATP and that the nucleotides move from one compartment to another down concentration gradients (Aprille and Austin, 1981; Austin and Aprille, 1984). The transport system is not affected by inhibitors of adenine nucleotide translocase if, under the conditions employed, the translocase is not essential to attain ratios of ATP and ADP within or without the mitochondria that permit net transport of ATP. These workers proposed that ATP is transported as the magnesium salt in exchange for divalent phosphate (Austin and Aprille, 1984; Nosek and Aprille, 1986).

A few years ago Dr. Helen Hamman and I carried out experiments to test the hypothesis that the increase in mitochondrial adenine nucleotides following glucagon administration may account for a number of changes in mitochondrial functions produced by the hormone. However, we were not able to demonstrate any stimulation of these functions resulting from an increased content of adenine nucleotides (Hamman and Haynes, 1983). In the course of these studies, William Zaks, then a student in the laboratory, made the observation that the net uptake of adenine nucleotides *in vitro* was blocked in the presence of EDTA, alerting us to the possibility that trace quantities of a metal ion, most likely calcium, might be required for net transport of the nucleotides. This was an idea that remained untested for several years, as our interest in net transfer of adenine nucleotides waned when we were unable to link this phenomenon to the other hormone-produced changes in mitochondrial functions.

When we found that glucagon, in near-physiological doses, increased cytosolic $[Ca^{2+}]$ in hepatocytes, we decided to test the long dormant hypothesis that calcium in very low concentrations was necessary for net transfer of adenine nucleotides. Our experimental approach was to incubate mitochondria at 30°C under air in a medium designed to approximate cytosol in ionic composition. It contained 150 mM KCl, 30 mM sucrose, 4 mM KH_2PO_4 (pH 7.1), 4.2 mM $MgCl_2$, 1 mM sodium succinate, 4 mM ATP, 5 mM EGTA, and variable amounts of $CaCl_2$ to provide $[Ca^{2+}]$ from 0.08 to 4 μM. After incubations, cold 0.3 M sucrose was added, the mitochondria sedimented and washed twice, followed by fixation and analysis of adenine nucleotides (Haynes *et al.*, 1986).

As is evident from Fig. 2, we found not only that submicromolar Ca^{2+} was indeed essential for accumulation by adenine nucleotides, but also that the approximate control level

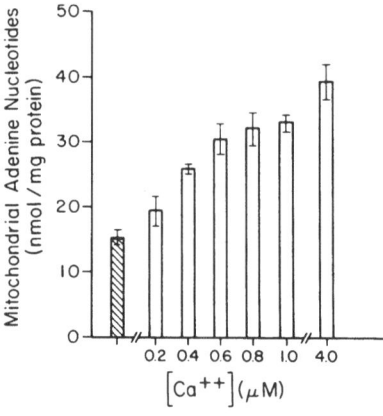

FIGURE 2. Accumulation of adenine nucleotides by mitochondria in relation to $[Ca^{2+}]$. Mitochondria were incubated for 15 min at 30°C in the presence of 4 mM ATP and then assayed for adenine nucleotide content as described by Haynes *et al.* (1986). The hatched bar represents nucleotide content of mitochondria not incubated. Means ± SE are from three experiments.

of $[Ca^{2+}]$ in the cytosol, 0.2 μM, was essentially ineffective while 0.6 μM $[Ca^{2+}]$, the peak reported to be reached with maximal stimulation by glucagon (Charest *et al.*, 1983; Sistare *et al.*, 1985), led to a 50% increase in mitochondrial adenine nucleotides in 15 min. We were curious to find out if the adenine nucleotides reach a steady-state level within the mitochondria that is determined by the concentration of extramitochondrial $[Ca^{2+}]$. Kinetic experiments indicated, however, that while there is an initial rapid uptake of nucleotides that is dependent on $[Ca^{2+}]$ concentration, the uptake then slows but continues at least for 35 min with the rate remaining dependent on $[Ca^{2+}]$ in the medium.

The accumulation of adenine nucleotides by mitochondria is insensitive to carboxy-atractyloside as anticipated from the work of Austin and Aprille (1984). Only a small increase in mitochondrial volume (0.73 ± 0.03 to 0.86 ± 0.03 $\mu l/mg$) was demonstrable when adenine nucleotides were accumulated. Inhibition of the electron transport chain by antimycin A and the proton translocating ATPase by oligomycin led to a greater accumulation of adenine nucleotides and a decreased membrane potential. The adenine nucleotides accumulated were found to be both ATP and ADP with AMP remaining essentially unchanged. However, in experiments with antimycin and oligomycin present in which increased quantities of nucleotides were taken up, the predominant species was ADP accompanied by some increase in AMP. Similarly, uptake in the absence of added phosphate led primarily to an accumulation of ADP and a slight increase in AMP.

The studies of Carafoli *et al.* (1965) suggested that mitochondrial accumulation of adenine nucleotides depends on uptake of calcium and phosphate. The dependence of adenine nucleotide uptake on submicromolar $[Ca^{2+}]$ is compatible with the hypothesis of Carafoli *et al.* (1965), but our experiments using several orders of magnitude less $[Ca^{2+}]$ than the earlier workers did not lend themselves readily to chemical analysis of calcium phosphate accumulation; furthermore, as mentioned above, we found no evidence of severe swelling of mitochondria loaded with adenine nucleotides under the influence of submicromolar $[Ca^{2+}]$. To approach the hypothesis that $[Ca^{2+}]$ uptake is a prerequisite for adenine nucleotide accumulation by mitochondria, we added ruthenium red to block electrogenic transport of Ca^{2+} in experiments in which uptake of adenine nucleotides was stimulated by 1 μM $[Ca^{2+}]$. We were surprised to find that ruthenium red in concentrations as high as 200 μM failed to inhibit adenine nucleotide accumulation. Our preparation of ruthenium red used at 4 μM completely inhibited the respiratory response of isolated mitochondria to 350 μM $[Ca^{2+}]$, so there is no question that our preparation of inhibitor was active. These experiments suggested that calcium may interact with the inner membrane of the mitochondria to activate the mechanism responsible for net transport of adenine nucleotides into the mitochondria. The dose response curve shown in Fig. 2 is compatible with control being exerted through such an interaction with a binding constant for $[Ca^{2+}]$ of approximately 0.6 μM. Another possible mechanism we suggested is that the calcium salt of ATP carrying a negative charge of two exchanges for divalent phosphate. This hypothesis derives from Austin and Aprille (1984), who proposed that the magnesium salt of ATP is the active species of transport. Activation of the transport system by $[Ca^{2+}]$ might be expected to lead to release of adenine nucleotides as well as stimulate their uptake. We therefore studied the effects of submicromolar $[Ca^{2+}]$ on release of adenine nucleotides. These experiments were made essentially as those studying uptake except ATP was omitted from the incubation. It was found that 0.2–4 μM $[Ca^{2+}]$ enhanced loss of adenine nucleotides in a dose-dependent fashion, a result compatible with an effect of $[Ca^{2+}]$ on the net transport system. In a single experiment, however, we also observed loss of pyridine nucleotides roughly proportional to loss of adenine nucleotides as $[Ca^{2+}]$ was increased. We feel that incubation of mitochondria in the presence of inorganic phosphate and no added adenine nucleotides is fraught with ambiguity inasmuch as these conditions can produce damage to mitochondrial membranes (Beatrice et al., 1980). Consequently, these experiments related to release of adenine nucleotides are only weak evidence for a calcium-activated adenine nucleotide transport system.

To help relate the effects of submicromolar $[Ca^{2+}]$ on adenine nucleotide uptake by isolated mitochondria to the hormonal effects, we carried out experiments using hormones and the calcium ionophore A-23187 on isolated hepatocytes. These experiments were designed to show that calcium-mobilizing hormones other than glucagon lead to elevations of mitochondrial adenine nucleotides, that cells partially depleted of calcium have a diminished response to glucagon, and that artificially increasing cytosolic calcium by the calcium ionophore imitates the effects of hormones. Hepatocytes were incubated in a modified Krebs–Ringer solution containing 2.5 mM or no calcium chloride. Hormones or the ionophore were added after 45 min of incubation and the cells incubated for an additional 15 min. The cells were subjected to digitonin fractionation, and the particulate fraction, assumed to consist primarily of mitochondria, was assayed for adenine nucleotides. It is evident from Fig. 3 that glucagon, epinephrine, and A-23187 increased the mitochondrial content of adenine nucleotides. It can also be seen that cells incubated with EGTA and no calcium chloride added to the medium had a lower mitochondrial content of adenine nucleotides, and the response to glucagon was attenuated. We feel, therefore, that the evidence supports our hypothesis that hormones acting to elevate mitochondrial adenine nucleotides do so by raising cytosolic $[Ca^{2+}]$.

The physiological significance of increased mitochondrial adenine nucleotides in the mature organ is not clear. Nevertheless, as mentioned earlier, the work of several investigators has established the critical nature of the rise in mitochondrial adenine nucleotides that takes place within an hour or two after birth in the rat (Nakazawa et al., 1973; Sutton and Pollak, 1978; Aprille and Asimakis, 1980; Aprille et al., 1981). Aprille and her colleagues have emphasized the importance of increased mitochondrial nucleotides during the neonatal period in the development of gluconeogenic capacity and the role that the neonatal surge in glucagon secretion probably plays in this phase of development. Recently, her group showed that maternal diabetes delays the postnatal uptake of adenine nucleotides into hepatic mitochondria (Aprille and Nosek, 1987). There appears to be no information on adenine nucleotide content of hepatic mitochondria from newborn infants. If human infants have a rapid hepatic mitochondrial maturation characterized by accumulation of adenine nucleotides as in the rat, it can be surmised that the hypoglycemia, occurring in some newborn infants of diabetic mothers, may be due in part to retardation of mitochondrial maturation.

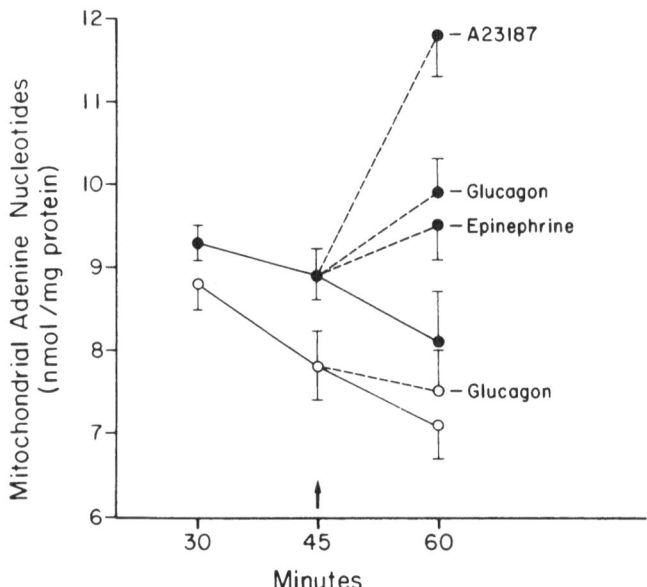

FIGURE 3. The effect of hormones, A-23187, and calcium depletion on adenine nucleotide content of mitochondria of isolated hepatocytes. Hepatocytes were incubated and fractionated with digitionin as described by Haynes et al. (1986). Analyses were made after 30, 45, and 60 min of incubation. Agonists or control vehicle were added at 45 min. Solid circles represent values from cells incubated in a medium containing 2.5 mM $CaCl_2$. Open circles represent values from cells incubated in the absence of added calcium and in the presence of 1 mM EGTA. Means ± SE are from seven experiments.

REFERENCES

Aprille, J. R., and Asimakis, G. K., 1980, Postnatal development of rat liver mitochondria: State 3 respiration, adenine nucleotide translocase activity, and the net accumulation of adenine nucleotide, Arch. Biochem. Biophys. 201:564–575.

Aprille, J. R. and Austin, J. (1981), Regulation of the mitochondrial adenine nucleotide pool size, Arch. Biochem. Biophys. 212:689–699.

Aprille, J. R. and Nosek, M. T. (1987), Neonatal hypoxia or maternal diabetes delays postnatal development of liver mitochondria, Ped. Res. 21:266–269.

Aprille, J. R., Yaswen, P., and Rulfs, J., 1981, Acute postnatal regulation of pyruvate carboxylase activity by compartmentation of mitochondrial adenine nucleotides, Biochim. Biophys. Acta 675:143–147.

Austin, J. and Aprille, J. R. (1984), Carboxyatractyloside sensitive influx and efflux of adenine nucleotides in rat liver mitochondria, J. Biol. Chem. 259:154–160.

Beatrice, M. L. Palmer, J. W. and Pfeiffer, D. R., 1980, The relationship between mitochondrial membrane permeability, membrane potential, and the retention of Ca^{2+} by mitochondria, J. Biol. Chem. 255:8663–8671.

Carafoli, E., Rossi, C. S. and Lehninger, A. L., 1965, Uptake of adenine nucleotide by respiring mitochondria during active accumulation of Ca^{2+} and phosphate, J. Biol. Chem. 240:2254–2261.

Charest, R., Blackmore, P. F., Berthon, B., and Exton, J. H., 1983, Changes in free cytosolic Ca^{2+} in hepatocytes following α_1-adrenergic stimulation, *J. Biol. Chem.* **258**:8769–8773.

Combettes, L., Berthon, B., Binet, A., and Claret, M., 1986, Glucagon and vasopressin interactions on Ca^{2+} movements in isolated hepatocytes, *Biochem. J.* **237**:675–683.

Connelly, P. A., Parker Botelho, L. H., Sisk, R. B., and Garrison, J. C., 1987, A study of the mechanism of glucagon-induced protein phosphorylation in isolated rat hepatocytes using (S$_p$)-cAMPS and (R$_p$-)cAMPS, the stimulatory and inhibitory diastereomers of adenosine cyclic 3',5'-phosphorothioate, *J. Biol. Chem.* **262**:4324–4332.

Hamman, H. C., and Haynes, R. C., Jr., 1983, Elevated intramitochondrial adenine nucleotides and mitochondrial function, *Arch. Biochem. Biophys.* **223**:85–94.

Haynes, R. C., Jr., Picking R. A., and Zaks, W. J., 1986, Control of mitochondrial content of adenine nucleotides by submicromolar calcium concentrations and its relationship to hormonal effects, *J. Biol. Chem.* **261**:16121–16125.

Keppens, S. and DeWulf, H., 1985, P$_2$-purinergic control of liver glycogenolysis, *Biochem. J.* **231**:797–799.

Meisner, H. and Klingenberg, M., 1968, Efflux of adenine nucleotides from rat liver mitochondria, *J. Biol. Chem.* **243**:3631–3639.

Nakazawa, T. Asami, K., Suzuki, H., and Vulowa, O., 1973, Appearance of energy conservation system in rat liver mitochondria during development. The role of adenine nucleotide translocation, *J. Biochem.* **73**:392–406.

Nosek, M. T., and Aprille, J. R., 1986, Divalent phosphate is a counter ion for carboxyatractyloside-insensitive adenine nucleotide transport in rat liver mitochondria, *Fed. Proc.* **45**:1924.

Sistare, F. D., Picking, R. A., and Haynes, R. C., Jr., 1985, Sensitivity of the response of cytosolic calcium in quin2-loaded rat hepatocytes to glucagon, adenine nucleosides and adenine nucleotides, *J. Biol. Chem.* **260**:12744–12747.

Sutton, R. and Pollak, J. K., 1978, The increasing adenine nucleotide concentration and the maturation of rat liver mitochondria during neonatal development, *Differentiation* **12**:15–21.

Williamson, J. R., Cooper, R. H., and Hoek, J. B., 1981, Role of calcium in the hormonal regulation of liver metabolism, *Biochim. Biophys. Acta* **639**:243–295.

38

Effect of Mitochondrial Ca²⁺ on Hepatic Aspartate Formation and Gluconeogenic Flux

ANNA STERNICZUK, STAN HRENIUK, RUSSELL SCADUTO, Jr., and KATHRYN F. LaNOUE

1. INTRODUCTION

According to current evidence, the mechanism of action of many gluconeogenic hormones involves a rise in intracellular free Ca^{2+} (Charest *et al.*, 1983; Sistare *et al.*, 1985). There is general agreement that hepatic gluconeogenesis is controlled at several sites in the pathway by Ca^{2+}-and cAMP-mediated phosphorylations of cytosolic enzymes (Hers and Hue, 1983; Pilkis *et al.*, 1986). Nevertheless, there is a perception that other Ca^{2+}-linked sites of action may be important; in particular, those which involve mitochondrial enzymes (Leverve *et al.*, 1986; McCormack, 1985; Staddon and Hansford, 1987). The purpose of the study was to evaluate the significance of the stimulation by Ca^{2+} of α-ketoglutarate dehydrogenase (McCormack, 1985) and pyruvate dehydrogenase (Oviasu and Whitton, 1984) which occurs following exposure of the liver to glucagon or phenylephrine. The dramatic decrease in α-ketoglutarate caused by these hormones (Siess *et al.*, 1977) is rather convincingly due to the stimulation of α-ketoglutarate dehydrogenase. Studies from this laboratory (LaNoue *et al.*, 1983; Schoolwerth and LaNoue, 1983) showed that α-ketoglutarate is a potent and physiologically important inhibitor of glutamate dehydrogenase. Stimulation of α-ketoglutarate dehydrogenase relieves inhibition of glutamate dehydrogenase by α-ketoglutarate and, thereby, may stimulate gluconeogenesis from amino acids. Glucagon and phenylephrine are also known to stimulate gluconeogenesis from lactate (Hutson *et al.*, 1976; Kneer *et al.*, 1979), and to stimulate alcohol oxidation (Ochs and Lardy, 1981) and the malate-aspartate cycle (Kneer *et al.*, 1979; Leverve *et al.*, 1986). These processes do not involve glutamate dehydrogenase but rather aspartate aminotransferase and the glutamate aspartate translocase (Rognstad and Katz, 1970; Williamson *et al.*, 1971). Aspartate amino-

ANNA STERNICZUK, STAN HRENIUK, RUSSELL SCADUTO, Jr., and KATHRYN F. LaNOUE
• Department of Physiology, Milton S. Hershey Medical Center, Pennsylvania State University, Hershey, Pennsylvania 17033.

transferase is subject to inhibition by α-ketoglutarate, competitive with oxalacetate (LaNoue *et al.*, 1987); the physiological relevance of this inhibition has not been previously evaluated. Since aspartate production by mitochondria is an irreversible process due to the electrophoretic nature of the glutamate aspartate translocase (LaNoue and Schoolwerth, 1979), and since it is an obligatory step in gluconeogenesis from lactate (Rognstad and Katz, 1970), it seems reasonable that it should increase when hormones stimulate glucose formation. However, the substrates of the mitochondrial aspartate aminotransferase decrease after the addition of gluconeogenic hormones to perfused livers (Williamson *et al.*, 1969; Parilla *et al.*, 1975) and to hepatocytes (Siess *et al.*, 1977). In order to quantitatively evaluate the significance of the fall of glutamate and oxalacetate, as well as the decrease in α-ketoglutarate after hormone addition, we have developed a rate equation to describe rates of aspartate formation from glutamate by isolated rat liver mitochondria (LaNoue *et al.*, 1987). This rate equation contains terms for the K_m values for oxalacetate and glutamate and the K_i for α-ketoglutarate. The equation was used to evaluate the rate of aspartate formation in perfused livers and these rates were then compared with observed rates of gluconeogenesis under a variety of physiological conditions.

2. EXPERIMENTAL PROCEDURES

2.1. Animals

Sprague–Dawley rats, 180–220 g in weight, were fasted 24–30 hr before being used in the experiments described.

2.2. Liver Perfusion

Livers were perfused in a flow-through system at 32°C at flow rates of 14 ml/min. Krebs–Ringer bicarbonate buffer containing 10% bovine red blood cells was oxygenated with a gas mixture of 95% O_2 plus 5% CO_2. The pH of the perfusion buffer after equilibration was 7.4. After 15 min of perfusion without substrates, 10 mM lactate and 1 mM pyruvate were added. In some cases at 25–30 min of perfusion in the presence of substrate, hormones were added and perfusion was continued for the next 30 min. Samples of perfusate were taken every 5 min and assayed for glucose (Trinder, 1969) and ketone bodies (Williamson and Corkey, 1969). The perfusions were terminated by freeze-clamping the livers. Neutralized perchloric acid (PCA) extracts (2 ml of 6% PCA/g tissue) were used for metabolite assays.

2.3. Assays and Calculations of Metabolite Concentrations and Fluxes

Aspartate, α-ketoglutarate, malate, glutamate, ATP, acetoacetate, and βOH butyrate were assayed enzymatically by standard spectrophotometric or fluorometric techniques (Williamson and Corkey, 1969).

In order to calculate free oxalacetate in the mitochondrial matrix from tissue malate, it was assumed that the mitochondrial malate dehydrogenase catalyzes an equilibrium reaction and that K_{eq} is 1.77×10^{-12} (Raval and Wolfe, 1962) and matrix pH $= 7.9$, as measured previously (Strzelecki *et al.*, 1984). The ratio of mitochondrial free NADH/NAD was calculated from the liver perfusate ratio of β-hydroxybutyrate to acetoacetate, again assuming that these metabolites are in equilibrium with NAD/NADH via β-hydroxybutyrate dehydrogenase and that K_{eq} is 2.81 × 10^{-9} (Krebs *et al.*, 1962).

In order to calculate the cytosolic concentration of glutamate and mitochondrial concentrations of malate and α-ketoglutarate, it was assumed that the tissue content of these compounds was evenly distributed in the 2 ml of intracellular H$_2$O/g dry wt (Tischler *et al.*, 1977). Moreover, the studies by other workers (Tischler *et al.*, 1977; Zurrendonk and Tager, 1974) using techniques of rapid cell fractionation show that gradients of malate and α-ketoglutarate across the mitochondrial membrane in intact cells are similar and vary from about 3 to 6 in different studies. It is reasonable that the gradient should be the same for the two metabolites since both metabolite gradients are subject to the same driving force (ΔpH) (LaNoue and Schoolwerth, 1979). Since oxalacetate, which is calculated from malate, and α-ketoglutarate compete for the catalytic site on aspartate aminotransferase, the absolute values of these two metabolites in the mitochondria matrix are not as important in determining flux as the ratio of free oxalacetate to α-ketoglutarate. Thus, varying the assumption about the metabolite gradient in the observed physiological range (3–6) has little influence ($\pm 5\%$) on the calculated fluxes. We assumed that the ratio was 4 in all flux determinations.

In prior studies using isolated mitochondria, we developed a kinetic expression for aspartate production by rat liver mitochondria (LaNoue *et al.*, 1987). This kinetic equation was empirically derived by studying the effects of varying the oxidation–reduction state of the mitochondria, as well as the concentration of added glutamate, malate, and α-ketoglutarate. The rate equation is:

$$ V = \frac{V_{max}}{\dfrac{K_m(A)}{A} \times \dfrac{K_m(B)}{B} \left(\dfrac{I}{K_i} + 1 \right) + \dfrac{K_m(A)}{A} + \dfrac{K_m(B)}{B} \left(\dfrac{1}{K_i} + 1 \right) + 1} $$

V_{max}, the maximal rate of aspartate formation by isolated rat liver mitochondria, is 100 nmol/min·mg mitochondrial protein. $K_m(A)$ is the K_m for glutamate (0.45 mM) and A is the concentration of cytosolic (media) glutamate. $K_m(B)$ is the K_m for oxalacetate (1.1 μM) and B is the concentration of matrix (free) oxalacetate. This was estimated from the measured matrix malate using the equilibrium constant of malate dehydrogenase. K_i is the inhibitor constant of α-ketoglutarate (0.55 mM), competitive with oxalacetate, and (I) is the concentration of matrix α-ketoglutarate.

In order to apply this equation to the intact perfused liver, some conversion factors must be applied and some assumptions made. We have previously determined that there is 423 mg mitochondrial protein per g dry weight of liver (LaNoue *et al.*, 1984). However, it is inconvenient to weigh livers directly either before or after the perfusion. Using 20 rats in the weight range 150–250 g, we found that there is 0.78 g dry weight liver/100 g body weight. One must convert μmol/hr·100 g body weight to nmol/min·mg mitochondrial protein by using the above factors.

To estimate the flux through pyruvate dehydrogenase, livers were perfused in the presence of [1-^{14}C]lactate (10 mM, 5 nCi/μmol) plus [1-^{14}C]pyruvate (1 mM, 5 nCi/μmol) and ^{14}CO$_2$ formation was measured. The amount of ^{14}CO$_2$ was assayed in perfusate samples before and after hormone addition by injecting 1-ml samples of perfusate into closed vials containing 1 ml of 2 N sodium acetate (pH 3), and a hyamine-hydroxide CO$_2$-trapping system. After injection of hyamine-hydroxide into the center well, samples were shaken for 1 hr at 37°C. A perfusate sample containing [^{14}C]lactate and [^{14}C]pyruvate was used as a blank to correct for nonenzymatic decarboxylation. This assay of pyruvate dehydrogenase activity is only a first approximation of pyruvate dehydrogenase flux since it also includes $^{14}CO_2$ production by the citric acid cycle following pyruvate carboxylation. Preliminary experiments suggest that this correction factor is small.

2. RESULTS AND DISCUSSION

The stoichiometric relationships between aspartate formation by liver mitochondria and rates of hepatic gluconeogenesis from lactate are shown in Fig. 1. The heavy arrows in the scheme show the three possible pathways that carbon may take after it leaves the mitochondria as aspartate. Cytosolic aspartate may be converted directly to glucose, or it may be converted to phosphoenolpyruvate and then back to pyruvate via pyruvate kinase. This recycling is controlled by the activity of pyruvate kinase. Pyruvate kinase activity is regulated both by calmodulin and cAMP-dependent phosphorylation (Schworer *et al.*, 1985; Pilkis *et al.*, 1986). Alternatively, cytosolic aspartate may be converted to malic acid and reenter the mitochondria in exchange for α-ketoglutarate. The latter results in utilization of cytosolic reducing equivalents generated by lactate dehydrogenase and should, therefore, be equivalent to the extra pyruvate formed from lactate which is not used for gluconeogenesis. This "extra" pyruvate is oxidized via pyruvate dehydrogenase and, therefore, the "malate recycling" can be approximated as the flux through pyruvate dehydrogenase. This flux was estimated as the rate of $^{14}CO_2$ generation from $[1\text{-}^{14}C]$lactate. Thus, according to the scheme shown in Fig. 1, aspartate efflux should be equal to the rate of gluconeogenesis times 2, plus the flux through pyruvate dehydrogenase, plus pyruvate kinase flux. With the exception of pyruvate kinase flux, these rates can be estimated directly. However, it is possible to estimate pyruvate

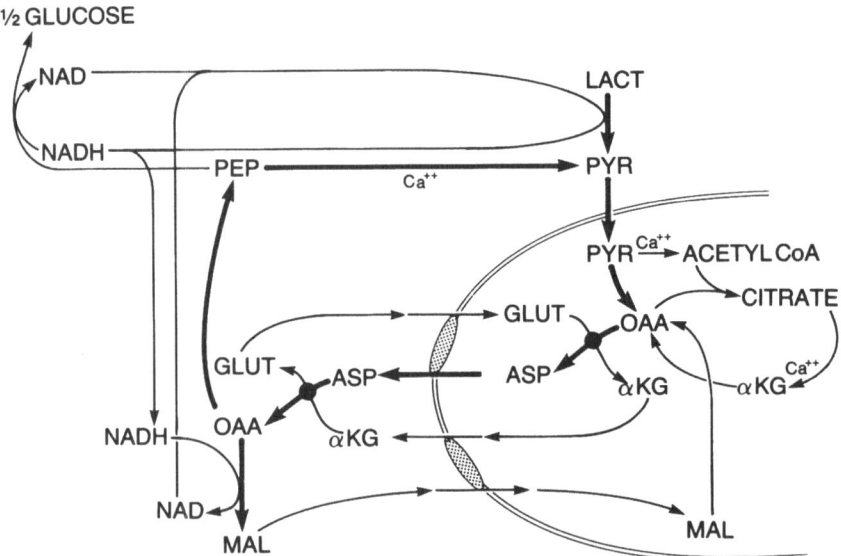

FIGURE 1. Relationships between mitochondrial aspartate formation and gluconeogenesis from lactate in rat liver.

kinase flux as the difference between aspartate efflux, as determined by the rate equation, and the sum of the rate of gluconeogenesis times 2 plus pyruvate dehydrogenase flux.

We have tested the effects of two hormones, glucagon and phenylephrine, on metabolite concentrations and calculated fluxes. Glucagon is thought to act mainly via a cAMP-dependent mechanism (Hers and Hue, 1983; Pilkis *et al.*, 1986), although it also induces a transient rise in cytosolic Ca^{2+} (Sistare *et al.*, 1985; Crompton and Golstone, 1986). Phenylephrine acts solely via Ca^{2+}, but it is not known which target(s) of Ca^{2+} action is physiologically important. Ca^{2+} stimulates calmodulin-dependent phosphorylation of pyruvate kinase (Pilkis *et al.*, 1986; Schworer *et al.*, 1985), thereby stimulating gluconeogenesis by an inhibition of pyruvate kinase. Ca^{2+} can also enter the mitochondria and stimulate α-ketoglutarate dehydrogenase, which could stimulate aspartate formation by lowering matrix levels of α-ketoglutarate. The relative importance of these possible targets of Ca^{2+} action were evaluated in isolated perfused rat livers.

Glucagon (1 μM) and phenylephrine (10 μM) were added at 30 min and the livers were freeze-clamped for metabolite assay at 60 min (Table I). To further modulate Ca^{2+} levels in the presence of hormones, the 12-myristate-13-acetate phorbol ester (1 μM) was added 15 min after the addition of hormones. Studies by other workers (Staddon and Hansford, 1986) showed that phorbol esters (presumably acting via protein kinase C) when added after Ca^{2+}-mobilizing hormones will diminish the free-Ca^{2+} levels in the liver toward control values. Selected metabolites were measured in perchloric acid extracts of the freeze-clamped livers. Recent studies (Staddon and Hansford, 1987) suggested a fairly direct relationship in hepatocytes between cytosolic free Ca^{2+} and mitochondrial redox state, probably due to the Ca^{2+} activation of pyruvate dehydrogenase and α-ketoglutarate dehydrogenase. Thus the data of Table I suggest that at the time the livers were clamped, cytosolic and mitochondrial Ca^{2+} were significantly higher in phenylephrine-treated livers compared to control livers. Phorbol ester diminished this hormonal effect but did not abolish it. The effect of glucagon on the redox state was minimal 30 min after hormone addition. Other studies have shown that there is a very significant effect of glucagon on cytosolic free Ca^{2+} and mitochondrial redox state at earlier times after addition of this hormone (Staddon and Hansford, 1986, 1987). When added after glucagon, phorbol ester had no effect on redox state. The assayed levels of α-ketoglutarate also reflect changes in Ca^{2+} levels in the mitochondria. However, these hormonal effects of Ca^{2+} on α-ketoglutarate metabolism appear to occur at much lower levels of Ca^{2+} than those which produce effects on mitochondrial redox state. There was a 59% decrease in α-ketoglutarate in the presence of glucagon and an 83% decrease in the presence of phenylephrine. Phorbol esters increased α-ketoglutarate only slightly when added after hormone. Aspartate levels were not significantly altered by hormones, while glutamate levels were decreased.

The data of Table I were used to determine aspartate flux in the perfused livers by means of the rate expression (see equation). The cytosolic concentration of glutamate and matrix concentrations of oxalacetate and α-ketoglutarate are expressed in Fig. 2 as a percentage of their respective K_m and K_i values. Since glutamate remains well above its K_m value, variations in glutamate appear unlikely to affect aspartate production. On the other hand, the levels of α-ketoglutarate and oxalacetate varied around the K_m and K_i values. Both gluconeogenic hormones caused a significant increase in the ratio of oxalacetate to α-ketoglutarate.

The concentrations of cytosolic glutamate, free mitochondrial oxalacetate, and α-ketoglutarate were inserted into the equation, and aspartate efflux was calculated and compared with rates of gluconeogenesis. [1-¹⁴C]lactate and [1-¹⁴C]pyruvate were added to the

TABLE I. Effect of Glucagon and Phenylephrine on Metabolite Concentrations (μmol/g dry wt) in Livers Perfused with Lactate and Pyruvate[a]

	Aspartate	Glutamate	Malate	α-Ketoglutarate	ATP	β-Hydroxybutyrate acetoacetate
Control	1.20 ± 0.32	5.67 ± 0.55	0.98 ± 0.09	1.41 ± 0.10	9.92 ± 0.42	0.26
Glucagon (1 μM)	1.48 ± 0.18	4.37 ± 0.41	1.19 ± 0.18	0.58 ± 0.07	8.98 ± 0.77	0.30
Glucagon + PMA(1 μM)	1.52 ± 0.15	4.84 ± 1.08	0.97 ± 0.05	0.77 ± 0.14	9.32 ± 0.51	0.33
Phenylephrine (10 μM)	1.74 ± 0.14	4.74 ± 0.46	1.42 ± 0.12	0.24 ± 0.02	9.79 ± 0.29	0.85
Phenylephrine + PMA	2.08 ± 0.04	2.78 ± 0.31	1.28 ± 0.07	0.33 ± 0.05	9.10 ± 0.68	0.69

[a] Fasted livers were perfused with 10 mM lactate and 1 mM pyruvate. Hormones were added to the perfusate 30 min after substrate addition. In some cases 12-myristate-13-acetate ether (PMA) was added 15 min after hormone addition. In all experimental conditions livers were freeze-clamped at 60 min after substrate addition. Metabolites were assayed in neutralized PCA extracts of tissues. β-Hydroxybutyrate/acetoacetate ratio was determined in the perfusate and represents the value for samples taken just prior to freeze-clamping.

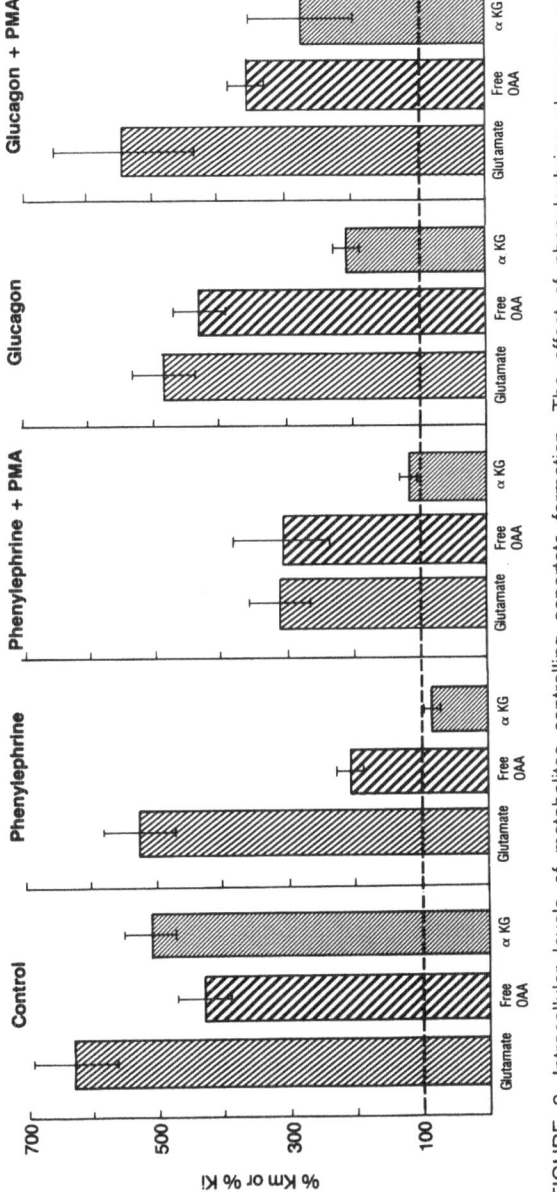

FIGURE 2. Intracellular levels of metabolites controlling aspartate formation. The effect of phenylephrine glucagon and 12-myristate-13-acetate phorbol ester (PMA). The cytosolic glutamate concentration and the mitochondrial concentrations of free oxaloacetate (OAA) and α-ketoglutarate (αKG) were calculated by using the factors described in Section 2. % of K_m or K_i was calculated on the basis of the following values: K_m for glutamate, 0.45 mM; K_m for oxaloacetate, 1.1 μM; K_i for α-ketoglutarate, 0.55 mM.

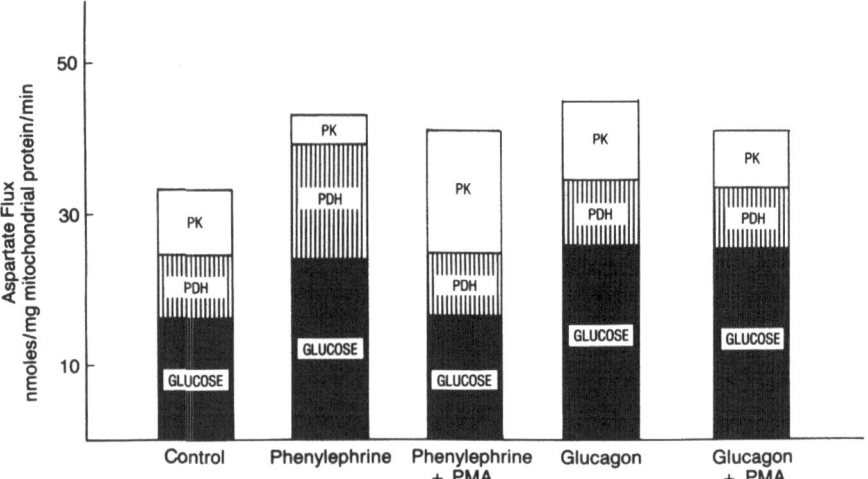

FIGURE 3. Effect of phenylephrine and glucagon on the contribution of each component influencing aspartate flux. Livers were perfused with 10 mM [1-^{14}C]lactate and 1 mM [1-^{14}C]pyruvate. Hormones and phorbol ester (PMA) were added according to the protocol described in the text. Aspartate flux (total height of the bars) was calculated from the flux equation (see Section 2). PDH, pyruvate dehydrogenase flux; PK, pyruvate kinase flux; glucose, two times the rate of measured glucose production.

perfusion in order to estimate flux through pyruvate dehydrogenase. In Fig. 3 the results are shown as bar graphs. The total height of the bar represents the calculated aspartate efflux under each experimental condition while gluconeogenic flux and the estimated values of pyruvate dehydrogenase flux are shown as fractions of the aspartate efflux. The remaining fraction of the bar is labeled as pyruvate kinase flux. This analysis allows for the evaluation of the relative importance of pyruvate kinase phosphorylation as well as the mitochondrial targets of Ca^{2+} action.

Flux through pyruvate kinase was relatively slow in the control state. Since aspartate flux was slow and the carbon drain created by excess production of NADH in the cytosol (pyruvate dehydrogenase flux) significant, pyruvate kinase is limited by substrate availability. This analysis suggests that inhibition of pyruvate kinase alone would be inadequate to increase the rate of gluconeogenesis in the absence of a factor to stimulate aspartate efflux.

In addition, this analysis suggests that glucagon stimulates gluconeogenesis primarily because of an increase in aspartate efflux. Phorbol esters when added after glucagon had no significant effect on gluconeogenesis or any of the other fluxes. Phenylephrine stimulated gluconeogenesis, pyruvate dehydrogenase, aspartate efflux, and inhibited pyruvate kinase. In opposition to glucagon, the phorbol ester added after phenylephrine inhibited gluconeogenesis to control values. Surprisingly, the phorbol ester had little effect on the oxalacetate/α-ketoglutarate ratio or on aspartate efflux, but it appeared to stimulate pyruvate kinase. These data imply that calmodulin-dependent phosphorylation of pyruvate kinase is an important aspect of the stimulation of gluconeogenesis by phenylephrine and that even quite small increases in cytosolic Ca^{2+} are sufficient to lower α-ketoglutarate enough to stimulate aspartate formation.

3. CONCLUSIONS

The data suggest that Ca^{2+}-mediated hormonal modulation of cell mitochondrial redox state and α-ketoglutarate levels may have profound metabolic effects. These effects include stimulation of aspartate efflux from mitochondria, which results in stimulation of gluconeogenesis. A kinetic expression for aspartate efflux from isolated rat liver mitochondria apparently provides reasonable estimates of aspartate efflux from mitochondria of perfused livers. The kinetic expression includes a term for the inhibitory influence of α-ketoglutarate. The levels of α-ketoglutarate in the tissue vary over a range in which they have an important influence on flux.

Calculation of mitochondrial aspartate formation from the kinetic expression using measured metabolite levels in perfused livers suggests that the gluconeogenic hormones would not stimulate this process were it not for the large decrease in α-ketoglutarate levels which occurs in the presence of both gluconeogenic hormones. Thus, Ca^{2+} stimulation of α-ketoglutarate dehydrogenase is a vital component of the mechanism of action of glucagon and phenylephrine.

Additional data, which include comparisons of the rate of glucose formation and estimates of pyruvate dehydrogenase flux under various hormonal conditions, suggest that stimulation of α-ketoglutarate dehydrogenase alone cannot stimulate gluconeogenesis, but that concomitant inhibition of pyruvate kinase is necessary for full expression of hormone action.

REFERENCES

Charest, R., Blackmore, P. F., Berthon, B., and Exton, J. H., 1983, Changes in free cytosolic Ca^{2+} in hepatocytes following α-adrenergic stimulation, *J. Biol. Chem.* **258**:8769–8773.

Crompton, M., and Goldston, T. P., 1986, The involvement of calcium in the stimulation of respiration in isolated rat hepatocytes by adrenergic agonists and glucagon, *FEBS Lett.* **204**:198–202.

Hers, H G., and Hue, L., 1983, Gluconeogenesis and related aspects of glycolysis, *Ann. Rev. Biochem.* **52**:617–653.

Hutson, N. J., Brumley, F. T., Assimocopoulos, F. D., Harper, S. C., and Exton, J. H., 1976, Studies on the α-adrenergic activation of hepatic glucose output, *J. Biol. Chem.* **251**:5200–5208.

Kneer, N. M., Wagner, M. J., and Lardy, H. A., 1979, Regulation by calcium of hormonal effects on gluconeogenesis, *J. Biol. Chem.*, **254**:12160–12168.

Krebs, H. A., Mellanby, J., and Williamson, D. H., 1962, The equilibrium constant of the β-hydroxybutyric-dehydrogenase system, *Biochem. J.* **82**:96–98.

LaNoue, K. F., and Schoolwerth, A. C., 1979, Metabolite transport in mitochondria, *Ann. Rev. Biochem.* **48**:871–922.

LaNoue, K. F., Schoolwerth, A. C., and Pease, A. J., 1983, Ammonia formation in isolated rat liver mitochondria, *J. Biol. Chem.* **258**:1726–1734.

LaNoue, K. F., Sterniczuk, A., Strzelecki, T., Hreniuk, S., Scaduto, R., Jr., 1987, The effect of mitochondrial Ca^{2+} on hepatic aspartate formation and gluconeogenic flux. In: *Integration of Mitochondrial Function* (J. J. Lemasters, C. R. Hackenbrock, R. G. Thurman, and H. V. Esterhoff, eds.), in press.

LaNoue, K. F., Strzelecki, T., and Finch, F., 1984, The effect of glucagon on hepatic respiratory capacity, *J. Biol. Chem.* **259**:4116–4121.

Leverve, X. M., Verhoeven, A. J., Groen, A. K., Meijer, A. J., and Tager, J. M., 1986, The malate/aspartate shuttle and pyruvate kinase as targets involved in the stimulation of gluconeogenesis by phenylephrine, *Eur. J. Biochem.* **155**:551–556.

McCormack, J. G., 1985, Studies on the activation of rat liver pyruvate dehydrogenase and 2-oxoglutarate dehydrogenase by adrenaline and glucagon, *Biochem. J.* **231:**597–608.

Ochs, R. S., and Lardy, H. A., 1981, Catecholamine stimulation of ethanol oxidation by isolated rat hepatocytes, *FEBS Lett.* **131:**119–121.

Oviasu, O. A., and Whitton, P. D., 1984, Hormonal control of pyruvate dehydrogenase activity in rat liver. *Biochem. J.* **224:**181–186.

Parrilla, R., Jimenez, I., Ayuso-Parrilla, M. S., 1975, Glucagon + insulin control of gluconeogenesis in the perfused isolated rat liver. Effects on cellular metabolite distribution, *Eur. J. Biochem.* **56:** 375–383.

Pilkis, S. J., Fox, E., Wolfe, L., Rothbarth, L., Colosia, A., Stewart, H. B., and El-Maghrabi, M. R., 1986, Hormonal modulation of key hepatic regulatory enzymes in the gluconeogenic/glycolytic pathway, *Ann. N.Y. Acad. Sci.* **478:**1–19.

Raval, D. N., and Wolfe, R. G., 1962, Malic dehydrogenase. II. Kinetic studies of the reaction mechanism, *Biochemistry* **1:**263–269.

Rognstad, R., and Katz, J., 1970, Gluconeogenesis in the kidney cortex, *Biochem. J.* **116:**483–491.

Schoolwerth, A. C., and LaNoue, K. F., 1983, Control of ammoniogenesis by α-ketoglutarate in rat kidney mitochondria, *Am. J. Physiol.* **244:**F399–408.

Schworer, C. M., E1-Maghrabi, M. R., Pilkis, S. J., and Soderling, T. R., 1985, Phosphorylation of L-type pyruvate kinase by a Ca^{2+}/calmodulin-dependent protein kinase, *J. Biol. Chem.* **260:** 13018–13022.

Siess, E. A., Brocks, D. G., Lattke, H. K., and Wieland, O. H., 1977, Effect of glucagon on metabolite compartmentation in isolated rat liver cells during gluconeogenesis from lactate, *Biochem. J.* **166:** 225–235.

Sistare, F. D., Picking, R. A., and Haynes, R. C. Jr., 1985, Sensitivity of the response of cytosolic calcium in Quin-2-loaded rat hepatocytes to glucagon, adenine nucleosides, and adenine nucleotides, *J. Biol. Chem.* **260:**12744–12747.

Staddon, J. M., and Hansford, R. G., 1986, 4β-phorbol 12-myristate 13-acetate attenuates the glucagon-induced increase in cytoplasmic free Ca^{2+} concentration in isolated rate hepatocytes, *Biochem. J.* **238:**737–743.

Staddon, J. M., and Hansford, R. G., 1987, The glucagon-induced activation of pyruvate dehydrogenase in hepatocytes is diminished by 4β-phorbol 12-myrisate 13-acetate, *Biochem. J.* **241:**729–735.

Strzelecki, T., Thomas, J. A., Koch, C. D., and Lanoue, K. F., 1984, The effect of hormones on proton compartmentation in hepatocytes, *J. Biol. Chem.* **259:**4122–4129.

Tischler, M. E., Hecht, P., and Williamson, J. R., 1977, Determination of mitochondrial/cytosolic metabolite gradients in isolated rat liver cells by cell disruption, *Arch. of Biochem. and Biophys.* **181:**278–292.

Trinder, P., 1969, *Ann. Clin. Biochem.* **6:**24.

Williamson, J. R., Browning, E. T., Thurman, R. G., and Scholz, R., 1969, Inhibition of glucagon effects in perfused liver by (+) decanoylcarnithine, *J. Biol. Chem.* **244:**5055–5064.

Williamson, J. R., and Corkey, B. E., 1969, Assays of intermediates of the citric acid cycle and related compounds by fluorometric enzyme methods, *Method Enzymol.* **13:**434–513.

Williamson, J. R., Jakob, A., and Refino, C., 1971, Control of the removal of reducing equivalents from the cytosol in perfused liver, *J. Biol. Chem.* **246:**7632–7641.

Zuurendonk, P. F., and Tager, J. M., 1974, Rapid separation of particular components and soluble cytoplasm of isolated rat liver cells, *Biochim. et Biophys. Acta* **333:**393–399.

39

Regulatory Effects of a Thromboxane A$_2$ Analogue on Hepatic Glycogenolysis and Vasoconstriction

RORY A. FISHER, MARK E. STEINHELPER,
and MERLE S. OLSON

INTRODUCTION

Many studies have indicated that the regulatory effects of hormones on glycogenolysis in the rat liver occur by at least two distinct mechanisms. While glucagon and β-adrenergic agonists stimulate hepatic glycogenolysis via cAMP-dependent protein kinase activation, α-adrenergic hormones as well as vasopressin and angiotension II mediate their effects by mechanisms independent of cAMP (Hems and Whitton, 1980; Exton et al., 1981). Convincing experimental evidence suggests that alterations in cellular calcium fluxes leading to increases in the cytosolic calcium concentration and subsequent activation of phosphorylase kinase are involved in the glycogenolytic actions of the cAMP-independent hormones (Blackmore et al., 1978; Murphy et al., 1980). Studies in several laboratories have indicated a possible second-messenger role of inositol-1,4,5-trisphosphate, a breakdown product of phosphatidylinositol-4,5-bisophosphate, in the calcium-mobilizing properties of these hormones (Berridge, 1984; Williamson et al., 1985). The glycogenolytic actions of both glucagon and calcium-mobilizing hormones are observed in isolated hepatocytes, which constitute approximately 90% of the liver volume and are the primary site of glycogen storage in the liver.

Several recent studies from our laboratory suggest that glycogenolysis in the perfused liver may be stimulated by mechanisms differing from those of cAMP-elevating or calcium-mobilizing hormones. Infusion of platelet-activating factor or IgG aggregates into perfused livers stimulates glucose release without increasing hepatic cAMP levels or causing the type of hepatic calcium efflux observed during infusion of calcium-mobilizing agonists (Buxton et al., 1986; Buxton et al., 1987). In isolated hepatocytes, platelet-activating factor and IgG aggregates do not stimulate either glucose release or glycogen phosphorylase activity (Fisher

RORY A. FISHER, MARK E. STEINHELPER, and MERLE S. OLSON • Department of Biochemistry, University of Texas Health Science Center, San Antonio, Texas 78284.

et al., 1984; Buxton *et al.*, 1987). However, coupled with the glycogenolytic action of these substances in the perfused liver, a pronounced constriction of the hepatic vasculature occurs which we have reasoned may stimulate hepatic glycogenolysis indirectly via redistribution of perfusate flow or by inducing hypoxia within the liver (Fisher *et al.*, 1986a; Buxton *et al.*, 1986, 1987). Both the glycogenolytic and vasoconstrictive actions of IgG aggregates in the perfused liver are substantially inhibited by coinfusion of indomethacin, suggesting that an arachidonic acid metabolite mediator also might be involved in the hepatic actions of substances like IgG aggregates.

Thromboxane A_2 is a potent vasoactive arachidonic acid metabolite which is produced in response to various stimuli in a number of cell types. Experimental evidence in a variety of systems has demonstrated the existence of a thromboxane-synthesizing system in the liver (Mahmud and Miura, 1981; Spolarics *et al.*, 1984; Bowers *et al.*, 1985). It was of great interest to examine whether thromboxanes might mediate the actions of substances such as IgG aggregates in the isolated perfused liver. In the present report, the effects of the thromboxane A_2 mimetic U-46619 on hepatic metabolism and vasoconstriction are described along with preliminary evidence for the production of thromboxanes during stimulation of the perfused rat liver with IgG aggregates.

2. HEPATIC RESPONSES TO U-46619

To examine potential metabolic and hemodynamic effects of thromboxanes in the liver, *in situ* perfused rat livers were stimulated with the thromboxane A_2 mimetic U-46619. Figure 1 illustrates various responses of the perfused liver during infusion of U-46619 (42 ng/ml). Upon infusion, U-46619 stimulated a prompt increase in glucose and lactate production, oxygen consumption, and perusate lactate/pyruvate and β-hydroxybutyrate/acetoacetate ratios, the latter indicative of reductions in the cytosolic and mitochondrial NADH/NAD$^+$ oxidation/reduction couples, respectively. During this same period, significant increases in hepatic portal pressure were observed, indicating that constriction of the hepatic vasculature occurred during stimulation with U-46619. The biphasic oxygen response at this dose of U-46619 likely represents a balance between stimulated oxygen consumption as a result of increased flux through the tricarboxylic acid cycle (i.e., glycogenolytic provision of substrate) and decreased consumption due to vasoconstriction-induced shutting down of regions of the liver. At concentrations of U-46619 which induced more moderate vasoconstriction (e.g., <20 ng/ml), hepatic oxygen consumption increased throughout agonist infusion whereas at higher doses (e.g., >63 ng/ml) a more pronounced vasoconstriction occurred and hepatic oxygen consumption decreased dramatically. The glycogenolytic and vasoconstrictive responses to U-46619 exhibited a similar dose dependence although portal pressure responses were shifted somewhat to the left of glycogenolytic responses (Fisher *et al.*, 1987).

Several experiments were performed to examine the possible involvement of cAMP and calcium in the mechanism of stimulation of hepatic glycogenolysis by U-46619 in the perfused liver. Time course experiments demonstrated that glycogen phosphorylase was activated without changes in hepatic cAMP levels during stimulation of perfused livers with 42 ng/ml U-46619, a concentration which stimulates glucose release from the perfused liver by more than 100% (Fisher *et al.*, 1987). Infusion of maximal concentrations of U-46619 (e.g., 1000 ng/ml) into perfused livers similarly activated glycogen phosphorylase without changes in cAMP at 30 sec of stimulation; however, modest increases in hepatic cAMP (i.e., <0.2 nmol/g wet wt) occurred at 1 and 2 min of stimulation. Glucagon (2.3 nM), in contrast, increased

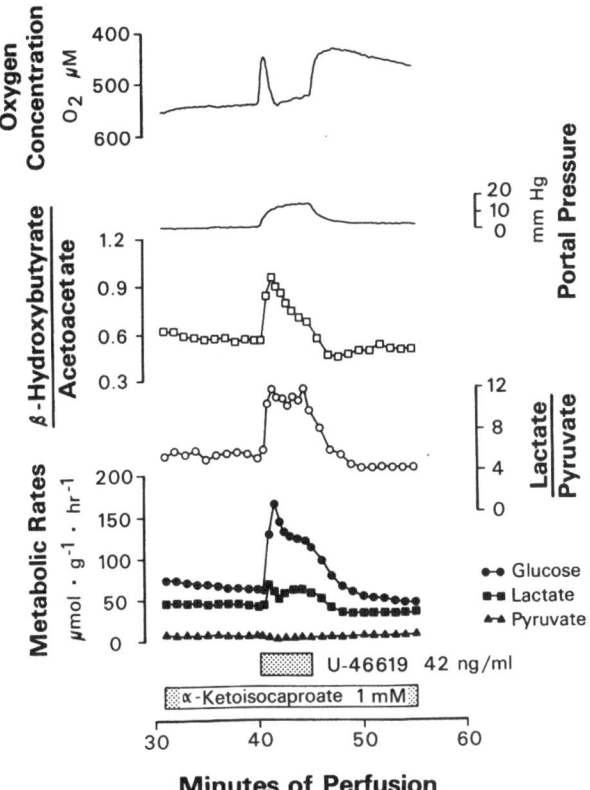

FIGURE 1. Metabolic responses of the perfused rat liver to 42 ng/ml U-46619 (for details, see reference below). [Reprinted by permission from: Fisher, R. A., Robertson, S. M., and Olson, M. S. *J. Biol. Chem.*, **262**:4631–4638 (1987).]

hepatic cAMP levels by more than 25 nmol/g wet wt. These results suggested that U-46619 stimulates hepatic glycogenolysis by a cAMP-independent mechanism in the perfused rat liver.

Calcium-mobilizing hormones cause a transient efflux of calcium from the perfused liver upon infusion which is thought to reflect mobilization of hepatocyte intracellular calcium stores (Blackmore *et al.*, 1978; Blackmore *et al.*, 1979). In Fig. 2, the effects of U-46619 and phenylephrine on perfusate calcium efflux and glycogenolysis is compared in experiments in which the perfusate calcium concentration was reduced to 50 μM to facilitate measurement of changes in effluent perfusate calcium. The glycogenolytic response to U-46619 was reduced significantly by reducing the perfusate calcium level to 50 μM (i.e., compare to Fig. 1) and was not associated with calcium efflux from the liver as seen with the α-adrenergic agonist phenylephrine. The importance of the perfusate calcium concentration in hepatic responses to U-46619 is further illustrated in Fig 3. Decreasing the perfusate calcium concentration reduced similarly the portal pressure and glucose output responses of the liver to U-46619. Approximately 25% of these hepatic responses to U-46619 appeared to be independent of

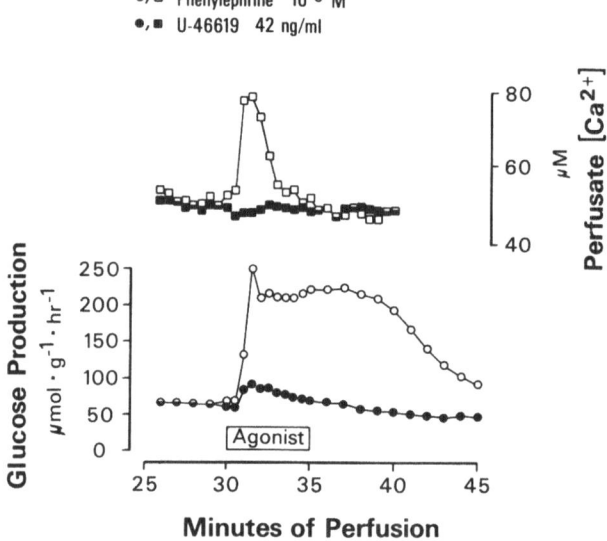

FIGURE 2. Comparison of effects of phenylephrine and U-46619 on stimulation of hepatic glucose output and calcium efflux at reduced perfusate calcium concentration (for details, see reference in Fig. 1). The perfusate calcium concentration was reduced to 50 μM in these liver perfusions. (Reproduced as described in Fig. 1.)

significant calcium concentrations in the perfusate since they occurred in the presence of EGTA which reduced perfusate calcium levels below 1 nM. The calcium requirement for these responses is likely extracellular in nature, since removal of perfusate calcium only 1 min prior to stimulation with U-46619 substantially reduced U-46619-stimulated glucose release and vasoconstriction (Fisher *et al.*, 1987). Thus, unlike calcium-mobilizing hormones, U-46619-induced glycogenolysis is largely dependent on extracellular calcium, is not as-

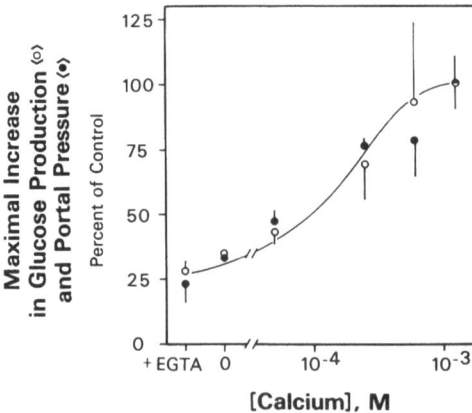

FIGURE 3. Effect of perfusate calcium concentration on hepatic responses to 42 ng/ml U-46619 (for details, see reference in Fig. 1). The maximal increase in glucose output and portal pressure in response to 42 ng/ml U-46619 at 1.25 mM perfusate calcium (i.e., control conditions) was 72 $\pm$ 7 μmol/g/hr and 13.9 mm Hg, respectively. (Reproduced as described in Fig. 1.)

sociated with net efflux of calcium from the liver, and is closely coupled to hepatic vaso-constriction.

Contrary to its stimulatory effects on glycogenolysis in perfused livers, U-46619 had no effect on glucose release or glycogen phosphorylase levels in isolated hepatocytes (Fisher *et al.*, 1987). Although it is possible that such a result was due to loss of receptors during hepatocyte preparation, it also provided evidence for an indirect mode of action of U-46619 on hepatocyte glycogenolysis in the perfused liver. Ischemia within the liver leads to rapid alterations in adenine nucleotides, conversion of glycogen phosphorylase b to glycogen phosphorylase a, and stimulation of glucose output (Theen *et al.*, 1970; Hems and Brosnan, 1970; Hems and Whitton, 1980). To examine whether U-46619-induced vasoconstriction might lead to ischemia within the liver, perfused livers were freeze-clamped and analyzed for adenine nucleotides at various times following infusion of U-46619. U-46619 stimulation of perfused livers resulted in rapid decreases in ATP and increases in ADP and AMP at maximal concentrations (i.e., 1000 ng/ml) and increases in ADP at lower concentrations (i.e., 42 ng/ml), suggesting that ischemia occurred within the liver under these conditions (Fisher *et al.*, 1987). The observed elevation in effluent perfusate lactate/pyruvate and β-hydroxybutyrate/acetoacetate ratios during U-46619-induced vasoconstriction (Fig. 1) as well as the decrease in hepatic oxygen consumption which occurs at higher concentrations are also in keeping with this idea. Thus, it is our contention that U-46619 exerts its gly-cogenolytic effect in the perfused liver by an indirect mechanism involving vasoconstriction-induced alterations in perfusate flow resulting in ischemia within the liver.

3. THROMBOXANE PRODUCTION IN PERFUSED LIVERS

In view of the observed stimulation of glycogenolysis in perfused livers by the throm-boxane A_2 mimetic U-46619, it was of interest to examine whether thromboxanes might serve mediator roles in the actions of substances such as IgG aggregates in the perfused liver. In previous studies we found that infusion of IgG aggregates into perfused livers resulted in a stimulation of glycogenolysis and vasoconstriction, both of which could be minimized by coinfusion of the cyclooxygenase inhibitor indomethacin (2 μM) (Buxton *et al.*, 1987). To examine the potential role of thromboxanes in IgG aggregate action in the liver, livers were stimulated with IgG aggregates (2 μg/ml) and effluent perfusate was collected for thromboxane extraction and measurement. In other experiments, IgG aggregate-stimulated livers were freeze-clamped for subsequent analysis of thromboxanes in the liver tissue. Arachidonic acid metabolites were extracted and isolated from perfusate or tissue samples essentially as described by Powell (1982) and thromboxane B_2 measurements per-formed on the resulting samples by radioimmunoassay. Recovery of thromboxane B_2 utilizing these extraction and separation methods was greater than 90%. Thromboxane B_2 levels in the effluent perfusate increased from 20 $\pm$ 4 to 159 $\pm$ 16 pg/min/g liver following infusion of IgG aggregates into the perfused liver for 2 min. In the freeze-clamped liver samples, thromboxane B_2 levels increased from 9 $\pm$ 1 to 100 $\pm$ 36 ng/g liver during this same period of stimulation with IgG aggregates. Zymosan, a potent stimulator of phagocytosis in various cells, also induces thromboxane B_2 production upon its infusion into the perfused liver (Fisher, Steinhelper, and Olson, unpublished observations). It is worth noting that the concentrations of thromboxane B_2 formed in response to IgG aggregate infusion into perfused livers are equivalent to the concentrations of U-46619 which exert pronounced glycogenolytic and vasoconstrictive actions in the perfused liver.

4. SUMMARY

Stimulation of hepatic glycogenolysis by the thromboxane mimetic U-46619 occurs by mechanisms clearly distinct from those of cAMP-elevating or calcium-mobilizing hormones. It seems likely that this agonist's effect on glucose release in the perfused liver is a consequence of its vasoactive properties which result in localized ischemia within the liver. The finding that thromboxanes are produced during stimulation of perfused livers with IgG aggregates suggests an autocoid role for thromboxanes in the liver. A possible scheme for regulatory influences of thromboxanes on hepatic glucose release is illustrated in Fig. 4. In this model, thromboxane A_2 interacts with cells in the hepatic sinusoids (e.g., endothelial cells) to produce vasoconstriction, which influences glycogenolysis in the hepatocyte via ischemia-induced alterations in AMP levels. It is possible also that the primary site of interaction is at the level of the Kupffer cell, which has been suggested to exert regulatory effects on flow through the hepatic sinusoid (McCuskey, 1966). Substances such as IgG aggregates or zymosan likely stimulate thromboxane and other mediator synthesis by interacting with reticuloendothelial cells, such as Kupffer cells, within the liver. Other mediators produced within the liver during such stimulation include platelet-activating factor and prostaglandin E_2 (Fisher, Steinhelper, and Olson, unpublished observations), both of which possess vasoactive and glycogenolytic properties in perfused livers (Buxton *et al.*, 1984, 1986; Fisher *et al.*, 1986b). The production of glycogenolytically active autocoid mediators within the liver during acute pathophysiological conditions such as systemic anaphylaxis, endotoxic shock, and serum sickness may be important in supplying extrahepatic tissues with glucose during such situations. The molecular mechanisms involved in thromboxane action at its primary site of interaction within the liver are not known; however, it seems likely that

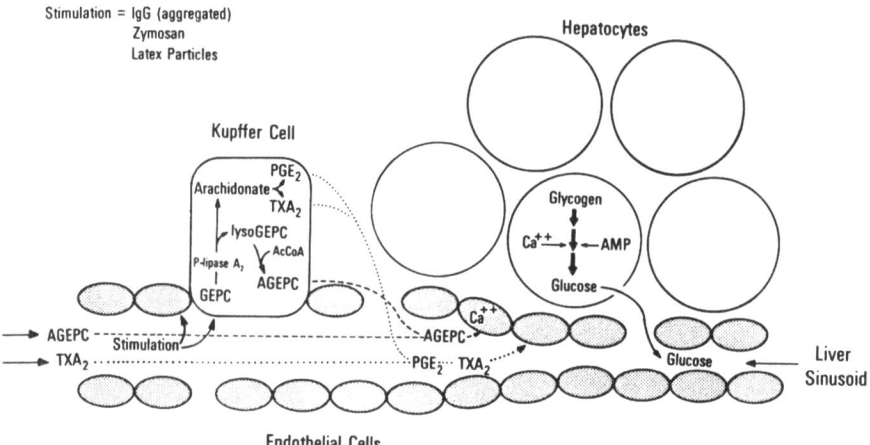

FIGURE 4. Autocoid production and regulation of glucose release in the perfused liver. Thromboxane A_2 (TXA₂), platelet-activating factor (AGEPC), prostaglandin E_2 (PGE₂).

calcium plays an important role in view of the calcium dependence of responses to U-46619 and IgG aggregates in the perfused liver.

REFERENCES

Berridge, M. J., 1984, Inositol trisphosphate and diacylglycerol as second messengers, *Biochem. J.* **220**:345–360.

Blackmore, P. F., Brumley, F. T., Marks, J. L., and Exton, J. H., 1978, Studies on α-adrenergic activation of hepatic glucose output: Relationship between α-adrenergic stimulation of calcium efflux and activation of phosphorylase in isolated rat liver parenchymal cells, *J. Biol. Chem.* **253**:4851–4858.

Blackmore, P. F., Dehaye, J.-P., and Exton, J. H., 1979, Studies on α-adrenergic activation of hepatic glucose output: The role of mitochondrial calcium release in α-adrenergic activation of phosphorylase in perfused rat liver, *J. Biol. Chem.* **254**:6945–6950.

Bowers, G. J., MacVittie, T. J., Hirsch, E. F., Conklin, J. C., Nelson, R. D., Roethel, R. J., and Fink, M. P., 1985, Prostanoid production by lipopolysaccharide-stimulated Kupffer cells, *J. Surg. Res.* **38**:501–508.

Buxton, D. B., Fisher, R. A., Hanahan, D. J., and Olson, M. S., 1986, Platelet-activating factor-mediated vasoconstriction and glycogenolysis in the perfused rat liver, *J. Biol. Chem.* **261**:644–649.

Buxton, D. B., Fisher, R. A., Briseno, D. L., Hanahan, D. J., and Olson, M. S., 1987, Glycogenolytic and haemodynamic responses to heat-aggregated immunoglobulin G and prostaglandin E₂ in the perfused rat liver, *Biochem. J.* **243**:493–498.

Buxton, D. B., Hanahan, D. J., and Olson, M. S., 1984, Stimulation of glycogenolysis and platelet-activating factor production by heat-aggregated immunoglobulin G in the perfused rat liver, *J. Biol. Chem.* **259**:13758–13761.

Exton, J. H., Blackmore, P. F., El-Refai, M. F., Dehaye, J.-P. Strickland, W. G., Cherrington, A. D., Chan, T. M., Assimacopoulos-Jeannet, F. D., and Chrisman, T.D., 1981, Mechanisms of hormonal regulation of liver metabolism, *Adv. Cyclic Nucleotide Res.* **14**:491–505.

Fisher, R. A., Kumar, R., Hanahan, D. J., and Olson, M. S., 1986a, Effects of β-adrenergic stimulation on 1-0-hexadecyl-2-acetyl-sn-glycero-3-phosphocholine-mediated vasoconstriction and glycogenolysis in the perfused rat liver, *J. Biol. Chem.* **261**:8817–8823.

Fisher, R. A., Kumar, R., Hanahan, D. J., and Olson, M. S., 1986b, Stimulation of glycogenolysis and platelet-activating factor synthesis by zymosan in the perfused rat liver, *Fed. Proc.* **45**:1838.

Fisher, R. A., Robertson, S. M., and Olson, M. S., 1987, Stimulation of glycogenolysis and vasoconstriction in the perfused rat liver by the thromboxane A₂ analogue U-46619, *J. Biol. Chem.* **262**:4631–4638.

Fisher, R. A., Shukla, S. D., Debuysere, M. S., Hanahan, D. J., and Olson, M. S., 1984, The effect of acetylglyceryl ether phosphorylcholine on glycogenolysis and phosphatidylinositol 4,5-bisphosphate metabolism in rat hepatocytes, *J. Biol. Chem.* **261**:8685–8688.

Hems, D. A., and Brosnan, J. T., 1970, Effects of ischemia on content of metabolites in rat liver and kidney in vivo, *Biochem. J.* **120**:105–111.

Hems, D. A., and Whitton, P. D., 1980, Control of hepatic glycogenolysis, *Physiol. Rev.* **60**:1–50.

Mahmud, I., and Miura, Y., 1981, Effects of stimulators and inhibitors on arachidonic acid metabolism in hepatoma, *Cell. Molec. Biol.* **27**:197–202.

McCuskey, R. S., 1966, A dynamic and static study of hepatic arterioles and sphincters, *Am. J. Anat.* **119**:455–477.

Murphy, E., Coll, K., Rich, T. L., and Williamson, J. R., 1980, Hormonal effects on calcium homeostasis in isolated hepatocytes, *J. Biol. Chem.* **255**:6600–6608.

Powell, W. J., 1982, Rapid extraction of arachidonic acid metabolites from biological samples using octadecylsilyl silica, in: *Methods in Enzymology* Vol. 86 (W.E.M. Lands and W. L. Smith, eds.), Academic Press, New York, pp. 467–477.

Spolarics, Z., Tanacs, B., Garzo, T., Mandl, J., Mucha, I., Antoni, F., Machovich, R., and Horvath, I., 1984, Prostaglandin and thromboxane synthesizing activity in isolated murine hepatocytes and non-parenchymal liver cells, *Prostaglandins Leukotrienes Med.* **16:**379–388.

Theen, J., Gilboe, D. P., and Nutall, F. Q., 1982, Liver glycogen synthase and phosphorylase changes in vivo with hypoxia and anaesthetics, *Am. J. Physiol.* **243:**E182–E187.

Williamson, J. R., Cooper, R. H., Joseph, S. K., and Thomas, A. P., 1985, Inositol trisphosphate and diacylglycerol as intracellular second messengers in liver, *Am. J. Physiol.* **248:**C203–C216.

40

Metabolic Regulation of Ca^{2+} Handling in Permeabilized Insulinoma Cells

BARBARA E. CORKEY, KEITH TORNHEIM,
JUDE T. DEENEY, M. CLAY GLENNON,
JANICE C. PARKER, FRANZ M. MATSCHINSKY,
NEIL B. RUDERMAN, and MARC PRENTKI

Signal generation in the pancreatic β cell requires metabolism of the stimulatory fuel and is accompanied by increases in oxygen consumption and intracellular free Ca^{2+} (Hedeskov, 1980; Matschinsky *et al.*, 1983; Meglasson and Matschinsky, 1986; Prentki and Matschinsky, 1987). We hypothesized that fuel phosphorylation decreased the cytosolic MgATP/MgADP ratio sufficiently to stimulate O$_2$ consumption, and simultaneously reduced the activity of Ca^{2+}-ATPase with resulting increases in free-Ca^{2+} levels. To explore this hypothesis, we varied the MgATP/MgADP ratio in permeabilized RINm5F insulinoma cells that maintain a low-medium Ca^{2+} concentration in the presence of MgATP. Either of the following was added: (1) creatine phosphokinase plus various fixed ratios of creatine/creatine phosphate, or (2) a cell-free extract of rat skeletal muscle that exhibits spontaneous oscillatory behavior of glycolysis, and linked oscillations in the MgATP/MgADP ratio when provided with glucose and a hexokinase (Tornheim, 1979). We found that the free-Ca^{2+} level maintained by the permeabilized cells varied inversely with the MgATP/MgADP ratio, regardless of the mechanism used to vary MgATP/MgADP. In addition, free Ca^{2+} was decreased by increasing levels of orthophosphate (Pi). Ca^{2+} levels oscillated in phase with glycolytic oscillations and correlated closely with the MgATP/MgADP ratio. Ca^{2+} oscillations were evoked by increasing glucose levels from 2.5 to 10 or 20 mM in the presence of glucokinase, whereas oscillations occurred at 2.5 mM glucose in the presence of hexokinase and were unaffected by increasing the glucose concentration. These results provide the first demon-

---

BARBARA E. CORKEY and KEITH TORNHEIM ● Department of Biochemistry and Division of Diabetes and Metabolism, Evans Department of Medicine, Boston University School of Medicine, Boston, Massachusetts 02118.   JUDE T. DEENEY and NEIL B. RUDERMAN ● Division of Diabetes and Metabolism, Evans Department of Medicine, Boston University School of Medicine, Boston, Massachusetts 02118.   M. CLAY GLENNON, JANICE C. PARKER, FRANZ M. MAT-SCHINSKY, and MARC PRENTKI ● Department of Biochemistry and Biophysics, University of Pennsylvania School of Medicine, Philadelphia, Pennsylvania 19014.

stration of a link between metabolite changes and free-Ca^{2+} levels, and suggest a mechanism by which fuel metabolism might be coupled to activation of the Ca^{2+} messenger system in pancreatic β cells.

1. EXPERIMENTAL PROCEDURES

1.1. Measurement of Free Ca^{2+}

Free-Ca^{2+} values were calculated from the fluorescence of Fura 2 signals at excitation wavelengths of 340 and 380 nm and emission at 510 nm (Grynkiewicz et al., 1985) using a time-sharing fluorometer designed and built by the Bio-Instrumentation Group of the University of Pennsylvania (Chance et al., 1975).

1.2. Fixed MgATP/MgADP Ratio and the Ca^{2+} Set Point

RINmF5 cells were cultured and prepared as described previously (Prentki et al., 1985). Different relatively stable values of the MgATP/MgADP ratio were achieved by adding to the media MgATP (4 mM) and creatine phosphate and creatine at different ratios with sufficient creatine phosphokinase to allow rapid equilibration. The total concentration of creatine plus creatine phosphate was kept constant at 10–12 mM. The values of metabolites were measured enzymatically under the actual experimental conditions. ADP levels were calculated using an equilibrium constant for creatine phosphokinase of 167 which was determined separately in the absence of cells under the specific incubation conditions used in these studies. Cells were permeabilized by addition of saponin. The Ca^{2+} set point is defined as the steady-state free-Ca^{2+} value maintained by permeabilized cells for periods of time in excess of 20–30 min in the presence of a continuous supply of ATP. Mild perturbations of the system by small additions of Ca^{2+} or EGTA did not alter the set point. Increasing the number of cells decreased the length of time required to achieve a set point but did not alter the value obtained. Set points were generally measured about 15 min after addition of saponin to permeabilize the cells.

1.3. Oscillating Glycolytic Extract

The gel-filtered, high-speed supernatant fraction of rat hindleg muscle was prepared as described previously (Tornheim and Lowenstein, 1973), except that tissue was pulverized using a Polytron homogenizer and EDTA was omitted from the gel filtration buffer. The extract was stored frozen prior to use; this largely inactivates adenylosuccinase, but adenylate deaminase, adenylosuccinate synthetase, and the glycolytic enzymes remain active. The occurrence of glycolytic oscillations under the conditions of the experiments was initially monitored spectrophotometrically by following changes in $A_{261}-A_{282}$ with an Aminco DW2A double-beam spectrophotometer (data not shown); such spectral changes are due to corresponding oscillations in the operation of the purine nucleotide cycle in response to changes in the MgATP/MgADP ratio associated with glycolytic oscillations (Tornheim and Lowenstein, 1974, 1975).

1.4. Metabolite Analysis

Samples were taken for metabolite analyses and deproteinized with perchloric acid. Nucleotides were measured by high-performance liquid chromatography using a reversed phase column and paired ion chromatography. The mobile phases were 100 mM KH_2PO_4, 2 mM PIC A (Waters Associates, Milford, MA) (pH 3.2), and 100 mM KH_2PO_4, 2 mM PIC A, 30% acetonitrile (pH 3.2). Nucleotides were eluted at a flow rate of 0.4 ml/min using a linear gradient from the first to the second buffer over 40 min. Elution times in minutes were as follows: AMP, 6; GMP, 8; IMP, 10; ADP, 14; NAD, 15; GDP, 17; ATP, 30; GTP, 32. Other metabolites were measured by enzymatic techniques described previously (Williamson and Corkey, 1969, 1979). Total nucleotide was conserved in these experiments and was equal to approximately 2 mM adenine (plus hypoxanthine) and 300 μM guanine nucleotide. The disappearance of phosphate from P_i and ATP was largely matched by the appearance of phosphorylated glycolytic intermediates and α-glycerophosphate (data not shown).

1.5. Calculation of Metal–Ligand Complexes

An iterative computer program was used to calculate free Mg^{2+} and the concentrations of free and bound ligands. Dissociation constants for magnesium, calcium, and potassium complexes of ATP, ADP, phosphate, and creatine phosphate were taken from O'Sullivan and Smithers (1979) and corrected for pH. Estimates of total Ca^{2+} and Mg^{2+} binding sites and their dissociation constants were taken from values obtained in rat liver (Corkey et al., 1986). Total values for Ca^{2+}, Mg^{2+}, and K^+ were measured by atomic absorption spectroscopy.

1.6. Materials

Glucokinase was prepared from rat liver as described by Meglasson et al (1983). RPMI 1640 tissue culture media were obtained from Flow Laboratories. Nucleotides were from PL Biochemicals and other biochemicals and enzymes were obtained from Boehringer Mannheim and Sigma. Fura 2 was purchased from Molecular Probes. Male Sprague–Dawley rats were obtained from Charles River Breeding Labs.

2. RESULTS AND DISCUSSION

2.1. Effect of MgATP/MgADP Ratio on the Ca^{2+} Set Point

As shown in Fig. 1A the Ca^{2+} set point varied inversely with the MgATP/MgADP ratio. The inverse relationship between the Ca^{2+} set point and the MgATP/MgADP ratio persisted with similar sensitivity when mitochondria were inhibited with antimycin A and oligomycin (Corkey et al., 1989). These data suggest that alterations in the MgATP/MgADP ratio alter free Ca^{2+}, primarily through effects on a nonmitochondrial, presumably endoplasmic reticulum (ER), Ca^{2+} pool.

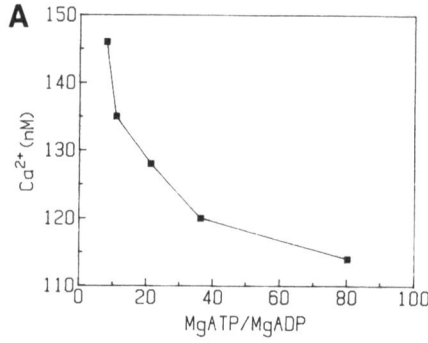

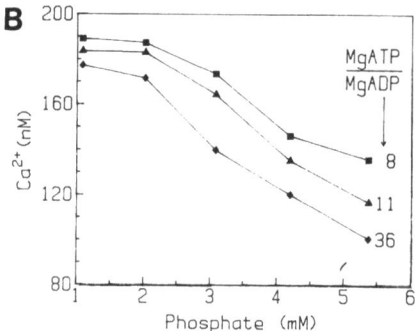

FIGURE 1. The effect of (A) the MgATP/MgADP ratio and (B) inorganic phosphate on the Ca^{2+} set point in permeabilized RINm5F insulinoma cells. Cells (0.75–0.85 mg/ml) were incubated at 30°C in buffer, pH 7.3, containing the following components: 100 mM KCl, 22 mM NaCl, 5 mM $KHCO_3$, 20 mM Hepes, 1 mM $MgCl_2$, 4 mM KH_2PO_4 (except where indicated), 4 mM Mg^{2+}-ATP, 12 mM creatine plus creatine phosphate, 50 μg/ml creatine phosphokinase, and 1 μM Fura 2.

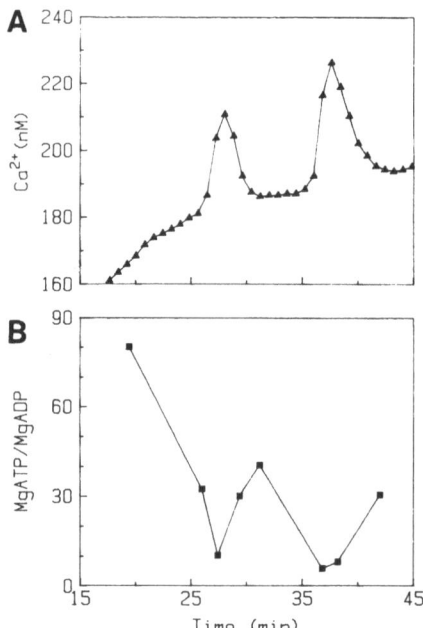

FIGURE 2. Oscillations in (A) free-Ca^{2+} levels and (B) the MgATP/MgADP ratio observed in saponin-permeabilized suspensions of cultured RINm5F insulinoma cells incubated in an extract of rat skeletal muscle exhibiting spontaneous glycolytic oscillations. The reaction mixture contained 1 μM Fura 2, 2 mM Na-ATP, 3 mM $MgCl_2$, 20 mM Hepes adjusted to pH 7.1 with KOH, 97 mM KCl, 5 mM $KHCO_3$, 6.5 mM potassium phosphate, 0.3 mM GTP, 4 mM sodium aspartate, 10 mM glucose, 30 μM NAD, 0.2 unit/ml yeast hexokinase (gel filtered in 20 mM Hepes), rat muscle extract equivalent to 1 mg protein/ml, 0.08 mg/ml of saponin, and RINm5F cells equivalent to 0.6 mg/ml of cell protein. The muscle extract contributed 47 mM of the KCl and 2.5 mM of the potassium phosphate to the reaction mixture plus 17 μM dithiothreitol.

2.2. Effect of Phosphate on the Ca^{2+} Set Point

The data in Fig. 1A are presented as MgATP/MgADP ratios rather than MgATP/MgADP·P$_i$ since the phosphate values were constant in this series of experiments. To determine whether the Ca^{2+} set point was being regulated by the MgATP/MgADP ratio or the phosphorylation potential, the effect of variations in P$_i$ on the Ca^{2+} set point at fixed ratios of MgATP/MgADP was investigated. An inverse relationship between the Ca^{2+} set point and medium phosphate was observed when concentrations of the latter were greater than 1 mM (Fig 1B). This occurred at all MgATP/MgADP ratios whether mitochondria were active or inhibited (Corkey et al., 1989). The relationship between phosphate and the Ca^{2+} set point was superimposed on the reciprocal relationship between the MgATP/MgADP ratio and free Ca^{2+}, since the free Ca^{2+} maintained at any given level of P$_i$ was lower at higher ratios of MgATP/MgADP. These data indicated that the correlation with Ca^{2+} (Fig. 1A) was a function of the MgATP/MgADP ratio and not the phosphorylation potential, which would have required an increase in Ca^{2+} set point with increasing P$_i$. The data suggested that phosphate influences the Ca^{2+} set point through an effect on a nonmitochondrial system and is consistent with a role for this anion in Ca^{2+} sequestration by the ER (Bygrave and Anderson, 1981).

2.3. Effect of Oscillations of Glycolysis and the MgATP/MgADP Ratio

When permeabilized RINm5F cells were incubated in a glycolyzing muscle extract, medium Ca^{2+} decreased initially. This was followed by a train of oscillations. The free-Ca^{2+} concentration oscillated between 180 and 230 nM, levels in the range reported to stimulate insulin release from permeabilized islets (Jones et al., 1985; Colca et al., 1985). Such oscillations in free Ca^{2+} were observed with eight separate cell preparations and three muscle extracts (data not shown). Subsequent removal of the insulinoma cells by centrifugation suppressed the Ca^{2+} oscillations but not the glycolytic oscillations (data not shown), indicating that the fluctuations in Ca^{2+} were not due to alterations in Ca^{2+}-ligand formation in the extract.

Samples were taken during the first two oscillations of medium Ca^{2+} (Fig. 2A) and analyzed for nucleotides, P$_i$, fructose-6-phosphate, and glucose-6-phosphate. There was a strong correlation between Ca^{2+} levels and the MgATP/MgADP ratio (Fig. 2B): a low MgATP/MgADP ratio was observed at times of high or rising Ca^{2+}, whereas a high MgATP/MgADP ratio was observed at times of low or falling Ca^{2+}. The results are consistent with modulation of the Ca^{2+}-ATPase activity of the endoplasmic reticulum by the MgATP/MgADP ratio, although other mechanisms such as effects on Ca^{2+}-releasing systems or mitochondrial Ca^{2+} carriers cannot be excluded. This interpretation is supported by the observations in Fig. 1 showing that steady-state Ca^{2+} levels maintained by endoplasmic reticulum in permeabilized RINm5F cells varied inversely with the MgATP/MgADP ratio, when the latter is maintained at values between 5 and 100 by a creatine phosphokinase buffering system, or when the MgATP/MgADP ratio is changed acutely (Prentki et al., 1985).

Other metabolite levels did not correlate with changes in free Ca^{2+} (data not shown). Orthophosphate declined nearly monotonically during the two oscillations. Levels of glucose-6-phosphate also fluctuated due to the glycolytic oscillations, but these fluctuations did not correspond with the changes in Ca^{2+}. Fructose-6-phosphate and glucose-1-phosphate were generally proportional to glucose-6-phosphate; lactate accumulated and the pH decreased monotonically.

2.4. Glucose Dependence of Glycolytic Ca^{2+} Oscillations

Pancreatic islets are stimulated by concentrations of glucose above about 5 mM. The glucose sensor of pancreatic islets has been proposed to be glucokinase (Meglasson and Matschinsky, 1983, 1986 and Matschinsky et al., 1986). Figure 3 illustrates the responsiveness of permeabilized RINm5F cells incubated in a glycolyzing muscle extract to glucose concentrations between 2.5 and 20 mM. In the presence of hexokinase, there was little change in the Ca^{2+} oscillations, since that enzyme would be saturated at all three glucose concentrations. However, when hexokinase was replaced with glucokinase, oscillations were not apparent at 2.5 mM glucose (physiologically, a nonstimulatory level), but oscillations appeared and increased in amplitude and frequency when the glucose concentration was increased to 10 and 20 mM.

Glucose initiation and modulation of glycolytic oscillations have been described in intact yeast (Hess and Boiteux, 1968) and ascites tumor cells (Ibsen and Schiller, 1967), but it is

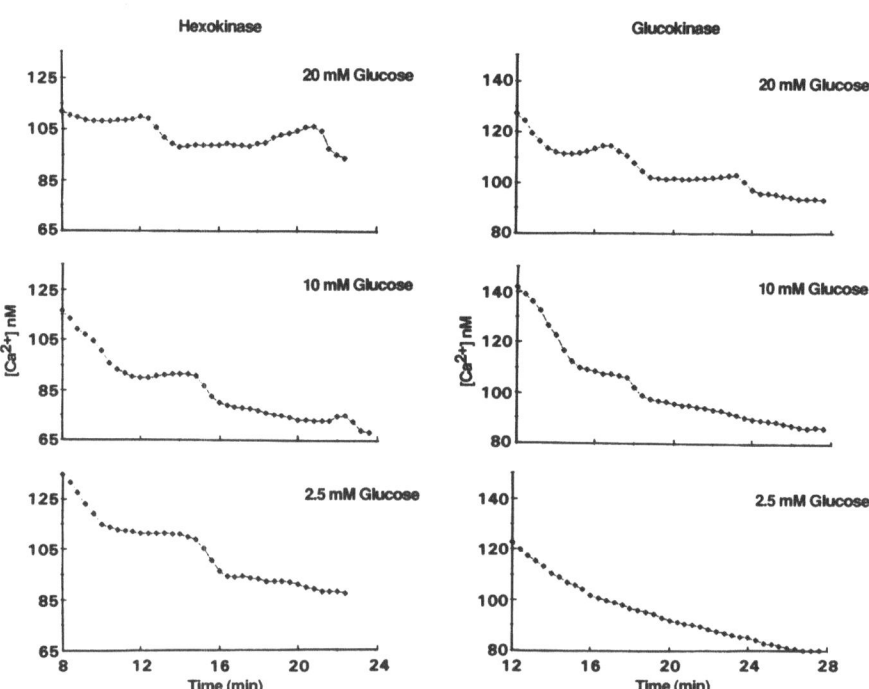

FIGURE 3. Dependence of free-Ca^{2+} oscillations on glucose concentrations when glucokinase was the initial glucose phosphorylating enzyme. RINm5F insulinoma cells were incubated with a muscle extract under the conditions described in Fig. 2, except that glucokinase (0.2 U/ml) replaced hexokinase in the incubations shown in the right panel and glucose was added at the concentrations indicated.

still unknown whether a similar phenomenon occurs in insulin-producing cells. On the other hand, it is noteworthy that insulin release *in vivo* is pulsatile (Jaspan *et al.*, 1986; Weigle, 1987) and that linked oscillations in insulin secretion, β-cell membrane potential, and Ca²⁺ concentration in the extracellular space within the islet have been shown in isolated single islets (Rosario *et al.*, 1986). It should be noted that although the cells within a single islet may be electrically and/or metabolically coupled, asynchrony of different islets may obscure oscillatory behavior in a large population of islets or β cells.

2.5. A Model of Glucose-Stimulated Insulin Secretion

The data suggest the following hypothesis linking glucose metabolism to β-cell Ca²⁺ signal transduction. This model is illustrated in Fig. 4. Stimulation of β cells with glucose would cause an initial fall in the MgATP/MgADP ratio due to phosphorylation of the sugar by glucokinase (Meglasson and Matschinsky, 1983, 1986). This would decrease the activity of Ca²⁺-ATPases in the endoplasmic reticulum and plasma membrane and result in the initial elevation in cytosolic Ca²⁺. Increased cytosolic Ca²⁺ would then activate mitochondrial enzymes such as α-glycerophosphate oxidase and pyruvate dehydrogenase. The decrease in the MgATP/MgADP ratio activates phosphofructokinase and stimulates flux through glycolysis. Increased glycolytic flux and mitochondrial energy production would then increase the MgATP/MgADP ratio. During sustained glucose stimulation we propose that alternate closing and opening of ATP-sensitive K⁺ channels, Ca²⁺ entry through voltage-dependent

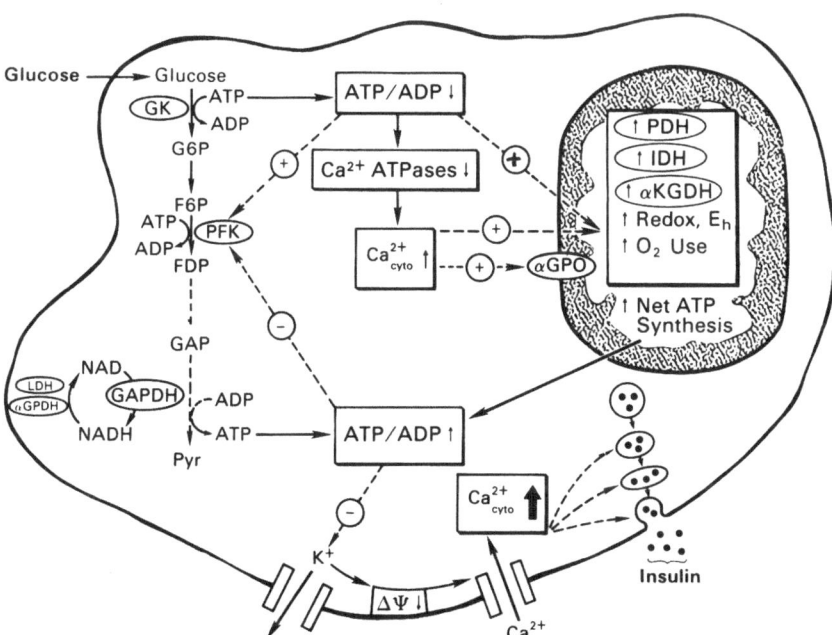

FIGURE 4. A model linking the early metabolic and ionic events in glucose-stimulated insulin secretion.

Ca^{2+} channels, and Ca^{2+} removal by Ca^{2+}-ATPases would be regulated in a coordinate manner by variations in the cytosolic MgATP/MgADP ratio. Further studies using intact cells are needed to evaluate the temporal relationships among the MgATP/MgADP ratio, cytosolic free-Ca^{2+} levels, and membrane potential in glucose-stimulated pancreatic islets.

ACKNOWLEDGMENTS. These studies were supported by NIH grants DK35914, DK19525, DK31559, NS17752, and HL26895.

REFERENCES

Bygrave, F. L., and Anderson, T. A., 1981, Ruthenium red-insensitive cation transport in ascites-sarcoma 180/TG cells, *Biochem. J.* **200**:343–348.

Chance, B., Legallais, V., Sorge, J., and Graham, N., 1975, A versatile time-sharing multichannel spectrophotometer, reflectometer and fluorometer, *Anal. Biochem.* **66**:498–514.

Colca, J. R., Wolf, B. A., Comens, P. G., and McDaniel, M. L., 1985, Protein phosphorylation in permeabilized pancreatic islet cells, *Biochem. J.* **228**:529–536.

Corkey, B. E., Duszynski, J., Rich, T. L., Matschinsky, B., and Williamson, J. R., 1986, Regulation of free and bound magnesium in rat hepatocytes and isolated mitochondria, *J. Biol. Chem.* **261**: 2567–2574.

Corkey, B. E., Tornheim, K., Deeney, J. T., Glennon, M. C., Matschinsky, F. M., and Prentki, M., 1989, Integrated mitochondrial and microsomal regulation of free Ca^{2+} in permeabilized insulinoma cells, in *Integration of Mitochondrial Function* (J. J. Lemasters, ed.), Plenum Press, New York: 543–550.

Grynkiewicz, G., Poenie, M., and Tsien, R. W., 1985, A new generation of Ca^{2+} indicators with greatly improved fluorescence properties, *J. Biol. Chem.* **260**:3440–3450.

Hedeskov, C. J., 1980, Mechanism of glucose-induced insulin secretion, *Physiol. Rev.* **60**:442–509.

Hess, B., and Boiteux, A., 1968, Mechanism of glycolytic oscillations in yeast, I, *Hoppe-Seyler's Z. Physiol. Chem.* **349**:1567–1574.

Ibsen, K. H., and Schiller, K. W., 1967, Inorganic ion concentrations in Ehrlich ascites carcinomal cells and fluid, *Biochim. Biophys. Acta* **131**:405–407.

Jaspan, J. B., Lever, E., Polonsky, K. S., and Van Cauter, E., 1986, In vivo pulsatility of pancreatic islet peptides, *Am. J. Physiol.* **251**:E215–E226.

Jones, P. M., Stutchfield, J., and Howell, S. L., 1985, Effects of Ca^{2+} and a phorbol ester on insulin secretion from islets of Langerhans permeabilized by high-voltage discharge, *FEBS Lett.* **191**: 102–106.

Matschinsky, F. M. M., Fertel, R., Kotler-Brajtburg, J., Stillings, S., Ellerman, J., Raybaud, F., and Holowach-Thurston, J., 1983, in *Eighth Midwest Conference on Endocrinology and Metabolism* (X. J. Mussachi, and R. P. Breitenbach, eds.), Univ. of Missouri Press, Columbia, MO.

Matschinsky, F. M., Meglasson, M. D., Ghosh, A., Appel, M., Bedoya, F., Prentki, M., Corkey, B., Shimizu, T., Berner, D., Najafi, H., and Manning, C., 1986, Biochemical design features of the pancreatic islet cell glucose-sensory system. *Adv. Exp. Med. Biol.* **211**:459–469.

Meglasson, M. D., and Matschinsky, F. M., 1983, Discrimination of glucose anomers by glucokinase from liver and transplantable insulinoma, *J. Biol. Chem.* **258**:6705–6708.

Meglasson, M. D., and Matschinsky, F. M., 1986, Pancreatic islet glucose metabolism and regulation of insulin secretion, *Diabetes/Metabolism Reviews* **2**:163–214.

Meglasson, M. D., Burch, P. T., Berner, D. K., Najafi, H., Vogin, A. P., and Matschinsky, F. M., 1983, Chromatographic resolution and kinetic characterization of glucokinase from islets of Langerhans, *Proc. Natl. Acad. Sci. U.S.A.* **80**:85–89.

O'Sullivan, W. J., and Smithers, G. W., 1979, *Methods Enzymol.* **63**:294–336.

Prentki, M., and Matschinsky, F. M., 1987, Ca^{2+}, cAMP and phospholipid-derived messengers in coupling mechanisms of insulin secretion, *Physiol. Rev.* **67**:1185–1247.

Prentiki, M., Corkey, B. E., and Matschinsky, F. M., 1985, Inositol 1,4,5-trisphosphate and the endoplasmic reticulum Ca^{2+} cycle of rat insulinoma cell line, *J. Biol. Chem.* **260**:9185–9190.

Rosario, L. M., Atwater I., and Scott, A. M., 1986, Pulsatile insulin release and electrical activity from single ob/ob mouse islets of Langerhans, *Adv. Exp. Med. Biol.* **211**:413–425.

Tornheim, K., 1979, Oscillations of the glycolytic pathway and the purine nucleotide cycle, *J. Theor. Biol.* **79**:491–541.

Tornheim, K., and Lowenstein, J. M., 1973, The purine nucleotide cycle. III. Oscillations in metabolite concentrations during the operation of the cycle in muscle extracts, *J. Biol. Chem.* **248**:2670–2677.

Tornheim, K., and Lowenstein, J. M., 1974, The purine nucleotide cycle. IV. Interactions with oscillations of the glycolytic pathway in muscle extracts, *J. Biol. Chem.* **249**:3241–3247.

Tornheim, K., and Lowenstein, J. M., 1975. The purine nucleotide cycle. V. Control of phosphofructokinase and glycolytic oscillations in muscle extracts, *J. Biol. Chem.* **250**:6304–6314.

Weigle, D. S., 1987, Pulsatile secretion of fuel-regulatory hormones, *Diabetes* **36**:764–775.

Williamson, J. R., and Corkey, B. E., 1969, Assays of intermediates of the citric acid cycle and related compounds by fluorimetric enzyme methods, *Methods Enzymol.* **13**:434–513

Williamson, J. R., and Corkey, B. E. 1979, Assay of citric acid cycle intermediates and related compounds—update with tissue metabolite levels and intracellular distribution, *Methods Enzymol.* **55**:200–222.

41

Calcium, Calmodulin, and Insulin Action in the Adipocyte

JAY M. McDONALD, JOSEPH P. LAURINO, and JERRY COLCA

1. INTRODUCTION

The cellular mechanism of insulin action remains enigmatic despite intensive research efforts. Over the past 20 years, considerable effort has focused on the signal transduction mechanisms that are responsible for linking the primary signal (insulin binding to its specific receptor on the cell surface) to the characteristic multicomponent pleiotypic cellular response (for reviews, see Czech, 1977, 1985). Although considerable evidence supports the concept that the final common pathway by which insulin regulates a variety of intracellular target pathways is the phosphorylation and dephosphorylation of key intracellular target proteins, the signal transduction processes remain unknown. Clearly, the complexity of the metabolic responses of the cell to insulin accompanied by our increasing knowledge about intracellular signals and mediators lend considerable support to the contention that the signal transduction process for insulin most assuredly involves multiple ''mediator'' pathways acting in a coordinated, orchestrated manner rather than a single mediator or messenger.

It is within this context that we will briefly review the potential role of Ca^{2+} as one of these multiple mediators in the cellular mechanism of insulin action. Specifically, we will review recently obtained data from our laboratory on the role of calmodulin and Ca^{2+} on insulin receptor function.

2. ROLE OF Ca^{2+} IN THE CELLULAR MECHANISM OF INSULIN ACTION

The potential role of Ca^{2+} in the insulin receptor effector mechanism is controversial. There have been considerable data both in support of and against the hypothesis that Ca^{2+}

JAY M. McDONALD and JOSEPH P. LAURINO ● Departments of Pathology and Medicine, Washington University School of Medicine, St. Louis, Missouri, 63110.   JERRY COLCA ● The Upjohn Company, Kalamazoo, Michigan, 49001.

plays an important role in insulin action. Since this topic has been reviewed extensively (Klip, 1984; McDonald et al., 1984; McDonald and Pershadsingh, 1985), this brief review will specifically focus on recent data which support the contention that Ca^{2+} is an important component of insulin receptor effector system in rat adipocytes. By limiting the scope of this review, controversies that may result from conclusions that are far too often generalized from one tissue, species, or even cultured cells to all relevant tissues and species will be avoided.

Before addressing this topic, it is important to understand that cytosolic Ca^{2+} is not evenly distributed especially during times of cellular activation. Dating back to the elegant studies of Rose and Lowenstein (1975), and more recently confirmed using many cell types (Williams et al., 1985), it is apparent that the cytosolic concentrations of Ca^{2+}, being continually regulated at various locations by the potent membrane Ca^{2+} mobilizing systems and the Ca^{2+}-buffering capacity of the cellular proteins, is not uniform (Matthews, 1980). Therefore, gradients for Ca^{2+} exist in the cytoplasm with the most dramatic changes occurring at or near the site of origin of primary or triggering signals. These concepts are especially important given the application of newer techniques using intracellular fluorescent probes to monitor "intracellular Ca^{2+} concentrations." Such probes are commonly employed with batches of cells and therefore, at best, yield an average signal from many cells at one time. However, the application of enhanced digital imaging with newer, more sensitive probes may ultimately permit us to monitor the more subtle subcompartmentalized Ca^{2+} transients that are likely to be important in many intracellular signal transduction processes.

Even with these present limitations, however, there are still considerable data which support the hypothesis that Ca^{2+} is important in insulin action. Using the intracellular Ca^{2+} probe, Fura 2, physiological concentrations of insulin increased the average cytoplasm free Ca^{2+} concentration in rat fat adipocytes (Draznin et al., 1987). In addition, we recently observed that loading rat adipocytes with the Ca^{2+} chelator, Quin2, inhibits insulin-stimulated glucose transport, glucose oxidation, and antilipolysis (Pershadsingh et al., 1987a). Importantly, the inhibiting effect of Quin2 on insulin-stimulated glucose transport is reversible by increasing Ca^{2+} with the Ca^{2+} ionophore, A23187. Other data, obtained using intact rat adipocytes include the following: (1) Trifluoperazine, a calmodulin inhibitor, blocks the ability of insulin to stimulate glucose transport without altering its ability to regulate lipolysis (Shechter, 1984). (2) Certain Ca^{2+} channel blockers (nicardipine, verapamil, and diltiazem) prolong the time course for insulin to stimulate glucose transport (Ishibashi and Kubo, 1984). We have found similar effects of these Ca^{2+} channel blockers (unpublished observations). (3) Certain activators of protein kinase C mimic the ability of insulin to stimulate glucose transport (Kirsch et al., 1985; Christensen et al., 1987). (4) Insulin causes a 40 ± 2% increase in [^{3}H]phorbol dibutyrate binding to adipocyte cytosol and a 27 ± 10% decrease in binding to plasma membranes (Pershadsingh et al., 1987b). These latter data contrast with the data of others who, using activity measurements, have been unable to detect insulin-induced changes in protein kinase C activation or distribution (Glynn et al., 1986; Vaartges et al., 1986).

Direct experimental support for the "Ca^{2+}-insulin action hypothesis" was obtained using subcellular preparations from adipocytes. These observations include (1) inhibition of the calmodulin-sensitive ($Ca^{2+} + Mg^{2+}$)-ATPase in plasma membranes by insulin (Pershadsingh and McDonald, 1979), (2) an increase in Ca^{2+} binding to adipocyte plasma membranes by insulin (McDonald et al., 1976), (3) an increase in calmodulin binding to adipocyte plasma membranes by insulin (Goewert et al., 1983), (4) the presence of a common 40-kDa phosphoprotein substrate in plasma membranes for insulin and activators of protein kinase C (Graves and McDonald, 1985a), (5) the presence of both Ca^{2+} (Williams and Turtle, 1984) and calmodulin (Graves *et al.*, 1985) binding sites on the insulin receptor, (6) identification of endogenous phosphatidylinositol kinase activity in the insulin receptor (Machicao and Wieland, 1984), (7) the ability of insulin to stimulate autophosphorylation of its receptor and associated tyrosine kinase activity being regulated (*in vitro*) by Ca^{2+} and calmodulin (Graves *et al.*, 1986), and (8) stimulation of the phosphorylation of calmodulin itself both *in vitro* (using solubilized enriched receptors) (Graves *et al.*, 1986) and *in vivo* by insulin (McDonald *et al.*, 1987; Colca *et al.*, 1987). It is these latter data that will be summarized here in more detail.

3. PHOSPHORYLATION OF CALMODULIN BY THE INSULIN RECEPTOR KINASE

Recently, Haring *et al.* (1985) and our group (Graves *et al.*, 1986), using similar assay systems employing enriched solubilized insulin receptors, reported that insulin phosphorylates calmodulin. The studies described by Haring *et al.* used receptors prepared from the Fao hepatoma cell line, whereas we used receptors prepared from rat adipocyte plasma membranes. Most of the remainder of the discussion will concentrate on the data derived from our laboratory employing rat adipocytes and adipocyte receptors with brief mention being made of more recent unpublished observations that utilize other biological preparations.

We originally observed that Ca^{2+} and calmodulin enhanced insulin-stimulated β-subunit autophosphorylation of the insulin receptor and associated tyrosine kinase activity as measured using the exogenous substrate Hf2B (Graves *et al.*, 1986). During the course of these studies, we demonstrated that calmodulin itself was phosphorylated. Some of the characteristics of insulin-stimulated calmodulin phosphorylation in this solubilized receptor preparation from rat adipocytes are summarized in Table I.

One of the most intriguing characteristics of this reaction is the absolute requirement for certain basic protein cofactors in the assay system to observe insulin-stimulated phosphorylation of calmodulin. The protein cofactors that have been shown to work are polylysine, protamine sulfate, and histone 2B. Interestingly, other compounds known to interact with calmodulin such as mellitin, myelin basic protein, chlorpromazine, trifluoperazine, substance P, glucagon, polyarganine, mastoparan, β-endorphin, spermine, spermidine, and putrescine were ineffective in supporting insulin-stimulated phosphorylation of calmodulin (Laurino *et al.*, 1988). Preliminary data employing insulin receptor preparations from different tissues

TABLE I. Characteristics of Insulin-Stimulated Phosphorylation of Calmodulin *in Vitro* Using Rat Adipocyte Receptors

$K_{0.5}$ (insulin)[a]	2.3×10^{-9} M
$K_{0.5}$ (calcium)[a]	1×10^{-7} M
Ions[a]	Requires Mg^{2+}; Mn^{2+} ineffective
$K_{0.5}$ ATP[b]	80–115 μM
Need for basic protein cofactors[b]	Polylysine, protamine sulfate, and histone Hf2B all effective
Phosphoaminoacid[b]	Tyrosine 99

[a] Data from Graves *et al.* (1986).
[b] Data from Laurino *et al.* (1987).

TABLE II. Various Kinases That Have Been
Reported to Phosphorylate Calmodulin

Insulin receptor preparations
Rat adipocyte (Graves et al., 1986)
Fao cultured hepatoma (Haring et al., 1985)
Rat liver[a]
Human placenta[a]
3T3 L1 adipocytes[a]
Other tyrosine kinases
SRC kinase (Fukami et al., 1986)
EGF kinase (Lin et al., 1987)
Intact cells and tissues
Rat adipocyte; insulin-dependent (Graves et al., 1986)
Skeletal muscle and brain (Plancke and Lazarides, 1983)
Brain (Nakajo et al., 1986)

[a] McDonald et al., unpublished observations.

(see Table II) indicate that the cofactors that are effective in the rat adipocyte receptor system are also effective in systems employing other insulin receptors. However, there appear to be subtle differences between the effectiveness of each of the cofactors in each system and in some cases certain other basic protein cofactors appear to uniquely support insulin-stimulated phosphorylation in one system but not in the adipocyte. Defining these protein interactions at the molecular level is clearly required before the significance of these findings can be elucidated.

The absolute requirement for Mg^{2+} instead of Mn^{2+} is interesting since the insulin receptor kinase is maximally stimulated in the presence of Mn^{2+}. Originally, we proposed that this may result from the inability of EGTA to chelate Ca^{2+} into the submicromolar range when Mn^{2+} is present in the assays due to the higher affinity of EGTA for Mn^{2+} than for Ca^{2+} (Graves et al., 1986). Alternatively, Mg^{2+} may interact in a unique way with calmodulin, the cofactor, and/or the receptor. Interestingly, calmodulin is phosphorylated solely on tyrosine 99 regardless of which cofactor is utilized (Laurino et al., 1988). Since tyrosine 99 is located in the third Ca^{2+} binding site, it is highly plausible to predict that the phosphorylation of this residue would alter the biological activity of calmodulin. For example, if the biological activity was decreased relative to nonphosphorylated calmodulin, this pathway would be a highly efficient way to decrease the activity of the vast spectrum of target enzymes of calmodulin.

Although the majority of the studies on tyrosine kinase phosphorylation of calmodulin have been performed using solubilized enriched insulin receptors prepared from rat adipocytes, this phenomenon is not confined to rat adipocytes (Table II). Although each preparation tested appears to be somewhat different in such aspects as cofactor concentration requirement, stoichiometry and Ca^{2+} sensitivity, calmodulin is phosphorylated in all. Haring et al. (1985) originally reported that calmodulin was phosphorylated in a system using receptors prepared from Fao cultured hepatoma cells. We were more recently able to demonstrate this phenomenon in insulin receptor preparations from rat liver, human placenta, and 3T3 L1 adipocytes. Furthermore, this phenomenon does not appear to be confined to the insulin receptor kinase, as other tyrosine kinases including the EGF kinase and a sarc-related (SRC) kinase have been reported to phosphorylate calmodulin.

TABLE III. Characteristics of *in Vivo* Insulin-Stimulated Phosphorylation of Calmodulin in Rat Adipocytes[a]

Insulin concentration dependency is similar to that for glucose transport $K_{0.5}$ (insulin) $\sim 1 \times 10^{-10}$ M.
Occurs rapidly; stimulated in <10 min.
Phosphorylation of calmodulin is resistent to alkali hydrolysis, suggesting the presence of tyrosine residues.

[a] Data from Colca *et al.* (1987).

Since the above observations were made using enriched or purified receptors, they could represent physiologically meaningless artefacts. Importantly, however, phosphorylated calmodulin has been identified and studied in intact cells and tissues. Phosphorylated calmodulin has been identified in three tissues: adipose, skeletal muscle, and brain (Table II). However, only in the adipocyte has its physiological regulation been addressed. As summarized in Table III, the ability of insulin to stimulate the phosphorylation of calmodulin has been investigated in intact adipocytes that had been preloaded with ^{32}Pi. At physiologically relevant concentrations, insulin stimulates the rapid phosphorylation of calmodulin on residues that are resistant to alkali treatment. This latter observation is consistent with the phosphorylation of tyrosine residues. In these studies, phosphocalmodulin was identified using a variety of criteria including appropriate migration on high-resolution two-dimensional gels, identity by the Ca^{2+} binding stain, Stains-All, and purification by W7 chromatography. To date no anticalmodulin or antityrosine antibodies have been effective immunoprecipitating reagents for phosphocalmodulin. Such reagents would obviously be invaluable for future studies.

4. SUMMARY AND CONCLUSIONS

The mechanism of insulin action remains enigmatic, and specifically the role of Ca^{2+} in this process remains controversial. Nevertheless, a number of recent observations have been summarized here that, with respect to one target cell, the adipocyte, support the concept that Ca^{2+} is a participant in the cellular mechanism of insulin. Some of the more interesting observations would appear to be those relating the phosphorylation of calmodulin to the function of the insulin receptor. Insulin phosphorylates calmodulin in a variety of *in vitro* systems and in intact adipocytes (Table II). *In vitro*, insulin stimulates the phosphorylation of calmodulin exclusively on tyrosine residue 99, located in the third Ca^{2+} binding pocket of the calmodulin molecule. It is therefore logical to predict that this phosphorylation will alter the Ca^{2+} binding properties and thus the function of calmodulin. Subsequently, the function of such ubiquitous Ca^{2+} and calmodulin-dependent enzymes and processes as the insulin receptor kinase itself, the insulin-sensitive $(Ca^{2+} - Mg^{2+})$ATPase, glycogen synthase, cAMP phosphodiesterase, and even ADP ribosylation (Graves and McDonald, 1985) may be regulated by this posttranslational modification of calmodulin. Beyond the concept that the phosphorylation of calmodulin may potentially be important in mediating the acute effects of insulin, it may also play a role in the mechanism of action of related growth factors. This is supported by reports that the SRC kinase and EGF phosphorylate calmodulin. Although such a posttransitional modification of a key regulatory protein as calmodulin permits speculation with regard to functional significance, considerable data remain to be gathered before its importance and underlying molecular mechanisms can be understood.

REFERENCES

Christensen, R. L., Shade, D. L., Graves, C. B., and McDonald, J. M., 1987, Evidence that protein kinase C is involved in regulating glucose transport in the adipocyte, *Int. J. Biochem.* **19:**259–265.

Colca, J. R., DeWald, D. B., Pearson, J. D., Palazuk, B. J., Laurino, J. P., and McDonald, J. M., 1987, Insulin stimulates the phosphorylation of calmodulin in intact adipocytes, *J. Biol. Chem.* **262:**11399–11402.

Czech, M. P. (ed.), 1985, *Molecular Basis of Insulin Action,* Plenum Press, New York.

Czech, M. P., 1977, Molecular basis of insulin action, *Ann. Rev. Biochem.* **46:**359–384.

Draznin, B., Kao, M., Sussman, K. E., 1987, Insulin and glyburide increase cytosolic free Ca^{2+} concentration in isolated rat adipocytes, *Diabetes* **36:**174–178.

Fukami, Y., Nakamura, T., Nakayama, A., and Kanehisa, T., 1986, Phosphorylation of tyrosine residues of calmodulin in Rous sarcoma virus-transformed cells, *Proc. Natl. Acad. Sci. USA* **83:**4190–4193.

Glynn, B. P., Collieton, J. W., McDermott, J. M., and Witters, L. A., 1986, Phorbol esters, but not insulin, promote depletion of cytosolic protein kinase C in rat adipocytes, *Biochem. Biophys. Res. Commun.* **135:**1119–1125.

Goewert, R. R., Klaven, N. B., and McDonald, J. M., 1983, Direct effect of insulin on the binding of calmodulin to rat adipocyte plasma membranes, *J. Biol. Chem.* **258:**9995–9999.

Graves, C. B., Gale, R. D., Laurino, J. P., and McDonald, J. M., 1986, The insulin receptor and calmodulin: Calmodulin enhances insulin-mediated receptor kinase activity and insulin stimulates phosphorylation of calmodulin, *J. Biol. Chem.* **261:**10429–10438.

Graves, C. B., Goewert, R. R., and McDonald, J. M., 1985, The insulin receptor contains a calmodulin binding domain, *Science* **230:**827–829.

Graves, C. B., and McDonald, J. M., 1985a, Insulin and phorbol ester stimulate phosphorylation of a 40-kDa protein in adipocyte plasma membranes, *J. Biol. Chem.* **260:**11286–11292.

Graves, C. B., and McDonald, J. M., 1985b, Effects of Ca^{2+} and calmodulin on endogenously catalyzed ADP-ribosylation in adipocyte plasma membranes, *Cell Calcium* **6:**491–501.

Haring, H. U., White, M. F., Kahn, C. R., Ahmed, Z., DePooli-Roach, A. A., and Roach, P., 1985, Interaction of the insulin receptor kinase with serine/threonine kinases in vitro, *J. Cell Biochem.* **28:**171–182.

Ishibashi, F., and Kubo, K., 1984, Inhibition of calcium antagonist of coupling of insulin binding and insulin action on glucose transport in isolated fat cells, *Hiroshima J. Med. Sci.* **33:**73–79.

Kirsch, D., Obermaier, B., and Haring, H. U., 1985, Phorbol esters enhance basal D-glucose transport but inhibit stimulation of D-glucose transport and insulin binding in isolated rat adipocytes, *Biochem. Biophys. Res. Commun.* **128:**824–832.

Klip, A., 1984, Is intracellular Ca^{2+} involved in insulin stimulation of sugar transport? Fact and prejudices, *Can. J. Biochem. Cell. Biol.* **62:**1228–1236.

Laurino, J. P., Colca, J. R., DeWald, D. B., Pearson, J. D., and McDonald, J. M., 1988, The in vitro phosphorylation of calmodulin by the insulin receptor tyrosine kinase, *Arch. Biochem. Biophys.* **265:**8–21.

Lin, P. H., Selinfreund, R., and Wharton, W., 1987, Calmodulin is phosphorylated on the tyrosine residue and binds to a different site from EGF on the EGF receptors, *Fed. Proc.* **46:**396.

Machicao, E., and Wieland, O. H., 1984, Evidence that the insulin receptor-associated protein kinase acts as a phosphatidylinositol kinase. *FEBS Lett.* **175:**113–116.

Matthews, E. K., 1980, in: *Secretory Mechanisms* (C. R. Hopkins and C. J. Duncan, eds.), Cambridge University Press, London, pp. 225–249.

McDonald, J. M., Bruns, D. E., and Jarett, L., 1976, Ability of insulin to increase calcium binding by adipocyte plasma membranes, *Proc. Natl. Acad. Sci. USA* **73:**1542–1546.

McDonald, J. M., Graves, C. B., and Christensen, R. L., 1984, in: *Calcium and Cell Function,* Vol. (W. Y. Cheung, ed.), Academic Press, New York, p. 209.

McDonald, J. M., Pershadsingh, H. A., and Colca, J., 1987, The role of calcium and calmodulin in insulin receptor function in the adipocyte, *Ann. N. Y. Acad. Sci.* **488:**406–418.

McDonald, J. M., and Pershadsingh, H. A., 1985, in: *Molecular Basis of Insulin Action* (M. P. Czech, ed.), Plenum Press, New York, p. 103.

Nakajo, S., Hayashi, K., Daimatsu, T., Tanaka, M., Nakaya, K., and Nakamura, Y., 1986, Phosphorylation of rat brain calmodulin in vivo and in vitro, *Biochem. Int.* **13:**687–693.

Pershadsingh, H. A., Shade, D. L., Delfert, D. M., and McDonald, J. M., 1987a, Chelation of intracellular calcium blocks insulin action in the adipocyte, *Proc. Natl. Acad. Sci. USA* **84:**1025–1029.

Pershadsingh, H. A., Shade, D. L., and McDonald, M., 1987b, Insulin-dependent alterations of phorbol ester binding to adipocyte subcellular constituents. Evidence for the involvement of protein kinase C in insulin action, *Biochem. Biophys. Res. Commun.* **145:**1384–1389.

Pershadsingh, H. A., and McDonald, J. M., 1979, Direct addition of insulin inhibits a high affinity of Ca^{2+}-ATPase in isolated adipocyte plasma membranes, *Nature* **281:**495–497.

Plancke, Y. D., and Lazarides, E., 1983, Evidence for a phosphorylated form of calmodulin in chicken brain and muscle, *Molec. Cell. Biol.* **3:**1412–1420.

Rose, B., and Lowenstein, W. R., 1975, Calcium ion distribution in cytoplasm visualized by aequorin: diffusion in cytosol restricted by energized sequestering. *Science* **190:**1204–1206.

Shechter, Y., 1984, Trifluoperazine inhibits insulin action on glucose metabolism in fat cells without affecting inhibition of lipolysis, *Proc. Natl. Acad. Sci. USA* **81:**327–331.

Vaartges, W. J., deHaas, C. C. M. and van der Bergh, S. O., 1986, Phorbol esters, but not epidermal growth factor or insulin, rapidly decrease soluble protein kinase C activity in rat hepatocytes, *Biochem. Biophys. Res. Commun.* **138:**1328–1333.

Williams, D. A., Fogarty, K. E., Tsien, R. Y., and Fay, F. S., 1985, Calcium radients in single smooth muscle cells revealed by the digital imaging microscope using fura 2, *Nature* **318:**558–561.

Williams, P. F., and Turtle, J. A., 1984, Terbium, a fluorescent probe for insulin receptor binding, *Diabetes* **33:**1106–1111.

42

Role of Calcium in the Regulation of Mammalian Lipoxygenases

BECKY M. VONAKIS and JACK Y. VANDERHOEK

1. INTRODUCTION

Arachidonic acid is the precursor to a variety of oxygenated metabolites that have been implicated as regulators of various cell functions. The prostanoids are one group of metabolites that are formed from arachidonic acid via the cyclooxygenase pathway. Arachidonic acid metabolism catalyzed by the lipoxygenase pathway, which produces metabolites such as hydroperoxyeicosatetraenoic acids (HPETEs), the corresponding hydroxy analogs (HETEs), leukotrienes (LTs), and lipoxins (Pace-Asciak and Smith, 1983), will be the focus of this paper. Our laboratory as well as others has shown that the HETEs can modulate cellular lipoxygenase, cyclooxygenase, and phospholipase activities (Vanderhoek, 1985). LTB_4 causes leukocyte adhesion, degranulation and chemotaxis of neutrophils while LTC_4, LTD_4, and LTE_4, which are released during an anaphylactic reaction, cause broncho- and vasoconstriction and an increase in vascular permeability (Samuelsson, 1983). Lipoxin A causes a contraction of bronchial smooth muscle while lipoxins A and B can inhibit natural killer cell function without preventing natural killer cell binding to target cells (Samuelsson, 1987).

Lipoxygenase enzymes catalyze the stereospecific incorporation of a molecule of oxygen into arachidonic acid to form a HPETE and are classified according to the carbon atom oxidized as the 5-, 12-, or 15-lipoxygenases. However, many cellular lipoxygenases exist in a relatively inactive or cryptic state, even in the presence of substrate, until properly stimulated. The extracellular stimulus leading to the release of arachidonic acid from cellular lipids as well as the conversion of an inactive lipoxygenase to the active enzyme are generally considered the key regulatory steps responsible for lipoxygenase activation (Vanderhoek and Bailey, 1985). It is the purpose of this report to discuss research conducted by our laboratory and other investigators concerning the role of calcium ions in the stimulation of cryptic cellular lipoxygenases and the ability of arachidonic acid and specific lipoxygenase metabolites to act as calcium release agents.

BECKY M. VONAKIS and JACK Y. VANDERHOEK ● Department of Biochemistry, The George Washington University School of Medicine and Health Sciences, Washington, D.C. 20037.

2. EFFECTS OF CALCIUM ON MAMMALIAN LIPOXYGENASES

Initially, Jakschik *et al.* (1980) and others (Parker and Aykent, 1982; Furukawa *et al.*, 1984) demonstrated that the inactive 5-lipoxygenase in a cell-free preparation of rat basophilic leukemia (RBL-1) cells was stimulated in a dose-dependent manner by calcium ions. Subsequently, the 5-lipoxygenases in broken cell preparations from casein-activated guinea pig and rat peritoneal polymorphonuclear leukocytes (PMNs) (Ochi *et al.*, 1983; Skoog *et al.*, 1986) and human and porcine peripheral blood leukocytes (Rouzer *et al.*, 1986; Ueda *et al.*, 1986) were shown to have a divalent cation dependence for activity in the presence of substrate with calcium ions being the most effective. The calcium-dependent 5-lipoxygenase activity is further enhanced by ATP (Ochi *et al.*, 1983; Furukawa *et al.*, 1984; Skoog *et al.*, 1986; Rouzer *et al.*, 1986; Ueda *et al.*, 1986).

The 5-lipoxygenase has been purified to near homogeneity from human and porcine leukocytes (Rouzer *et al.*, 1986; Ueda *et al.*, 1986), RBL-1 cells (Goetze *et al.*, 1985; Hogaboom *et al.*, 1986), and a murine bone marrow-derived mast cell line (Shimizu *et al.*, 1986). Unexpectedly, these highly purified 5-lipoxygenase preparations also contain the LTA_4 synthase activity which converts 5-HPETE to LTA_4 (Rouzer *et al.*, 1986; Ueda *et al.*, 1986; Hogaboom *et al.*, 1986; Shimizu *et al.*, 1986). Both enzymatic activities reside on the same polypeptide chain and require a millimolar concentration of calcium ions for activity.

The 12-lipoxygenase activity in a 105,000g fraction of RBL-1 cells was stimulated over fivefold by 1 mM $CaCl_2$ versus a metal ion-free preparation as measured by 12-HETE formation (Hamasaki and Tai, 1984) and a 12-lipoxygenase purified from the microsomal fraction (160,000g pellet) of human uterine cervix was stimulated approximately twofold by the addition of 1 mM $CaCl_2$ and 2 mM ATP (Flatman *et al.*, 1986). However, the cytosolic fraction of rat platelets showed a decrease in 12-HETE formation when incubated with 1–5 mM $CaCl_2$ versus calcium-free incubations (Hamasaki and Tai, 1984).

The 105,000g supernatant from glycogen-activated rabbit peritoneal PMNs contains a 15-lipoxygenase that is stimulated 10-fold by 5 mM $CaCl_2$ versus untreated controls (Narumiya *et al.*, 1981). Upon purification, the calcium dependency disappeared suggesting that another part of the activation mechanism rather than the enzyme itself has a calcium requirement. A 15-lipoxygenase purified from the cytosolic fraction of human PMNs exhibited maximal activity in the presence of 1 mM $CaCl_2$ although it retained 22% of its activity in the presence of 5 mM EGTA (Soberman *et al.*, 1985).

The enhancement of lipoxygenase activity by calcium ions in broken cell preparations clarifies the mechanism by which the divalent cation ionophore A23187 stimulates intact cellular lipoxygenases to metabolize exogenous arachidonic acid. Since the phospholipases that release arachidonic acid from cellular lipid pools have a calcium requirement, the use of exogenously added arachidonic acid should bypass the need for substrate release and allow the calcium requirements of other enzymes in the activation pathway to be probed. In mouse peritoneal macrophages, phagocytosis of particulate agonists is accompanied by the activation of both the lipoxygenase and cyclooxygenase pathways as evidenced by the formation of LTC_4 and thromboxane B_2, prostaglandin E_2, and 6-keto-prostaglandin $F_1\alpha$, respectively (Tripp *et al.*, 1985). However, stimulation of the macrophages with the soluble agonist phorbol myristate acetate (PMA) or exogenous arachidonic acid leads to prostaglandin metabolism only. When the cells were stimulated with PMA and A23187, a 50-fold increase in LTC_4 production was seen over the quantity of LTC_4 formed by stimulation with PMA or A23187 alone. Furthermore, macrophages stimulated with arachidonic acid and A23187 showed a 10-fold increase in LTC_4 formation versus product formation due to substrate or

A23187 alone. In either case (PMA or arachidonic acid), addition of A23187 led to no change in prostaglandin formation. Therefore, it appears that the addition of this calcium ionophore released the cryptic 5-lipoxygenase from its inactive state and led to leukotriene synthesis from exogenous substrate. Apparently, PMA and exogenous arachidonic acid cannot provide the proper stimulus for activation of the 5-lipoxygenase pathway until the intracellular calcium ion concentration is raised by A23187. Rabbit PMNs stimulated with A23187 produce 5-HETE and 5,12-diHETE in calcium-containing media (Borgeat and Samuelsson, 1979b; Walker and Parish, 1981). Depletion of calcium ions by 60-min incubation in 10 mM EGTA followed by stimulation with A23187 led to a time-dependent decrease in the formation of both 5-HETE and 5,12-diHETE from rabbit PMNs. Replenishment of calcium to PMNs previously depleted of calcium by EGTA treatment followed by stimulation with A23187 led to the restoration of 5-lipoxygenase activity. Casein-activated guinea pig PMNs incubated with 10 μM A23187 produce 5-HETE from exogenous arachidonic acid (Ochi *et al.*, 1983). A cloned murine mast cell line (MC 9) contains a cryptic 5-lipoxygenase which produces 5,12-diHETE and LTC_4 upon incubation with 0.5 μM A23187 and exogenous arachidonic acid (Musch *et al.*, 1985). Activation of the 5-lipoxygenase pathway was only partially dependent on extracellular calcium ions since 225 pmol/10^8 cells of 5-HETE was formed upon addition of A23187 and substrate in calcium-free media. Increasing the extracellular calcium ion concentration from 0 to 500 μM caused a twofold increase in 5-HETE formation. Production of LTC_4 by ionophore-stimulated cells was independent of extracellular calcium ion concentration. In another study, human peripheral blood PMNs from 12 of 18 donors showed cryptic 5-lipoxygenase activity while 6 of 18 donors showed weak 5-lipoxygenase activity upon incubation with substrate (Borgeat and Samuelsson, 1979a). When the cryptic lipoxygenase-containing cells were stimulated with 20 μM A23187, a 5- to 20-fold increase in 5-HETE formation was detected. The noncryptic cells showed no increase in 5-HETE production when incubated with ionophore. The strongest stimulation of 5-lipoxygenase activity was seen with PMNs receiving 20 μM A23187 plus exogenous arachidonic acid such that 20% of the added substrate was utilized.

Finally, a fourfold enhancement in 15-lipoxygenase activity was observed when human PMNs were treated with A23187 (6 μM), 12-HETE (42 μM), and arachidonic acid (16 μM) relative to that obtained when the cells were incubated only with 12-HETE and arachidonic acid (Vanderhoek *et al.*, 1985). This suggests that an influx of calcium ion is an essential requirement in the stimulation process of the cellular 15-lipoxygenase by exogenous HETEs.

3. ROLE OF CALCIUM IN THE 15-HETE-INDUCED STIMULATION OF THE CRYPTIC 5-LIPOXYGENASE IN PT-18 CELLS: CURRENT STUDIES

We reported previously that 15-HETE can activate the cryptic 5-lipoxygenase enzyme in the murine mast/basophil cell line PT-18 (Vanderhoek *et al.*, 1982). When PT-18 cells were incubated with [^{14}C]arachidonic acid alone, very little metabolism to ^{14}C-labeled products was observed (Fig. 1). However, when the cells were preincubated with the 15-lipoxygenase product 15-HETE followed by addition of [^{14}C]arachidonic acid, a dose-dependent stimulation of a cryptic 5-lipoxygenase was observed as measured by production of [^{14}C]5-HETE and [^{14}C]5,12-diHETEs (including LTB_4) (Fig. 1). 12-HETE was ineffective relative to

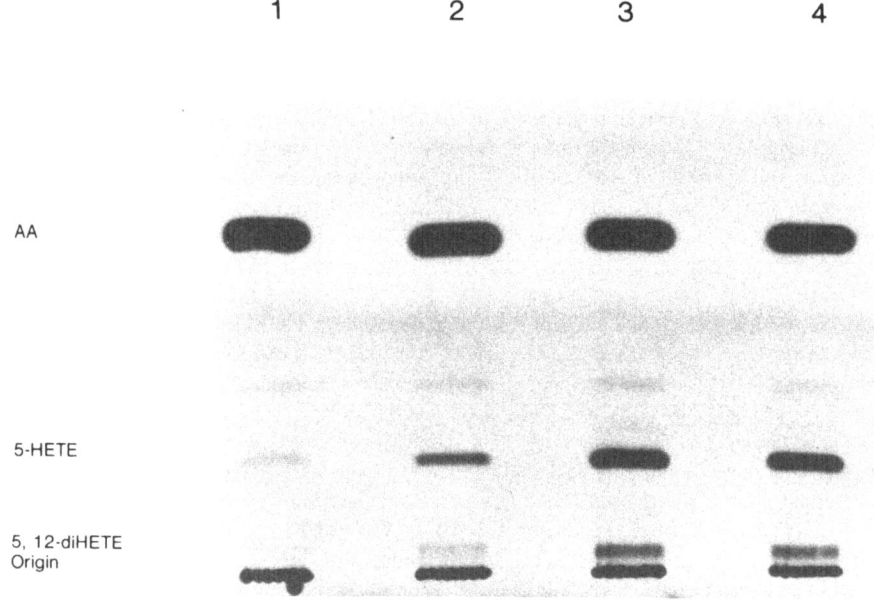

FIGURE 1. 15-HETE-induced activation of 5-lipoxygenase activity in PT-18 mast/basophil cells. Autoradiograph of a thin-layer chromatogram of arachidonic acid metabolites formed from PT-18 cells (7 × 10⁶/ml) and [¹⁴C]arachidonic acid (16 μM) in the absence (lane 1) and presence (lane 2, 3 μM; lane 3, 11 μM; lane 4, 17 μM) of 15-HETE. TLC solvent system, petroleum ether/ether/acetic acid (50:50:1). HETE, hydroxyeicosatetraenoic acid; AA = arachidonic acid.

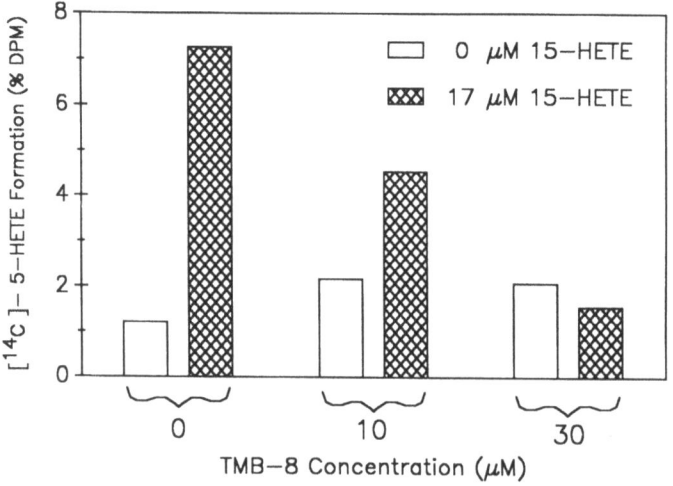

15-HETE in stimulating 5-lipoxygenase activity (Vanderhoek *et al.*, 1982). Similarly, neither A23187 (Vanderhoek and Pluznik, 1985) nor ionomycin (see Fig. 4) was able to appreciably activate the cryptic 5-lipoxygenase. In view of the previous discussion on the role of calcium in stimulating 5-lipoxygenases, we examined whether calcium was involved in the 15-HETE-induced activation of the cryptic 5-lipoxygenase in PT-18 cells. When PT-18 cells were pretreated with 3,4,5-trimethoxybenzoic acid-8-(diethylamino)octyl ester (TMB-8), an intracellular calcium antagonist (Chiou and Malagodi, 1975), the 15-HETE-induced activation of the 5-lipoxygenase was not observed (Fig. 2). This result suggests the involvement of calcium in the activation process and is similar to a report that TMB-8 reduced the ricinoleic acid-induced stimulation of [^{14}C]LTB$_4$ and [^{14}C]5-HETE formation from [^{14}C]arachidonic acid in rat isolated intestine (Capasso *et al.*, 1985). [However, since TMB-8 was shown to inhibit mitochondrial ATP production in rat thymocytes (Brand and Felber, 1984), we are currently examining whether this alternative mechanism is operating in the PT-18 system.]

Based on the results with TMB-8, we hypothesized that perhaps 15-HETE stimulates the 5-lipoxygenase by modulating intracellular calcium levels. In order to measure intracellular calcium, PT-18 cells were loaded with the calcium indicator Fura 2 (Grynkiewicz *et al.*, 1985). Various fatty acids were then tested for their ability to affect cytosolic calcium levels. As shown in Fig. 3, 15 μM arachidonic acid was able to induce significant increases in cytosolic calcium ion levels while 15-HETE or 12-HETE at this concentration was relatively ineffective. When the fatty acid concentration was increased to 30 μM, 12-HETE and 15-HETE became more effective enhancers of cytosolic calcium (12-HETE > 15-HETE) but both HETE isomers were still much less potent than arachidonic acid. Although both arachidonic acid (Wolf *et al.*, 1986; Chan and Turk, 1987) and 12-HETE (Naccache *et al.*, 1981) were previously reported to alter cytosolic calcium levels in different systems, the present results indicate the relative order of potencies of these agents in the same system, i.e., PT-18 cells. In order to correlate intracellular calcium changes with stimulation of the 5-lipoxygenase, Fura 2-loaded PT-18 cells were pretreated with either the calcium ionophore ionomycin, 12-HETE, or 15-HETE, followed by the addition of [^{14}C]arachidonic acid. The results in Fig. 4 show that both ionomycin and arachidonic acid induced increases in intracellular calcium levels but there was no corresponding enhancement of the 5-lipoxygenase activity. On the other hand, 15-HETE caused a minimal change in cytosolic calcium concentration but was quite effective in stimulating the cryptic 5-lipoxygenase. When Fura-loaded PT-18 cells were treated with 12-HETE, increases in both cytosolic calcium levels (Fig. 3) and the cryptic 5-lipoxygenase activity ([^{14}C]5-HETE formation (% DPM) was 8.3 $\pm$ 3.1 ($n = 2$) for 12-HETE-treated cells and 1.5 $\pm$ 0.17 ($n = 3$) for control cells were observed. This result with 12-HETE contrasts with that obtained with non-Fura 2-loaded cells where 12-HETE does not stimulate the PT-18 5-lipoxygenase (Vanderhoek *et al.*, 1982) and it appears that loading PT-18 cells with Fura 2 changes the specificity of the

---

FIGURE 2. TMB-8 prevents 15-HETE-induced 5-lipoxygenase activation in PT-18 cells. PT-18 cells (7 × 10^6/ml) were preincubated with either ethanol (control) or 17 μM 15-HETE for 1 min, followed by the addition of 0, 10, or 30 μM TMB-8. After 1 min, [^{14}C]arachidonic acid (16 μM) was added and the reaction was terminated 10 min later. Products were extracted, chromatographed on TLC, and the radioactive band with the same retention time as authentic 5-HETE was isolated, counted, and used as an indicator of 5-lipoxygenase activity. Results are representative of at least three experiments and are expressed as % DPM (DPM in the 5-HETE band × 100/total TLC plate DPM).

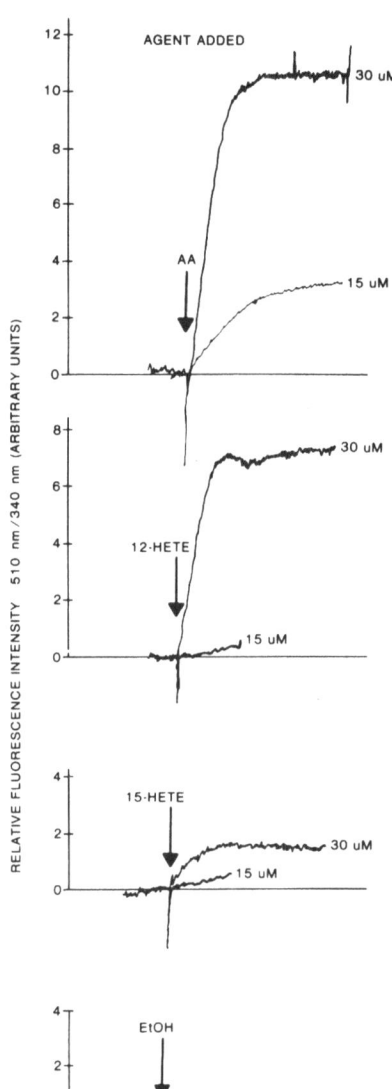

FIGURE 3. Effects of arachidonic acid, 12-HETE, and 15-HETE on cytosolic calcium ion concentration in Fura 2-loaded PT-18 cells. PT-18 cells (7 × 10^6 cells/ml) were loaded with 5 μM Fura 2 and incubated with the indicated concentrations of arachidonic acid, 12-HETE, 15-HETE, or ethanol (control). The excitation wavelength was 340 nm and the fluorescence emission was monitored at 510 nm. The results are representative of at least four experiments.

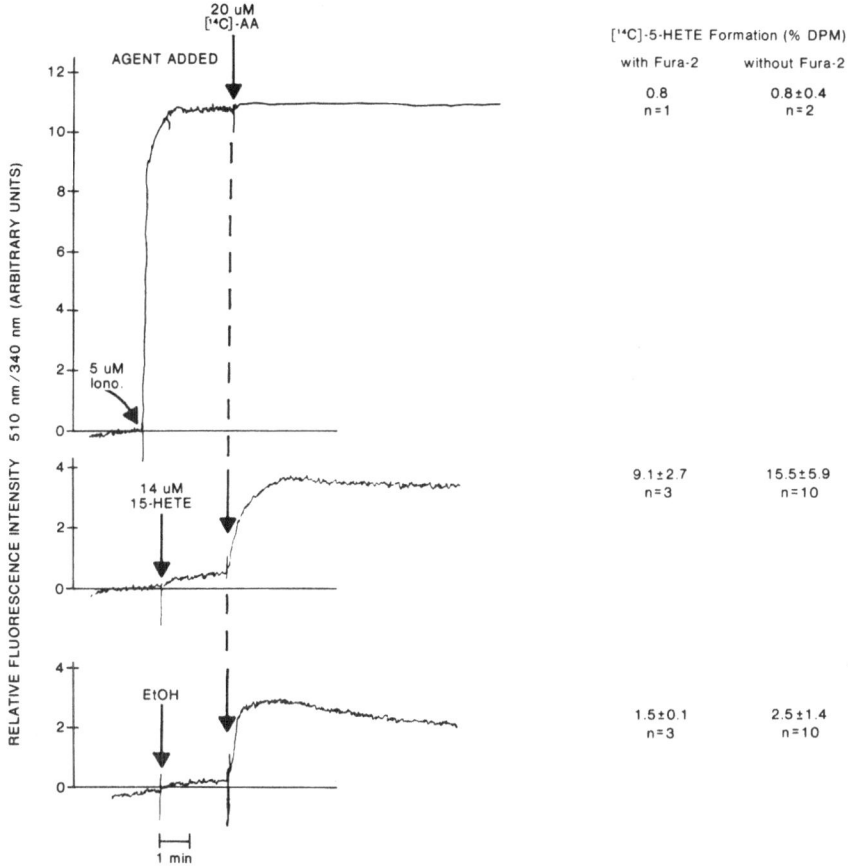

FIGURE 4. Comparison of the effects of ionomycin and 15-HETE on cytosolic calcium levels and activation of the cryptic 5-lipoxygenase in PT-18 cells. PT-18 cells (7 × 10⁶/ml) previously loaded with 5 μM Fura 2 were incubated with ionomycin (5 μM), 15-HETE (14 μM), or ethanol (control). The samples were excited at 340 nm and the fluorescence emission at 510 nm was monitored throughout the entire run. After 2 min, [¹⁴C]arachidonic acid (20 μM) was added and the reaction was stopped 10 min later. 5-Lipoxygenase activity was determined as described in the legend of Fig. 2.

activation requirements such that either 12-HETE or 15-HETE is effective. In view of these results, we propose that cytosolic calcium mobilization does not appear to play a role in the 15-HETE-induced activation of the cryptic 5-lipoxygenase in intact PT-18 cells.

4. CONCLUSIONS

Evidence gathered to date indicates calcium is a required cofactor for the activity of 5-lipoxygenase enzymes in a variety of mammalian species and cells, as well as several

12- and 15-lipoxygenases. This calcium requirement may be met by the steady-state cytoplasmic calcium concentration and as such may not be part of the regulatory mechanism which holds unstimulated lipoxygenases (in intact cells) in the cryptic state. Studies with intact cells stimulated by the divalent cation ionophore A23187 in calcium-containing media indicate that an influx of calcium into the cytoplasm of mouse peritoneal macrophages, rabbit and guinea pig PMNs, and human peripheral blood PMNs releases the 5-lipoxygenase enzymes to metabolize exogenous arachidonic acid. On the other hand, the murine mast cell lines PT-18 and MC9 do not seem to require extracellular calcium ions for 5-lipoxygenase activation. In the activation of the PT-18 enzyme, the inability of calcium ionophores (A23187 and ionomycin) and arachidonic acid to stimulate the 5-lipoxygenase indicates that an influx of extracellular calcium ions into the cytosol is not sufficient to release the enzyme from its cryptic state. This observation is corroborated by the studies conducted with Fura 2-loaded PT-18 cells. Hence, these results suggest that the 15-HETE-induced activation of the cryptic 5-lipoxygenase in PT-18 cells may not be regulated by calcium. In view of the recent study by Hansson *et al.* (1986) regarding activation of protein kinase C by 15-HETE, it is tempting to speculate that protein kinase C may play a role in the 15-HETE-induced stimulation of the cryptic 5-lipoxygenase in PT-18 cells.

ACKNOWLEDGMENT. This work was supported by grants from the National Institutes of Health.

REFERENCES

Borgeat, P., and Samuelsson, B., 1979a, Arachidonic acid metabolism in polymorphonuclear leukocytes: Effects of ionophore A23187, *Proc. Natl. Acad. Sci. USA* **76:**2148–2152.

Borgeat, P., and Samuelsson, B., 1979b, Transformation of arachidonic acid by rabbit polymorphonuclear leukocytes, *J. Biol. Chem.* **254:**2643–2646.

Brand, M. D., and Felber, S. M., 1984, The intracellular calcium antagonist TMB-8 [8-(NN-diethylamino)octyl-3,4,5-trimethoxybenzoate] inhibits mitochondrial ATP production in rat thymocytes, *Biochem. J.* **224:**1027–1030.

Capasso, F., Tavares, I. A., Tsang, R., and Bennett, A., 1985, The role of calcium in eicosanoid production induced by ricinoleic acid or the calcium ionophore A23187, *Prostaglandins* **30:**119–124.

Chan, K. M., and Turk, J., 1987, Mechanism of arachidonic acid-induced Ca^{2+} mobilization from rat liver microsomes, *Biochim. Biophys. Acta* **928:**186–193.

Chiou, C. Y., and Malagodi, M. H., 1975, Studies on the mechanism of action of a new Ca^{2+} antagonist, 8-(N,N,diethylamino)octyl 3,4,5-trimethoxybenzoate hydrochloride in smooth and skeletal muscles, *Br. J. Pharmac.* **53:**279–285.

Flatman, S., Hurst, J. S., McDonald-Gibson, R. G., Jonas, G. E. G., and Slater, T. F., 1986, Biochemical studies on a 12-lipoxygenase in human uterine cervix, *Biochim. Biophys. Acta* **883:**7–14.

Furukawa, M., Yoshimoto, T., Ochi, K., and Yamamoto, S., 1984, Studies on arachidonate 5-lipoxygenase of rat basophilic leukemia cells, *Biochim. Biophys. Acta* **795:**458–465.

Goetze, A. M., Fayer, L., Bouska, J., Bornemeier, D., and Carter, G., 1985, Purification of a mammalian 5-lipoxygenase from rat basophilic leukemia cells, *Prostaglandins* **29:**689–701.

Grynkiewicz, G., Poenie, M., and Tsien, T., 1985, A new generation of Ca^{2+} indicators with greatly improved fluorescence properties, *J. Biol. Chem.* **260:**3440–3450.

Hamasaki, Y. and Tai, H., 1984, Calcium stimulation of a novel 12-lipoxygenase from rat basophilic leukemia (RBL-1) cells, *Biochim. Biophys. Acta* **793:**393–398.

Hansson, A., Serhan, C., Haeggstrom, J., Ingelman-Sundberg, M., and Samuelsson, B., 1986, Activation of protein kinase C by lipoxin A and other eicosanoids: Intracellular action of oxygenation products of arachidonic acid, *Biochem. Biophys. Res. Commun.* **134**:1215–1222.

Hogaboom, K. G., Cook, M., Newton, J. F., Varrichio, A., Shorr, R. G. L., Sarau, H. M., and Crooke, S. T., 1986, Purification, characterization, and structural properties of a single protein from rat basophilic leukemia (RBL-1) cells possessing 5-lipoxygenase and leukotriene A_4 synthetase activities, *Molec. Pharm.* **30**:510–519.

Jakschik, B. A., Sun, F. F., Lee, L., and Steinhoff, M. M., 1980, Calcium stimulation of a novel lipoxygenase, *Biochem. Biophys. Res. Commun.* **95**:103–109.

Maclouf, J., de la Baume, H., Levy-Toledano, S., and Caen, J., 1982, Selective stimulation of human platelet lipoxygenase product 12-hydroxy-5,8,10,14-eicosatetraenoic acid by chlorpromazine and 8-(N,N-diethylamino)-octyl-3,4,5-trimethoxybenzoate, *Biochim. Biophys. Acta* **711**:377–385.

Musch, M. W., Bryant, R. W., Coscolluela, C., Myers, R. F., and Siegel, M. I., 1985, Ionophore-stimulated lipoxygenase activity and histamine release in a cloned murine mast cell, MC9, *Prostaglandins* **29**:405–430.

Naccache, P. H., Sha'afi, R. I., Borgeat, P., and Goetzl, E. J., 1981, Mono- and dihydroxyeicosatetraenoic acids alter calcium homeostasis in rabbit neutrophils, *J. Clin. Invest.* **67**:1584–1587.

Narumiya, S., Salmon, J., Cottee, F., Weatherley, B., and Flower, R., 1981, Arachidonic acid 15-lipoxygenase from rabbit peritoneal polymorphonuclear leukocytes: Partial purification and properties, *J. Biol. Chem.* **259**:9583–9592.

Ochi, K., Yoshimoto, T., Yamamoto, S., Taniguchi, K., and Miyamoto, T., 1983, Arachidonate 5-lipoxygenase of guinea pig peritoneal polymorphonuclear leukocytes: Activation by adenosine 5'-triphosphate, *J. Biol. Chem.* **258**:5754–5758.

Pace-Asciak, C. R., and Smith, W. L., 1983, Enzymes in the biosynthesis and catabolism of the eicosanoids: Prostaglandins, thromboxanes, leukotrienes and hydroxy fatty acids, in: *The Enzymes*, Vol. 16, Academic Press, New York, 1983, pp. 543–603.

Parker, C. W., and Aykent, S., 1982, Calcium stimulation of the 5-lipoxygenase from RBL-1 cells, *Biochem. Biophys. Res. Commun.* **109**:1011–1016.

Rouzer, C. A., Matsumoto, T., and Samuelsson, B., 1986, Single protein from human leukocytes possesses 5-lipoxygenase and leukotriene A_4 synthase activities, *Proc. Natl. Acad. Sci. USA* **83**:857–861.

Samuelsson, B., 1983, Leukotrienes: Mediators of immediate hypersensitivity reactions and inflammation, *Science* **220**:568–575.

Samuelsson, B., 1987, An elucidation of the arachidonic acid cascade: Discovery of prostaglandins, thromboxane and leukotrienes, *Drugs* **33**:2–9.

Shimizu, T., Izumi, T., Seyama, Y., Tadokoro, K., Radmark, O., and Samuelsson, B., 1986, Characterization of leukotriene A_4 synthase from murine mast cells: Evidence for its identity to arachidonate 5-lipoxygenase, *Proc. Natl. Acad. Sci. USA* **83**:4175–4179.

Skoog, M. T., Nichols, J. S., and Wiseman, J. S., 1986, 5-Lipoxygenase from rat PMN lysate, *Prostaglandins* **31**:561–576.

Soberman, R. J., Harper, T. W., Betteridge, D., Lewis, R. A., and Austen, K. F., 1985, Characterization and separation of the arachidonic acid 5-lipoxygenase and linoleic acid n-6 lipoxygenase (arachidonic acid 15-lipoxygenase) of human polymorphonuclear leukocytes, *J. Biol. Chem.* **260**:4508–4515.

Tripp, C. S., Mahoney, M., and Needleman, P., 1985, Calcium ionophore enables soluble agonists to stimulate macrophage 5-lipoxygenase, *J. Biol. Chem.* **260**:5895–5898.

Ueda, N., Kaneko, S., Yoshimoto, T., and Yamamoto, S., 1986, Purification of arachidonate 5-lipoxygenase from porcine leukocytes and its reactivity with hydroperoxyeicosatetraenoic acids, *J. Biol. Chem.* **261**:7982–7988.

Vanderhoek, J. Y., Tare, N. S., Bailey, J. M., Goldstein, A. L., and Pluznik, D. H., 1982, New role for 15-hydroxyeicosatetraenoic acid: Activator of leukotriene biosynthesis in PT-18 mast/basophil cells, *J. Biol. Chem.* **257**:12191–12195.

Vanderhoek, J. Y., 1985, Biological effects of hydroxy fatty acids, in *Biochemistry of Arachidonic Acid Metabolism* (W. E. M. Lands, ed.), Martinus Nijhoff, Boston, pp. 213–226.

Vanderhoek, J. Y., and Bailey, J. M., 1985, Postphospholipase activation of lipoxygenase/leukotriene systems, in *Prostaglandins, Leukotrienes, and Lipoxins* (J. M. Bailey, ed.), Plenum Press, New York, pp. 133–146.

Vanderhoek, J. Y., Karmin, M. T., and Ekborg, S. L., 1985, Endogenous hydroxyeicosatetraenoic acids stimulate the human polymorphonuclear leukocyte 15-lipoxygenase pathway, *J. Biol. Chem.* **260:**15482–15487.

Vanderhoek, J. Y., and Pluznik, D. H., 1985, Structural requirements in hydroxyeicosanoids for the activation of the 5-lipoxygenase in PT-18 mast/basophil cells, *Biochim. Biophys. Acta* **837:**119–122.

Walker, J. R., and Parish, H. A., 1981, Metabolic requirements for rabbit polymorphonuclear leucocyte lipoxygenase activity, *Inter. Archs. Allergy Appl. Immun.* **66:**83–90.

Wolf, B. A., Turk, J., Sherman, W. R., and McDaniel, M. L., 1986, Intracellular Ca^{2+} mobilization by arachidonic acid: Comparison with myo-inositol 1,4,5-triphosphate in isolated pancreatic islets, *J. Biol. Chem.* **261:**3501–3511.

43

Phosphorylation and Modulation of Enzymic Activity of 3-Hydroxy-3-Methylglutaryl Coenzyme A Reductase by Multiple Kinases

ZAFARUL H. BEG, JOHN A. STONIK, and
H. BRYAN BREWER, JR.

1. INTRODUCTION

3-Hydroxy-3-methylglutaryl coenzyme A reductase (HMG-CoA reductase, EC 1.1.1.34) is the rate-limiting enzyme which regulates the synthesis of cholesterol and other polyisoprenoid compounds (Rodwell *et al.*, 1976; Goldstein and Brown, 1977; Brown and Goldstein, 1980; Ingebritsen and Gibson, 1980; Beg and Brewer, 1981, 1982; Kennelly and Rodwell, 1985; Beg *et al.*, 1987). In mammalian cells, HMG-CoA reductase is a transmembrane glycoprotein with its active site facing the cytosol and a carbohydrate-containing site oriented toward the luminal surface of the endoplasmic reticulum. The protomer of HMG-CoA reductase is an approximately 100,000-D protein (Chin *et al.*, 1984). Proteolysis (Ness *et al.*, 1981) of the native protein results in a 53,000 molecular weight fragment that contains the active site of the enzyme (Chin *et al.*, 1984).

Modulation of the enzymic activity of HMG-CoA reductase and the formation of mevalonate are of major importance in cellular metabolism since mevalonate serves as a precursor for four separate metabolic pathways including the formation of cholesterol, ubiquinones, dolichols, and isopentenyl tRNAs. Two major modes of control of HMG-CoA reductase activity and mevalonate formation have been elucidated: (1) *Long-term* regulation involves the modulation of HMG-CoA reductase activity by changes in enzyme concentration through transcriptional and posttranscriptional modifications as well as enzyme degradation. (2) *Short-term* regulation involves phosphorylation and dephosphorylation of HMG-CoA reductase. Recently, we identified three separate systems for the regulation of HMG-CoA reductase activity by short-term covalent modification. These three systems are mediated by different

ZAFARUL H. BEG, JOHN A. STONIK, and H. BRYAN BREWER, Jr. • Molecular Disease Branch, National Heart, Lung, and Blood Institute, National Institutes of Health, Bethesda, Maryland 20892.

kinases including reductase kinase, Ca^{2+}-activated and phospholipid-dependent protein kinase (protein kinase C), and a Ca^{2+}, calmodulin-dependent protein kinase. This report summarizes our current understanding of the mechanisms involved in the short-term regulation of HMG-CoA reductase activity and mevalonate formation.

2. BICYCLIC CASCADE SYSTEM OF RAT AND HUMAN HEPATIC HMG-CoA REDUCTASE

2.1. In Vitro Phosphorylation of HMG-CoA Reductase and Reductase Kinase

We and others have reported previously that the enzymic activity of both microsomal and protease-cleaved purified HMG-CoA reductase from rat and human liver (Beg et al., 1984a) can be modulated in vitro by reversible phosphorylation (Ingebritsen and Gibson, 1980; Kennelly and Rodwell, 1985; Beg et al., 1987). The ATP-dependent phosphorylation and inactivation of HMG-CoA reductase are catalyzed by a specific cAMP-independent protein kinase, designated reductase kinase. Dephosphorylation and concomitant reactivation of inactivated HMG-CoA reductase requires incubation with a phosphoprotein phosphatase. Both reductase kinase and the phosphatases are present in microsomes as well as in cytosol. The native HMG-CoA reductase enzyme ($M_r = 100,000$) was also phosphorylated and inactivated by reductase kinase (Beg et al., 1987).

Both rat and human hepatic reductase kinase, like HMG-CoA reductase, exist in active and inactive forms with the phosphorylated form of reductase kinase being catalytically active. Dephosphorylation of active reductase kinase is associated with inactivation of the enzyme (Beg et al., 1979, 1984a; Ingebritsen et al., 1979). The phosphorylation and activation of inactive reductase kinase is catalyzed by a second protein kinase, designated reductase kinase kinase.

Based on the combined in vitro results it was proposed that the enzymic activity of both rat and human hepatic HMG-CoA reductase can be modulated in a bicyclic cascade system involving reversible phosphorylation of HMG-CoA reductase and reductase kinase (Fig. 1, I). In this system the regulation of HMG-CoA reductase consists of an open type of bicyclic cascade. The bicyclic cascade is composed of a reductase kinase cycle and a HMG-CoA reductase cycle (Fig. 1, I). The initial cascade involves the conversion of inactive or less active reductase kinase to the more active phosphorylated form. This conversion is catalyzed by reductase kinase kinase. The active form of reductase kinase catalyzes the phosphorylation of HMG-CoA reductase which is associated with the conversion of an active form to an inactive or a less active form of HMG-CoA reductase. The activation of phosphorylated (inactive) HMG-CoA reductase and inactivation (dephosphorylation) of the phosphorylated (active) form of reductase kinase are catalyzed by HMG-CoA reductase phosphatase and reductase kinase phosphatase, respectively (Fig. 1, I).

2.1.1. In Vivo Phosphorylation of HMG-CoA Reductase

Several studies have now been performed to evaluate the regulation of HMG-CoA reductase in vivo by reversible phosphorylation.

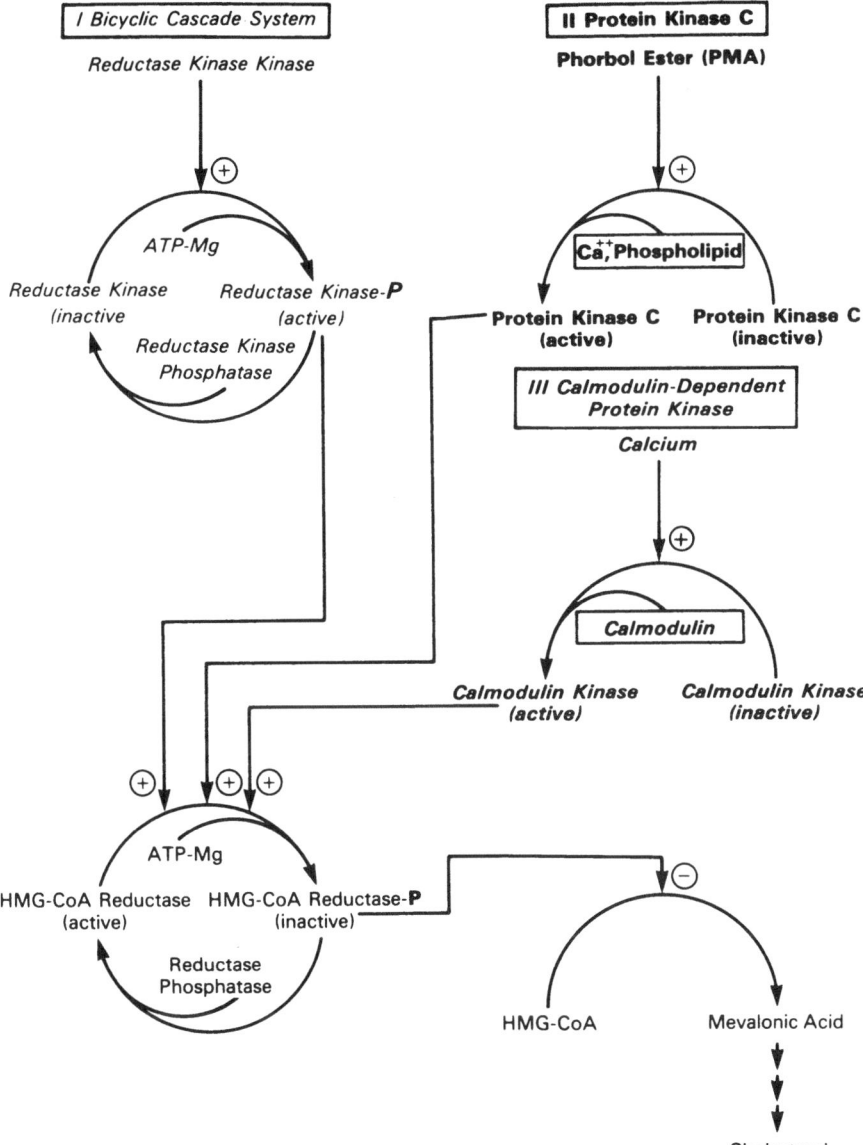

FIGURE 1. Schematic conceptual representation of three kinase systems involved in the modulation of the enzymic activity of HMG-CoA reductase. In this model the enzymic activity of HMG-CoA reductase and mevalonate formation is regulated by three different kinase systems. The first is a bicyclic cascade system composed of HMG-CoA reductase, reductase kinase, reductase kinase kinase, and phosphatases. The second system includes protein kinase C, which is activated by Ca^{2+}, phospholipid, and phorbol-12-myristate-13-acetate (PMA). The third kinase system involved in the phosphorylation of HMG-CoA reductase is dependent on Ca^{2+} and calmodulin for its enzymic activity. The phosphorylated enzymes are designated P.

2.1.1a. Regulation of HMG-CoA Reductase Activity and Cholesterol Synthesis in Cultured Fibroblasts from Normal and Familial Hypercholesterolemic (FH) Subjects. Recently, it was demonstrated that HMG-CoA reductase activity in normal and receptor negative-cultured human fibroblasts is modulated by a short-term mechanism involving reversible phosphorylation of both HMG-CoA reductase and reductase kinase (Beg *et al.*, 1986).

Low-density lipoprotein, 25-hydroxycholesterol, 7-ketocholesterol, and mevalonolactone reduced HMG-CoA reductase activity in cultured human fibroblasts by reversible phosphorylation (Beg *et al.*, 1986). 25-Hydroxycholesterol, which does not require receptor binding, phosphorylated and inhibited HMG-CoA reductase activity in FH cells (Beg *et al.*, 1986). Quantitation of the short-term effects of inhibitors on the rate of cholesterol synthesis from acetate revealed that HMG-CoA reductase phosphorylation was responsible for rapid suppression of sterol synthesis while long-term regulation of synthesis was related to reduction in total enzyme activity (Beg *et al.*, 1986).

2.1.1b. Modulation of HMG-CoA Reductase Enzymic Activity by Glucagon. The short-term administration of glucagon to rats was associated with an increase in hepatic cAMP content and an increase in the degree of phosphorylation with a concomitant decrease in enzymic activity of HMG-CoA reductase (Beg *et al.*, 1980; Beg and Brewer, 1982). In addition to a decrease in the enzymic activity of HMG-CoA reductase, changes were also observed in the catalytic activity of reductase kinase. The inactivation of HMG-CoA reductase by microsomal reductase kinase was twofold greater in glucagon-treated rats than in controls. Phosphoprotein phosphatase treatment of the reductase kinase from both glucagon-treated and control rats was associated with a significant loss of enzymic activity and ability to phosphorylate HMG-CoA reductase (Beg *et al.*, 1980).

The combined results from these studies represented the first *in vivo* correlation of the modulation of the enzymic activity of HMG-CoA reductase with changes in the degree of phosphorylation of the enzyme. Short-term regulation of the enzymic activity of HMG-CoA reductase by polypeptide hormones such as glucagon is therefore achieved by an increase in the degree of phosphorylation of reductase kinase, providing a mechanism for the phosphorylation and inactivation of HMG-CoA reductase. These results are consistent with the findings that both HMG-CoA reductase and reductase kinase activities are regulated in freshly isolated rat hepatocytes by glucagon and insulin (Ingebritsen *et al.*, 1979).

2.1.1c. Modulation of HMG-CoA Reductase Activity by Cholesterol. Cholesterol biosynthesis is known to be regulated by modulation of HMG-CoA reductase activity by negative feedback inhibition (Rodwell *et al.*, 1976; Goldstein and Brown, 1977; Brown and Goldstein, 1980). Feeding of cholesterol to rats was shown to be associated with a rapid decrease in the active form of the HMG-CoA reductase. This decrease was a consequence of increased phosphorylation of HMG-CoA reductase (Arebalo *et al.*, 1981, 1982; Beg and Brewer, 1981, 1982). Dephosphorylation of the inactive enzyme was associated with complete restoration of enzymic activity (Arebalo *et al.*, 1981; Beg and Brewer, 1982). The reduced level of HMG-CoA reductase following long-term (2–2.5 hr) feeding of cholesterol was not due to an increase in phosphorylated (inactive) enzyme, suggesting that the decline in enzyme activity was due to a decrease in enzyme concentration (Arebalo *et al.*, 1981; Beg and Brewer, 1981, 1982).

2.1.1d. Modulation of HMG-CoA Reductase Activity by Mevalonolactone. Mevalonic acid, a key intermediate in the biosynthesis of cholesterol, is synthesized from HMG-CoA by the catalytic action of HMG-CoA reductase. Like cholesterol, mevalonolactone also modulated the activity of HMG-CoA reductase by two mechanisms (Erickson *et al.*, 1980; Arebalo *et al.*, 1980, 1982; Beg and Brewer, 1981, 1982; Beg *et al.*, 1984b). The first mechanism involved short-term regulation by changes in the enzymic activity mediated by covalent phosphorylation. The long-term effect on HMG-CoA reductase activity was an irreversible process and independent of phosphorylation (Arebalo *et al.*, 1981; Beg and Brewer, 1984).

In a series of recent experiments we delineated the mechanism(s) involved in the phosphorylation of HMG-CoA reductase following a short-term (20-min) administration of mevalonolactone to rats (Beg *et al.*, 1984b). The results of this study indicated that the increase in *in vivo* phosphorylation (inactivation) of HMG-CoA reductase after 20 min of mevalonolactone administration was achieved by two separate mechanisms:

1. *Increased activity of reductase kinase kinase.* An increase in the catalytic activity of reductase kinase kinase was associated with an increase in the phosphorylation and activation of reductase kinase, which in turn inhibited the enzymic activity of HMG-CoA reductase by increasing the quantity of inactive HMG-CoA reductase without any change in total enzyme concentration (Beg *et al.*, 1984b). These studies represented the initial demonstration that the catalytic activity of reductase kinase kinase can be significantly induced *in vivo*, thus providing a mechanism for the regulation of the activities of both reductase kinase (increased phosphorylation) and HMG-CoA reductase (increased phosphorylation).

2. *Inhibition of enzymic activity of HMG-CoA reductase phosphatase and reductase kinase phosphatase.* A rapid decline in the activity of the phosphoprotein phosphatase(s) significantly reduced the dephosphorylation of both reductase kinase (inactivation) and HMG-CoA reductase (activation). Thus, the administration of mevalonolactone to rats produced an increase in the steady-state level of the phosphorylated forms of both reductase kinase and HMG-CoA reductase, which resulted in a net decline in HMG-CoA reductase activity and mevalonate formation (Beg *et al.*, 1984b).

3. PHOSPHORYLATION OF HMG-CoA REDUCTASE AND MODULATION OF ITS ENZYMIC ACTIVITY BY PROTEIN KINASE C

In addition to the well-characterized reductase kinase–reductase kinase kinase system for the modulation of the enzymic activity of HMG-CoA reductase, a Ca^{2+}-activated and phospholipid-dependent protein kinase (protein kinase C) was identified which modulates HMG-CoA reductase activity. Protein kinase C purified from rat brain cytosol is able to inactivate and phosphorylate both microsomal native and protease-cleaved soluble, purified HMG-CoA reductase (Fig. 2). Incubation of ^{32}P-phosphorylated and inactivated HMG-CoA reductase with cytosolic purified phosphoprotein phosphatase was associated with a time-dependent reactivation (dephosphorylation) of enzymic activity and loss of ^{32}P radioactivity (Fig. 2). Autoradiography of the phosphorylated purified HMG-CoA reductase following NaDodSO$_4$–gel electrophoresis revealed a single band of radioactivity in the electrophoretic position of the 53,000-dalton form of HMG-CoA reductase, which was lost following incubation and phosphatase (Fig. 2). Autoradiography of the ^{32}P-phosphorylated native mi-

ZAFARUL H. BEG *et al.*

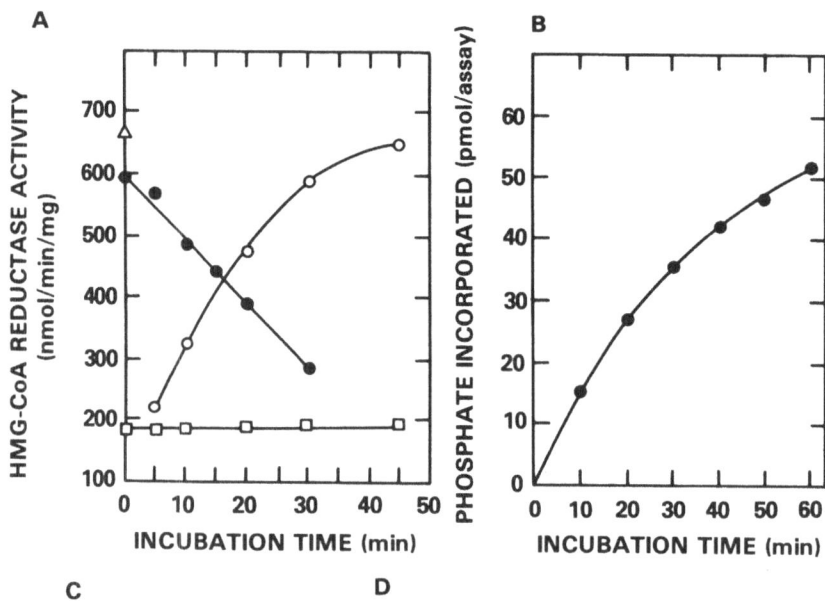

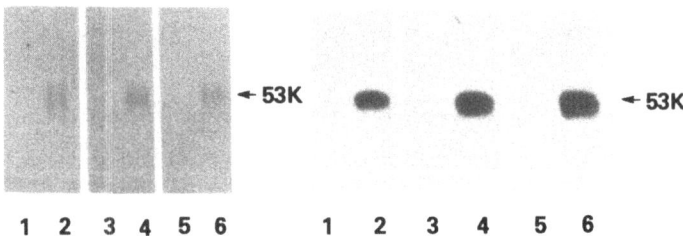

FIGURE 2. Reversible phosphorylation of HMG-CoA reductase by protein kinase C. (A) Protein kinase C-mediated inactivation (●) and reductase phosphatase-mediated reactivation (○) of purified (M_r 53,000) HMG-CoA reductase. △, control + phosphatase; □, inactivated HMG-CoA reductase + inactive phosphatase. (B) Time course (10–60 min) of protein kinase C-mediated phosphorylation of purified HMG-CoA reductase. (C) Coomassie blue-stained gel of aliquots of [^{32}P]HMG-CoA reductase at 20, 40, and 60 min of incubation (samples from panel B). (D) Autoradiogram of the gel (panel C) showing the position of phosphorylated HMG-CoA reductase (see arrow). Lanes 1, 3, and 5 in panels C and D are control samples. (E) Autoradiogram of

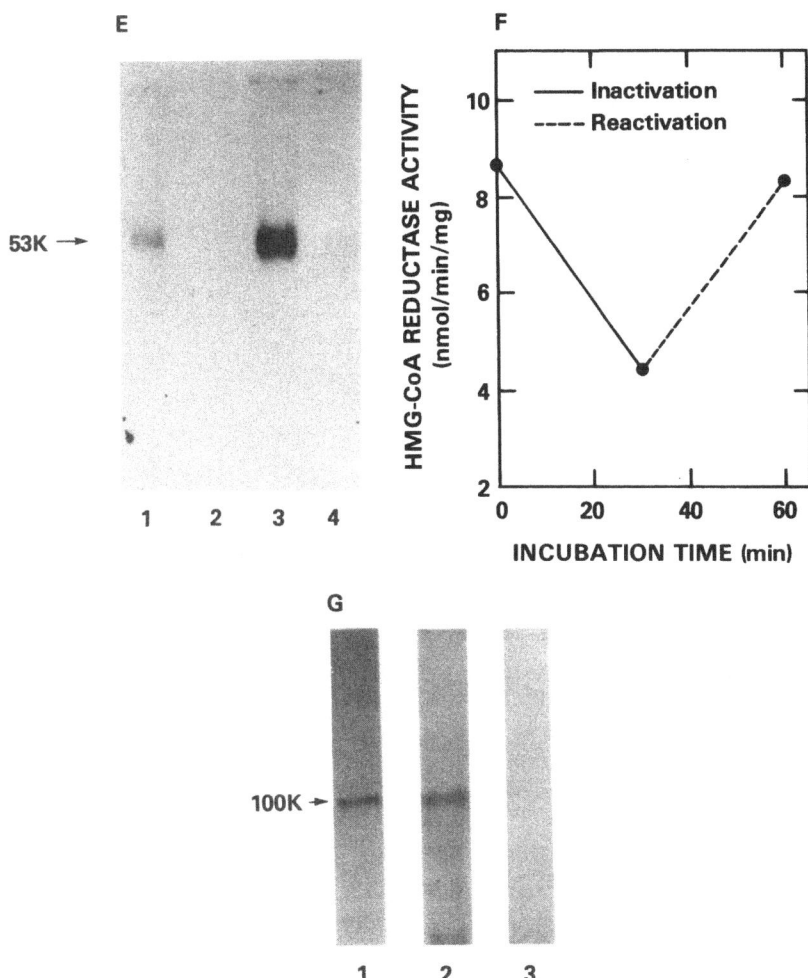

NaDodSO$_4$–gel electrophoretic analysis of HMG-CoA reductase phosphorylation by protein kinase C (lanes 1 and 3) and subsequent dephosphorylation with phosphatase (lanes 2 and 4; 7.5-min incubation). (F) Inactivation (—) and reactivation (---) of the native form (M_r = 100,000) of HMG-CoA reductase. (G) Protein kinase C-mediated phosphorylation of the native form of HMG-CoA reductase. Lane 1 represents the immunodetection of native HMG-CoA reductase (M_r = 100,000). The immunoblot was used for autoradiogram (lane 2). The position of the native subunit of HMG-CoA reductase is indicated by an arrow. The autoradiogram of the immunoblot from a control sample is shown in lane 3.

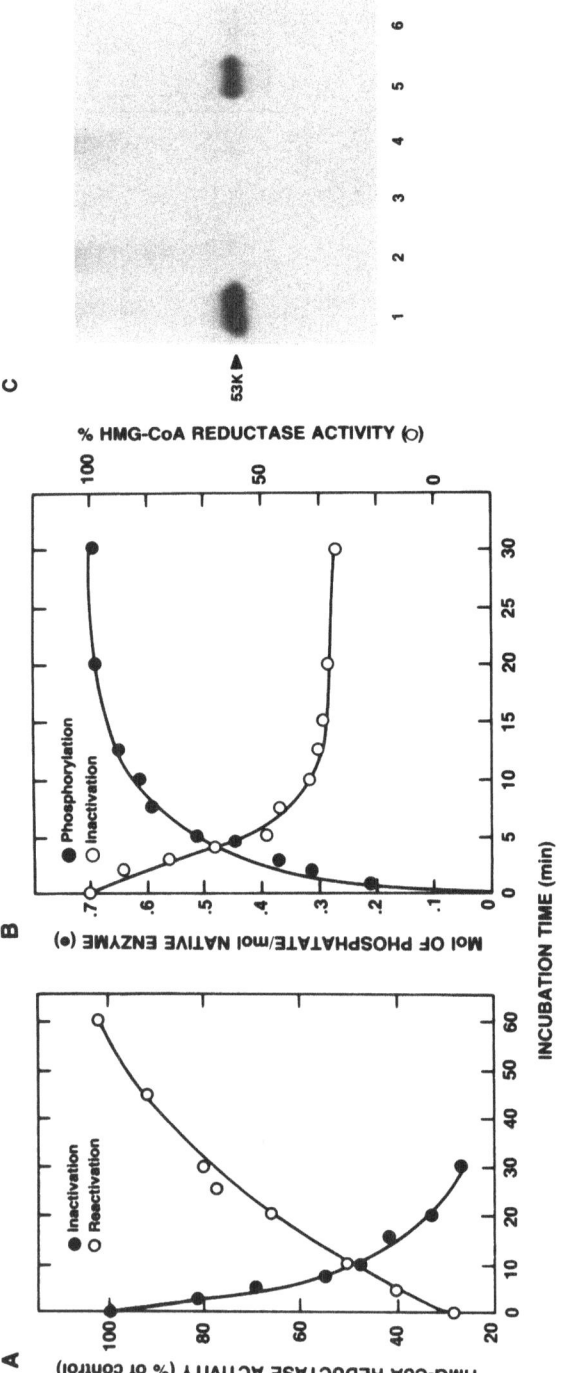

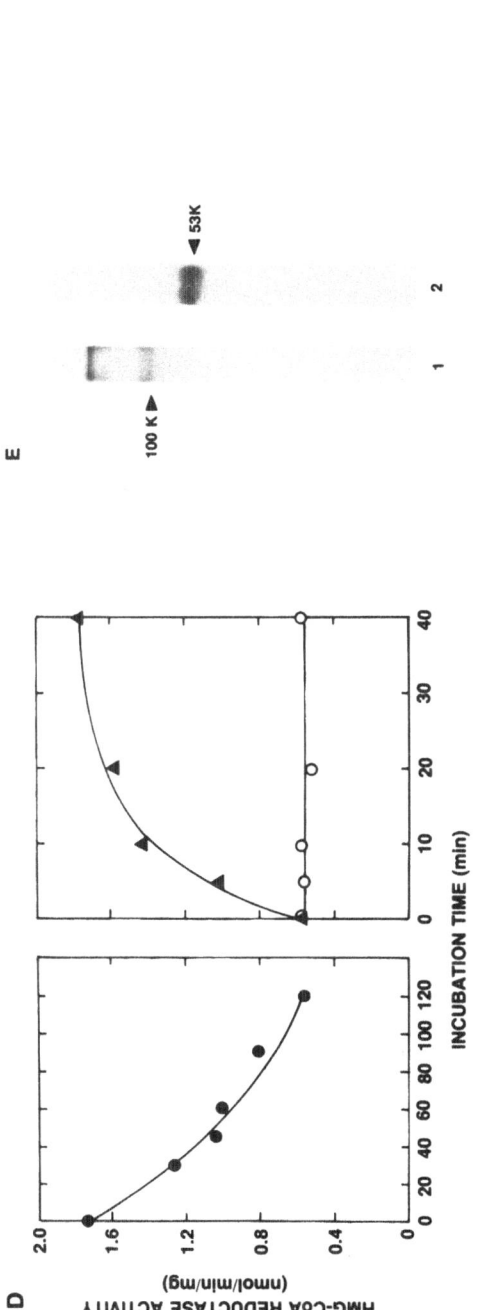

FIGURE 3. Analysis of Ca^{2+}, calmodulin-dependent kinase-mediated phosphorylation of HMG-CoA reductase. (A) Ca^{2+}, calmodulin-dependent protein kinase-mediated inactivation (●) and reductase phosphatase-mediated reactivation (○) of purified HMG-CoA reductase (M_r = 53,000). (B) Rate of phosphorylation (●) and concomitant inactivation (○) of purified HMG-CoA reductase. (C) Autoradiogram of phosphorylated and dephosphorylated HMG-CoA reductase after NaDodSO$_4$–gel electrophoresis. Lane 1 represents the position of [^{32}P]HMG-CoA reductase (M_r = 53,000) phosphorylated in the presence of Ca^{2+}, calmodulin-dependent protein kinase (complete system). Lanes 2, 3, and 4 had EGTA + calmodulin, Ca^{2+} alone, and EGTA (− Ca^{2+}, − calmodulin), respectively. Lane 6 represents the dephosphorylated [^{32}P]HMG-CoA reductase after incubation with phosphatase (45 min) whereas lane 5 contains phosphorylated HMG-CoA reductase + fluoride-inactivated phosphatase. (D) Calmodulin-dependent kinase-mediated inactivation (●) and phosphatase-mediated reactivation (▲) of native lubrol-solubilized microsomal HMG-CoA reductase. Inactivated HMG-CoA reductase containing inactive phosphatase is also shown (○). (E) Autoradiogram of native (lane 1) and protease-cleaved soluble purified (lane 2) HMG-CoA reductase following phosphorylation by Ca^{2+}, calmodulin-dependent kinase, immunoprecipitation, and NaDodSO$_4$-urea–gel electrophoresis.

crosomal lubrol-solubilized HMG-CoA reductase following NaDodSO$_4$–gel electrophoresis and immunoblotting revealed a radiolabeled band which corresponded to the 100,000-dalton molecular form of the enzyme (Fig. 2).

It was reported previously that the tumor-promoting phorbol esters could replace diacylglycerol in modulating the activity of protein kinase C (Castagna *et al.*, 1982; Niedel *et al.*, 1983). Phorbol-12-myristate-13-acetate (PMA) also stimulated the protein kinase C-catalyzed phosphorylation of HMG-CoA reductase (Beg *et al.*, 1985), suggesting a possible *in vivo* protein kinase C-mediated mechanism for the regulation of HMG-CoA reductase activity (Fig. 1, II). Although protein kinase C is known to be present in soluble and particulate fractions of rat liver (Niedel *et al.*, 1983), definitive conclusions regarding the physiological significance of the protein kinase C-mediated phosphorylation of hepatic HMG-CoA reductase remains to be investigated (Fig. 1, II).

4. CALCIUM, CALMODULIN-DEPENDENT PROTEIN KINASE-MEDIATED PHOSPHORYLATION OF HMG-CoA REDUCTASE

A second recently reported mechanism for the modulation of HMG-CoA reductase activity by reversible phosphorylation involves a third kinase completely dependent on calmodulin and Ca^{2+} (Fig. 1, III). A low molecular weight Ca^{2+} calmodulin- dependent protein kinase ($M_r = 110,000$) which phosphorylates HMG-CoA reductase, histone H$_1$, and synapsin I as major substrates has been purified to homogeneity from rat brain cytosol. The other previously characterized larger M_r 600,000 calmodulin-dependent kinase from brain and liver revealed a significantly different substrate specificity when compared to M_r 110,000 kinase. The initial characterization and properties of the low molecular weight ($M_r = 110,000$) calmodulin-dependent kinase have been recently described (Beg *et al.*, 1987). The relationship of this low molecular weight ($M_r = 110,000$) calmodulin-dependent kinase to other kinases purified from rat forebrain and cerebellum is as yet unknown. Nevertheless, the isolation of a M_r 110,000 calmodulin-binding enzyme complex provides additional insights into the mechanisms of the calcium-dependent regulation of phosphorylation of protein substrates.

The *in vitro* phosphorylation and concomitant inactivation of purified HMG-CoA reductase mediated by the purified calmodulin-dependent kinase was completely dependent on Ca^{2+} and calmodulin (Fig. 3). No phosphorylation of purified HMG-CoA reductase was observed in the absence of calmodulin and/or Ca^{2+}. Incubation of phosphorylated and inactivated HMG-CoA reductase with rat liver cytosolic purified phosphoprotein phosphatase was associated with the reactivation of inactive HMG-CoA reductase. The calmodulin-dependent protein kinase that catalyzed inactivation of purified HMG-CoA reductase was directly correlated with the incorporation of [^{32}P]phosphate. A 72% inactivation of enzyme activity followed incorporation of 0.35 mol of phosphate per mol of the M_r 53,000 fragment of HMG-CoA reductase (Fig. 3B). Maximal phosphorylation of purified HMG-CoA reductase by Ca^{2+}, calmodulin-stimulated HMG-CoA reductase kinase revealed stoichiometry of approximately 0.5 mol of phosphate/mol of the M_r 53,000 enzymic fragment with near-complete inactivation of enzyme activity. Autoradiography of purified [^{32}P]HMG-CoA reductase following NaDodSO$_4$–gel electrophoresis revealed a radiolabeled electrophoretic band corresponding to the 53,000 molecular weight form of the cleaved enzyme (Fig. 3C). Dephosphorylation of ^{32}P-phosphorylated HMG-CoA reductase with hepatic phosphoprotein phosphatase was associated with the loss of radioactivity and disappearance of the ^{32}P-radiolabeled HMG-CoA reductase band in the autoradiogram. Microsomal native HMG-CoA reductase (M-100,000) was also phosphorylated and inactivated in a time-dependent manner following incubation with Ca^{2+}, calmodulin, and ATP-Mg; dephosphorylation was catalyzed by the purified phosphoprotein phosphatase enzyme, resulting in a time-dependent restoration of HMG-CoA reductase activity. Phosphorylation of the native microsomal form of HMG-CoA reductase was confirmed by immunoprecipitation of the ^{32}P-radiolabeled native enzyme with HMG-CoA reductase antiserum and analysis by $NaDodSO_4$-urea–polyacrylamide gel electrophoresis (Fig. 3). The *in vivo* physiological role of Ca^{2+}, calmodulin-dependent kinase in the regulation of HMG-CoA reductase activity via reversible phosphorylation remains to be established (Fig. 1, III).

5. CONCLUSION

The enzymic activity of HMG-CoA reductase in rat, human, and cultured fibroblasts of human skin has been established as modulated by a bicyclic cascade system involving phosphorylation and dephosphorylation of both HMG-CoA reductase and reductase kinase. These systems are associated with significant signal amplification since both positive and negative effectors such as glucagon and mevalonate can produce maximal responses (by affecting either kinases and/or phosphatases in each cycle) with minimal changes in their concentration (Fig. 1, I).

Recently, protein kinase C was shown to be of major importance in mediating the phosphorylation of several different proteins. Under physiological conditions protein kinase C may be activated by diacylglycerol, which is transiently formed in membranes from phosphatidylinositol when cells are stimulated by a variety of ligands including cholinergic and α-adrenergic stimulators, peptide hormones, and growth factors which are known to alter phosphatidylinositol turnover. Further importance of protein kinase C in cellular metabolism is suggested by the modulation of the *in vitro* phosphorylation of HMG-CoA reductase and the activation of protein kinase C by PMA (Fig. 1, II).

In addition to cAMP, recent evidence indicates that Ca^{2+} may serve as a second messenger for phosphorylation of endogeneous proteins (substrates) in cellular metabolism. Ca^{2+}-dependent phosphorylation of proteins has primarily involved the multifunctional Ca^{2+} binding protein, calmodulin, which confers a Ca^{2+} sensitivity on specific protein kinases by serving as either an integral subunit or an obligatory cofactor for the enzymes. A similar mechanism may be involved in the phosphorylation of HMG-CoA reductase by the Ca^{2+}-calmodulin-dependent protein kinase (Fig. 1, III).

The conceptual scheme outlined in Fig. 1 summarizes the current mechanisms known to modulate the reversible phosphorylation of HMG-CoA reductase. Knowledge of the molecular mechanisms involved in the regulation of HMG-CoA reductase may provide new insight into the coordinate control of cholesterol biosynthesis and may ultimately lead to the development of new modulators of HMG-CoA reductase for use in patients with hypercholesterolemia and premature cardiovascular disease (Beg *et al.*, 1987).

6. SUMMARY

This report summarizes the current findings and concepts regarding the short-term modulation of HMG-CoA reductase activity in rat and human liver. Three separate systems for the short-term control of hepatic HMG-CoA reductase involving covalent phosphorylation have been described. These systems involve three separate specific kinases including reductase kinase, protein kinase C, and a Ca^{2+}, calmodulin-dependent kinase. The conceptual schemes outlined in this report and the elucidation of multiple kinase systems for the modulation of HMG-CoA reductase activity may provide new insights into the molecular mechanisms involved in the complex and multifaceted regulation of this key enzyme in the biosynthetic pathway of cholesterol.

REFERENCES

Arebalo, R. E., Hardgrave, J. E., and Scallen, T. J., 1981, The in vivo regulation of rat liver 3-hydroxy-3-methylglutaryl coenzyme A reductase: phosphorylation of the enzyme as an early regulatory response following cholesterol feeding. *J. Biol. Chem.* **256**:571–574.

Arebalo, R. E., Hardgrave, J. E., Roland, B. J., and Scallen, T. J., 1980, In vivo regulation of rat liver 3-hydroxy-3-methylglutaryl-coenzyme A reductase: Enzyme phosphorylation as an early regulatory response after intragastric administration of mevalonolactone, *Proc. Natl. Acad. Sci. USA* **77**:6429–6433.

Arebalo, R. E., Tormanen, C. C., Hardgrave, J. E., Nolan, B. J., and Scallen, T. J., 1982, In vivo regulation of rat liver 3-hydroxy-3-methylglutaryl-coenzyme A reductase: Immunotitration of the enzyme after short-term mevalonate or cholesterol feeding, *Proc. Natl. Acad. Sci. USA* **79**:51–55.

Beg, Z. H., and Brewer, H. B., Jr., 1981, Regulation of liver 3-hydroxy-3-methylglutaryl-CoA reductase, in *Current Topics in Cellular Regulation*, Vol. 20 (B. L. Horecker and E. R. Stadtman, eds.), Academic Press, New York, pp. 139–184.

Beg, Z. H., and Brewer, H. B., Jr., 1982, Modulation of rat liver 3-hydroxy-3-methylglutaryl-CoA reductase activity by reversible phosphorylation, *Fed. Proc.* **41**:2634–2638.

Beg, Z. H., Stonik, J. A., and Brewer, H. B., Jr., 1985, Phosphorylation of hepatic 3-hydroxy-3-methylglutaryl coenzyme A reductase and modulation of its enzymic activity by calcium-activated and phospholipid-dependent protein kinase, *J. Biol. Chem.* **260**:1682–1687.

Beg, Z. H., Stonik, J. A., and Brewer, H. B., Jr., 1984a, Human hepatic 3-hydroxy-3-methylglutaryl coenzyme A reductase: Evidence for the regulation of enzymic activity by a bicyclic phosphorylation cascade, *Biochem. Biophys. Res. Commun.* **119**:488–498.

Beg, Z. H., Stonik, J. A., and Brewer, H. B., Jr., 1979, Characterization and regulation of reductase kinase, a protein kinase that modulates the enzymic activity of 3-hydroxy-3-methylglutaryl coenzyme A reductase, *Proc. Natl. Acad. Sci. USA* **76**:4375–4379.

Beg, Z. H., Reznikov, D. C., and Avigan, J., 1986, Regulation of 3-hydroxy-3-methylglutaryl coenzyme A reductase activity in human fibroblasts by reversible phosphorylation: Modulation of enzymic activity by low density lipoprotein, sterols and mevalonolactone, *Arch. Biochem. Biophys.* **244**:310–322.

Beg, Z. H., Stonik, J. A., and Brewer, H. B., Jr., 1980, In vitro and in vivo phosphorylation of rat liver 3-hydroxy-3-methylglutaryl coenzyme A reductase and its modulation by glucagon, *J. Biol. Chem.* **255**:8541–8545.

Beg, Z. H., Stonik, J. A., and Brewer, H. B., Jr., 1984b, In vivo modulation of rat liver 3-hydroxy-3-methylglutaryl-coenzyme A reductase, reductase kinase, and reductase kinase kinase by mevalonolactone, *Proc. Natl. Acad. Sci. USA* **81**:7293–7297.

Beg, Z. H., Stonik, J. A., and Brewer, H. B., Jr., 1987, Modulation of the enzymic activity of 3-hydroxy-3-methylglutaryl coenzyme A reductase by multiple kinase systems involving reversible phosphorylation: A review, *Metabolism* **36**:900–917.

Brown, M. S., and Goldstein, J. L., 1980, Multivalent feedback regulation of HMG-CoA reductase, a control mechanism coordinating isoprenoid synthesis and cell growth, *J. Lipid. Res.* **21**:505–517.

Castagna, M., Takai, Y., Kaibuchi, K., Sano, K., Kikkawa, U., and Nishizuka, Y., 1982, Direct activation of calcium-activated, phospholipid-dependent protein kinase by tumor-promoting phorbol esters, *J. Biol. Chem.* **257**:7847–7851.

Chin, D. J., Gil, G., Russell, D. W., Liscum, L., Luskey, K. L., Basu, S. K., Okayama, H., Berg, P., Goldstein, J. L., and Brown, M. S., 1984, Nucleotide sequence of 3-hydroxy-3-methylglutaryl coenzyme A reductase, a glycoprotein of endoplasmic reticulum, *Nature* **308**:613–617.

Erickson, S. K., Shrewsbury, A., Gould, R. G., Cooper, A. D., 1980, Studies on the mechanisms of the rapid modulation of 3-hydroxy-3-methylglutaryl coenzyme A reductase in intact liver by mevalonolactone and 25-hydroxycholesterol, *Biochim. Biophys. Acta.* **620**:70–79.

Goldstein, J. L., and Brown, M. S., 1977, The low density lipoprotein pathway and its relation to atherosclerosis, *Ann. Rev. Biochem.* **46**:897–930.

Ingebritsen, T. S., Geelen, M. J. H., Parker, R. A., Evenson, K. J., Gibson, D. M., 1979, Modulation of hydroxymethylglutaryl-CoA reductase activity, reductase kinase activity, and cholesterol synthesis in rat hepatocytes in response to insulin and glucagon, *J. Biol. Chem.* **254**:9986–9989.

Ingebritsen, T. S., and Gibson, D. M., 1980, Reversible phosphorylation of hydroxymethylglutaryl CoA reductase, in *Recently Discovered Systems of Enzyme Regulation by Reversible Phosphorylation*, Vol. 1. (P. Cohen, ed.), Elsevier, Amsterdam, pp. 63–93.

Kennelly, P. J., and Rodwell, V. W., 1985, Regulation of 3-hydroxy-3-methylglutaryl coenzyme A reductase by reversible phosphorylation-dephosphorylation, *J. Lipid Res.* **26**:903–914.

Ness, G. C., Way, S. C., and Wickham, P. S., 1981, Proteinase involvement in the solubilization of 3-hydroxy-3-methylglutaryl coenzyme A reductase, *Biochem. Biophys. Res. Commun.* **102**:81–85.

Niedel, J. E., Kuhn, L. J., and Vanenbark, G. R., 1983, Phorbol diester receptor copurifies with protein kinase C, *Proc. Natl. Acad. Sci. USA* **80**:36–40.

Rodwell, V. W., Nordstrom, J. L., and Mitschelen, J. J., 1976, Regulation of HMG-CoA reductase, *Adv. Lipid Res.* **14**:1–74.

44

Origin and Role of Calcium in Platelet Activation–Contraction–Secretion Coupling

GUNDU H. R. RAO, JONATHAN M. GERRARD,
ISAAC COHEN, CARL J. WITKOP, Jr.,
and JAMES G. WHITE

1. INTRODUCTION

Agonist interaction with platelet surface receptors initiates a complex but concerted series of events inducing alterations in morphology, biochemistry, and physiology (White, 1968; Marcus, 1978; Holmsen and Weiss, 1979; Zucker and Nachmias, 1985). Intracellular movements of calcium ions accompany and may play a critical role in regulating these biochemical and physiological processes (Statland *et al.*, 1969; Feinman and Detwiler, 1974; Hathaway and Adelstein, 1979; Feinstein, 1980; Gerrard *et al.*, 1981; Fox *et al.*, 1983; Menahsi *et al.*, 1984). For lack of precise methods to monitor intracellular free calcium, earlier studies relied heavily on indirect methods to demonstrate agonist-induced calcium mobilization. However, synthesis of novel calcium indicators has made it possible to demonstrate agonist-induced calcium mobilization concurrent with cell activation (Rink *et al.*, 1982; Grynkiewicz *et al.*, 1985; Rao *et al.*, 1985a,b).

Although it is presumed that cytosolic calcium mobilization is the final common pathway for cell activation, the sources and exact molecular mechanisms involved in this process are not clearly understood (Massini *et al.*, 1978; Detwiler *et al.*, 1978; Owen and LeBreton, 1981; Brass and Shattil, 1982; Hallam *et al.*, 1984; Thompson and Scrutton, 1985). Chief sources of calcium in the platelet environment include the surrounding plasma, mitochondria, membranes covering the cell surface, lining channels of the open canalicular system and the

GUNDU H. R. RAO and JAMES G. WHITE • Department of Laboratory Medicine and Pathology, University of Minnesota, Minneapolis, Minnesota 55455.   JONATHAN M. GERRARD • Department of Pediatrics, University of Manitoba, Winnepeg, Canada. R3E OW1   ISAAC COHEN • Artherosclerosis Research Laboratory, Northwestern University, Chicago, Illinois 60611.   CARL J. WITKOP, Jr. • Department of Human and Oral Genetics, University of Minnesota School of Dentistry, Minneapolis, Minnesota 55455.

dense tubular system (Charo *et al.*, 1976; Zucker and Grant, 1978; White, 1981). Several previous studies evaluated the role of calcium from these various sources, but the results were not conclusive (Murer, 1985).

In an earlier report (Rao *et al.*, 1986), we demonstrated the need for free cytosolic calcium in the platelet release reaction and irreversible aggregation using a specific intracellular calcium chelator, 2-methyl-6-methoxy-8-nitroquinoline tetraacetoxymethyl ester (Quin-2 AM). In the present study, we employed three different calcium chelators to probe the role of extracellular, membrane-associated, and cytosolic calcium in the platelet release reaction, irreversible aggregation, and clot retraction. Results of our study suggest that chelation of calcium from any one of these sources can prevent single agonist-induced platelet secretion and irreversible aggregation.

2. MATERIALS AND METHODS

2.1. Reagents

Quin-2, tetraacetoxymethyl ester (Quin-2 AM), and Quin-2 free acid (Quin-2 FA) were purchased from Lancaster Synthesis Ltd., White Lund (Lancastershire, England). Fura 2 AM and its free acid were from Molecular Probes, Inc. (Junction City, OR). Arachidonic acid as the sodium salt was from NuChek Prep (Elysian, MN); injectable adrenaline and bovine thrombin were from the Parke-Davis Company (Detroit, MI). Calcium ionophore A23187 and ionomycin were purchased from Calbiochem/Behring (Los Angeles, CA). Unless otherwise mentioned, all other chemicals were obtained from Sigma Chemical Company (St. Louis, MO)

2.2. Cell Preparation and Measurement of Platelet Function

Blood for these studies was obtained from adult volunteers who had not taken any medication for at least 2 weeks prior to venesection. Blood was also provided by patients wiht Hermansky–Pudlak syndrome (White and Witkop, 1972) and gray platelet syndrome (Gerrard *et al.*, 1982) for special studies. It was mixed immediately with citrate–citric acid–dextrose (CCD) anticoagulant (citrate 0.1 M, citric acid 7 mM, dextrose 0.14 M, pH 6.5) in a ratio of 9 parts blood to 1 part anticoagulant (Rao *et al.*, 1985a,b). Platelet-rich plasma (PRP) was obtained by centrifugation of whole blood at room temperature for 20 min at 100g. In the majority of these studies, the platelet count was adjusted to approximately 300,000/mm^3 with platelet-poor plasma (PPP). Platelet aggregation studies were carried out on a Payton dual-channel aggregometer interfaced with an Apple IIe microcomputer. Release of adenosine triphosphate was monitored using a Chronolog lumiaggregometer (Rao *et al.*, 1985a,b).

2.3. Preparations of Platelet-Rich Plasma Clots and Measurement of Retraction

Cylindrical clots of PRP and PPP were made in siliconized glass tubes (10 × 0.5 cm I.D.). After sealing one end with parafilm, 1 ml of PRP or PPP was transferred into the tube with a long-tipped Pasteur pipette (Cohen *et al.*, 1982). Fifty microliters of thrombin (20 U/ml) in saline was added to the PRP or PPP (final concentration of thrombin: 1 U/ml), gently mixed, and allowed to form clots for 9 min at room temperature. After the incubation period, clots were poured into a plastic dish containing ice-cold Hank's balanced salt solution (HBSS, pH 7.4) by gently tapping the tube onto the bottom of petri dishes so that the clot slid into the ice-cold buffer.

For measurement of tension due to retraction, both ends of the clot were tied with cotton thread and attached to a wire holder and force displacement transducer (Universal Transducer, Model SC1001, Gould Instrument Company, Minneapolis, MN). The clot was immersed in a tissue bath containing modified Tyrode's solution maintained at 37°C. The tension of clot retraction was monitored and quantitated according to the method described in an earlier report (Cohen *et al.*, 1982).

2.4. Measurement of Cytosolic Calcium

For measurement of intracellular calcium both Quin-2 and Fura-2 were employed. Quin-2 was loaded into platelets in plasma (10 μM, 10 min, 37°C), washed, and reconstituted in Hepes buffer containing 145 mM NaCl, 5 mM KCl, 1 mM MgSO$_4$, 1 mM Na$_2$HPO$_4$, 1 mM CaCl$_2$, 10 mM Hepes, and 5 mM glucose (pH 7.4). Fura-2 was loaded into washed platelets for 10 min at a concentration of 1 μM and the cells were washed again to remove excess dye. Fluorescence measurements were performed with a Fluorolog scanning fluorometer as described in earlier reports (Rink *et al.*, 1982; Grynkiewicz *et al.*, 1985).

2.5. Effect of Inhibitors on Cell Function

To evaluate the role of extracellular calcium and cytosolic calcium, Quin-2 free acid was employed. Quin-2 acetoxymethyl ester permeates the cell membrane and is hydrolyzed in the cytosol to free acid by the action of esterases. The free acid has a high affinity for free calcium (K_d = 115 nM). Since Quin-2-treated platelets accumulate the free acid in their cytosol, Quin-2 AM in various concentrations was employed to evaluate the role of cytosolic calcium in platelet function. Quin-2 free acid, which will not permeate cell membranes, was used to buffer extracellular calcium. For comparison, equimolar ethylenediaminetetraacetic acid (EDTA) (pH 7.4) was used in some studies. To complex membrane-associated calcium, the antibiotic chlortetracycline was utilized (Schneider *et al.*, 1983). Inhibitors were added to platelets in plasma and incubated for 30 min at room temperature before challenging with agonists. The same protocol was employed for evaluating the effect of inhibitors on platelet secretion, irreversible aggregation, and thrombin-induced clot formation and retraction. To evaluate mitochondria as a source of calcium for the cytosolic rise in divalent cation, platelets were incubated with two different mitochondrial poisons (Takeo and Sakanashi, 1985; McGill, 1980), sodium azide (0.5 mM) and ethidium bromide (100 μM). All studies were repeated several times, and the data presented represent the average findings in at least three different experiments.

3. RESULTS

3.1. Role of Calcium Sequestered in Platelet Organelles

Platelet dense bodies contain a substantial store of releasable calcium. To evaluate the role of releasable calcium from dense granules or α granules on agonist-induced elevation of intracellular calcium levels, platelets from patients with Hermansky–Pudlak syndrome (HPS) or gray platelet syndrome (GPS) were utilized and compared with platelets obtained from normal donors. Earlier studies from our laboratory demonstrated that the platelets of the patient with HPS contain only rare dense bodies (White and Witkop, 1972; White, 1983; White et al., 1971). Platelets from the patient with GPS contained normal numbers of dense bodies but lacked α granules (Gerrard et al., 1980). In the present study, HPS platelets stirred with arachidonate aggregated irreversibly but did not release detectable ATP, confirming the absence of a releasable pool of adenine nucleotides (Fig. 1). Normal control and GPS platelets, upon stimulation with arachidonate, aggregated and released significant quantities of ATP (Fig. 1).

Platelets from normal controls as well as those from HPS and GPS patients were loaded with Quin-2 or Fura-2, and the thrombin-induced increase of cytosolic calcium followed. As shown in Fig. 2, thrombin caused significant elevation of cytoplasmic calcium levels (>800 nM) in HPS platelets, suggesting noninvolvement of dense body calcium in the agonist-induced calcium flux. Similarly, GPS platelets stirred with thrombin also showed significant elevation in cytosolic calcium, indicating that α granules are not the source for increased cytosolic calcium in activated platelets (Fig. 3).

3.2. Mitochondria as a Source of Ionized Calcium

Mitochondria also sequester calcium and therefore could contribute to the elevation of cytosolic calcium. Platelets exposed to two different mitochondrial toxicants (Takeo and Sakanashi, 1985; McGill, 1980) were evaluated for their response to the action of thrombin (Fig. 4). Exposure of platelets to sodium azide (0.5 mM) or ethidium bromide (100 μM) did not block thrombin-induced elevation of cytosolic calcium, thereby suggesting that the contribution of calcium from mitochondria is minimal.

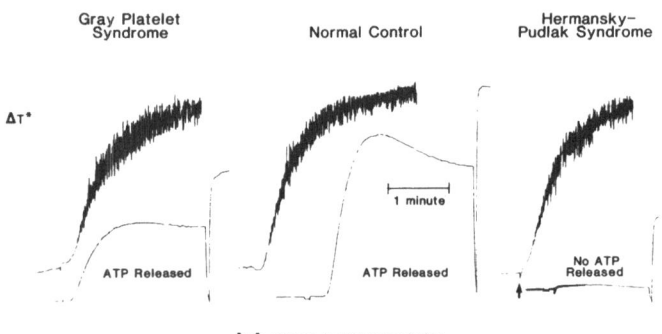

FIGURE 1. Platelets from patients as well as from normal donors aggregated irreversibly when stirred with arachidonate (AA, 0.45 mM). However, platelets from patients with Hermansky–Pudlak syndrome did not release detectable ATP, whereas platelets from normal control donors as well as patients with gray platelet syndrome released significant quantities of ATP.

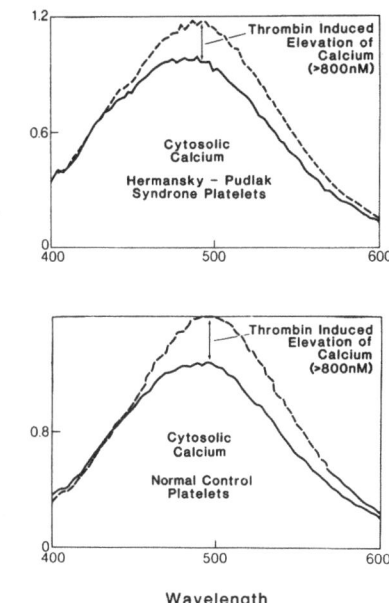

FIGURE 2. Platelets from normal controls as well as those from Hermansky–Pudlak syndrome patients loaded with Quin-2 (10 μM) were exposed to thrombin (0.5 U/ml) and alterations in the level of cytosolic calcium followed. Thrombin caused significant elevation of cytoplasmic calcium in HPS platelets, suggesting noninvolvement of dense body calcium in the agonist-induced calcium flux.

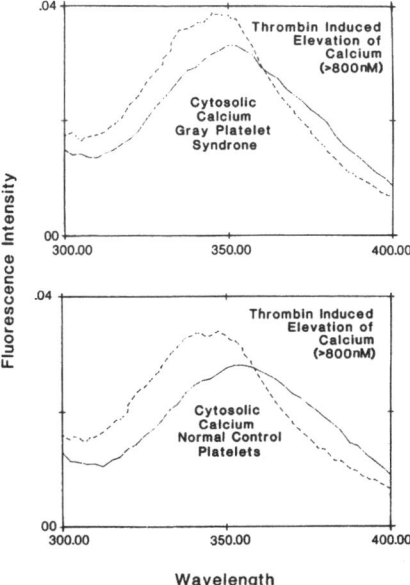

FIGURE 3. Platelets from normal controls and from gray platelet syndrome patients loaded with Fura-2 (1 μM) were exposed to thrombin (0.5 U/ml) and alterations in the level of cytosolic calcium followed. Thrombin caused significant elevation of cytosolic calcium in GPS platelets, suggesting noninvolvement of α-granule calcium in the agonist-induced calcium.

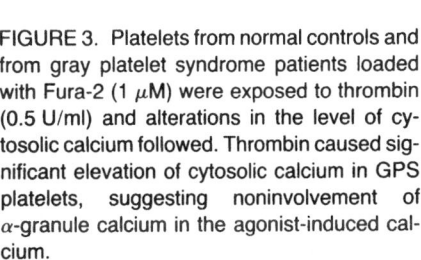

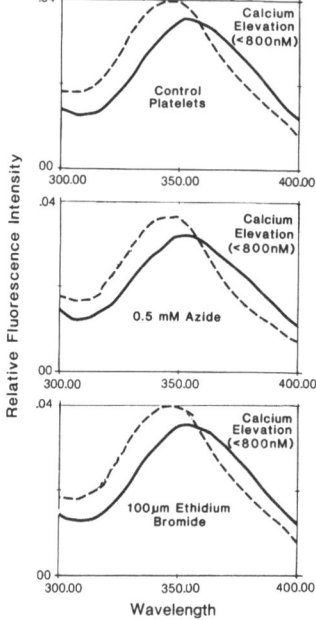

FIGURE 4. Fura-2-loaded normal control platelets were incubated with the mitochondrial toxicants, sodium azide, and ethidium bromide, and thrombin-induced alterations in cytosolic calcium followed. Thrombin (0.5 U/ml) induced significant elevation of cytosolic calcium in drug-treated platelets, suggesting that mitochondria may not contribute significantly to the agonist-induced elevation of cytosolic calcium.

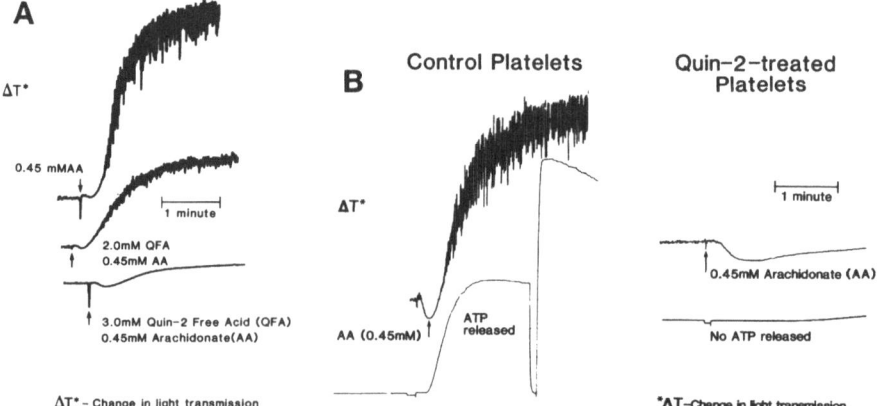

FIGURE 5. (A) Quin-2 ester (QAM), (B) its free acid (QFA), and (C) chlortetracycline (CTC) were used as intra- and extracellular and membrane-associated calcium chelators. All three drugs inhibited arachidonate-induced aggregation and secretion.

3.3. Role of Extracellular, Membrane-Associated, and Cytosolic Calcium in the Platelet Release Reaction and Irreversible Aggregation

To evaluate the role of extracellular and cytosolic calcium, Quin-2 free acid (Quin-2 FA) and Quin-2 tetraacetoxymethyl ester (Quin-2 AM) were used. For the chelation of membrane-associated calcium, chlortetracycline was employed. All three drugs inhibited arachidonic acid-induced aggregation as well as secretion of granule contents (Fig. 5a–c). They blocked secondary aggregation induced by agonists such as epinephrine (5 μM) and adenosine diphosphate (3 μM). At the effective inhibitory concentration of these drugs, platelets treated with epinephrine (5 μM) regained their sensitivity to the action of arachidonate (0.45 mM), aggregated irreversibly, and released significant quantities of ATP (Fig. 6a–c).

To see if chelation of calcium from two out of three critical sources would affect function, platelets were exposed to a combination of Quin-2 AM (40 μM) + Quin-2 FA (3 mM), or Quin-2 AM (40 μM) + chlortetracycline (1 mM), or Quin-2 FA (3 mM) + chlortetracycline (1 mM). All combinations employed effectively blocked the release reaction and aggregation induced by several agonists, including epinephrine, which failed to restore sensitivity to the action of other agonists when a combination of chelators was employed.

3.4. Role of Calcium in Thrombin-Induced Clot Formation and Retraction

Quin-2-loaded (40 μM) platelets in plasma exposed to 1.0 U/ml of thrombin formed normal clots. Prolonged exposure of platelets to Quin-2 resulting in the accumulation of large quantities of Quin-2 free acid in the cytosol inhibited tension development dramatically (Fig. 7). Normal PRP clots developed a tension of 140 mg/cm^2/min, whereas platelets exposed to 80 μM Quin-2 AM for 30 min developed a tension of only 41 mg/cm^2/min (>70% inhibition). As seen in Fig. 8, Quin-2 caused a dose dependent inhibition in clot

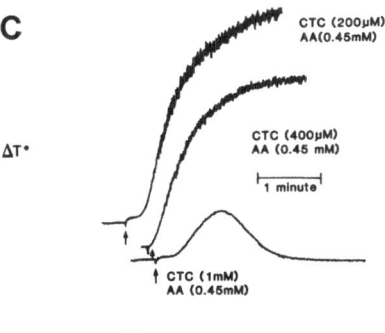

C

CTC (200µM)
AA(0.45mM)

ΔT*

CTC (400µM)
AA (0.45 mM)

1 minute

CTC (1mM)
AA (0.45mM)

ΔT*–Change in light transmission

FIGURE 5. (*Continued*)

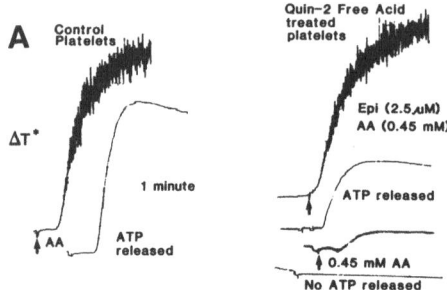

ΔT*-Change in light transmission

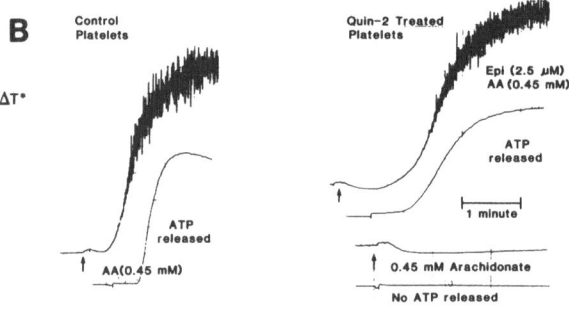

ΔT* -Change in light transmission

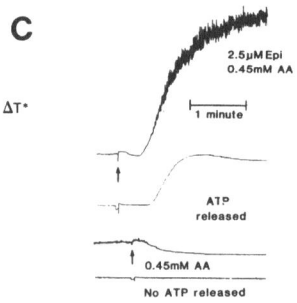

ΔT* -Change in light transmission

FIGURE 6. (A) Quin-2 free acid (QFA)(3 mM), (B) Quin-2 ester (QAM)(40 μM), and (C) chlor-tetracycline (CTC) (1 mM) inhibited arachidonate-induced platelet aggregation. Exposure of drug-treated refractory platelets to epinephrine restored the sensitivity of platelets to the action of arachidonate and caused irreversible aggregation.

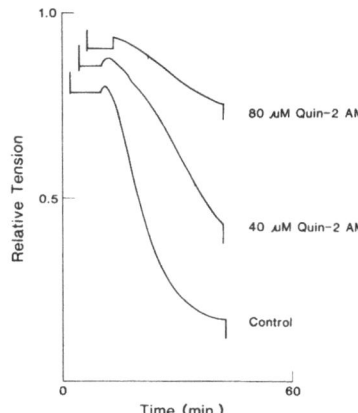

FIGURE 7. Normal control platelets as well as Quin-2 AM-loaded platelets formed clots when exposed to thrombin (1 U/ml). However, when clots were incubated at 37°C and allowed to retract, Quin-2 AM caused a dose-dependent inhibition in clot retraction.

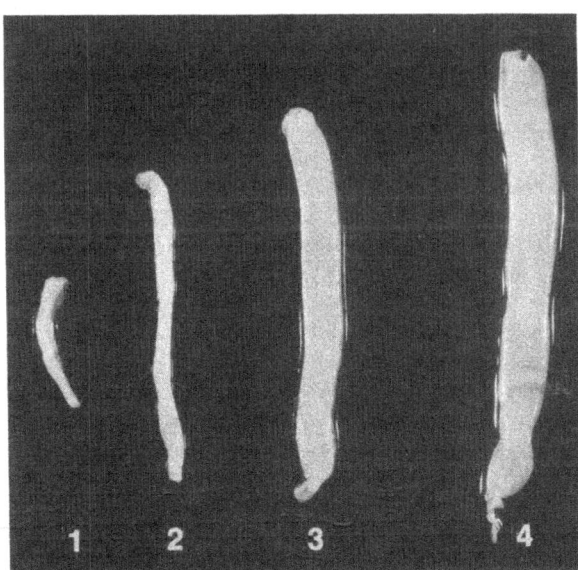

FIGURE 8. Quin-2 AM exerted a dose-dependent inhibition of thrombin-induced retraction of platelet-rich plasma clots. (1) Normal control clot after complete retraction, (2) influence of 40 μM Quin-2 AM on clot retraction, (3) influence of 80 μM Quin-2 AM on clot retraction, and (4) thrombin-induced clot of platelet-poor plasma.

retraction. To evaluate the effect of various sources of calcium on clot retraction, studies were done with Quin-2 free acid, Quin-2 ester, and chlortetracycline. Chelation of extracellular calcium or membrane calcium alone did not have a significant inhibitory effect on thrombin-induced clot formation and retraction, whereas chelation of intracellular calcium with Quin-2 had a dramatic effect on clot retraction.

4. DISCUSSION

The source of calcium that contributes to the agonist-induced elevation of cytosolic free calcium and specific roles of the divalent cation in platelet function have not been clearly defined (Feinman and Detwiler, 1974; Massini *et al.*, 1978; Feinstein, 1980; Gerrard *et al.*, 1981; Rink *et al.*, 1982; Zucker and Nachmias, 1985; Grynkiewicz *et al.*, 1985; Thompson and Scrutton, 1985; Murer, 1985). As part of a continuing investigation, we examined the various sources of calcium available and their contribution to platelet activation and function. The possible sources of calcium in the platelet environment are extracellular plasma, various membrane systems of the cell, mitochondria, dense bodies, and α granules. Although there is a 10,000-fold higher concentration in the extracellular plasma, the cytosolic level of calcium is maintained at an approximate 100 nM concentration. Platelets do not take up and concentrate calcium avidly, even though they lack an effective calcium extrusion pump (Steiner and Luscher, 1985). They have an efficient means of sequestering calcium in dense bodies, mitochondria, and a complex of membranes referred to as the dense tubular system, which may provide the cations in response to agonist-induced stimulation (White *et al.*, 1974; White, 1981).

Apart from the surrounding plasma, the richest source of calcium in platelets are the dense bodies. Calcium stored in this organelle complexed with other components is nonexchangeable, and is probably greater than 4 M in concentration (Costa, 1977). Hermansky–Pudlak syndrome platelets have only rare dense bodies and, upon stimulation, do not secrete ATP (White *et al.*, 1971; White and Witkop, 1972). However, thrombin-induced stimulation of HPS platelets in the present study resulted in an elevation of cytosolic calcium comparable to that obtained by stimulating normal control platelets. Platelets from patients with gray platelet syndrome are deficient in α granules and are larger than normal cells (Gerrard *et al.*, 1980). However, they contain normal-appearing membrane systems, dense bodies, mitochondria, and other organelles (White, 1979). They respond in the same manner as normal cells when stimulated with agonists. Gray platelet syndrome platelets stimulated with thrombin developed as high a level of cytosolic calcium as the normal control platelets. These observations contrast with those of Hardisty *et al.* (1985) presented in a preliminary form in which they report slightly impaired thrombin-induced calcium flux in GPS platelets. Results of our studies with platelets from HPS and GPS further confirm earlier suggestions that the α granules and dense bodies do not contribute significantly to the cytosolic ionized pool of calcium during agonist-induced activation of platelets (Holmsen and Weiss, 1979; Gerrard *et al.*, 1980, 1981; Thompson and Scrutton, 1985).

Previous works by Feinstein and Walenga (1981), Purdon and colleagues (1984), and Luscher *et al.* (1980) suggested that mitochondria are not an important source of calcium for physiological platelet activation. Their suggestions were based on indirect evidence, and not by direct measurement of ionized cytosolic calcium following exposure to agents that rapidly release the divalent cation from mitochondria. In the present study we have used ethidium bromide, which destroys mitochondria without affecting other platelet organelles or membrane systems (McGill, 1980). Platelets treated with sodium azide (0.5 mM) or ethidium bromide (100 μM) did not show elevated levels of basal calcium, suggesting that no significant leak or release of calcium occurred following such treatment. In addition, drug-treated platelets developed the same levels of cytosolic calcium after exposure to thrombin as untreated control platelets. Based on electron probe analysis of mitochondria in a variety of cells, Somlyo et al. (1978) concluded that normal mitochondria contained very little endogenous calcium, and increases in calcium content were due to damage. Our results and the findings of others (Feinstein, 1980; Purdon et al., 1984) support the concept that mitochondria do not serve as important repositories for calcium flux into the cytosol during agonist-induced platelet activation, although they may very well serve as a sink for excess calcium present in the cytosol (Statland et al., 1969; Somlyo et al., 1978; Luscher et al., 1980; Feinstein and Walenga, 1981; Purdon et al., 1984).

If one excludes mitochondria, dense bodies, and α granules as possible sources of releasable calcium, then calcium from the surrounding plasma or the platelet membrane systems could serve as the primary source for agonist-induced elevation of divalent cations. Zucker and Grant demonstrated that prolonged exposure of platelets to EDTA or EGTA resulted in loss of their ability to aggregate in response to agonist-induced stimulation (Zucker and Grant, 1978). Recent studies have shown that membrane glycoproteins IIb and IIIa exist as a Ca^{2+}-dependent complex and treatment with EDTA irreversibly dissociates the complex (Jennings and Phillips, 1982).

In the present study both EDTA and Quin-2 free acid were used for chelating extracellular calcium. Platelets exposed to these agents failed to aggregate when stimulated by agonists. However, exposure of Quin-2 free acid-treated platelets of epinephrine restored the sensitivity of the refractory cells to the action of agonists. Epinephrine failed to reverse the inhibitory effect of EDTA. The difference in the results may be related to the ability of EDTA to chelate membrane calcium in addition to extracellular calcium (Zucker and Grant, 1978). Since epinephrine can potentiate arachidonate-induced aggregation and secretion in the presence of 3 mM Quin-2 free acid, it is reasonable to suggest that potent agents (thrombin, ionomycin, ionophore A23187) that can mobilize membrane calcium will also be able to induce elevation of cytosolic calcium in the absence of extracellular calcium.

Chlortetracycline is considered a specific complexing agent for membrane-associated calcium (Owen and LeBreton, 1981; Schneider et al., 1983). Thompson and Scrutton analyzed agonist-induced CTC-fluorescence changes and the secretory response in platelets, and concluded that chlortetracycline could be used as a specific probe to monitor mobilization of membrane-associated calcium (Thompson and Scrutton, 1985). However, the method does not distinguish between calcium that is associated with the surface and channels of the open canalicular system or the membranes of the dense tubular system. Earlier studies demonstrated the existence of an internal membrane system (dense tubular system) (Behnke, 1967) capable of sequestering calcium in an energy-dependent manner (Statland et al., 1969; Menashi et al., 1982, 1984). The data obtained by biochemical, ultrastructural, and physiological studies support the concept that these membrane systems are the major source of a releasable pool of calcium that contributes to the agonist-induced elevation of cytosolic calcium.

In the present study, we used chlortetracycline to monitor alterations in the membrane calcium level, as well as a chelator for the cation. Chlortetracycline at high concentrations ($>$1 mM) specifically blocked second-wave aggregation and secretion in response to agonists without preventing the formation of thromboxane via the cyclooxygenase pathway. Earlier studies suggested that epinephrine can mobilize membrane calcium in CTC-treated platelets

(Owen *et al.*, 1980; Owen and LeBreton, 1981). Indeed, in the present study, epinephrine was able to restore the sensitivity of chlortetracycline-treated platelets to the action of arachidonate.

Quin-2 AM, used as an intracellular chelator, blocked the second-wave aggregation of platelets to the action of agonists such as epinephrine and ADP. The inhibition was associated with absence of secretion. Potent agonists such as thrombin and ionophore, although able to induce release of arachidonic acid, were unable to cause aggregation of platelets. Since Quin-2 loading does not prevent arachidonic acid conversion to thromboxane A_2 (Rao *et al.*, 1986), the inhibition of aggregation by agonists such as thrombin, ionophore, and arachidonate may be due to the inability of the newly formed thromboxane to induce platelet stickiness. However, exposure of Quin-2-loaded platelets to epinephrine restored the sensitivity of refractory platelet to the action of all agonists tested (Rao *et al.*, 1981). Therefore, it is reasonable to speculate that Quin-2 AM loading will concentrate Quin-2 free acid in the cytoplasm, and this free acid by buffering calcium probably antagonizes the glycoprotein IIb and IIIa complex similar to what happens when the free acid is added to the exterior of the cells. Epinephrine, by facilitating availability of calcium for this glycoprotein complex, restores the sensitivity of the membrane to the action of agonists.

In view of the dynamic role played by calcium in platelet activation–contraction–secretion coupling, we evaluated the effect of these three chelators on thrombin-induced PRP clot formation and retraction. Several earlier studies demonstrated a critical role for extracellular calcium in this process (Budtz-Olsen, 1951; Zucker and Borelli, 1981). Ethylenediaminetetraacetic acid can block clot retraction, but the concentration used in this study to inhibit agonist-induced aggregation (2.5 mM) did not prevent retraction. Similarly, Quin-2 free acid (3 mM) and chlortetracycline (1.5 mM) alone also were unable to prevent thrombin-induced clot formation and retraction. Quin-2 AM, however, when loaded into platelets inhibited clot retraction in a dose-dependent manner. A combination of any two of the three chelators (Quin-2 FA, Quin-2 AM, or CTC) prevented clot retraction probably by blocking thrombin-induced contraction and secretion.

In summary, results of our study suggest that the presence of extracellular calcium, the ability to mobilize membrane pools of calcium, and the availability of free cytosolic calcium are essential to achieve complete activation of platelets leading to contraction, secretion, and clot retraction. Calcium from dense bodies, mitochondria, and α granules do not contribute significantly to the cytosolic pool of calcium mobilized by agonists. Chelation of calcium from extracellular, intracellular, or membrane pools significantly inhibited agonist-induced aggregation and secretion. However, epinephrine-induced membrane modulation restored the sensitivity of refractory platelets to the action of agonists. A combination of any two chelators blocked secretion and clot retraction, and prevented correction by epinephrine-induced membrane modulation. Results of our present study and those of others suggest that membranes of the surface and channel systems are the primary source of calcium for platelet activation, with channels of the dense tubular system being the major reservoir.

5. SUMMARY

Changes in intracellular calcium concentration appear to play a critical role in platelet activation–contraction–secretion coupling. However, the specific source and precise function of the divalent cation involved in platelet physiology remain obscure. In the present study, we used the calcium-sensitive fluorophore Quin-2 ester (QAM), its free acid (QFA), and chlortetracycline (CTC), along with EDTA as intra- and extracellular calcium chelating agents to assess the origin and function of calcium in normal platelets and cells from patients with storage pool deficiency (Hermansky–Pudlak syndrome, HPS) and α-granule deficiency (gray platelet syndrome, GPS). Rises in cytosolic calcium levels following exposure to thrombin in the presence of EDTA were as great in HPS and GPS as in control platelets, indicating that dense bodies and α granules are not major sources of cytosolic calcium flux. Treatment of platelets with sodium azide or ethidium bromide to poison or destroy mitochondria had no effect on resting cytosolic calcium levels or the elevation after exposure to thrombin. In order to complex surface membrane, intracellular and extracellular calcium, respectively, CTC, QAM and QFA were employed. Chelation at any site was sufficient to block single agonist-induced secretion and irreversible aggregation. However, sensitivity of the refractory cells to agonists was restored by pretreatment with epinephrine. A combination of any two chelators (CTC + QAM, CTC + QFA, or QAM + QFA) completely blocked secretion, aggregation, and clot retraction, which could not be restored by membrane modulation. The results indicate that membranes and membrane systems are the primary source of calcium for platelet activation, with channels of the dense tubular system as the major reservoir. Elevation of cytoplasmic calcium appears essential for contractile events, including secretion and clot retraction.

ACKNOWLEDGMENTS. The authors wish to thank Ms. Janet Peller for her expert technical assistance and Ms. Susan Schwarze for her secretarial skills in the preparation of this manuscript. Supported by USPHS grants HL11880, CA-21737, HL-30217, March of Dimes (MOD-1-886), CRC (PR-400), MMF (HRDF 38-85; 47-87).

REFERENCES

Behnke, O., 1967, Electron microscopic observations on the membrane systems of the rat blood platelet, *Anat. Rec.* **158**:121–137.

Brass, L. F., and Shattil, S. J., 1982, Changes in surface-bound and exchangeable calcium during platelet activation, *J. Biol. Chem.* **257**:14000–16005.

Budtz-Olsen, O. E., 1951, *Clot Retraction,* Charles C Thomas, Springfield, IL.

Charo, I. F., Feinman, R. D., Detwiler, T. C., 1976, Inhibition of platelet secretion by an antagonist of intracellular calcium, *Biochem. Biophys. Res. Commun.* **72**:1462–1467.

Cohen, I., Gerrard, J. M., and White, J. G., 1982, Ultrastructure of clots during isometric contraction, *J. Cell Biol.* **93**:775–787.

Costa, J. L., Detwiler, T. C., Feinman, R. D., Murphy, D. L., Patlak, C. S., and Pettigrew, K. D., 1977, Quantitative evaluation of the loss of human platelet dense bodies following stimulation by thrombin or A23187, *J. Physiol.* **264**:297–306.

Detwiler, T. L., Charo, I. F., and Feinman, R. D., 1978, Evidence that calcium regulates platelet function, *Thromb. Haemost.* **404**:207–211.

Feinman, R. D., and Detwiler, T. C., 1974, Platelet secretion induced by divalent cation ionophores, *Nature* **249**:172–173.

Feinstein, M. D., 1980, Release of intracellular membrane-bound calcium precedes the onset of stimulus induced exocytosis in platelets, *Biochem. Biophys. Res. Commun.* **93**:593–600.

Feinstein, M. B., and Walenga, R. W., 1981, The role of calcium in platelet activation, In.: *Biochemistry of the Acute Allergic Reactions* (E. Becker, A. Simon, and K. Austen, eds.), Alan R. Liss, New York, pp. 279–306.

Fox, J. E. B., Reynolds, C. C., Phillips, D. R., 1983, Calcium-dependent proteolysis occurs during platelet aggregation, *J. Biol. Chem.* **258**:9973–9981.

Gerrard, J. M., Peterson, D. A., and White, J. G., 1981, Calcium mobilization, in: *Platelets in Biology and Pathology* (J. Gordon, ed.), Elsevier Oxford, pp. 407–436.

Gerrard, J. M., Phillips, D. R., Rao, G. H. R., Plow, E. F., Walz, D. A., Ross, R., Harker, L. A., and White, J. G., 1980, Biochemical studies of two patients with the gray-platelet syndrome–selective deficiency of platelet alpha granules, *J. Clin. Invest.* **66**:102–109.

Grynkiewicz, G., Poenie, M., and Tsien, R. Y., 1985, A new generation of Ca^{2+} indicators with greatly improved fluorescence properties, *J. Biol. Chem.* **260**:3440–3450.

Hallam, T. J., Sanchez, A., and Rink, T. J., 1984, Stimulus–response coupling in human platelets, *Biochem. J.* **218**:819–827.

Hardisty, R. M., Powling, J., Nokes, T. J. C., Patrick, A. D., and Srivastava, P. C., 1985, Gray platelet syndrome: New biochemical and functional studies, *Thromb. Haemostas.* **54**:73.

Hathaway, B. R., and Adelstein, R. S., 1979, Human platelet myosin light chain kinase requires the calcium-binding protein calmodulin for reactivity, *Proc. Natl. Acad. Sci. USA* **76**:1653–1657.

Holmsen, H., and Weiss, H. J., 1979, Secretable storage pools in platelets, *Ann. Rev. Med.* **30**:119–134.

Jennings, L. K., and Phillips, D. R., 1982, Purification of glycoproteins IIb and IIIa from human platelet plasma membranes and characterization of a calcium dependent glycoprotein IIb IIIa complex, *J. Biol. Chem.* **257**:10466–10658.

Luscher, E. F., Massini, P., and Kaser-Glanzmann, R., 1980, The role of calcium in the induction of platelet activities, in: *Cellular Mechanisms and Their Biological Significance* (A. Rotman, F. A. Meyer, C. Giltler, A. Sildberg, eds.), J. Wiley and Sons, New York, London, pp. 67–77.

Marcus, A. J., 1978, The role of lipids in platelet function with particular reference to the arachidonic acid pathway, *J. Lipid Res.* **19**:793–826.

Massini, P., Kaser-Glanzmann, R., and Luscher, E. F., 1978, Movement of calcium ions and their role in the activation of platelets, *Thromb. Haemost.* **40**:212–218.

McGill, M., 1980, Inhibition of mitochondrial-specific protein synthesis in human lymphocytes and platelets is dependent upon the site of cellular differentiation, *Cytogenet. Cell Genet.* **26**:117–126.

Menashi, S., Authi, K. S., Carey, F., and Crawford, N., 1984, Characterization of the calcium sequestering process associated with human platelet intracellular membranes isolated by free flow electrophoresis, *Biochem. J.* **222**:413–417.

Menashi, S., Davis, C., and Crawford, N., 1982, Calcium uptake associated with human platelet intracellular membranes isolated by free flow electrophoresis, *FEBS Lett.* **140**:298–302.

Murer, E. H., 1985, The role of platelet calcium, *Sem. Hematol.* **22**:313–323.

Owen, N. E., Feinberg, H., and LeBreton, G. C., 1980, Epinephrine induces Ca^{2+} uptake in human blood platelets, *Am. J. Physiol.* **239**:H483–H488.

Owen, N. E., LeBreton, G. C., 1981, Ca^{2+} mobilization in blood platelets as visualized by chlortetracycline fluorescence, *Am. J. Physiol.* **241**:H613–H619.

Purdon, D. A., Daniel, J. L., Stewart, G. J., and Holmsen, H., 1984, cytoplasmic free calcium concentration in porcine platelets. Regulation of an intracellular nonmitochondria calcium pump and increase after thrombin stimulation, *Biochem. Biophys. Acta* **800**:178.

Rao, G. H. R., and White, J. G., 1985a, Role of arachidonic acid metabolism in human platelet activation and irreversible aggregation, *Am. J. Hematol.* **19**:339–347.

Rao, G. H. R., and White, J. G., 1985b, Disaggregation and reaggregation of "irreversibly" aggregated platelets: A method for more complete evaluation of anti-platelet drugs, *Agents and Actions* **16**:425–434.

Rao, G. H. R., Peller, J. D., and White, J. G., 1985, Measurement of ionized calcium in blood platelets with a new generation of calcium indicator, *Biochem. Biophys. Res. Commun.* **132**:652–657.

Rao, G. H. R., Peller, J. D., Semba, C. P., and White, J. G., 1986, Influence of the calcium sensitive fluorophore Quin 2 on platelet function, *Blood* **67**:356–361.

Rao, G. H. R., Reddy, K. R., and White, J. G., 1981, Modification of human platelet response to sodium arachidonate by membrane modulation, *Prost. Med.* **6**:75–90.

Rink, T. J., Smith, S. W., and Tsien, R. Y., 1982, Intracellular free calcium in platelet shape change and aggregation, *J. Physiol.* **324**:53P–54P.

Schneider, A. S., Herz, R., and Sonenberg, M., 1983, Chlortetracycline as a probe of membrane associated calcium and magnesium: Interaction with red cell membranes, phospholipids, and proteins monitored by fluorescence and circular dichroism, *Biochemistry* **22:**1680–1686.

Somlyo, A. P., Somlyo, A. V., Schulman, H., Sloane, B., and Scrapa, A., 1978, Electron probe analysis of calcium compartments in cryosections of smooth and striated muscles, *Ann. N.Y. Acad. Sci.* **307:**523–566.

Statland, B., Heagan, B., and White, J. G., 1969, Uptake of calcium by platelet relaxing factor, *Nature* **223:**521.

Steiner, B., Luscher, E. F., 1985, Evidence that the platelet plasma membranes do not contain a $(Ca^{2+} + Mg^{2+})$-dependent ATPase, *Biochim. Biophys. Acta* **818:**299–309.

Takeo, S., and Sakanashi, M., 1985, Calcium accumulating ability of mitochondria from bovine coronary artery: Comparison with aortic mitochondria, *Jpn. Heart J.* **26:**91–103.

Thompson, N. T., and Scrutton, M. C., 1985, Intracellular calcium fluxes in human platelets, *Eur. J. Biochem.* **147:**421–427.

White, J. G., 1968, Fine structural alterations induced in platelets by adenosine diphosphate, *Blood* **31:**604–622.

White, J. G., 1979, Ultrastructural studies of the gray platelet syndrome, *Am. J. Pathol.* **95:**445–462.

White, J. G., 1981, Is the canalicular system the equivalent of the muscle sarcoplasmic reticulum? *Haemostasis* **4:**185–191.

White, J. G., 1983, The morphology of platelet function, in: *Methods in Hematology, Series 8: Measurements of Platelet Function* (W. A. Harker and T. S. Zimmerman, eds.), Churchill-Livingstone, New York, pp. 1–25.

White, J. G., and Witkop, C. J., Jr., 1972, Effects of normal and aspirin platelets on defective secondary aggregation in the Hermansky–Pudlak syndrome: A test for storage pool deficient platelets, *Am. J. Pathol.* **68:**57–66.

White, J. G., Rao, G. H. R., and Gerrard, J. M., 1974, Effects of the ionophore A23187 on blood platelets. I. Influence on aggregation and secretion, *Am. J. Pathol.* **77:**135–149.

White, J. G., Edson, J. R., Desnick, S. J., Witkop, C. J., 1971, Studies of platelets in a variant of the Hermansky–Pudlak syndrome. *Am. J. Pathol.* **63:**319–332.

Zucker, M. B., and Borelli, J., 1981, Some effects of divalent cations on the clotting mechanisms and the platelets of EDTA blood, *J. Appl. Physiol.* **12:**453–460.

Zucker, M. B., and Grant, R. A., 1978, Nonreversible loss of platelet aggregability induced by calcium deprivation, *Blood* **52:**505–514.

Zucker, M. B., and Nachmias, V. T., 1985, Platelet activation, *Arteriosclerosis* **5:**1–18.

45

Calmodulin-like Ca^{2+}-Binding Proteins of Smooth Muscle

GWYNETH DE VRIES, JOHN R. McDONALD, and
MICHAEL P. WALSH

1. INTRODUCTION

Ca^{2+} ions serve as intracellular messengers mediating the effects of a variety of extracellular signals (hormones, neurotransmitters, growth factors, etc.) in eliciting appropriate physiological responses (Carafoli and Penniston, 1985). For example, the neurotransmitter acetylcholine causes an elevation of cytosolic [Ca^{2+}] in smooth muscle leading to contraction. The effects of Ca^{2+} ions are mediated by a number of Ca^{2+}-binding proteins (Kretsinger, 1980). Calmodulin is one such Ca^{2+}-binding protein which responds to physiological [Ca^{2+}] transients by binding Ca^{2+} and undergoing a conformational change (Klee, 1977) which includes exposure of a hydrophobic site(s) (LaPorte et al., 1980; Tanaka and Hidaka, 1980). In this altered conformation, calmodulin can interact with a target enzyme, e.g., myosin light-chain kinase of smooth muscle (Walsh, 1985). Usually, such interaction converts the target enzyme from an inactive to an active state, triggering a cascade of biochemical reactions (often protein phosphorylations) and leading ultimately to the desired physiological response.

Calmodulin exhibits a number of unusual properties which have been exploited in the development of purification procedures, e.g., Ca^{2+}-dependent exposure of a hydrophobic site(s) (Gopalakrishna and Anderson, 1982), heat stability in the presence of bound Ca^{2+} (Cheung, 1971), and a high content of acidic amino acid residues (pI = 4.0) (Watterson et al., 1976). We demonstrated the presence of several proteins in bovine brain which exhibit properties similar to those of calmodulin yet are clearly distinct proteins (Walsh et al., 1984). These include a 21-kDa Ca^{2+}-binding protein of unknown function (McDonald and Walsh, 1985a; McDonald et al, 1985, 1987a) and 17- and 12-kDa protein inhibitors of the Ca^{2+}- and phospholipid-dependent protein kinase (McDonald and Walsh, 1985a,b; McDonald et al., 1987b). These studies were significantly facilitated by the demonstration by Maruyama et al. (1984) that some high-affinity Ca^{2+}-binding proteins recover their ability to bind Ca^{2+} following sodium dodecyl sulfate/polyacrylamide gel electrophoresis (SDS–PAGE) and

GWYNETH DE VRIES, JOHN R. MCDONALD, and MICHAEL P. WALSH • Department of Medical Biochemistry, University of Calgary, Calgary, Alberta, T2N 4N1, Canada.

transblotting onto nitrocellulose sheets. Such proteins can be readily identified by incubating the nitrocellulose sheet in ^{45}Ca-containing solution followed by autoradiography.

Here we report the initial results of a long-term study aimed at the isolation and structural and functional characterization of calmodulin-like Ca^{2+}-binding proteins of smooth muscle. These proteins share the following properties with calmodulin: (1) Ca^{2+}-dependent interaction with a hydrophobic matrix (phenyl-Sepharose); (2) heat stability in the presence of Ca^{2+}; (3) ability to bind $^{45}Ca^{2+}$ in the transblot-^{45}Ca autoradiographic procedure; and (4) a high content of acidic amino acids. In addition, some of these proteins exhibit a Ca^{2+}-dependent electrophoretic mobility shift similar to that of calmodulin (Burgess *et al.*, 1980). We thus identified seven calmodulin-like Ca^{2+}-binding proteins in chicken gizzard smooth muscle.

2. MATERIALS AND METHODS

2.1. Materials

[γ-^{32}P]ATP (10–40 Ci/mmol) and ^{45}CaCl$_2$ (10–40 mCi/mg of Ca) were purchased from Amersham (Oakville, Ontario). Phenyl-Sepharose CL-4B and DEAE-Sephacel were purchased from Pharmacia (Mississauga, Ontario). Nitrocellulose membranes, M_r marker proteins, and electrophoresis reagents were purchased from Bio-Rad Laboratories (Richmond, CA). General laboratory reagents used were of analytical grade or better and were purchased from Fisher Scientific (Calgary, Alberta).

2.2. Protein Purification

Calmodulin, the 21-kDa Ca^{2+}-binding protein, and cAMP phosphodiesterase were purified from bovine brain as described by Walsh *et al.* (1984), McDonald *et al.* (1987a), and Sharma *et al.* (1983), respectively. Myosin light-chain kinase was purified from chicken gizzard according to Ngai *et al.* (1984). Chicken gizzard calmodulin-like Ca^{2+}-binding proteins were prepared as follows. All procedures were carried out at 4°C. Frozen chicken gizzards were chopped and minced to yield 250 g of muscle mince and homogenized in a Waring blender for 2 × 30 sec in 4 volumes of 0.25 M sucrose, 20 mM Tris-HCl (pH 7.5), 2 mM EDTA, 10 mM EGTA, and 1 mM DTT. The homogenate was centrifuged at 30,000g for 15 min. The supernatant was filtered through glass wool and centrifuged at 100,000g for 60 min. The supernatant was again filtered through glass wool. Solid (NH$_4$)$_2$SO$_4$ was added with stirring to 75% saturation (i.e., 480 g/liter), stirred for 30 min, and centrifuged at 30,000g for 45 min. The pellet was redissolved in a minimum volume (~120 ml) of buffer A [20 mM Tris-HCl (pH 7.5), 1 mM EDTA, 1 mM EGTA, 0.2 mM DTT] and dialyzed overnight versus two changes (10 liters each) of buffer A. The dialyzate was clarified by centrifugation at 30,000g for 30 min. Solid MgCl$_2$ and CaCl$_2$ were added to the supernatant with stirring to final concentrations of 2 mM each. The sample was again clarified by centrifugation at 30,000g for 30 min prior to loading on a column (1.5 × 40 cm) of phenyl-Sepharose CL-4B previously equilibrated with buffer B [20 mM Tris-HCl (pH 7.5), 0.1 mM CaCl$_2$, 0.2 mM DTT]. The column was washed with buffer B until A$_{280}$ returned to baseline and then with buffer B containing 1 M NaCl to elute any proteins electrostatically bound to the column. Finally, bound proteins were eluted with 20 mM Tris-HCl (pH 7.5), 1 mM EGTA, and 0.2 mM DTT. Fractions (4 ml) were collected at a flow rate of 20 ml/h. Calmodulin-containing fractions (pool I) were combined and $CaCl_2$ was added to a final concentration of 2 mM. The sample was heated on a boiling water bath to 80°C, the temperature was maintained at 80–85°C for 2 min, and the sample was placed on ice. Denatured proteins were removed by centrifugation at 48,000g for 30 min. The supernatant was dialyzed overnight versus two changes (10 liters each) of buffer C [20 mM Tris-HCl (pH 7.5) 1 mM EGTA, 0.1 mM NaCl). The dialyzate was applied to a column (1 × 40 cm) of DEAE-Sephacel previously equilibrated with buffer C. The column was washed with buffer C until A_{280} returned to baseline and bound proteins were eluted with a linear gradient made from 150 ml each of buffer C and buffer C containing 0.4 M NaCl. Fractions (4 ml) were collected at a flow rate of 15 ml/hr.

2.3. Other Methods

Myosin light-chain kinase and cAMP phosphodiesterase activities were measured as previously described (Teo *et al.*, 1973; Walsh *et al.*, 1983). Polyclonal antibodies to the brain 21-kDa Ca^{2+}-binding protein and the isolated 20-kDa light chain of chicken gizzard myosin were raised in rabbits and purified as previously described (Ngai *et al.*, 1987). Immunoblotting was carried out as described by Ngai *et al.* (1987). Electrophoretic procedures, transblotting, and $^{45}Ca^{2+}$ autoradiography were also described by Ngai *et al.* (1987). When Ca^{2+}-dependent electrophoretic mobility shifts were examined, samples were made 10 mM in either $CaCl_2$ or EGTA prior to electrophoresis. Amino acid analysis was performed as described previously (McDonald *et al.*, 1987b).

3. RESULTS

In this study we focused on the set of proteins which interacts with a hydrophobic matrix in a Ca^{2+}-dependent manner. The cytosolic fraction of chicken gizzard smooth muscle was applied to a phenyl-Sepharose column in the presence of Ca^{2+} at low ionic strength. After washing off all unbound proteins, those proteins which were electrostatically bound to the resin were eluted with 1 M NaCl in the presence of Ca^{2+}. Finally, specifically bound proteins were eluted by chelation of Ca^{2+} with EGTA. The results are depicted in Fig. 1. Comparison of lanes 3 and 4 (Fig. 1A) indicates that almost all the proteins applied to the column were recovered in the flow-through. This emphasizes the usefulness of Ca^{2+}-dependent hydrophobic interaction chromatography. Very little protein was eluted in the high [NaCl] wash (lane 5). The remainder of Fig. 1 indicates the specific elution of a large number of proteins upon chelation of Ca^{2+}. Many of these could only be visualized by applying large amounts of column fractions to the gel (see legend to Fig. 1). The major protein component of the EGTA eluate (lanes 9 and 10) was subsequently identified as calmodulin (see below).

Selected fractions from the phenyl-Sepharose column were subjected to SDS–PAGE transblotted onto nitrocellulose, incubated with $^{45}Ca^{2+}$, and autoradiographed (Fig. 2A,B). Owing to the transblotting step, the autoradiogram and stained transblot in Fig. 2 have the opposite orientation of the gels shown in Fig. 1. From the autoradiogram (Fig. 2A), six high-affinity Ca^{2+}-binding proteins in addition to calmodulin were observed. These are denoted 1–6 in order of decreasing M_r. They have the following M_r values: 1 = 24.6 kDa, 2 = 21.8 kDa, 3 = 20.1 kDa, 4 = 17.5 kDa, 5 = 14.1 kDa, 6 = 12.4 kDa. The Amido black-stained transblot (Fig. 2B) verified the efficiency of protein transfer and indicated that

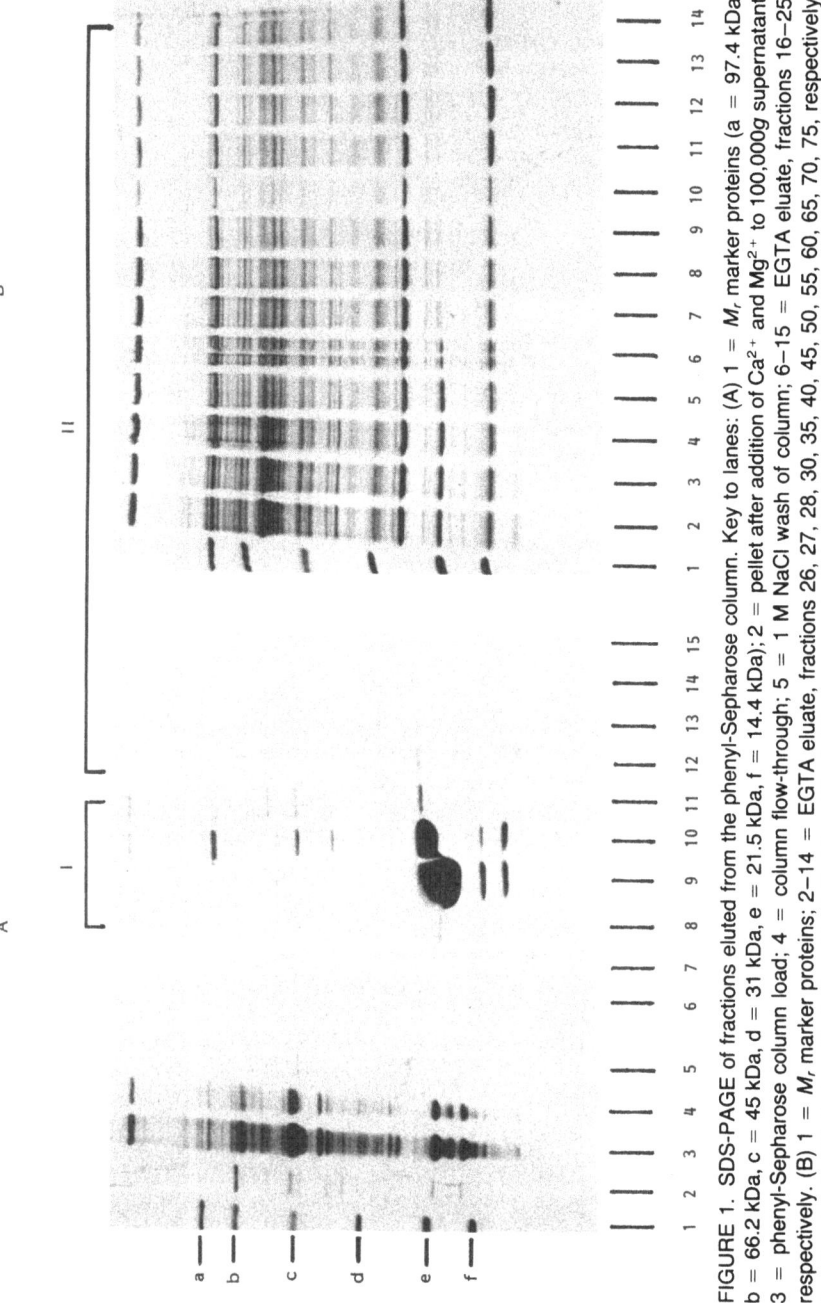

FIGURE 1. SDS-PAGE of fractions eluted from the phenyl-Sepharose column. Key to lanes: (A) 1 = M_r marker proteins (a = 97.4 kDa, b = 66.2 kDa, c = 45 kDa, d = 31 kDa, e = 21.5 kDa, f = 14.4 kDa); 2 = pellet after addition of Ca^{2+} and Mg^{2+} to 100,000g supernatant; 3 = phenyl-Sepharose column load; 4 = column flow-through; 5 = 1 M NaCl wash of column; 6–15 = EGTA eluate, fractions 16–25, respectively. (B) 1 = M_r marker proteins; 2–14 = EGTA eluate, fractions 26, 27, 28, 30, 35, 40, 45, 50, 55, 60, 65, 70, 75, respectively. Loading levels: (A) 1–4 = 5 µl; 5 = 10 µl; 6–15 = 20 µl. (B) 1 = 5 µl; 2–14 = 200 µl. Fractions were pooled as indicated: I and II.

A

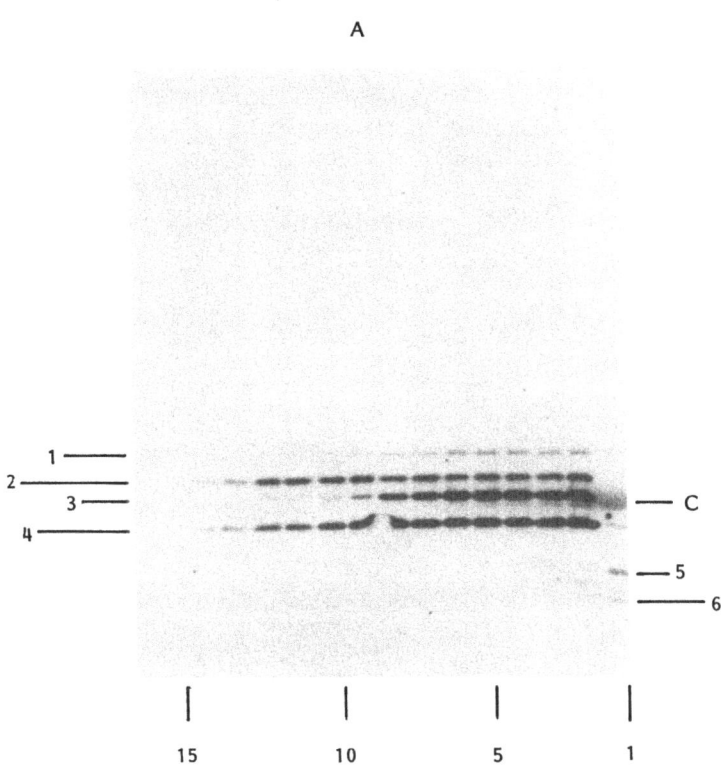

FIGURE 2. Analysis of Ca^{2+}-binding proteins partially purified by Ca^{2+}-dependent hydrophobic interaction chromatography. Selected fractions from the phenyl-Sepharose column (Fig. 1) were electrophoresed, transblotted onto nitrocellulose, incubated with $^{45}CaCl_2$, and autoradiographed (A). The blot was stained with Amido black (B). Key to lanes in A and B: 1 = pool I (fractions 18–21) (20 μl); 2–16 = EGTA eluate, fractions 25–30, 35, 40, 45, 50, 55, 60, 65, 70, 75, respectively (200 μl each). Numbers at left and right indicate the Ca^{2+}-binding proteins discussed in the text; "C" = calmodulin. (C) Pool II was heat-treated in the presence of Ca^{2+}, electrophoresed in the presence (lane 1) and absence (lane 2) of Ca^{2+}, transblotted onto nitrocellulose, incubated with $^{45}CaCl_2$, and autoradiographed. (D) Heat-treated pool II (lanes 3 and 4) and brain 21-kDa Ca^{2+}-binding protein (lanes 1 and 2) were electrophoresed in the presence (lanes 1 and 3) and absence (lanes 2 and 4) of Ca^{2+}, transblotted onto nitrocellulose, and treated with anti-21-kDa brain Ca^{2+}-binding protein. (E) Immunoblot using anti-chicken gizzard myosin LC_{20}. Key to lanes: 1 = intact gizzard myosin; 2 = isolated light chains of gizzard myosin; 3 = heat-treated pool II + Ca^{2+}; 4 = heat-treated pool II − Ca^{2+}.

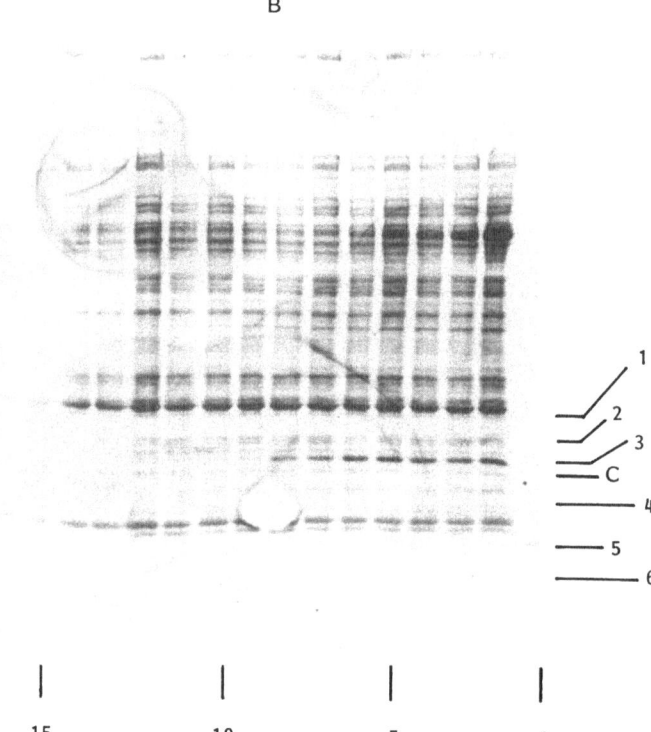

FIGURE 2. (Continued)

proteins 5 and 6 corresponded to the two major low-M_r proteins migrating faster than calmodulin observed in pool I (Fig. 1A). Of the other four Ca^{2+}-binding proteins observed in Fig. 2A, only protein 3 comigrated with a major stained band. The others therefore represent relatively minor protein components of the EGTA eluate. Calmodulin is visible, albeit weakly, in the autoradiogram (Fig. 2A, lane 1). In agreement with the observations of others, we consistently find that calmodulin does not bind quantitatively to nitrocellulose. This is apparent from the stained blot (Fig. 2B, lane 1). Coomassie blue staining of the SDS gel after transblotting revealed that calmodulin was quantitatively electroeluted (data not shown).

Proteins 1–4 were found to be heat-stable in the presence of Ca^{2+} and exhibited a Ca^{2+}-dependent electrophoretic mobility shift (Fig. 2C). Proteins 5 and 6 were also heat-stable in the presence of Ca^{2+} but did not exhibit a Ca^{2+}-dependent electrophoretic mobility shift (data not shown).

In initial attempts to identify some of these Ca^{2+}-binding proteins, we subjected pool II from the phenyl-Sepharose column to heat treatment in the presence of Ca^{2+} followed by SDS–PAGE and immunoblotting using antibodies against either the isolated brain 21-kDa Ca^+-binding protein or the 20-kDa light chain of gizzard myosin. The immunoblot in Fig. 2D indicates that heat-treated pool II contains a doublet of M_r ~21 kDa which cross-reacts

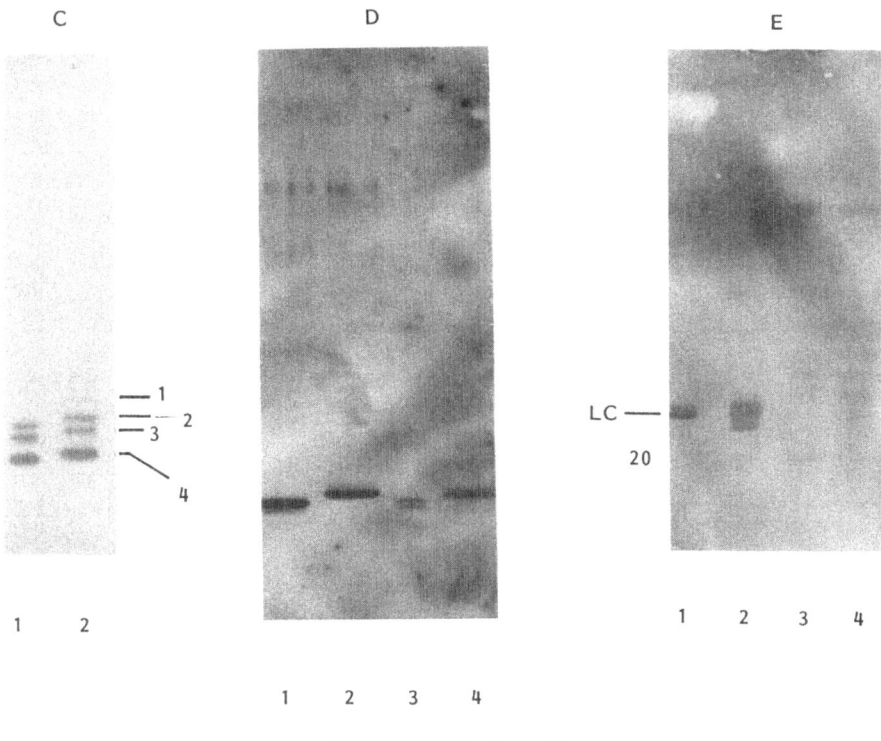

FIGURE 2. (*Continued*)

with anti-brain 21-kDa Ca^{2+}-binding protein. This corresponds to protein 2. On the other hand, pool II does not contain the 20-kDa light chain of myosin (Fig. 2E).

The Ca^{2+}-binding proteins in pool I from phenyl-Sepharose were further purified by heat treatment in the presence of Ca^{2+} followed by ion exchange chromatography (Fig. 3). Heat treatment removed essentially all the high-M_r components (compare Fig. 3, lane 2 with Fig. 1, pool I), leaving predominantly calmodulin and proteins 5 and 6. Ion exchange chromatography effected the separation of protein 5 (pool A), protein 6 (pool B), and calmodulin (pool D). Calmodulin was identified by its ability to activate cAMP phosphodiesterase in a Ca^{2+}-dependent manner. In the presence of bovine brain calmodulin (29.7 nM), the cAMP phosphodiesterase exhibited an initial rate of hydrolysis of cAMP of 11.8 μmol/min/mg in the presence of Ca^{2+} and 1.6 μmol/min/mg in the absence of Ca^{2+}. In the presence of the same concentration of gizzard pool D, the observed rates were 11.8 μmol/min/mg in the presence of Ca^{2+} and 2.1 μmol/min/mg in the absence of Ca^{2+}. Similarly, gizzard pool D was equally effective as bovine brain calmodulin in Ca^{2+}-dependent activation of smooth-muscle myosin light-chain kinase (data not shown).

Protein 6 (pool B) and calmodulin (pool D) were electrophoretically homogeneous (Fig. 3). In addition, another protein of M_r very similar to that of protein 6 was obtained in electrophoretically homogeneous form from the ion exchange column: pool C, denoted protein 6′. Amino acid analysis of proteins 6 (M_r = 12.4 kDa) and 6′ (M_r = 12.9 kDa) revealed that they have very similar compositions (Table I).

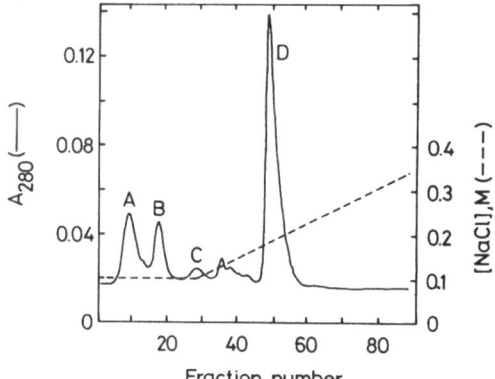

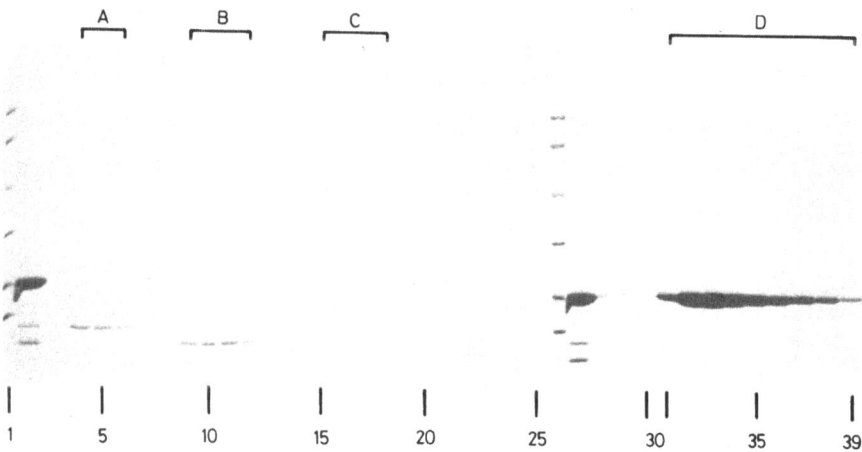

FIGURE 3. Ion exchange chromatography of pool I from phenyl-Sepharose. Pool I was heat-treated in the presence of Ca^{2+}, centrifuged, and the supernatant loaded on a DEAE-Sephacel column equilibrated with buffer containing 0.1 M NaCl. After washing unbound proteins from the column, bound proteins were eluted with a linear 0.1–0.4 M NaCl gradient (---). Protein elution (A_{280}, —) was monitored continuously. Selected fractions were examined by SDS-PAGE. Key to lanes: 1 = M_r marker proteins; 2 = column load; 3–25 = fractions 7, 9, 10, 11, 13, 15, 17, 18, 19, 20, 22, 24, 26, 28, 29, 30, 32, 34, 35, 36, 37, 38, 40, respectively; 26 = M_r marker proteins; 27 = column load; 28–39 = fractions 42, 44, 46, 48, 49, 50, 51, 52, 53, 54, 55, 57, respectively. Fractions were pooled as follows: A = 8–12 (14.1 kDa); B = 17–20 (12.4 kDa); C = 28–32 (12.9 kDa); D = 48–60 (calmodulin).

TABLE I. Comparison of Amino Acid
Compositions of Proteins 6 and 6'

Amino acid	Composition (mol of residue/mol)	
	Protein 6	Protein 6'
Lys	13.0	10.4
His	1.7	1.6
Arg	1.8	1.9
Asp	15.0	11.7
Thr	3.6[a]	3.5[a]
Ser	5.9[a]	5.4[a]
Glu	20.0	17.6
Pro	3.6	3.2
Gly	10.8	17.6
Ala	12.1	10.4
Cys	0.2	N.D.[b]
Val	4.6	4.1
Met	1.8	2.0
Ile	6.4	5.6
Leu	18.1	15.6
Tyr	5.2	4.6
Phe	4.0	3.6
Trp	0.0	N.D.[b]

[a] Extrapolated to zero time of hydrolysis.
[b] N.D., not detected.

4. DISCUSSION

The physiological regulatory effects of Ca^{2+} ions are exerted via a series of Ca^{2+}-binding proteins. A subclass of Ca^{2+}-binding proteins has been identified on the basis of detailed structural characterization and analysis of Ca^{2+}-binding parameters. This family of Ca^{2+}-binding proteins includes calmodulin, troponin C, parvalbumins, myosin light chains, S-100 proteins, calpain, and calbindin. In recent years, several other Ca^{2+}-binding proteins were identified which may belong to this calmodulin family, e.g., a 21-kDa brain Ca^{2+}-binding protein (McDonald and Walsh, 1985a) and CAB-18 (Manalan and Klee, 1984). In addition, a variety of Ca^{2+}-binding proteins which are structurally unrelated to calmodulin have been identified (see, e.g., Geisow and Walker, 1986). The great diversity of physiological processes subject to regulation by Ca^{2+} suggests that many more Ca^{2+}-binding proteins remain to be identified. We have initiated a series of studies aimed at the identification and structural and functional characterization of calmodulin-like Ca^{2+}-binding proteins from smooth muscle, specifically chicken gizzard. This tissue was chosen because it is a pure smooth muscle and is readily available in large quantities suitable for biochemical analysis. Furthermore, avian gizzard has been the system most utilized in biochemical studies of the Ca^{2+}-mediated regulation of smooth-muscle contraction.

In this initial study, we have demonstrated, in addition to calmodulin, the existence of seven calmodulin-like Ca^{2+}-binding proteins in chicken gizzard. These proteins, like calmodulin, interact with a hydrophobic matrix (phenyl-Sepharose) in a Ca^{2+}-dependent manner, are heat-stable in the presence of Ca^{2+}, and exhibit direct Ca^{2+} binding by the transblot-^{45}Ca autoradiographic technique. They have M_r values, as determined by SDS–PAGE in the absence of Ca^{2+}, of 24.6, 21.8, 20.1, 17.5, 14.1, 12.9, and 12.4 kDa. Amino acid analysis indicates that the 12.9- and 12.4-kDa proteins have very similar compositions and therefore may represent isoforms of a single Ca^{2+}-binding protein. The 21.8-kDa Ca^{2+}-binding protein cross-reacts with an antibody to a well-characterized bovine brain Ca^{2+}-binding protein (Walsh *et al.*, 1984; McDonald and Walsh, 1985a; McDonald *et al.*, 1985, 1987a). The function of this protein, however, is unknown. Immunoblotting with anti-gizzard myosin LC_{20} indicated that the 20.1-kDa Ca^{2+}-binding protein is not the 20-kDa light chain of myosin.

As indicated in this chapter, the 12.9- and 12.4-kDa Ca^{2+}-binding proteins have been purified to electrophoretic homogeneity. Future efforts will be aimed at purification of the other five Ca^{2+}-binding proteins and extensive physicochemical characterization of all these calmodulin-like proteins with a view to the elucidation of their respective roles in Ca^{2+}-mediated regulation of cellular functions.

ACKNOWLEDGMENTS. This work was supported by grants from the Medical Research Council of Canada and Hässle Pharmaceutical, Mölndal, Sweden. G. de V. was recipient of a summer studentship from the Alberta Heritage Foundation for Medical Research. J. R. McD. is recipient of a postdoctoral fellowship from the A.H.F.M.R. M. P. W. is a recipient of a Medical Research Council of Canada development grant and A.H.F.M.R. scholarship.

REFERENCES

Burgess, W. H., Jemiolo, D. K., and Kretsinger, R. H., 1980, Interaction of calcium and calmodulin in the presence of sodium dodecyl sulfate, *Biochim. Biophys. Acta* **623**:257–270.

Carafoli, E., and Penniston, J. T., 1985, The calcium signal, *Sci. Am.* **253**(5):70–78.

Cheung, W. Y., 1971, Cyclic 3′, 5′-nucleotide phosphodiesterase. Evidence for and properties of a protein activator, *J. Biol. Chem.* **246**:2859–2869.

Geisow, M. J., and Walker, J. H., 1986, New proteins involved in cell regulation by Ca^{2+} and phospholipids, *Trends Biochem. Sci.* **11**:420–424.

Gopalakrishna, R., and Anderson, W. B., 1982, Ca^{2+}-induced hydrophobic site on calmodulin: Application for purification of calmodulin by phenyl-Sepharose affinity chromatography, *Biochem. Biophys. Res. Commun.* **104**:830–836.

Klee, C. B., 1977, Conformational transition accompanying the binding of Ca^{2+} to the protein activator of 3′, 5′-cyclic adenosine monophosphate phosphodiesterase, *Biochemistry* **16**:1017–1024.

Kretsinger, R. H., 1980, Structure and evolution of calcium-modulated proteins, *CRC Crit. Rev. Biochem.* **8**:119–174.

LaPorte, D. C., Wierman, B. M., and Storm, D. R., 1980, Calcium-induced exposure of hydrophobic surface on calmodulin, *Biochemistry* **19**:3814–3819.

Manalan, A. S., and Klee, C. B., 1984, Purification and characterization of a novel Ca^{2+}-binding protein (CBP-18) from bovine brain, *J. Biol. Chem.* **259**:2047–2050.

Maruyama, K., Mikawa, T., and Ebashi, S., 1984, Detection of calcium binding proteins by ^{45}Ca autoradiography on nitrocellulose membrane after sodium dodecyl sulfate gel electrophoresis, *J. Biochem.* **95**:511–519.

McDonald, J. R., and Walsh, M. P., 1985a, Ca^{2+}-binding proteins from bovine brain including a potent inhibitor of protein kinase C, *Biochem. J.* **232**:559–567.

McDonald, J. R., and Walsh, M. P., 1985b, Inhibition of the Ca^{2+}- and phospholipid-dependent protein kinase by a novel M_r 17,000 Ca^{2+}-binding protein, *Biochem. Biophys. Res. Commun.* **129**:603–610.

McDonald, J. R., Walsh, M. P., McCubbin, W. D., Oikawa, K., and Kay, C. M., 1985, Ca^{2+}-binding proteins from bovine brain including a potent inhibitor of protein kinase C, *Biochem. J.* **232:** 569–575.

McDonald, J. R., Walsh, M. P., McCubbin, W. D., and Kay, C. M., 1987a, Isolation and characterization of novel 21 kDa Ca^{2+}-binding protein from bovine brain, *Methods Enzymol.* **139:**88–105.

McDonald, J. R., Gröschel-Stewart, U., and Walsh, M. P., 1987b, Properties and distribution of the protein inhibitor (M_r 17000) of protein kinase C, *Biochem. J.* **242:**695–705.

Ngai, P. K., Carruthers, C. A., and Walsh, M. P., 1984, Isolation of the native form of chicken gizzard myosin light-chain kinase, *Biochem. J.* **218:**863–870.

Ngai, P. K., Scott-Woo, G. C., Lim, M. S., Sutherland, C., and Walsh, M. P., 1987, Activation of smooth muscle myosin Mg^{2+}-ATPase by native thin filaments and actin/tropomyosin, *J. Biol. Chem.* **262:**5352–5359.

Sharma, R. K., Taylor, W. A., and Wang, J. H., 1983, Use of calmodulin affinity chromatography for purification of specific calmodulin-dependent enzymes, *Methods Enzymol.* **102:**210–219.

Tanaka, T., and Hidaka, H., 1980, Hydrophobic regions function in calmodulin-enzyme(s) interactions, *J. Biol. Chem.* **255:**11078–11080.

Teo, T. S., Wang, T. H., and Wang, J. H., 1973, Purification and properties of the protein activator of bovine heart cyclic adenosine 3′,5′-monophosphate phosphodiesterase, *J. Biol. Chem.* **248:** 588–595.

Walsh, M. P., 1985, Calcium regulation of smooth muscle contraction, in *Calcium and Cell Physiology* (D. Marmé, ed.), Springer-Verlag, Berlin, pp. 170–203.

Walsh, M. P., Hinkins, S., Dabrowska, R., and Hartshorne, D. J., 1983, Smooth muscle myosin light chain kinase, *Methods Enzymol.* **99:**279–288.

Walsh, M. P., Valentine, K. A., Ngai, P. K., Carruthers, C. A., and Hollenberg, M. D., 1984, Ca^{2+}-dependent hydrophobic-interaction chromatography. Isolation of a novel Ca^{2+}-binding protein and protein kinase C from bovine brain, *Biochem. J.* **224:**117–127.

Watterson, D. M., Harrelson, W. G., Jr., Keller, P. M., Sharief, F., and Vanaman, T. C., 1976, Structural similarities between the Ca^{2+}-dependent regulatory proteins of 3′:5′-cyclic nucleotide phosphodiesterase and actomyosin ATPase, *J. Biol. Chem.* **251:**4501–4513.

V

Mechanisms of Cell Injury

46

Role of Ion Regulation in Cell Injury, Cell Death, and Carcinogenesis

BENJAMIN F. TRUMP and IRENE K. BEREZESKY

1. INTRODUCTION

The understanding of acute and chronic cell injury constitutes one of the major problems in medical science, involving such issues as the mechanism of cell death and the control of cell growth and differentiation. Disease represents the summation of the effects of responses to cell injury in multicellular forms. Cell injury can be classified into those injuries that lead to cell death preceded by a reversible phase and those injuries that result in altered steady states compatible with cell survival, though in a different form (Trump and Arstila, 1971). We have observed that the responses to acute as well as to chronic cell injuries are accompanied by changes in cell ion content, including cytosolic sodium ($[Na^+]_i$), potassium ($[K^+]_i$), chloride ($[Cl^-]_i$), magnesium ($[Mg^{2+}]_i$), and calcium ($[Ca^{2+}]_i$) (Trump et al., 1971). In the present paper, we will continue to explore our hypothesis that deregulation of free, ionized cytosolic $[Ca^{2+}]_i$ is directly or indirectly related to many of the events that follow acute cell injury as well as to control of cell division and differentiation (Trump and Berezesky, 1985, 1987a–c; Trump et al., 1984).

Changes of cell injury leading to acute cell death can be separated into those that occur prior to and those that occur after the point of cell death, commonly termed ''the phase of necrosis.'' The sequence of such stages has been classified by our group into stages I–V of prelethal and lethal cell injury (Trump and Ginn, 1969). To briefly summarize these stages, the changes occurring prior to the point of cell death or ''the point of no return'' include cytoplasmic blebbing at the cell surface, mitochondrial condensation, clumping of nuclear chromatin, and dilatation of the endoplasmic reticulum (ER). The changes that occur following cell death (stage V) include breaks in the plasmalemma and massive swelling of mitochondria which contain one or two types of densities: flocculent densities and/or Ca phosphate precipitates. In fact, from observing the morphological changes in stage V, some characteristics of the initial injury can be deduced. That is, mitochondrial Ca phosphate

BENJAMIN F. TRUMP and IRENE K. BEREZESKY ● Department of Pathology, University of Maryland School of Medicine, and The Maryland Institute for Emergency Medical Services Systems, Baltimore, Maryland 21201.

accumulation is an active process and therefore agents that interfere with such active accumulation (e.g., anoxia, ischemia, and other areas of mitochondrial function) kill cells without Ca precipitates, whereas others such as complement, oxidative stress, and $HgCl_2$ can modify cell membrane Ca permeability to the point where the mitochondrial buffering systems will accumulate Ca, resulting in intramitochondrial Ca phosphate precipitates as the cells die (Gritzka and Trump, 1968). This is not a new observation as pathologists have recognized for many years that accumulation of insoluble Ca precipitates, usually Ca phosphate, occurs in dead and dying cells. At the same time, it is now recognized that such precipitates typically represent markers of cell death rather than act as participants in the initiation process that leads to cell killing. It was this putative relationship that led us some years ago to begin the systematic examination of the role of ion deregulation, especially that of $[Ca^{2+}]_i$, in the processes of cell injury and cell death (Gritzka and Trump, 1968; Laiho and Trump, 1974).

2. ACUTE CELL INJURY AND ION DEREGULATION

2.1. Alterations in $[Ca^{2+}]_i$

The role of $[Ca^{2+}]_i$ deregulation was studied in our laboratory in cultured rabbit and rat proximal tubular epithelial (PTE) cells following injury by 50 μM $HgCl_2$, 250 μM N-ethylmaleimide (NEM), 5 μM ionomycin, 10 μM A23187, 1 μM carbonyl cyanide m-chlorophenylhydrazone (CCCP), 5 μM potassium cyanide (KCN), 1 mM p-chloromercuribenzene sulfonic acid (PCMBS), 1 mM p-chloromercuribenzene (PCMB) and anoxia in the presence or absence of 1.37 mM extracellular Ca ($[Ca^{2+}]_e$). For spectrofluorometry, monolayers were preloaded with 2 μM Fura 2/AM in a modified Hank's buffered salt solution (pH 7.2) for 1 hr at 25°C and cells then suspended by trypsinization. The injurious agents were added and $[Ca^{2+}]_i$ measurements made on cell suspensions in a spectrofluorometer. Resting $[Ca^{2+}]_i$ in normal cells was approximately 100 nm. For each type of injury studied, there was a rise in $[Ca^{2+}]_i$ which varied in extent and in degree of buffering (Smith et al., 1987). Injury after $HgCl_2$ treatment showed an early, rapid rise in $[Ca^{2+}]_i$ of approximately 10-fold which was not dependent on $[Ca^{2+}]_e$ and was not related to cell killing. The slower second $[Ca^{2+}]_i$ rise did correlate with cell killing and was dependent on $[Ca^{2+}]_e$. Injuries resulting from NEM, PCMB, PCMBS, and anoxia showed $[Ca^{2+}]_i$ elevations which were most probably related to redistribution of $[Ca^{2+}]_i$ from intracellular compartments (e.g., mitochondria and ER) rather than from $[Ca^{2+}]_e$. Cytoplasmic blebbing was seen in all cases, varied in time of appearance, and depended on the agent and the amount of increase in $[Ca^{2+}]_i$ (Phelps et al., 1989). Blebbing was particularly dramatic after treatment with $HgCl_2$ in the presence of $[Ca^{2+}]_e$ where some blebs were observed by 3 min and many by 6–8 min. Most blebs appeared to pinch off, float free into the medium, and to increase in size as a function of time. In the case of NEM, small blebs were seen by 30 sec both with and without $[Ca^{2+}]_e$, indicating a relationship between bleb formation and deregulation of $[Ca^{2+}]_i$. Our data showed that cytoplasmic blebbing may involve changes in the arrangement of cytoskeletal components including actin and tubulin and/or their association with the plasma membrane, and that bleb formation appears to be triggered when $[Ca^{2+}]_i$, regardless of its source, passes a threshold of approximately 300 nm.

Using $HgCl_2$, Ca entry blockers did not prevent blebbing and/or the early rise in $[Ca^{2+}]_i$ (Phelps et al., 1989), further supporting a role for $[Ca^{2+}]_i$ redistribution. This deregulation is associated with cytoplasmic blebbing during the reversible phase of cell injury, whereas the later stages that lead to cell death are believed to result from Ca-mediated membrane damage, possibly through Ca-activated phospholipases and/or proteases. These and our other studies on *in vivo* models have thus provided direct evidence for the role of $[Ca^{2+}]_i$ deregulation in the process of cell injury.

Imaging studies were performed on rabbit PTE cell cultures grown on coverslips and loaded with 1 μM Fura 2/AM for 30 min at 37°C. After an initial control period of observation, $HgCl_2$ (50 μm) was added at 37°C. All treatments were performed under continual live observation and were recorded using digital imaging fluorescence microscopy (DIFM). In preliminary results, $[Ca^{2+}]_i$ increased in individual PTE cells in response to $HgCl_2$ treatment with variation between cells and between regions within individual cells (Fig. 1). Quantitation and the spatial pattern of $[Ca^{2+}]_i$ increase may possibly have been inaccurate because deesterification and compartmentation of the Fura 2/AM were not tested. In addition, spontaneous changes in $[Ca^{2+}]_i$ in untreated cells were not examined in any detail.

In another series of preliminary experiments using DIFM, blebs not only began to form at the cell periphery very quickly after the addition of $HgCl_2$, but also appeared to contain uniform distributions of Fura 2. Using pixel-by-pixel analysis of $[Ca^{2+}]_i$, rapid increases were seen, often more prominently at the cell periphery, and corresponding to the blebbed regions. After approximately 2–3 min, overall $[Ca^{2+}]_i$ concentrations stabilized at levels of approximately 1400 nM. At later time periods, as cell death occurred, propidium iodide staining of nuclei was observed; these dead cells showed no Fura 2 fluorescence. With $HgCl_2$, rhodamine 123 fluorescence showed changes beginning within 2–3 min. These changes began with alterations in the conformation of mitochondrial staining and later, a striking diminution in fluorescence intensity; again, dead cells showed little or no rhodamine 123 fluorescence.

2.2. Mechanisms of Bleb Formation

Observations on a variety of types of cell injury from our laboratory suggest that formation of cytoplasmic blebs at the cell surface correlates with an increase of $[Ca^{2+}]_i$ (Phelps *et al.*, 1989). It also appears that bleb formation occurs at the cell surface as $[Ca^{2+}]_i$ passes a threshold of approximately 300–500 nM and is associated with altered patterns of actin distribution; furthermore, very similar or identical blebs can be produced by compounds such as cytochalasin or vinblastine. It is therefore likely that blebbing involves changes in the cell membrane, the cytoskeleton, or in cytoskeletal-membrane interactions. Although in rabbit PTE, Ca entry blockers did not modify Ca entry or bleb formation, some modification of bleb formation did appear to occur with calmodulin inhibitors (such as TFP), suggesting that the process may be dependent on calmodulin. In our studies, especially with $HgCl_2$, the blebs commonly enlarge, seal off, detach, and drift away into the culture medium without the occurrence of cell death. This observation seems to correspond to *in vivo* studies on both rat PTE following ischemia (Glaumann *et al.*, 1975) and hepatocytes following injury in which blebs form, detach, and pass into the proximal tubule lumen or hepatic sinusoid, respectively. However, Lemasters *et al.* (1987), studying hepatocytes treated with KCN plus iodoacetic acid *in vitro*, reported bleb formation without an increase in $[Ca^{2+}]_i$ and furthermore suggested that cell death is related to bleb rupture. This contrasts with our observation on $HgCl_2$-treated rabbit PTE *in vitro* and rat PTE following total renal ischemia both *in vivo* and *in vitro*.

Also, in the $HgCl_2$ model studied in our laboratory on rabbit or rat PTE, elimination of $[Ca^{2+}]_e$ from the medium greatly retards the formation of blebs, suggesting that bleb

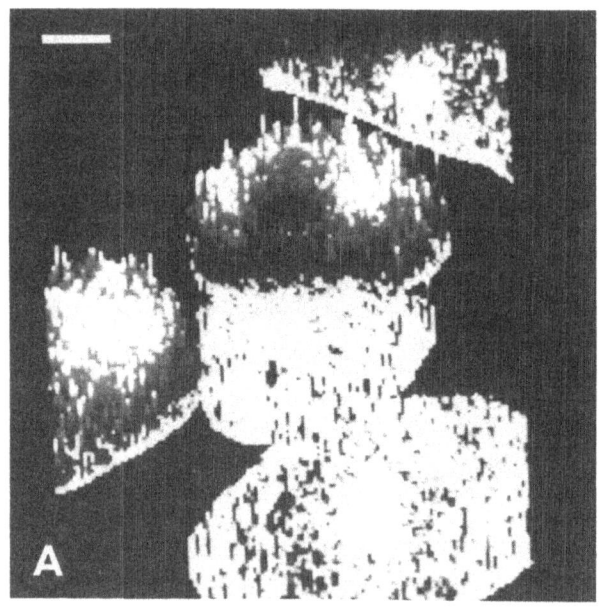

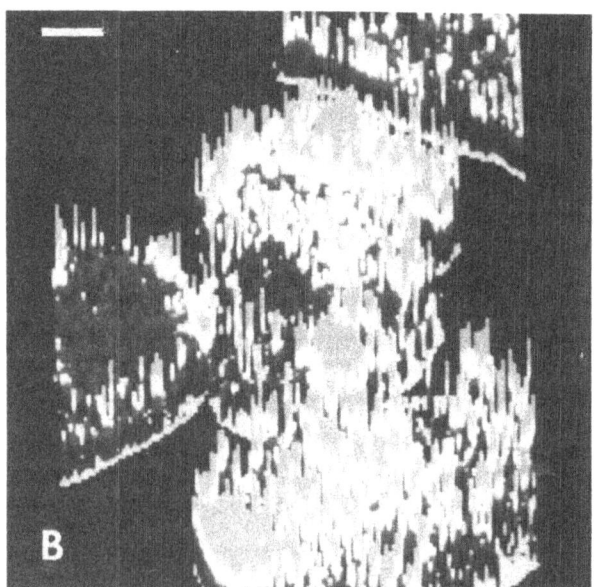

FIGURE 1. Digitized Fura 2 fluorescence ratioed images illustrating $[Ca^{2+}]_i$ in rabbit proximal tubular epithelial (PTE) cells in primary cultures. Cultured PTE cells were grown on no. 1 cover slips which had been mounted over a 1/2-in. hole drilled in the bottom of 35-mm plastic Petri dishes, loaded with 1 μM Fura 2/AM for 30 min at 37°C, and rinsed. The specially prepared Petri dishes with cultures were then transferred to a Leitz Diavert inverted microscope equipped with epifluorescence and quickly scanned using a 40X Nikon UV-FLuor objective. Following the

genesis is related to influx of $[Ca^{2+}]_e$. At the same time, increased $[Ca^{2+}]_i$ induced by inhibitors presumed to release Ca^{2+} from nonmitochondrial sources even in the absence of $[Ca^{2+}]_e$ results in bleb formation, indicating that the source of $[Ca^{2+}]_i$ associated with blebbing can either be from extracellular influx or intracellular redistribution.

2.3. Mechanisms of Cell Death

Our general hypothesis is illustrated in Fig. 2. Note that there are three principal mechanisms through which $[Ca^{2+}]_i$ deregulation may occur: modification of cell membrane function; modification of mitochondrial function; or modification of ER function. It is recognized that the ER is not homogeneous and that subsets of this organelle may have particular Ca-regulating capacities. As mentioned above, increases in $[Ca^{2+}]_i$ can occur with each of these three types of injury and can mediate a number of processes that could relate to cell killing. Furthermore, we noted that modification of $[Ca^{2+}]_e$ can modify cell killing (Laiho et al., 1983; Smith et al., 1987). These processes include stimulation of membrane peroxidation, activation of phospholipases (e.g., phospholipase C and phospholipase A_2), activation of Ca-dependent proteases, and activation of Ca-activated nucleases. In ischemic rat kidney in vitro, early reversible changes in mitochondrial membrane phospholipids have been noted (Smith et al., 1980) and these changes gradually progress as cell death occurs. Therefore, it is conceivable that activation of Ca-dependent phospholipases could eventually result in irreparable membrane damage. On the other hand, Nicotera et al. (1986), in their studies of isolated hepatocytes following cystamine-induced hepatotoxicity, found that inhibition of Ca-activated proteases, not phospholipases, could result in protection against cell death. Nicotera (personal communication) also observed that other types of injury were apparently not protected by protease inhibitors. Another possibility, although less likely in our view, which may relate to cell death but perhaps more likely to alterations of cell division and differentiation is that Ca activation of nucleases is capable of producing strand breaks in DNA. These breaks, if unrepaired, could result in chromosomal or genetic abnormalities. Increases of $[Ca^{2+}]_i$ can also mediate cell damage in cells adjacent to the target cell. For example, a recent study from our laboratory suggests that the respiratory burst in neutrophils induced by zymosan binding can be blocked by the Ca entry blocker verapamil (Nishihira et al., 1987a). Other concurrent studies from our laboratory indicate that it is possible that both in vivo and in vitro hepatocyte damage in sepsis or endotoxin injury is related to leukocyte–hepatocyte interactions (Nishihira et al., 1987b). The effect of $[Ca^{2+}]_i$ modification is also related to cytosolic pH; experiments in our laboratory have shown that reduction of medium pH is protective for several types of cell injury in various cells (Pentilla and Trump, 1974).

---

selection of an area, cells were excited at 340- and 380-nm wavelengths (emission 500 nm) and video images obtained via an ISIT camera (Dage, Inc.). An Epyx board (Silicon Video) in an IBM/AT computer was used to digitize the video images. Calculations for $[Ca^{2+}]_i$ were performed according to the method of Grynkiewicz et al. (1985). (A) Zero time control. (B) 18 min following $HgCl_2$ treatment. Note the increases in $[Ca^{2+}]_i$ after treatment with $HgCl_2$ as indicated by taller pixel heights and whiteness in color of the cells. Scale = 20 μm.

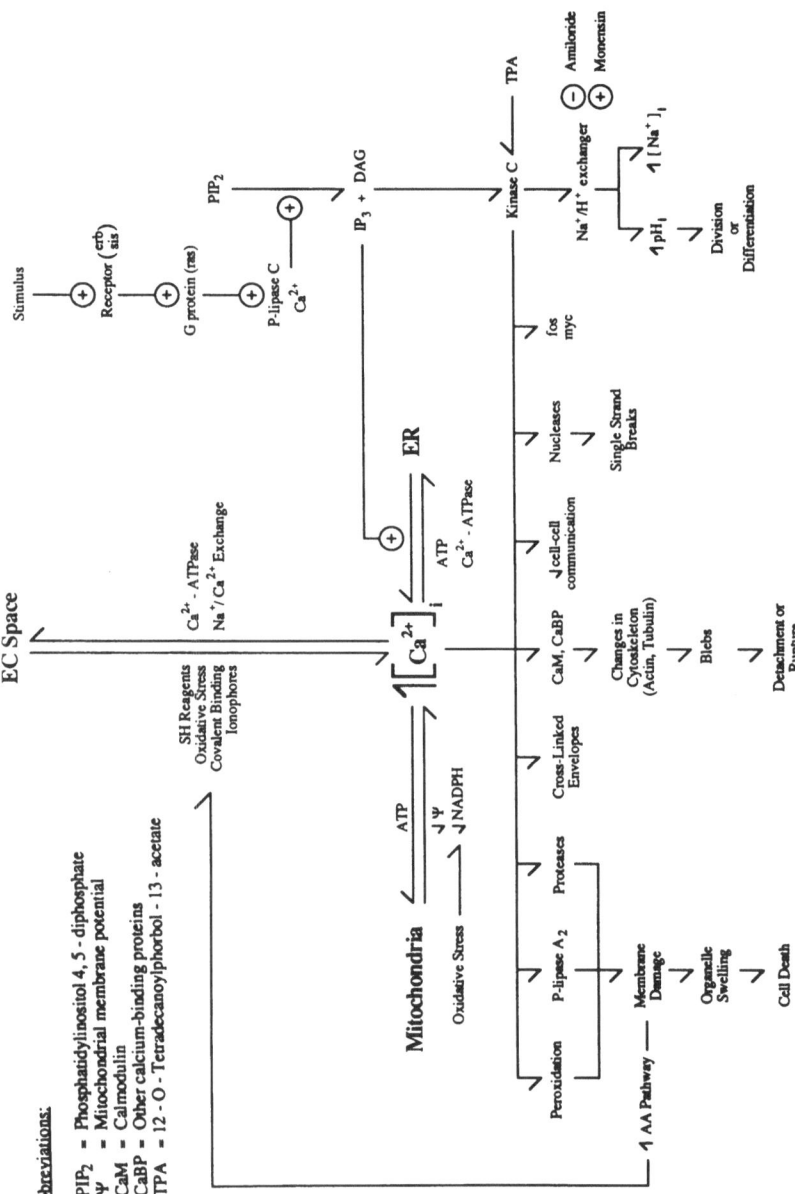

FIGURE 2. Diagram of our hypothesis illustrating the relationships between ion deregulation, cell injury, and carcinogenesis. (Reprinted with permission from B. F. Trump and I. K. Berezesky, 1987c.)

2.4. Control of Differentiation and Gene Expression

The relationship between cell toxicity and carcinogenesis is undisputed as virtually all complete carcinogens given at a carcinogenic dose *in vivo* produce acute toxicity and cell death during the process of conversion from normal to neoplastic cells. This process has been generally overlooked by all but a few investigators such as Farber and Sarma (1987). However, at the same time, it is known from *in vivo* studies, e.g., with CCl_4 toxicity in the liver or $HgCl_2$ toxicity in the kidney, that the reversibly altered cells at the edge of the necrotic zone in the hepatic parenchyma or kidney tubule, respectively, enter the mitotic cycle, divide, and repair the defect. In both cases, the issue becomes the identification of the stimulus for cell division, on the one hand, and the loss of cell division control on the other. As indicated in Fig. 2, there are many potential relationships between $[Ca^{2+}]_i$ deregulation, cell differentiation, and cell division. These involve the PI pathway which can result both in release of Ca from the ER by IP_3 and modification of Ca entry by IP_4 (Berridge *et al.*, 1983). Current data relate the activation of protein kinase C by the tumor promoter TPA, the induction of Ca-dependent nucleases capable of producing DNA strand breaks, and the relationship between elevated $[Ca^{2+}]_i$ and formation of keratinization and crosslinked envelopes to cell division and differentiation.

It was recently noted that changes in $[Ca^{2+}]_e$ can have opposite effects in normal as compared to transformed cells. For example, in both normal cultured mouse skin keratinocytes (Hennings *et al.*, 1980) and the human tracheobronchial epithelium (Lechner, 1984), 1–2 mM concentrations of $[Ca^{2+}]_e$ induce cell terminal differentiation with formation of crosslinked envelopes while transformed cells, cells treated with carcinogens, or tumor cell lines do not respond to this stimulus and continue to divide even with normal $[Ca^{2+}]_e$. It is still not known how this stimulus is transduced; however, it is reasonable to postulate an involvement of $[Ca^{2+}]_i$ either directly or through pathways such as that of the PI hydrolysis. Trosko and colleagues (Trosko and Chang, 1984; Wade *et al.*, 1986) observed that diminished cell–cell communication, presumably through altered gap junctions, is a very common response to tumor promoters *in vitro*. Loewenstein (1981) clearly showed that increased $[Ca^{2+}]_i$ concentrations can modify gap junctions. All of the above suggest that $[Ca^{2+}]_i$ levels in the cytosol may be deregulated in tumor cells. At the same time, however, several studies have indicated that increased $[Ca^{2+}]_i$ can produce rapid expression of genes related to cell division such as *fos* and *myc*. Moreover, an increasing number of receptor-related genes involving G proteins, phospholipase C, and PI hydrolysis including *erb*, *sis*, and *ras* are capable of modulating $[Ca^{2+}]_i$.

The ultimate mechanisms of induction of DNA synthesis and cell division remain undisclosed; however, both in fertilization of marine eggs and in 3T3 cells *in vitro,* division is preceded by alkalinization of the cytoplasm which can simulate the effects of growth factors such as EGF and fibroblasts (Moolenaar, 1986). This, in turn, seems to relate to activation of a Na^+–H^+ exchanger which is sometimes, at least, amiloride-sensitive at the cell membrane. In this context, it is important to realize that data from Cameron and Smith (1982) strongly indicate that neoplastic cells typically have increased levels of $[Na^+]_i$, and current studies utilizing DIFM indicate regional changes in $[Ca^{2+}]_i$ following growth factor stimulation (Tucker and Loats, 1986).

3. SUMMARY

Our studies show a significant involvement of $[Ca^{2+}]_i$ in both acute and chronic cell injury. Furthermore, changes in $[Ca^{2+}]_i$ concentrations may provide an important link between the two, including changes in cell division and differentiation. These relationships are currently being investigated using a number of new technologies, such as fluorescent probes and digital imaging fluorescence microscopy. Modification of $[Ca^{2+}]_i$ and/or $[Ca^{2+}]_e$ can clearly modify the course of cell injury in a number of models. Furthermore, modification of either $[Ca^{2+}]_i$ or $[Ca^{2+}]_e$ can also apparently affect the course of cellular differentiation.

ACKNOWLEDGMENT. This is contribution no. 2488 from the Pathobiology Laboratory. Supported by NIH grant AM15440.

REFERENCES

Berridge, M. J., Dawson, R. M. C., Downes, C. P., Hislop, J. P., and Irvine, R. F., 1983, Changes in the levels of inositol phosphates after agonist-dependent hydrolysis of membrane phosphoinositides, *Biochem. J.* **212**:473–482.

Cameron, I. L., and Smith, N. K. R., 1982, Energy dispersive spectroscopy in the study of the ionic regulation of growth in normal and tumor cells, in: *Ions, Cell Proliferation, and Cancer* (A. L. Boynton, W. L. McKeehan, and J. F. Whitfield, eds.), Academic Press, New York, pp. 13–40.

Farber, J. L., and Sarma, D. S. R., 1987, Biology of disease. Hepatocarcinogenesis: A dynamic cellular perspective, *Lab. Invest.* **56**:4–22.

Glaumann, B., Glaumann, H., Berezesky, I. K., and Trump, B. F., 1975, Studies on the pathogenesis of ischemic cell injury. II. Morphological changes of the pars convoluta (P_1 and P_2) of the proximal tubule of the rat kidney made ischemic in vivo, *Virchows Arch. B Cell Pathol.* **19**:281–302.

Gritzka, T. L., and Trump, B. F., 1968, Renal tubular lesions caused by mercuric chloride. Electron microscopic observations: Degeneration of the pars recta, *Am. J. Pathol.* **52**:1225–1277.

Grynkiewicz, G., Poenie, M., and Tsien, R. Y., 1985, A new generation of Ca^{2+} indicators with greatly improved fluorescence properties, *J. Biol. Chem.* **260**:3440–3450.

Hennings, H., Michaels, D., Cheng, C., Steinert, K., Holbrook, K., and Yuspa, S. H., 1980, Calcium regulation of growth and differentiation of mouse epidermal cells in culture, *Cell* **19**:145–154.

Laiho, K. U., and Trump, B. F., 1974, Relationship of ionic, water, and cell volume changes in cellular injury of Ehrlich ascites tumor cells. *Lab. Invest.* **31**:207–215.

Laiho, K. U., Berezesky, I. K., and Trump, B. F., 1983, The role of calcium in cell injury. Studies in Ehrlich ascites tumor cells following injury with anoxia and organic mercurials, *Surv. Synth. Pathol. Res.* **2**:170–183.

Lechner, J. F., 1984, Interdependent regulation of epithelial cell replication by nutrients, growth factors, and cell density, *Fed. Proc.* **43**:116–121.

Lemasters, J. J., DiGuiseppi, J., Nieminen, A-L., and Herman, B., 1987, Blebbing, free Ca^{2+} and mitochondrial membrane potential preceding cell death in hepatocytes, *Nature* **325**:78–81.

Loewenstein, W. R., 1981, Junctional intercellular communication: The cell-to-cell membrane, *Physiol. Rev.* **61**:829–913.

Moolenaar, W. H., 1986, Effects of growth factors on intracellular pH regulation, *Ann. Rev. Physiol.* **48**:363–376.

Nicotera, P., Hartzell, P., Baldi, C., Svensson, S-A., Bellomo, G., and Orrenius, S., 1986, Cystamine induces toxicity in hepatocytes through elevation of cytosolic Ca^{2+} and the stimulation of a nonlysosomal proteolytic system, *J. Biol. Chem.* **261**:14628–14635.

Nishihira, T., Kobayashi, N., Berezesky, I. K., and Trump, B. F., 1987a, Effect of verapamil on rat neutrophil respiratory burst, *J. Cell. Biol.* **105**:186a.

Nishihira, T., Sato, T., Koizuma, K., Berezesky, I. K., and Trump, B. F., 1987b, Neutrophil-mediated cell injury in cultured rat hepatocytes, *Circ. Shock* **21**:330.

Pentilla, A., and Trump, B. F., 1974, Extracellular acidosis protects Ehrlich acites tumor cells and rat renal cortex against anoxic injury, *Science* **185**:277–278.

Phelps, P. C., Smith, M. W., and Trump, B. F., 1989, Cytosolic ionized calcium and bleb formation following acute cell injury of cultured rabbit renal tubules cells, *Lab. Invest.* (in press.).

Smith, M. W., Collan, Y., Kahng, M. W., and Trump, B. F., 1980, Changes in mitochondrial lipi of rat kidney during ischemia, *Biochim. Biophys. Acta* **618:**192–201.

Smith, M. W., Ambudkar, I. S., Phelps, P. C., Regec, A. L., and Trump, B. F., 1987, HgCl$_2$-induced changes in cytosolic Ca^{2+} of cultured rabbit renal tubular cells, *Biochim. Biophys. Acta* **931:** 130–142.

Trosko, J. E., and Chang, C. C., 1984, Adaptive and nonadaptive consequences of chemical inhibition of intercellular communication, *Pharmacol. Rev.* **36:**1375–1445.

Trump, B. F., and Arstila, A. U., 1971, Cell injury and cell death, in: *Principles of Pathobiology* (M. F. LaVia, and R. B. Hill, Jr., eds.), Oxford University Press, New York, pp. 9–95.

Trump, B. F., and Berezesky, I. K., 1985, The role of calcium in cell injury and repair: A hypothesis, *Surv. Synth. Pathol. Res.* **4:**248–256.

Trump, B. F., and Berezesky, I. K., 1987a, Calcium regulation and cell injury: A heuristic hypothesis, *N. Y. Acad. Sci.* **494:**280–282.

Trump, B. F., and Berezesky, I. K., 1987b, Cell injury, ion regulation, and tumor promotion, in: *Nongenotoxic Mechanisms in Carcinogenesis* (B. E. Butterworth and T. J. Slaga, eds.), Banbury Report 25. Cold Spring Harbor, New York, p. 69.

Trump, B. F., and Berezesky, I. K., 1987c, Ion regulation, cell injury and carcinogenesis, *Carcinogenesis* **8:**1027–1031.

Trump, B. F., and Ginn, F. L., 1969, The pathogenesis of subcellular reaction to lethal injury, in: *Methods and Achievements in Experimental Pathology* (E. Bajusz and G. Jasmin, eds.), Karger, Basel, pp. 1–29.

Trump, B. F., Croker, B. P., Jr., and Mergner, W. J., 1971, The role of energy metabolism, ion, and water shifts in the pathogenesis of cell injury, in: *Cell Membranes: Biological and Pathological Aspects* (G. W. Richter and D. G. Scarpelli, eds.), Williams and Wilkins, Baltimore, pp. 84–128.

Trump, B. F., Berezesky, I. K., Sato, T., Laiho, K. U., Phelps, P. C., and DeClaris, N., 1984, Cell calcium, cell injury and cell death, *Env. Health Persp.* **57:**281–297.

Tucker, R. W., and Loats, H., 1986, Analysis of localized increases in free cytosolic (Ca$_i$) induced by different growth factors in human fibroblasts, *J. Cell Biol.* **103:**452a.

Wade, M. H., Trosko, J. E., and Schindler, M., 1986, A fluorescence photobleaching assay of gap junction-mediated communication between human cells, *Science* **232:**525–528.

47

Role of Calcium in Oxidative Cell Injury

STEN ORRENIUS, DAVID J. McCONKEY, and PIERLUIGI NICOTERA

1. INTRODUCTION

Tissue necrosis has long been known to be associated with the accumulation of large amounts of calcium in the necrotic tissue, and it has been proposed that the calcium ion may play a critical role in the development of lethal alterations of cell structure and function (Campbell, 1983). Although the generality of this hypothesis has been questioned (Starke et al., 1986), the involvement of Ca^{2+} in the development of lethal cell injury is now supported by a large number of observations. For example, Shanne and associates (1979) found that removal of extracellular Ca^{2+} exerted a protective effect against the toxicity of a number of agents in cultured hepatocytes. This and similar observations led to the proposal that an influx of extracellular Ca^{2+} could play a predominant role in the development of irreversible cell damage (Chien et al., 1979; Farber, 1981). However, subsequent studies showed that an influx of extracellular Ca^{2+} is not necessarily required for cytotoxicity and that a disruption of intracellular Ca^{2+} homeostasis may also result in irreversible cell damage.

Studies in this and other laboratories demonstrated that the toxicity caused by compounds whose metabolism generates oxidative stress in the liver is associated with the inhibition of Ca^{2+} translocases involved in the maintenance of intracellular Ca^{2+} homeostasis (Orrenius and Bellomo, 1986, and references therein). This may lead to a sustained elevation of cytosolic Ca^{2+} concentration which, in turn, can cause the activation of various Ca^{2+}-dependent degradative enzymes, including phospholipases, proteases, and endonucleases, whose activities may combine to cause irreversible damage. This "calcium hypothesis" of toxic cell injury is summarized in Fig. 1. The following sections will be concerned with the mechanisms by which a perturbation of intracellular Ca^{2+} homeostasis can cause cytotoxicity and cell death.

STEN ORRENIUS, DAVID J. McCONKEY, and PIERLUIGI NICOTERA • Department of Toxicology, Karolinska Institutet, S-104 01 Stockholm, Sweden.

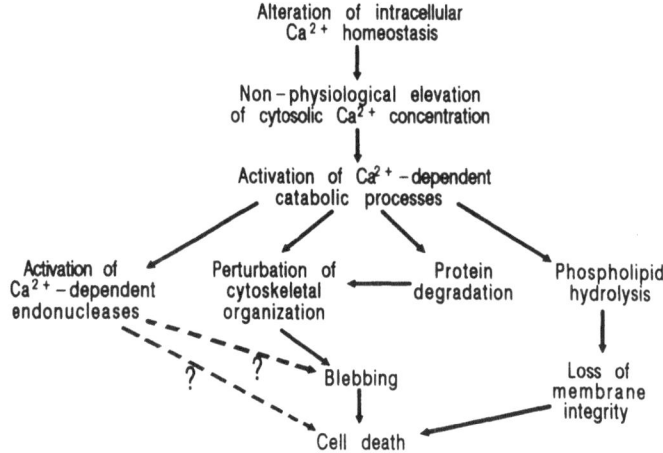

FIGURE 1. The hypothetical role of calcium in lethal cell injury.

2. CALCIUM COMPARTMENTATION IN HEPATOCYTES

The cytosolic concentration of free Ca^{2+} in hepatocytes is normally very low (0.1–0.2 μM) compared to that present in extracellular fluids (about 1.3 mM). Indeed, changes in the cytosolic Ca^{2+} level underlie a large number of physiological processes, such as cell growth and differentiation, cytosolic transport and secretion, and control of cell motility and structure (Campbell, 1983).

The low resting concentration of Ca^{2+} in the cytosol of hepatocytes is maintained by active compartmentation processes and by Ca^{2+} binding to specific proteins, including calmodulin and various calcium-dependent enzymes. The mitochondria and the endoplasmic reticulum represent the predominant sites of intracellular Ca^{2+} sequestration. They can exchange Ca^{2+} with the cytosolic compartment and can thereby buffer the effects of normal Ca^{2+} utilization. The plasma membrane Ca^{2+} translocase then plays the major role in the maintenance of the concentration gradient existing between the extra- and intracellular environments by actively extruding Ca^{2+} from the cell.

Mitochondrial Ca^{2+} homeostasis is regulated by a cyclic mechanism, involving Ca^{2+} uptake by an energy-dependent pathway, and Ca^{2+} release which, in liver, is probably mediated by a Ca^{2+}/H^+ antiporter. The latter appears to be regulated by the redox level of intramitochondrial pyridine nucleotides (Lehninger *et al.*, 1978), although membrane-bound protein thiols may also be important in modulating mitochondrial Ca^{2+} fluxes. The active transport of calcium ions through the endoplasmic reticular and plasma membranes is mediated by Ca^{2+}-stimulated, Mg^{2+}-dependent ATPases which appear to be dependent on free sulfhydryl groups for their activity (Moore *et al.*, 1975; Bellomo *et al.*, 1983).

Thus, it seems that the major mechanisms responsible for Ca^{2+} movements across cellular membranes are dependent on the redox status of protein thiol groups. It is therefore conceivable that oxidation of and/or binding to thiol groups may result in the inactivation of the pumps involved in the regulation of intracellular Ca^{2+} compartmentation, leading to an alteration of the level of cytosolic Ca^{2+}.

3. ROLE OF GLUTATHIONE AND PROTEIN THIOL MODIFICATION

Damage to biological systems caused by generation of active oxygen species is often referred to as "oxidative stress" (Sies, 1986). Since aerobic cells possess antioxidant systems to trap and/or inactivate such reactive species, oxidative stress is the result of an imbalance between the generation of active oxygen species and their inactivation by protective systems. Glutathione (GSH) plays a unique role in cellular defense, inasmuch as it functions as a reductant in the metabolism of hydrogen peroxide and organic hydroperoxides, and can also bind electrophilic metabolites (Orrenius and Moldéus, 1984). During the glutathione peroxidase-catalyzed metabolism of hydroperoxides, GSH serves as an electron donor and the glutathione disulfide (GSSG) formed in the reaction is subsequently reduced back to GSH by glutathione reductase at the expense of NADPH. Under conditions of oxidative stress, when the cell must cope with large amounts of H_2O_2 and/or organic hydroperoxides, the glutathione reductase is unable to match the rate of glutathione oxidation, and GSSG accumulates. In an apparent effort to avoid the detrimental effects of increased intracellular levels of GSSG, e.g., formation of mixed disulfides with protein thiols, the cell actively excretes the disulfide, which can lead to a depletion of the intracellular glutathione pool.

Other thiol groups with much biological relevance are those present in the structure of cellular proteins. In general, thiol groups are highly reactive and participate in several different types of reactions, such as arylation, oxidation, and thiol–disulfide exchange. All of these reactions may be involved in the modification of protein thiols as the consequence of their interaction with the reactive intermediates formed during the metabolism of toxic chemicals. Such modification of protein thiols may be particularly important during oxidative stress (Di Monte *et al.*, 1984).

4. MENADIONE-INDUCED OXIDATIVE STRESS

During the past several years our laboratory has been actively engaged in the study of the mechanism of cell killing by oxidative stress in isolated hepatocytes. A model compound extensively used in our studies is menadione (2-methyl-1,4-naphthoquinone). In isolated liver cells menadione metabolism involves both one- and two-electron reduction pathways which differ in their contributions to cytotoxicity (Thor *et al.*, 1982). One-electron reduction of quinones results in the formation of semiquinone radicals which can rapidly reduce dioxygen, forming the superoxide anion radical (O_2^-), and regenerating the parent quinone. Dismutation of O_2^- to hydrogen peroxide and the production of other highly reactive species quickly lead to a condition of oxidative stress as redox cycling of the quinone continues.

However, quinones can also undergo two-electron reduction, forming hydroquinones without the production of free semiquinone intermediates. This reaction, catalyzed by NAD(P)H quinone reductase (Ernster, 1967), may serve an important protective function by competing with the one-electron reduction pathway, since hydroquinones can be more easily excreted by the cell than can semiquinone radicals or their parent quinones.

Exposure of hepatocytes to toxic concentrations of menadione results in the extensive formation of O_2^- and H_2O_2 and the oxidation of GSH and pyridine nucleotides. Following GSH depletion, a dose- and time-dependent loss of protein thiols precedes cell death in hepatocytes exposed to menadione (Di Monte *et al.*, 1984). Although a general decrease in protein thiols has been observed in our experiments in which cells were exposed to cytotoxic levels of menadione, this does not imply that all protein thiols are equally important for cell survival. It appears that the fraction of critical thiols may be quite small. As mentioned above, the proper functioning of intracellular Ca^{2+} translocases is dependent on thiol groups in specific proteins, and the disruption of Ca^{2+} homeostasis resulting from their modification appears to be ultimately related to the onset of cytotoxicity. Other critical thiols affected by menadione metabolism may be located on microfilament proteins, whose modification may result in the disruption of the structural integrity of the plasma membrane and the onset of widespread bleb formation.

5. APPEARANCE OF SURFACE BLEBS IN HEPATOCYTES EXPOSED TO TOXIC AGENTS

Incubation of hepatocytes with menadione results in an alteration of surface morphology characterized by a loss of microvilli and the appearance of multiple blebs on the surface of the hepatocytes (Thor et al., 1982). Many other toxic agents cause similar alterations in surface structure (Jewell et al., 1982), indicating that plasma membrane blebbing is a common event in the progression of toxic injury. Blebs usually appear before any sign of increased plasma membrane permeability is observed and seem to be initially reversible (i.e., they often disappear when the toxic agent is removed from the incubation). The formation of surface blebs does not seem to be related to cell swelling caused by increased plasma membrane permeability since it is not affected by changes in the osmolarity of the incubation medium.

Cell surface morphology is thought to be determined by the organization of cortical microfilaments associated with the plasma membrane (Cheung, 1980, and references therein). This assumption is supported by the finding that two classes of compounds, the cytochalasins and phalloidins, which disrupt cortical microfilament structure, cause bleb formation on the surface of hepatocytes similar to that observed with other toxic agents.

The polymerization of monomeric to filamentous actin is dependent on ATP; 1 mole of bound ATP is converted to ADP for every monomeric actin subunit polymerized. Thus, one would expect that, in the hepatocyte, ATP depletion would result in actin depolymerization, breakdown of the actomyosin network, and plasma membrane blebbing. Indeed, inhibition of ATP synthesis by treatment of isolated hepatocytes with antimycin A is associated with extensive plasma membrane blebbing which precedes loss of cell viability. However, these alterations in surface morphology occur before ATP depletion does and are better correlated with alterations in intracellular Ca^{2+} distribution following antimycin A-induced release of mitochondrial Ca^{2+} (Smith et al., 1984).

As will be discussed below, during oxidative stress a perturbation of intracellular Ca^{2+} homeostasis is intimately related to a modification of the intracellular thiol status. Both disruption of intracellular thiol and Ca^{2+} homeostasis may result in alteration of cytoskeletal structures. Thiol oxidation can alter cytoskeletal assembly by causing a depolymerization of filamentous actin and through S-S crosslinking of cytoskeletal proteins. A change in intracellular Ca^{2+} distribution can affect cytoskeletal structure because the calcium ion and its associated binding proteins play a pivotal role in regulating the cytoskeletal network (Cheung, 1980). Alternatively, Ca^{2+}-activated catabolic enzymes, such as cytosolic proteases, can cleave cytoskeletally associated proteins (Collier and Wang, 1982) and thus cause alteration of the microfilament network and blebbing (Nicotera et al., 1986b). It is therefore conceivable that both thiol modification and Ca^{2+}-dependent processes may combine to produce cytoskeletal alterations during oxidative stress. The relative contribution of each of the different mechanisms remains to be established.

6. DISRUPTION OF INTRACELLULAR Ca^{2+} HOMEOSTASIS BY MENADIONE METABOLISM

The availability of noninvasive techniques to measure Ca^{2+} content in intracellular compartments has made it possible to monitor alterations in Ca^{2+} distribution during the development of toxicity in hepatocytes. Thus, in studies with isolated hepatocytes, we have been able to demonstrate that menadione-induced oxidative stress causes the mobilization of Ca^{2+} from both mitochondria and the endoplasmic reticulum (Thor et al., 1982; Jewell et al., 1982). Further studies with isolated subcellular fractions showed that menadione metabolism impairs the ability of mitochondria to take up and retain Ca^{2+} by causing the oxidation of pyridine nucleotides (Moore et al., 1986). Menadione also inhibits Ca^{2+} uptake by rat liver microsomes through a mechanism which seems to involve oxidation/arylation of thiol group(s) necessary for Ca^{2+}-ATPase activity. The latter study further revealed that arylation, oxidation, and formation of mixed disulfides with protein thiols can all inhibit microsomal Ca^{2+} sequestration (Thor et al., 1985). The incubation of hepatocytes with menadione therefore causes the release into the cytosol of Ca^{2+} which cannot be resequestered. Normally, this would result in only a transient rise in the cytosolic Ca^{2+} concentration because the plasma membrane Ca^{2+}-ATPase would remove this Ca^{2+} from the cell and the Ca^{2+} concentration would return to its usual, very low level. However, recent studies in our laboratory showed that the hepatic plasma membrane Ca^{2+}-ATPase is inhibited by agents, including menadione, which oxidize membrane protein thiols (Nicotera et al., 1985). Our ability to restore menadione-inhibited Ca^{2+}-ATPase activity by treatment with reducing agents, such as dithiothreitol (DTT), strongly suggests that this inhibition is due to protein thiol oxidation. The effects of menadione and some thiol reagents on the microsomal and plasma membrane Ca^{2+} translocases are shown in Tables I and II.

The redox cycling of menadione can therefore inhibit the Ca^{2+} translocases present in the mitochondria, endoplasmic reticulum, and plasma membrane of hepatocytes, leading to

TABLE I. Inhibition of Rat Liver Microsomal Ca^{2+} Sequestration by Agents which Modify Protein Thiols[a]

	Microsomal Ca^{2+} sequestration	
	nmol $\times$ mg protein^{-1}	% of control
No addition	7.6	100
Diamide (1 mM)	4.5	55
Cystamine (5 mM)	2.1	29
p-Chloromercuribenzoic acid (50 μM)	0	0
Menadione (400 μM) + NADPH (1 mM) after 60 min	4.6	57
Menadione + NADPH + GSH (1 mM) after 60 min	7.4	97

[a] For experimental details, see Thor et al. (1985).

TABLE II. Inhibition of Ca^{2+}-ATPase Activity in Rat Liver Plasma Membrane Fragments by Agents which Modify Protein Thiols

	Plasma membrane Ca^{2+}-ATPase activity[a]	
	nmol $P_i \times hr^{-1}$ $\times$ mg protein^{-1}	% of control
No addition	0.20	100
Diamide (2.5 mM)	0.07	35
Cystamine (1 mM)	0.04	20
N-ethylmaleimide (1 mM)	0	0
Menadione (200 μM) for 60 min	0.02	10
Menadione (200 μM) for 60 min followed by dithiothreitol (1 mM)	0.20	100

[a] Values were determined at pCa = 8.6. For experimental details, see Nicotera *et al.* (1985).

a sustained rise in cytosolic Ca^{2+}. Under physiological conditions, the cytosolic Ca^{2+} concentration in hepatocytes is modulated by various hormones and other Ca^{2+}-mobilizing agents. Thus, to investigate the difference between physiological and toxicological fluctuations in cytosolic Ca^{2+} levels, we compared the changes induced by an α-adrenergic agonist, phenylephrine, and by nontoxic and toxic concentrations of menadione in Quin2-loaded hepatocytes. Our study showed that both the extent and the duration of the rise in cytosolic Ca^{2+} differ between nontoxic and toxic conditions. Toxic concentrations of menadione caused a sustained elevation of cytosolic Ca^{2+}, which reached micromolar concentrations and fully saturated the binding capacity of the trapped dye (Fig. 2). Furthermore, in view of the ability of reducing agents, such as DTT, to prevent protein thiol modification and to protect hepatocytes from menadione toxicity, we investigated the effect of DTT on the increase in cytosolic Ca^{2+} caused by menadione. Treatment of hepatocytes with DTT prevented both

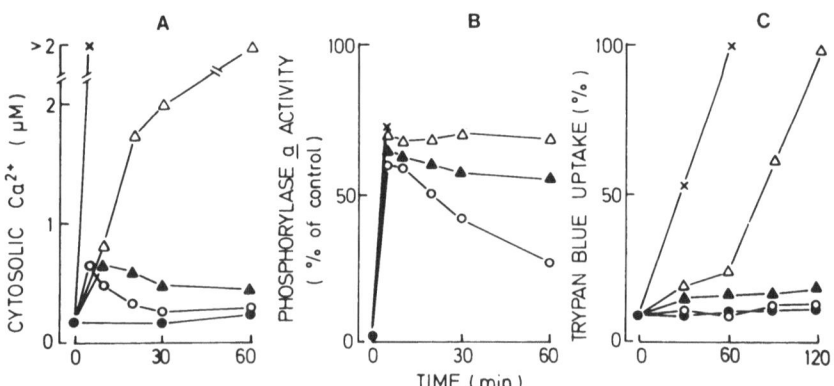

FIGURE 2. Relationship between cytosolic Ca^{2+} concentration (A), phosphorylase a activity (B), and cell viability (C) in hepatocytes exposed to toxic or nontoxic agents. Hepatocytes loaded with Quin2 were incubated in the absence (O) or presence of 100 μM menadione (▲), 200 μM menadione (△), 30 μM A23187 (X), or 1 μM phenylephrine (O). Results of one experiment typical of three. (For experimental details, see Nicotera *et al.*, 1986a.)

the increase in cytosolic Ca^{2+} and the onset of cytotoxicity in hepatocytes exposed to menadione (Nicotera *et al.*, 1988).

7. MECHANISMS OF CALCIUM-MEDIATED CYTOTOXICITY

The discovery of a relationship between a sustained increase in cytosolic Ca^{2+} concentration and the toxicity of menadione and a number of other agents in hepatocytes has led to a search for a mechanism by which the increased cytosolic Ca^{2+} could trigger cytotoxicity. It has been proposed that a sustained increase in Ca^{2+} concentration may cause abnormal stimulation of physiological processes, such as those dependent on calmodulin. It is conceivable that calmodulin may mediate some of the cytotoxic effects of Ca^{2+}, although direct evidence for this is still lacking. On the other hand, Ca^{2+} can activate catabolic enzymes, e.g., lipases, proteases, and endonucleases (cf. Fig. 1). Calcium-activated phospholipases are widely distributed in mammalian cells, and it has been proposed that an enhanced rate of phospholipid hydrolysis may result in irreversible cell damage. This assumption is supported by the observation that inhibitors of phospholipase activity prevent ischemic cell death in liver and heart, and by the finding that phospholipid breakdown is enhanced during tissue injury (Chien *et al.*, 1979). We recently investigated whether phospholipase activation might contribute to the toxicity of the reactive disulfide cystamine in hepatocytes (Nicotera *et al.*, 1986a). Indeed, exposure to cystamine did increase the rate of phospholipid hydrolysis, and this effect followed the elevation of cytosolic Ca^{2+} and preceded loss of cell viability. However, in our study phospholipase inhibitors failed to prevent cystamine toxicity.

Intracellular protein degradation is also known to be stimulated by a rise in cytosolic Ca^{2+}, and a group of cytosolic, thiol-sensitive proteases, whose activity depends on micromolar concentrations of Ca^{2+}, seems to be involved in this process (Murachi, 1983). These proteases function in the cleavage of membrane receptors, cytoskeletal proteins, and protein kinase C. Exposure of hepatocytes to toxic levels of cystamine resulted in a stimulation of calcium-dependent, nonlysosomal proteases which appeared to be responsible for cystamine toxicity. Moreover, Ca^{2+}-activated proteases also seem to be involved in the toxicity of other agents, i.e., extracellular ATP and ionophore A23187, which cause a sustained increase in cytosolic Ca^{2+} by selectively interfering with the normal Ca^{2+} influx–efflux balance (Nicotera *et al.*, 1986b). However, it remains to be established whether this mechanism of toxicity may also apply to other toxic compounds and different cell systems.

A third catabolic process found to be sensitive to a nonphysiological elevation of cytosolic Ca^{2+} concentration is that of an endogenous endonuclease. Indigenous to a number of different tissues, including liver, its activity has been associated with "apoptosis," or programmed cell death (Wyllie, 1980). Thus, both glucocorticoid hormones (Wyllie, 1980; Cohen and Duke, 1984) and the calcium ionophore A23187 (Wyllie *et al.*, 1984) stimulate apoptosis in isolated mouse thymocytes. Characteristic morphological changes accompany the activation of the chromatin degradation process. First is the appearance of widespread plasma and nuclear membrane perturbations (blebbing). This is followed by activation of the endonuclease, which is associated with extensive condensation of the nucleus.

As noted in Section 5, we found that plasma membrane blebbing can often be linked to a disruption of Ca^{2+} homeostasis and that it can be mediated by the activation of catabolic processes (Nicotera *et al.*, 1986b). In fact, the blebbing observed in hepatocytes treated with ionophore A23187 has been shown to involve activation of calcium-dependent proteases.

The appearance of blebs on apoptotic cells could therefore be an indication that a disruption of Ca^{2+} homeostasis accompanies this process. Since menadione metabolism creates such a disruption, we studied its effects on endogenous endonuclease activity in hepatocytes.

Figure 3 shows a typical dose-response pattern for the menadione activation of the endonuclease. With a toxic dose of menadione (200 μM), cell death occurred before induction of the endonuclease could be detected. However, in the presence of lower concentrations of menadione the activation of the enzyme was apparent after about 2 hr; maximal DNA cleavage was seen with 100 μM menadione after 4 hr of incubation. As also shown in Fig. 3, activation of the endonuclease preceded cell death by an appreciable period of time and was associated with the appearance of plasma membrane blebbing, characteristic of menadione-induced oxidative stress.

Indeed, comparison of Figs. 2 and 3 reveals that a significant increase in cytosolic Ca^{2+} preceded both membrane blebbing and induction of endonuclease activity in hepatocytes exposed to 100 μM menadione. This increase was sustained longer than the increase in Ca^{2+} observed in the presence of phenylephrine. It appears, therefore, that an increase in cytosolic Ca^{2+} of nonphysiological duration is linked both to the membrane perturbations and to the stimulation of endonuclease activity in hepatocytes exposed to 100 μM menadione.

Interestingly, morphological patterns observed as the consequence of menadione treatment of hepatocytes follow those observed with apoptotic thymocytes quite closely. The blebbing has already been mentioned, but chromatin condensation is also seen and a close similarity exists between cleavage products isolated from apoptotic thymocytes (Wyllie,

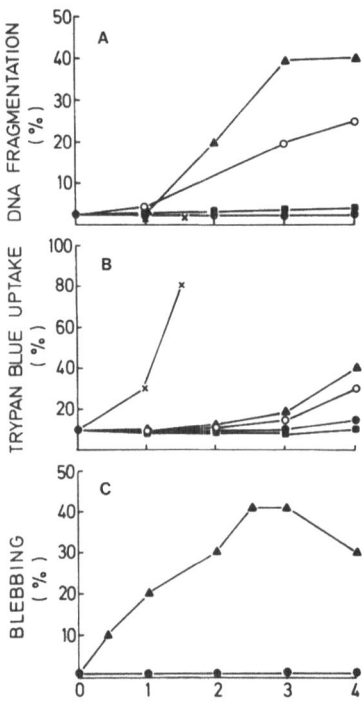

FIGURE 3. Comparison of the effects of menadione on endonuclease activity (A), cell viability (B), and plasma membrane blebbing (C) in isolated hepatocytes. Endonuclease activity was measured as described by Wyllie (1980). Hepatocytes were incubated in the absence (●) or presence of 25 μM (■), 50 μM (○), 100 μM (▲), or 200 μM (X) menadione. Samples were taken at indicated time points for measurement of DNA fragmentation.

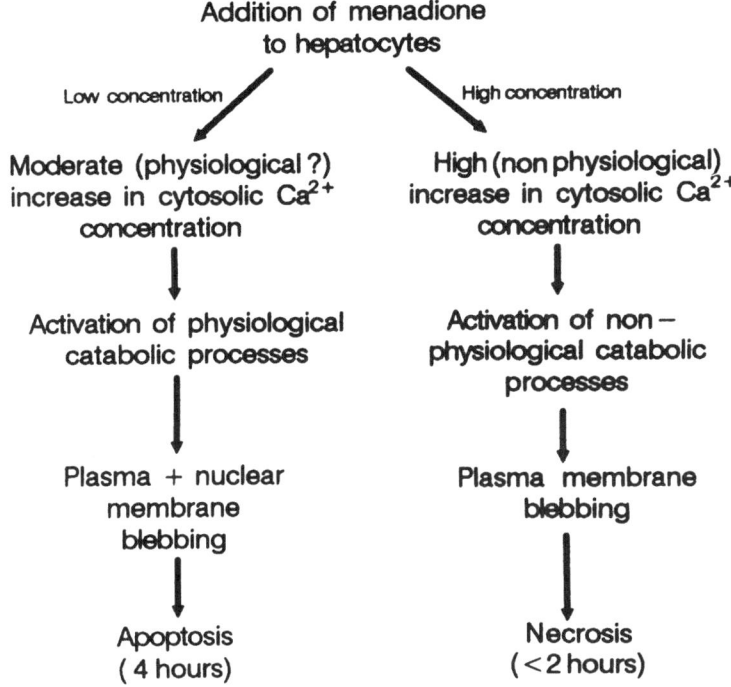

FIGURE 4. Two mechanisms of cell death induced by menadione through the elevation of cytosolic Ca^{2+}.

1980) and those isolated from menadione-treated hepatocytes (McConkey *et al.*, 1988). Gel electrophoresis of these products has shown that host chromatin is preferentially cleaved into oligonucleosome length fragments by the endonuclease in each case. Thus, it appears that the activation of hepatocyte endonuclease activity is not only a critical step in the sequence of events involved in programmed cell death in the liver but can also occur during the development of oxidative cell injury. Figure 4 shows how the results of this study could fit into a model for two modes of menadione-induced cell death.

8. CONCLUDING REMARKS

A series of potentially toxic mechanisms are activated when a cell is exposed to agents which cause oxidative stress. They include a perturbation of the normal redox balance, thiol depletion, and disruption of intracellular Ca^{2+} homeostasis. The pathways through which a perturbation of intracellular Ca^{2+} homeostasis may cause irreversible damage appear to involve the activation of catabolic enzymes such as phospholipases, proteases, and endonucleases. Although additional mechanisms may contribute to cell killing, our findings suggest that a disruption of Ca^{2+} homeostasis is a critical event in the development of lethal cell injury.

ACKNOWLEDGMENTS. These studies were supported by grants from the Swedish Medical Research Council (03X-2471) and the Swedish Council for the Planning and Coordination of Research.

REFERENCES

Bellomo, G., Mirabelli, F., Richelmi, P., and Orrenius S., 1983, Critical role of sulfhydryl group(s) in the ATP-dependent Ca^{2+}-sequestration by the plasma membrane fraction from rat liver, *FEBS Lett.* **163**:136–139.

Campbell, A. K., 1983, Intracellular Ca^{2+}: Its universal role as regulator, John Wiley and Sons Ltd., London.

Cheung, W. Y., 1980, Calmodulin plays a pivotal role in cellular regulation, *Science* **217**:1257–1259.

Chien, K. R., Pfau, R. G., and Farber, J. L., 1979, Ischemic myocardial cell injury. Prevention by chlorpromazine of an accelerated phospholipid degradation and associated membrane dysfunction, *Am. J. Pathol.* **97**:505–530.

Cohen, J. J., and Duke, R. L., 1984, Glucocorticoid-activation of a calcium-dependent endonuclease in thymocyte nuclei leads to cell death, *J. Immunol.* **132**:38–42.

Collier, N. L., and Wang, K., 1982, Purification and properties of human platelets P235, *J. Biol. Chem.* **257**:6937–6943.

Di Monte, D., Bellomo, G., Thor, H., Nicotera, P., and Orrenius, S., 1984, Menadione-induced cytotoxicity is associated with protein thiol oxidation and alteration in intracellular Ca^{2+} homeostasis, *Arch. Biochem. Biophys.* **235**:343–350.

Ernster, L., 1967, DT-diaphorase, *Meth. Enzymol.* **10**:309–321.

Farber, J. L., 1981, The role of Ca^{2+} in cell death, *Life Sci.* **29**:1289–1295.

Jewell, S. A., Bellomo, G., Thor, H., Orrenius, S., and Smith, M. T., 1982, Bleb formation in hepatocytes during drug metabolism is caused by disturbances in thiol and calcium ion homeostasis, *Science* **214**:1257–1259.

Lehninger, A. L., Vercesi, A., and Bababunmi, E., 1978, Regulation of Ca^{2+} release from mitochondria by the oxidation-reduction state of pyridine nucleotides, *Proc. Natl. Acad. Sci. USA* **75**:1690–1694.

McConkey, D., Nicotera, P., Hartzell, P., Wyllie, A. H., and Orrenius, S., 1988, Stimulation of endogenous endonuclease activity in hepatocytes exposed to oxidative stress, *Toxicol. Lett.* **42**:123–130.

Moore, G. A., O'Brien, J. P., and Orrenius, S., 1986, Menadione (2 methyl-1,4-naphthoquinone)-induced Ca^{2+} release from rat-liver mitochondria is caused by NAD(P)H oxidation, *Xenobiotica* **16**:873–882.

Moore, L., Chen, T., Knapp, H. R., and Landon, E., 1975, Energy-dependent Ca^{2+} sequestration activity in rat liver microsomes, *J. Biol. Chem.* **250**:4562–4568.

Murachi, T., 1983, Intracellular Ca^{2+} proteases and its inhibitor protein: Calpain and calpastatin, in: Ca^{2+} and Cell Function, Vol. 4 (W. Y. Cheung, ed.), Academic Press, Orlando, pp. 376–410.

Nicotera, P., Moore, M., Mirabelli, F., Bellomo, G., and Orrenius, S., 1985, Inhibition of hepatocyte plasma membrane Ca^{2+}-ATPase activity by menadione metabolism and its restoration by thiols, *FEBS Lett.* **181**:149–153.

Nicotera, P., Hartzell, P., Baldi, C., Svensson, S. Å., Bellomo, G., and Orrenius, S., 1986a, Cystamine induces toxicity in hepatocytes through the elevation of cytosolic Ca^{2+} and the stimulation of a non-lysosomal proteolytic system, *J. Biol. Chem.* **261**:14628–14635.

Nicotera, P., Hartzell, P., Davis, G., and Orrenius, S., 1986b, The formation of plasma membrane blebs in hepatocytes exposed to agents that increase cytosolic Ca^{2+} is mediated by the activation of a non-lysosomal proteolytic system, *FEBS Lett.* **209**:139–144.

Nicotera, P., McConkey, D., Svensson, S. Å., Bellomo, G., and Orrenius, S., 1988, Correlation between cytosolic Ca^{2+} concentration and cytotoxicity in hepatocytes exposed to oxidative stress, *Toxicology* **52**: 55–63.

Orrenius, S., and Bellomo, G., 1986, Toxicological implications of perturbation of Ca^{2+} homeostasis in hepatocytes, in: Ca^{2+} and Cell Function, Vol. 6 (W. Y. Cheung, ed.), Academic Press, Orlando, pp. 186–204.

Orrenius, S., and Moldéus, P., 1984, The multiple roles of glutathione in drug metabolism, Trends Pharmacol. Sci. 5:432–635.

Sies, H., 1986, Biochemistry of oxidative stress, Angew. Chem. Int. Ed. Engl. 25:1058–1071.

Shanne, F. A., Kane, A. B., Young, E. E., and Farber, J. L., 1979, Calcium dependence of toxic cell death: A final common pathway, Science 206:700–702.

Smith, M. T., Thor, H., Jewell, S. A., Bellomo, G., Sandy, M. S., and Orrenius, S., 1984, Free radical-induced changes in the surface morphology of isolated hepatocytes, in: Free Radicals in Molecular Biology, Aging and Disease, Vol. 27 (D. Armstrong, R. S. Sohal, R. G. Cutler, and T. F. Slater, eds.), Raven Press, New York, pp. 103–118.

Starke, P. E., Hoek, J. B., and Farber, J. L., 1986, Calcium-dependent and calcium-independent mechanisms of irreversible cell injury in cultured hepatocytes, J. Biol. Chem. 261:3006–3012.

Thor, H., Smith, M. T., Hartzell, P., Bellomo, G., Jewell, S. A., and Orrenius, S., 1982, The metabolism of menadione (2-methyl-1,4-naphthoquinone) by isolated hepatocytes: A study of implications of oxidative stress in intact cells, J. Biol. Chem. 257:12419–12424.

Thor, H., Hartzell, P., Svensson, S. Å., Orrenius, S., Mirabelli, F., Marinoni, V., and Bellomo, G., 1985, On the role of thiol groups in the inhibition of liver microsomal Ca^{2+}-sequestration by toxic agents, Biochem. Pharmacol. 34:3717–3723.

Wyllie, A. H., 1980, Glucocorticoid-induced thymocyte apoptosis is associated with endogenous endonuclease activation, Nature 284:555–556.

Wyllie, A. H., Morris, R. G., Smith, A. C., and Dunlop, D., 1984, Chromatin cleavage in apoptosis: Associated with condensed chromatin morphology and dependence on macromolecular synthesis, J. Pathol. 14:67–77.

48

Cytosolic Free Calcium and Cell Injury in Hepatocytes

JOHN J. LEMASTERS, ANNA-LIISA NIEMINEN,
GREGORY J. GORES, BARNABY E. WRAY, and
BRIAN HERMAN

1. INTRODUCTION

Formation of cell surface blebs is an early event in hypoxic and toxic injury to liver (Lemasters *et al.*, 1981, 1983; Jewell *et al.*, 1982). Many authors have proposed that a rise of cytosolic free Ca^{2+} is the stimulus for bleb formation and the initiating factor in a sequence of events leading to irreversible injury and cell death (see Schanne *et al.*, 1979; Trump *et al.*, 1980; Bellomo and Orrenius, 1985). Recently, we applied the technique of digitized video microscopy to quantitate changes in cytosolic free Ca^{2+} in relation to blebbing and other cellular parameters during "chemical hypoxia" with metabolic inhibitors in single cultured hepatocytes (Lemasters *et al.*, 1987). Here we present new data examining the relation of bleb formation to the onset of cell death during cellular injury induced by a variety of toxic chemicals. We also determine whether changes in cytosolic free Ca^{2+} occur during cellular injury which might initiate bleb formation and lead to cell death. The results indicate that the onset of cell death is a very rapid event initiated by rupture of a large plasma membrane bleb leaving the cell in a hyperpermeable state. An increase in cytosolic free Ca^{2+} is not a necessity for the progression of bleb formation or the onset of cell death.

2. MATERIALS AND METHODS

Rat hepatocytes were isolated by collagenase digestion and cultured for 24 hr on collagen-coated coverslips at 37°C with 5% CO_2 in Waymouth media MB-752/2 containing 100 nM insulin and 1% fetal calf serum. Coverslips were incubated for 30 min with Fura-2/acetoxymethyl ester (2.5–5 μM), washed three times with Hank's medium, and

JOHN J. LEMASTERS, ANNA-LIISA NIEMINEN, GREGORY J. GORES, BARNABY E. WRAY, and BRIAN HERMAN • Laboratories for Cell Biology, Department of Cell Biology and Anatomy, School of Medicine, University of North Carolina at Chapel Hill, Chapel Hill, North Carolina 27599-7090.

transferred to Krebs–Henseleit bicarbonate medium buffered with 20 mM Na-Hepes, pH 7.4, containing propidium iodide (50–500 nM), and mounted in a chamber on the stage of a digitized video microscope. Described in detail elsewhere (DiGuiseppi *et al.*, 1985), the digitized video microscope consisted of an inverted fluorescence microscope equipped with UV-transmitting optics and light source, a low-light intensified silicon-intensified target (ISIT) television camera, and computer components to digitize, store, and analyze video images. The output of the camera was also recorded using a time-lapse video cassette recorder.

Cytoplasmic esterases serve to hydrolyze acetoxymethyl ester bonds and release free Fura-2 into the cytosol. Excited at 350 or 340 nm, Fura-2 exhibits a Ca^{2+}-dependent fluorescence with an apparent K_d for Ca^{2+} of about 225 nM, whereas at an excitation wavelength of 365 nm, Fura-2 fluorescence is independent of Ca^{2+} concentration (Grynkiewicz *et al.*, 1985). Therefore, we used the ratio of the fluorescence intensities at the Ca^{2+}-dependent and Ca^{2+}-independent wavelengths to determine cytosolic free Ca^{2+} quantitatively with a spatial resolution of 1 μm within individual hepatocytes. The ratio method corrected for variable loading of Fura-2 and differential volume fractions in the two-dimensional image. Ca^{2+} concentrations so calculated were displayed as a grey scale map or averaged for the total cell. The other fluorescent probe employed, propidium iodide, labeled nuclei of non-viable cells where propidium iodide labeling occurs identically with trypan blue labeling (Lemasters *et al.*, 1986). The high sensitivity of the ISIT camera together with the frame-averaging capability of our system allowed us to work at very low levels of excitation energy and fluorescent probe concentration, thus preventing photobleaching, photodamage, and any consequent disruption of cellular functions.

Three types of toxic regimes were employed. KCN (2.5–5 mM) and iodoacetate (0.5–10 mM) were employed to inhibit ATP production by oxidative phosphorylation and glycolysis. This treatment mimics ATP depletion which occurs during oxygen deprivation; hence we term it "chemical hypoxia." $HgCl_2$ (50 μM) was used to disrupt plasma membrane transport mechanisms (see Zalme *et al.*, 1976). Cystamine (5 mM) was employed to cause oxidative stress. Cystamine is membrane-impermeant and acts by forming mixed disulfides with plasma membrane protein thiols (Nicotera *et al.*, 1986).

3. RESULTS AND DISCUSSION

3.1. Progression of Bleb Formation after Chemical Hypoxia

Bleb formation after chemical hypoxia with KCN plus iodoacetate was followed using time-lapse video recording (Fig. 1). Bleb formation occurred in three stages. Stage I involved the formation of numerous small surface blebs (arrowheads, Fig. 1A) within 15 min of addition of the metabolic inhibitors. The blebs were always clear and excluded any granular organelles. In stage II, beginning about 20–25 min after induction of chemical hypoxia, the small blebs enlarged and coalesced by fusion (arrowheads, Fig. 1B and C) until one to three large terminal blebs per cell remained (Fig. 1C; see also Fig. 2). Bleb enlargement was a gradual process, but bleb fusion resembled the fusion of soap bubbles and was complete within consecutive video frames (33 msec) (data not shown). The third and final stage of bleb development was initiated by rupture of one of the terminal blebs (Fig. 1D). Subsequent to this lysis, further bleb enlargement ceased. Progression through stages I and II to stage III required 30–60 min during chemical hypoxia.

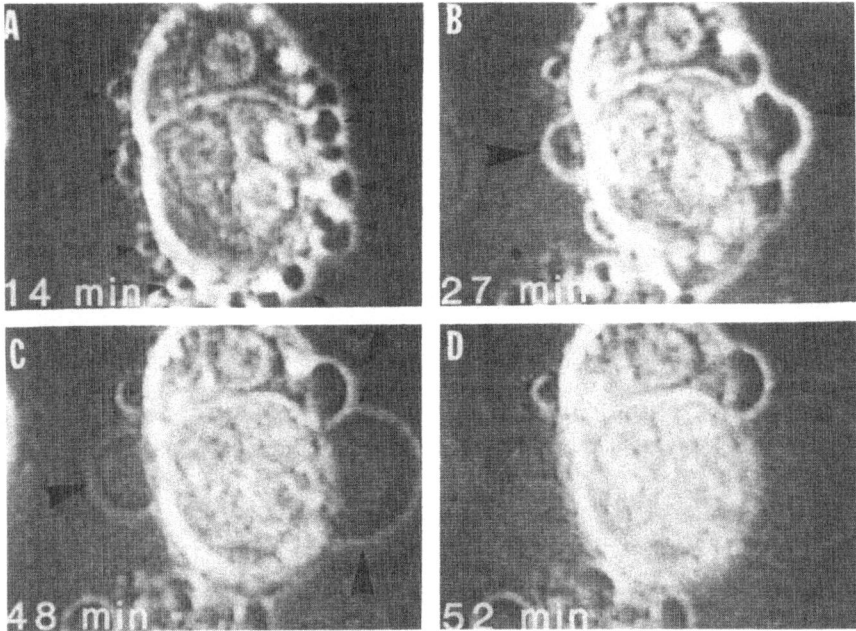

FIGURE 1. Three stages of bleb development. (A) Stage I. (B) Early stage II. (C) Late stage II. (D) Stage III. Cultured hepatocytes were subjected to chemical hypoxia with KCN plus iodoacetic acid. The elapsed time after addition of the toxins is shown in each frame. Arrowheads identify examples of blebs. In B arrowheads identify blebs formed by fusion of smaller blebs. Note the disappearance in D of the lower left bleb seen previously in panel C.

3.2. Bleb Rupture and Propidium Iodide Uptake

Rupture of a terminal bleb was very rapid (within 33 msec) and resembled a soap bubble bursting. Figure 2 shows the time course of propidium iodide uptake following bleb rupture. After about 50 min of chemical hypoxia, a cultured hepatocyte had developed three large terminal blebs on its surface (Fig. 2A). A second later, one of the blebs ruptured and disappeared from view (Fig. 2B), and after 3 sec, propidium iodide began to enter the cell to produce diffuse staining of the cytoplasm (Fig. 2C). Two minutes later, propidium iodide had accumulated to a high concentration in the nucleus and was visible even with the phase lamp on (Fig. 2C). In other experiments, we observed that trapped anionic cytosolic probes such as Fura-2 or a fluorescein derivative rapidly exited the cell following bleb rupture (data not shown). These results indicate that at the time of bleb rupture, the cell is left in a hyperpermeable state which permits entry of propidium iodide into the cytoplasmic and nuclear compartments, exit of cytoplasmic trapped dyes, and presumably collapse of ionic gradients and loss of metabolic intermediates and enzymes. Thus, bleb rupture is the initiating factor in the abrupt onset of cell death.

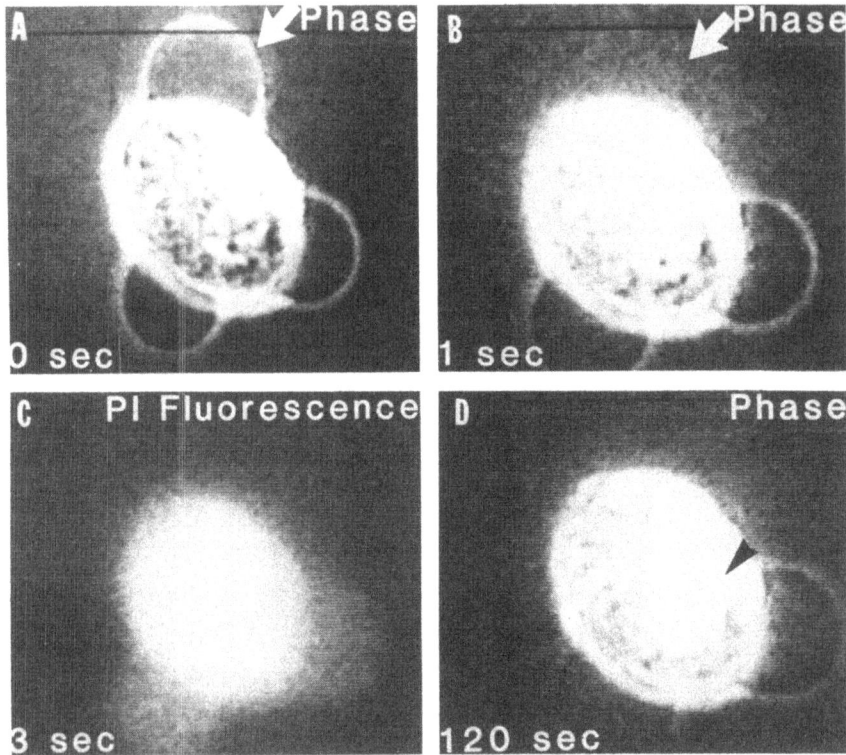

FIGURE 2. Time course of propidium iodide labeling after bleb rupture. Cultured hepatocytes were subjected to chemical hypoxia with propidium iodide in the medium. After about 50 min, a hepatocyte showed three large terminal blebs (A). One of these blebs ruptured (B, arrowhead) and propidium iodide began to enter the cell within seconds (C). Nuclear uptake of propidium iodide was evident after 2 min (D, arrowhead). The elapsed time from when the image in A was acquired is shown in the bottom left corner of each panel. In D, the hepatocyte was simultaneously illuminated with the phase and fluorescence light sources.

3.3. Cytosolic Free Ca^{2+} during Cell Injury

Ratio imaging of Fura-2 fluorescence was used to determine cytosolic free Ca^{2+} after toxic injury with KCN plus iodoacetate, $HgCl_2$, and cystamine (Fig. 3). After KCN plus iodoacetate, no change of cytosolic free Ca^{2+} preceded formation of blebs or the transition to irreversible injury (Fig.-3C). In these experiments, all cells initially contained Fura-2 fluorescence and were counted; however, as the experiment progressed individual cells died and Fura-2 leaked out of these dead cells. Once this began to happen (as indicated by the dotted lines in Fig. 3), average free Ca^{2+} for the remaining cells was calculated since free Ca^{2+} in the dead cells could no longer be determined. After 55 min of chemical hypoxia, all cells had lost viability and their content of Fura-2. Several control experiments were performed which demonstrated that the Fura-2 signal was responsive to the effects of alterations in cytosolic free Ca^{2+}. Various Ca^{2+}-mobilizing hormones (e.g., phenylephrine,

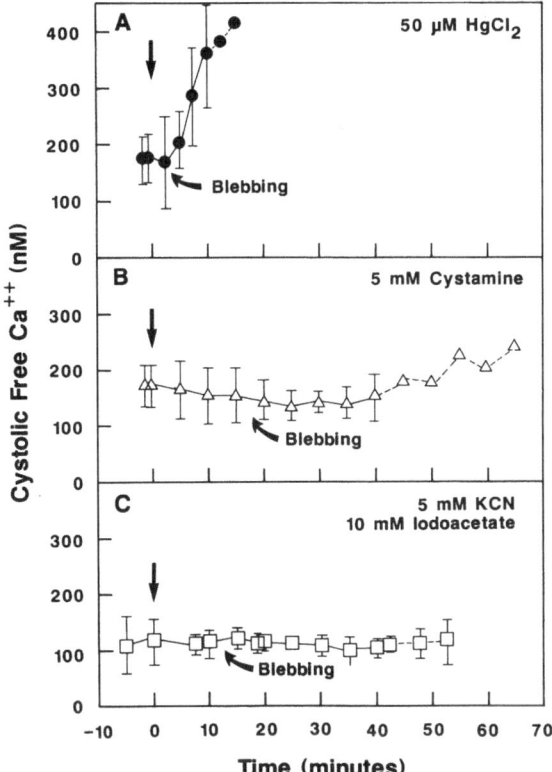

FIGURE 3. Cytosolic free Ca^{2+} after toxic injury to cultured hepatocytes. Hepatocytes were labeled with Fura-2 and incubated with 50 μM HgCl$_2$, 5 mM cystamine, or 5 mM KCN plus 10 mM iodoacetic acid. Cytosolic free Ca^{2+} was determined by Fura-2 ratio imaging on a pixel-by-pixel basis and averaged for all cells in a field. Dotted lines indicated the first loss of Fura-2 in individual experiments. Curved arrows indicate the time when blebbing was first observed.

vasopressin, and epidermal growth factor) increased cytosolic free Ca^{2+} by one- to threefold as measured by Fura-2 ratio imaging. Conversely, EGTA added to lower extracellular Ca^{2+} also lowered cytosolic free Ca^{2+}. Thus Fura-2 was competent to report both increases and decreases in cytosolic free Ca^{2+}. Interestingly, despite the increases of cytosolic free Ca^{2+} produced by the hormones, cell surface blebbing did not occur and viability was not lost.

Following HgCl$_2$, blebbing occurred very rapidly and more quickly than during chemical hypoxia. Cell viability was lost within 20 min. Unlike chemical hypoxia, average cytosolic free Ca^{2+} increased from an initial average value of about 180 nM to more than 400 nM (Fig. 3A). Grey scale maps showed that before addition of HgCl$_2$ acytosolic free Ca^{2+} was evenly distributed within the cells (Fig. 4A). However, after 16 min exposure to HgCl$_2$, free Ca^{2+} distribution was inhomogeneous, and highest concentrations exceeding 1 μM were observed inside the blebs. Since blebs contained endoplasmic reticulum but not mitochondria (see Lemasters et al., 1981, 1983), the lower free Ca^{2+} of the central portion of the cell may be the consequence of Ca^{2+} uptake by mitochondria, suggestive that at this stage of

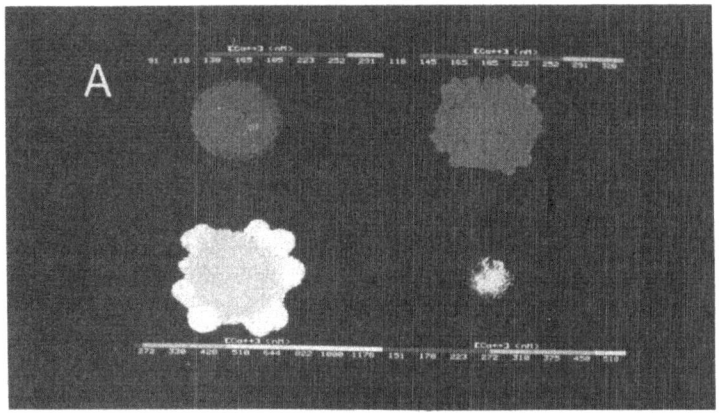

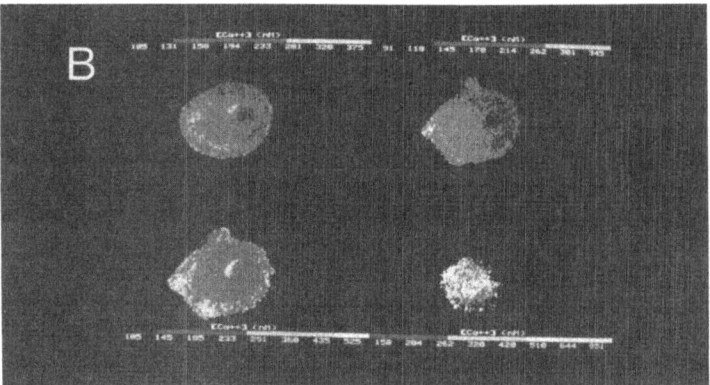

FIGURE 4. Grey scale maps of cytosolic free Ca²⁺ after toxic injury to cultured hepatocytes. In A, a hepatocyte was exposed to 50 μM HgCl₂. In B, a hepatocyte was exposed to 5 mM cystamine. Time of exposure: in A top left is baseline, top right is 8.0 min, bottom left is 16.0 min, and bottom right is 28.4 min; in B, top left is baseline, top right is 38.4 min, bottom left is 45.1 min, and bottom right is 77.3 min. The grey scale is the same for all experiments. The grey scale key bar below each panel indicates upper and lower range of cytosolic free Ca²⁺ for each image. The bottom right panels in A and B were taken after the cells had died and leaked Fura-2. The images record the weak autofluorescence of the nucleus. Cytoplasmic fluorescence above background was absent.

injury the endoplasmic reticulum is overloaded or otherwise impaired in Ca²⁺ uptake. Although bleb formation and the increase of free Ca²⁺ were quite marked after HgCl₂, the onset of bleb formation actually preceded the increase in free Ca²⁺ as shown in the upper right panel of Fig. 4B.

The dissociation of bleb formation from an elevation of cytosolic free Ca²⁺ was also found in cystamine toxicity. Cystamine was less toxic in terms of killing time than HgCl₂ or KCN plus iodoacetate. Blebbing occurred after between 10 and 25 min, and loss of viability ensued after between 40 and 70 min. Bleb formation again preceded cell death

(Fig. 4B) but as in chemical hypoxia was not accompanied by an increase of cytosolic free Ca^{2+} (Fig. 3B). In three experiments, cytosolic free Ca^{2+} had an average maximal change of 3 ± 35% (SD).

The absence of an increase of cytosolic free Ca^{2+} after cystamine appears inconsistent with the recent work by Nicotera et al. (1986), who reported an increase of Ca^{2+} after exposure of freshly isolated hepatocytes to cystamine. Recent preliminary experiments performed by us suggest an explanation for this discrepancy. Freshly isolated hepatocytes were plated on collagen-coated coverslips and incubated for 2–24 hr. In cells incubated only 2 hr, cytosolic free Ca^{2+} did increase after addition of KCN and iodoacetate. The magnitude of the increase in cytosolic free Ca^{2+} declined as the length of incubation increased. Hepatocytes incubated for 12 hr or more did not display an increase in cytosolic free Ca^{2+} after KCN plus iodoacetate. Basal cytosolic free Ca^{2+} levels also decreased as the length of time the cells were cultured increased. These preliminary results suggest that the plasma membranes of freshly isolated cells are relatively leaky to Ca^{2+}; during incubation in culture medium, the membranes lose this leakiness. This observation may serve to explain the discrepancies regarding levels of cytosolic free Ca^{2+} following cystamine between our data in 24-hr cultured cells and those of Nicotera et al. (1986) in freshly isolated hepatocytes.

4. CONCLUSION

In conclusion, cell surface blebs developed rapidly in cultured hepatocytes following "chemical hypoxia" and after exposure to toxic chemicals. Bleb formation occurred in three stages. Stage I was characterized by numerous small blebs. In stage II, blebs grew and fused until one to three terminal blebs remained. The onset of stage III occurred when a terminal bleb ruptured. Bleb formation in these models was not the consequence of an increase in cytosolic free Ca^{2+}, although free Ca^{2+} may rise after exposure of cells to certain toxic agents. Even when cytosolic free Ca^{2+} increased, bleb formation preceded rather than followed the change in ion concentration. Since cell morphology did not change after addition of Ca^{2+}-mobilizing hormones, an increase in cytosolic free Ca^{2+} was neither necessary nor sufficient to cause bleb formation or cell death. The rapid entry of propidium iodide following bleb rupture points to this event as the precipitating event in cell death.

ACKNOWLEDGMENTS. This work was supported in part by grants DK37034, HL35490, and AG07218 from the National Institutes of Health and grant-in-aid 86-1299 from the American Heart Association. J. J. L. is an established investigator of the American Heart Association.

REFERENCES

Bellomo, G., and Orrenius, S., 1985, Altered thiol and calcium homeostasis in oxidative hepatocellular injury, Hepatology 5:876–882.
DiGuiseppi, J., Inman, R., Ishihara, I., Jacobson, K., and Herman, B., 1985, Applications of digitized fluorescence microscopy to problems in cell biology, BioTechniques 3:394–403.
Grynkiewicz, G., Poenie, M., and Tsien, R. Y., 1985, A new generation of Ca^{2+} indicators with greatly improved fluorescence properties, J. Biol. Chem. 260:3440–3450.
Jewell, S. A., Bellomo, G., Thor, H., and Orrenius, S., 1982, Bleb formation in hepatocytes during drug metabolism is caused by disturbances in thiol and calcium ion homeostasis, Science 217:1257–1259.

Lemasters, J. J., Ji, S., and Thurman, R. G., 1981, Centrilobular injury following hypoxia in isolated, perfused rat liver, *Science* **213**:661–663.

Lemasters, J. J., Stemkowski, C. J., Ji, S., and Thurman, R. G., 1983, Cell surface changes and enzyme release during hypoxia and reoxygenation in the isolated, perfused rat liver, *J. Cell Biol.* **97**:778–786.

Lemasters, J. J., Fleishman, K. E., Nieminen, A.-L., Shah, S., and Herman, B., 1986, Continuous measurement of cell viability in suspensions of isolated hepatocytes by propidium iodide fluorometry, *Hepatology* **6**:1177.

Lemasters, J. J., DiGuiseppi, J., Nieminen, A.-L., and Herman, B., 1987, Blebbing, free Ca^{2+} and mitochondrial membrane potential preceding cell death in hepatocytes, *Nature* **325**:78–81.

Nicotera, P., Hartzell, P., Baldi, C., Svensson, S.-A., Bellomo, G., and Orrenius, S., 1986, Cystamine induces toxicity in hepatocytes through the elevation of cytosolic Ca^{2+} and the stimulation of a nonlysosomal proteolytic system, *J. Biol. Chem.* **261**:14628–14635.

Schanne, F. A. X., Kane, A. B., Young, E. E., and Farber, J. L., 1979, Calcium dependence of toxic cell death: A final common pathway, *Science* **206**:700–702.

Trump, B. F., Berezesky, I. K., Laiho, K. U., Osornio, A. R., Mergner, W. J., and Smith, M. W., 1980, The role of calcium in cell injury. A review, in *Scanning Electron Microscopy* (A. M. R. O'Hare, ed.), SEM, Inc., Chicago, pp. 437–492.

Zalme, R. C., McDowell, E. M., Nagle, R. B., McNeil, J. S., Flamenbaum, W., and Trump, B. F., 1976, Studies on the pathophysiology of acute renal failure II. A histochemical study of the proximal tubule of the rat following administration of mercuric chloride, *Virchows Arch. B. Cell Pathol.* **22**:197–216.

49

Mechanisms of Anoxic Injury to Transport and Metabolism of Proximal Renal Tubules

LAZARO J. MANDEL, WILLIAM R. JACOBS, RICK SCHNELLMANN, MARIA SGAMBATI, ANN LeFURGEY, and PETER INGRAM

1. INTRODUCTION

The pathological effects of ischemia have been extensively studied in the kidney (e.g., Venkatachalam *et al.*, 1978; Trump *et al.*, 1982; Johnston *et al.*, 1984). By subjecting the kidney to ischemia of varying duration, investigators have been able to describe the many morphological, physiological, and biochemical alterations that occur in the kidney as a function of time. Upon deprivation of oxygen, proximal tubules rapidly lose K, gain Na, and decrease their ATP content. Within 15 min of clamp-induced ischemia, structural alterations are clearly observed in the form of cellular and mitochondrial swelling and irregular brush borders. At the same time, mitochondrial dysfunctions are clearly discernible (Kahng *et al.*, 1978).

Most renal ischemia studies have been performed in animals *in vivo* by clamping the renal artery for varying amounts of time, after which reflow is initiated. Thus, anoxia (or hypoxia) is produced simultaneously with a hemodynamic blockage which eliminates tubule perfusion, inhibiting both substrate support and removal of metabolic products. Therefore, when experimental maneuvers have been applied which ameliorate the effects of ischemia, it has been difficult to separate hemodynamic from tubular effects. In the present study, we utilized a suspension of proximal tubules obtained from the rabbit kidney (Balaban *et al.*, 1980), which allowed the investigation of the effects of anoxia independently from the other consequences of clamp-induced ischemia. The suspension was subjected to anoxia of short

LAZARO J. MANDEL, WILLIAM R. JACOBS, RICK SCHNELLMANN, and MARIA SGAMBATI ● Duke University Medical Center, Durham, North Carolina 27710.   ANN LeFURGEY ● Department of Physiology, Duke University Medical Center, Durham, North Carolina 27710.   PETER INGRAM ● Research Triangle Institute, Research Triangle Park, North Carolina 27709.

duration (10–40 min) with subsequent reoxygenation *in vitro,* and the effects of these perturbations were evaluated on a number of cellular transport and respiratory parameters.

Cellular respiration was measured upon addition of nystatin (Nys) to maximally stimulate
Na pump activity (Harris *et al.,* 1981). This is the equivalent of a cellular "stress test,"
since the increase in pump activity requires more ATP, stimulating respiration to the maximal
mitochondrial respiratory rate (Harris *et al.,* 1981). The cellular content of K was measured
as an index of ionic integrity and the release of lactate dehydrogenase (LDH) was measured
as an index of plasma membrane damage.

In attempting to elucidate the mechanisms of anoxic cell injury, the roles of three
variables were particularly examined. These cellular variables were ATP, calcium, and
glutathione. In each case, factors that determined reversibility versus irreversibility of the
injury were stressed.

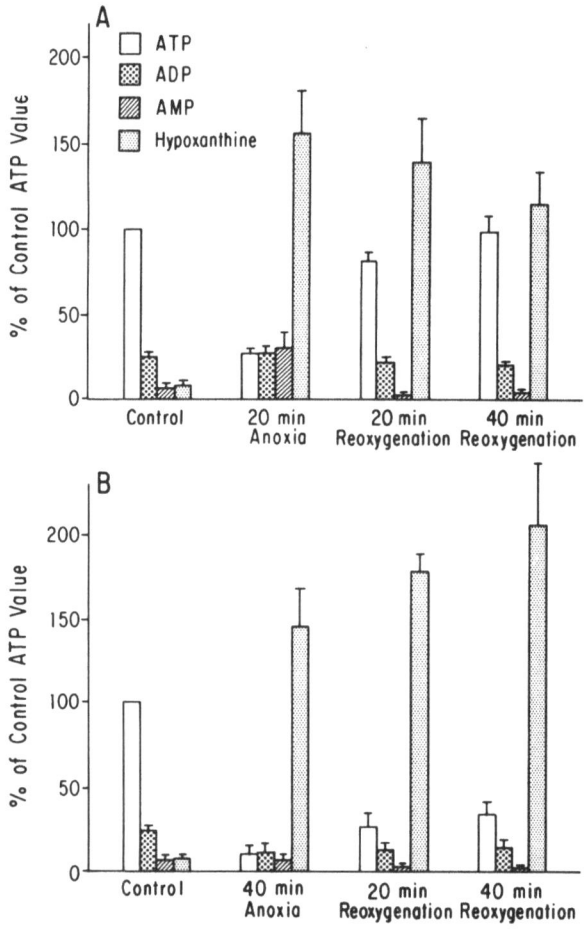

FIGURE 1. Cellular content of ATP, ADP, AMP, and HX when the tubules were subjected to
either (A) 20 min of anoxia or (B) 40 min of anoxia, followed by 40 min of reoxygenation (*n* =
4). [Reproduced with permission from Mandel *et al.* (1988).]

2. ROLE OF ATP

Results from our laboratory (Takano *et al.*, 1985) show that 10 and 20 min of anoxia partially inhibited Nys-stimulated respiration and partially decreased the K content, but these effects were largely reversible after 20 min of reoxygenation. After 40 min of anoxia and 20 min of reoxygenation, all these variables remained irreversibly inhibited: Nys-stimulated respiration by 54%, K content by 42%, and LDH release was 40% of total.

ATP levels decreased rapidly during the first 10 min of anoxia and continued decreasing by a total of 80–90% during 40 min of anoxia. Similarly, total adenine nucleotide content decreased, being inhibited by 75% after 40 min of anoxia. Recovery from anoxia was examined in paired experiments in which tubules were exposed to 20 or 40 min of anoxia (Mandel *et al.*, 1988). As shown in Fig. 1A, the ATP levels increased from 27% of control immediately after 20 min of anoxia to 76% of control following 20 min of oxygenated recovery. The decrease in AMP levels (to almost zero) accounted stoichiometrically for most of the ATP recovery, the rest probably due to resynthesis from other sources which included hypoxanthine (HX). Complete ATP recovery was achieved after 40 min of reoxygenation, and this was accompanied by a small but significant decline in HX.

The pattern of recovery was very different after 40 min of anoxia, as shown in Fig. 2B. Reoxygenation for 20 min produced recovery of ATP from 10 to 27% of control and an additional 20 min of reoxygenation only increased this value to 34% of control. Again, much of the ATP recovery could be attributed to the decrease in AMP levels during this period; however, the total nucleotide content remained inhibited. The HX level was already 45% larger than the control ATP value after 40 min of anoxia. During recovery, the HX level continued to *increase*, in contrast to the behavior observed during the recovery from 20 min of anoxia (Fig. 2A). The continued production of HX suggests that the cells may be undergoing a generalized deterioration, leading to irreversible cell injury.

A series of paired experiments was performed to determine the effects of Mg-ATP addition on adenine nucleotides and their breakdown products during 40 min of anoxia

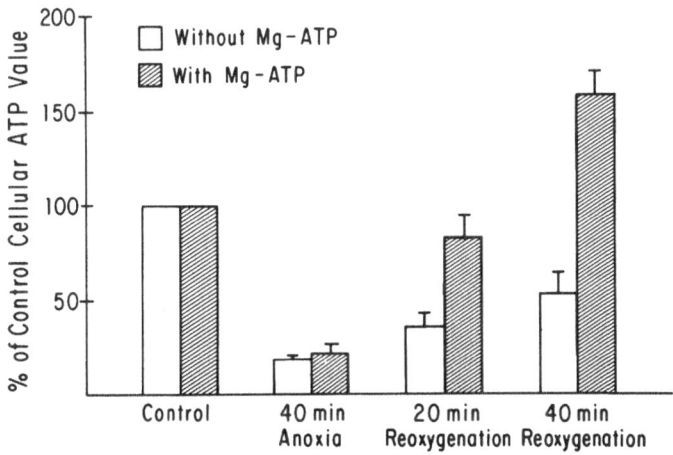

FIGURE 2. Cellular ATP contents of tubules exposed to 40 min of anoxia followed by 40 min of reoxygenation. Paired studies (n = 5) comparing the effects of external Mg-ATP addition during anoxia. [Reproduced with permission from Mandel *et al.* (1988).]

followed by 40 min of reoxygenation (Mandel et al., 1988). Mg-ATP was added in five aliquots, each adding a concentration of 0.2 mM to the solution at 10-min intervals during the 40 min of anoxia. The presence of external Mg-ATP during anoxia protected the plasma membrane, as measured by the decrease in LDH release from 30 to 40% without additions to only 10% when Mg-ATP was present (Takano et al., 1985). As seen in Fig. 2, addition of extracellular Mg-ATP did not change the measured value of intracellular ATP after 40 min of anoxia. This result suggests that the extracellular presence of ATP or its breakdown products during anoxia may stabilize the plasma membrane and help preserve cellular integrity. During reoxygenation there were significant increases in the ATP recoveries obtained after 20 and 40 min. The addition of Mg-ATP during anoxia produced an $82 \pm 11\%$ recovery of ATP level after 20 min and $157 \pm 12\%$ recovery following 40 min of reoxygenation. Other experiments (Mandel et al., 1988) suggest that this replenishment of ATP content during reoxygenation proceeds in three states: breakdown of external ATP to AMP and adenosine (ADO), transport of ADO into the cells, and resynthesis of intracellular ATP.

These results emphasize the role of ATP and plasma membrane damage in anoxic renal injury. The large amount of membrane damage observed after 40 min of anoxia is associated with poor recovery of cellular ATP. The enhanced recovery from anoxia obtained upon external addition of Mg-ATP occurs through two steps: (1) protection of plasma membrane breakdown during anoxia, and (2) supply of adenine nucleotide precursors for ATP resynthesis during reoxygenation.

3. ROLE OF CALCIUM

Previous results (Takano et al., 1985) showed that decreasing the extracellular calcium concentration from 1 mM to 2.5 μM during anoxia drastically reduced the percentage of LDH loss during anoxia. Since previous investigators have suggested that the intracellular free-calcium concentration (Ca_f) increases during anoxia (Farber et al., 1988; Trump et al., 1988), this protection of LDH release could have occurred through an amelioration of the postulated increase in Ca_f. These considerations motivated the direct measurement of Ca_f during anoxia in proximal renal tubules using Fura 2 (Jacobs and Mandel, 1987b).

In a series of preliminary experiments, Ca_f was measured as a function of time in anoxia. Under normoxic conditions, Ca_f averaged 100 ± 15 nM ($n = 6$) and was unchanged after 5, 15, and 40 min of anoxia ($n = 3$), or during reoxygenation. This result is identical to that recently obtained by Lemasters et al. (1987) in hepatocytes subjected to anoxia, but differs from that obtained by Borle et al. (1986) and Goligorsky et al. (1987), who observed increases in Ca_f during anoxia in cultured renal cells. The variability in the response of Ca_f may reflect differences in plasma membrane calcium permeability or cytoplasmic calcium-buffering capacity between freshly isolated and cultured renal cells. Another aspect that needs to be evaluated concerns the different methods used: aequorin by Borle et al. (1986), Fura 2 (340:380 nm ratio) by Goligorsky et al. (1987), and our method (Jacobs and Mandel, 1987a), which permeabilizes the cells with digitonin to calibrate the Fura 2 signal. Nevertheless, our results suggest that no increase in Ca_f occurs under conditions that cause demonstrable damage to the renal cell. Therefore, changes in cytosolic Ca may not be necessary to initiate the cascade of events that lead to the damage. Conversely, when Ca_f does increase, it may reflect differences in cellular calcium handling rather than a causal relationship to cellular damage.

Another possibility that was explored concerned alterations in intracellular calcium compartmentation during anoxia. This was approached by use of quantitative electron probe X-ray microanalysis. This technique allows the measurement of Ca content within subcellular compartments in thin cryosections obtained from biological samples (Somlyo *et al.*, 1978). The application of these techniques to proximal renal tubules has been described in detail elsewhere (LeFurgey *et al.*, 1986). A typical spectrum obtained from a mitochondrion of a healthy proximal renal tubule cell is shown in Fig. 3. It shows low Na and Cl contents and high K, which are typical of a healthy cell. Its Ca content is extremely low and has to be calculated by a deconvolution of the K-Kβ peak from the overlapped Ca-Kα peak (Kitazawa *et al.*, 1983). The great majority of cells in the suspension display a K/Na ratio of 3:4 and Ca content of 3–4 mmol/g dry wt (4–5 nmol/mg protein) in both the cytoplasm and the mitochondria. Preliminary results suggest that the cytoplasmic calcium content of anoxic cells may be unchanged or slightly elevated, whereas the mitochondrial calcium content is either unchanged or slightly decreased. These results, analyzed in conjunction with the Ca$_f$ measurements, indicate that calcium homeostasis is maintained fairly well during anoxia. Further work is needed to determine whether more subtle redistributions of calcium occur under these conditions.

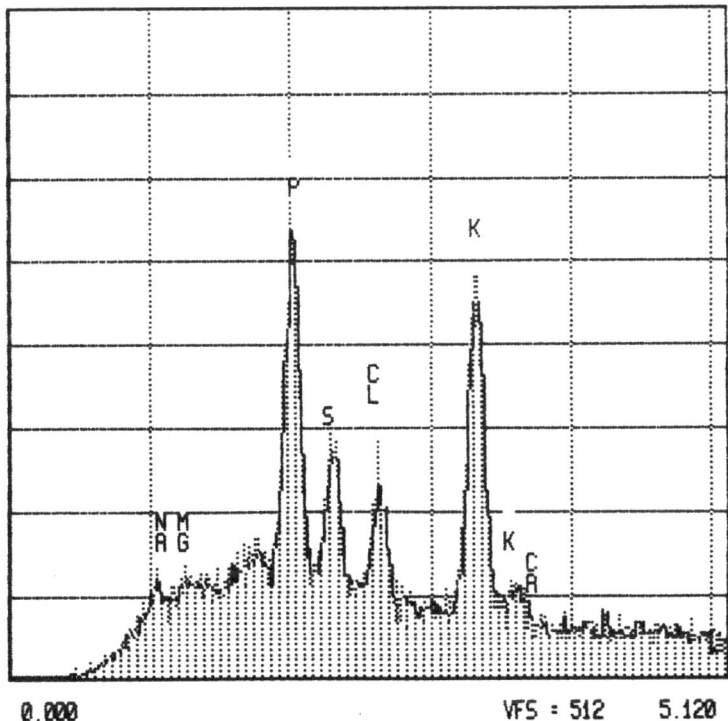

0.000   VFS = 512   5.120

FIGURE 3. Representative energy dispersive X-ray spectrum from a mitochondrion of a renal proximal tubule. For experimental details, see LeFurgey *et al.* (1986).

4. ROLE OF GLUTATHIONE

Glutathione is found in millimolar concentrations in the kidney (Meister and Anderson, 1983). Although the exact physiological role of glutathione in the kidney remains to be determined, this compound has been reported to be protective against a variety of toxic insults (Meister and Anderson, 1983). Addition of 1 mM glutathione to renal cells during 40 min of anoxia reduced LDH release from $29 \pm 10\%$ to $8 \pm 2\%$, and significantly enhanced the recovery of cellular ATP and K contents after 40 min of reoxygenation. Similar protection from anoxia was obtained by extracellular addition of 1 mM cysteine. The mechanism of this protective action is presently unknown. It is unclear whether intracellular glutathione is involved since preliminary experiments suggest that the cellular glutathione content is unaffected by these external additions during anoxia. Glutathione content rapidly falls during anoxia from 5–8 to about 2–3 nmol/mg protein, and this value is also unaffected by other protective maneuvers, such as the addition of ATP or ADP during anoxia. Our own unpublished observations show that decreasing the cellular glutathione content from about 5 to less than 1 nmol/mg protein causes no observable deleterious effects during normoxia. Therefore, the fall in cellular glutathione observed during anoxia would not be by itself deleterious to the cell.

5. SUMMARY AND CONCLUSIONS

The possible role of three variables was considered in anoxic cell injury to the renal proximal tubule. Cellular ATP content decreases rapidly during anoxia, initiating a continuous catabolic process that progresses with time in anoxia. The disruption of the plasma membrane, as measured by LDH release, seems to be closely associated with irreversible cell injury. The detailed steps that lead to plasma membrane breakdown are currently unknown; however, an increase in cytosolic free calcium does not appear to be a prerequisite for cellular damage. Three extracellular maneuvers performed during anoxia significantly ameliorated this injury: removal of extracellular calcium, addition of adenine nucleotides, and addition of glutathione or cysteine. It is presently unclear what their mechanisms of action are. Based on their protection of plasma membrane integrity from the extracellular medium, it is hypothesized that they act by stabilizing cytoskeletal–plasma membrane interactions.

REFERENCES

Balaban, R. S., Soltoff, S., Storey, J. M., and Mandel, L. J., 1980, Improved renal cortical tubule suspension: Spectrophotometric study of O₂ delivery, Am. J. Physiol. **238**:F50–F59.

Borle, A. B., Freudenrich, C. C., and Snowdowne, K. W., 1986, A simple method for incorporating aequorin into mammalian cells, Am. J. Physiol. **251**:C323–C326.

Farber, J. L., 1988, Calcium and ischemic cell injury, this volume.

Goligorsky, M. S., Huskey, M., Loftus, D. J., and Hruska, K. A., 1987, Anoxia and re-oxygenation: Cytosolic calcium in individual proximal tubular cells. Abstracts, Cell Calcium Metabolism 87 (G. Fiskum, ed.), no. 77.

Harris, S. I., Balaban, R. S., Barrett, L., and Mandel, L. J., 1981, Mitochondrial respiratory capacity and Na⁺- and K⁺-dependent adenosine triphosphatase-mediated ion transport in the intact renal cell, J. Biol. Chem. **256**:10319–10328.

Jacobs, W. R., and Mandel, L. J., 1987a, Fluorescent measurements of intracellular free calcium in isolated toad urinary bladder epithelial cells, *J. Membrane Biol.* **97**:53–62.

Jacobs, W. R., and Mandel, L. J., 1987b, Role of cytosolic free calcium in renal tubule damage during anoxia, Abstract, 20th Annual Meeting, Am. Soc. Nephrol.

Johnston, P. A., Rennke, H., and Levinsky, N. G., 1984, Recovery of proximal tubular function from ischemic injury, *Am. J. Physiol.* **246**:F159–F166.

Kahng, M. W., Berezesky, I. K., and Trump, B. F., 1978, Metabolic and ultrastructural response of rat kidney cortex to in vitro ischemia, *Exp. Mol. Pathol.* **29**:183–198.

Kitazawa, T., Shuman, H., and Somlyo, A. P., 1983, Quantitative electron probe analysis: Problems and solutions, *Ultramicroscopy* **11**:251–262.

LeFurgey, A., Ingram, P., and Mandel, L. J., 1986, Heterogeneity of calcium compartmentation: Electron probe analysis of renal tubules. *J. Membrane Biol.* **94**:191–196.

Lemasters, J. J., DiGuiseppi, J., Nieminen, A.-L., and Herman, B., 1987, Blebbing, free Ca^{2+} and mitochondrial membrane potential preceding cell death in hepatocytes, *Nature* **325**:78–81.

Mandel, L. J., Takano, T., Soltoff, S. P., and Murdaugh, S., 1988, Mechanisms whereby exogenous adenine nucleotides improve rabbit renal proximal function during and after anoxia, *J. Clin. Invest.* **81**:1255–1264.

Meister, A., and Anderson, M. E., 1983, Glutathione, *Ann. Rev. Biochem.* **52**:711–760.

Somlyo, A. P., Somlyo, A. V., Shuman, H., Sloane, B., and Scarpa, A., 1978, Electron probe analysis of calcium compartments in cryosections of smooth and striated muscle, *Ann. N. Y. Acad. Sci.* **307**:523–544.

Takano, T., Soltoff, S. P., Murdaugh, S., and Mandel, L. J., 1985, Intracellular respiratory dysfunction and cell injury in short-term anoxia of rabbit renal proximal tubules, *J. Clin. Invest.* **76**:2377–2384.

Trump, B. F., and Berezesky, I. K., 1988, The role of ion regulation in cell injury, cell death and carcinogenesis, Chap. 46 in this volume.

Trump, B. F., Berezesky, I. K., and Cowley, R. A., 1982, The cellular and subcellular characteristics of acute and chronic injury with emphasis on the role of calcium, in: *Pathophysiology of Shock, Anoxia, and Ischemia* (R. A. Cowley and B. F. Trump, eds.), Williams & Wilkins, Baltimore, pp. 6–46.

Venkatachalam, M. A., Bernard, D. B., Donohoe, J. F., and Levinsky, N. G., 1978, Ischemic damage and repair in the rat proximal tubule: Differences among the S_1, S_2 and S_3 segments, *Kidney Int.* **14**:31–49.

50

X-Ray Microprobe Analysis of Subcellular Elemental Distribution in Normal and Injured Peripheral Nerve Axons

RICHARD M. LOPACHIN, Jr., VICKI R. LOPACHIN, and
ALBERT J. SAUBERMANN

1. INTRODUCTION

Studies involving X-ray microanalysis of liver, heart, and other nonneuronal cells have contributed significantly to our understanding of submembrane distribution of elements (e.g., Na, K, Ca) and their role in cellular injury (Trump *et al.*, 1979). However, with respect to elemental distribution in nerve cells, there is little direct information available from either X-ray microprobe analysis or biochemical studies. Relevant data from the few published microprobe studies are difficult to interpret due to certain methodological problems. In some cases the results were not quantitated (Rick *et al.*, 1976; Schlote *et al.*, 1981), whereas in other studies the tissue was chemically fixed and/or cytochemical markers such as pyroantimonate were used (Hillman and Llinas, 1974; Oschman *et al.*, 1974; Duce and Keen, 1978; Ellisman *et al.*, 1979; Chan *et al.*, 1984). These chemical procedures can produce artefactual translocation of elements (Morgan, 1979; Somlyo, 1985). To provide information concerning the subcellular distribution of elements in nerve cells, electron probe X-ray microanalysis was used in this study. The microprobe data are generated by visually identifying a cellular compartment, followed by manual placement of the microprobe electron beam over the compartment and subsequent analysis. However, subtle patterns of elemental distribution which might exist within a compartment cannot be determined conveniently because the beam is placed manually and the large amount of time required to produce the necessary data. To circumvent this problem, digital X-ray imaging with computer-controlled beam placement was used. This method can analyze a cellular region point by contiguous point and thereby provide a chemical map of elemental distribution. In the present study, selected compartments of normal rat sciatic nerve axons were examined by both X-ray microprobe

RICHARD M. LOPACHIN, Jr., VICKI R. LOPACHIN, and ALBERT J. SAUBERMANN • Department of Anesthesiology, State University of New York Health Sciences Center, Stony Brook, New York 11794-8480.

analysis and digital imaging. In addition, we determined how elemental distribution was altered in axons injured by hypoxia or transection. To prevent artefactual translocation of elements, frozen, unfixed sections of nerves were analyzed. The results from this study indicate that each axonal compartment examined displayed a characteristic distribution of elemental content and that following injury this distribution was altered.

2. MATERIALS AND METHODS

2.1. Preparation of Sciatic Nerves

Male Sprague–Dawley rats (275–299 g) were sacrificed by decapitation. Longitudinal mid-thigh incisions were made to expose the sciatic nerves. Nerves were freed from surrounding connective tissue and were rapidly frozen *in situ* using copper pliers which had been cooled in liquid nitrogen (Hagler and Buja, 1984). This procedure removes a piece of frozen nerve approximately 15 mm in length and 0.3 mm thick. Samples were placed immediately in liquid nitrogen (LN_2) and stored under LN_2 until analyzed.

2.2. Axonal Injury

To induce sciatic nerve hypoxia, rats were sacrificed by decapitation, exsanguinated, and 15 min later segments from both sciatic nerves were frozen *in situ*. In a second experiment, rats were anesthetized with pentobarbitone, and axotomy was performed by making a mid-thigh incision followed by transection of the sciatic nerve with a scalpel. The wound was closed and 16 hr later the distal nerve stump was frozen *in situ*, removed, and stored in LN_2 for later analysis. The sections used for analysis were removed from a region 3–5 mm behind the distal stump.

2.3. Cryoultramicrotomy

Frozen nerve segments were mounted for cryosectioning using a LN_2-cooled vise-type flat tissue holder. After a 20-min thermal equilibration period, 200-nm (normal thickness) sections were cut at $-55°C$ with a glass knife at extremely slow cutting speeds using a Sorvall-MT2B microtome equipped with a specially designed cryochamber (Burlington Scientific). Both cross-sectional and longitudinal sections of nerve samples were cut. Frozen sections were transferred to a precooled carbon-coated, nylon-covered beryllium grid located on a specially designed beryllium specimen holder (Saubermann *et al.*, 1981a,b). Since analysis was to be performed on frozen dried sections, transferred frozen hydrated sections were dehydrated in the Cameca vacuum column through sublimation by raising the cold-stage temperature to $-60°C$ for 30 min before measurements were taken.

2.4. X-Ray Microanalysis

X-ray spectra were obtained with a Tracor Northern 2100 energy-dispersive X-ray analysis system with a 10-mm², 148-eV resolution Si(Li) detector (working distance < 15 mm). A 20-keV accelerating voltage, 1.0-nA probe current, and 100-sec live counting time were used for all measurements. Digital filtering was used to suppress continuum X-ray counts (Saubermann *et al.*, 1981a,b). Characteristic X-rays for each element were integrated over the following energy ranges: sodium, 0.96–1.12 keV; phosphorus, 1.92–2.08 keV; chlorine, 2.52–2.68 keV; potassium, 3.24–3.40 keV; and calcium, 3.60–3.76 keV. Continuum radiation generation rate, used as a measurement of tissue mass, was obtained by integrating total X-ray counts minus extraneous background counts from the region between 4.60 and 6.00 keV. Individual elements were quantified using the continuum normalization technique (Hall *et al.*, 1973). By determining the ratio of characteristic counts to continuum (Rx) for each element and relating this value to elemental standards, absolute mass fraction for each element can be ascertained. Elemental analysis for each region is expressed as mmol element/kg dry wt.

The neuronal compartments analyzed were identified using morphological information obtained from scanning transmission electron microscopy (STEM) images. Four internodal compartments were analyzed: mitochondria, axoplasm, myelin, and extra-axonal space. Analysis was performed by rastering the electron beam within each compartment. Identification of cellular compartments was confirmed by transmission electron microscopy (TEM) of conventionally fixed tissues (200 nm nominal thickness).

2.5. Quantitative Digital X-Ray Imaging

Cryosections of rat sciatic nerve were prepared as above. Based on morphology visible in STEM image, a region of the frozen dried nerve cryosection was selected. The selected region was then analyzed, point by point, with a computer-controlled electron beam (1.4-nA probe current) according to a 64 × 64 matrix design. A 4-sec beam dwell time was used per point (pixel) and the resulting X-ray spectrum for each point was collected, processed by top-hat digital filtering (Saubermann and Heyman, 1987; Heyman and Saubermann, 1987), and stored on computer disks. To display the X-ray image, stored values for a given element were retrieved to form the image where each pixel represents an absolute mass fraction of element (i.e., mmol/kg dry wt). A grey scale was chosen to represent a specific range of elemental concentrations. Accordingly, the particular shade of grey associated with an individual pixel reflects a specific amount of element.

2.6. Statistics

One-way analysis of variance and the least significant difference (LSD) method of multiple comparisons were used to determine differences among means. The two-sided Student's t test was used to determine significance of difference between two means.

3. RESULTS

3.1. Analysis of Morphological Compartments in Normal Sciatic Nerve

The distribution of selected elements in normal rat sciatic nerve axons was quantitated in four morphological compartments: axoplasm, mitochondria, myelin, and extra-axonal space. These measurements were made using six different nerve segments ($n = 6$ rats). The axoplasmic compartment is probably composed of several components including smooth endoplasmic reticulum as well as protein constituents of cytoskeleton. This compartment

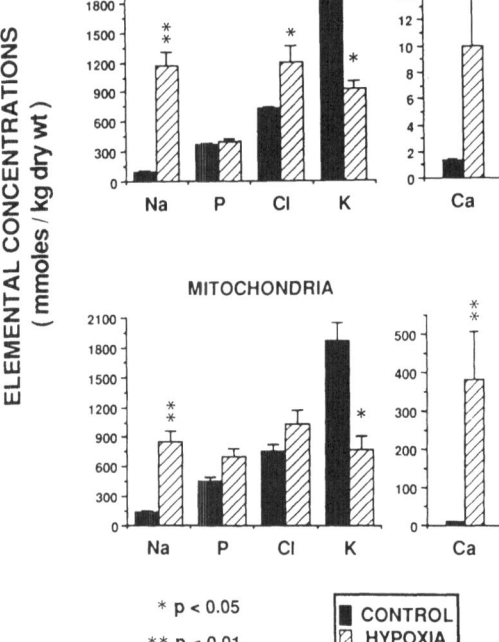

FIGURE 1. Histograms showing elemental content of axoplasm and mitochondria from hypoxic and control sciatic nerve axons ($n = 4$ nerves, 30–50 analyses per compartment). Ultrathin sections were obtained from sciatic nerve samples which had been frozen *in situ*. Sections were unfixed, frozen, and analyzed in the dehydrated state. Data are expressed as mmol element/kg dry wt $\pm$ SEM. Data from myelin and extra-axonal space are not shown.

includes areas adjacent to mitochondria as well as other areas located throughout the axoplasm. In STEM images, mitochondria appeared as electron-dense, ovoid, or elongated structures (depending on the plane of section), although cristae were not visible. Elemental analysis (Figs. 1 and 2) of mitochondria and axoplasm in normal nerves indicated that these compartments were alike with respect to Na, K, Cl, and Ca elemental mass fractions. However, mitochondria exhibited P and Cl concentrations which were significantly higher than those of the axoplasm. As noted above, both mitochondria and axoplasm of normal nerves had comparable levels of K, and these intracellular concentrations of element were approximately eightfold greater than that of extra-axonal or myelin compartments.

The myelin sheath of normal axons was similar to axoplasm with respect to Na and Ca mass fractions (data not shown). In contrast, the myelin compartment had the lowest levels of Cl and K and the highest concentration of P when compared to other compartments. Elemental analysis of extra-axonal space encompassed extracellular fluid and, possibly, portions of small unmyelinated fibers and collagen. The extra-axonal space had a Na level which was approximately 10-fold greater than that of any other compartment analyzed. (data not shown). Moreover, Cl and Ca concentrations in this compartment were significantly higher than in other neuronal regions. Concentration of K in extra-axonal space was about twice that of myelin, whereas extra-axonal P mass fraction was lower than any of the compartments examined.

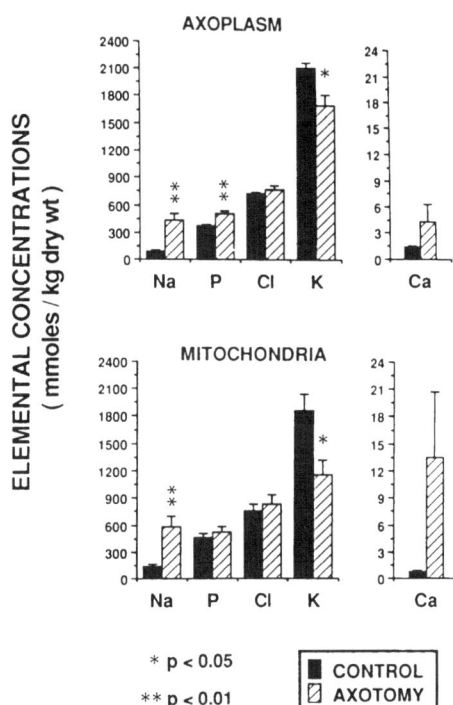

FIGURE 2. Histograms showing elemental content of axoplasm and mitochondria from transected and control sciatic nerves ($n = 4$ nerves, 50–70 analyses per compartment). Ultrathin sections were obtained from sciatic nerve samples which had been frozen *in situ*. Sections were unfixed, frozen, and analyzed in the dehydrated state. Data are expressed as mmol element/kg dry wt $\pm$ SEM. Data from myelin and extra-axonal space are not shown.

3.2. Digital X-Ray Imaging of Normal Sciatic Nerve Axons

Elemental distribution for normal axons determined by microanalysis was confirmed by quantitative digital X-ray imaging. Figure 3a is a STEM image of small-, medium-, and large-diameter internodal myelinated fibers from normal sciatic nerve. The respective quantitative digital images for K(b), P(c), and Na(d) are also presented. Since these chemical maps are superimposable with the STEM image (Fig. 3a), the spatial as well as quantitative relationships of elemental distribution and morphological compartmentation are immediately evident. The digital images clearly demonstrate cellular localization for each element: Na is mainly extracellular, K is almost entirely intracellular, and P is predominantly associated with myelin.

3.3. Analysis of Morphological Compartments in Injured Sciatic Nerve

Injured axons displayed characteristic changes in elemental content when compared to normal axons. Sciatic nerves made hypoxic *in situ* did not exhibit gross morphological changes; however, numerous swollen mitochondria were seen at the subcellular level. Microprobe analysis (Fig. 1) of hypoxic nerves showed marked increases in axoplasmic Na and Cl in conjunction with a loss of K. Of the axons analyzed, only a few demonstrated dramatic increases in axoplasmic Ca. This accounts for the large but nonsignificant increase

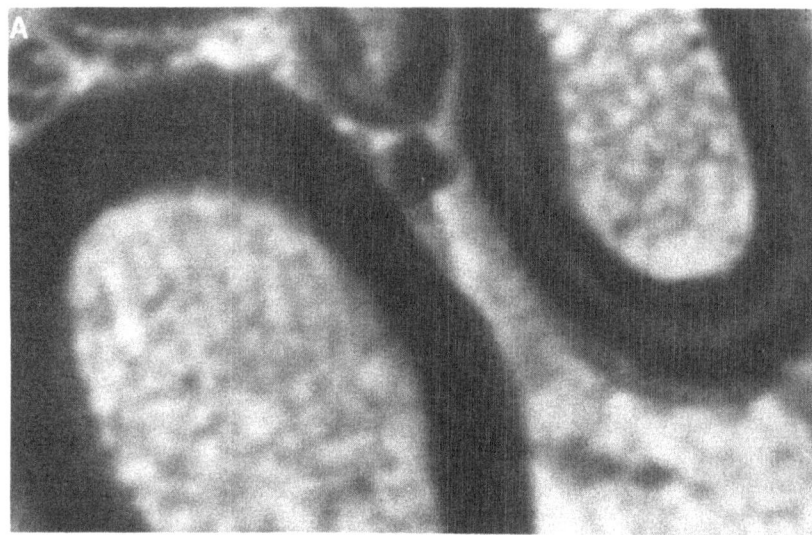

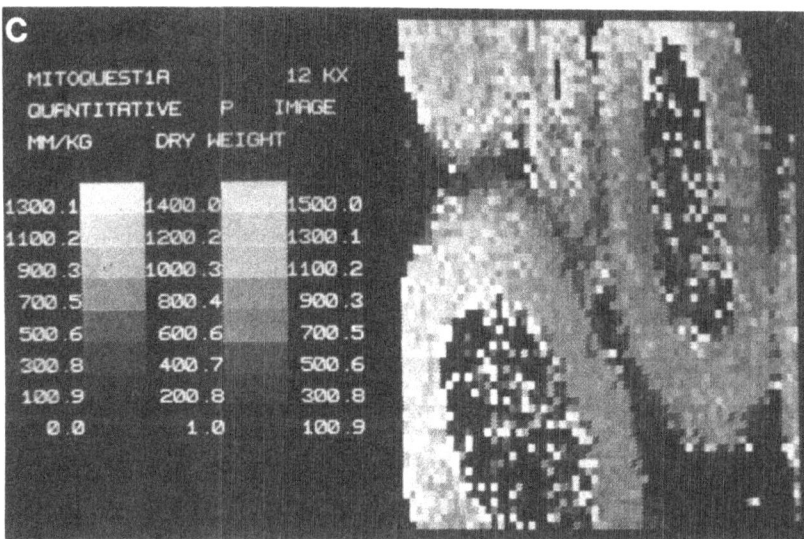

FIGURE 3. The top left-hand figure (A) is a scanning transmission electron micrograph (STEM, X7800) of a frozen, dehydrated section from a control sciatic nerve. This section shows three myelinated axons of varying sizes and was analyzed by computer-controlled digital imaging. The remaining figures are corresponding digital images for K (B), P (C), and Na (D). Each pixel of the image represents a specific elemental concentration according to the corresponding quantitative grey scale. The individual digital images are superimposable on the STEM image (A).

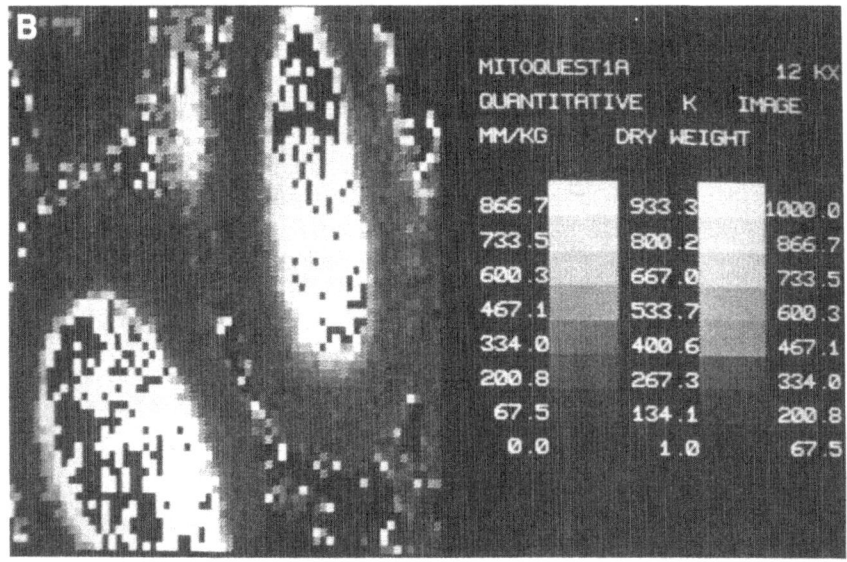

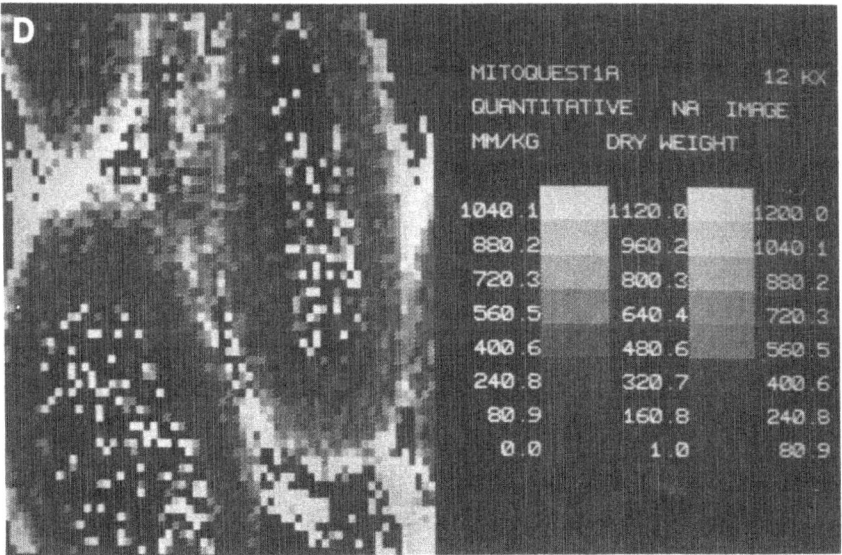

FIGURE 3. (*Continued*)

(Fig. 1) which occurred when all axoplasmic Ca data were included in the analysis. Phosphorus levels were not affected by hypoxia. Mitochondria from hypoxic nerves showed increases in Na and Cl concentrations and a decrease in K. These changes were similar to those observed in axoplasm. However, unlike this compartment, mitochondrial Ca and P levels were consistently increased. These alterations in elemental distribution occurred in myelinated axons regardless of their diameter. In contrast, hypoxia did not perturb the elemental content of either myelin or extra-axonal space.

The effects of axotomy on elemental distribution in nerve fibers were also studied. Gross morphological analysis of axotomized distal stumps at 16 hr post-transection did not demonstrate striking changes other than edema. At the subcellular level the axoplasm appeared normal with only a few swollen mitochondria evident. The quantitative results (Fig. 2) indicate that axoplasmic Na levels increased while K concentrations were reduced. Significant changes in elemental concentration occurred most frequently in small myelinated fibers (<3 μm diameter). Also noted was a trend toward raised Ca content which was not statistically significant, since only 13% of the axoplasms examined showed changes in Ca. However, the data indicate that those axons exhibiting increases in Ca also had the highest axoplasmic Na values.

In mitochondria from axotomized nerves, there was a marked increase in Na with a corresponding loss of K (Fig. 2). Although not significant, there was a trend toward increased mitochondrial Ca. As was the case with axoplasm, this trend reflects a subpopulation (33%) of selectively affected mitochondria which displayed increased Ca concentrations in association with the largest changes in Na. For both axoplasm and mitochondria, P and Cl were not altered. Moreover, axotomy did not affect elemental distribution in either myelin or extra-axonal space.

4. DISCUSSION

The results of our microprobe study demonstrate that a characteristic compartmentalization of elements exists in normal peripheral nerves. As is the case with other cell types, Na, Cl, and Ca are predominant elements of the extracellular compartment whereas K is almost exclusively localized within intracellular compartments. By providing detailed maps of elemental distribution within internodal compartments, digital X-ray imaging (Fig. 3) confirms and extends our nonimaging microprobe data. The distribution of Na, Cl, and K in sciatic nerve determined directly by the present study substantiates that estimated by electrophysiological techniques and biochemical methods (Nichols and Kuffler, 1965; Walker and Brown, 1977; Caldwell-Violich and Requena, 1979; Deitmer and Schlue, 1979). Moreover, the compartmentalization of elements identified in this study probably reflects differentiation and specialization of corresponding axonal regions (Saubermann and Scheid, 1985). Microprobe analysis and X-ray imaging indicate that in axoplasm K is the most abundant element of those measured. This is consistent with previous microprobe studies of axoplasmic K levels in peripheral nerves (Rick *et al.*, 1976; Schlote *et al.*, 1981) and of cytoplasmic K concentrations in nonneuronal cells (Somlyo *et al.*, 1977; Buja *et al.*, 1983; Millette *et al.*, 1985). The high intracellular K concentrations reported in the present study appear to be a unique attribute of axons since these levels are much higher than those determined by microprobe analysis of rat dorsal root ganglion cell bodies and sciatic nerve Schwann cell bodies (LoPachin *et al.*, unpublished data, 1986) and of nonneuronal cell bodies (Somlyo *et al.*, 1977; Millette *et al.*, 1985).

The subcellular elemental distribution observed in normal axons is disturbed following sciatic nerve injury. In both mitochondria and axoplasm, hypoxia causes large increases in Na and Cl coupled with a loss of K. In addition, mitochondria display selective increases in Ca and P which reflect their ability to sequester Ca and subsequently store it as calcium phosphate. These changes in intra-axonal elements induced by hypoxia are similar to those alterations observed in cytoplasm of nonneuronal cells damaged by several different methods (Trump *et al.*, 1979; James-Kracke *et al.*, 1980). Moreover, mitochondrial changes in Ca and P are consistent with those reported for mitochondria from hypoxic myocytes (Buja *et al.*, 1983) and ischemic myocardium (Trump *et al.*, 1979). These mitochondrial alterations occurring under hypoxia, an energy-limiting condition, appear to contradict the finding that mitochondrial calcium accumulation is an energy-dependent process (Lehninger *et al.*, 1978). However, these changes are occurring after only a short period of hypoxia; therefore, they might be taking place once an increase in membrane permeability has been initiated but prior to total mitochondrial energy inhibition. Although morphology was not studied, the early onset and magnitude of changes observed suggest an involvement of Na, Cl, and other elements in mediating the effects of hypoxia on nerve cells.

In both axoplasm and mitochondria from axotomized nerves, Na levels were increased substantially whereas K concentrations were reduced. In contrast to hypoxia, these intra-cellular compartments did not exhibit significant changes in the concentrations of either Ca or Cl. This pattern of distribution in transected axons occurs at a time point (16 hr) which coincides or precedes previously reported early morphological alterations (Schlote *et al.*, 1981; Lee, 1963). Morphological studies showed that the majority of initial neuropathological changes are taking place in small- to medium-diameter myelinated fibers (Lee, 1963). Correspondingly, it is these fiber types which exhibit the majority of elemental changes. The temporal nature of these results and the correlation between morphological and elemental changes suggests that shifts in subcellular distribution of Na and K play a role in mediating the structural and functional alterations observed in transected axons.

The generalized perturbation of elemental distribution caused by hypoxia suggests a loss of membrane integrity associated with an influx of extracellular ions and efflux of intracellular K. However, in axotomized nerves, selective changes in the labile elements Na and K imply a more subtle mechanism of injury. In a previous microprobe study of axons from crushed rat sciatic nerve, Schlote *et al.* (1981) found changes in axoplasmic Na and K at 36-hr post-crush which were similar to those detected in this study. These data and those of the present study suggest a mechanism of injury involving an initial impairment of the Na, K-ATPase membrane pump possibly caused by developing axonal energy deficit (Schleapfer, 1974; Schlote *et al.*, 1981). As a result of compromised Na pump activity, a loss of os-moregulation might occur (Ginn *et al.*, 1968). The subsequent expansion of axolemma and increase in membrane permeability could promote an influx of Ca in addition to ongoing loss of axoplasmic K and gain of Na. This suggestion is supported by results from the present study showing that levels of intra-axonal Ca were increased only in those axons showing the largest changes in Na and K. Thus, an influx of Na and loss of K which occur early in the process of axotomy-induced injury are followed by a later developing entry of Ca. Schleapfer (1977) proposed that following nerve transection, Ca influx with subsequent activation of a Ca-dependent neutral protease is responsible for the breakdown of neurofilaments and their dissociation from microtubules. This dissolution of the cytoskeleton represents axoplasmic degranulation which is characteristic of Wallerian degeneration.

Two types of nerve cell injury were examined in this study: hypoxia, which represents injury caused by metabolic inhibition, and axotomy, which produces injury by mechanical trauma. The corresponding elemental changes are characteristic for each type of damage and they are detected during early stages of the injury process. In the case of axotomy, alterations of elemental homeostasis are noted mainly in small- and medium-diameter myelinated fibers. This selectivity of effect correlates with the earliest morphological changes which also occur in these fiber types following nerve transection. Furthermore, instead of focusing on a single ion, microprobe analysis has provided simultaneous information concerning several different elements. This inclusive approach indicates that more than one element is affected and that elemental changes probably occur according to a sequential pattern (i.e., Na influx, K efflux followed later by a Ca influx). Thus, a comprehensive picture of the primary role played by elements in nerve cell injury is developing. However, a complete understanding of the involvement of elements requires continued microprobe research which will necessarily encompass studies at various points during the developing nerve damage.

ACKNOWLEDGMENTS. We wish to thank John Stockton, Robert Heyman, and Joan Lowery for their excellent technical help. We also wish to thank Dr. Ruth Bulger for her comments and criticisms. This work was supported by a Pharmaceutical Manufacturer's Association starter grant and a NIH grant (ES 03830) to R. M. L. and a NIH grant (NS 21455-01A1) to A. J. S.

REFERENCES

Buja, L. M., Burton, K. P., Hagler, H. K., and Wilkerson, J. T., 1983, Quantitative X-ray microanalysis of the elemental composition of individual myocytes in hypoxic rabbit myocardium, *Lab. Invest.* **68:**872–882.

Caldwell-Violich, M., and Requena, J., 1979, Magnesium content and net fluxes in squid giant axons, *J. Gen. Physiol.* **74:**739–752.

Chan, S. Y., Ochs, S., and Jersild, R. A., Jr., 1984, Localization of calcium in nerve fibers, *J. Neurol. Biol.* **15:**89–108.

Deitmer, J. W., and Schlue, W. R., 1979, Measurements of the intracellular potassium activity of Retzius cells in the leech central nervous system, *J. Exp. Biol.* **91:**87–101.

Duce, I. R., and Keen, P., 1978, Can neuronal smooth endoplasmic reticulum function as a calcium reservoir? *Neuroscience* **3:**837–848.

Ellisman, M. H., Friedman, P. L., and Hamilton, W. J., 1979, Cytochemical localization of cations in myelinated nerve using TEM, HVEM, SEM and electron probe microanalysis, *Scan. Electron Microsc.* **2:**793–800.

Ginn, F. L., Shelburne, J. D., and Trump, B. J., 1968, Disorders of cell volume regulation, *Lab. Invest.* **53:**1041–1071.

Hagler, H. K., and Buja, L. M., 1984, New techniques for the preparation of thin freeze dried cryosections for X-ray microanalysis, in: *Science of Biological Specimen Preparation* (J. P. Revel, ed.), AMF O'Hare, IL: SEM Inc., pp. 161–166.

Hall, T. A., Anderson, H. C., and Appleton, T., 1973, The use of thin specimens for X-ray microanalysis in biology, *J. Microsc.* **99:**177–182.

Hillman, D. E., and Llinas, R., 1974, Calcium-containing electron-dense structures in the axons of the squid giant synapse, *J. Cell Biol.* **61:**146–155.

Heyman, R. B., and Saubermann, A. J., 1987, Multifunctional mini-computer program providing quantitative digital X-ray microanalysis of cryosectioned biological tissue for the inexperienced analyst, *J. Electron Microsc. Tech.* **5:**315–345.

James-Kracke, M. R., Sloane, B. F., Shuman, H., Karp, R., and Somlyo, A. P., 1980, Electron probe analysis of cultured vascular smooth muscle, *J. Cell Physiol.* **103:**313–322.

Lee, J. C.-Y., 1963, Electron microscopy of Wallerian degeneration, *J. Comp. Neurol.* **120**:65–71.

Lehninger, A. L., Reynafarje, B., Vercesi, A., and Tew, W. P., 1978, Transport and accumulation of calcium in mitochondria, *Ann. N.Y. Acad. Sci.* **307**:160–172.

Millette, J. R., Allenspach, A. L., Clarke, P. J., McCauley, P. T., and Washington, I. S., 1985, X-ray microanalysis of calcium, potassium, and phosphorus in liver mitochondria stressed by carbon tetrachloride, *J. Anal. Toxicol.* **8**:145–151.

Morgan, A. J., 1979, Non-freezing techniques of preparing biological specimens for electron microprobe X-ray microanalysis, *Scan. Electron Microsc.* **2**:635–648.

Nicholls, J. G., and Kuffler, S. W., 1965, Na and K content of glial cells and neurons determined by flame photometry in the central nervous system of the leech, *J. Neurophysiol.* **28**:519–525.

Oschman, J. L., Hall, T. A., Peters, P. D., and Wall, B. J., 1974, Association of calcium with membranes of squid giant axon, *J. Cell Biol.* **61**:156–165.

Rick, R., Dorge, A., and Tippe, A., 1976, Elemental distribution of Na, P, Cl and K in different structures of myelinated nerve of *Rana esculenta, Experientia* **32**:1018–1019.

Saubermann, A. J., Echlin, P., Peters, P. D., and Beeuwkes, R., 1981a, Application of scanning electron microscopy to X-ray analysis of frozen hydrated sections. I. Specimen handling techniques, *J. Cell Biol.* **88**:257–267.

Saubermann, A. J., Beeuwkes, R., and Peters, P. D., 1981b, Application of scanning electron microscopy to X-ray analysis of frozen-hydrated sections. II. Analysis of standard solutions and artificial electrolyte gradients, *J. Cell Biol.* **88**:268–273.

Saubermann, A. J., and Scheid, V. L., 1985, Elemental composition and water content of neuron and glial cells in the central nervous system of the North American medicinal leech (*Macrobdella decora*), *J. Neurochem.* **44**:825–834.

Saubermann, A. J., and Heyman, R. B., 1987, Quantitative digital X-ray imaging using frozen hydrated and frozen dried tissue sections, *J. Microsc.* **146**(2):169–182.

Schlaepfer, W. W., 1974, Calcium-induced degeneration of axoplasm in isolated segments of rat peripheral nerve, *Brain Res.* **69**:203–215.

Schlaepfer, W. W., 1977, Structural alterations of peripheral nerve induced by the calcium ionophore, A23187, *Brain Res.* **136**:1–9.

Schlote, W., Wolburg, H., and Wendt-Gallitelli, M. F., 1981, Ionic shifts in myelinated nerve fibers during early stages of Wallerian degeneration, *Acta Neuropathol.* **7**:31–35.

Somlyo, A. P., 1985, Cell calcium measurement with electron probe and electron energy loss analysis, *Cell Calcium* **6**:197–212.

Somlyo, A. V., Shuman, H., and Somlyo, A. P., 1977, Elemental distribution in striated muscle and the effects of hypertonicity, *J. Cell Biol.* **74**:828–854.

Trump, B. F., Berezesky, I. K., Change, S. H., Pendergrass, R. E., and Mergner, W. J., 1979, The role of ion shift in cell injury, *Scan. Electron Microsc.* **3**:1–14.

Walker, J. L., and Brown, H. M., 1977, Intracellular ionic activity measurements in nerve and muscles, *Physiol. Rev.* **57**(4):729–775.

51

Prevention of Ischemic Brain Mitochondrial Injury by Lidoflazine

ROBERT E. ROSENTHAL, GARY FISKUM, and
FOZIA HAMUD

1. INTRODUCTION

Viability of the brain is dependent on maintenance of ionic and electrochemical gradients within neurons as well as within subcellular components. Of particular interest is neuronal handling of calcium. Under aerobic, steady-state conditions intracellular Ca^{2+} concentration is maintained at basal levels of approximately 0.1 μM by ATP-dependent ionic pumps, as compared to extracellular levels of 1.0 mM. In the brain, more than 80% of ATP is generated through mitochondrial oxidative phosphorylation (Fiskum, 1983, 1985). If oxidative phosphorylation were interrupted, ATP synthesis would be greatly impaired, thus preventing maintenance of ionic gradients, with potentially catastrophic consequences. Loss of calcium homeostasis has been linked to the initiation of various degradative processes including activation of phospholipases (Fiskum et al., 1985), lipid peroxidation (McCord, 1985), vascular spasm (Borgers et al., 1983), induction of coagulative necrosis (Farber, 1982), and eventually cell death (Borgers et al., 1983).

One key event which has repeatedly been shown to compromise mitochondrial oxidative phosphorylation is ischemia. Various models (Mela, 1979; Hamud and Fiskum, 1985; White et al., 1985) have demonstrated that prolonged episodes of cerebral ischemia markedly impair the ability of brain mitochondria to effectively synthesize ATP. In fact, the supply of ATP in the brain is expended within several minutes from the onset of ischemic anoxia. As suggested above, without adequate ATP to fuel ionic pumps, calcium homeostasis is rapidly lost (Farber, 1982) with intracellular concentration rapidly approaching extracellular levels of 1.0 mM. As previously noted, such elevations of intracellular Ca^{2+} concentration have been shown to have potentially serious effects. Brain mitochondria, however, demonstrate

ROBERT E. ROSENTHAL • Department of Emergency Medicine, George Washington University School of Medicine, Washington, D.C. 20037 . GARY FISKUM • Departments of Biochemistry and Emergency Medicine, George Washington University School of Medicine and Health Sciences, Washington, D.C. 20037 . FOZIA HAMUD • Department of Biochemistry, George Washington University, Washington, D.C. 20037 .

an impressive capacity to recover the ability to synthesize ATP in a limited number of studies of complete cerebral ischemia (Rehncrona et al., 1979; Mela, 1979; Hillered et al., 1984). Similar recovery is not noted following prolonged incomplete ischemia (Hillered et al., 1984).

In addition to being the primary generator of ATP, respiring brain mitochondria are capable of actively accumulating large amounts of Ca^{2+}, thereby buffering the Ca^{2+} concentration of the cytosol (Fiskum et al., 1985). Mitochondrial calcium influx occurs via a ruthenium red-sensitive electrophoretic influx pathway in response to the negative inside membrane potential generated by proton efflux catalyzed by the electron transport chain (Fiskum and Lehninger, 1982). While clearly not the primary determinant of the cytosolic free-Ca^{2+} concentration during steady-state conditions, mitochondrial transport may be extremely important during abnormal rises in cellular Ca^{2+} concentration (Siesjo, 1981). If this capacity of mitochondria were damaged by prolonged ischemia (Hamud and Fiskum, 1985), this form of injury might limit the capability of the reoxygenated tissue to establish normal intracellular Ca^{2+} concentrations during reperfusion.

It was recently suggested that calcium antagonist drugs may play a role in the prevention of neurological injury during cerebral ischemia and recirculation (Winegar et al., 1983; Steen et al., 1985; Edmonds et al., 1985). Despite various hypotheses, the method of neuroprotection afforded by this class of drugs remains unclear. While calcium antagonists have been shown to prevent ischemic mitochondrial injury to myocardium (Nayler et al., 1980; Wolkowicz et al., 1983), little work on the effect of calcium antagonists on ischemic brain mitochondria has been published to date. Following is a brief description of our findings on the effect of the calcium antagonist lidoflazine on brain mitochondrial injury following prolonged complete ischemia (Rosenthal et al., 1987).

2. EXPERIMENTAL DESIGN

Cardiac arrest and resuscitation in dogs was used as the model for complete cerebral ischemia and recirculation for our experiments. Twenty-three adult beagles were separated into four experimental groups: (1) nonischemic controls ($n = 5$); (2) 10 min cardiac arrest, no resuscitation ($n = 5$); (3) lidoflazine (1 mg/kg I.V.) followed 20 min later by 10 min of cardiac arrest, no resuscitation ($n = 5$); and (4) 10 min cardiac arrest followed by 100 min of spontaneous circulation ($n = 8$).

3. MITOCHONDRIAL RESPIRATION

At the conclusion of the appropriate experimental period, a 2-g sample of left parietal cortex was surgically removed, immediately cooled to 4°C, and processed for mitochondrial studies using a modified Clark–Nicklas (1970) procedure, designed to recover both free and synaptosomal mitochondria. Mitochondrial oxygen consumption was measured polarographically at 30°C with an oxygen electrode. The measurements were performed at pH 7.0 in a thermostatically controlled glass chamber in a medium containing concentrations of K^+, Mg^+, and phosphate that are similar to those present in cytosol.

A typical oxygen electrode measurement of oxygen consumption by a suspension of normal isolated brain mitochondria is described in Fig. 1. Mitochondria were added at a concentration of 0.5 mg protein per ml to the respiratory medium which contained the

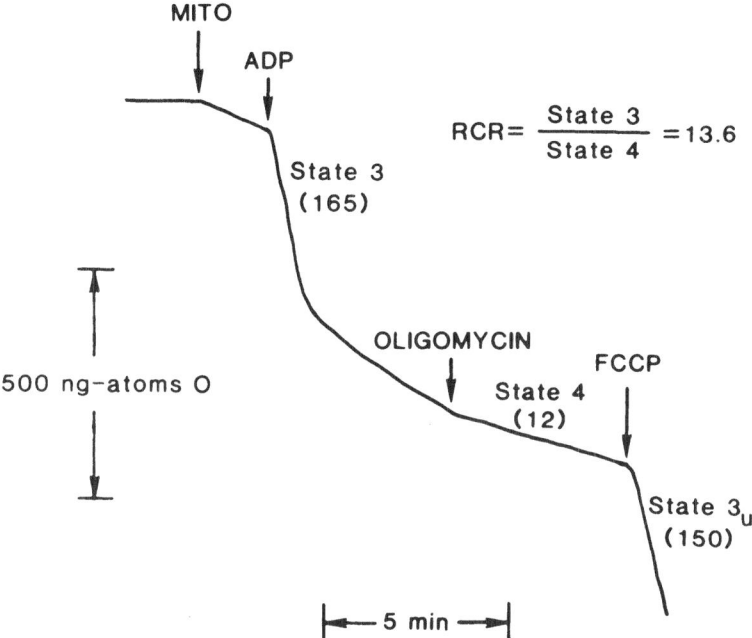

FIGURE 1. Oxygen electrode tracings describing methods for measuring respiration and the respiratory control ratio of isolated brain mitochondria. Brain mitochondria were added to a medium which contained 125 mM KCl, 1 mM MgCl$_2$, 2 mM KH$_2$PO$_4$, 5 mM malate, 5 mM glutamate, and 5 mM Hepes (pH 7.0). Phosphorylating (state 3) respiration was initiated by the addition of 0.26 mM ADP. A true resting rate of respiration (state 4) was established by adding the ATP synthetase inhibitor oligomycin (2 μg/ml). Approximately 5 min later, an optimal concentration of the proton ionophore uncoupler FCCP (0.05 μg/ml) was added to induce uncoupled (state 3$_u$) respiration. Other conditions were as described in text. The respiratory control ratio (RCR) was defined and calculated as the rate of state 3 respiration divided by the rate of state 4 respiration in the presence of oligomycin. The numbers in parentheses are the rates of oxygen consumption in ng-atoms O/min/mg protein.

oxidizable substrates malate and glutamate. Phosphorylating (state 3) respiration was initiated by the addition of 0.25 mM ADP to the mitochondrial suspension. A slower rate of oxygen consumption followed state 3 respiration once the added ADP had been converted into ATP. The post-state 3 rate of respiration is limited by the turnover of ATP due to the activity of contaminating ATP hydrolases. In order to achieve a true resting rate of respiration (state 4) which is limited by the proton permeability of the mitochondrial inner membrane, the turnover of ATP must be eliminated. This was achieved by blocking ATP synthesis by the addition of the mitochondrial ATP synthetase inhibitor oligomycin. After a short period of resting respiration, mitochondrial oxygen consumption was stimulated by the addition of the respiratory uncoupler FCCP (carbonylcyanide p-trifluoromethoxyphenylhydrazone). This agent maximizes the proton permeability of the mitochondrial inner membrane and provides a rate of uncoupled respiration (state 3$_u$) that is limited by the rate of electron flow through the mitochondrial electron transport chain. Theoretically, this rate should be greater than or equal to the rate of ADP-stimulated respiration. In these mitochondrial suspensions, concentrations of FCCP that are optimal for respiratory uncoupling apparently also result in partial respiratory inhibition. The degree to which the mitochondrial respiration is coupled to oxidative phosphorylation is reflected by the respiratory control ratio (RCR), which is defined as the rate of state 3 respiration divided by the rate of state 4 respiration. The RCR of 13.6 obtained for these mitochondria indicates that their energy-coupling efficiency was high.

A comparison of the different rates of respiration and respiratory control ratios for brain mitochondria isolated from the different groups of animals is shown in Table I. There was a 35% decrease in the rate of mitochondrial oxidative phosphorylation following 10 min of complete cerebral ischemia, as demonstrated by the decline in state 3 respiration from 170 ± 12 ng-atoms O/min/mg to 111 ± 6 ($p < 0.001$). By 100 min after recirculation, state 3 respiration had returned to control. Those animals pretreated with lidoflazine showed significantly less decline in state 3 respiration, 143 ± 7 ng-atoms O/min/mg versus 111 ± 6 nontreated ($p < 0.01$). Resting state 4 respiration ($p < 0.01$) as well as uncoupled state 3_u respiration ($p < 0.01$) both showed significant declines following 10 min of complete cerebral ischemia. Pretreatment with lidoflazine significantly protected uncoupled respiration when compared to nontreated animals ($p < 0.025$) while the rate of resting respiration for pretreated dogs subjected to 10 min of cerebral ischemia did not differ statistically from either control or ischemic nontreated animals. No significant decline in the mitochondrial respiratory control ratio was observed for any of the fibrillated or resuscitated groups of animals, as a result of simultaneous decline of state 3 and state 4 respiration during ischemia. The decrease in state 3 respiration in the absence of a decrease in the RCR implies that damage to the systems involved in oxidative phosphorylation occurs in the absence of generalized disruption of the mitochondrial inner membrane. Uncoupler-induced respiration (state 3_u) was also significantly depressed after 10 min of complete cerebral ischemia. This finding indicates that the mitochondrial electron transport chain per se was damaged during the ischemic episode. Since the percent inhibition of state 3_u respiration is comparable to that of state 3 respiration, it is likely that the respiratory chain is also the site of inhibition of phosphorylating respiration under the conditions used in these experiments.

Interestingly, by 100 min after resuscitation the RCR was significantly greater than values obtained for control, fibrillated, or pretreated dogs. While mechanisms of mitochondrial recovery from ischemic insults are poorly understood, this study demonstrates that the brain possesses an impressive ability to reverse the process of mitochondrial damage, at least for a short period following 10 min of complete ischemia.

TABLE I. Mitochondrial Respiration and Ca^{2+} Accumulation[a]

	Respiration[b]				Ca^{2+} accumulation[c]
	State 3	State 3_u	State 4	RCR	
Control ($n = 5$)	170 ± 12	134 ± 50	15 ± 4	12 ± 2	504 ± 66
V-Fib ($n = 5$)	111 ± 6	63 ± 25	10 ± 2	11 ± 2	452 ± 95
Lidoflazine ($n = 5$)	143 ± 7	118 ± 30	14 ± 1	11 ± 1	400 ± 114
ROSC ($n = 8$)	165 ± 21	104 ± 29	12 ± 3	14 ± 2	712 ± 191

[a] Values are represented as mean $+/-$ one standard deviation.
[b] ng-atoms O/min/mg mitochondrial protein.
[c] nmol/mg mitochondrial protein.

As previously mentioned, dogs pretreated with the calcium antagonist lidoflazine showed significantly less decline in mitochondrial respiration following 10 min of complete cerebral ischemia when compared to nontreated animals. This effect may be secondary to lidoflazine's inhibition of Ca^{2+} influx into ischemic nervous tissue, thereby ameliorating the cellular injury known to be associated with abnormally high levels of intracellular calcium. In the present study, the protective activity of lidoflazine may be attributed to the indirect inhibition of calcium-activated degradative enzymes, e.g., phospholipases and proteases. Alternatively, by inhibiting the efflux of adenine nucleotide metabolites (Van Belle *et al.*, 1986), lidoflazine may lower the rate of depletion of cellular ATP and possibly decrease residual glycolysis and associated intracellular lactic acidosis.

4. MITOCHONDRIAL CALCIUM UPTAKE

The ability of mitochondria to sequester calcium was measured with a Ca^{2+} electrode at 30°C in the same medium used for the respiratory experiments supplemented with an additional 3 mM $MgCl_2$ and 3 mM Na_2ATP. The pH of the medium was continuously monitored and adjusted to pH 7.0. The response of the electrode was calibrated against known free-calcium concentrations using a calcium–EGTA buffer system (Portzehl *et al.*, 1964).

A representative Ca^{2+} electrode determination of respiration-dependent Ca^{2+} accumulation by a suspension of dog brain mitochondria is shown in Fig. 2. The addition of isolated mitochondria to the test medium, which contained respiratory substrates (malate plus glutamate) and ATP, resulted in little change in the free-Ca^{2+} concentration of the medium. The addition of 44 nmol $CaCl_2$ per mg of mitochondrial protein caused a large increase in the Ca^{2+} concentration followed by a rapid decline to a steady-state "buffer

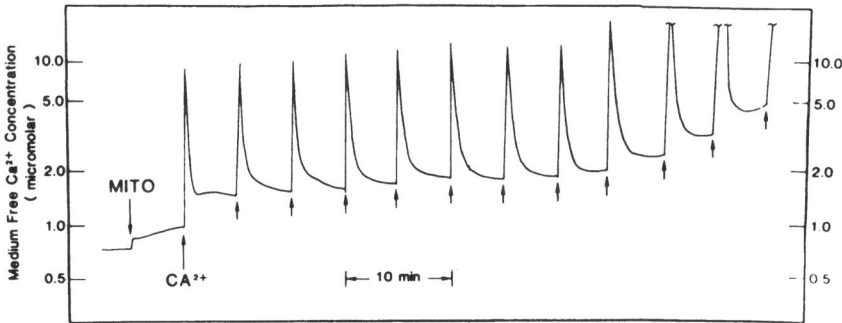

FIGURE 2. Calcium electrode tracings describing method for measuring the Ca^{2+} uptake capacity of isolated brain mitochondria. Mitochondria were added to a medium similar to that described in Fig. 1 except that 4 mM $MgCl_2$ and 3 mM Na_2ATP were present. Exogenous $CaCl_2$ was added at a concentration of 44 nmol/mg protein every 5 min until the mitochondria were unable to sequester any more Ca^{2+}. An addition of 5 mM malate plus 5 mM glutamate was made every 20 min to ensure a continuous source of respiratory substrates. The response of the Ca^{2+} electrode was calibrated against known free-Ca^{2+} concentrations using a Ca^{2+}–EGTA buffer system.

point'' of approximately 2 M. Multiple subsequent additions of calcium were followed by complete reuptake of Ca^{2+} by the mitochondria. At a total added calcium level of 352 nmol per mg protein (eight additions), the mitochondrial buffer point for medium free Ca^{2+} began to rise. At a total level of 484 nmol added calcium per mg protein (11 additions), the mitochondria were still able to accumulate nearly all of the added Ca^{2+}. This was taken as their maximum uptake capacity since the subsequent addition of Ca^{2+} was not followed by further accumulation.

The average maximum capacity for Ca^{2+} accumulation by brain mitochondria isolated from different experimental groups of animals is summarized in Table I. Mitochondrial calcium uptake appears to be relatively insensitive to ischemia. In the presence of ATP plus oxidizable substrate there appears to be a small decrease in calcium accumulation in mitochondria subjected to 10 min ischemia either without (452 ± 95 nmol/mg) or with lidoflazine pretreatment (400 ± 114 nmol/mg) when compared to control dogs (504 ± 66 nmol/mg). These differences were not, however, statistically significant. By 100 min after resuscitation, there is a significant rise in the ability to sequester calcium (712 ± 191 nmol/mg) when compared to either control ($P < 0.02$), pretreated ($P < 0.001$), or fibrillated groups ($P < 0.01$).

This study demonstrates that 10 min of fibrillatory arrest substantially impairs the ability of dog brain mitochondria to carry out oxidative phosphorylation. This injury to brain mitochondria may logically contribute to the cellular damage occurring following cardiac arrest and resuscitation. Without sufficient ATP to fuel cellular processes, degradative biochemical events, such as cellular Ca^{2+} overload with activation of Ca^{2+}-dependent degradative enzymes, ensue. While mitochondrial ability to carry out oxidative phosphorylation is markedly impaired by ischemia, it appears during identical periods of ischemia that mitochondrial capacity for Ca^{2+} uptake is preserved.

Administration of the calcium antagonist lidoflazine has been shown to preserve the ability of brain mitochondria to carry out oxidative phosphorylation during prolonged complete ischemia (Rosenthal et al., 1987). While pretreatment with this agent is not feasible for all victims of cardiac arrest, some degree of neuroprotection appears to be afforded with prophylactic treatment, perhaps extending the viability of cells exposed to prolonged periods of ischemia. If a subset of patients prone to cardiac arrest were identified, e.g., patients with recurrent ventricular tachycardia or survivors of sudden death, oral pretreatment with calcium antagonists could result in improved neurological recovery should a patient undergo cardiac arrest. Pretreatment with calcium antagonists for potential victims of cardiac arrest appears to merit further investigation.

5. CONCLUSIONS

1. Ten minutes of complete cerebral ischemia causes a substantial (35%) decline in the maximum rate of phosphorylating respiration by isolated cortical brain mitochondria.
2. Cerebral ischemia causes damage to the mitochondrial electron transport chain but apparently does not result in respiratory uncoupling, i.e., an increase in the mitochondrial membrane H^+ permeability.
3. Pretreatment of dogs with lidoflazine results in a greater than 50% reduction in ischemia-induced respiratory damage.
4. Mitochondrial damage is completely reversed following 100 min of restored cerebral circulation.

5. The ability of mitochondria to actively accumulate and retain Ca^{2+} is relatively insensitive to the adverse effects of ischemia.
6. Pretreatment with lidoflazine for potential victims of cardiac arrest merits further investigation as a means of extending neuronal viability during ischemia.

REFERENCES

Borgers, M., Thone, F., Van Reempts, J., and Verheyen, F., 1983, The role of calcium in cellular dysfunction, *Am. J. Emerg. Med.* **2**:154–161.

Clark, J. B., and Nicklas, W. J., 1970, The metabolism of rat brain mitochondria, *J. Biol. Chem.* **245**: 4724–4731.

Edmonds, H. L., Wauquier, A., Melis, W., Van Den Broeck, W. A. E., Van Loon, J., and Janssen, P. A. J., 1985, Improved short-term neurological recovery with flunarizine in a canine model of cardiac arrest, *Am. J. Emerg. Med.* **3**:150–155.

Farber, J. L., 1982, Biology of disease: Membrane injury and calcium homeostasis in the pathogenesis of coagulative necrosis, *Lab. Invest.* **47**:114–123.

Fiskum, G., and Lehninger, A. L., 1982, Mitochondrial regulation of intracellular calcium, in: *Calcium and Cell Function*, Vol. 2 (W. Y. Chung, ed.), Academic Press, New York, pp. 39–79.

Fiskum, G., 1983, Involvement of mitochondria in ischemic cell injury and in regulation of intracellular calcium, *Am. J. Emerg. Med.* **2**:147–153.

Fiskum, G., 1985, Mitochondrial damage during cerebral ischemia, *Ann. Emerg. Med.* **14**:810–815.

Fiskum, G., Pfeiffer, D. R., Broekemeir, K. M., and Baroody, B., 1985, Calcium buffering characteristics and phospholipase activities of rat brain mitochondria (Abstract), *Biophys. J.* **47**:443a.

Hamud, F., and Fiskum, G., 1985, Loss of maximal capacities for Ca^{2+} accumulation and oxidative phosphorylation by rat brain mitochondria during cerebral ischemia (Abstract), *Biophys. J.* **47**: 415a.

Hillered, L., Siesjo, B. K., and Arfors, K. E., 1984, Mitochondrial response to transient forebrain ischemia and recirculation in the rat, *J. Cereb. Blood Flow. Metabol.* **4**:438–446.

McCord, J. M., 1985, Oxygen-derived free radicals in postischemic tissue injury, *N. Engl. J. Med.* **312**:159–163.

Mela, L., 1979, Reversibility of mitochondrial metabolic response to circulatory shock and tissue ischemia, *Circ. Shock Suppl.* **1**:61–67.

Nayler, W. G., Ferrari, R., and Williams, A., 1980, Protective effect of pretreatment with verapamil, nifedipine and propranolol on mitochondrial function in the ischemic and reperfused myocardium, *Am. J. Cardiol.* **46**:242–248.

Portzehl, H., Caldwell, P. C., and Ruegg, J. C., 1964, The dependence of contraction and relaxation of muscle fibres from the crab maia squinado on the internal concentration of free calcium ions, *Biochim. Biophys. Acta* **79**:581–591.

Rehncrona, S., Mela, L., and Siesjo, B. K., 1979, Recovery of brain mitochondrial function in the rat after complete and incomplete cerebral ischemia, *Stroke* **10**:437–446.

Rosenthal, R. E., Hamud, F., Fiskum, G., Varghese, P. J., and Sharpe, S., 1987, Cerebral ischemia and reperfusion: Prevention of brain mitochondrial injury by lidoflazine, *J. Cereb. Blood Flow Metab.* **7**:752–758.

Siesjo, B. K., 1981, Cell damage in the brain: A speculative synthesis, *J. Cereb. Blood Flow Metab.* **1**:155–185.

Steen, P. A., Gisvold, S. E., Milde, J. H., Newberg, L. A., Scheithauer, B. W., Lanier, W. L., and Michenfelder, J. D., 1985, Nimodipine improves outcome when given after complete cerebral ischemia in primates, *Anesthesiology* **62**:406–414.

Van Belle, H., Wynants, J., Xhonneux, R., and Flameng, W., 1986, Changes in creatine phosphate, inorganic phosphate, and the purine pattern in dog hearts with time of coronary artery occlusion and effect thereon of mioflazine, a nucleoside transport inhibitor, *Cardiovasc. Res.* **20**:658–664.

White, B. C., Hildebrandt, J. F., Evans, A. T., Aronson, L., Indrieri, R. J., Hoehner, T., Fox, L., Huang, R., and Johns, D., 1985, Prolonged cardiac arrest and resuscitation in dogs: Brain mitochondrial function with different artificial perfusion methods, *Ann. Emerg. Med.* **14**:383–388.

Winegar, C. D., Henderson, O., White, B. C., Jackson, R. E., O'Hara, R., Krause, G. S., Vigor, D. N., Kontry, R., Wilson, W., and Shelby-Lane, C., 1983, Early amelioration of neurologic deficits by lidoflazine after 15 minutes of cardiopulmonary arrest in dogs, *Ann. Emerg. Med.* **12:** 470–476.

Wolkowicz, P. E., Michael, L. H., Lewis, R., and McMillin-Wood, J., 1983, Sodium–calcium exchange in dog heart mitochondria: Effects of ischemia and verapamil, *Am. J. Physiol.* **244**:H644–H651.

52

Role of Phosphoinositides in the Response of Mammalian Cells to Heat Shock

STUART K. CALDERWOOD and MARY ANN STEVENSON

1. INTRODUCTION

All cellular organisms possess a profound response to temperature shock which consists of the expression of a group of phylogenetically well-preserved genes, the heat shock genes (Lindquist, 1986; Pelham, 1986). They are among the most highly conserved of all known genes. The response appears to be homeostatic in nature leading to heat resistance (Lindquist, 1986). In animal cells, the heat shock response leads to profound changes at the level of transcription (Parker and Topol, 1984; Pelham, 1986), RNA processing (Toot and Lindquist, 1986), translation (Duncan et al., 1987), and intracellular protein transport (Napolitano et al., 1987). The biochemical regulation of these effects is not understood.

In the present study we have examined the possibility that physiological pathways of transmembrane signal transduction are involved in the induction of the heat shock response. This was suggested by our earlier studies indicating heat shock-induced increases in Ca^{2+}, diacylglycerol, and phosphoinositide turnover (Stevenson et al., 1986). Here we have shown that heat shock induces inositol trisphosphate (IP_3) release with a similar time–temperature dependency to heat shock protein induction. Heat-induced IP_3 release is observed in four different species and other inducers of the heat shock response induce IP_3 release. Heat shock-induced IP_3 release is GTP-dependent.

2. MATERIALS AND METHODS

2.1. Cell Culture

Cells were maintained in monolayer tissue culture at 37°C using standard techniques. HA-1 chinese hamster ovary fibroblasts were maintained in minimal essential medium. Hela

STUART K. CALDERWOOD and MARY ANN STEVENSON • Joint Center for Radiation Therapy and Dana Farber Cancer Institute, Harvard Medical School, Boston, Massachusetts 02115.

cells, NIH 3T3 fibroblasts, and Balbc 3T3 fibroblasts were grown in Dulbecco's medium containing 4.5 mg/liter glucose. The above cells and HA-1 fibroblasts were cultured in the presence of 5% CO_2/air and in 10% fetal calf serum. PC-12 pheochromocytoma cells were grown in Dulbecco's minimal essential medium (MEM) with 10% fetal calf serum, 5% horse serum, and 10% CO_2/air.

2.2. Inositol Phosphates

Methods were devised from those described by Berridge (1983) and Irvine (1986). Inositol phosphates were measured after metabolic labeling of cells with [³H]myoinositol. Confluent monolayers of cells (10^7) were thus labeled for 24 hr in inositol-free Dulbecco's MEM supplemented with 5 μM myoinositol and 0.2% fetal calf serum (medium A) containing 4.0 μCi/ml of [³H]myoinositol. Before experiment, cells were washed in medium A. Experiments were carried out either in lithium-free conditions or with 15 μM LiCl. LiCl is an inhibitor of IP_1 phosphatase (Berridge, 1983) and its use permits the measurement of overall stimulation of phospholipase C without the complicating effects of breakdown by phosphatases.

Experiments were terminated by adding ice-cold 80% trichloracetic acid to a final concentration of 8.0%. Acid extracts were clarified by centrifugation and then extracted four times with 5 volumes of ethyl ether. Samples were then separated by chromatography on 0.7-ml Dowex-1 columns as described previously (Irvine, 1986). Inositol trisphosphate is reported as the fraction eluted between 0.1 M formic acid/0.4 M ammonium formate and 0.1 M formic acid/0.8 M ammonium formate (10 ml). In control experiments, 72–75% of standard [³H]inositol-1,4,5-trisphosphate was eluted in this fraction. The recommendations of Berridge (1983) and Irvine (1986) were adhered to in all analytical procedures.

2.3. Cell Permeabilization

Cells were washed and overlaid with buffered salt solution which approximated physiological intracellular fluid. The buffer contained KCl (115 mM), NaCl (5 mM), KH_2PO_4 (1 mM), $NaHCO_3$ (5 mM), $MgCl_2$ (300 μM), $CaCl_2$ (100 μM), and ethylene glycol-bis (β aminoethyl ether) NNN'N' tetraacetic acid (182 μM). The buffer also contained 15 mM LiCl. Cells were made permeable by adding digitonin (2.5 mg/ml in ethanol) to a concentration of 15 μg/ml (Wolf et al., 1985).

3. RESULTS

The effects of temperature on IP_3 production in Hela cells are indicated in Fig. 1A. Levels of IP_3 were not increased by temperature up to 39°C. At 41°C a gradual increase in IP_3 levels was observed and this increase was more marked at 42°C. IP_3 release was extremely rapid at temperatures above 42°C. Previous studies indicate that at 45°C, IP_3 release occurs on a time scale of seconds (Stevenson et al., 1986), which is characteristic of receptor-mediated release (Berridge, 1983; Wolf et al., 1985). Examination of proteins synthesized after heat shock by SDS–PAGE and autoradiography indicates that a similar time–temperature relationship governs heat shock protein (hsp) synthesis. No evidence of increases in heat shock proteins 70 or 89 (hsp 70, hsp 89) is observed by 120 min at 39°C. Hsp 70 and 89 bands are observed after 120 min at 41°C and the bands are more intense after a similar

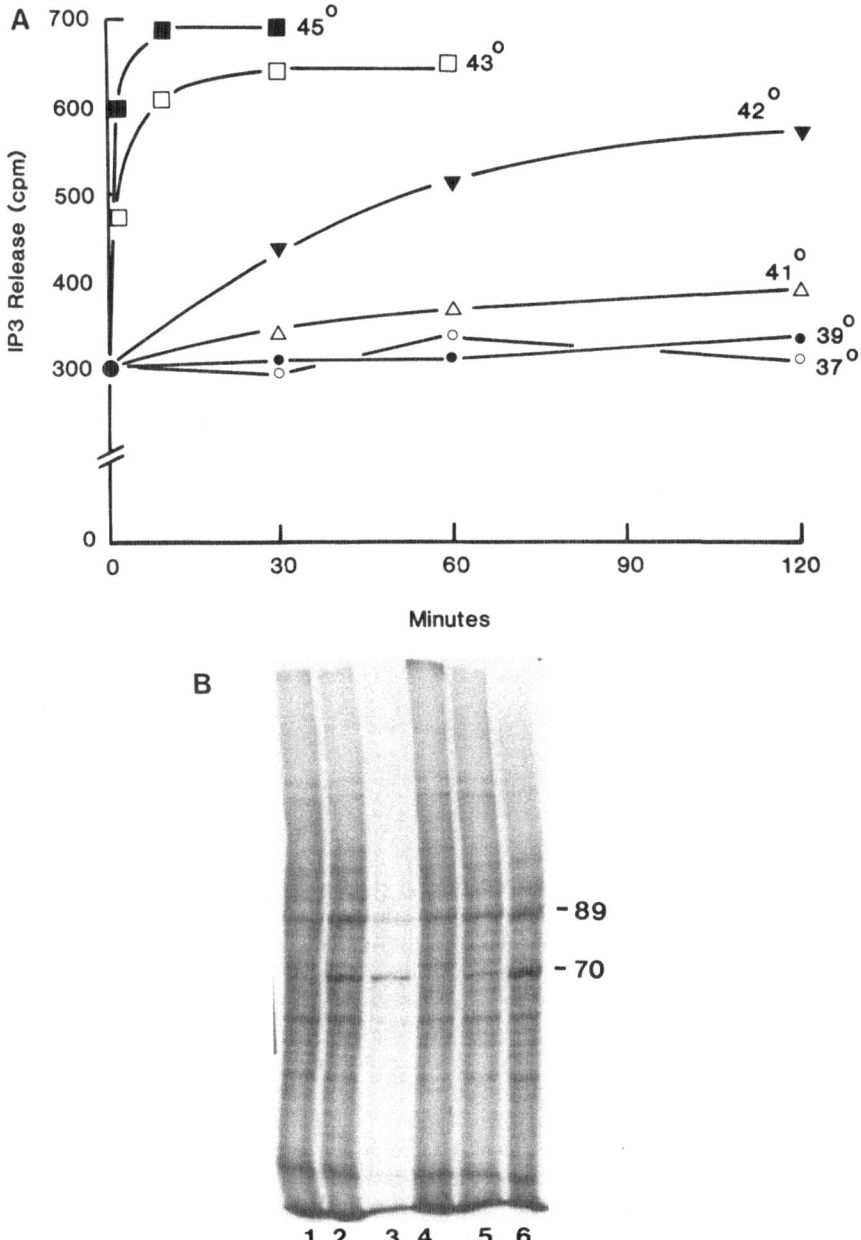

FIGURE 1. (A) Effect of heat shock on IP_3 release in Hela cells. Cells were heated in the presence of LiCl (15 mM). (B) Autoradiography of Hela cell proteins metabolically labeled with [^{35}S]methionine and separated by SDS–PAGE (7.0% gel). Cells were treated: (1) control, (2) 30 min at 43°C, (3) 30 min at 45°C, (4) 120 min at 39°C, (5) 120 min at 41°C, and (6) 120 min at 42°C. Datum points are means of duplicate assays. Experiment was performed twice, yielding consistent data.

42°C heating. Hsp 70 and 89 are induced by 30 min at 43 or 45°C. At 45°C there is a major inhibition of normal protein synthesis as described previously (Lindquist, 1986).

Stimulation of inositol phosphate release by heat shock (45°C) in cell lines comprising four different species (human, rat, mouse, hamster) is shown in Fig. 2. The response thus seems to be a general one in heat-shocked cells.

Heat shock gene expression is induced by a small, disparate group of agents (Ashburner and Bonner, 1979; Lindquist, 1986). We examined in HA-1 cells the effect on IP_3 release of five such agents, which are ethanol, arsenite, arsenate, $ZnSO_4$, and $CdSO_4$ (Fig. 3). The agents all caused IP_3 release at levels similar to that of the heat-induced response. In the figure we also show the effects on IP_3 concentration of vanadate and fluoroaluminate ($AlCl_3/NaF$), agents which cause activation of G proteins (Uhing et al., 1986). Both compounds induce IP_3 release in HA-1 cells, vanadate most effectively. Physiological agonists were more effective inducers of IP_3 release. Thrombin in particular caused an almost 500% increase in IP_3 levels (compared to 180% in heat-shocked). The effects of mitogenic stimulation with serum appeared to be additive with heat as observed previously (Stevenson et al., 1986). The effects of heat shock on IP_3 release were GTP-dependent. Figure 4 shows IP_3 release from digitonin-permeabilized HA-1 cells. Experiments were carried out at 37° or 45°C at a range of GTPγS concentrations. In the absence of GTPγS, the degree of IP_3 release was similar at 37 or 45°C. Heat shock led to IP_3 release at lower concentrations of GTPγS (10^{-7}–10^{-6} M), which were not effective in inducing IP_3 release at 37°C. As the concentration of GTPγS increased to 10^{-5} M, IP_3 release increased to a maximum. At this concentration the differential between 37°C and 45°C was not observed. The effect of heat shock was thus to cause stimulation of phospholipase C at concentrations of GTPγS that were nonstimulatory or only faintly effective in controls.

4. DISCUSSION

The data indicate that heat shock induces both IP_3 release and heat shock protein expression with similar dose–response relationships (Fig. 1). We observed this response in each of the five cell lines so far examined, four of which are shown in Fig. 2. As these cell lines encompass four different species, this response to heat appears to be phylogenetically well conserved. Further evidence linking phospholipase C action to the heat shock response is that five disparate agents, connected by their induction of hsp expression (Lindquist, 1986; Ashburner and Bonner, 1979), each cause IP_3 release (Fig. 3). The effect of heat appears to require the activation of a GTP-binding protein (Gp) involved in the regulation of phospholipase C action (Ui, 1986). However, our preliminary data indicate that pertussis toxin (10–1000 μg/ml for 3 hr) does not inhibit heat-induced IP_3 release in HA-1, Hela, and NIH 3T3 cells. As pertussis acts by inhibiting receptor–G protein interactions (Ui, 1986), its failure to block the effects of heat shock may indicate a direct physical effect of heat on G proteins not involving receptors or may indicate that the G protein involved in transducing the effects of heat is not a pertussis substrate.

Since the heat shock response is ubiquitous in cellular organisms, any general mechanism proposed for its induction must be equally widespread. The G protein gene family is observed in all species so far investigated (Stryer and Bourne, 1986; Ahnn et al., 1986). G proteins would seem possible candidates as transducing agents required to convert the physical effects of heat shock into a molecular response. Further evidence which may support such a sug-

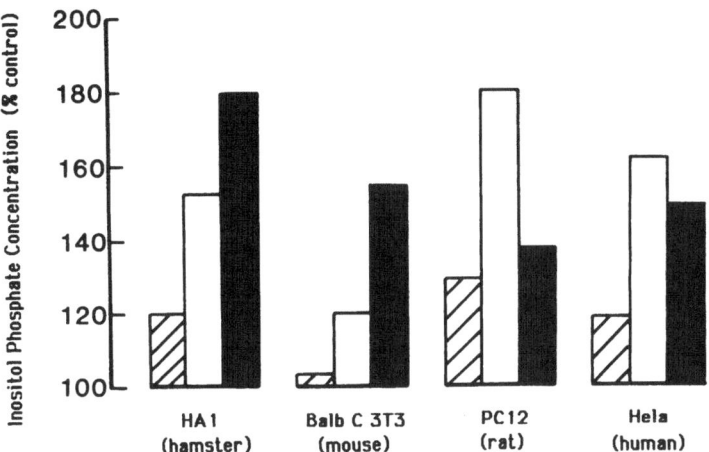

FIGURE 2. Levels of inositol phosphates in HA-1 (hamster), Balb C 3T3 (mouse), PC-12 (rat) and Hela (human) cells after 5 min at 45°C. Experiments were carried out in the absence of Li$^+$. Levels of IP$_1$ (▨), IP$_2$ (▢), and IP$_3$ (■) are indicated. Experiments were carried out in duplicate on at least three occasions.

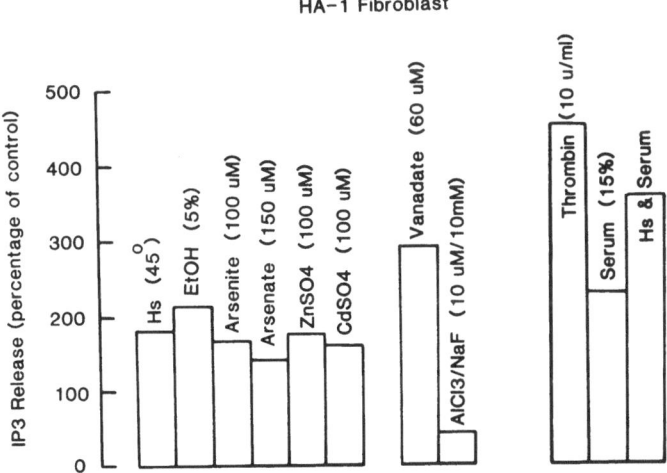

FIGURE 3. Effects of heat shock inducers, vanadate, fluoroaluminate, and mitogens on IP$_3$ release in HA-1 fibroblasts. Cells were treated for 30 min in the presence of LiCl with heat shock gene inducers: 45°C heat shock, ethanol (5%), arsenite (100 μM), arsenate (150 μM), ZnSO$_4$ (100 μM), and CdSO$_4$ (100 μM). These agents were shown to induce hsp expression under the conditions used (Ashburner and Bonner, 1979). Vanadate and AlCl$_3$/NaF were previously shown to stimulate phospholipase C activity under similar conditions to those used in this experiment (Paris et al., 1987). For treatment with mitogens, serum, and thrombin, cells were pretreated for 48 hr under low serum conditions (0.2% fetal calf serum) in order to accumulate them in a mitogen-sensitive quiescent state. Data are means of triplicate assays. Experiments were performed two to five times.

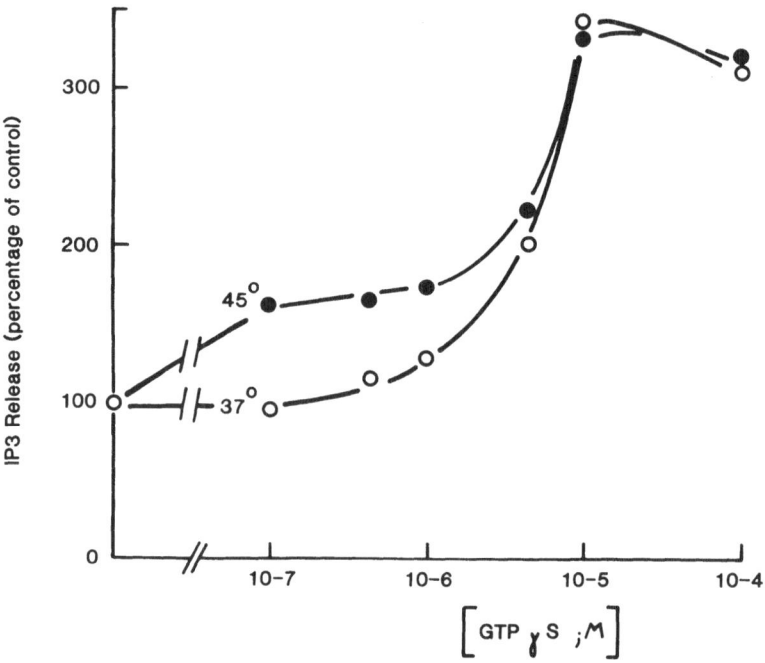

FIGURE 4. IP$_3$ release from HA-1 cells at 37°C and 45°C at a range of GTPγS concentrations. Experiments were carried out in cells permeabilized with digitonin. Datum points are means of duplicate assays; experiments were carried out on two occasions, yielding consistent results.

gestion is that heat shock activates adenylate cyclase (Calderwood *et al.*, 1985) and phospholipase A$_2$ (S. K. Calderwood, unpublished data). Both enzymes are regulated by guanine nucleotide-binding proteins (Ui, 1986). Thus, the suggestion that heat shock acts by directly activating G proteins would appear to be a good working hypothesis to direct further study of processes involved in induction of the heat shock response.

5. SUMMARY

All cellular organisms exhibit a response to temperature shock which results in *de novo* expression of heat shock proteins (hsp), molecules thought to mediate heat resistance. The mechanisms involved in induction of the heat shock genes are not known. We have examined the potential role of inositol phosphates in the response. Heat shock causes IP$_3$ release in human, rat, mouse, and hamster cells. The time–temperature relationships for IP$_3$ release are similar to those for hsp synthesis. Other inducers of the heat shock response (ethanol, arsenite, arsenate, ZnSO$_4$, CdSO$_4$) also cause IP$_3$ release. Stimulation of IP$_3$ release by heat shock is observed only in the presence of GTP. This suggests a role for G proteins in heat-induced IP$_3$ release. The results overall indicate that heat shock activates G proteins and phospholipase C, and suggest a role for these events in the heat shock response.

REFERENCES

Ahnn, J., March, P. E., Takiff, H. E., and Inouye, M., 1986, A GTP binding protein of Escherichia coli has homology to yeast RAS proteins, *Proc. Natl. Acad. Sci. USA* **83**:8849–8853.

Ashburner, M., and Bonner, J. J., 1979, The induction of gene activity in drosophila by heat shock, *Cell* **17**:241–254.

Calderwood, S. K., Stevenson, M. A., and Hahn, G. M., 1985, Cyclic AMP and the heat shock response in chinese hamster ovary cells, *Biochem. Biophys. Res. Commun.* **126**:911–916.

Duncan, R., Milburn, S. C., and Hershey, J. W. B., 1987, Regulated phosphorylation and low abundance of Hela cell initiation factor eIF-4F suggests a role in translational control, *J. Biol. Chem.* **262**: 380–388.

Irvine, R. F., 1986, The structure, metabolism, and analysis of inositol lipids and inositol phosphates, in: *Phosphoinositides and Receptor Mechanisms*, Vol. 7, *Receptor Biochemistry and Methodology*, (J. C. Venter and L. C. Harrison, eds.), Alan R. Liss, New York, pp. 89–109.

Lindquist, S. L., 1986, The heat shock response, *Ann. Rev. Biochem.* **55**:563–582.

Napolitano, E. W., Pachter, J. S., and Liem, R. K. H., 1987, Intracellular distribution of mammalian stress proteins, *J. Biol. Chem.* **262**:1493–1504.

Paris, S., Chambard, J. C., and Pouyssegur, J., 1987, Coupling between phosphoinositide breakdown and mitogenic events in fibroblasts, *J. Biol. Chem.* **262**:1977–1983.

Parker, C. S., and Topol, J., 1984, A drosophila RNA polymerase II transcription factor binds to the regulatory site of an hsp 70 gene, *Cell* **37**:273–283.

Pelham, H. R. B., 1986, Speculations on the functions of the major heat shock and glucose regulated proteins, *Cell* **46**:959–961.

Stevenson, M. A., Calderwood, S. K., and Hahn, G. M., 1986, Rapid increases in inositol trisphosphate and intracellular Ca^{2+} after heat shock, *Biochem. Biophys. Res. Commun.* **137**:826–833.

Stryer, L., and Bourne, H. R., 1986, G proteins: A family of signal transducers, *Ann. Rev. Cell Biol.* **2**:391–419.

Toot, H. J., and Lindquist, S. L., 1986, RNA splicing is interrupted by heat shock and is rescued by heat shock protein synthesis, *Cell* **45**:185–193.

Uhing, R. J., Prpic, V., Hang, J., and Exton, J. H., 1986, Hormone stimulated polyphosphoinositide breakdown in rat liver plasma membranes, *J. Biol. Chem.* **261**:2140–2146.

Ui, M., 1986, Pertussis toxin as a probe of receptor coupling to inositol lipid metabolism, in: *Phosphoinositides and Receptor Mechanisms;* Vol. 7, *Receptor Biochemistry and Methodology* (J. C. Venter and L. C. Harrison, eds.), Alan R. Liss, New York, pp. 163–197.

Wolf, B. A., Comens, P. G., Ackermann, K. E., Sherman, W. R., and McDaniel M. L., 1985, The digitonin-permeabilized pancreatic islet model, *Biochem. J.* **227**:965–969.

53

Cell Calcium in Malignant Hyperthermia
Skeletal Muscle Contracture and Ca^{2+} Regulation by Sarcoplasmic Reticulum

T. E. NELSON

1. INTRODUCTION

Malignant hyperthermia (MH) is a genetic disease of man and various animal species that predisposes to a hypermetabolic syndrome triggered by certain anesthetic agents (Gronert, 1965; Nelson and Flewellen, 1983). Susceptibility to the MH syndrome is prevalent in swine and this also predisposes to a stress-induced syndrome in this species (Nelson, 1973). Potent volatile anesthetics such as halothane and depolarizing skeletal muscle relaxants such as succinylcholine are triggering agents for MH in susceptible man and animals. The state of anesthesia alone is not responsible for MH as evidenced by the fact that barbiturates, nitrous oxide, and narcotics can be used to provide anesthesia for the MH patient without triggering the syndrome. The primary defect for MH appears to be in skeletal muscle and a working hypothesis is that anesthetic agents which cause MH do so by producing a sustained increase in Ca^{2+} inside the muscle cell. Increases in myoplasmic Ca^{2+} below contracture threshold can increase oxygen consumption (Bianchi *et al.*, 1975), and if unabated this can deplete energy necessary to maintain Ca^{2+} below contracture threshold. Further increases in myoplasmic Ca^{2+} can then activate the contractile elements, producing the clinical features of skeletal muscle rigidity and lactic acidosis that are observed in the malignant phases of the syndrome.

In this paper we describe experiments that show (1) dependence of abnormal MH muscle responses on Ca^{2+} content of the muscle cell; (2) the effect of dantrolene, a direct-acting skeletal muscle relaxant, on the abnormal contracture of MH muscle; and (3) properties of the Ca^{2+}-regulating sarcoplasmic reticulum membrane that may be defective in MH skeletal muscle.

T. E. NELSON ● Department of Anesthesiology, University of Texas Health Science Center at Houston, Houston, Texas 77030.

2. THE MALIGNANT HYPERTHERMIA SYNDROME

Malignant hyperthermia is a genetic disease of man and pigs. Human MH pedigrees indicate an autosomal-dominant form of inheritance although multigenic modes have not been excluded. In the pig an autosomal-recessive inheritance is evident (Smith and Bampton, 1977; Archibald and Imlak, 1985). Induction of the anesthetized state in man and animals results in depression of the central thermoregulatory mechanisms and the homeotherm is pharmacologically transformed to a poikilotherm. Consequently, most body temperatures decrease following anesthesia unless active measures are taken to prevent body heat loss. It is usually abnormal and unexpected to observe increases in patients' body temperature following general anesthesia. In its most malignant form the MH syndrome can produce heat at a rate that increases body temperature by 0.5°C every 5 min and final temperatures as high as 45°C. This hypermetabolic storm is produced by peripheral rather than central mechanisms as demonstrated by the experiment of Harrison (1973). In an MH-susceptible pig prepared by spinal cord transection and extracorporeal perfusion of the hind limbs, exposure to halothane produced all clinical manifestations of the syndrome. In the intact pig, the first symptoms of the MH syndrome are tachycardia and tachypnea secondary to increased oxygen consumption. Following this, metabolic acidosis occurs followed by increases in body temperature and global skeletal muscle rigidity. If untreated, the syndrome progresses unabated and produces severe acidosis, electrolyte disturbances, and other losses of homeostasis that ultimately produce cardiac failure and death. All of these symptoms can be prevented or effectively treated by the direct-acting skeletal muscle relaxant dantrolene sodium (Flewellen and Nelson, 1980; Kolb et al., 1982). This effect of dantrolene on the MH syndrome lends credence to the concept that MH is a consequence of abnormality in skeletal muscle.

3. ABNORMALITY IN SKELETAL MUSCLE METABOLISM

The occurrence of skeletal muscle rigidity during the MH syndrome raised suspicion that a defect existed in skeletal muscle. The first direct evidence for this was the observation by Harrison et al. (1969) showing that skeletal muscle from MH pigs consumed more ATP in vitro when exposed to halothane than muscle from normal pigs (Harrison et al., 1969). Subsequent studies showed that halothane also produced abnormal lactate production in MH skeletal muscle (Nelson et al., 1972) and that production of lactate is proportional to the concentration of Ca^{2+} (0–1.0 mM) in the in vitro media (Nelson, 1978). Several studies were performed to determine if a metabolic enzyme defect exists in genetically predisposed MH skeletal muscle (Kastenschmidt et al., 1968; Berman et al., 1972) but no conclusive results were obtained. We propose that the hypermetabolic state of MH muscle is secondary to abnormal increases in myoplasmic Ca^{2+}.

4. ABNORMALITY IN SKELETAL MUSCLE CONTRACTILITY

4.1. Skeletal Muscle Contraction

Under normal physiological control the electromechanical coupling pathway is thought to involve depolarization of the sarcolemma by the action of acetylcholine on the endplate receptors, propagation of the depolarizing signal along the sarcolemma and down the transverse tubule, and by some unknown mechanism the signal is transduced from the transverse tubule to the terminal cisternae of the sarcoplasmic reticulum (SR). The terminal cisternae structure of the SR is the main Ca^{2+} storage site and is thought to contain Ca^{2+} channels that are opened to release Ca^{2+} for the activation of the contractile elements. A postulate for MH that has directed our studies is based on the supposition that one or more defects exist along this electromechanical coupling pathway in MH skeletal muscle that produce an abnormal sustained increase in myoplasmic Ca^{2+} which triggers the MH syndrome. Clinically, the MH-susceptible individual has no discernible manifestations and most lead relatively normal lives until confronted with the triggering anesthetic agents. If a defect did exist along the electromechanical coupling pathway, then it may be possible to detect such a defect by measuring rates of contraction and relaxation and the time from depolarization of the sarcolemma to the onset of mechanical tension, i.e., excitation–contraction coupling time interval. A study in MH-susceptible pigs showed that the E—C coupling time interval was about 1 msec longer than for normal muscle but the kinetics for contraction and relaxation did not differ from normal values (Nelson et al., 1983). A subsequent study (Quinlan et al., 1986) failed to confirm the earlier results but did show that multiple stimuli to dantrolene (3 mg/kg I.V.)-treated pigs produced different contraction kinetics between MH and normal skeletal muscle.

In normal human skeletal muscle the resting myoplasmic Ca^{2+} level is about 1 × 10^{-7} M (Wood et al., 1982), i.e., pCa^{2+} = 7.0, and the optimal Ca^{2+} concentration for contraction is between pCa^{2+} = 6 to 5. However, increases in Ca^{2+} below the contraction threshold can stimulate metabolism and oxygen consumption (Bianchi et al., 1975). Thus, any abnormal increase in myoplasmic Ca^{2+} could stimulate metabolic and/or contractile events. The significant discovery by Kalow et al. (1970) that skeletal muscle from MH-susceptible patients produced abnormal in vitro contracture responses to caffeine and halothane implied that these drugs produced increases in myoplasmic Ca^{2+} at much lower concentrations in MH muscle. Previous studies had shown the effect of caffeine on isolated sarcoplasmic reticulum membranes to be one of producing a net effect of releasing Ca^{2+} from the membrane stores (Weber and Hertz, 1968). The sarcoplasmic reticulum (SR) membrane serves at least three different functions in regulation of skeletal muscle cell Ca^{2+}. The SR is the terminal membrane in transduction of signal from depolarization of the sarcolemma to release of activator Ca^{2+} for contraction and the terminal cisternae is the site for Ca^{2+} storage and release. Second, the SR must contain some mechanism for release of Ca^{2+} stores into the myoplasm. This role is described for putative Ca^{2+} channels in the SR terminal cisternae. Finally, the SR regulates myoplasmic Ca^{2+} by way of the Mg^{2+}-dependent ATPase Ca^{2+} pump that actively transports Ca^{2+} back inside the SR membrane. Although other organelles may play a role in regulation of myoplasmic Ca^{2+}, i.e., mitochondria, for the most part these seem to be of secondary importance to functions of the SR in regulation of contraction–relaxation functions.

The final step in excitation–contraction coupling is activation of contraction by binding of Ca^{2+} to troponin. The pCa^{2+} versus tension curves do not differ between MH and normal muscle (Britt et al., 1982; Wood et al., 1982; Endo et al., 1983) and halothane is without effect, excluding this as a etiological mechanism for MH.

The SR is a likely target for a defect in MH skeletal muscle. Attempts to compare Ca^{2+} transport function in SR from MH muscle have been difficult to assess due to small numbers of patients and to extreme variations in methodology. We have compared Ca^{2+} transport rates in heavy and light SR fractions from MH and control pigs and from MH and control humans, and concluded that no defect occurs in this function of the SR membrane from MH skeletal muscle.

Using skinned single fibers from MH patients, Takagi (1976) and Endo *et al.* (1983) concluded that Ca^{2+}-induced Ca^{2+} release was abnormal and that the hypercontracture effects of caffeine and halothane on MH muscle could be produced through the Ca^{2+}-induced Ca^{2+} release pathway. Subsequent studies on Ca^{2+}-induced Ca^{2+} release from SR isolated from MH pig skeletal muscle tend to support the initial observations on human muscle (Nelson, 1983; Ohnishi and Gronert, 1983; Kim *et al.*, 1984; Mickelson *et al.*, 1986).

In conclusion, there are considerable data to support the hypothesis that the hypermetabolic syndrome malignant hyperthermia is caused primarily by a defect in skeletal muscle. Further, this defect appears to be manifested by a loss in regulation of muscle cell Ca^{2+}, and the consequent sustained increase in muscle cell free Ca^{2+} produces the hypermetabolic state and associated skeletal muscle rigidity. Anesthetic agents that trigger the MH syndrome do so not by blocking the sarcoplasmic reticulum Ca^{2+} pump relaxing mechanism, but more probably by increasing the probability of Ca^{2+} channel openings. The Ca^{2+} channel openings produce a rise in muscle cell Ca^{2+} which triggers metabolic and contracture events. This loss in regulation of Ca^{2+} at the skeletal muscle cell level produces a life-threatening syndrome, illustrating the importance of regulation of cell Ca^{2+} to normal homeostasis of the organism.

REFERENCES

Archibald, A. L., and Imlak, P., 1985, The halothane sensitivity locus and its linkage relationships, *An. Bld. Groups. Biochem. Gen.* **16**:253–256.

Berman, M. C., Conradie, P. J., and Kench, J. E., 1972, The mechanism of accelerated skeletal muscle glycogenolysis during malignant hyperthermia in swine, *S. Afr. Med. J.* **46**:1810.

Bianchi, C. P., Narayan, S., and Lakshminarayanaiah, N., 1975, Mobilization of muscle calcium and oxygen uptake in skeletal muscle, in: *Calcium Transport in Contraction and Secretion* (E. Carafoli, F. Clementi, W. Drabikowski, and A. Margreth, eds.), Elsevier, New York, pp. 503–515.

Britt, B.A., Frodis, W., Scott, E., Clements, M. J., and Endrenyi, L., 1982, Comparison of the caffeine skinned fibre tension (CSFT) test with the caffeine-halothane contracture (CHC) test in the diagnosis of malignant hyperthermia, *Can. Anaesth. Soc. J.* **29**(6):550–562.

Endo, M., Yagi, S., Ishizuka, T., Horiuti, K., Koga, Y., and Amaha, K., 1983, Changes in the Ca-induced Ca release mechanism in the sarcoplasmic reticulum of the muscle from a patient with malignant hyperthermia, *Biomed. Res.* **4**(1):83–92.

Flewellen, E. H., and Nelson, T. E., 1980, Dantrolene dose response in malignant hyperthermia-susceptible (MHS) swine: Method to obtain prophylaxis and therapeusis, *Anesthesiology* **52**(4):303–308.

Gronert, G. A., 1980, Malignant hyperthermia, *Anesthesiology* **53**:395–423.

Harrison, G. G., 1973, Recent advances in understanding of anaesthetic-induced malignant hyperpyrexia, *Anaesthetist* **22**:373–376.

Harrison, G. G., Saunders, S. J., Biebuyck, J. F., Hickman, R., Dent, D. M., Weaver, V., and Terblanche, J., 1969, Anaesthetic-induced malignant hyperpyrexia and a method for its prediction, *Br. J. Anaesthiol.* **41**:844–854.

Kalow, W., Britt, B. A., Terreau, M. E., and Haist, C., 1970, Metabolic error of muscle metabolism after recovery from malignant hyperthermia, *Lancet* **2**:895–898.

Kastenschmidt, L. L., Hoekstra, W. G., and Briskey, E. J., 1968, Glycolytic intermediates and co-factors in "fast-" and "slow-glycolyzing" muscles of the pig, *J. Food Service* **33**:151–158.

Kim, D. H., Sreter, F. A., Ohnishi, T., Ryan, J. F., Roberts, J., Allen, P. D., Meszaros, L. G., and Antoniu, B., 1984, Kinetic studies of Ca^{2+} release from sarcoplasmic reticulum of normal and malignant hyperthermia susceptible pig muscles, *Biochim. Biophys. Acta* **775**(3):320–327.

Kolb, M. E., Horne, M. L., and Martz, R., 1982, Dantrolene in human malignant hyperthermia, *Anesthesiology* **56**:254–262.

Mickelson, J. R., Ross, J. A., Reed, B. K., and Louis, C. F., 1986, Enhanced Ca^{2+}-induced calcium release by isolated sarcoplasmic reticulum vesicles from malignant hyperthermia susceptible pig muscle, *Biochim. Biophys. Acta* **862**:318–328.

Nelson, T. E., 1973, Porcine stress syndromes, in: *International Symposium on Malignant Hyperthermia* (R. A. Gordon, B. A. Britt, and W. Kalow, eds.), Charles C Thomas, Springfield, pp. 191–197.

Nelson, T. E., 1977, Excitation–contraction coupling: A common etiologic pathway for malignant hyperthermia susceptible muscle, in: *Malignant Hyperthermia* (J. A. Aldrete and B. A. Britt, eds.), Grune and Stratton, New York, pp. 23–36.

Nelson, T. E., 1983, Abnormality in calcium release from skeletal sarcoplasmic reticulum of pigs susceptible to malignant hyperthermia, *J. Clin. Invest.* **72**:862–870.

Nelson, T. E., Flewellen, E. H., 1983, Current concepts, The malignant hyperthermia syndrome, *N. Engl. J. Med.* **309**:416–418.

Nelson, T. E., Jones, E. W., Venable, J. H., and Kerr, D. D., 1972, Malignant hyperthermia of Poland China swine, *Anesthesiology* **36**:52–56.

Nelson, T. E., Flewellen, E. H., and Arnett, D. W., 1983, Prolonged electromechanical coupling time intervals in skeletal muscle of pigs susceptible to malignant hyperthermia, *Muscle Nerve* **6**:263–268.

Ohnishi, S. T., and Gronert, G. A., 1983, The sarcoplasmic reticulum of swine with malignant hyperthermia has abnormal calcium-induced and halothane-induced calcium release phenomena, *Biophy. J.* **41**:167A.

Quinlan, J. G., Iaizzo, P. A., Gronert, G. A., and Taylor, S. R., 1986, Use of dantrolene plus multiple pulses to detect stress-susceptible porcine muscle, *J. Appl. Physiol.* **60**:1313–1320.

Smith, C., and Bampton, P. R., 1977, Inheritance of reaction to halothane anaesthesia in pigs, *Genet. Res. Camb.* **22**:287–292.

Takagi, A., 1976, Abnormality of sarcoplasmic reticulum in malignant hyperpyrexia, *Adv. Neurol. Sci.* **20**:109–113.

Weber, A., and Hertz, R., 1968, The relationship between caffeine contracture of intact muscle and the effect of caffeine on reticulum, *J. Gen. Physiol.* **52**:750–759.

Wood, D. S., Willner, J. H., and Solviati, G., 1982, Malignant hyperthermia: The pathogenesis of abnormal caffeine contracture, in *Disorders of the Motor Unit* (D. L. Schotland, ed.), John Wiley and Sons, New York, pp. 597–610.

54

Pathogenetic Roles of Intracellular Calcium and Magnesium in Membrane-Mediated Progressive Muscle Degeneration in Duchenne Muscular Dystrophy

SYAMAL K. BHATTACHARYA, ALICE J. CRAWFORD, JAY H. THAKAR, and PATTI L. JOHNSON

1. INTRODUCTION

Duchenne muscular dystrophy (DMD) in humans is a progressively crippling X-linked recessive neuromuscular disease with no effective treatment (Rowland, 1980; Moser, 1984). It is characterized by profound biochemical (Kar and Pearson, 1976; Bertorini et al., 1982; Bhattacharya and Crawford, 1985), electrocardiographic (Sanyal and Johnson, 1982), histopathological (Bodensteiner and Engel, 1978; Emery and Burt, 1980; Bertorini et al., 1982, 1984), and ultrastructural (Mokri and Engel, 1975; Oberc and Engel, 1977) abnormalities of skeletal and cardiac muscle, and a 70–80% reduced life expectancy. Although a "vascular hypothesis" implicating abnormal microvasculature has been presented in the past to explain many aspects of the dystrophic pathophysiology, the most tenable mechanism for the classical muscle degeneration in DMD is now widely attributed to a generalized functional and structural defect(s) in the plasma membrane integrity of myofibers (Mokri and Engel, 1975; Schotland et al., 1977) and erythrocytes (Araki and Mawatari, 1971).

This "membrane hypothesis" suggests a sarcolemmal abnormality which results in an increased cellular ingress of Ca^{2+} and Na^+, and a leakage of Mg^{2+}, K^+, creatine kinase (CK), and lactate dehydrogenase (LD) to the extracellular compartment. Cumulative evidence indicates that among the multitude of discernible pathogenetic events which transpire between the initiation of membranous changes and ultimate toxic cell death in DMD (Carpenter and Karpati, 1979; Schanne et al., 1979), membrane-mediated excessive intracellular Ca ac-

SYAMAL K. BHATTACHARYA, ALICE J. CRAWFORD, JAY H. THAKAR, and PATTI L. JOHNSON • Edward Dana Mitchell Surgical Research Laboratories, Departments of Surgery, Anatomy, and Neurobiology, University of Tennessee Medical Center, Memphis, Tennessee 38163.

cumulation (EICA) represents one of the earliest detectable biochemical events (Emery and Burt, 1980; Bhattacharya and Crawford, 1985). Excessive intracellular Ca accumulation appears to be the most crucial factor in the pathogenesis of muscular dystrophy in both DMD and animal models (Wrogemann and Pena, 1976; Brambati et al., 1980; Bertorini et al., 1984; Bhattacharya et al., 1984).

In this chapter, we will briefly discuss (1) the roles of intracellular Ca and Mg in the pathogenesis of membrane-mediated progressive muscle degeneration in DMD and in animal models of muscular dystrophy, (2) the interrelationships of EICA with other pathobiological aspects of progressive muscle degeneration, and (3) the potential pharmacological regulation of EICA in degenerating cardiac and skeletal muscles affected by hereditary muscular dystrophy.

2. PATHOGENETIC ROLES OF CALCIUM AND MAGNESIUM IN PROGRESSIVE MUSCLE DEGENERATION IN HEREDITARY MUSCULAR DYSTROPHY

The biological roles of Ca and plasma membranes in cellular and subcellular functions are well characterized in numerous physiological processes (Ebashi and Sugita, 1979); however, the intricacies of their abnormalities in a variety of pathogenetic conditions are far from completely understood (Carpenter and Karpati, 1979; Chizzonite and Zak, 1981). In the mammalian system, extracellular $[Ca^{2+}]$ is 10^{-3} M, whereas that in the intracellular milieu is approximately 10^{-7} M. This 10,000-fold $[Ca^{2+}]$ gradient is maintained by active transport, and is augmented by the functional and structural integrity of the plasma membrane. Likewise, $[Ca^{2+}]$ in the intracellular compartments is regulated by the mitochondrial (MIT) and sarcoplasmic reticulum membranes which maintain the intraorganelle $[Ca^{2+}]$ at 10^{-4} and 10^{-6} M, respectively.

These subcellular organelles participate in the regulation of intracellular Ca to meet specialized cellular functions which are vital for the interaction and synchronization of many biochemical and physiological processes within the cell. The MIT primarily function as a cellular power house for ATP synthesis, serve as a reservoir for intracellular Ca^{2+}, and regulate the intracellular $[Ca^{2+}]$ within a narrow physiological threshold during acute and transient Ca accumulation. Sarcoplasmic reticulum (SR), on the other hand, is involved in the release and uptake of Ca^{2+} during excitation–contraction coupling of myofibers in response to neurotransmitter-mediated stimuli for muscle contraction.

There are many mechanisms by which cells maintain electrolyte equilibrium. These mechanisms include passive diffusion; membrane-bound ATPases such as Na^+, K^+-ATPase, Mg^{2+}-ATPase, and Ca^{2+}-ATPase which are present in sarcolemmal, mitochondrial, and SR membranes; slow inward Ca^{2+} channels which allow the influx of Ca^{2+}; and the $2Na^+ \leftrightarrow Ca^{2+}$ exchange channels (Ebashi and Sugita, 1979). A normal cell also has at least two distinct pools of Ca. The intracellular free Ca exists in the ionized form and is freely transported between the intra- and extracellular compartments, whereas the bound Ca may remain complexed with ATP, lipids, phosphates, proteins, or other biomolecules. The initial abnormalities in degenerating muscle presumably are expressed in the cellular and subcellular membranes, and may be characterized by a decreased transmembrane potential and increased membrane permeability. Under a partially depolarized state, uncontrolled amounts of extracellular Ca^{2+}, Na^+, and water can passively diffuse into the cells, while K^+ and Mg^{2+} leak out along with intracellular enzymes such as CK and LD (Fig. 1). Thus, whenever cells

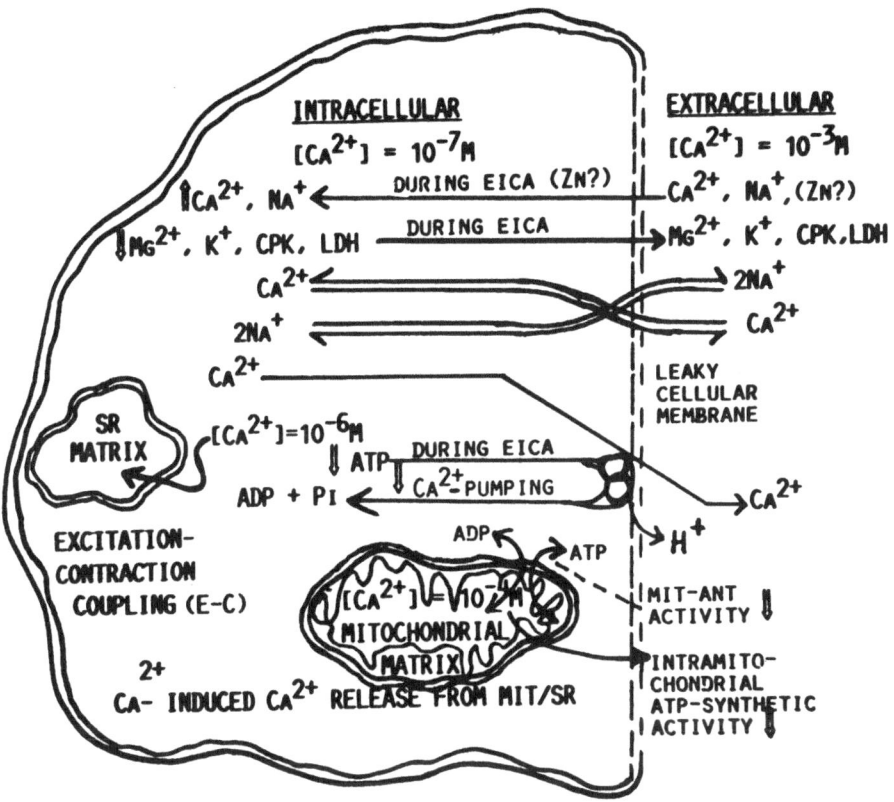

FIGURE 1. Cellular events of membrane hypothesis in muscle degeneration resulting in electrolyte and enzymatic imbalances.

experience chronic membrane abnormalities, the intracellular free [Ca] increases with a consequent elevation of the total cellular [Ca^{2+}] to a critical level which initiates myofibrillar necrosis (Wrogemann and Pena, 1976; Carpenter and Karpati, 1979) (Fig. 2).

Progressive muscle degeneration in a variety of human diseases such as muscular dystrophy and malignant hyperthermia is therefore thought to be membrane-mediated. Numerous independent reports and the preceding arguments have led to the development of our central hypothesis that the resting intracellular free [Ca^{2+}], and cellular and subcellular exchangeable Ca^{2+} pools, are increased in degenerating muscle with EICA, and that the mitigation of EICA can improve cellular structure and function.

Since EICA and low cellular energy charge with reduced [ATP] are the invariable fate of dystrophic myofibers, the impairment of intracellular energy-producing machinery and the excitation–contraction coupling system seems inevitable. During EICA, cellular [ATP] is depleted due to reduced mitochondrial ATP synthesis, the increased demand of pumping out excess intracellular Ca^{2+}, and the activation of the Na^+, K^+-ATPase system to normalize Na^+-K^+ imbalances. In response to reduced cellular [ATP], the mitochondrial adenine

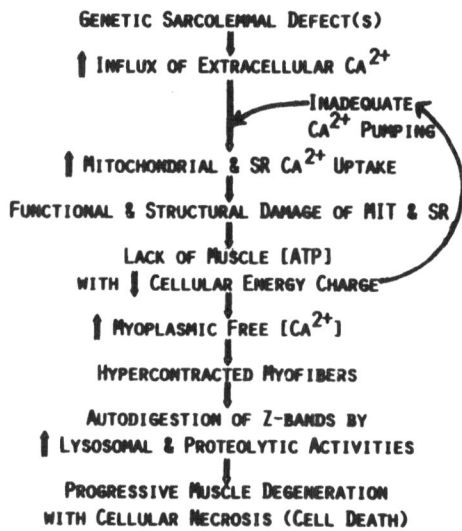

FIGURE 2. Sequence of pathogenetic events in membrane-mediated excessive intracellular calcium accumulation.

nucleotide translocase (ANT) activity, which regulates the influx and efflux of adenine nucleotides across the MIT membrane, is stimulated and the cyclic AMP level is suppressed. As more Ca^{2+}, Na^+, and H_2O enter the cell, swelling occurs in the cell as a whole as well as in the MIT and SR. A further diminished rate of ATP synthesis ensues with reduced mitochondrial ANT activity and increased long-chain free fatty acid accumulation within the MIT (Fig. 1). Thus, the inability of the damaged MIT to maintain a normal ATP synthesis rate leads to a further increase in accumulation of Ca and inorganic phosphates (P_i) within the cell and MIT. Ultimately, mitochondrial $[Ca^{2+}]$ and $[P_i]$ increase at a steady rate until their solubility product is exceeded, resulting in deposition of granular $Ca_3(PO_4)_2$ within the MIT matrix. These extreme intramitochondrial alterations lead to decreased cellular metabolic capabilities, and initiate a final common pathway to cellular destruction via EICA and enhanced lysosomal and neutral protease activities (Fig. 2) (Wrogemann and Pena, 1976; Duncan, 1978; Schanne *et al.*, 1979).

Cellular magnesium, on the other hand, appears to play an antagonistic role in EICA during membrane-mediated acute and chronic muscle degeneration. Magnesium is a vital component of several crucial enzymatic reactions and is involved in ATP synthesis and utilization. Thus, either low intracellular magnesium and/or excess calcium seems to be detrimental to cell survival. A lower Mg-to-Ca ratio is probably a necessary prerequisite to initiate membrane hyperpermeability impairing several membrane-associated biochemical processes.

3.1. Excessive Intracellular Ca Accumulation in Duchenne Muscular Dystrophy and Other Neuromuscular Disorders in Humans

Excessive intracellular calcium accumulation has been demonstrated in degenerating muscle from DMD patients by histochemical (Bodensteiner and Engel, 1978) and ultrastructural (Oberc and Engel, 1977) techniques, as well as by X-ray fluorescence spectroscopy (Maunder-Sewry *et al.*, 1980). However, until recently, because of the lack of a sensitive analytical technique to measure Ca and Mg in needle biopsy specimens, no quantitative data on EICA in DMD were available (Bhattacharya *et al.*, 1979).

In our study of muscle biopsies from patients with various neuromuscular disorders including congenital myopathies, DMD, peripheral neuropathies, spinal muscular atrophy, polymyositis, dermatomyositis, limb-girdle muscular dystrophy, and fascioscapulohumeral muscular dystrophy, we provided a clear evidence for EICA (Bertorini *et al.*, 1982). Although muscle [Ca] was significantly elevated in all neuromuscular diseases ($p < 0.0001$), it was highest in the muscles of DMD patients. On the other hand, muscle [Mg] was diminished by 44% in DMD patients ($p < 0.0001$), with corresponding lesser depletion in patients with other neuromuscular diseases (Table I). When DMD patients were subgrouped according to their ambulatory status and age, the older (>8 years) nonambulatory patients revealed a significantly lower muscle [Mg] than younger (≤ 8 years) ambulatory patients whose muscle was found to be already Mg-deficient relative to normal subjects ($p < 0.0001$), indicating a progressive depletion of muscle [Mg] and probably cell necrosis in DMD. However, the fact that the elevated [Ca] in dystrophic muscle remained stable irrespective of age and ambulatory status of the patients suggested an early pathogenetic role of Ca in DMD. Histopathologically, biopsies from the DMD patients revealed sporadic necrosis, centronucleated and hypercontracted myofibers, marked variation in the fiber size and shape, and an abundance of Ca^{2+}-positive and opaque fibers which are characteristic of profound dystrophic degeneration (Bertorini *et al.*, 1982).

3.2. Evidence of EICA in Neonates and Human Fetuses at Risk of DMD

Although Ca^{2+}-positive and opaque fibers have been reported to be present in abundance in muscles of male fetuses at risk of DMD obtained from definite carriers of this incurable disease, no quantitative data on fetal muscle [Ca] are available (Brambati *et al.*, 1980; Emery and Burt, 1980). To evaluate the role of Ca and Mg in the pathogenesis of DMD during fetal development, we measured Ca and Mg in biceps and quadriceps muscle of aborted normal and DMD male fetuses at the gestational age of 10–22 weeks. These studies were conducted in collaboration with Dr. Cornelio from Milan, Italy and Dr. Emery of Edinburgh, Scotland.

A three- to sixfold increase in muscle [Ca] was noted in the dystrophic fetal tissues (Table II), which correlated well with the EICA reported in the older DMD patients. On the other hand, normal human fetal muscle revealed strikingly similar levels of [Ca] and [Mg] compared to the older normal subjects. An 18–57% increase in muscle [Mg], however, was found in the fetuses at risk of DMD as compared to a 44% depletion of muscle [Mg] in DMD patients (Tables I and II). A portion of each of the muscle samples was subjected to histological analysis using calcium-specific alizarine red-S, pentahydroxyflavone, and Van Kossa stains (Bertorini *et al.*, 1984; Bhattacharya *et al.*, 1984). No necrotic fibers could be detected in any of the fetal muscle tissues or in the muscle from a premature infant. Thus, it is possible that the elevated Mg content of the tissue inhibited precipitation of excess calcium, preventing fiber damage and necrosis. It may be of interest to note that studies such as those reported here may be performed on tissue samples obtained using fetoscopy. Gustavii *et al.* (1983) showed that such an antenatal diagnostic procedure is feasible and involves minimal risk to fetal viability.

TABLE I. Ca, Mg, and Noncollagen Nitrogen Levels in Human Muscle with DMD and Other Neuromuscular Disorders[a]

Group (N)	Calcium[b]	Magnesium[b]	Mg/Ca	Noncollagen nitrogen[c] (N)
A. Normal subjects (22)	8.49 ± 0.36	75.18 ± 3.69	9.06 ± 0.49	25.53 ± 1.37 (10)
B. Duchenne muscular dystrophy (27)	24.63 ± 0.84	42.36 ± 2.45	1.75 ± 0.11	13.14 ± 1.22 (25)
C. Denervating diseases[d] (14)	16.12 ± 1.07	60.07 ± 3.61	3.98 ± 0.44	14.11 ± 2.04 (9)
D. Polymyositis–dermatomyositis (11)	15.77 ± 1.74	60.80 ± 3.37	4.55 ± 0.71	19.28 ± 3.34 (5)
E. Other neuromuscular diseases[e] (11)	15.61 ± 1.82	77.26 ± 5.34	5.41 ± 0.51	17.87 ± 4.99 (5)
Significance (P)				
A versus B	<0.0001	<0.0001	<0.0001	<0.0001
A versus C	<0.0001	<0.01	<0.0001	<0.0001
A versus D	<0.0001	<0.02	<0.0001	N.S.
A versus E	<0.0001	N.S.	<0.0001	N.S.
B versus C, D, and E	<0.0001	<0.005	<0.001	N.S.

[a] Data presented in mean ± SEM. N.S. = not significant. Figures in parentheses represent the number of individuals studied. (From Bertorini et al., 1982, with permission from Harcourt Brace Jovanovich, Cleveland, OH.)

[b] meq/kg fat-free dry tissue.

[c] μg/mg fresh muscle.

[d] Peripheral neuropathies and spinal muscular atrophy.

[e] Congenital myopathies, limb-girdle muscular dystrophy, and fascioscapulohumeral muscular dystrophy.

TABLE II. Muscle Calcium and Magnesium Concentrations in Male Human Fetuses and Neonates at Risk of DMD[a]

Serial ID	Gestational age[b]	Muscle type	Calcium[c]	Magnesium[c]	Mg/Ca ratio
TB-1 (neonate)	28	Quadriceps	35.1	—[d]	—[d]
AE-1	22	Quadriceps	18.1	109.7	6.0
FC-1	20	Quadriceps	25.9	85.1	3.3
FC-2	10	Quadriceps	35.9	106.7	3.0
AE-2	20	Quadriceps	19.7	102.5	5.2
FC-3	10	Biceps	54.5	113.0	2.1
Fetuses at risk of DMD	10–22	Quadriceps/biceps	30.8 ± 6.7 (5)	103.4 ± 4.9 (5)	3.9 ± 0.7 (5)
Normal fetuses	10–22	Quadriceps/biceps	8.5 ± 0.5 (5)	71.8 ± 3.6 (5)	8.4 ± 0.4 (5)
Significance (P)			<0.01	<0.001	<0.001

[a] Group data presented in mean ± SEM (N). (From Bertorini et al., 1984; and Bhattacharya et al., 1984, with permission from Harcourt Brace Jovanovich, Cleveland, OH.)
[b] Gestational age is given in weeks.
[c] meq/kg fat-free dry tissue.
[d] Quantity was insufficient for Mg determination.

3.3. Evidence of EICA in Human Myoblasts of Duchenne Origin

Ca and Mg were also measured in human myoblasts collected from apparently normal fetuses and those at risk of DMD, and were cultured in 80% Dulbecco's medium fortified with 10% horse serum and 10% fetal calf serum. Myoblasts were harvested at equivalent stages of growth and washed twice with Ca-Mg free normal saline. The calcium content was found to be significantly elevated in DMD myoblasts, while the Mg levels were lower than normal (Bhattacharya and Crawford, 1985). Also, the Mg/Ca ratio was found to be severely depressed in dystrophic myoblasts. Although our observation of myoblast [Ca] correlated well with the previously reported EICA in DMD patients and fetuses at risk of DMD (Tables I and II), there was an apparent contradiction between the muscle Mg concentrations. This Mg paradox may be attributed to inherent differences between the prenatal development of human fetuses *in utero* and growth of human myoblasts *in vitro*.

3.4. Comparative Studies of EICA in Dystrophic Animal Models

During the past several years, BIO-14.6 and CHF-146 strains of dystrophic hamsters (DH), the C57BL/6J-dy^{2J} strain of dystrophic mice (DM), and the University of California, Davis line 413 dystrophic chickens (DC) have been studied to elucidate the degree of EICA, other biochemical abnormalities, cardiac malfunctions, and histological aberrations associated with degenerating muscles. These studies were crucial to identify the animal model which most closely represents the multiple lesions present in degenerating muscles of DMD patients. Although all species demonstrated the presence of significant voluntary muscle weakness and elevated plasma CK activity, classical histological changes and EICA were not found in the cardiac muscles of DC and DM, but were present in the myocardium of both strains of DH. Skeletal muscles such as rectus femoris and extensor digitorum longus revealed similar histological abnormalities and significant EICA in all species. Correlation of these and other biochemical, mechano- and electrophysiological, functional, and histopathological findings indicated that DH were the most suitable animal model to study membrane-mediated progressive muscle degeneration in muscular dystrophy. Also, DH exhibit both skeletal and cardiac muscle impairment (Table III) as well as other classical histological, biochemical, and functional abnormalities present in DMD (Bertorini *et al.*, 1982; Bhattacharya *et al.*, 1982; Bhattacharya, Crawford, and Pate, 1987a; Riedel *et al.*, 1988).

3.4.1. Electrocardiographic Abnormalities in Dystrophic Hamsters

Because of the profound cardiac involvement in DMD and BIO-14.6 strain DH, the electrocardiographic (EKG) changes in 7-month-old DH with advanced hypertrophic cardiomyopathy were investigated to determine whether the EKG changes commonly seen in patients with DMD were also present in DH. Abnormally tall R-I and R-aVL amplitudes, deep S-III and S-aVR waves, and elongated PR-I, QT-I, and QRS-I intervals ($p < 0.0001$) were observed in DH when compared to age- and sex-matched normal F1B strain hamsters (NH) (Bhattacharya *et al.*, 1987a). These EKG changes were strikingly similar to those reported in DMD patients, who frequently die from congestive heart failure. It appears that longitudinal EKG evaluation could be a useful noninvasive technique for monitoring the progressive cardiac changes in DH, and could be used as preliminary predictor of the efficacy of pharmacological regulation, which is capable of mitigating membrane-mediated EICA and associated pathobiology in the dystrophic myocardium.

TABLE III. Muscle Calcium and Magnesium Concentrations in Various
Dystrophic Animal Models[a]

Group (N)	Cardiac muscle		Skeletal muscle[b]	
	Calcium	Magnesium	Calcium	Magnesium
Normal hamsters (5)	9.1 ± 1.4	96.4 ± 3.5	16.6 ± 0.8	77.6 ± 4.7
Dystrophic hamsters (6)	324.6 ± 19.0[c]	118.9 ± 11.5	26.2 ± 5.7[c]	77.1 ± 4.3
Normal mice (8)	13.4 ± 0.8	87.6 ± 2.2	15.7 ± 0.5	90.1 ± 2.0
Dystrophic mice (7)	14.8 ± 0.9	83.2 ± 2.3	16.8 ± 0.8	86.9 ± 2.4
Normal chickens (6)	10.6 ± 0.7	83.9 ± 1.8	7.2 ± 0.4	98.9 ± 6.0
Dystrophic chickens (6)	11.0 ± 0.7	82.7 ± 2.0	11.8 ± 0.1[c]	80.7 ± 3.2[c]

[a] Data presented in meq/kg fat-free dry tissue, mean ± SEM. Figures in parentheses represent number of animals studied. (From Bhattacharya et al., 1982, 1987a, with permission of John Wiley and Sons, New York; Riedel et al., 1988.)
[b] Skeletal muscle data obtained from the rectus femoris region only.
[c] Significantly ($p < 0.05$) different from normal counterparts.

3.4.2. EICA-Dependent Voluntary Muscle Weakness in Dystrophic Muscle

In a mechanophysiological study using BIO-14.6 strain DH, we observed a strong inverse correlation ($r = -0.9974$) between contractile properties, such as twitch and tetanus tensions of soleus (slow) and extensor digitorum longus (fast) muscles, and EICA. The soleus muscle of 35-day-old, male DH and NH were compared in vivo at 37°C with supra-maximal nerve stimulation using standard procedures. Dystrophic hamsters generated significantly smaller maximal twitch tension ($p < 0.01$) and tetanus tension ($p < 0.005$) not only in absolute terms, but also when expressed per unit wet weight of muscle. Prolongation of the half-relaxation time ($p < 0.01$) and contraction time were also observed in DH compared to sex- and age-matched NH (Law et al., 1983). These findings are of great clinical significance in that they demonstrate a direct relationship between voluntary muscle weakness and EICA, which has been shown to be a fundamental pathogenetic event in muscular dystrophy.

3.4.3. Evidence of Increased Intracellular Free-Calcium and Sodium and Decreased Potassium Ion Concentrations in Dystrophic Myofibers

Not only does the "membrane hypothesis" suggest that dystrophic muscle may acquire increased intracellular free [Ca], but we also reported an increased [Na] and decreased [K] in the muscle of DH (Bhattacharya and Crawford, 1986). Therefore, an attempt was made to characterize the intracellular free-Ca ($[Ca^{2+}]_i$), Na ($[Na^+]_i$), and K ($[K^+]_i$) concentrations in 75-day-old male CHF-146 strain DH and CHF-148 strain NH. The resting $[Ca^{2+}]_i$ and $[Na^+]_i$ in DH were significantly higher than those in NH. On the other hand, the intracellular free-potassium concentration was significantly lower in DH muscle (Bhattacharya et al., 1987b).

Our above results suggest that a generalized defect in the cellular ion-gating mechanism at the plasmalemmal level is responsible for the derangement of $[Ca^{2+}]_i$, $[Na^+]_i$, and $[K^+]_i$ in DH. These observations represent the first direct demonstration of significantly increased intracellular free [Ca] and [Na] associated with decreased [K] in dystrophic myofibers, and support our central hypothesis that disruption in the equilibrium of these essential intracellular cations in muscular dystrophy is a fundamental membrane-mediated pathogenetic event.

4. REGULATION OF EICA IN DEGENERATING CARDIAC AND SKELETAL MUSCLES

Cumulative evidence suggests that muscle degeneration in DMD and DH is mediated by impaired membrane integrity which results in EICA. Therefore, it is plausible that EICA may be mitigated by one or more of the following means: (1) preventing or inhibiting further entry of Ca^{2+} through the slow inward Ca^{2+} current by administration of Ca channel blocking agents such as diltiazem (DTZM), (2) improving the membrane stability and intracellular Ca pump efficiency by exogenous ATP-MgCl$_2$ (Bhattacharya and Crawford, 1986), or (3) administering a combination of these agents to accomplish multiple beneficial effects. Although many Ca channel blockers are available, some are tissue-specific and others produce considerable deleterious side effects (Emery *et al.*, 1982). Recently, we evaluated the comparative efficacies of DTZM, nifedipine, and verapamil in the mitigation of EICA in cardiac and skeletal muscles of DH. Our results showed that DTZM was the most effective and best tolerated agent tested (Crawford and Bhattacharya, 1987). Furthermore, both ATP-MgCl$_2$ and DTZM can be used individually in DH without adverse side effects. These agents appear to lower plasma CK activity and significantly reduce EICA in the heart, diaphragm, and rectus femoris muscles of DH, probably by blocking Ca^{2+} entry, stabilizing the cell membrane, and improving the energy supply of these degenerating muscle cells (Bhattacharya *et al.*, 1982; Bhattacharya and Crawford, 1986).

DTZM acts by inhibiting the slow inward Ca^{2+} current without blocking the β-adrenergic receptors. Bhattacharya and co-workers first studied the chronic effects of oral DTZM in regulating EICA and improving histology of the heart, diaphragm, and rectus femoris muscles of DH. Excessive intracellular calcium accumulation was reduced by 73% in the heart, 61% in the diaphragm, and 48% in the rectus femoris compared to saline-treated DH. Histologically, DTZM significantly reduced cellular Ca deposition in the heart and variable fiber size in the rectus femoris, and also significantly reduced the elevated plasma CK activity in DH. However, it neither changed the plasma [Ca] and [Mg] in DH and NH, nor modified any other biochemical parameters tested in NH (Bhattacharya *et al.*, 1982).

These findings suggest that DTZM may play a role in the regulation of membrane-mediated EICA and associated biochemical, histological, and ultrastructural changes in muscular dystrophy. Since orally administered DTZM (half-life of 26 min) is readily metabolized and its metabolites are pharmacologically inactive, it is possible that the therapeutic efficacy of DTZM can be improved by infusing it subcutaneously via an osmotic minipump.

5. ROLE OF MAGNESIUM IN EICA

Several studies, including those from our laboratory, provided convincing evidence that EICA is a fundamental pathogenetic event in the membrane-mediated progressive muscle degeneration in muscular dystrophy. It is also apparent from the membrane hypothesis that attempts to reduce EICA and replenish muscle Mg deficiency in DMD by administration of Ca channel blocking agents or membrane-stabilizing agents such as ATP-MgCl$_2$ could have profound therapeutic significance in the intervention of this hitherto uncontrollable chronic muscle degeneration in muscular dystrophy. It has been well documented that a progressive Mg deficiency occurs in the presence of stable EICA. Also, Mg depletion and EICA occur in dystrophic myoblasts of human origin. Even though EICA is present in muscles of fetuses at risk of DMD, increased [Mg] in these muscles probably prevents cellular necrosis.

It seems, therefore, from the aforementioned evidence that Mg acts as an endogenous Ca antagonist and prevents cellular necrosis during fetal development partly by maintaining EICA in ionic form. Also, muscle Mg anomalies noted in patients, fetuses, and myoblasts of DMD origin may be related to inherent differences between the prenatal development of human fetuses *in utero* and growth of human myoblasts *in vitro,* and the ongoing postnatal degenerative dystrophic processes. It is also equally noteworthy that the *in vitro* studies with Ca^{2+}-loaded normal and dystrophic mitochondria demonstrated that the early addition of Mg^{2+} into the incubation media protects mitochondria from Ca^{2+}-induced functional impairment, probably by preventing the loss of intramitochondrial NADH and stabilizing the mitochondrial membrane (Thakar *et al.,* 1973; Wrogemann *et al.,* 1975). This protective role of Mg in mitochondrial viability remains to be further elucidated.

6. CONCLUSION

Membrane-mediated EICA represents a central pathway leading to cell death in animal models of muscular dystrophy and DMD, as well as in many other acute, chronic, and hereditary pathological conditions. Excessive intracellular calcium accumulation is associated with altered permeability of the cell membrane, and it results in depletion of cellular energy (ATP). Pharmacological agents that inhibit the influx of Ca and stabilize the structure and function of the cellular and subcellular membranes or promote energy synthesis and utilization may improve the survivability of degenerating cells. These latter avenues seem to have significant therapeutic potential in the control of acute and chronic membrane-mediated cellular degeneration.

ACKNOWLEDGMENTS. We would like to thank Drs. James W. Pate and Louis G. Britt for their encouragement and critical review of this chapter. This work was supported in part by grants from MDA, USPHS (NIAMS), AHA–Memphis Chapter, and the Varian Instrument Group of America.

REFERENCES

Araki, S., and Mawatari, S., 1971, Ouabain and erythrocyte-ghost adenosine triphosphatase. Effects in human muscular dystrophies, *Arch. Neurol.* **24**:187–190.

Bertorini, T. E., Bhattacharya, S. K., Palmieri, G. M. A., Chesney, C. M., Pifer, D., and Baker, B., 1982, Muscle calcium and magnesium content in Duchenne muscular dystrophy, *Neurology* **32**: 1088–1092.

Bertorini, T. E., Cornelio, F., Bhattacharya, S. K., Palmieri, G. M. A., Dones, I., Dworzak, F., and Brambati, B., 1984, Calcium and magnesium content in fetuses at risk and prenecrotic Duchenne muscular dystrophy, *Neurology* **34**:1436–1440.

Bhattacharya, S. K., and Crawford, A. J., 1985, Preliminary evidence of magnesium depletion and excessive calcium accumulation in human myoblasts of dystrophic origin, *Soc. Neurosci. Abstr.* **11**:1302.

Bhattacharya, S. K., and Crawford, A. J., 1986, Beneficial effects of adenosine triphosphate-magnesium chloride administration in muscular dystrophy, *Soc. Neurosci. Abstr.* **12**:264.

Bhattacharya, S. K., Crawford, A. J., and Emery, A. E. H., 1984, Quantitation of total calcium in fetal muscle: A differential antenatal diagnosis for Duchenne muscular dystrophy, *Clin. Res.* **32**: 288.

Bhattacharya, S. K., Crawford, A. J., and Pate, J. W., 1987a, Electrocardiographic, biochemical, and morphologic abnormalities in dystrophic hamsters with cardiomyopathy, *Muscle Nerve* **10:**168–176.

Bhattacharya, S. K., Lopez, J. R., Sanchez, V., Crawford, A. J., Vergara, J. L., and Sreter, F., 1987b, Direct evidence of membrane-mediated cellular degeneration by changes in the intracellular free calcium, potassium, and sodium concentrations in dystrophic hamsters, *Soc. Neurosci. Abstr.* **13:** 1680.

Bhattacharya, S. K., Palmieri, G. M. A., Bertorini, T. E., and Nutting, D. F., 1982, The effects of diltiazem in dystrophic hamsters, *Muscle Nerve* **5:**73–78.

Bhattacharya, S. K., Williams, J. C., and Palmieri, G. M. A., 1979, Determination of calcium and magnesium in cardiac and skeletal muscle by atomic absorption spectroscopy using stoichiometric nitrous oxide-acetylene flame, *Anal. Lett.* **12:**1451–1475.

Bodensteiner, J. B., and Engel, A. G., 1978, Intracellular calcium accumulation in Duchenne dystrophy and other myopathies: A study of 567,000 muscle fibers in 114 biopsies, *Neurology* **28:**439–446.

Brambati, B., Cornelio, F., Dworzak, F., and Dones, I., 1980, Calcium-positive muscle fibres in fetuses at risk for Duchenne muscular dystrophy, *Lancet* **2:**969–970.

Carpenter, S., and Karpati, G., 1979, Duchenne muscular dystrophy: Plasma membrane loss initiates muscle cell necrosis unless it is repaired, *Brain* **102:**147–161.

Chizzonite, R. A., and Zak, R., 1981, Calcium-induced cell death: Susceptibility of cardiac myocytes is age-dependent, *Science* **213:**1508–1510.

Crawford, A. J., and Bhattacharya, S. K., 1987, Regulation of membrane-mediated chronic muscle degeneration by Diltiazem, Nifedipine and Verapamil in dystrophic hamsters, *Soc. Neurosci. Abstr.* **13:**1680.

Duncan, C. J., 1978, Role of intracellular calcium in promoting muscle damage: A strategy for controlling the dystrophic condition, *Experientia* **34:**1531–1535.

Ebashi, S., and Sugita, H., 1979, The role of calcium in physiological and pathological processes of skeletal muscle, in: *Current Topics in Nerve and Muscle Research,* ICS No. 455 (A. J. Aguayo and G. Karpati, eds.), Excerpta Medica, Amsterdam, pp. 73–84.

Emery, A. E. H., and Burt, D., 1980, Intracellular calcium and pathogenesis and antenatal diagnosis of Duchenne muscular dystrophy, *Br. Med. J.* **280:**355–357.

Emery, A. E. H., Skinner, R., Howden, L. C., and Matthews, M. B., 1982, Verapamil in Duchenne muscular dystrophy, *Lancet* **1:**559.

Gustavii, B., Loefberg, L., and Enriksson, K. G., 1983, Fetal muscle biopsy, *Acta Obstet. Gynecol. Scand.* **62:**369–371.

Kar, N. C., and Pearson, C. M., 1976, A calcium-activated neutral protease in normal and dystrophic human muscle, *Clin. Chim. Acta* **73:**293–297.

Law, P. K., Luther, R. W., Goodwin, T. G., and Bhattacharya, S. K., 1983, Comparative mechano-physiologic studies on normal and dystrophic hamsters soleus muscles correlated with muscle calcium content, *Clin. Res.* **31:**718.

Maunder-Sewry, C. A., Gorodetsky, R., Yarom, R., and Dubowitz, V., 1980, Element analysis of skeletal muscle in Duchenne muscular dystrophy using X-ray fluorescence spectrometry, *Muscle Nerve* **3:**502–508.

Mokri, B., and Engel, A. G., 1975, Duchenne dystrophy: Electron microscopic findings pointing to a basic or early abnormality in the plasma membrane of the muscle fiber, *Neurology* **25:**1111–1120.

Moser, H., 1984, Duchenne muscular dystrophy: Pathogenetic aspects and genetic prevention, *Hum. Genet.* **66:**17–40.

Oberc, M. A., and Engel, W. K., 1977, Ultrastructural localization of calcium in normal and abnormal skeletal muscle, *Lab. Invest.* **36:**566–577.

Riedel, D. M., Entrikin, R. K., and Bhattacharya, S. K., 1988, Relevance of dystrophic chickens as an experimental model for hereditary muscular dystrophy in humans, *Soc. Neurosci. Abstr.* **14:** 828.

Rowland, L. P., 1980, Biochemistry of muscle membranes in Duchenne muscular dystrophy, *Muscle Nerve* **3:**3–20.

Sanyal, S. K., and Johnson, W. W., 1982, Cardiac conduction abnormalities in children with Duchenne's progressive muscular dystrophy: Electrocardiographic features and morphologic correlates, *Circulation* **66**:853–863.

Schanne, F. A. X., Kane, A. B., Young, E. E., and Farber, J. L., 1979, Calcium dependence of toxic cell death: A final common pathway, *Science* **206**:700–703.

Schotland, D. L., Bonilla, E., and Van Meter, M., 1977, Duchenne dystrophy: Alteration in muscle plasma membrane structure, *Science* **196**:1005–1007.

Thakar, J. H., Wrogemann, K., and Balnehaer, M. C., 1973, Effect of ruthenium red on oxidative phosphorylation and the calcium and magnesium content of skeletal muscle mitochondria of normal and BIO 14.6 dystrophic hamsters, *Biochim. Biophys. Acta* **314**:8–14.

Wrogemann, K., Blanchaer, M. C., Thakar, J. H., and Mezon, B. J., 1975, On the role of mitochondria in the hereditary cardiomyopathy of the Syrian hamster, in: *Recent Advances in Studies on Cardiac Structure and Metabolism,* Vol. 6 (A. Fleckenstein and G. Rona, eds.), University Park Press, Baltimore, pp. 231–241.

Wrogemann, K., and Pena, S. D. J., 1976, Mitochondrial calcium overload: A general mechanism for cell-necrosis in muscle diseases, *Lancet* **1**:672–673.

VI

Cardiovascular Pathophysiology

55

Characteristics and Functional Implications of Spontaneous Sarcoplasmic Reticulum-Generated Cytosolic Calcium Oscillations in Myocardial Tissue

EDWARD G. LAKATTA, MAURIZIO C. CAPOGROSSI, HAROLD A. SPURGEON, and MICHAEL D. STERN

The process by which electrical excitation of cardiac muscle cells leads to contraction is incompletely understood. It is clear, however, that in all mammalian hearts, release of Ca^{2+} from the sarcoplasmic reticulum (SR) contributes to the activation of the myofilaments. Spontaneous SR oscillatory Ca^{2+} release (Fig. 1), given the required conditions, appears to be a universal phenomenon in mammalian preparations (see Lakatta *et al.*, 1985 for review). While spontaneous chaotic cellular contractions, a mechanical sequalae of SR Ca^{2+} oscillations (CaOsc), was observed over 70 years ago (see Capogrossi *et al.*, 1986b), the universality of CaOsc had not generally been recognized, and its multiple effects on myocardial function have not generally been considered collectively.

1. SPONTANEOUS Ca^{2+} OSCILLATIONS (CaOsc) IN MYOCARDIAL CELLS DEVOID OF SARCOLEMMAL FUNCTION

A revived interest in spontaneous cardiac CaOsc within the last two decades began with studies of spontaneous phasic contractions in cardiac myocytes from which the sarcolemma had been mechanically or chemically removed (Bloom, 1970; Fabiato and Fabiato, 1972;

EDWARD G. LAKATTA, MAURIZIO C. CAPOGROSSI, HAROLD A. SPURGEON, and MICHAEL D. STERN ● Laboratory of Cardiovascular Science, Gerontology Research Center, National Institute on Aging, National Institutes of Health, and Johns Hopkins Medical Institutions, Baltimore, Maryland 21224.

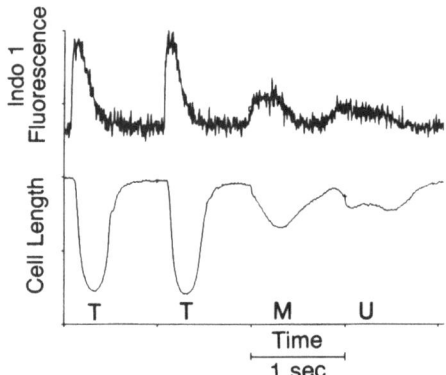

FIGURE 1. Changes in cell length (lower tracing) and Ca_i (upper tracing) in an indo-1-loaded adult rat myocyte during an electrically stimulated twitch (T) and during spontaneous multifocal (M) and unifocal (U) SR Ca^{2+} release following the twitch. The system for measuring cytosolic Ca^{2+} transients uses the fluorescent probe indo-1. Indo-1 fluorescence is excited by epi-illu-mination with 10-μsec flashes of 350 $\pm$ 5 nm light at repetition rates of up to 333 Hz. The emitted light is collected by paired photomultipliers to measure simultaneously spectral windows of 411 $\pm$ 20 nm and 481 $\pm$ 25 nM selected by custom-designed bandpass interference filters. The fluorescence emission from each flash is collected by a pair of fast integrator sample-and-hold circuits of custom design under the control of a VAX 11/730 computer which calculates the ratio of indo-1 emission at the two wavelengths as a measure of Ca_i with a time precision of better than 20 μsec. Cell length is measured from the bright-field image of the cell by an optical edge tracking method using a video edge detector (or a photodiode array when milli-second time resolution is required). By using red light (650–700 nm) for the bright-field image and a dichroic mirror to transmit the fluorescent light (380–550 nm) and reflect the red light, and taking advantage of the low-duty cycle of the fluorescence excitation, length, and Ca_i measurements are obtained simultaneously without cross-talk. The membrane potential may also be monitored simultaneously with patch electrodes. In this figure the fluorescence of indo-1 is the ratio of that measured at 410 nm to that measured at 490 nm and was collected over the entire cell area. The signals are unfiltered and unaveraged. That the fluorescence underlying U and M is actually instantaneously present in only 15–30% of cell (area within the contractile band; see Fig. 2) indicates that the localized increase in Ca_i is of the order of magnitude of that during the stimulated twitch. (From Capogrossi and Lakatta, 1988).

Dani *et al.*, 1979; Chiesi *et al.*, 1981; Fabiato, 1985). The modulation of CaOsc characteristics in response to step changes in bathing [Ca^{2+}] and to intermittent application of caffeine made it clear that the activating Ca^{2+} came from the SR; from aequorin luminescence measurements the peak free [Ca^{2+}] in the cytosolic space during spontaneous contractions was estimated to be 3–4 μM (i.e., equivalent to that triggered by an action potential). This process has been termed "spontaneous Ca^{2+} release" (Fabiato, 1985), distinguishing it from "Ca^{2+}-induced Ca^{2+} release" (CICR), manifested by the same preparations, which is be-lieved by many to be a step in normal excitation–contraction coupling.

The threshold bathing [Ca^{2+}] required for skinned rat preparations to exhibit spontaneous (CaOsc) is quite low (about 100 nM), i.e., approximately at the threshold for SR Ca^{2+} pumping (Dani *et al.*, 1979), and is less than most estimates of cytosolic [Ca^{2+}] in intact resting cardiac preparations. Thus, it is not surprising that *unstimulated* intact rat cells and muscles exhibit CaOsc when bathed in solutions containing physiological [Ca^{2+}]. The CaOsc frequency, which can vary from <0.1 to 7–8 Hz (see Lakatta *et al.*, 1985 for review), depends on the bathing free [Ca^{2+}] in the range 0.1–10 μM (Chiesi *et al.*, 1981; Fabiato, 1985). The probability for CaOsc to occur in chemically skinned rat cells is dependent on species, e.g., about 20% of rat myocytes devoid of sarcolemmal function exhibit CaOsc when bathed in pCa of 7.0, and this increases to 100% at a pCa of 6.8; in contrast, rabbit myocytes do not begin to exhibit CaOsc until the myoplasmic [Ca^{2+}] approaches this level (Chiesi *et al.*, 1981).

2. CaOsc IN INTACT CARDIAC MUSCLE

In "resting," i.e., unstimulated, rat cardiac muscle with intact sarcolemmal function, we observed the occurrence of spontaneous fluctuations in the intensity of laser light scattered (Lakatta and Lappe, 1981); the frequency of these fluctuations depends on the intracellular Ca^{2+} loading of the muscle and is correlated with the presence of a Ca^{2+}-dependent resting (diastolic) tone. In the unstimulated rat muscle, the phenomenon occurs spontaneously under conditions when the bathing [Ca^{2+}] is in the physiological range and the muscles have apparently normal Ca^{2+} loading. Subsequently, we demonstrated (Stern *et al.*, 1983) that the underlying cause of these scattered intensity light fluctuations (SLIF) was a persistent subcellular chaotic "squirming" motion which depended on the presence of an intact SR sufficiently loaded with Ca^{2+}. In the edges of thin papillary muscles, this motion could be directly seen via high magnification to consist of random waves of contraction, considerably shorter than the length of a cell, propagating back and forth in a partially periodic manner. By means of spectral analysis of fluctuations in aequorin luminescence, it was shown that the chaotic motion which produces SLIF in whole muscle is associated with quasi-periodic oscillations in myoplasmic free [Ca^{2+}] (see Lakatta *et al.*, 1985 for review). This chaotic motion, with its accompanying SLIF, could be produced in the steady rested state in any mammalian cardiac muscle by maneuvers which increase intracellular Ca^{2+}, e.g., high extracellular [Ca^{2+}], low extracellular [Na$^+$] or digitalis glycosides (Kort and Lakatta, 1984); changes in SLIF frequency in a given muscle during changes in cell Ca loading correlate with changes in contractile wave frequency in single cells (Kort *et al.*, 1985a). In intact isolated hearts SLIF can be detected in the backscatter of laser light (Stern *et al.*, 1985). The Ca^{2+}, drug, and species dependence of SLIF have proven to be exactly the same as those of the "spontaneous Ca^{2+} release" observed in mechanically skinned cells (Fabiato and Fabiato, 1978; Lakatta *et al.*, 1985); in particular, they are abolished by ryanodine or high concentrations of caffeine, which deplete the Ca^{2+} load of the SR and thus abolish Ca^{2+} release.

Most studies of spontaneous CaOsc in cardiac preparations (other than those of the rat) with intact sarcolemma have been implemented during experimental "Ca^{2+} overload" because this exaggerates the CaOsc and renders them more easily measurable (see Lakatta *et al.*, 1985 review). Under these conditions the cytosolic [Ca^{2+}] reached during a CaOsc has been estimated to be as high as 4–40 μM (Orchard *et al.*, 1983; Kort *et al.*, 1985b).

3. CaOsc IN SINGLE CARDIAC MYOCYTES WITH INTACT SARCOLEMMAL FUNCTION

It is our opinion that spontaneous CaOsc may be most effectively observed in isolated intact adult cardiac single myocytes with intact sarcolemma function (Kort *et al.*, 1985a).

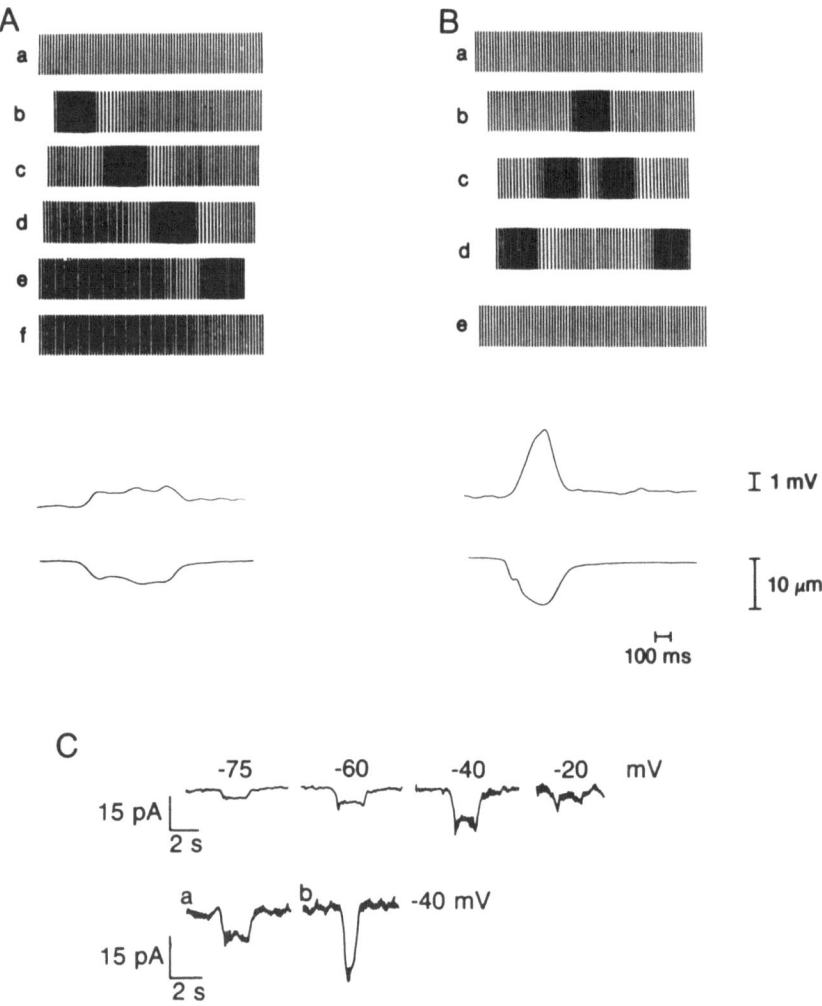

FIGURE 2. (A and B) Schematic representations of the localized myofilament shortening; contractile manifestation of localized spontaneous Ca^{2+} release from the SR into the myoplasmic space. The localized myofilament shortening produces a contractile band which propagates presumably by diffusion of Ca^{2+} causing CICR at the wave front with SR repumping of Ca^{2+} within the wake of the wave (Stern et al., 1984). Spontaneous Ca^{2+} release from the SR can be defined as either unifocal if at a given time it is present in only a single localized area within the cell, and thus causes only a single band of contracted sarcomeres (panel A), or multifocal if at a given time the localized increase in Ca_i is present at more than one locus, giving rise to two or more bands of contracted sarcomeres simultaneously (panel B). Note that the presence of a contractile band within the cell causes the cell length to decrease from the resting level (see lower tracings). Multifocal SR Ca^{2+} release leads to a greater total area of myofilament activation and thus greater cell shortening than the unifocal type (see also Fig. 4). The occurrence of the localized increase in cytosolic $[Ca^{2+}]$ leads to a depolarization of the sarcolemma (panels

In these single cells, it takes the form of spontaneous contractions which are spatially inhomogeneous within the cell. Local contraction of a band of sarcomeres is initiated at one point, usually at one of the ends of the cell, and then propagates as a solitary wave of contraction (Fig. 2), traveling at a velocity of the order of 100 μM/sec (Stern et al., 1984; Kort et al., 1985a), with relaxation of sarcomeres occurring in the wake of the wave. The localized increase in cytosolic [Ca²⁺] within the contractile band has been estimated to be up to greater than 1 μM (Cobbold and Bourne, 1984). The initiation of these waves occurs roughly, but not exactly, periodically, with a frequency that depends on the extracellular [Ca²⁺], and on the conditions that promote Ca²⁺ entry into the cell (Stern et al., 1984; Kort et al., 1985a). Under conditions of high Ca²⁺ loading, when the contractile waves become very frequent, multifocal waves (Capogrossi and Lakatta, 1985; Capogrossi et al., 1987) may occur (Figs. 1 and 2); in this case the waves annihilate if they collide. Microelectrode measurements of sarcolemmal membrane potential and current (Fig. 2) show that contractile waves in rat myocyte bathed in 1.0 mM Ca²⁺ are accompanied by minute inward currents that produce a small (2–3 mV) depolarization under most conditions (Capogrossi et al., 1987, Orchard et al., 1987), indicating that the intracellular [Ca²⁺] release is the primary event in these contractions.

4. ADRENERGIC MODULATION OF CaOsc

Stimulation of adrenergic receptors in intact myocytes or the application of biochemical messengers that result from receptor stimulation, i.e., cAMP or inositol trisphosphate (IP₃) in "skinned" preparations, and diacylglycerol or phorbol esters in intact myocytes are known to alter the spontaneous Ca²⁺ oscillation characteristics.

The application of cAMP to cardiac preparations in which the sarcolemma has been mechanically removed or rendered hyperpermeable causes an increase in the spontaneous oscillation frequency (Fabiato and Fabiato, 1975; Chiesi et al., 1981) and enhances the SR Ca²⁺ load, which can result from the well-documented effect of cAMP to modulate SR Ca²⁺ pumping via phospholamban phosphorylation. Presently, there is no evidence for direct cAMP modulation of the SR Ca²⁺ release channel.

The effect of β-adrenergic agonists on CaOsc in cardiac myocytes with intact sarcolemmal function depends on the membrane potential. In unstimulated but depolarized cells, isoproterenol increases the frequency of spontaneous CaOsc (Lehto et al., 1983), as it does

---

A and B, lower traces) which is greater in the multifocal versus unifocal case (see also Fig. 4A). (From Capogrossi et al., 1987.) (C) Inward membrane current (I) during a singe wave (W) was measured in voltage-clamped single myocytes (150 mM KCl patch electrode, whole-cell configuration). In 10 such cells (Hepes buffer pH 7.4, 1 mM Ca²⁺, 23°C) membrane potential was -74 ± 2.1 mV, W frequency was 3.9 ± 0.57 min⁻¹, and I_w during the occurrence of a single W was 11 ± 2.0 pA. I_w varies biphasically with holding potential (V_h) (upper tracings). I_w did not reverse for $V_h = -90$ to $+40$ mV. Lower tracings show that when a multifocal W occurs (b) the peak I_w produced is increased by a factor of 2 and its duration decreased about one-half compared to a unifocal W (a). These experiments show that (1) I_w is small in magnitude and (2) since I_w can summate during multifocal W, it likely represents a localized current rather than synchronously involving the entire sarcolemma. (From Talo et al., 1986.)

in skinned preparations. In regularly electrically stimulated myocytes high concentrations of β-agonists can cause the *de novo* appearance of spontaneous CaOsc between twitches or increase the frequency of preexisting CaOsc and increase the likelihood for it to be multifocal (Capogrossi and Lakatta, 1985). These effects can be attributed to cAMP modulation of voltage-dependent L-type Ca^{2+} channels to effect a net cell Ca gain. With cessation of stimulation in myocytes with normal membrane potentials, however, this effect dissipates and in the absence of further stimulation spontaneous CaOsc frequency decreases (Capogrossi and Lakatta, 1985). In the continued presence of receptor stimulation, since cAMP levels remain high even in the absence of electrical stimulation, the above result suggests that cAMP exerts its effects on CaOsc via modulation of the amount of Ca^{2+} within the cell and thus available to the SR. In unstimulated individual cardiac myocytes (Williford *et al.*, 1987) isoproterenol decreases the "resting" myoplasmic $[Ca^{2+}]$. In addition to changes in SR pumping of Ca^{2+}, this result may be explained via enhanced Ca^{2+} efflux via stimulation of the sarcolemmal Ca^{2+} pump or via changes of myoplasmic Na^+ concentration.

α-Adrenergic stimulation leads to a degradation of phosphatidylinositol within the plasma membrane and results in the production of IP_3 and diacylglycerol. The application of IP_3 to cardiac preparations in which the sarcolemma has been chemically skinned causes oscillatory SR Ca^{2+} release to occur (Nosek *et al.*, 1986). In contrast, diacylglycerol or the phorbol ester PMA decreases CaOsc frequency in cardiac myocytes with intact sarcolemmal function (Capogrossi *et al.*, 1986c). This effect, like that of cAMP, may be due to a reduction in Ca^{2+} available to participate in the CaOsc; a reduction of cytosolic $[Ca^{2+}]$ in resting rat myocytes by PMA was recently demonstrated (Uglesity *et al.*, 1987).

In summary, neurotransmitter substances or their intracellular mediators can modulate CaOsc. While IP_3 may have a direct effect on the SR to initiate spontaneous CaOsc or to enhance their frequency, cAMP and probably diacylglycerol appear to express their effects via changes in the cell Ca load. Thus, all factors considered, Ca^{2+} itself seems to be the most important modulator of spontaneous Ca^{2+} release.

5. MECHANISMS UNDERLYING SPONTANEOUS SARCOPLASMIC RETICULUM CaOsc

While the Ca^{2+} dependence of spontaneous SR CaOsc has clearly been established, the mechanism of this Ca^{2+} release is not known. Three mechanisms of Ca^{2+} release need to be considered: (1) the mechanism by which the contractile wave is periodically initiated, (2) the mechanism by which the wave of Ca^{2+} release propagates through the cell (Stern *et al.*, 1984), and (3) the normal mechanism of SR Ca^{2+} release during excitation—contraction coupling following an action potential. These mechanisms may or may not all be the same or related.

A unifying hypothesis would be to attribute all three processes to Ca^{2+}-induced Ca^{2+} release (CICR). CICR has been demonstrated in mechanically skinned cardiac cells under conditions in which these preparations behave as "SR *in situ*" (Chiesi *et al.*, 1981; Fabiato, 1985) and in isolated SR vesicles (Rousseau *et al.*, 1986), and is widely believed to be the process by which Ca^{2+} entering the cell during the action potential triggers the release of Ca^{2+} from within the SR during normal excitation–contraction coupling. Since Ca^{2+}, presumably in the cytosolic space, can trigger the release of more Ca^{2+} into this space, the process involves positive feedback and is intrinsically unstable.

Without a detailed kinetic knowledge of the CICR channel, it is impossible to model the process quantitatively, but it must be subject to the constraints that there exists a stable resting equilibrium with low cytosolic [Ca^{2+}] and that the cell must return to this equilibrium following a release. By mathematical modeling it was found that all models satisfying these constraints give rise to spontaneous CaOsc when provided with a sufficient source of Ca^{2+} (Stern, personal communication). As the Ca^{2+} in the SR rises, a point is reached at which the loop gain of the positive feedback process exceeds unity. Traces of Ca^{2+} released from the SR will then trigger the release of further Ca^{2+} in an explosive process, resulting in a rapid spontaneous release of large amounts of Ca^{2+} (Stern *et al.*, 1984). Following release, the SR will reload with Ca^{2+}, and the process will repeat itself with a frequency that depends on the availability of Ca^{2+} to the cell. The regeneration of CICR thus provides an attractive mechanism for the spontaneous initiation of periodic contractile waves; indeed, it is difficult to see how they could be avoided if a CICR mechanism exists in the intact cell.

Because of inevitable inhomogeneities, the spontaneous Ca^{2+} release initiated by the cycle postulated above would occur first in one area of the cell. When this happens, diffusion of Ca^{2+} from the site of release can trigger release at adjacent sites, causing the release to propagate (Fig. 1) as a wave (Stern *et al.*, 1984). Model calculations show that, under conditions in which spontaneous release is initiated, such a wave is always self-sustaining. The propagation velocity calculated from such models agrees with the experimentally observed velocity (Stern *et al.*, 1984).

From these theoretical studies we conclude that regeneration of CICR is a plausible mechanism for both the initiation and propagation of spontaneous Ca^{2+} release, agreeing with the facts within the rather broad limits set by our ignorance of the details of CICR itself. It is important to point out, however, that there exists no direct experimental evidence excluding the alternative hypotheses: that spontaneous release is initiated directly by "overload" of the SR and that it propagates by means of diffusion of some second messenger other than Ca^{2+}, or by electrical depolarization of some intracellular membrane structure of very high longitudinal resistance and parallel capacitance. The direct experimental demonstration of the mechanism of spontaneous Ca^{2+} release remains an important unsolved problem.

6. FREQUENCY-DEPENDENT MODULATION OF SPONTANEOUS CONTRACTION AND DEPOLARIZATION AMPLITUDES BY CaOsc

Regardless of the exact mechanism of spontaneous Ca^{2+} release, it has important consequences for the function of cardiac cells and cardiac tissue. When cells are regularly electrically stimulated, spontaneous CaOsc can be suppressed as long as the stimulation interval is shorter than the period of spontaneous release (Capogrossi and Lakatta, 1985; Capogrossi *et al.*, 1986a). Under conditions which load cells with Ca^{2+}, spontaneous release, having been overdriven by regular stimulation, can then reappear between stimuli (Capogrossi and Lakatta, 1985; Capogrossi *et al.*, 1986a).

6.1. Effect of CaOsc on Diastolic Tonus

The presence of a contractile wave causes a decrease in cardiac myocyte cell length (Fig. 2); this cell displacement leads to force production. Spontaneous Ca^{2+} release of an

A

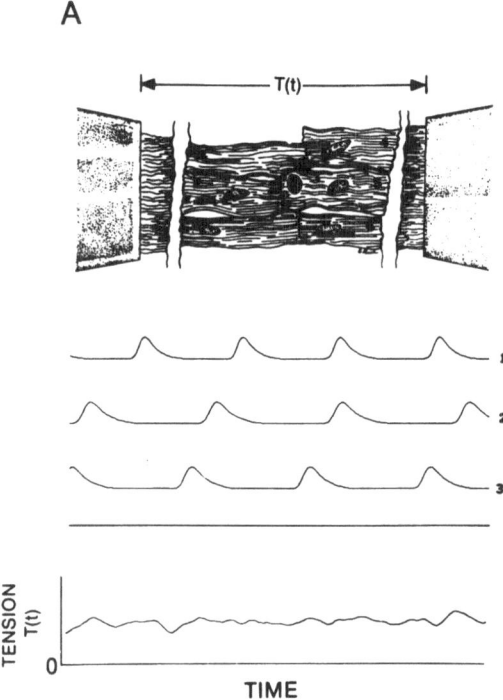

FIGURE 3. (A) A schematic of the model of independent spontaneous CaOsc within cells of intact cardiac tissue at rest, in the absence of stimulation. The sum of independent periodic asynchronous CaOsc in cells 1–3 summates to cause a fluctuating tension $T(t)$, measured at the ends of the preparation. The curve shown as $T(t)$ in the lower part of the panel was computer-simulated by adding up 100 curves similar to curves in 1–3 but with arbitrary phases and periods normally distributed about an average value. Note that this average Ca^{2+}-dependent "tone" is nonzero and thus contributes to the overall diastolic force. An increase in CaOsc frequency due to enhanced cell Ca loading will augment this Ca^{2+}-dependent tonus through partial synchro-

ensemble of myocytes results in random asynchronous contraction within bulk muscle during diastole (Stern *et al.*, 1983). The aggregate effect of this, averaged over the many cells, is to produce an active diastolic tone (Fig. 3A), which may affect diastolic filling function in the working heart, particularly under conditions of pathological Ca^{2+} overload. The effect of spontaneous Ca^{2+} release on diastolic tone in whole muscle depends on the distribution of frequencies and phases of spontaneous Ca^{2+} release cycles throughout the muscle at any instant of stimulation (Stern *et al.*, 1983; Kort *et al.*, 1985b; Capogrossi *et al.*, 1988). This distribution depends, in turn, on the interval since the last stimulus, referred to as the "delay interval" (Capogrossi *et al.*, 1986a). Myocardial cells that are stimulated during high Ca^{2+} loading states, e.g., in the presence of glycosides, enhanced bathing $[Ca^{2+}]$, or catechol-amines, exhibit not only a decrease in the delay interval for spontaneous Ca^{2+} release to occur following the prior twitch, leading to the appearance of CaOsc in the diastolic period, but also an enhanced probability of CaOsc to occur in more than a single focus (Figs. 2B

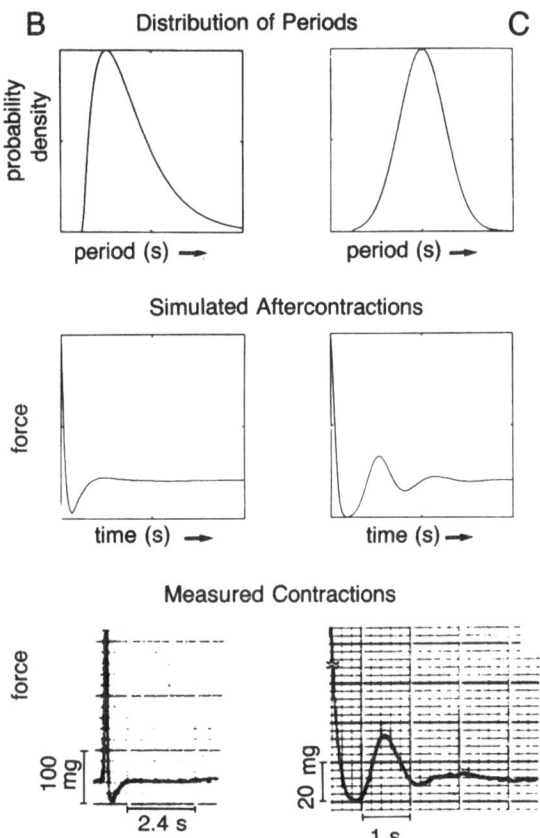

nization of the individual oscillations. (B and C) Simulated force transients generated by numerical integration of an infinite number of oscillators whose intrinsic periods are distributed with a probability density $P(T)$. If $f(t)$ describes the periodic force produced by one oscillator with unit period, then the predicted total force at time t after synchronization by an action potential, i.e., a twitch, will be $F(f) = \int_0^\infty P(t)(t/T)dT$. Synchronization of oscillations with a skewed distribution (panel B, upper tracing) gives rise to a simulated force transient (twitch) with a prominent hyperrelaxation and a subtle nonperiodic "aftercontraction" (middle tracing), resembling a measured transient (lower tracing) from a rat papillary muscle in $[Ca^{2+}]_0$ of 1 mM. A symmetrical (Gaussian) distribution of periods (panel C, upper tracing) gives rise to a periodic series of aftercontractions (middle tracing) resembling the measured aftercontractions (lower tracing) from a rat papillary muscle in $[Ca^{2+}]_0$ of 2.5 mM and caffeine concentration of 2.5 mM (From Stern et al., 1983).

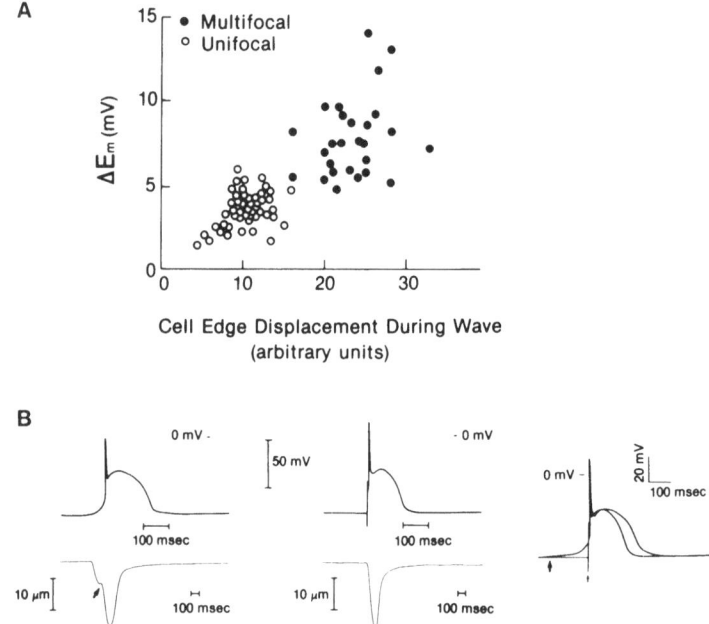

FIGURE 4. (A) Multifocal Ca^{2+} release from the SR causes greater myocyte shortening and depolarization than unifocal release within the same cardiac myocyte. (B, upper left tracings). The depolarization induced by the spontaneous increase in Ca_i of the multifocal type can be sufficient to reach threshold to trigger an action potential. An electrically stimulated action potential and twitch are shown for comparison (middle tracings). Note the slow changes in membrane potential and cell length related to spontaneous localized and multifocal SR Ca^{2+} release that precede the action potential and twitch. The spontaneous and stimulated action potentials are superimposed to contrast their initiating events (right tracings: spontaneous depolarization, large arrow; stimulus artifact for stimulated twitch by the small arrow). (From Capogrossi *et al.*, 1987.)

and 4). Both Ca^{2+}-dependent phenomena—an increase in the frequency of spontaneous release and an increase in the probability for it to be multifocal—have a summation effect on CaOsc contractile amplitude during diastole (see Figure 3A), because the individual oscillations become partially synchronized (Stern *et al.*, 1983; Kort *et al.*, 1985b). While a complete quantitative analysis of these phenomena would require a greater knowledge of the kinetics of spontaneous Ca^{2+} release than we now possess, a crude model can be constructed by assuming that stimulated and spontaneous Ca^{2+} release both draw on the same pool of stored Ca^{2+}, which is spontaneously released when it exceeds a certain size. Partial synchronization following stimulation in such a model (Fig. 3B) predicts the development of hyperrelaxation oscillatory restitution, and aftercontractions in states of high Ca^{2+} loading (Fig. 3C) (Stern *et al.*, 1983; Capogrossi *et al.*, 1988). These phenomena, which are all well known in intact muscle, are not observed in single cells, and are best understood as a statistical effect of the partially synchronized occurrence of spontaneous Ca^{2+} release throughout the muscle (Fig. 3).

6.2. Effect of CaOsc to Cause Spontaneous Sarcolemmal Depolarization

Important electrophysiological consequences of spontaneous SR Ca^{2+} release have also been recognized. It is well known that an increase in the cytosolic [Ca^{2+}] release following an action potential or voltage clamp step under conditions of high cell Ca loading leads to a so-called transient depolarization mediated via a "transient inward current" (Lederer and Tsien, 1976; Kass and Tsien, 1982), due to Ca^{2+} activation of nonspecific ion channels, or of electrogenic Na–Ca^{2+} exchange. In the single intact cell, such currents usually produce only a small depolarization (Figs. 2A, B lower panels and 4). In a tissue consisting of many asynchronously oscillating cells, electrotonic spread of current will occasionally permit the local development of large depolarizations. A model of this process can be found even in one cell as noted above because, as cell Ca loading is increased, spontaneous Ca^{2+} release becomes multifocal even in a single cell (Figs. 1 and 2). Since the myocyte length is not greater than 150 μm, the Ca^{2+} modulation of sarcolemmal conductances resulting from the synchronized loci of spontaneous Ca^{2+} release within a myocyte summates to produce an augmented depolarization (Figs. 2 and 4) compared to that which accompanies a unifocal release (Capogrossi et al., 1987). When this happens, the degree of diastolic depolarization is correlated with the extent of contractile edge displacement (Fig. 4A), suggesting that it depends on the area of sarcolemma simultaneously exposed to high cytosolic [Ca^{2+}]. This augmented depolarization can be sufficient to trigger an action potential, even when the initial membrane potential is normal (Fig. 4B). Additionally, the rapid application of a high concentration of caffeine to single ventricular myocytes at the normal membrane potential, which causes a relatively synchronous release of Ca^{2+} from the SR into the myoplasmic space, mimics the ability of multifocal spontaneous SR Ca^{2+} release to result in a depolarization sufficient in magnitude to elicit an action potential from normal resting membrane potential (Capogrossi et al., 1987).

Spontaneous Ca^{2+} release can therefore be expected to play an important role in the initiation of arrhythmias of the "abnormal automaticity" variety (Capogrossi et al., 1987) and may be important in some cases of reentry due to dispersion of refractory periods. The "after depolarization" that underlies "triggered activity" can be considered to result from a partial synchronization of spontaneous SR Ca^{2+} release following action potential-mediated SR Ca^{2+} release to cause a twitch (Kass and Tsien, 1982). Thus, a role for spontaneous SR Ca^{2+} release also seems highly likely for arrhythmias due to digitalis intoxication and for multifocal atrial tachycardia. Whether spontaneous Ca^{2+} release is also important in the genesis of arrhythmias due to ischemia and reperfusion is an important unanswered question. Even in the absence of a spontaneous action potential, the characteristics of the stimulated (or conducted) action potential are modified by the presence of spontaneous Ca^{2+} release (see Fig. 5C) because the effects of cytosolic [Ca^{2+}] to inactivate the sarcolemmal Ca^{2+} channel or to activate the TI current or Na–Ca^{2+} exchange affect the action potential configuration.

6.3. Effect of Diastolic CaOsc on Twitch Amplitude

Spontaneous SR Ca^{2+} release also has an important effect on systolic function. The ultimate modulation of contractile amplitude by SR Ca^{2+} release is for it to occur in a totally synchronous mode in response to an action potential, i.e., as a twitch. Assuming that the Ca^{2+} released during the spontaneous diastolic wave arises from the same source as the

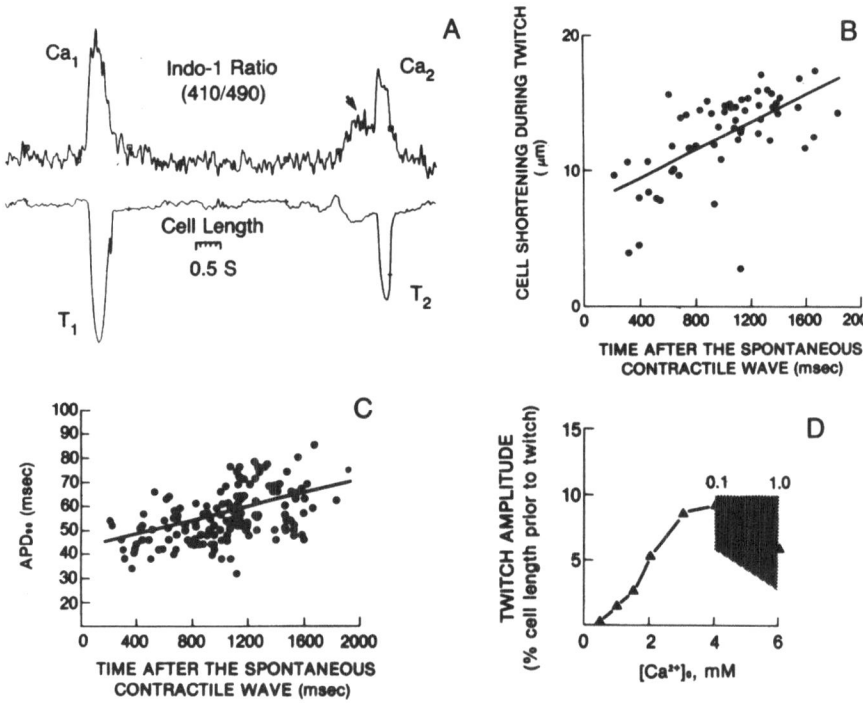

FIGURE 5. The appearance of spontaneous Ca^{2+} release in the diastolic interval between stimulated twitches has a negative effect on the ensuing stimulated twitches. (A) Two twitches (T_1 and T_2) in a single rat ventricular myocyte loaded with indo-1 AM, bathed in Ca_0 of 3 mM, and stimulated at 0.2 Hz at 23°C. The upper tracing shows indo-1 fluorescence and the lower tracing depicts cell length both measured as in Fig. 1. The appearance of spontaneous SR Ca^{2+} release in the diastolic interval (arrow) leads to a diminution in the ensuing cytosolic $[Ca^{2+}]$ transient (Ca_2) and decrease in the amplitude of the following twitch (T_2). (B and C) Respectively, the time dependence of the effect of spontaneous SR Ca^{2+} release on the ensuing twitch amplitude, and associated action potential repolarization time (90% repolarization) in a representative rat myocyte bathed in 1.8 mM Ca_0 at 35°C and regularly stimulated to twitch at 0.4 Hz via the impaling microelectrode. (From Capogrossi et al., 1986a.) (D) Representative example of an individual myocyte stimulated at 1.0 Hz in varying Ca_0. The line is the *average* twitch amplitude. The shaded area indicates the range of twitch amplitudes in a given Ca_0 when spontaneous contractile waves appeared. Numbers above the shaded area are the average number of waves between two consecutive twitches at each Ca_0. Twitch amplitude saturate as diastolic waves appear. (From Capogrossi et al., 1988.)

Ca^{2+} released during the twitch (Capogrossi et al., 1986a), it might be expected that the occurrence of a spontaneous release immediately prior to a twitch would reduce the magnitude of the Ca^{2+} transient evoked by an action potential, and therefore reduce twitch amplitude. This proves to be the case, as shown in Fig. 5. In bulk muscle, as noted above, spontaneous diastolic CaOsc occur asynchronously, so that at any time when an action potential stimulates a twitch, SR Ca^{2+} stores throughout the muscle will be in various phases of spontaneous

Ca^{2+} release and reuptake. The result will be a diminution of the twitch amplitude, compared to the value it would have had in the absence of spontaneous Ca^{2+} release (Capogrossi *et al.*, 1988). The diminution is most marked when the CaOsc in the prior diastole are synchronized, i.e., during an aftercontraction. In addition, the asynchronous nature of CaOsc in muscle means that sarcomere lengths will be inhomogeneous at the time of the synchronous contraction, and that those sarcomeres which produce maximal force will be in series with others which are "weak" and provide an effective series compliance. This effect will further dissipate the force production during a twitch (Kort and Lakatta, 1984). We therefore anticipate that the onset of spontaneous Ca^{2+} release during diastole limits the effectiveness of positive inotropic interventions (Capogrossi *et al.*, 1985, 1988).

These effects can be directly demonstrated in single cells, in which we find that maneuvers which increase cell Ca loading produce increased twitch amplitude up to the point at which spontaneous release begins to occur (Fig. 5D). For still higher Ca^{2+} loading, average twitch amplitude actually decreases. Under a wide variety of conditions, it appears that spontaneous Ca^{2+} release limits the inotropic effect of interventions that act by increasing intracellular Ca^{2+} (Capogrossi *et al.*, 1988). This relationship can be demonstrated directly in bulk muscle by SLIF measurements (Lakatta and Lappe, 1981) or the chemiluminescent aequorin (Allen *et al.*, 1985).

From the above considerations it becomes clear that spontaneous SR Ca^{2+} release can have important and even dominant effects in determining the contractile and electrophysiological properties of cardiac muscle under conditions of Ca^{2+} loading sufficient to permit occurrence of CaOsc between stimulated contractions. It should be noted that "high Ca^{2+} loading" is somewhat of a tautology in this context, since it refers to those conditions in which spontaneous Ca^{2+} release can occur between stimuli. When spontaneous CaOsc do occur, however, it is likely that their aforementioned functional consequences *collectively* contribute to the common clinical manifestations of diverse pathophysiological etiology of heart failure.

7. SUMMARY

Small increases in cytosolic $[Ca^{2+}]$ loading of mammalian myocardial cells can lead to spontaneous oscillatory amplifications of cytosolic $[Ca^{2+}]$, i.e., CaOsc. These are generated by SR and the increases in cytosolic $[Ca^{2+}]$ produced are on the order of that triggered by an action potential but are localized rather than diffuse. In the intact cardiac muscle CaOsc occur asynchronously among the ensemble of cells comprising that tissue. The occurrence of CaOsc in the interval between twitches leads to (1) localized myofilament Ca^{2+} activation which summates to result in an increased resting "tonus," (2) localized Ca^{2+} modulation of sarcolemmal ionic conductances resulting in a small inward current leading to small oscillatory depolarizations, and (3) transient Ca^{2+}-dependent impairment of specific excitation–contraction coupling mechanisms, e.g., SR Ca^{2+} loading and release and inhomogeneities in sarcomere and cell length, each of which can limit the twitch amplitude in response to a subsequent action potential. The probability for these spontaneous CaOsc to occur is modulated by Ca^{2+} itself: an increase in cell (and therefore SR) Ca loading increases the frequency of CaOsc occurrence in a given locus within the cell and makes it more likely for CaOsc to occur simultaneously at more than a single locus. Intracellular messengers of adrenergic receptor stimulation also modulate the CaOsc frequency: cAMP and IP_3 increase it and diacylglycerol decreases it. The effects of cAMP and diacylglycerol on CaOsc frequency, however, appear to be secondary to their changing cell Ca loading and thus changing Ca^{2+} available to the SR. An increase in CaOsc frequency per se produces a partial "synchronization" of the localized CaOsc occurring asynchronously within and among cells and leads to a summation of their individual effects, and thus to more marked functional sequalae, i.e., a further increase in resting tone and the occurrence of aftercontractions, "after-depolarizations" and an increased likelihood for spontaneous action potentials to occur, and a decline in myocardial contractile state from its optimum. These adverse functional effects of spontaneous SR CaOsc may be a cause of the triad of clinical symptoms common to diverse etiologies of heart failure, i.e., an increase in diastolic pressure, compromised systolic function, and an increased probability for the occurrence of spontaneous arrhythmias.

REFERENCES

Allen, D. G., Eisner, D. A., Pirolo, J. S., and Smith, G. L.,1985, The relationship between intracellular calcium and contraction in calcium overloaded ferret papillary muscles, *J. Physiol.*, **364**:169–182.

Bloom, S., 1970, Spontaneous rhythmic contraction of separated heart muscle cells, *Science* **167**: 1727–1729.

Capogrossi, M. C., and Lakatta, E. G., 1985, Frequency modulation and synchronization of spontaneous oscillations in cardiac cells, *Am. J. Physiol.* **248**:H412–H418.

Capogrossi, M. C., and Lakatta, E. G., 1988, Intracellular calcium and activation of contraction as studied by optical techniques, in: *Isolated Adult Cardiomyocytes*, Volume II (H. M. Piper and G. Isenberg, eds.), CRC Press, Boca Raton, FL (in press).

Capogrossi, M. C., Suarez-Isla, B. A., and Lakatta, E. G., 1986a, The interaction of electrically stimulated twitches and spontaneous contractile waves in single cardiac myocytes. *J. Gen. Physiol.* **88**:615–633.

Capogrossi, M. C., Kort, A. A., Spurgeon, H. A., and Lakatta, E. G., 1986b, Single adult rabbit and rat cardiac myocytes retain the Ca^{2+}- and species-dependent systolic and diastolic contractile properties of intact muscle, *J. Gen. Physiol.* **88**:589–613.

Capogrossi, M. C., Kaku, T., Filburn, C. H., Pelto, D. J., Hansford, R. G., and Lakatta, E. G., 1986c, Phorbol ester stimulates membrane association of protein kinase C and inhibits spontaneous Ca^{2+} dependent sarcoplasmic reticulum Ca^{2+} release in rat cardiac cells, *Fed. Proc.* **45**:210 (Abstr.).

Capogrossi, M. C., Houser, S. R., Bahinski, A., and Lakatta, E. G., 1988, Synchronous occurrence of spontaneous localized calcium release from the sarcoplasmic reticulum generates action potentials in rat cardiac ventricular myocytes at normal resting membrane potential, *Circ. Res.* **61**:498–503.

Capogrossi, M. C., Stern M. D., Spurgeon H. A., and Lakatta E. G., 1988, Spontaneous Ca^{2+} release from the sarcoplasmic reticulum limits Ca^{2+}-dependent twitch potentiation in individual cardiac myocytes. *J. Gen. Physiol.* **91**:133–155.

Chiesi, M., Ho, M. M., Inesi, G., Somlyo, A. V., and Somlyo, A. P., 1981, Primary role of sarcoplasmic reticulum in phasic contractile activation of cardiac myocytes with shutted myolemma, *J. Cell. Biol.* **91**:728–742.

Cobbold, P. H., and Bourne, P. K., 1984, Aequorin measurements of free calcium in single heart cells, *Nature* **312**:444–446.

Dani, A. M., Cittadini, A., and Inesi, G., 1979, Calcium transport and contractile activity in dissociated mammalian heart cells, *Am. J. Physiol.* **237**:C147–C155.

Fabiato, A., 1985, Rapid ionic modifications during the aequorin-detected calcium transient in a skinned canine cardiac Purkinje cell. *J. Gen. Physiol.* **85**:189–246.

Fabiato, A., and Fabiato, F., 1972, Excitation-contraction coupling of isolated cardiac fibers with disrupted or closed sarcolemma. Calcium dependent cyclic and tonic contractions. *Circ. Res.* **31**: 293–307.

Fabiato, A., and Fabiato, F., 1975, Relaxing and inotropic effects of cyclic AMP on skinned cardiac cells. *Nature* **253**:556–558.

Fabiato, A., and Fabiato, F., 1978, Calcium induced release of calcium from the sarcoplasmic reticulum of skinned cells from adult human, dog, cat, rabbit, rat, and frog hearts and from fetal and newborn rat ventricules. *Ann. N.Y. Acad. Sci.* **307**:491–522.

Kass, R. S., and Tsien, R. W., 1982, Fluctuations in membrane current driven by intracellular calcium in cardiac Purkinje fibers, *Biophys. J.* **38**:259–269.

Kort, A. A., and Lakatta, E. G., 1984, Calcium-dependent mechanical oscillations occur spontaneously in unstimulated mammalian cardiac tissues. *Circ. Res.* **54**:396–404, 1984.

Kort, A. A., Capogrossi, M. C., and Lakatta, E. G., 1985a, Frequency, amplitude, and propagation velocity of spontaneous Ca^{2+}-dependent contractile waves in intact adult rat cardiac muscle and isolated myocytes. *Circ. Res.* **57**:844–855.

Kort, A. A., Lakatta, E. G., Marban, E., Stern, M. D., and Wier, W. G., 1985b, Fluctuations in intracellular calcium concentrations and their effect on tonic tension in canine cardiac Purkinje fibres. *J. Physiol.* **367**:391–308.

Lakatta, E. G., and Lappe, D. L., 1981, Diastolic scattered light fluctuations, resting force and twitch force in mammalian cardiac muscle, *J. Physiol.* **315**:369–394.

Lakatta, E. G., Capogrossi, M. C., Kort, A. A., and Stern, M. D., 1985, Spontaneous myocardial Ca oscillations: overview with emphasis on ryanodine and caffeine. *Fed. Proc.* **44**:2977–2983.

Lederer, W. J., and Tsien, R. W., 1976, Transient inward current underlying arrhythmogenic effects of cardiotonic steroids in Purkinje fibres, *J. Physiol.* **263**:73–100.

Lehto, H., Talo, A., Tirri, R., and Vornanen, 1983, Membrane potential oscillations in enzymatically isolated rat myocardial cells, *Acta Physiol. Scand.* **118**:385–391.

Nosek, T. M., Williams, M. F., and Zeigler, S.T., 1986, Inositol trisphosphate enhances calcium release in skinned cardiac and skeletal muscle, *Am. J. Physiol.* **250**:C807–811.

Orchard, C. H., Eisner, D. A., and Allen, D. G., 1983, Oscillations of intracellular Ca^{2+} in mammalian cardiac muscle, *Nature* **304**:735–738.

Orchard, C. H., Houser, S. R., Kort, A. A., Bahinski, A., Capogrossi, M. C., and Lakatta, E. G., 1987, Acidosis facilitates spontaneous sarcoplasmic reticulum Ca^{2+} release in rat myocardium, *J. Gen. Physiol.* **90**:145–165.

Rousseau, E., Smith, J. S., Henderson, J. S., and Meissner, G., 1986, Single channel and ^{45}Ca^{2+} flux measurements of the cardiac sarcoplasmic reticulum calcium channel, *Biophys. J.* **50**:1009–1014.

Stern, M. D., Capogrossi, M. C., and Lakatta, E. G., 1984, Propagated contractile waves in single cardiac myocytes modeled as regenerative calcium-induced calcium release from the sarcoplasmic reticulum, *Biophys. J.* **45**:94 (Abstr.).

Stern, M. D., Kort, A. A., Bhatnagar, G. M., and Lakatta, E. G., 1983, Scattered-light intensity fluctuations in diastolic rat cardiac muscle caused by spontaneous Ca^{++}-dependent cellular mechanical oscillations, *J. Gen. Physiol.* **82**:119–153.

Stern, M. D., Weisman, H. F., Renlund, D. G., Gerstenblith, G., and Lakatta, E. G., 1985, Cellular calcium oscillations in intact perfused hearts detected by laser light scattering: cellular mechanism for diastolic tone, *Circulation* **72**(3):196 (Abstr.).

Talo, A., McIvor, M. E., Spurgeon, H. A., and Lakatta, E., 1986, Membrane currents during spontaneous contractile waves in rat cardiac myocytes, *Fed. Proc.* **45**:769 (Abstr.).

Uglesity, A., Sharma, V. K., and Sheu, S.-S., 1987, Effect of protein kinase C activation on the inotropic response induced by α-adrenoceptor stimulation in rat myocardium, *Biophys. J.* **51**:264a (Abstr.).

Williford, D. J., Sharma, V. K., Walton, M. K., and Sheu, S.-S., 1987, Isoproterenol reduces cytosolic calcium concentration measured with fura 2 in resting single isolated rat ventricular myocytes, *Biophys. J.* **51**:262a (Abstr.).

56

Phospholipid–Calcium Relations at the Sarcolemma of the Cardiac Cell
Their Possible Role in Control of Contractility

G. A. LANGER

1. INTRODUCTION

The mammalian heart controls its force development by two general mechanisms: (1) The Frank–Starling response to changing preload which involves a change in diastolic fiber length, and (2) a change of contractile state in which force changes without a primary change in the relation of sarcomeric myofilaments to each other. There is evidence (Allen and Kentish, 1985) that the Frank–Starling response might involve changes in intracellular calcium (Ca) distribution and/or myofilament sensitivity that may affect force, but the primary event is the length change. In contrast, the second mechanism, change in contractile state, is primarily dependent on alterations in cellular Ca flux and exchange. The characteristics of this flux and exchange have been the focus of my laboratory at UCLA for the past 20 years. This paper will present a brief review of the current state of our knowledge with emphasis on the possible role of phospholipid in the sarcolemmal membrane in the regulation of Ca binding and exchange.

2. CHARACTERISTICS OF CELLULAR Ca EXCHANGE

It is clear that in the heart transsarcolemmal Ca flux must occur if contraction is to be maintained. The experiment illustrated in Fig. 1 demonstrates a number of important characteristics about the excitation–contraction coupling (ECC) sequence in the myocardial cell. The record represents the contractile excursion of a single cell from an adult rabbit heart. The cell is adhered to the bottom of a serum-treated culture dish and remains fixed in place despite superfusion at a flow rate of >50 ml/min. With direction of this flow over the immediate region to which the cell is adherent, the perfusion medium is completely exchanged

G. A. LANGER • Cardiovascular Research Laboratories, University of California, Los Angeles School of Medicine, Center for the Health Sciences, Los Angeles, California 90024-1760.

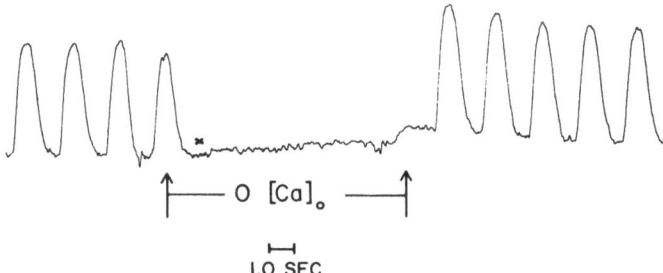

$$\longmapsto \quad 0 \ [Ca]_o \quad \longrightarrow$$

$$\longmapsto$$
1.0 SEC

FIGURE 1. Contraction amplitude of a single rabbit ventricular cell continuously stimulated at 0.4 Hz and perfused at >50 ml/min. The perfusion medium is completely exchanged in <200 msec. Removal of Ca at first arrow completely abolishes contraction at the next excitation 1 sec later, marked by (x). Ca-free perfusion continued for 10 sec and 1 mM $[Ca]_o$ was reinstituted at second arrow. With the next excitation, 1 sec following reinstitution of 1 mM $[Ca]_o$ perfusion, contractile force has returned to 100% control level. (Reproduced with permission of the *Can. J. Physiol. Pharmacol.*)

in <200 msec. The contractile excursion is recorded with a photocell mounted over the image of the end of the cell on a video screen. The cell is stimulated at 0.4 Hz by electrodes spanning the cell. At the first arrow (peak of fourth contraction) the perfusate is switched by a solenoid-controlled valve from one containing 1 mM $[Ca]_o$ to one containing <10 μM $[Ca]_o$. Note that by the time of the next excitation, indicated by (x) 1 sec after the solution change, contraction has been *completely abolished*. The "zero" $[Ca]_o$ perfusion is continued for 10 sec and is then replaced by the original 1.0 mM $[Ca]_o$ solution at the second arrow. By the time of the next excitation 1 sec later, contraction is *100% of control*. The experiment indicates that a pool of Ca, absolutely critical in the support of contraction, can be depleted and repleted in <1 sec. Recently completed experiments indicate that the depletion and repletion can be accomplished in <200 msec. This indicates that this rapidly exchangeable Ca pool could turn over completely in the course of a single contractile cycle.

The Ca in the rapidly exchangeable pool may simply be that in free solution bathing the external surface of the cell. This does not, however, seem to be the case. The Ca most immediately at the surface will occupy, in part, the so-called diffuse double layer. This is the region directly adjacent to the sarcolemma where the ionic composition is a direct function of the membrane surface potential. Since the latter is quite strongly negative (due to fixed anionic sites), cations, including Ca, will accumulate in the solution at the cellular surface. When external sodium concentration $[Na]_o$ is reduced, Ca becomes a major counterion in the diffuse double layer. There is a series of relatively large organic cations which can be used to displace inorganic cations, quite specifically, from the diffuse double layer without affecting cations actually bound to the sarcolemma. Such a compound is dimethonium (ethanebistrimethylammonium) (McLaughlin *et al.*, 1983). Dimethonium can be used in association with reduction in $[Na]_o$ to examine the possibility that Ca in free solution at the surface of the cell represents the rapidly exchangeable pool (Fintel *et al.*, 1985).

A reduction of $[Na]_o$ produces a large increase in contractile force, probably mediated through an effect on Na–Ca exchange. It also produces a large increase of Ca in the diffuse double layer (as well as in the cell). Application of dimethonium after low $[Na]_o$ perfusion displaces a major portion of Ca from the diffuse double layer but has no effect on contractile force in the isolated adult rat or rabbit cardiac cell. This indicates that rapidly exchangeable Ca actually bound or located within the cell is the Ca of prime importance in support of contraction.

That this Ca is bound, at least in part, to the sarcolemma is confirmed by membrane studies which showed that 0.5 mM cadmium ion (Cd^{2+}) displaced eight times the quantity of Ca from the sarcolemma as compared to 0.5 mM magnesium (Mg^{2+}) (Langer and Nudd, 1983). The nonhydrated radius of Cd is 0.97 Å and of Mg 0.66 Å with Ca at 0.99 Å. Divalent cations of intermediate radius, e.g., manganese (Mn^{2+}) at 0.80 Å, produced displacements proportional to the proximity of their radii to that of Ca. If this "displacement" by the cation series were from the diffuse double layer and not from membrane-binding sites, all divalent cations would be equipotent. Such was clearly not the case. In addition, it was demonstrated that the ability of a cation to displace Ca from either whole cell or isolated sarcolemma is proportional to its ability to uncouple excitation from contraction (Bers and Langer, 1979). The kinetics of Ca displacement by the competitive cations in the intact cell indicate that the binding is at sarcolemmal sites.

A series of studies using ^{45}Ca as a tracer confirms the existence of a large cellular pool of rapidly exchangeable Ca (Langer and Frank, 1972; Langer et al., 1979, 1987). Using monolayers of cultured cardiac cells and a unique measurement system, ^{45}Ca uptake and washout can be followed continuously second by second and on line for hours. The technique can detect changes of ± 125 μmol/kg dry wt with half-times in the range of 10–12 sec. It is the only technique capable of measuring a component of the rapidly exchangeable Ca pool in the intact cell. This is in contrast to methods which employ the technique of immersion of tissue in a ^{45}Ca bath followed by a number of seconds of ^{45}Ca-free wash to remove non-cell-associated "adherent" ^{45}Ca before cellular washout is commenced (Barry and Smith, 1982). The initial wash removes, along with "adherent" ^{45}Ca, a large component of the rapidly exchangeable cell-associated ^{45}Ca.

Using the technique described above, it was shown in cultured cardiac cells from neonatal (Langer et al., 1979) and adult (Langer et al., 1987) rat that between 80 and 90% of the cell's exchangeable Ca exchanges as rapidly as the flow cell system can be exchanged, i.e., half-time of 10–12 sec. The exchange is likely to be more rapid than measured because the system is perfusion-limited for this rapidly exchangeable Ca pool. It is probable that the exchange rate approaches that indicated by the loss of contractile force shown above to be <1 sec. The remainder of the exchangeable Ca in these cells, 10–20%, exchanges much more slowly with a half-time of 30–50 min. This pool cannot be directly related to contractile control since it is so slow. These exchangeable pools, rapid and slow, account for approximately 45% of the total cellular Ca. The remaining 55% is termed "inexchangeable" in that it demonstrates no significant exchange after 60 min isotopic perfusion.

The sarcolemmal location of the rapidly exchangeable pool is indicated by experiments which demonstrate that a quantity of Ca equal to the content of the pool is displaced from the cell by the Ca competitor lanthanum (La^{3+}) (Langer and Nudd, 1983). La is a trivalent rare earth which does not penetrate deeper in the cell than the sarcolemmal bilayer. Since it displaces virtually all of the rapid Ca pool, this indicates that this pool is sarcolemmal in location. La, in addition to its Ca-displacing ability, completely uncouples excitation from contraction.

In summary, the functional studies on single cells and the ^{45}Ca exchange studies on cultured cells using the unique on-line, continuous-perfusion technique give evidence that a rapidly exchangeable cellular Ca pool is localized within the sarcolemma and that this pool is important in the cell's regulation of force development.

3. CALCIUM BINDING SITES IN THE SARCOLEMMA

In a study using highly purified sarcolemmal vesicles, Philipson et al. (1980) found that, at 1.0 mM $[Ca]_0$, 80–85% of the sarcolemmal bound Ca was bound to the phospholipids (PL) of the membrane.

The class of PL sites important in binding is indicated by a study using polymyxin B (PXB), an antibiotic with a highly charged cationic head and a lipophilic tail. It binds only to membranes that contain negatively charged PL or PL zwitterions that have a negative dipole (Sixl and Galla, 1981). At low concentration (1.0 mM or less) PXB reversibly displaces large amounts of Ca from isolated sarcolemma and myocardial cells (Burt and Langer, 1983; Burt et al., 1983). PXB is also a potent EC uncoupler. These results confirm what was to be expected, i.e., that the negatively charged PL serve as the sites for binding.

Though the evidence supported the concept of a large fraction of cell Ca bound to negatively charged PL sites in the sarcolemma, there was no connection established between this bound Ca and contractile force developed by the cell. This relationship was explored as follows: (1) Phospholipase D (PLD) cleaves the nitrogenous base from PL with the production of phosphatidic acid. This increases the net anionic charge on the membrane. Phospholipase D induces a 40–60% increase in Ca binding to isolated sarcolemma (Langer and Nudd, 1983) and to whole myocardial cells in tissue culture (Burt et al., 1984). The enzyme applied to whole ventricular tissue reversibly increased contractile force by 70–150% (Langer and Rich, 1985). Therefore, an alteration of endogenous membranous neutral PL to produce anionic PL was associated with a marked increase in force development. (2) Next we inserted negatively charged amphiphile by adding the amphiphile exogenously (Philipson et al., 1985). We chose dodecyl sulfate (DDS) at concentrations below which disruptive micelle formation did not occur (<100 μM). The lipophilic dodecyl tail inserts into the membrane and the negatively charged hydrophilic sulfate headgroup can interact with counterion at the sarcolemmal surface. Dodecyl sulfate augmented Ca binding to sarcolemmal vesicles (Philipson et al., 1985) to gas-dissected sarcolemmal membranes and cultured cells (Langer and Rich, 1986) by 80% and increased contractile force by 60% in intact rabbit papillary muscle. Therefore these experiments demonstrated that when sarcolemmal anionic PL was increased, both Ca binding and force increased. It was also shown that insertion of cationic amphiphile (dodecyltrimethylamine) caused Ca displacement and a decrease of contractile force (Philipson et al., 1985).

Subsequent studies in which pH was varied and the effect on sarcolemmal Ca binding and force measured suggested that the Ca binding of PL may not be a simple one-site binding. A large increase in binding and force occurs as pH is increased from 6.0 to 8.0 (Langer, 1985). The putative PL-binding sites (carboxyl and phosphate) have apparent pK values in the range 3.7–4.0. At pH 6.0–8.0 these sites would already be fully ionized and no further effect of pH in this range would be expected. We proposed (Langer, 1985) that a second set of sites with pK approximately 7.5 act to control "access" of Ca to the more acidic phosphate and/or carboxyl groups on the PL. It is proposed that these groups of higher pK interact with the phosphate and carboxyl groups when these are positively charged. This would block access of Ca^{2+} to the PL groups. As pH is increased through the 6.0–8.0 range, these blocking groups are neutralized and the more acidic and fully ionized phosphate groups are open to bind Ca^{2+}. The "access-controlling" groups might be amine groups (NH_3^+ $\rightarrow$ NH_3 + H^+ as pH increases over the 6.0–8.0 range).

The PL that are predicted to play the major role in Ca binding are the anionic phosphatidylserine (PS) and phosphatidylinositol (PI) and the zwitterionic phosphatidylethanolamine (PE). Work in progress in cooperation with the Biochemistry Laboratories, University of Utrecht, The Netherlands has defined the distribution of the PL in the sarcolemmal bilayer of the cultured neonatal rat cells, the model used for much of the experimentation reviewed above. One hundred percent of PS and PI and 75% of the PE are located at the *inner* or cytoplasmic side of the sarcolemmal bilayer. The amounts of these components are clearly sufficient to bind the quantity of Ca (approximately 140 nmol Ca/mg sarcolemmal protein) measured.

4. CONCLUSIONS

The possibility that the large amount of the myocardial cell's exchangeable Ca that is bound to the sarcolemma has little or nothing to do with regulation of the cell's force development has to be considered. This would, however, seem unlikely in view of the fact that alterations in binding cannot as yet be dissociated from the expected alterations in force. We thought, until the location of the anionic PL was defined, that the binding sites might be at the outer bilayer and the Ca in this outside location would "feed" or interact with the two systems responsible for transsarcolemmal movement of Ca—the Ca channel and the sodium–calcium (Na–Ca) exchanger. In was visualized that Ca would bind and subsequently be moved inward by one or both of the systems. With the likelihood that most of the binding is at the inner leaflet, this concept is no longer tenable.

There is clear evidence, however, that at least two of the three (Ca channel, Na–Ca exchanger, sarcolemmal Ca pump) sarcolemmal Ca systems may interact with the anionic PL. In sarcolemmal vesicles Caroni *et al.* (1983) showed that addition of PS augments the activity of the sarcolemmal Ca pump by twofold. Philipson and Nishimoto (1984) showed that addition of PLD to their vesicular system to generate anionic phosphatidic acid increased Na–Ca exchange by three- to fourfold. Possible interaction of inner leaflet Ca-binding sites with the channel is less clear. It was shown, however, that influx of Ca feeds back to produce closure of the channel with diminution of Ca current (Lee *et al.*, 1985). It is possible that the inner leaflet sites might be involved in this process. The inside location of the large quantity of Ca poses a problem: The K_d for the sarcolemmal sites which account for most of the binding is high, in the range of 1.2–1.5 mM Ca (Bers and Langer, 1979). The intracellular free-Ca concentration $[Ca]_i$ is $<10^{-7}$ M during diastole and no more than 5×10^{-6} M during systole. If the binding sites were in equilibrium with $[Ca]_i$, very little Ca would be bound to the low-affinity sites—at most a few percent of that actually found. In order to achieve the binding that is measured, there would have to be a rapidly exchangeable compartment with a high Ca concentration in equilibrium with the inner surface of the sarcolemma. The intracellular compartment that comes to mind is, of course, the junctional sarcoplasmic reticulum (JSR). Very little is presently known about the region between the JSR and the sarcolemma and less about the actual mechanisms by which Ca moves between the JSR and the sarcolemma in the course of the contractile cycle. The problem of further definition of the model for EC coupling focuses on this region of the cell.

REFERENCES

Allen, D. G., and Kentish, J. C., 1985, The cellular basis of the length-tension relation in cardiac muscle, *J. Mol. Cell. Cardiol.* **17**:821–840.

Barry, W. H., and Smith, T. W., 1982, Mechanisms of transmembrane calcium movement in cultured chick embryo ventricular cells, *J. Physiol.* **325:**243–260.

Bers, D. M., and Langer, G. A., 1979, Uncoupling cation effects on cardiac contractility and sarcolemmal Ca^{2+} binding, *Am. J. Physiol.* **237:**H332–H341.

Burt, J. M., and Langer, G. A., 1983, Ca^{2+} displacement of polymyxin B from sarcolemma isolated by "gas dissection" from cultured neonatal rat myocardial cells, *Biochem. Biophys. Acta.* **729:** 44–52.

Burt, J. M., Duenas, C. J., and Langer, G. A., 1983, Influence of polymyxin B, a probe for anionic phospholipids, on calcium binding and calcium and potassium fluxes of cultured cardiac cells, *Circ. Res.* **53:**679–687.

Burt, J. M., Rich, T. L., and Langer, G. A., 1984, Phospholipase D increases cell surface Ca^{2+} binding and positive inotropy in rat heart, *Am. J. Physiol.* **247:**H880–H885.

Caroni, P., Zurini, M., Clark, A., and Carafoli, E., 1983, Further characterization and reconstitution of the purified Ca^{2+}-pumping ATPase of heart sarcolemma, *J. Biol. Chem.* **258:**7305–7310.

Fintel, M., Langer, G. A., Rohloff, J. C., and Jung, M., 1985, Contribution of myocardial diffuse double layer calcium to contractile function, *Am. J. Physiol.* **249:**H989–H994.

Langer, G. A., and Frank, J. S., 1972, Lanthanum in heart cell culture, *J. Cell. Biol.* **54:**441–455.

Langer, G. A., 1985, The effect of pH on cellular and membrane calcium binding and contraction of myocardium. A possible role for sarcolemmal phospholipid in EC coupling, *Circ. Res.* **57:**374–382.

Langer, G. A., and Nudd, L. M., 1983, Effects of cations, phospholipases and neuraminidase on calcium binding to "gas dissected" membranes from cultured cardiac cells, *Circ. Res.* **53:**482–490.

Langer, G. A., and Rich, T. L., 1985, Phospholipase D produces increased contractile force in rabbit ventricular muscle, *Circ. Res.* **56:**146–149.

Langer, G. A., and Rich, T. L., 1986, Augmentation of sarcolemmal Ca by anionic amphiphile: Contractile response of three ventricular tissues, *Am. J. Physiol.* **250:**H247–H254.

Langer, G. A., Frank, J. S., and Nudd, L. M., 1979, Correlation of calcium exchange, structure, and function in myocardial tissue culture, *Am. J. Physiol.* **237:**H239–H246.

Langer, G. A., Frank, J. S., Rich, T. L., and Orner, F. B., 1987, Calcium exchange, structure, and function in cultured adult myocardial cells, *Am. J. Physiol.* **252:**H314–H324.

Lee, K. S., Marban, E., and Tsien, R. W., 1985, Inactivation of calcium channels in mammalian heart cells: joint dependence on membrane potential and intracellular calcium, *J. Physiol.* **364:**395–411.

McLaughlin, A., Eng, W., Vaio, G., Wilson, T., and McLaughlin, S., 1983, Dimethonium, a divalent cation that exerts only a screening effect on the electrostatic potential adjacent to negatively charged phospholipid bilayer membranes, *J. Membrane Biol.* **76:**183–193.

Philipson, K. D., and Nishimoto, A. Y., 1984, Stimulation of $Na^{+}-Ca^{2+}$ exchange in cardiac sarcolemmal vesicles by phospholipase D, *J. Biol. Chem.* **259:**16–19.

Philipson, K. D., Bers, D. M., and Nishimoto, A. Y., 1980, The role of phospholipids in the Ca^{2+} binding of isolated cardiac sarcolemma, *J. Mol. Cell. Cardiol.* **12:**1159–1173.

Philipson, K. D., Langer, G. A., and Rich, T. L., 1985, Charged amphiphiles regulate heart contractility and sarcolemmal Ca^{2+} interactions, *Am. J. Physiol.* **249:**H147–H150.

Sixl, F., and Galla, H. J., 1981, Polymyxin interaction with negatively charged lipid bilayer membranes and the competitive effect of Ca^{2+}, *Biochim. Biophys. Acta.* **643:**626–635.

57

Role of Membrane Dysfunction and Altered Calcium Homeostasis in the Pathogenesis of Irreversible Myocardial Injury

L. MAXIMILIAN BUJA and JAMES T. WILLERSON

1. HYPOTHESES REGARDING IRREVERSIBLE MYOCARDIAL INJURY

The initial consequences of impaired coronary blood flow in the ischemic myocardium are decreased oxygen-dependent energy metabolism, decreased high-energy phosphate content, acidosis, accumulation of lactate and other metabolites, K^+ efflux, electromechanical uncoupling, and depressed myocardial function. A major issue is how the initial metabolic alterations, if they persist, lead to irreversible myocardial injury or necrosis. A large body of evidence supports the conclusions that progressive membrane damage, including damage to the sarcolemma and organellar membranes, is the essential factor in the conversion from reversible to irreversible injury in myocardial ischemia and related conditions, and that altered calcium homeostasis may have an important role in the pathogenesis of the membrane damage (Nayler, 1981).

Several observations document that membrane dysfunction and damage develop in association with lethal myocardial ischemic and hypoxic injury. Pathological calcium accumulation occurs in infarcted myocardium with collateral perfusion or reperfusion and allows for the detection of the infarcts with Tc-99m pyrophosphate (Buja et al., 1977). Ischemic myocardium exhibits abnormal volume regulation in vitro (Buja and Willerson, 1981). In response to hypoxia or ischemia in isolated myocardial preparations, a membrane permeability defect to multivalent ions develops in association with irreversible contractile depression, as demonstrated by a lanthanum probe technique (Burton et al., 1981). Hypoxia also may be associated with the accumulation of calcium and phosphorus in mitochondria of severely affected myocytes to levels shown to cause severe dysfunction of the isolated organelles (Buja et al., 1983).

L. MAXIMILIAN BUJA and JAMES T. WILLERSON ● Department of Pathology, University of Texas Southwestern Medical Center at Dallas, Dallas, Texas 75235-9072.

TABLE I. Mechanisms of Membrane
Injury in Myocardial Ischemia

Progressive phospholipid (PL) degradation
  activation of phospholipase(s)
  imbalance in deacylation-reacylation of PL
  accumulation of PL degradation products

Amphiphile-induced membrane injury
  free fatty acids
  acyl CoA, acyl carnitine
  lysophospholipids

Toxic effects of altered intracellular milieu
  elevated Ca^{2+}
  metabolites

Free-radical-induced lipid peroxidation

Physical effects on membranes
  activation of proteases
  loss of cytoskeleton–membrane connections
  effects of cell swelling

Observations also have been made in support of a number of mechanisms of membrane injury in lethal myocardial damage (Table I). These mechanisms are (1) progressive phospholipid degradation, (2) amphiphile-induced membrane injury, (3) toxic effects of an altered intracellular environment, including an elevated cellular calcium level, (4) free-radical-induced membrane lipid peroxidation, and (5) physical effects on membranes. Several of these phenomena are encompassed within the calcium-induced phospholipid degradation hypothesis, and others may act in an additive manner to worsen membrane damage.

According to the calcium-induced phospholipid degradation hypothesis, the metabolic alterations of ischemia are postulated to result in discrete alterations in membrane transport systems leading to altered calcium flux across the sarcolemma and release of calcium from deenergized membranes of the sarcoplasmic reticulum and mitochondria. The result would be a small but significant increase in cytosolic free calcium (Ca^{2+}). This Ca^{2+} elevation in energy-deficient myocytes would lead to activation of catabolic enzymes, including phospholipases and proteases. There would follow a number of catabolic reactions, including accelerated phospholipid degradation. This would lead to a vicious cycle of progressive ATP depletion, impaired membrane integrity, and further calcium overloading, which eventuates in irreversible injury. It is important to indicate that two stages of altered calcium homeostasis are postulated, namely, (1) an initial early change in Ca^{2+} and (2) a late stage of progressive calcium overloading, the magnitude of which is largely dependent on the availability of Ca^{2+} from extracellular sources, which in turn is influenced by the extent of collateral blood flow or reperfusion. According to this hypothesis, membrane injury is progressive with potentially reversible and irreversible phases. Early changes in membranes may lead to altered function of membrane proteins, including increased expression of adrenergic receptors (Buja *et al.,* 1985b). Such changes may contribute to arrhythmogenesis and the progression of ischemic cell injury. Observations related to the pathogenesis of membrane injury are summarized in the following sections.

2. ALTERATIONS OF PHOSPHOLIPIDS AND FATTY ACIDS IN ISCHEMIC MYOCARDIUM

Phospholipid levels were measured in lipid extracts of control and ischemic canine myocardium (Chien et al., 1981). After 1 hr of coronary occlusion there was a 3% decrease in total phospholipids with a 5% decrease in phosphatidylcholine in the ischemic myocardium. After 3 hr there was a significant 10% decrease in total phospholipids with 11% decreases in phosphatidylcholine and phosphatidylethanolamine. Measurements also were made of free fatty acids in lipid extracts of ischemic myocardium. Over the first 60 min of coronary occlusion, progressive increases in saturated and unsaturated free fatty acids on the order of 100–400% were measured, including a significant increase in free arachidonic acid (37 versus 11 nmol/mg DNA) (Chien et al., 1984). These findings confirmed the similar observations made earlier by Weglicki et al. (1972). Since in normal tissues nearly all arachidonic acid is contained in membrane phospholipids, accumulation of free arachidonate provides evidence of accelerated phospholipid degradation in ischemic myocardium. Furthermore, the measurement of free fatty acid accumulation relative to a low basal level is a more sensitive index of accelerated phospholipid degradation than is a decrease in total phospholipids. Similar increases in free fatty acids have been observed in isolated hearts subjected to ischemia and reperfusion (Burton et al., 1986). In addition to free fatty acids, ischemic myocardium also may accumulate long-chain acyl CoA and acylcarnitine; however, accumulation of acylcarnitine is blunted in severely ischemic canine myocardium (Chien et al., 1983; Corr et al., 1984). Although an increase in lysophospholipids may occur during the first few minutes of myocardial ischemia in some models, a sustained increase in lysophospholipids is not observed in severely ischemic canine myocardium, probably because of degradation of lysophospholipids by lysophospholipases (Chien et al., 1981; Corr et al., 1984). Thus, myocardial ischemia of sufficient duration to result in the transition from reversible to irreversible injury is associated with perturbation of membrane phospholipids without gross phospholipid depletion.

In order to evaluate the potential significance of accelerated phospholipid degradation in ischemic myocardium, membrane vesicles from normal myocardium were prepared and loaded with ^{45}Ca^{2+} (Chien et al., 1981). Efflux of ^{45}Ca^{2+} was measured under control conditions and after incubation with exogenous phospholipase C or A (10 μg ml). Phospholipase exposure resulted in significantly increased ^{45}Ca^{2+} efflux indicative of a significant Ca^{2+} permeability defect. Importantly, this permeability defect was associated with only an 8% decrease in phosphatidylcholine, which was comparable to the decrease observed in ischemic myocardium.

3. STUDIES IN CULTURED CARDIAC MYOCYTES

In order to more precisely define factors involved in the pathogenesis of cellular and membrane injury, we employed a cultured neonatal rat cardiac myocyte model. This is a system in which the effects of individual components of ischemia can be directly evaluated in a uniform population of myocytes without the complexities of in vivo ischemia, which involves multiple cell types, regional variation in collateral blood flow, and the influence of the autonomic nervous system. Briefly, cultures were prepared by enzymatic digestion of 2- to 3-day old rat ventricles followed by myocyte enrichment using either a differential attachment or a discontinuous density gradient technique, thereby yielding preparations containing colonies of beating myocytes with few nonmyocytic cells after 3 days of culture. These preparations have been used to investigate relationships among adenosine triphosphate (ATP) depletion, calcium overloading, and membrane phospholipid degradation in response to various perturbations (Buja *et al.*, 1985a; Chien *et al.*, 1985; Gunn *et al.*, 1985).

Release of tritiated arachidonic acid into the medium was used as an index of phospholipid degradation. This was accomplished by labeling myocytes with tritiated arachidonic acid 24 hr before the experiments were performed. ATP was measured by high-pressure liquid chromatography. Whole-cell calcium content was measured by atomic absorption spectrophotometry. Values were normalized to cellular protein content. Elemental content of subcellular compartments of individual myocytes was measured by electron probe X-ray microanalysis. For these studies, the myocytes were grown on collagen gels so that specimens could be freeze-clamped and cryosections prepared for electron probe microanalysis. For measurement of intracellular free-calcium concentration, $[Ca^{2+}]_i$, the Fura 2 technique was used. Myocytes were grown on coverslips and incubated in medium containing 2–3 μM Fura 2/AM for 1 hr, washed, and then experiments performed. Bulk measurements were made by placing the coverslips in cuvettes for spectrofluorometric measurements. $[Ca^{2+}]_i$ was calculated according to the method of Grynkiewitz *et al.* (1985).

Inhibition of high-energy phosphate metabolism was induced by a number of metabolic inhibitors, including deoxyglucose, cyanide, oligomycin, and iodoacetate, alone or in combination for various periods of time (Fig. 1). These studies established the relationship between ATP reduction and accelerated arachidonate release in the model. Accelerated arachidonate release did not occur until ATP was reduced below 50% of control, following which there was an exponential increase in arachidonate release as ATP levels decreased to very low levels (Gunn *et al.*, 1985). In the case of treatment with the glycolytic inhibitor and alkylating agent, iodoacetic acid (30 μM), ATP levels were mildly reduced after 1 hr followed by a marked decrease in ATP levels at 2 hr and thereafter (Buja *et al.*, 1985a; Chien *et al.*, 1985). Arachidonate release remained at control level for 1 hr followed by progressive increase in arachidonate release thereafter. $[Ca^{2+}]_i$ in control myocytes was 80 ± 6.6 nM (n = 5). There was no change in $[Ca^{2+}]_i$ after 1 hr of iodoacetate, whereas

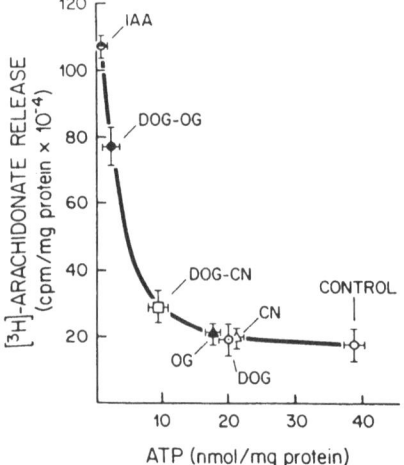

FIGURE 1. Relationship between cellular ATP content and [³H]arachidonic acid released into the medium following exposure of cultured neonatal rat cardiac myocytes to various metabolic inhibitors for 12 hr. Results are expressed at mean ± SEM for n = 3 or greater. DOG, 20 mM deoxyglucose; IAA, 30 μM iodoacetate; OG, 0.1 μg/ml oligomycin; CN, 100 μM cyanide; DOG-CN, deoxyglucose and cyanide; DOG-OG, deoxyglucose and oligomycin. [From Gunn *et al.* (1985).]

after 2 hr, $[Ca^{2+}]_i$ increased over fivefold to 496 ± 116 nM ($n = 5$) (Morris et al., 1987). The electron probe microanalysis measurements showed normal levels of electrolytes in cytoplasm and mitochondria of most myocytes after 1 hr of iodoacetate treatment, whereas after 1.5–2 hr, marked elemental changes occurred including increases in sodium, chlorine, and calcium and decreases in potassium and magnesium in cytoplasm and mitochondria (Buja et al., 1985a). The mitochondria showed marked calcium accumulation consistent with marked uptake into these organelles (Fig. 2). The electrolyte changes were associated with marked morphological alterations characterized by hypercontracture and bleb formation. In experiments in which myocytes were treated with iodoacetate for various periods of time and then returned to control medium to test for potential recovery, myocytes were capable of regenerating ATP to normal levels 24 hr after exposure to 1 hr of iodoacetate, whereas ATP levels remained markedly depressed 24 hr after exposure to iodoacetate for 2 or more hr (Buja et al., 1985a; Chien et al., 1985).

Murphy et al. (1985) reported that cultured embryonic chick cardiac myocytes were relatively resistant to calcium overloading induced by Na$^+$, K$^+$ pump inhibition, which leads to Na$^+$ accumulation and activation of Na$^+$–Ca^{2+} exchange. We performed a series of studies comparing the effects of different forms of calcium overloading on calcium homeostasis, high-energy phosphate metabolism, and the ability of myocytes to recover from calcium overloading. Specifically, we compared metabolic inhibition with iodoacetic acid to Na$^+$, K$^+$ pump inhibition induced by incubation in 0 mM K$^+$ medium or 10^{-3} M ouabain (Buja et al., 1986; Morris et al., 1986, 1987).

Measurements of whole-cell calcium content showed a marked increase in calcium following Na$^+$, K$^+$ pump inhibition at 1 hr and a persistent increase at 2 hr (Fig. 3). The

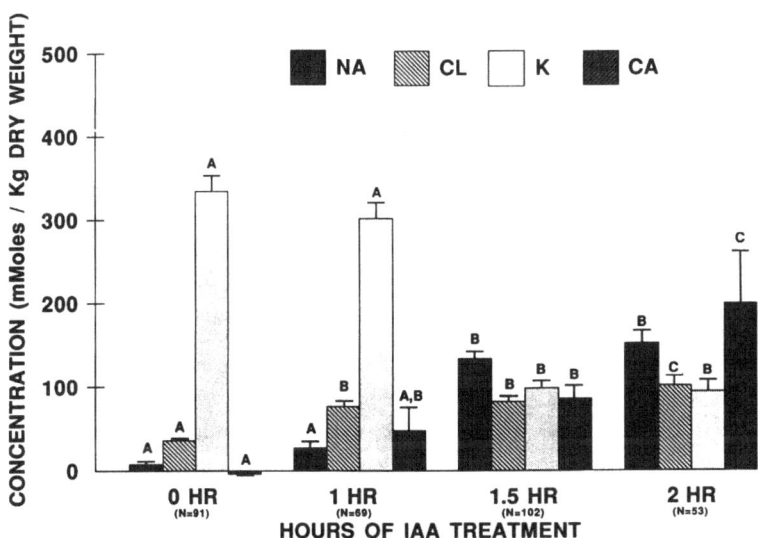

FIGURE 2. Elemental concentrations in mitochondria of cultured neonatal rat cardiac myocytes under control conditions and after exposure to 30 μM iodoacetic acid (IAA) for 1, 1.5, and 2 hr. Data were obtained by electron probe X-ray microanalysis of freeze-dried cryosections of myocytes (N = numbers of spectra analyzed). [Adapted from Table 3 of Buja et al. (1985a).]

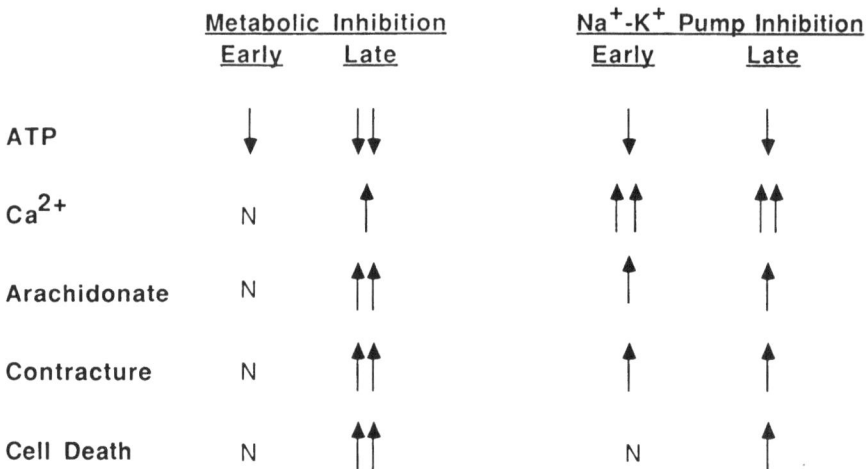

FIGURE 3. Summary of changes produced in cultured neonatal rat cardiac myocytes by metabolic inhibition and Na$^+$, K$^+$ pump inhibition. Arrows indicate increases and decreases. N indicates normal or no change.

rapidity of the calcium overloading and the magnitude of the increase were greater than that observed with iodoacetic acid. However, electron probe microanalysis showed individual cell variability in the magnitude of calcium accumulation. The cultures subjected to Na$^+$, K$^+$ pump inhibition exhibited a rapid onset of increased arachidonic acid release, with a significant increase observed at 60 min and a further increase thereafter. ATP was only moderately decreased at 1–3 hr. Return of cultures previously exposed to Na$^+$, K$^+$ pump inhibition to normal medium showed complete recovery of ATP levels 24 hr following 1 hr of Na$^+$, K$^+$ pump inhibition but incomplete recovery after 3 hr of pump inhibition. After 1 hr of Na$^+$, K$^+$ pump inhibition followed by return to normal medium, electron microscopy showed that most myocytes had normal ultrastructure, whereas after 3 hr, there was evidence of a mixed population of myocytes with some showing relatively normal ultrastructure and others showing features of irreversible injury.

Comparison of the effects of metabolic inhibition and Na$^+$, K$^+$ pump inhibition indicated significant differences in the response of myocytes to these interventions (Fig. 3). Exposure to metabolic inhibitors resulted in progressive decline in ATP level to very low levels. When ATP was reduced to a critical level, myocytes developed calcium overloading, marked arachidonate release, contracture, and cell death. In contrast, Na$^+$, K$^+$ pump inhibition was associated with a rapid onset of marked calcium accumulation associated with progressive arachidonate release, moderate decrease in ATP, and contracture. In spite of the calcium accumulation and arachidonate release, injury was largely reversible after 1 hr, but after 3 hr, irreversible injury occurred in significant numbers of myocytes.

The following conclusions are drawn from these studies. First, the mechanism of calcium overloading appears to be more important than its magnitude in influencing the potential for recovery following calcium overloading. Second, an increase in Ca^{2+} has been directly shown to cause accelerated phospholipid degradation with release of arachidonate acid. Third, the level of ATP reduction at the time of calcium overloading has an important influence on the severity of injury. Fourth, the degree of impaired energy metabolism—specifically, the ability to regenerate ATP—is an important determinant of the potential for recovery following calcium overloading. Thus, these studies predict an important synergism between impaired energy metabolism and calcium accumulation in the progression of membrane and cellular injury.

To further test the importance of membrane phospholipid degradation and altered Ca^{2+} homeostasis in cell injury, we evaluated the effects of agents with phospholipase inhibitory properties. Experiments have been performed with mepacrine, an alkylacridine, which has been reported to inhibit phospholipase A_2, possibly through a calcium calmodulin-mediated effect, and with U26384 (Upjohn Pharmaceutical Company), a steroidal diamine which inhibits phospholipase activities from human platelets, neutrophils, and hog pancreas *in vitro* (Jones *et al.*, 1987). Treatment of cultured cardiac myocytes with these agents protected against iodoacetate-induced arachidonate release as well as changes in elemental alterations, including calcium accumulation. Morphologically, the treated cells maintained normal structure, and hypercontracture and blebbing were prevented. Previously, a protective effect of chlorpromazine pretreatment was observed in isolated perfused myocardium subjected to ischemia (Burton *et al.*, 1981). The specificity and mechanism of action of mepacrine, U26384, and chlorpromazine have not been fully determined. Nevertheless, these experiments indicate that, in the setting of impaired energy metabolism, preservation against progressive release of fatty acids from membrane phospholipids is associated with protection against electrolyte shifts, including inhibition of calcium overloading, and preservation of cellular integrity.

4. SYNTHESIS AND PERSPECTIVES

Normal phospholipid homeostasis involves a balance between degradation, including deacylation (removal of fatty acids), and synthesis, including reacylation. Our data indicate that an increase in Ca^{2+} leads to accelerated phospholipid degradation with release of free fatty acids. This may be due to activation of a phospholipase A or other phospholipases, but other mechanisms, including direct effects of calcium on phospholipids, may also be operative. Evidence has been presented for accelerated phospholipid degradation in *in vivo* myocardial ischemia and *in vitro* models. Reacylation of membrane phospholipids involves at least one ATP-dependent step (Chien *et al.*, 1984). Thus, when ATP levels are partially preserved, reacylation of phospholipids may occur and membrane integrity is preserved. This may explain why normal myocytes can withstand a period of Ca^{2+} overloading secondary to Na^+, K^+ pump inhibition without initially developing irreversible injury. However, when metabolically inhibited myocytes become energy-depleted and develop a prominent increase in cytosolic Ca^{2+}, we postulate that accelerated phospholipid degradation occurs in association with an impaired ability to reacylate membrane phospholipids. The result is net loss of phospholipids, impaired membrane integrity, the potential for severe calcium overloading, total ATP depletion, and cell death. Interventions which block this vicious cycle by preventing accelerated phospholipid degradation are associated with preservation of cellular viability.

Membranes also may be injured by free-radical-induced lipid peroxidation during myocardial ischemia and reperfusion (Burton, 1985; Werns *et al.*, 1986). The relative importance of free-radical-induced lipid peroxidation and phospholipase-induced phospholipid degradation and the exact interaction between these two mechanisms require further investigation (Douglas *et al.*, 1986; Weglicki *et al.*, 1984; von Kuijk *et al.*, 1987). The final stage of membrane injury is characterized by large physical defects ("holes") in the membrane. Steenbergen *et al.* (1985) postulated that an important mechanism of formation of these defects involves damage to the cytoskeletal filaments anchoring the membrane coupled with cell swelling, with these phenomena leading to blebbing and multiple ruptures of the untethered membrane. However, it is likely that this stage of damage occurs in a membrane with an altered phospholipid composition (Table I). Furthermore, Ca^{2+} may be involved in this process by its role in activation of proteases which degrade cytoskeletal filaments. Further studies are needed to determine the specific phospholipases and proteases involved in membrane injury and the factors contributing to their activation.

ACKNOWLEDGMENTS. This work was supported by Ischemic Heart Disease SCOR Grant HL17669 and the Harry S. Moss Heart Fund, Dallas, Texas.

REFERENCES

Buja, L. M., and Willerson, J. T., 1981, Abnormalities of volume regulation and membrane integrity in myocardial tissue slices after early ischemic injury in the dog: Effects of mannitol, polyethylene glycol and propranolol, *Am. J. Pathol.* **103**:79–95.

Buja, L. M., Tofe, A. J., Kulkarni, P. V., Mukherjee, A., Parkey, R. W., Francis, M. D., Bonte, F. J., and Willerson, J. T., 1977, Sites and mechanisms of localization of technetium-99m phosphorus radiopharmaceuticals in acute myocardial infarcts and other tissues, *J. Clin. Invest.* **60**:724–740.

Buja, L. M., Burton, K. P., Hagler, H. K., and Willerson, J. T., 1983, Alterations of elemental composition of individual myocytes in hypoxic rabbit myocardium: A quantitative x-ray microanalytical study, *Circulation* **68**:872–882.

Buja, L. M., Hagler, H. K., Parsons, D., Chien, K., Reynolds, R. C., and Willerson, J. T., 1985a, Alterations of ultrastructure and elemental composition in cultured neonatal rat cardiac myocytes following metabolic inhibition with iodoacetic acid, *Lab. Invest.* **53**:397–412.

Buja, L. M., Muntz, H. K., Rosenbaum, T., Haghani, Z., Buja, D. K., Sen, A., Chien, K. R., and Willerson, J. T., 1985b, Characterization of a potentially reversible increase in beta adrenergic receptors in isolated neonatal rat cardiac myocytes with impaired energy metabolism, *Circ. Res.* **57**:640–645.

Buja, L. M., Williams, P. K., Buja, D. K., Chien, K. R., and Willerson, J. T., 1986, Comparative effects of cardiac myocyte injury induced by inhibition of the Na^+, K^+ pump and intermediary metabolism, *Clin. Res.* **34**:627A.

Burton, K. P., 1985, Superoxide dismutase enhances recovery following myocardial ischemia, *Am. J. Physiol.* **248**:H637–H643.

Burton, K. P., Hagler, H. K., Willerson, J. T., and Buja, L. M., 1981, Relationship of abnormal intracellular lanthanum accumulation to progression of ischemic injury in isolated perfused myocardium: Effect of chlorpromazine, *Am. J. Physiol.* **241**:H714–H723.

Burton, K. P., Buja, L. M., Sen, A., Willerson, J. T., and Chien, K. R., 1986, Accumulation of arachidonate in triacylglycerols and unesterified fatty acids during ischemia and reflow in the isolated rat heart: Correlation with the loss of contractile function and the development of calcium overload, *Am. J. Pathol.* **124**:238–245.

Chien, K. R., Reeves, J. P., Buja, L. M., Bonte, F. J., Parkey, R. W., and Willerson, J. T., 1981, Phospholipid alterations in canine ischemic myocardium: Temporal and topographical correlations with Tc-99m-PPi accumulation and an in vitro sarcolemmal Ca^{2+} permeability defect, *Circ. Res.* **48**:711–719.

Chien, K. R., Sen, A., Buja, L. M., and Willerson, J. T., 1983, Fatty acylcarnitine accumulation and membrane injury in ischemic canine myocardium, *Am. J. Cardiol.* **52**:893–897.

Chien, K. R., Han, A., Sen, A., Buja, L. M., and Willerson, J. T., 1984, Accumulation of unesterified arachidonic acid in ischemic canine myocardium: Relationship to a phosphatidylcholine deacylation-reacylation cycle and depletion of membrane phospholipids, *Circ. Res.* **54:**313–322.

Chien, K. R., Sen, A., Reynolds, R. C., Chang, A., Kim, Y., Gunn, M. D., Buja, L. M., and Willerson, J. T., 1985, Release of arachidonate from membrane phospholipids in cultured myocardial cells during ATP depletion: Correlation with the progression of cell injury, *J. Clin. Invest.* **75:**1770–1780.

Corr, P. B., Gross, R. W., and Sobel, B. E., 1984, Amphiphatic metabolites and membrane dysfunction in ischemic myocardium, *Circ. Res.* **55:**135–154.

Douglas, C. E., Chan, A. C., Choy, P. C., 1986, Vitamin E inhibits platelet phospholipase A$_2$, *Biochim. Biophys. Acta.* **876:**639–645.

Grynkiewitz, G., Poenie, M., and Tsien, R. Y., 1985, A new generation of Ca^{2+} indicators with greatly improved fluorescence properties, *J. Biol. Chem.* **260:**3440–3450.

Gunn, M. D., Sen, A., Chang, A., Willerson, J. T., Buja, L. M., and Chien, K. R., 1985, Mechanisms of accumulation of arachidonic acid in cultured myocardial cells during ATP depletion, *Am. J. Physiol.* **249:**H1188–H1194.

Jones, R. L., Miller, J. C., Williams, P. K., Chien, K. R., Willerson, J. T., and Buja, L. M., 1987, The relationship between arachidonate release and calcium overloading during ATP depletion in cultured neonatal rat cardiac myocytes, *Fed. Proc.* **46:**1152.

Morris, A. C., Williams, P. K., Lattanzio, F. A., Bellotto, D. J., Hagler, H. K., Willerson, J. T., and Buja, L. M., 1986, Altered calcium homeostasis induced by Na$^+$, K$^+$ pump inhibition in cultured ventricular myocytes, *Circulation* **74** (Suppl II):418 (Abstr.).

Morris, A. C., Williams, P. K., Bellotto, D. J., Hagler, H. K., Willerson, J. T., and Buja, L. M., 1987, Effects of Na$^+$, K$^+$ pump inhibition and metabolic inhibition on intracellular Ca concentration and localization in cultured neonatal rat cardiac myocytes, *Fed. Proc.* **46:**1095.

Murphy, E., Jacob, R., and Lieberman, M., 1985, Cytosolic free calcium in chick heart cells: Its role in cell injury, *J. Mol. Cell Cardiol.* **17:**221–231.

Nayler, W. G., 1981, The role of calcium in the ischemic myocardium, *Am. J. Pathol.* **102:**262–270.

Steenbergen, C., Hill, M. L., and Jennings, R. B., 1985, Volume regulation and plasma membrane injury in aerobic, anaerobic, and ischemic myocardium in vitro: Effects of osmotic cell swelling on plasma membrane integrity, *Circ. Res.* **57:**864.

von Kuijk, F. J. G. M., Sevanian, A., Handelman, G. J., and Dratz, E. A., 1987, A new role for phospholipase A$_2$: Protection of membranes from lipid peroxidation damage, *TIBS* **12:**31–34.

Weglicki, W. B., Owens, K., Urschel, C. W., Serur, J. R., and Sonnenblick, E. H., 1972, Hydrolysis of myocardial lipids during acidosis and ischemia, in: *Recent Advances in Studies on Cardiac Structure and Metabolism,* Vol. 3 (Dhalla, N.S., ed.), University Park Press, Baltimore, pp. 781–793.

Weglicki, W. B., Dickens, B. F., and Mak, I. T., 1984, Enhanced lysosomal phospholipid degradation and lysophospholipid production due to free radicals, *Biochem. Biophys. Res. Commun.* **124:**229–235.

Werns, S. W., Shea, M. J., and Lucchesi, B. R., 1986, Free radicals and myocardial injury: Pharmacologic implications, *Circulation* **74:**1–5.

58

Ischemic Alterations in Mitochondrial Calcium Transport Kinetics
An Indicator of Matrix Sulfhydryl Redox State

JEANIE B. MCMILLIN, DANIEL F. PAULY, and KIMINORI KAJIYAMA

1. ROLE OF CALCIUM IN MYOCARDIAL ISCHEMIA: CELLULAR CORRELATES

Cardiac function and energetics are intimately linked to an extracellular source of activator calcium. The levels of cytosolic free calcium available to the contractile proteins are precisely regulated by a variety of ion pump and channel proteins present in the cell membrane and sarcoplasmic reticulum. Low to moderate levels of calcium entry across the sarcolemma results in positive inotropy with increases in systolic pressure. However, further increase in calcium influx has only modest effects on systolic function, and diastolic pressure becomes elevated. High concentrations of cell calcium reduce systolic pressure from its optimal levels and increase diastolic pressure even more. It is now believed that the calcium-dependent "diastolic tonus" results from calcium loading and spontaneous release from the sarcoplasmic reticulum (Lakatta *et al.*, 1986). Therefore, calcium appears to play a central role in aberrant cardiac function. In model studies of cardiac disease, accumulation of cellular calcium following 30 min of ischemia with reflow is also associated with a large change in diastolic pressure and low rates of systolic pressure development (Burton *et al.*, 1986). Both conditions are diagnostic of a high calcium load. The sequence of events which leads to postischemic calcium entry is not well defined; however, an observed ischemic increase in intracellular sodium is thought to activate sarcolemmal sodium–calcium exchange with large increases in cellular calcium influx (Grinwald, 1982). Decreases in sarcolemmal Na^+,K^+-ATPase activity in early ischemia (Bersohn *et al.*, 1982) could be the basis for the increases in intracellular Na^+. However, other investigators observed only small changes in activity

JEANIE B. MCMILLIN, DANIEL F. PAULY, and KIMINORI KAJIYAMA • Department of Medicine, Division of Cardiovascular Disease, University of Alabama at Birmingham, University Station, Birmingham, Alabama 35294, and Department of Medicine, Section of Cardiovascular Sciences, Baylor College of Medicine, Houston, Texas 77030.

associated with irreversible cell injury (Schwartz et al., 1973). Calcium overload during reperfusion of reversibly injured, acutely ischemic tissue could explain the prolonged contractile abnormalities which accompany the reperfusion period (McMillin-Wood et al., 1979).

On the other hand, tissue calcium accumulation is also an indicator of irreversible ischemic injury. Ultrastructural studies provided the first evidence for sarcolemmal defects associated with cellular calcium deposition (Herdson et al., 1963). The onset of irreversible damage is characterized by biochemical and pathological data which suggest that degradation of cellular membrane phospholipid is causally related to increased cell calcium content (Chien et al., 1984). This is consistent with histological and chemical evidence for increased cellular lipid deposition in peripheral and border zones of experimental myocardial infarcts. The appearance of arachidonate in triacylglycerol and in unesterified free fatty acid further implicates membrane phospholipid breakdown in the ischemic response (Burton et al., 1986). Increased cellular calcium which occurs on reflow of ischemic hearts has been proposed to activate membrane phospholipase A_2 with subsequent release of arachidonic acid and lysophospholipids, the latter compounds having been shown to have toxic effects on myocardial cells (Corr et al., 1984). However, there has been no direct evidence for increased ischemic tissue calcium in the absence of reflow, although lipid droplets are still observed and arachidonate content in triacylglycerol is tripled (Burton et al., 1986). These data suggest that some phospholipid degradation may occur independent of calcium-activated mechanisms, although redistribution of subcellular calcium stores is possible. The subcellular site of arachidonate release from the phospholipid bilayer is not known, although a sarcolemmal origin has been presumed. However, studies on the phospholipid and cholesterol content of sarcolemma isolated from ischemic heart demonstrate no significant changes (Bersohn et al., 1982). Therefore, our studies have centered on the role of mitochondria in the ischemic alterations in lipid metabolism and the implication for increases in cellular calcium in the mitochondrial response to injury.

2. MITOCHONDRIAL METABOLISM IN ISCHEMIA

The transport of calcium across the mitochondrial membrane is a high-capacity but low-affinity mechanism which precludes participation of the mitochondria in the beat-to-beat control of cytosolic calcium in the heart. Alternatively, a role for the mitochondrial uptake and efflux pathways in buffering of cytosolic calcium has been proposed (Fiskum and Lehninger, 1982). The ability to maintain cytosolic calcium accurately depends on saturation of the efflux pathway, i.e., Na^+-Ca^{2+} antiport, with matrix calcium (Hansford, 1985). Half-saturation of the efflux carrier with calcium can occur only if mitochondrial calcium is several fold higher than can be detected using electron probe microanalysis of mitochondria in situ (Somlyo et al., 1981). The mitochondrial calcium concentration is within a sufficiently low concentration range consistent with regulation of calcium-sensitive dehydrogenase activities in the mitochondrial matrix (Hansford, 1985). Under normal conditions in the cardiac muscle cell, activation of mitochondrial dehydrogenase occurs in response to the cytosolic calcium signal which transmits to the mitochondria the need for increased ATP production to match increased mechanical work.

The participation of the mitochondria in the cellular response to calcium at moderate to high contractile states in the normal heart is becoming more clearly defined. Likewise, we have some information concerning the response of the mitochondria to increased cellular calcium loads in the expression of irreversible cell injury. A role for mitochondria in the onset of cell death was first proposed by Jennings *et al.* (1967). Ischemic injury is also associated with dramatic alterations in mitochondrial ultrastructure *in situ* where irreversible cellular damage is associated with mitochondrial membrane rupture and fragmentation of the cristae. These latter changes may be accompanied by the appearance of dense bodies in the mitochondria which have been identified as calcium precipitates believed to accumulate during the reperfusion interval with increased calcium flux into the cell. Despite the dramatic effect of reperfusion on cellular architecture in the irreversibly injured myocardium, it is likely that the event(s) responsible for the ultimate expression of irreversible injury occurs prior to the reintroduction of coronary flow. The basis for this belief is the fact that pre-treatment of ischemic myocardium with protective agents, e.g., calcium channel antagonists, preserves mitochondrial ultrastructure as well as isolated mitochondrial function *in vitro* both during the ischemic interval and following reperfusion (Pinsky *et al.*, 1981; Kloner *et al.*, 1982; Yoon *et al.*, 1985). Conversely, verapamil provides no protection when present only during the reperfusion period (Grinwald, 1982). These results suggest that protection of the myocardium during the ischemic interval is necessary to prevent expression of cell damage during reflow. The protective action on ischemic membrane disruption by the calcium channel antagonists may be related to the calcium dependence of membrane phospholipid hydrolysis (Van der Vusse *et al.*, 1982; Chien *et al.*, 1984, 1985; Burton *et al.*, 1986). Although elevation of myocardial calcium has not been demonstrated unequivocally during the ischemic period, accumulation of arachidonate, a fatty acid species present entirely in membrane phospholipid in the normal myocyte, has been observed during ischemia (Van der Vusse *et al.*, 1982). In the cell model of ischemia, however, release of arachidonate may be closely related to an increase in calcium uptake in cultured myocardial cells during ATP depletion (Chien *et al.*, 1985).

Relatively little is known concerning the effects of calcium on mitochondrial metabolism during the reversible period of ischemic injury and reflow. In two recent studies employing P-31 nuclear magnetic resonance to monitor high-energy phosphate metabolism in the intact heart, cellular calcium was increased by perfusion with either low sodium (Renlund *et al.*, 1985) or low potassium and ouabain (Hoerter *et al.*, 1986). In both cases, a persistent increase in oxygen consumption concomitant with increased intracellular calcium was dissociated from systolic function and a decline in cellular high-energy phosphates. Since the observed metabolic alterations are abolished by ryanodine, the calcium-dependent decline in high-energy phosphates probably results from the futile cycling of calcium across the sarcoplasmic reticulum during calcium overload. Alternatively, it is also possible that at increased levels of cytosolic calcium, the mitochondrial calcium uniport may begin to take priority over oxidative phosphorylation. It is likely, however, that the increased calcium oscillatory response, diagnostic of elevations in myoplasmic calcium, is communicated directly to the mitochondria. The relevance of these observations to the reversible period of ischemic contractile dysfunction is provocative.

3. STUDIES ON ISCHEMIC MITOCHONDRIA: CLUES TO THE NATURE OF MYOCARDIAL INJURY

Studies from our laboratory (McMilli n-Wood *et al.*, 1979; Pinsky *et al.*, 1981; Yoon *et al.*, 1985) consistently demonstrated that the degree of respiratory impairment of mito-chondria isolated from ischemic myocardium parallels the extent to which contractile function recovers during the subsequent reflow period. Consistent with this observation, we also showed that reversal of depressed mechanical activity by calcium channel antagonists is accompanied by a similar improvement in mitochondrial respiration. Two important observations can be made from these types of experiments. First, the depression in mitochondrial respiration is similar whether mitochondria are isolated from ischemic myocardium or from ischemic myocardium which has undergone a period of reflow. These results suggest either that reperfusion of ischemic heart has no additional effects on the population of mitochondria which are isolated or that respiratory function *per se* is not a sensitive criteria for detection of reperfusion injury. Second, pretreatment of the ischemic myocardium with the calcium channel antagonist verapamil significantly protects the respiratory function of mitochondria isolated from ischemic myocardium and completely restores respiration to control values following a period of reperfusion. Therefore, one might anticipate that the mitochondrial impairment(s) which is present following either reflow and/or ischemia and which may affect mitochondrial function is absent in mitochondria prepared from verapamil-pretreated hearts.

A sensitive indicator of ischemic dysfunction in isolated mitochondria is alteration in the kinetics of calcium uptake and release. In heart mitochondria, calcium uptake proceeds by a potential-dependent, uniport mechanism and the efflux pathway is primarily that of an electroneutral $Na^+ - Ca^{2+}$ exchange. However, preparations of mitochondria which are exposed to increased calcium loading will demonstrate enhanced susceptibility to spontaneous

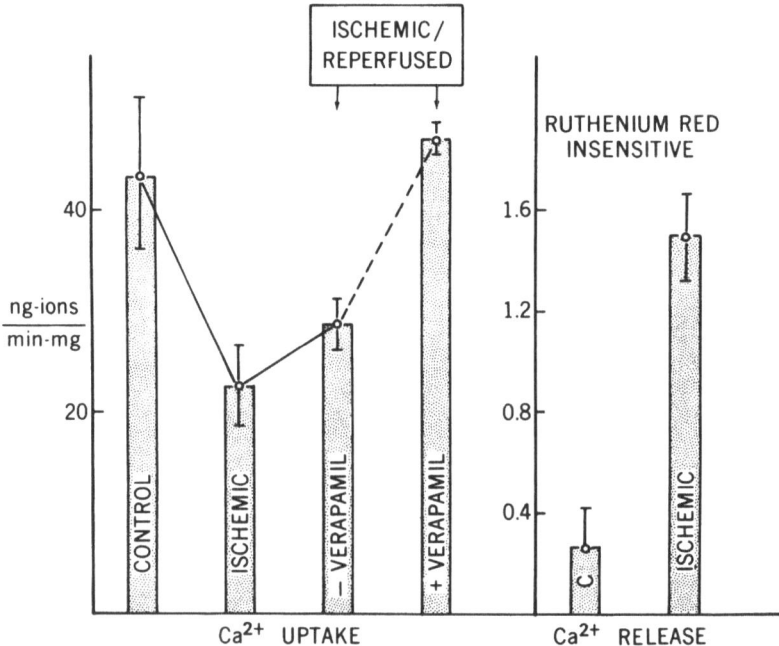

FIGURE 1. Ischemic changes in calcium uptake and release in cardiac mitochondria. Calcium uptake and release rates were measured following addition of 10 μM $CaCl_2$ employing dual-wavelength spectrophotometry and the calcium-sensitive dye antipyrylazo III (Kajiyama *et al.*, 1987).

calcium release which has been demonstrated to occur as a result of membrane swelling and associated alterations in mitochondrial permeability (Jurkowitz et al., 1983). Mitochondria isolated from ischemic hearts and exposed to low levels of exogenous calcium will also demonstrate decreased rates of calcium uptake as well as premature release of the accumulated calcium (Pinsky et al., 1981). Moreover, the decreased rates of calcium uptake and the appearance of a spontaneous release phase are reversed by verapamil pretreatment of the ischemic hearts (Pinsky et al., 1980; Wolkowicz et al., 1983). Similar to rates of phosphorylating respiration, calcium uptake velocities are also diminished to approximately the same extent whether the mitochondria are isolated from ischemic or ischemic-reperfused tissue (Fig. 1). Control mitochondria retain calcium in the absence of operation of a release pathway, following addition of ruthenium red to inhibit calcium uptake on the uniport. However, in mitochondria from ischemic hearts, a significant Na^+-independent ruthenium red-insensitive calcium release is observed, suggesting that these membranes have become more permeable to calcium (Fig. 1). Consistent with this suggestion is the dramatic drop in the rate of Na^+-Ca^{2+} exchange from 5.66 ± 0.94 in control mitochondria to 1.33 ± 0.91 ng-atoms/min/mg in ischemic mitochondria.

Changes in mitochondrial membrane permeability to calcium (as well as to other matrix cofactors including adenine nucleotides and magnesium) can be induced by a variety of compounds which include palmitoyl-CoA and inorganic phosphate (Wolkowicz and Mc-Millin-Wood, 1980) and sulfhydryl agents (Beatrice et al., 1984). The nonspecific permeability increase in mitochondrial membranes has been proposed to result from changes in matrix glutathione and thiol group reduction by a reaction coupled to the energy-linked pyridine nucleotide transhydrogenase (Beatrice et al., 1984). The metabolic basis of this permeability change is thought to arise from perturbation of a phospholipid deacylation–reacylation cycle by inhibition of the sulfhydryl-sensitive enzyme acyl-CoA: lysophospholipid acyltransferase, concomitant with activation of mitochondrial phospholipase A_2 by calcium. Since ischemic mitochondria consistently demonstrate a permeability change to calcium, it seemed reasonable to anticipate that the basis for this change might be related to calcium-dependent activation of phospholipase A_2 simultaneous with inhibition of lysophospholipid reacylation.

It has been proposed that reperfusion of the ischemic myocardium results in the generation of oxygen free radicals leading to sarcolemmal membrane disruption and explosive cell swelling (Jolly et al., 1984; Przyklenk and Kloner, 1986). Detoxification of oxygen free radicals present in the mitochondria would proceed by activation of glutathione peroxidase and subsequent rereduction of the oxidized glutathione (GSSG) by glutathione reductase. In ischemia, reconversion of GSSG to reduced glutathione (GSH) may be limited by a decreased supply of reducing equivalents to the mitochondrial dehydrogenase and transhydrogenase as a result of mitochondrial substrate deprivation. In ischemia, a decrease in reduced glutathione with a rise in the levels of oxidized glutathione is observed (Fig. 2). These alterations in glutathione levels produce an 85% decrease in the glutathione redox level with ischemia (Fig. 2). Reperfusion of the ischemic heart partially reverses the loss in GSH and returns the levels of GSSG to control values. No statistical differences between GSH or GSSG or in the redox ratio between control and verapamil-pretreated ischemic groups can be observed (data not shown).

It has been suggested that lipid peroxidation products play a role in the toxicity of oxygen free radicals (Kellogg and Fridovich, 1975). The present results suggest that mitochondrial glutathione peroxidase may be acting to detoxify free radicals which are present in the inner membrane matrix space. Therefore, selected peroxidation species of arachidonate

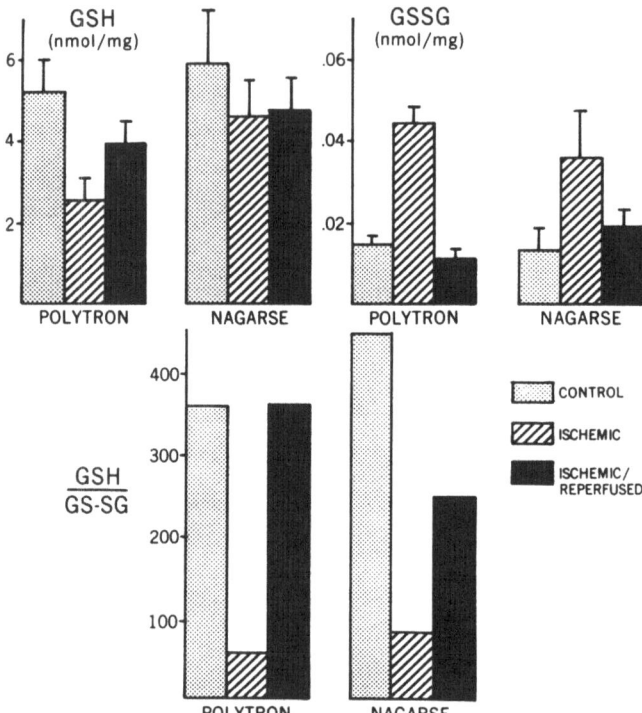

FIGURE 2. Changes in mitochondrial glutathione content during ischemia and reperfusion. Mitochondria were isolated following polytron homogenization and nagarse digestion. Mitochondrial glutathione contents in each mitochondrial preparation was determined as described by Kajiyama et al. (1987).

(11-, 12-, and 15-hydroxy 20:4) were measured in lipid extracts of mitochondria from control, ischemic, and ischemic reperfused heart (Kajiyama et al., 1987). Ischemia had no significant effect on the levels of hydroxyarachidonate measured in these experiments. However, significant decreases in hydroxyarachidonate were seen in mitochondria isolated from ischemic reperfused hearts. The lack of elevation in the levels of hydroxyarachidonate with ischemia does not exclude the possibility that mitochondrial lipid peroxidation occurs during ischemia and reperfusion. It is probable that these mitochondria may exhibit competent rates of detoxification and excision via enhanced phospholipase A_2 activity.

If phospholipase A_2 activity is enhanced due to elevations in both intracellular and mitochondrial calcium upon reflow of the ischemic cells, and if reacylation activity is depressed as a consequence of the sulfhydryl sensitivity of the acyl-CoA lysophospholipid acyltransferase, then mitochondria isolated from ischemic and/or ischemic reperfused hearts should demonstrate changes in the phospholipid content and composition when compared to control mitochondria. Indeed, total phospholipid content is decreased with ischemia due to net losses in phosphatidylcholine and cardiolipin (Fig. 3). No significant effect of ischemia was seen on the mitochondrial content of phosphatidylethanolamine (Fig. 3). With reperfusion, the total mitochondrial content of phospholipids decreased even further and the major

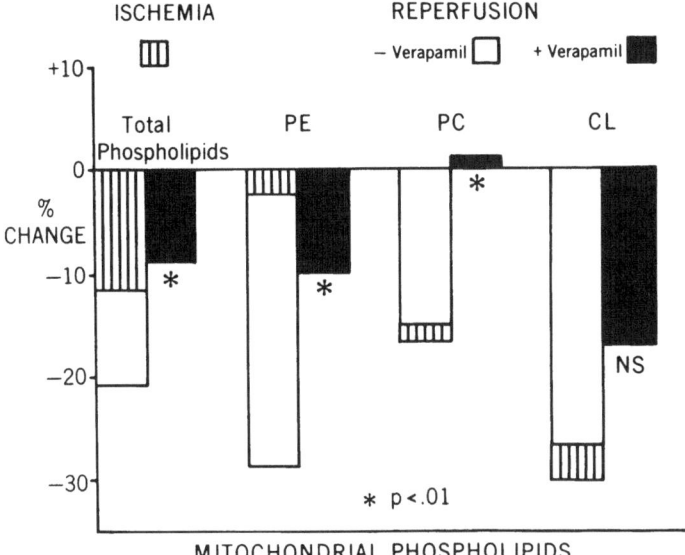

FIGURE 3. Losses in mitochondrial phospholipids during ischemia and reperfusion. Changes in total mitochondrial phospholipid, phosphatidylethanolamine (PE), phosphatidylcholine (PC), and cardiolipin (CL) following coronary occlusion and reperfusion were determined by thin-layer chromatography of extracted lipids as described by Kajiyama et al. (1987).

portion of this loss was due to a reperfusion-specific excision of phosphatidylethanolamine (Fig. 3). Verapamil pretreatment protected against loss of mitochondrial phospholipids [with the exception of cardiolipin, which is a relatively poor substrate for phospholipase (Okuyama et al., 1965)]. The decrease in mitochondrial content of phospholipids was not accompanied by any change in lysophospholipid levels. Since high lysophospholipase activities are present in cardiac tissue (Gross et al., 1984), rapid degradation of lysophospholipids may occur which obscures their production.

Although the mole percentage of phospholipid fatty acid constituents did not change significantly with ischemia, upon reperfusion there was a significant loss in C18:0 and C20:4, the major constituent fatty acids of phosphatidylethanolamine. Concomitantly, there is an enrichment in the mole percent of C18:1 which may relate to the increased fractional content of phosphatidylcholine with reperfusion (via specific loss of phosphatidylethanolamine from the mitochondrial membranes).

4. CONCLUSIONS

The purpose of these studies was twofold. First, since reperfusion of ischemic heart has a profound effect on cellular architecture and contractile function, and since mitochondria isolated from ischemic hearts reflect the impact of injury to the myocyte, identification of a perfusion-specific defect in the mitochondria could provide some insight into the metabolic state of the cells during reperfusion. Second, the initiating event for changes in both metabolic events and membrane composition may be generated during the ischemic interval, so that flow deprivation sets the stage for reperfusion injury. Alternatively, this event may be unique to the reperfusion setting only. In the latter regard, it is interesting that the sulfhydryl redox state of the mitochondrial matrix is dramatically oxidized with ischemia and becomes partially reduced upon reperfusion. The excision of phospholipids during both the ischemic and reperfusion interval suggests that the oxidation of glutathione is an important precursor of reperfusion injury. The finding that phosphatidylcholine and cardiolipin (but not phosphotidylethanolamine) are depleted from mitochondria during ischemia and in the absence of reperfusion is surprising. Although it is possible that both of these phospholipids are lost through phospholipase A_2 activation, it is also possible that this loss of phospholipid is not through a calcium-mediated pathway. Other investigators have found net increases in phospholipid degradation in isolated ischemic hearts prior to reperfusion and during a time when no increases in tissue calcium can be observed (Burton *et al.*, 1986).

The lack of an effect of specific loss of phosphatidylethanolamine on mitochondrial respiration or calcium transport may be related to a lack of respiratory and enzymatic dependency on pools of phosphatidylethanolamine which are readily accessible to phospholipase A_2. However, the dramatic loss of this phospholipid species (together with stearic and arachidonic acid during reperfusion) establishes that the mitochondria constitute an intracellular membrane source of arachidonic acid which contributes to the fatty acid pool that is liberated with reperfusion of the ischemic heart.

ACKNOWLEDGMENT. This work was supported by NIH grants HL23161 and HL38455.

REFERENCES

Beatrice, M. C., Stiers, D. L., and Pfeiffer, D. R., 1984, The role of glutathione in the retention of Ca^{2+} by liver mitochondria, *J. Biol. Chem.* **259:**1279–1287.

Bersohn, M. M., Philipson, K. D., and Fukushima, J. Y., 1982, Sodium–calcium exchange and sarcolemmal enzymes in ischemic rabbit hearts, *Am. J. Physiol.* **242:**C288–C295.

Burton, K. P., Buja, L. M., Willerson, J. T., and Chien, K. R., 1986, Accumulation of arachidonate in triacylglycerols and unesterified fatty acids during ischemia and reflow in the isolated rat heart, *Am. J. Pathol.* **124:**238–245.

Chien, K. R., Han, A., Sen, A., Buja, L. A., and Willerson, J. T., 1984, Accumulation of unesterified arachidonic acid in ischemic canine myocardium: Relationship to a phosphatidylcholine deacylation–reacylation cycle and the depletion of membrane phospholipids, *Circ. Res.* **54:**313–322.

Chien, K. R., Sen, A., Reynolds, R., Chang, A., Kim, Y., Gunn, M. D., Buja, L. M., and Willerson, J. T., 1985, Release of arachidonate from membrane phospholipids in cultured neonatal rat myocardial cells during adenosine triphosphate depletion. Correlation with the progression of cell injury, *J. Clin. Invest.* **75:**1770–1780.

Corr, P. B., Gross, R. W., and Sobel, B. E., 1984, Amphipathic metabolites and membrane dysfunction in ischemic myocardium, *Circ. Res.* **55:**135–154.

Fiskum, G., and Lehninger, A. L., 1982, Mitochondrial regulation of intracellular calcium, in: *Calcium and Cell Function,* Vol. 2 (W. Y. Cheung, ed.), Academic Press, New York, pp. 39–80.

Grinwald, P. M., 1982, Calcium uptake during post-ischemic reperfusion in the isolated rat heart: Influence of extracellular sodium, *J. Mol. Cell. Cardiol.* **14:**359–365.

Gross, R. W., Ahumada, G. G., and Sobel, B. E., 1984, Cytosolic lysophospholipase in cardiac myocytes and its inhibition by 1-palmitoylcarnitine, *Am. J. Physiol.* **15:**C266–C270.

Hansford, R. G., 1985, Relation between mitochondrial calcium transport and control of energy metabolism, in: *Rev. Physiol. Biochem. Pharmacol.*, Vol. 102, Springer-Verlag, New York, pp. 1–72.

Herdson, P. B., Sommers, N. H., and Jennings, R. B., 1963, A comparative study of the fine structure of normal and ischemic dog myocardium with special reference to early changes following temporary occlusion of a coronary artery, *Am. J. Pathol.* **46**:367–386.

Hoerter, J. A., Miceli, M. V., Renlund, D. G., Jacobus, W. E., Gerstenblith, G., and Lakatta, E. G., 1986, A phosphorus-31 nuclear magnetic resonance study of the metabolic, contractile and ionic consequences of induced calcium alterations in the isovolumic rat heart, *Circ. Res.* **58**:539–551.

Jennings, R. B., Kaltenbach, J. P., and Summers, H. M., 1967, Mitochondrial metabolism in ischemic injury, *Arch. Pathol.* **84**:15–19.

Jolly, S. R., Kane, W. J., Bailie, M. B., Abrams, G. D., and Lucchesi, B. R., 1984, Canine myocardial reperfusion injury: Its reduction by the combined administration of superoxide dismutase and catalase, *Circ. Res.* **54**:277–285.

Jurkowitz, M. S., Geisbuhler, T., Jung, D. W., and Brierley, G. P., 1983, Ruthenium red-sensitive and-insensitive release of Ca^{2+} from uncoupled heart mitochondria, *Arch. Biochem. Biophys.* **223**:120–128.

Kajiyama, K., Pauly, D. F., Hughes, H., Yoon, S. B., Entman, M. L., and McMillin-Wood, J., 1987, Protection by verapamil of mitochondrial glutathione equilibrium and phospholipid changes during reperfusion of ischemic canine myocardium, *Circ. Res.* **61**:301–310.

Kellogg, E. W., and Fridovich, I., 1975, Superoxide hydrogen peroxide and singlet oxygen in peroxidation by a xanthine oxidase system, *J. Biol. Chem.* **250**:8812–8817.

Kloner, R. A., DeBoer, L. W. V., Carlson, N., and Braunwald, E., 1982, The effect of verapamil on myocardial ultrastructure during and following release of coronary artery occlusion, *Exp. Mol. Pathol.* **36**:277–286.

Lakatta, E. G., Capogrossi, M. C., Kort, A. A., and Stern, M. D., 1987, Functional sequelae of diastolic sarcoplasmic reticulum Ca^{2+} release in the myocardium, in: *Diastolic Relaxation of the Heart*, Proceedings of an International Symposium (W. Grossman and B. H. Lorell, eds.), pp. 49–64, Martinez Nijhass, Boston, Mass.

McMillin-Wood, J. B., Entman, M. L., Hanley, H. G., Lewis, R. M., Busch, U., Chang, C. H., Swain, J. A., Morgan, W. J., and Schwartz, A., 1979, Biochemical and morphological correlates of acute experimental myocardial ischemia. III. Energy producing mechanisms, *Circ. Res.* **44**:52–61.

Okuyama, H., and Nojimi, S., 1965, Studies on hydrolysis of cardiolipin by snake venom phospholipase A, *J. Biochem. (Tokyo)* **57**:529–538.

Pinsky, W. W., Lewis, R. M. McMillin-Wood, J. B., Hara, H., Gillette, P. C., and Entman, M. L., 1981, Myocardial protection from ischemic arrest: Potassium and verapamil cardioplegia, *Am. J. Physiol* **240**:H326–H355.

Przyklenk, R., and Kloner, R. A., 1986, Superoxide dismutase plus catalase improve contractile function in the canine model of the "stunned myocardium," *Circ. Res.* **58**:148–156.

Renlund, D. G., Lakatta, E. G., Mellits, E. D., and Gerstenblith, G., 1985, Calcium-dependent enhancement of myocardial diastolic tone and energy utilization dissociates systolic work and oxygen consumption during low sodium perfusion, *Circ. Res.* **57**:876–888.

Schwartz, A., McMillin-Wood, J., Allen, J. C., Bornet, E. P., Entman, M. L., Goldstein, M. A., Sordahl, L. A., and Suzuki, M., 1973, Biochemical and morphological correlates of cardiac ischemia, *Am. J. Cardiol* **32**:46–61.

Somlyo, A. P., Somlyo, A. V., Shuman, H., Scarpa, A,. Endo, M., and Inesi, G., 1981, Mitochondria do not accumulate significant Ca^{2+} concentrations in normal cells, in: *Calcium and Phosphate Transport across Biomembranes* (F. Bronner and M. Peterlik, eds.), Academic Press, New York, pp. 87–93.

Van der Vusse, G. J., Roemen, Th. H. M., Pringen, F. W., Coumans, W. A., and Reneman, R. S., 1982, Uptake and tissue content of fatty acids in dog myocardium under normoxic and ischemic conditions, *Circ. Res.* **50**:538–546.

Wolkowicz, P. E., and McMillin-Wood, J., 1980, Dissociation between mitochondrial calcium ion release and pyridine nucleotide oxidation, J. Biol. Chem. 255:10348–10353.

Wolkowicz, P. E., Michael, L. H., Lewis, R. M., and McMillin-Wood, J., 1983, Sodium–calcium exchange in dog heart mitochondria: Effects of ischemia and verapamil, Am. J. Physiol. 244: H644–H651.

Yoon, S. B., McMillin-Wood, J. B., Michael, L. H., Lewis, R. M., and Entman, M. L., 1985, Protection of canine cardiac mitochondrial function by verapamil cardioplegia during ischemic arrest, Circ. Res. 56:704–708.

59

Structural, Biochemical, and Elemental Correlates of Injury in Cultured Cardiac Cells

ANN LEFURGEY, ELIZABETH MURPHY,
BERNHARD WAGENKNECHT, PETER INGRAM,
and MELVYN LIEBERMAN

1. INTRODUCTION

Several investigators (Trump *et al.*, 1976; Nayler, 1981; Nayler and Grinwald, 1981; J. G. Murphy *et al.*, 1987) have suggested that the movement and redistribution of cellular ion contents, especially calcium, play a key role in the pathophysiology of cardiac ischemia. Since 1982 (Murphy *et al.*) our investigations have focused on the study of subcellular ionic mechanisms underlying this pathophysiology. We have utilized a model system of heart cells grown in culture to obtain basic information about (1) the regulation of ion transport in cardiac cells (Lieberman *et al.*, 1984) and (2) the relationship between maintenance of ionic homeostasis, metabolic integrity, and the onset of irreversible cell injury (Lieberman *et al.*, 1985).

One of our findings suggested that elevation of the total cell content of sodium (Na_i) or calcium (Ca_t) *per se* is not directly responsible for irreversible cell injury (Murphy *et al.*, 1983). In this study the observed increases in cell Na_i and Ca_t were not associated with elevated lactate dehydrogenase (LDH) release, indicative of lethal disruption of the sarcolemma. Additionally, a correlation between elevated cytoplasmic free calcium (Ca_i) and LDH release existed only in cells with similar (or comparable) adenosine triphosphate (ATP) levels (Murphy *et al.*, 1985). In continuing efforts to determine the mechanisms responsible for cell injury, we have now followed the time course of changes in ATP, ion (Na, K, Ca) content, LDH release, and morphology of heart cells subjected to hypoxia, inhibition of

ANN LEFURGEY, BERNHARD WAGENKNECHT, and MELVYN LIEBERMAN • Department of Physiology, Duke University Medical Center, Durham, North Carolina 27710.   ELIZABETH MURPHY • Laboratory of Molecular Biophysics, National Institute of Environmental Health Sciences, Research Triangle Park, North Carolina 27709.   PETER INGRAM • Research Triangle Institute, Research Triangle Park, North Carolina 27709.

glycolytic and/or oxidative metabolism by various means, and Na,K pump inhibition (Wagenknecht et al., 1987; E. Murphy et al., 1987). The experiments were conducted in parallel with high-resolution electron microscopy and X-ray microanalysis to quantitate changes in heart cell ultrastructure and elemental compartmentation in organelles such as mitochondria, nuclei, or sarcoplasmic reticulum. By combining information on subcellular structure and composition with biochemical data on metabolic function and total ion content, we are continuing to evaluate the changes which occur prior to the onset of irreversible injury as determined by LDH release and morphological disruption.

2. GENERAL EXPERIMENTAL METHODS

Embryonic chick heart cells were grown as monolayers (Murphy et al., 1983) for biochemical analyses or spherical aggregates (Ebihara et al., 1980) for electron probe X-ray microanalysis (EPXMA) as previously described. Total cell Ca, Na, and K were measured by atomic absorption spectroscopy (AA). ATP was measured fluorimetrically via NADPH-linked reactions and LDH release was determined spectrophotometrically. Details of biochemical methodologies have been presented in Murphy et al. (1983, 1985, 1986). Hypoxia experiments were performed in a commercially available anaerobic chamber equilibrated with nitrogen to a final pO_2 of $<0.1\%$ at 37°C.

Ultrastructure of the monolayer and aggregate preparations was assessed by conventional transmission electron microscopy as previously described (Murphy et al., 1986; LeFurgey et al., 1986). EPXMA was performed on rapidly frozen, cryosectioned aggregates (LeFurgey et al., 1986, 1987). Energy-dispersive spectra were obtained at 80 kV and -100°C for 500 sec each from cellular regions (0.1×0.1 μm raster or smaller) using a JEOL 1200EX analytical electron microscope equipped with a Tracor Northern 30-mm^2 Si(Li) detector and multichannel analyzer interfaced to a VAX 11-750 computer. The Hall continuum normalization method (Hall, 1979) as implemented by Shuman et al. (1976) was used for quantitation. Ca measurements were optimized by correlating the Ca counts with the first and second derivatives of K peak counts in order to identify and correct for detector calibration shifts (Kitazawa et al., 1983). Ca content was also measured (1) in solutions of Ca standards by AA and (2) in air-dried films of the same solutions by EPXMA; the correlation coefficient for AA-EPXMA was 0.998.

3. RESULTS

3.1. Hypoxia and Cardiac Cell Injury

Hypoxia was established by maintaining heart cell cultures in an anaerobic chamber for 1–6 hr in substrate (glucose)-free media (Wagenknecht et al., 1987). Over the initial 4–5 hr of hypoxia, Na_i increased by approximately fourfold; concurrently K_i decreased to 20% of control. Over the same time course ATP fell to less than 5 nmol/mg protein from a control value of ≈35 nmol/mg protein. Ca_t also increased by 250–400%; however, the increase in Ca_t occurred after the increase in Na_i. In addition, LDH release was minimal at 3 hr, beyond which time the level of Na_i did not change. At 4 hr, although LDH release was only 25% of total cell LDH, ATP had declined to <5 nmol/mg protein, and the increase in Na_i and decrease in K_i had reached the maximum level. These data clearly demonstrate that during

ROTENONE + IAA

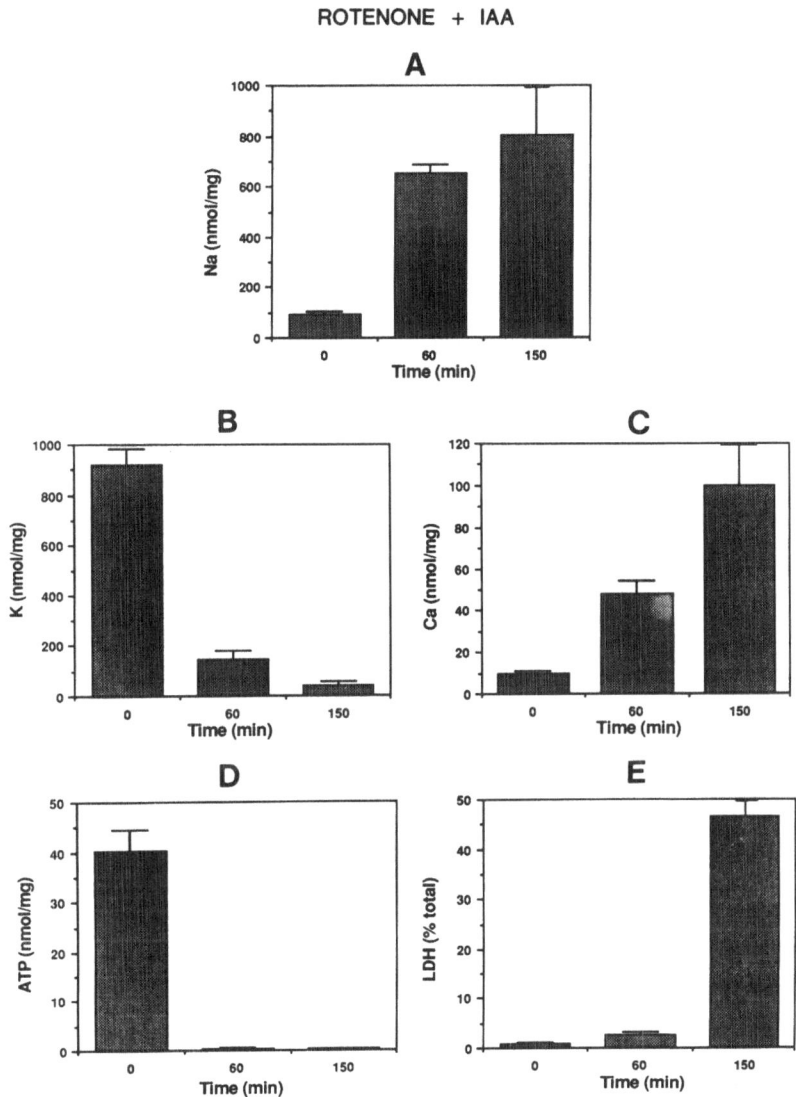

FIGURE 1. Effects of rotenone (10^{-4} M) and iodoacetic acid (10^{-3} M) on cell Na, K, Ca, and ATP contents and LDH release. (A) Na_i, (B) K_i, (C) Ca_t, (D) ATP, (E) LDH.

hypoxia the fall in ATP is accompanied by a rise in Na_i and a fall in K_i. The fact that the large increase in Na_i precedes the increase in Ca_t is consistent with the involvement of Na–Ca exchange (Murphy *et al.*, 1986). The disparity between the time courses for development of maximum Na_i increase and for development of maximum LDH release support the conclusion that the Na_i increase is not due to a nonselective permeability change associated with membrane disruption. This hypothesis was also confirmed by the following morphological observations.

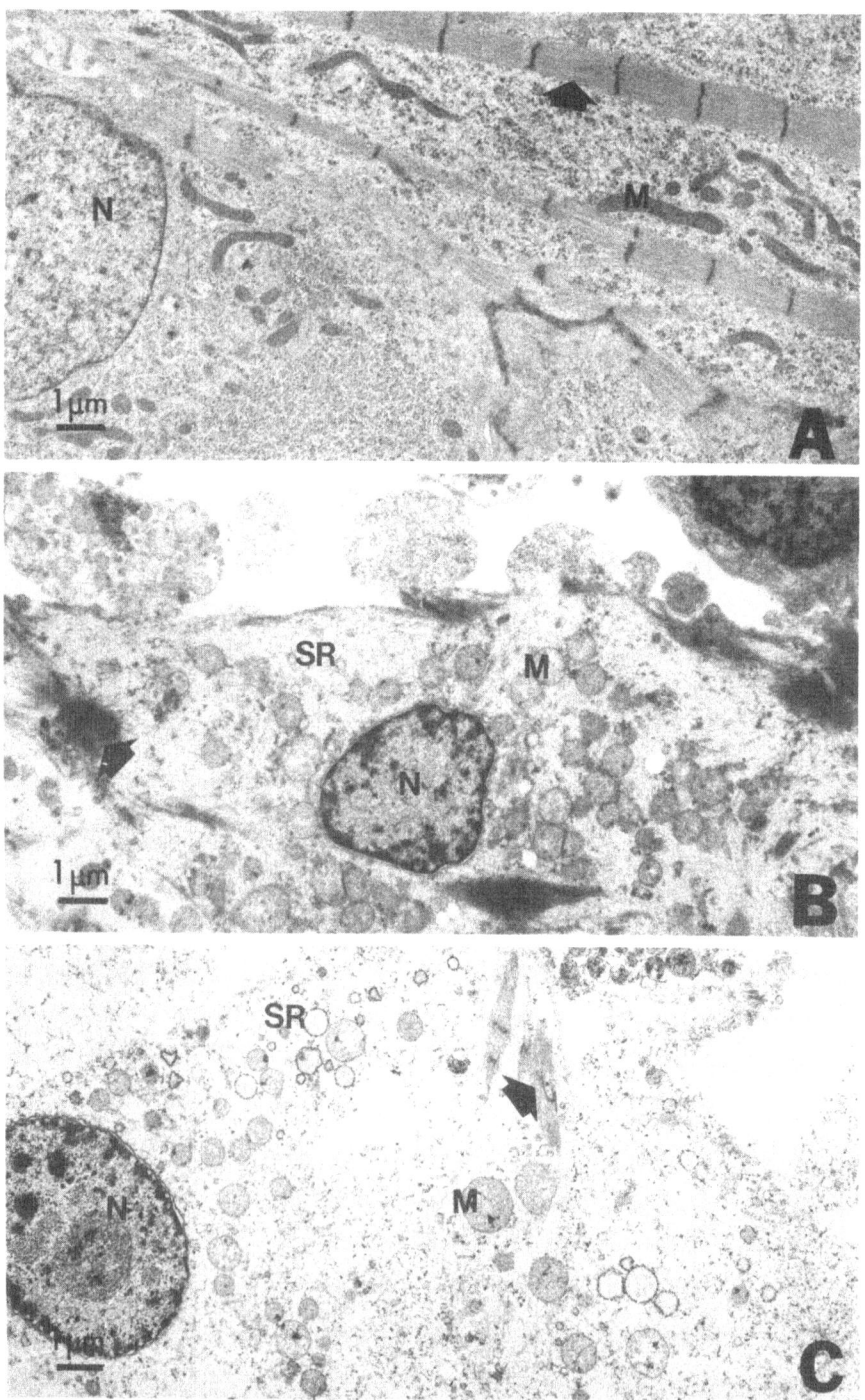

To assess the time course for structural changes in membrane, organelle, and overall cell integrity, electron microscopy was performed in parallel with measurement of ATP, LDH, Na, K, and Ca content. The fall in ATP to <5 nmol/mg protein following 4–5 hr of hypoxia was associated with swelling of mitochondria and sarcoplasmic reticulum, myofibrillar contraction, karyolysis, plasmalemmal blebbing, and disruption (Fig. 1A, B). This ultrastructural damage occurred prior to maximum LDH release and after the onset of decreases in total cell K/Na ratios.

These long-term (0–6 hr) time-dependent changes resulting from hypoxia alone are consistent with our earlier observations of ionic, metabolic, and morphological alterations induced by hypoxia (and reoxygenation) and substrate deprivation (Murphy *et al.*, 1982). The time course of the observed changes suggests that generation of myocardial cell injury by severe hypoxia coincides with the maximal decline in ATP content.

3.2. Inhibition of Oxidative Phosphorylation and Glycolysis and Cardiac Cell Injury

Cultured heart cell metabolism is capable of ATP synthesis by both glycolytic and oxidative pathways (Doorey and Barry, 1983; Hasin and Barry, 1984). Murphy *et al.* (1983) showed that inhibition of mitochondrial oxidative phosphorylation at site I by rotenone (10^{-4} M, substrate-free) reduces cell ATP by 50% within 10 min. In subsequent studies rotenone alone, as well as in combination with iodoacetic acid (IAA), an inhibitor of glycolysis, was applied to the cultured heart cells to determine the time course for development of myocardial cell injury as assessed by LDH release and morphological changes (E. Murphy *et al.*, 1987). The changes in ion content, ATP, and LDH release that occurred in the presence of rotenone alone were analogous to those seen during hypoxia and followed a similar time course. Simultaneous treatment with rotenone (10^{-4} M) and IAA (10^{-3} M, substrate-free) accelerated the time course of ionic and metabolic changes, since in the presence of IAA, endogenous glycogen cannot supply ATP (Fig. 2). Within 1 hr of inhibition of both glycolysis and oxidative phosphorylation, Na_i increased to $\simeq 800$ nmol/mg protein, K_i decreased to <100 nmol/mg protein, and ATP decreased to <5 nmol/mg protein. LDH release did not rise above 3% until 2.5 hr. This time disparity between the ATP fall and LDH release did not occur during hypoxia; therefore the role of metabolic inhibition in myocardial cell injury may depend not only on reduction in cell ATP to minimum levels but also on the occurrence of reduced ATP levels for a sustained period of time.

Significant ultrastructural changes were observed as early as 1 hr in the presence of both rotenone and IAA (Fig. 1C). Disruption of mitochondrial, nuclear, and plasma membranes occurred, as well as loss of cell junctions, contraction and disruption of myofibrils,

---

FIGURE 2. Transmission electron micrographs (TEM) of cultured heart cells preserved in 2.5% glutaraldehyde in 0.1 M sodium cacodylate buffer, poststained with osmium tetroxide and embedded in epoxy. (A) TEM of control heart cell preparations in modified Earle's balanced salt solution (MEBSS). Myofibrils (arrow), mitochondria (M), and nucleus are all normal in appearance ($\times 6650$). (B) TEM of heart cells exposed to hypoxia for 6 hr. Myofibrils are contracted (arrow), mitochondria are swollen (M), and nuclear karyolysis is visible (N) ($\times 6300$). (C) TEM of heart cells treated with rotenone (10^{-4} M) and IAA (10^{-3} M) in MEBSS for 2.5 hr. Cells are severely disrupted with swollen mitochondria (N), contracted myofibrils (arrow), and swollen sarcoplasmic reticulum (SR) ($\times 6300$).

and swelling of the sarcoplasmic reticulum. These structural changes were similar to those observed in the heart cells following 3–5 hr of hypoxia. As this early occurrence of reduced ATP and structural changes suggests, LDH release may be a relatively late event in the multifactorial process leading to irreversible cell injury (E. Murphy *et al.*, 1987).

3.3. Active Transport (Na,K Pump) Inhibition and Cardiac Cell Injury

The severe ATP depletion which occurs during ischemia (Gunn *et al.*, 1985; Chien *et al.*, 1985) should inhibit ATP-dependent membrane transport processes, such as the Na,K pump, with subsequent changes in ion content. To evaluate the relationship between the ionic changes which accompany ischemia and mechanisms of cell injury, heart cell preparations were exposed to ouabain (10^{-4} M), a specific inhibitor of the Na,K pump which induces similar ionic alterations without depleting ATP. Following 60 min in ouabain, cell Na_i increased $\approx$3- to 4-fold; K_i decreased by $\approx$40 to 50%; and Ca_t increased $\approx$5- to 10-fold (Fig. 3). This decline in Na_i could be modulated by extracellular calcium (Ca_0), with decreasing Ca_0 leading to greater increases in Na_i. As we previously showed with this and other studies (E. Murphy *et al.*, 1983, 1986, 1987), Na–Ca exchange is, in part, the mechanism by which Ca_t is elevated during Na,K-ATPase inhibition. However, during the initial hour in which Na_i, K_i, and Ca_t showed significant alterations, ATP remained at 75% of control and LDH release was insignificant ($\approx$2% of total LDH release).

In contrast to the ultrastructural disruption which occurred during metabolic inhibition, the morphological changes occurring during active transport were slight and were restricted primarily to mitochondria, which developed electron-dense deposits (Fig. 4A). Because of these morphological results and earlier biochemical studies which had demonstrated that much of the increased Ca_t was FCCP-releasable mitochondrial calcium (Murphy *et al.*, 1983, 1986), cells were prepared for EPXMA following incubation with ouabain. Analysis of densities in mitochondria of rapidly frozen, cryosectioned heart cell aggregates demonstrated the calcium content to be $\approx$500–900 mmol Ca/kg dry wt (Fig. 4B), in comparison with 2.3 ± 1.2 (SEM) mmol Ca/kg dry wt ($n = 25$) measured in control mitochondria. This massive accumulation in the presence of ouabain was reversible, as demonstrated in the cultured heart cells (Murphy *et al.*, 1983) and in other intact preparations such as portal vein (Broderick *et al.*, 1986). Although ouabain caused total cell ionic changes similar to those observed during metabolic inhibition, ionic alterations *per se* did not lead to cell injury.

4. DISCUSSION

Many investigators have suggested that elevation of LDH release is an appropriate index of cell injury (for review, see Trump *et al.*, 1976; Jennings *et al.*, 1981). Consequently, changes in LDH levels may correlate with other factors associated with cell damage, such as elevation in Ca_t or decrease in ATP. In cells exposed to metabolic inhibition, a positive relationship exists between Ca_t and LDH release, and significant ultrastructural alterations occur. However, when Ca_t is elevated by Na,K pump inhibition, neither an increase in LDH nor a change in morphology occurs. With Na,K pump inhibition, ATP is maintained at

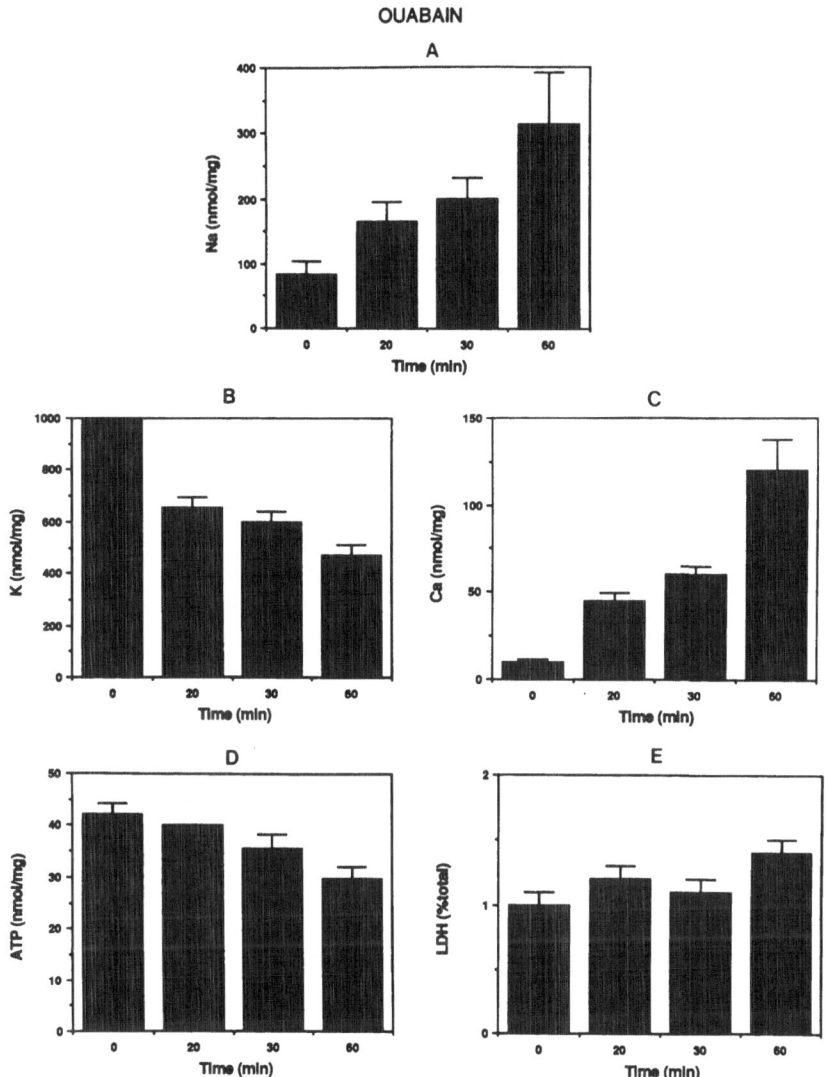

FIGURE 3. Effects of ouabain (10^{-4} M) on cell Na, K, Ca, and ATP contents and LDH release. (A) Na_i, (B) K_i, (C) Ca_t, (D) ATP, (E) LDH.

≥ 30 nmol/mg protein, no increase in LDH, and no structural alterations occur. These facts would suggest that an elevation in Ca_t alone does not lead to irreversible cell injury. With metabolic inhibition, as ATP decreases to below 5–10 nmol/mg protein, LDH release increases, and morphological disruption takes place. Thus these comparative results demonstrate that a correlation exists between ATP, LDH release, and ultrastructural integrity.

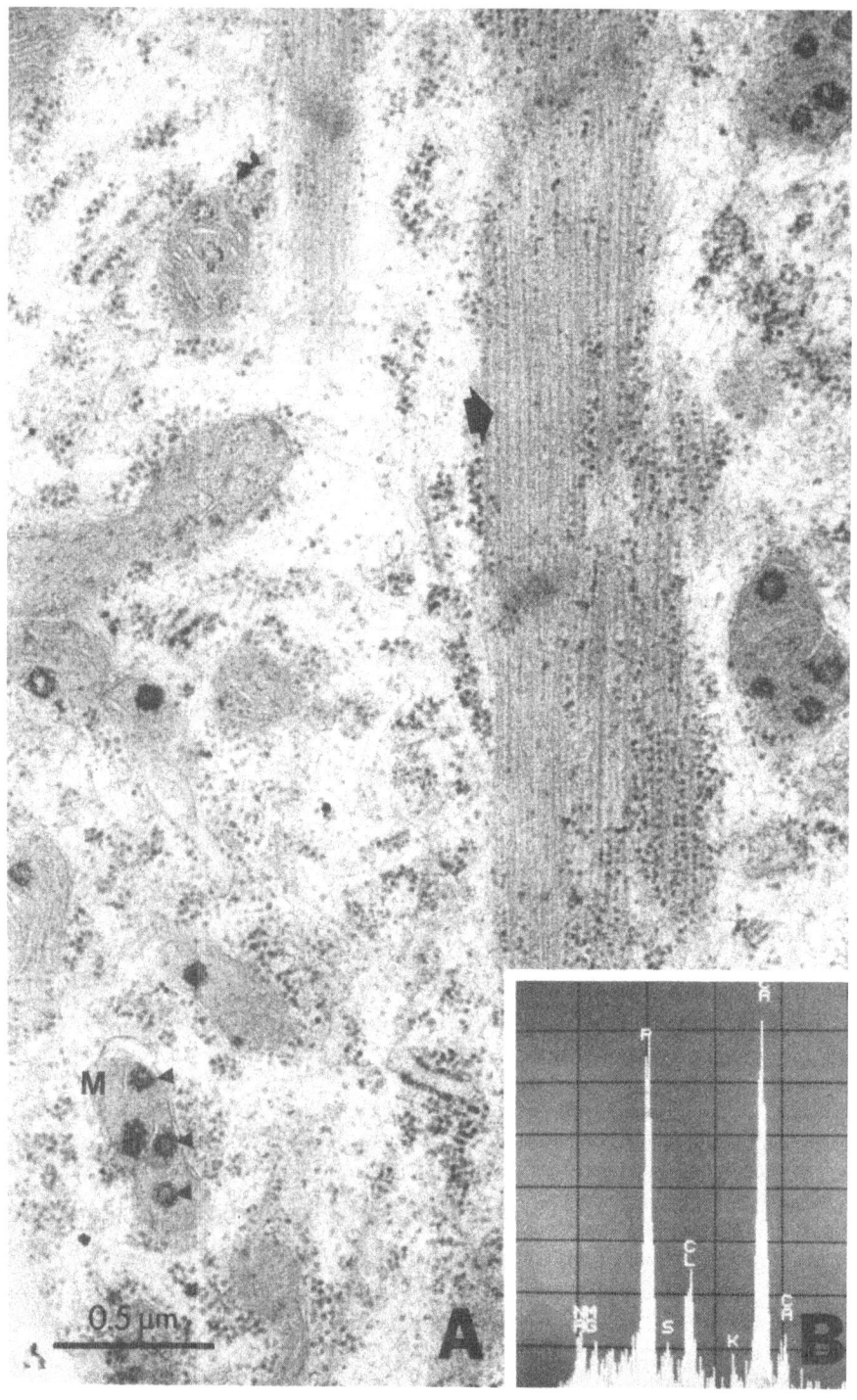

5. SUMMARY AND CONCLUSIONS

Metabolic inhibition by hypoxia, inhibition of oxidative phosphorylation, or inhibition of oxidative phosphorylation and glycolysis induces decreases in K_i and ATP; increases in Ca_t, Na_i, and LDH release; and morphological disruption in a time-dependent manner. Active transport (Na,K pump) inhibition generates similar changes in Na_i, Ca_t, and K_i but does not decrease ATP, increase LDH release, or cause morphological damage.

In these studies and others utilizing cultured heart cell preparations (Buja *et al.*, 1985; Chien *et al.*, 1985; Gunn *et al.*, 1985), the decline in ATP correlates with cell injury as indicated by LDH release and/or cell ultrastructure. However, some investigations have dissociated ATP from cell injury. For example, Kane *et al.* (1985) demonstrated that ATP depletion was not accompanied by loss of viability in either hepatocytes or macrophages. In intact myocardium Neely and Grotyohann (1984) described a major role for anaerobic glycolytic products (lactate, hydrogen ion, or NADH) in cardiac ischemic damage, which was unrelated to loss of adenine nucleotides, e.g., ATP. As our data suggest, cell injury may be due to interactive effects of decreased ATP with increased Ca_t. Increased Ca_t may activate calpain-like proteases which play a role in many irreversible Ca^+-dependent cellular processes (Kishimoto *et al.*, 1983); however, adequate ATP levels might allow for cell repair or recovery.

ACKNOWLEDGMENTS. The authors acknowledge the excellent technical assistance of Mr. Larry Hawkey, Ms. Marisa Menold, Ms. Kathleen Mitchell, and the skilled preparation of the manuscript by Ms. Gay Blackwell. The research was supported in part by NIH HL07101, HL17670, HL27105, and by DFG-Wa576/1-1 to B.W.

REFERENCES

Broderick, R., Wasserman, A. J., Fujimori, T., and Somlyo, A. P., 1986, Mitochondrial Ca^{2+} uptake during massive cellular Na^+ efflux and its reversibility in situ: An electron probe study, *J. Gen. Physiol.* **88:**13a.

Buja, L. M., Hagler, H. K., Parsons, D., Chien, K., Reynolds, R. C., and Willerson, J. T., 1985, Alterations of ultrastructure and elemental composition in cultured neonatal rat cardiac myocytes after metabolic inhibition with iodoacetic acid, *Lab. Invest.* **53:**397–412.

Chien, K. R., Sen, A., Reynolds, R., Chang, A., Kim, Y., Gunn, M. D., Buja, L. M., and Willerson, J. T., 1985, Release of arachidonate from membrane phospholipids in cultured neonatal rat myocardial cells during ATP depletion. Correlation with the progression of cell injury, *J. Clin. Invest.* **75:**1770–1780.

Doorey, A. J., and Barry, W. H., 1983, The effects of inhibition of oxidative phosphorylation and glycolysis on contractility and high energy phosphate content in cultured chick heart cells, *Circ. Res.* **53:**192–201.

Ebihara, L., Shigeto, N., Lieberman, M., and Johnson, E. A., 1980, The initial inward current in spherical clusters of chick embryonic heart cells, *J. Gen. Physiol.* **75:**437–456.

FIGURE 4. (A) TEM of cultured heart cells prepared as described in Fig. 2 after treatment for 1 hr with ouabain (10^{-4} M) in MEBSS. Mitochondria (M) contain large electron-dense inclusions (arrows); myofibrils (arrow), nucleus, and other structures are normal in appearance ($\times 43,050$). (B) EPXMA spectrum from mitochondrial-dense inclusion with large Ca peak visible. Cells for EPXMA were rapidly frozen, cryosectioned, freeze-dried, and analyzed as described in Section 2.

Gunn, M. D., Sen, A., Chang, A., Willerson, J. T., Buja, L. M., and Chien, K. R., 1985, Mechanisms of accumulation of arachidonic acid in cultured myocardial cells during ATP depletion, *Am. J. Physiol.* **249**:(*Heart Circ. Physiol.* **18**) H1188–H1194.

Hall, T. A., 1979, Biological x-ray microanalysis, *J. Microsc.* **117**:145–163.

Hasin, Y., and Barry, W. H., 1984, Myocardial metabolic inhibition and membrane potential, contraction, and potassium uptake, *Am. J. Physiol.* **247**:(*Heart Circ. Physiol.* **16**):H322–H329.

Jennings, R. B., and Reimer, K. A., 1981, Lethal myocardial ischemic injury, *Am. J. Pathol.* **102**: 241–255.

Kane, A. B., Petrovich, D. R., Stern, R. O., and Farber, J. L., 1985, ATP depletion and loss of cell integrity in anoxic hepatocytes and silica-treated P388D macrophages, *Am. J. Physiol.* **249** (*Cell Physiol.* **18**):C256–C266.

Kishimoto, A., Kajikawa, N., Shiota, M., and Nioshizuka, Y., 1983, Proteolytic activation of calcium-activated, phospholipid-dependent kinase by calcium dependent neutral protease, *J. Biol. Chem.* **258**:1156–1164.

Kitazawa, T., Shuman, H., and Somlyo, A. P., 1983, Quantitative electron probe analysis: Problems and solutions, *Ultramicroscopy* **11**:251–262.

LeFurgey, A., Hawkey, L. A., Lieberman, M., and Ingram, P., 1987, Na-Ca compartmentation in cultured heart cells, *Microbeam Analysis* **1987**:267–268.

LeFurgey, A., Liu, S., Lieberman, M., and Ingram, P., 1986, Quantitative elemental characterization of cultured heart cells by electron probe x-ray microanalysis and ion selective electrodes, *Microbeam Analysis* **1986**:205–208.

Lieberman, M., Horres, C. R., Jacob, R., Murphy, E., Piwnica-Worms, D., and Wheeler, D. M., 1984, Physiologic criteria for electrogenic transport in tissue-cultured heart cells, in: *Electrogenic Transport: Fundamental Principles and Physiological Implications* (M. P. Blaustein and M. Lieberman, eds.), Raven Press, New York, pp. 181–191.

Lieberman, M., LeFurgey, A., Murphy, E., and Liu, S., 1985, Cultured heart cells as a model for studying myocardial ischemia, in: *Pathology of Cardiovascular Injury* (H. S. Stone and W. B. Weglicki, eds.), Martinus Nijhoff, Boston, pp. 145–155.

Murphy, E., Aiton, J. F., Horres, C. R., and Lieberman, M., 1983, Calcium elevation in cultured heart cells: Its role in cell injury, *Am. J. Physiol.* **245**:(*Cell Physiology 14*)C316–C321.

Murphy, E., LeFurgey, A., and Lieberman, M., 1987, Biochemical and structural changes in cultured heart cells induced by metabolic inhibition, *Am. J. Physiol.* **253**:(*Cell Physiology 22*)C700–C706.

Murphy, E., LeFurgey, A., Horres, C. R., and Lieberman, M., 1982, Hypoxia and calcium in cultured heart cell injury, *Fed. Proc.* **41**:1273 (abstr.).

Murphy, E., Jacob, R., and Lieberman, M., 1985, Cytosolic free calcium in chick heart cells: Its role in cell injury, *J. Mol. Cell. Cardiol.* **17**:221–231.

Murphy, E., Wheeler, D. M., LeFurgey, A., Jacob, R., Lobaugh, L. A., and Lieberman, M., 1986, Coupled sodium-calcium transport in cultured chick heart cells, *Am. J. Physiol.* **250** (*Cell Physiol.* **19**):C442–C452.

Murphy, J. G., Marsh, J. D., and Smith, T. W., 1987, The role of calcium in ischemic injury, *Circulation* **75** (suppl. V):V15–V24.

Nayler, W. G., and Grinwald, P., 1981, Calcium entry blockers and myocardial function, *Fed. Proc.* **40**:2855–2861.

Nayler, W. G., 1981, The role of calcium in the ischemic myocardium, *Am. J. Pathol.* **102**:262–270.

Neely, J. R., and Grotyohann, L. W., 1984, Role of glycolytic products in damage to ischemic myocardium. Dissociation of adenosine triphosphate levels and recovery of function of reperfused ischemic hearts. *Circ. Res.* **55**:816–824.

Shuman, H., Somlyo, A. V., and Somlyo, A. P., 1976, Quantitative electron probe microanalysis of biological thin sections: Methods and validity, *Ultramicroscopy* **1**:317–339.

Trump, B. F., Mergner, W. J., Kahng, M. W., and Saladino, A. J., 1976, Studies on the subcellular pathophysiology of ischemia, *Circulation* **53**:Suppl. I17–I26.

Wagenknecht, B., LeFurgey, A., and Lieberman, M., 1987, Response of cultured embryonic chick heart cells to hypoxia, *Fed. Proc.* **46**(3):529.

60

Interaction of Volatile Anesthetics with Calcium-Sensitive Sites in the Myocardium

T. J. J. BLANCK and E. S. CASELLA

1. INTRODUCTION

There are a number of inhaled agents that can produce general anesthesia. They are known as volatile anesthetics (VA). Three of the most commonly used VA are halothane ($CF_3CHBrCl$), enflurane ($CFHClCF_2OCHF_2$), and isoflurane ($CF_3CHClOCHF_2$). They are effective and safe anesthetics, but their major side effect is a concentration-dependent depression of cardiac contractility (Brown and Crout, 1971). The depression of contractility, like anesthesia itself, is amazingly reversible but can have serious hemodynamic consequences in patients with cardiac disease, vascular abnormalities, and other organ system malfunctions. In these cases, the depression of cardiac contractility can result in dangerous metabolic imbalances.

There are two important reasons to study the effects of VA on cardiac contractility. The first is to determine their site(s) and mechanism(s) of action so that one can pharmacologically intervene to either reverse or prevent the cardiac depression and be able to predict the pharmacological interactions that might occur in patients who are receiving other cardioactive drugs. An awareness of site and mechanism of action of the anesthetic would allow rational therapy to be administered to those patients who are taking such drugs as β blockers, Ca^{2+} antagonists, antiarrhythmics, and cardiac glycosides. The second reason is to use this well-described biochemical system, i.e., the contractile process, as a model for anesthetic action in other excitable tissue.

The depression of cardiac contractility by VA is believed to be related to an interference with the availability or recognition of Ca^{2+} by Ca^{2+}-sensitive sites within the myocardial cell (Merin *et al.*, 1974). Several reports have noted alterations in sarcolemmal, sarcoplasmic reticulum (SR), and myofibrillary proteins in response to the VA (Ohnishi *et al.*, 1980; Casella *et al.*, 1987). These current data suggest that the VA have multiple sites of action within the myocardial cell.

T. J. J. BLANCK and E. S. CASELLA ● Division of Cardiac Anesthesia, Department of Anesthesiology and Critical Care Medicine, Johns Hopkins University, Baltimore, Maryland 21205.

Our investigations examined the interaction of the VA with subcellular elements of the myocardium, in particular the SR[2], in order to clarify the mechanisms and sites of action of the VA and their effect on Ca^{2+} sensitivity. In this chapter we will present data comparing the effects of the VA on Ca^{2+} transport from the SR of cardiac muscle and the effects of VA on binding of the Ca^{2+} antagonist nitrendipine to voltage-dependent Ca^{2+} channels in a cardiac membrane preparation.

2. EFFECT OF VOLATILE ANESTHETICS ON THE pH DEPENDENCE OF CALCIUM UPTAKE

Investigations of the effects of VA on the uptake of Ca^{2+} by the SR have resulted in reports of increased Ca^{2+} uptake (Blanck and Thompson, 1981, 1982), decreased Ca^{2+} uptake (Su and Kerrick, 1980; Blanck and Thompson, 1981), and no effect on Ca^{2+} uptake (Lain et al., 1968). These results are most likely due to variations in Ca^{2+} uptake assay conditions, including the method of isolating SR pH, incubation temperature, and concentration of ATP and Ca^{2+}. The uptake of Ca^{2+} by the SR has a strong pH dependence, with increased Ca^{2+} uptake at higher concentrations of H^+ and decreased uptake at lower H^+ concentrations (Nakamara and Schwartz, 1970; Fabiato and Fabiato, 1978; Dunnett and Nayler, 1979; Tate et al., 1980, 1981). Preliminary data suggested that different pH conditions may alter the response of the SR Ca^{2+} uptake function to the VA and may be responsible for the discrepant results. In order to describe this relationship fully, we examined the effect of the VA (halothane, enflurane, and isoflurane) on Ca^{2+} uptake by cardiac SR at pH 6.6–7.6.

We used rabbit cardiac SR, prepared by a modification of the method of Harigaya and Schwartz (1969), for the Ca^{2+} uptake experiments. The reaction medium was placed in 31-ml glass vials and consisted of 100 mM KCl, 5 mM $MgCl_2$, 5 mM sodium azide, 5 mM potassium oxalate, 5 mM ATP, 25 mM Hepes, (4-2-hydroxyethyl-1-piperazine sulfonic acid), 25 mM Pipes [piperazine-N,N'-bis-2-ethanesulfonic acid], 0.1 mM EGTA, and 0.1 mM $CaCl_2$ with $^{45}CaCl_2$ (150,000 cpm/sample). The temperature of the reaction mixture was brought to 37°C in a water bath, and then the reaction was started with the addition of 0.025–0.050 mg of SR, resulting in a total volume of 1 ml.

All experiments were performed with and without ATP to show that it was an ATP-dependent process. Those experiments carried out without ATP had results similar to those obtained measuring background, i.e., no uptake was observed.

Experiments were performed in the absence of anesthetic (control state) and in the presence of each VA (1.3% halothane, 1.8% enflurane, 1.2% isoflurane), separately. Volatile anesthetic concentrations were determined by infrared spectroscopy as previously described (Blanck, 1981). The anesthetics were added with a Hamilton microliter syringe at the same time the SR was added via a micropipette. Immediately after the VA and SR were added, the vials, sealed with teflon caps, were incubated at 37°C for either 2 or 20 min. The added anesthetic, equilibrated in the sealed 31-ml vial, resulted in a constant and reproducible anesthetic concentration in the reaction medium. Experiments were done in triplicate at each pH ranging from 6.6 to 7.6 in increments of 0.2 pH unit, for both incubation times. The pH, checked prior to and after each experiment, varied less than 0.04 pH unit during the course of the incubation. The incubation was terminated by filtration of an aliquot of the reaction medium into a glass fiber filter on a Millipore filtration apparatus. The filter was immediately rinsed with 15 ml of 10 mM $CaCl_2$ and counted in a scintillation vial with 5 ml of scintillation-counting cocktail.

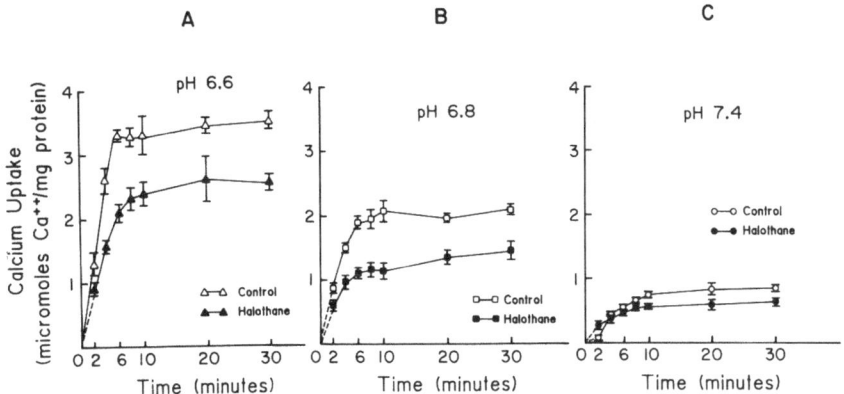

FIGURE 1. Time dependence of calcium uptake using $^{45}CaCl_2$ at (A) pH 6.6, (B) pH 6.8, and (C) pH 7.4, in the control state (open symbols) and in the presence of 1.3% halothane (blackened symbols). The initial uptake of calcium is linear up to 3 min and the reaction reaches a plateau at 20 min. Halothane decreases both the calcium uptake rate and plateau at pH 6.6–6.8. At pH 7.4, halothane increases the calcium uptake rate but decreases the plateau. The plateau represents the maximal uptake of calcium by the SR. Each data point is the mean of three experimental determinations. Error bars represent standard error of the mean.

The time dependence of Ca^{2+} uptake was examined at each pH using $^{45}CaCl_2$ as a tracer in order to determine the initial rate of Ca^{2+} uptake and the maximal uptake of Ca^{2+} by the SR. Figure 1 is an example of the time dependence of Ca^{2+} uptake at pH 6.6, 6.8, and 7.4. The uptake of Ca^{2+} is linear up to about 3 min. The reaction reaches a plateau at 20 min. The rate of Ca^{2+} uptake was approximated in subsequent experiments from a 2-min incubation, and maximal Ca^{2+} uptake (at a Ca^{2+} concentration of 10^{-6} M) from a 20-min incubation.

The computer program of Fabiato and Fabiato (1979) was used to estimate the free Ca^{2+} concentration at each pH. The ability of EGTA to buffer Ca^{2+} is known to vary with pH, with greater binding of Ca^{2+} by EGTA at more alkaline pH (Bartfai, 1979). A Ca^{2+} electrode (Madiera, 1975) was used to measure the differences in Ca^{2+} ion concentration at different pH levels. $CaCl_2$ was added to achieve a constant free-Ca^{2+} concentration at 10^{-6} M for each pH in the control state. Ca^{2+} uptake by cardiac SR, as measured by the calcium electrode at a constant initial free-Ca^{2+} concentration, was performed at pH 6.6–7.6 to verify that the effects observed with $^{45}CaCl_2$ measurements were not due to small changes in free Ca^{2+} in the reaction mixture. The pH dependence of Ca^{2+} uptake, determined by this method, was identical to that found using $^{45}CaCl_2$ (Fig. 2). In the $^{45}CaCl_2$ experiments, the free-Ca^{2+} concentration ranged from 0.8 to 8 μM.

2.1. Effect on Initial Rate of Ca^{2+}

The initial rate of Ca^{2+} uptake was examined from pH 6.6 to 7.6 by estimating the rate of uptake from a 2-min incubation. As can be seen in Fig. 3 and verified statistically by three-way analysis of variance, there is a highly significant ($p < 0.0001$) dependence of

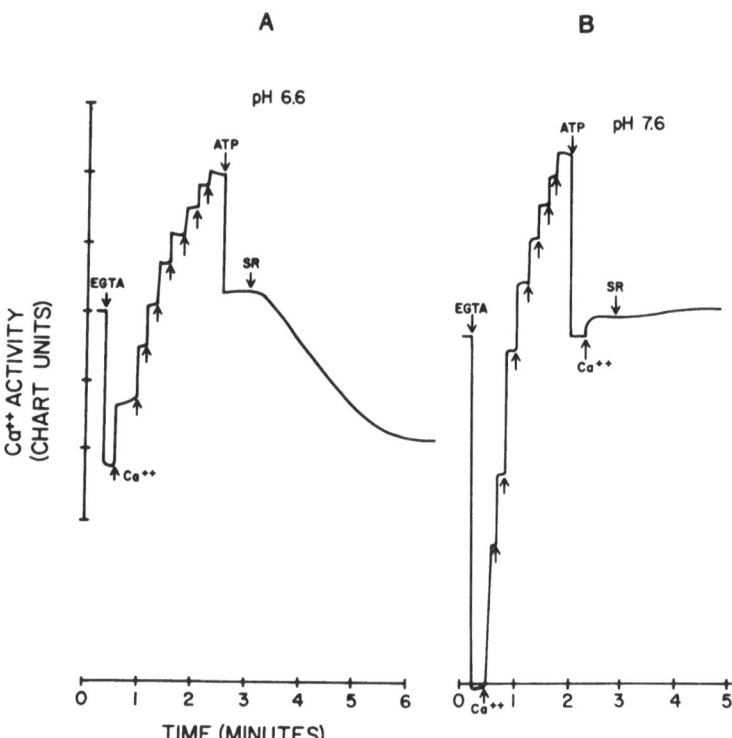

FIGURE 2. Representative drawing of the calcium electrode tracing which measured calcium uptake by cardiac SR at (A) pH 6.6 and (B) pH 7.6. Calcium uptake is measured in terms of chart units over time in minutes. Calcium is added to the reaction medium after the EGTA in increments of 50 nmol. ATP is then added (final concentration 5 mM) and is seen to bind some of the calcium, as would be expected. The SR is then added and calcium uptake is seen by the relative decrease in the amount in the reaction medium. At pH 7.6, additional calcium (100 nmol) was added after the ATP in order to keep the calcium ion concentration constant because EGTA binds more calcium at higher pH levels.

Ca^{2+} uptake on pH in the absence of VA (control state). However, there is a remarkable difference in the response of the SR to pH in the presence of VA when compared to control. All three VA significantly ($p < 0.0001$) stimulate the rate of uptake, relative to control, at pH 7.2–7.6. The exposure to VA results in a 52–78% increase in Ca^{2+} uptake by the SR in this pH range.

The response of the SR Ca^{2+} uptake function is also different from control at pH 6.6 and 6.8. Ca^{2+} uptake is inhibited by the VA at these pH levels relative to control. At pH 7.0, the VA have essentially no effect on the Ca^{2+} uptake function.

In summary, the effect of the VA on the rate of Ca^{2+} uptake by the SR is to alter the pH dependence, such that the effect of H^+ concentration on Ca^{2+} uptake in the control state is lost in the presence of VA.

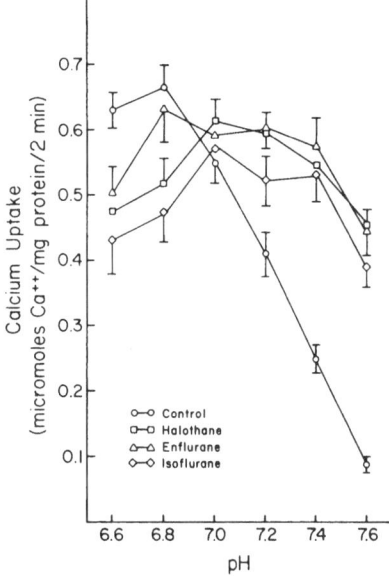

FIGURE 3. The pH dependence of the rate of calcium uptake (2-min incubation) in the control state (circles) and in the presence of the volatile anesthetics: 1.3% halothane (squares), 1.7% enflurane (triangles), and 1.2% isoflurane (diamonds). Calcium uptake was measured using $^{45}CaCl_2$. Each point is the mean of nine experimental determinations performed in duplicate ($n = 18$). Error bars represent standard error of the mean.

2.2. Effect on Maximal Ca^{2+} Uptake

Figure 1 shows that the Ca^{2+} uptake process reaches a plateau at approximately 20 min. We therefore estimated the maximal Ca^{2+} uptake by the SR from a 20-min incubation and examined the effect of the VA on maximal uptake. As seen in Fig. 4 and verified statistically by three-way analysis of variance, there is a highly significant ($p < 0.0001$) difference in the maximal amount of calcium uptake in control SR compared to the VA-treated SR. Despite the quantitative difference in the maximal amount of Ca^{2+} uptake between control and anesthetic-treated SR, both respond to pH in a similar fashion, i.e., maximal Ca^{2+} uptake decreases with increasing pH. Halothane- and isoflurane-treated SR responded identically to changes in pH. However, enflurane produced significantly less depression of Ca^{2+} uptake.

The results we obtained after the 2- and 20-min incubations can be discussed in light of previous findings by other investigators. The results for the initial rate of Ca^{2+} uptake (2-min incubation) by cardiac SR in the control state concur with previous findings which show that the rate of oxalate-supported, ATP-dependent Ca^{2+} uptake by cardiac SR of several species in the absence of VA is pH-dependent (Nakamaru and Schwartz, 1970; Fabiato and Fabiato, 1978; Dunnett and Nayler, 1979; Tate *et al.*, 1980, 1981; Blanck and Thompson, 1981). The highest initial Ca^{2+} uptake rate was reported at pH 6.2–6.8 and inhibition of uptake at pH 7.2–7.6.

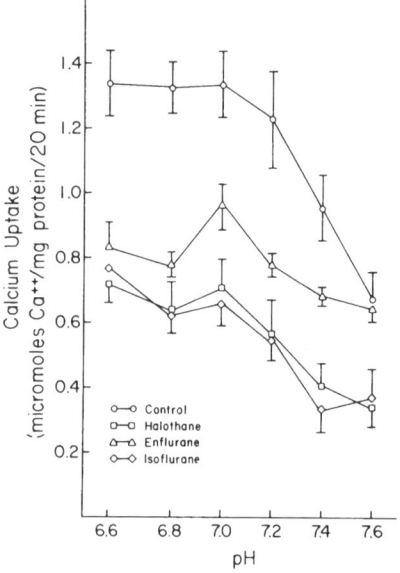

FIGURE 4. The pH dependence of the maximal uptake of calcium by the SR (20-min incubation) in the control state (circles) and in the presence of the volatile anesthetics: 1.3% halothane (squares), 1.7% enflurane (triangles), and 1.2% isoflurane (diamonds). Calcium uptake is measured using $^{45}CaCl_2$. Each point is the mean of nine experimental determinations performed in duplicate ($n = 18$). Error bars represent standard error of the mean.

It has previously been shown that preincubation of cardiac SR with Ca^{2+} prevents the inhibition of Ca^{2+} uptake at pH 7.2–7.6 (Tate *et al.*, 1981). Our experiments show that the VA can also prevent the inhibition of the initial rate of Ca^{2+} uptake in the same pH range. These data suggest that both the VA and Ca^{2+} preincubation change the H^+ dependence of the Ca^{2+} uptake process. It suggests that the anesthetics alter the intramembrane environment of the Ca^{2+} pump protein, either by altering the structure of the membrane or directly changing the conformation of the pump protein itself.

Our results concur with previous results measuring Ca^{2+} uptake by canine cardiac SR at 5 mM ATP and 37°C, which showed increased Ca^{2+} uptake at pH 7.3 in the presence of 1.0–1.13% halothane (Blanck and Thompson, 1981), as well as in the presence of 0.64–2.82% enflurane and 0.82–1.99% isoflurane (Blanck and Thompson, 1982). The findings of a decrease in the maximal uptake of Ca^{2+} by the SR in the presence of enflurane at pH 7.0 and halothane at pH 6.9 are also supported by previous studies, even though the concentration used in those studies (2.5–7.5% enflurane and 1.75% halothane) are higher than those used in this study (Su and Kerrick, 1980; Blanck and Thompson, 1981).

Using two different methods of measurement, the intramyocardial pH of the perfused rabbit heart was found to be 7.18 (Gonzalez and Clancy, 1975; Jacobus *et al.*, 1978; Flaherty *et al.*, 1982). At that pH, all three VA cause an increase in the velocity of Ca^{2+} uptake and a decrease in the maximal Ca^{2+} uptake when compared to control.

In comparing the effects of the VA on Ca^{2+} uptake by the SR, the results show that halothane and isoflurane effect the SR in an almost identical fashion. Enflurane, on the other hand, effects the SR somewhat differently in that it has less of a depressant effect on the maximal Ca^{2+} uptake by the SR at all H^+ concentrations studied. Enflurane also has less of a depressant effect on the rate of Ca^{2+} uptake by the SR, most notably at pH 6.8.

In summary, the VA decrease the rate of Ca^{2+} uptake by cardiac SR at pH 6.6–6.8, have little effect on pH 7.0, and significantly increase the rate of Ca^{2+} uptake at pH 7.2–7.6. Overall, the VA depress the maximal amount of Ca^{2+} uptake by the SR from pH 6.6 to 7.6 when compared to the control state.

3. EFFECT OF HALOTHANE ON THE BINDING OF NITRENDIPINE

The plateau phase of the action potential of ventricular muscle, which reflects the movement of Ca^{2+} through voltage-dependent channels of the sarcolemma, is depressed by halothane (Lynch, 1986). Recently, Bosnjak and Kampine (1986) demonstrated by aequorin luminescence that halothane decreases intracellular free Ca^{2+} in papillary muscle during contraction, presumably by inhibiting Ca^{2+} influx into the cell. The above evidence suggests that halothane might decrease the influx of Ca^{2+} by altering the function of the Ca^{2+} channels in the heart. In this study, we used the radioligand-binding technique to examine this possibility.

In order to determine whether or not halothane alters Ca^{2+} channel binding, we tested the effect of halothane on the binding of nitrendipine (NTP), a competitive voltage-dependent Ca^{2+} channel antagonist. Binding of [^{3}H]NTP was used as an index of the alteration of a site involved in Ca^{2+} translocation.

Two types of membrane preparations were used in this study. The first was a standard rat cardiac membrane preparation used by other investigators to characterize voltage-dependent Ca^{2+} channels (Gould et al., 1984). The other was the same SR preparation that we employed for the studies in Section 2. This SR preparation has been shown by Besch et al. (1976) to contain a significant amount of sarcolemma. We have not yet established the relative proportion of SR to sarcolemma, but it has been shown that not only the sarcolemma but also the junctional SR contain specific, high-affinity Ca^{2+} channel antagonist-binding sites (William and Jones, 1983; Sarmeinto et al., 1983; Brandt, 1985). Due to the impurity of our membrane preparations, we cannot unequivocally say which subcellular binding site for [^{3}H]NTP binding is altered by halothane.

Nitrendipine-binding experiments were performed in the same sealed system as used for the SR Ca^{2+} uptake studies in Section 2. The reaction medium contained 5 nM [^{3}H]NTP (70 μCi/ml) in 50 mM tris-HCl (pH 7.7) and 0.150 mg of cardiac membranes. Incubations were performed in the dark. The rat membranes were incubated in the presence or absence of 1.0 μM nifedipine and the rabbit membranes in the presence or absence of 0.5 μM nitrendipine to define nonspecific binding. The different competing agents, nifedipine and nitrendipine, were chosen because of availability. Both nitrendipine and nifedipine readily displace [^{3}H]NTP from binding sites in the heart (Ehlert et al., 1982). Nonspecific binding is defined as that binding of [^{3}H]NTP which occurs in the presence of a high concentration of unlabeled ligand. Specific binding, which indicates binding to the voltage-dependent Ca^{2+} channel, is the difference between total binding and nonspecific binding. Following incubation, the reaction mixture was added to Whatman GF/C filters, rinsed, and then counted in a liquid scintillation counter.

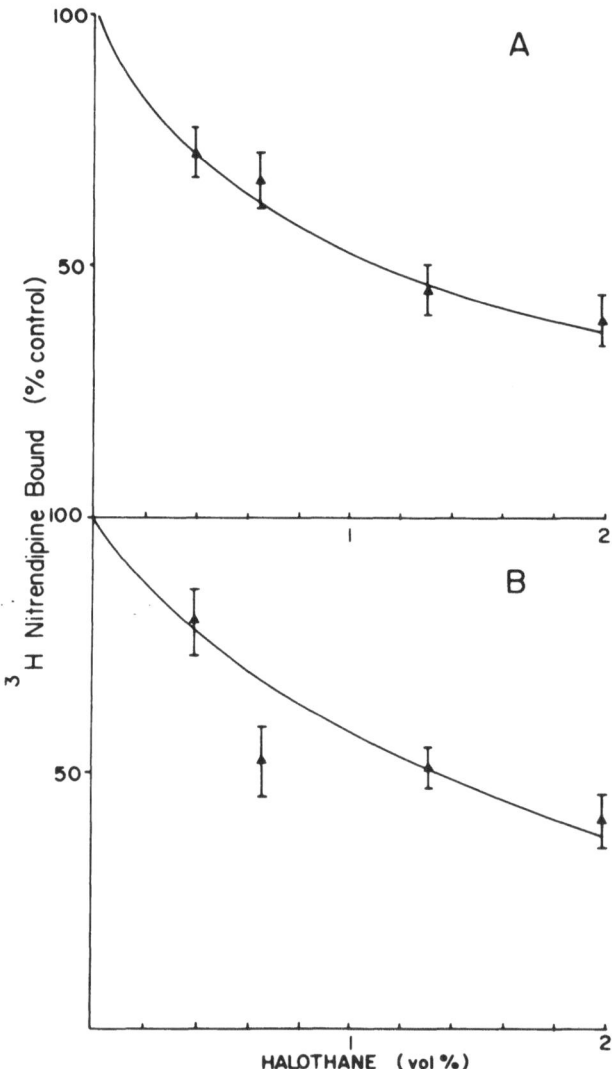

FIGURE 5. Amount of specifically bound [³H]NTP as a function of halothane concentration. Incubations were performed at 25°C for 90 min. The incubation medium included 5 nM [³H]NTP in 5 mM Tris-HCl at pH 7.7. Data are expressed as percent of control (no halothane). (A) Rat cardiac membranes; each point is the mean of 19 experimental observations ± the standard error of the mean. (B) Rabbit cardiac membranes; each point is the mean of nine experimental observations ± standard error of mean.

The dependence of Ca^{2+} channel-specific [^{3}H]NTP binding on vapor phase halothane concentration is shown in Fig. 5A for the rat and 5B for the rabbit membranes. The data are presented as percent control, i.e., the ratio of [^{3}H]NTP binding decreases with increasing halothane concentration. The specific binding data were analyzed by a two-way analysis of variance, which demonstrated a statistically significant ($p < 0.001$) inverse relationship between specific [^{3}H]NTP binding and halothane concentration for both rat and rabbit cardiac membranes.

At this time, little direct evidence is available to localize the negative inotropic effect of halothane to one specific subcellular site. Lynch (1986) demonstrated that the plateau phase of isoproterenol-stimulated slow-action potentials is decreased by halothane; this observation suggests that the movement of Ca^{2+} through the voltage-dependent Ca^{2+} channels is decreased by halothane. The voltage-dependent Ca^{2+} channels involved in the slow-action potential are located in the sarcolemma and are one of the sites at which [^{3}H]NTP specifically binds. Bosnjak and Kampine (1986) showed, by the use of Ca^{2+}-sensitive aequorin luminescence, that intracellular Ca^{2+} is decreased by halothane during papillary muscle contraction. They could not unequivocally attribute the decrease in intracellular Ca^{2+} to either decreased influx of Ca^{2+} through the sarcolemma Ca^{2+} channels or decreased release of Ca^{2+} from the SR. They noted that verapamil, a Ca^{2+} antagonist which binds to the voltage-dependent Ca^{2+} channels, caused a marked decrease in intracellular Ca^{2+}. These observations, together with our observation that the specific binding of [^{3}H]NTP to voltage-dependent Ca^{2+} channels is decreased by halothane, suggest that the sarcolemma is one of the important negative inotropic sites of action of halothane.

It has long been debated whether the VA act by altering membrane structure due to their high lipid solubility or by specific actions on certain protein targets in the cell. Our data could be interpreted by invoking either mechanism. The decrease in [^{3}H]NTP binding to membrane Ca^{2+} channel antagonist-binding sites by halothane has implications not only for the mechanism of depression of cardiac contractility but also for the mechanism of general anesthesia in the central nervous system.

4. SUMMARY

We have demonstrated that (1) halothane, enflurane, and isoflurane alter the pH dependence of Ca^{2+} uptake by the SR and decrease the total amount of Ca^{2+} accumulated by the SR; (2) halothane decreases the binding of the voltage-dependent Ca^{2+} channel antagonist nitrendipine. These data suggest that the halothane-induced decrease in intracellular Ca^{2+} noted by Bosnjak and Kampine (1986) can probably be attributed to effects at the SR, with an increased rate of Ca^{2+} at pH 7.2, as well as at the sarcolemma, which could result in a decrease in maximal Ca^{2+} uptake by the SR due to less Ca^{2+} entering the sarcolemma.

One can hypothesize at this time that halothane alters sarcolemmal voltage-dependent Ca^{2+} channels and all three VA cause a subsequent decrease in the amount of Ca^{2+} available in the SR for release for contraction. These biochemical experiments yield information about the possible mechanism of action by which the VA effect Ca^{2+} transport. Further experiments, in more physiological preparations, will be needed to verify these hypotheses.

REFERENCES

Bartfai, T., 1979, Preparation of metal-chelate complexes and the design of steady-state kinetic experiments involving metal nucleotide complexes, in: *Advances in Cyclic Nucleotide Research,* Vol. 10 (G. Brooker, P. Greengard, and G. A. Robison, eds.), Raven Press, New York, pp. 219–242.

Besch, H. R., Jones, L. R., and Watanabe, A. M., 1976, Intact vesicles of canine cardiac sarcolemma: Evidence from vectorial properties of Na^+K^+-ATPase, *Circ. Res.* **39**:586–595.

Blanck, T. J. J., 1981, A simple closed system for performing biochemical experiments at clinical concentration of volatile anesthetics, *Anesth. Analg.* **60**:435–332.

Blanck, T. J. J., and Thompson, M., 1981, Calcium transport by cardiac sarcoplasmic reticulum: modulation of halothane action by substrate concentration and pH, *Anesth. Analg.* **60**:390–394.

Blanck, T. J. J., and Thompson, M., 1982, Enflurane and isoflurane stimulate calcium transport by cardiac sarcoplasmic reticulum, *Anesth. Analg.* **61**:142–145.

Bosnjak, Z. J., and Kampine, J. P., 1986, Effects of halothane on transmembrane potentials, Ca^{2+} transients, and papillary muscle tension in the cat, *Am. J. Physiol.* **251** (*Heart Circ. Physiol.* **20**): H374–H381.

Brandt, N., 1985, Identification of two populations of cardiac microsomes with nitrendipine receptors: Correlation of the distribution of dihydropyridine receptors with organelle specific markers, *Arch. Biochem. Biophys.* **242**:306–319.

Brown, B. R., and Crout, J. R., 1971, A comparative study of the effects of five general anesthetics on myocardial contractility, *Anesthesiology* **34**:236–245.

Casella, E. S., Suite, N. D. A., Fisher, Y. I., and Blanck, T. J. J., 1987, The effect of volatile anesthetics on the pH dependence of calcium uptake by cardiac sarcoplasmic reticulum, *Anesthesiology* **67**:98–102.

Dunnett, J., and Nayler, W. G., 1979, Effect of pH on calcium accumulation and release by isolated fragments of cardiac and skeletal muscle sarcoplasmic reticulum, *Arch. Biochem. Biophys.* **198**: 434–438.

Ehlert, F. J., and Itoga, E., 1982, The interaction of [^{3}H] nitrendipine with receptors for calcium antagonists in the cerebral cortex and heart of rats, *Biochem. Biophys. Res. Commun.* **104**(3): 937–943.

Fabiato, A., and Fabiato, F., 1978, Effects of pH on the myofilaments and the sarcoplasmic reticulum of skinned cells from cardiac and skeletal muscles, *J. Physiol.* (Lond.) **276**:233–253.

Fabiato, A., and Fabiato, F., 1979, Calculator programs for computing the composition of the solutions containing multiple metals and ligands used for experiments in skinned muscle cells, *J. Physiol.* (Paris) **75**:463–505.

Flaherty, J. T., Weisfeldt, M. L., Bulkley, B. H., Gardner, T. J., Gott, V. L., and Jacobus, W. E., 1982, Mechanisms of ischemic myocardial cell damage assessed by phosphorus-31 nuclear magnetic resonance, *Circulation* **65**:561–571.

Gonzalez, N. C., and Clancy, R. L., 1975, Inotropic and intracellular acid-base changes during metabolic acidosis, *Am. J. Physiol.* **228**:1060–1064.

Gould, R. J., Murphy, K. M. M., and Snyder, S. H., 1984, Tissue heterogeneity of calcium channel antagonist binding sites labeled by [^{3}H]nitrendipine, *Mol. Pharmacol.* **25**:235–244.

Harigaya, S., and Schwartz, A., 1969, Rate of calcium binding and uptake in normal animal and failing human cardiac muscle, *Circ. Res.* **25**:781–794.

Jacobus, W. E., Pores, I. H., Taylor, G. E., Nunnally, R. L., Hollis, D. P., and Weisfeldt, M. L., 1978, Tight coupling of intracellular pH and ventricular performance, *J. Mol. Cell. Cardiol.* **10**: 39.

Lain, R. F., Hess, M. L., Gertz, E. W., Briggs, F. N., 1968, Calcium uptake activity of canine myocardial sarcoplasmic reticulum in the presence of anesthetic agents, *Circ. Res.* **23**:597–604.

Lynch, C., 1986, Differential depression of myocardial contractility by halothane and isoflurane in vitro, *Anesthesiology* **64**:620–631.

Madiera, V. M. C., 1975, A rapid and ultrasensitive method to measure Ca^{++} movements across biological membranes, *Biochem. Biophys. Res. Commun.* **64**:870–876.

Merin, R. G., Kumasawa, T., and Honig, C. R., 1974, Reversible interaction between halothane and calcium on cardiac actomyosin ATPase: Mechanisms and significance, *J. Pharmacol. Exp. Ther.* **190:**1–14.

Nakamaru, Y., and Schwartz, A., 1970, Possible control of intracellular calcium metabolism by [H$^+$]: sarcoplasmic reticulum of skeletal and cardiac muscle, *Biochem. Biophys. Res. Commun.* **41:** 830–836.

Ohnishi, S. T., DiCamillo, C. A., Singer, M., and Price, H. L., 1980, Correlation between halothane-induced myocardial depression and decreases in La^{3+}-displaceable calcium in cardiac muscle cells, *J. Cardiovasc. Pharmacol.* **2:**67–75.

Sarmeinto, J., Janis, R. A., Colvin, R. A., Triggle, D. J., and Katz, A. M., 1983, Binding of the calcium channel blocker nitrendipine to its receptor in purified sarcolemma from canine cardiac ventricle, *J. Mol. Cell. Cardiol.* **15:**135–137.

Su, J. Y., and Kerrick, G. L., 1980, Effects of enflurane on functionally skinned myocardial fibers from rabbits, *Anesthesiology* **52:**385–389.

Tate, C. A., Van Winkle, W. B., and Entman, M. L., 1980, Time-dependent resistance to alkaline pH of oxalate-supported calcium uptake by sarcoplasmic reticulum, *Life Sci.* **27:**1453–1464.

Tate, C. A., Chu, A., McMillin-Wood, J., Van Winkle, W. B., and Entman, M. L., 1981, Evidence for a calcium-sensitive factor which alters the alkaline pH sensitivity of sarcoplasmic reticulum calcium transport, *J. Biol. Chem.* **256:**2934–2939.

Williams, L. T., and Jones, L. R., 1983, Specific binding of the calcium antagonist [^{3}H]Nitrendipine to subcellular fractions isolated from canine myocardium, *J. Biol. Chem.* **258:**5344–5347.

61

Cardiac Protection by Halothane Following Ischemia and Calcium Paradox

ZELJKO J. BOSNJAK, SUMIO HOKA,
LAWRENCE TURNER, and JOHN P. KAMPINE

1. INTRODUCTION

Calcium ions play a crucial role in the regulation of cardiac function and their movements may be profoundly affected in various pathophysiological conditions. Specifically, reperfusion of the severely ischemic myocardium is associated with a rapid net gain in intracellular calcium, largely in the mitochondria, and the development of contraction band necrosis (Shen and Jennings, 1972, Nayler *et al.*, 1985). The extent of calcium accumulation on reperfusion of the ischemic heart is related to the severity and duration of ischemia and the degree of mechanical recovery (Henry *et al.*, 1977; Bourdillon and Poole-Wilson, 1981). A marked calcium accumulation is also produced by exposure to calcium-containing solutions following a brief period of exposure to calcium-free perfusate, a phenomenon known as the "calcium paradox" (Ruigrok *et al.*, 1985). While the mechanisms underlying myocardial injury under these conditions are controversial, several interventions limiting depletion of high-energy phosphate stores and calcium accumulation have been shown to have a protective effect on the heart (Bourdillon and Poole-Wilson, 1982). Pretreatment with calcium-blocking agents before calcium-free perfusion or the onset of ischemia is reported to reduce the degree of ultrastructural damage and enhance recovery of contractile function (Watts *et al.*, 1980; Ohhara *et al.*, 1982). This potential beneficial intervention may have relevance to the clinical problems of reducing cardiac injury and restoring ventricular function following transient ischemic episodes and the use of cardioplegic solutions in patients undergoing myocardial revascularization.

Although not chemically related to other drugs which alter myocardial calcium fluxes, the negative inotropic actions of the volatile anesthetic halothane are often used to advantage in limiting myocardial oxygen demand and the occurrence of ischemic episodes in patients undergoing myocardial revascularization (Hug, 1979). Lynch *et al.* (1981) initially dem-

ZELJKO J. BOSNJAK, SUMIO HOKA, LAWRENCE TURNER, and JOHN P. KAMPINE • Department of Anesthesiology, Medical College of Wisconsin and USVA Medical Center, Milwaukee, Wisconsin 53295.

onstrated that halothane depressed the slow-action potential responses, and by implication inward calcium ion fluxes, of potassium-depolarized ventricular myocardial fibers. Studies in this laboratory have shown that the direct negative inotropic action of halothane is associated with inhibition of the intracellular calcium ion transient (Bosnjak and Kampine, 1986) as measured with the calcium-sensitive photoprotein aequorin. Figure 1 illustrates the effect of halothane on the transmembrane potential, contractile force, and intracellular Ca^{2+} transient in a papillary muscle preparation from cat. Halothane produced a dose-dependent decrease in the calcium transients and contractile force with little change in the action potential configuration. Recently, Ikemoto et al. (1985), utilizing voltage-clamping techniques, demonstrated that clinically relevant concentrations of halothane inhibit the slow inward current of single isolated ventricular myocytes. In vivo halothane is reported to reduce infarct size following experimental coronary artery ligation (Davis et al., 1983) and to decrease the incidence of fibrillation in a canine acute occlusion/reperfusion model (Kroll and Knight, 1984). Studies in our laboratory are continuing to test the hypothesis that pretreatment with halothane may inhibit myocardial calcium accumulation associated with postischemic reperfusion and the calcium paradox.

2. CALCIUM UPTAKE

Spontaneously beating guinea pig hearts were perfused retrograde through the aorta at constant pressure (40 mm Hg) using a nonworking Langendorff preparation perfused with modified Krebs solution containing 2.5 mmol calcium (Hoka et al., 1987).

Calcium accumulation into the myocardium was determined by measuring tissue ^{45}Ca (New England Nuclear) which was loaded for 40 or 60 min followed by a 20-min washout period. The loading solution included 0.5 μCi/ml of ^{45}Ca. At the end of each experiment,

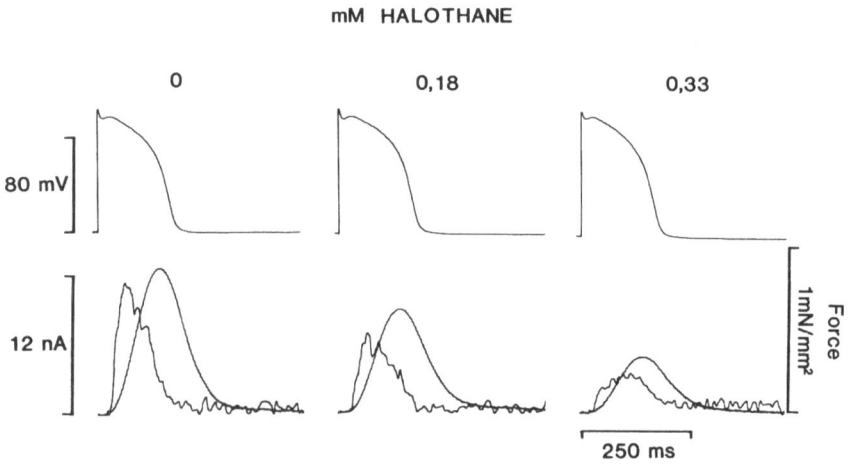

FIGURE 1. Influence of halothane (0.18 and 0.33 mM) on transmembrane potentials, Ca^{2+} transients, and tension development of cat papillary muscle. Stimuli were delivered at 1-sec intervals. Aequorin light intensity expressed as averaged photomultiplier current. (Modified from Bosnjak and Kampine, 1986.)

the hearts were dismounted and quickly dipped into an inactive Krebs solution to wash away superficially adherent radioactivity. Tissue samples (20–80 mg each) were removed, gently blotted, and weighed. Following overnight solubilization in NCS solution (tissue solubilizer, Amersham Corp.) at 55°C, tissue ^{45}Ca was determined using a liquid scintillation spectrometer (Packard Tri-Carb). The^{45}Ca content of the tissue was calculated as cpm/100 mg wet weight. The relative ^{45}Ca content of an individual preparation was expressed as tissue [^{45}Ca] divided by the loading solution [^{45}Ca] × 100.

Halothane (1%) was vaporized in a gas mixture (97% O_2/3% CO_2) using a Fluotec Mark III vaporizer. A dial setting of 1% produced a concentration of 0.40 mmol in the Krebs solution as measured using a gas chromatograph (Perkin Elmer) with a flame ionization detector.

The effects of 1% halothane alone on the^{45}Ca uptake was tested and the data are summarized in Fig. 2. There was no significant difference in the calcium uptake as compared to the control hearts. Therefore, halothane alone did not influence the overall calcium uptake over time in hearts which performed no external work, although its influence on a per beat basis could not be determined.

3. MYOCARDIAL ISCHEMIA

Regional myocardial ischemia was produced by occlusion of the left anterior descending (LAD) coronary artery at a point 2–4 mm from its origin for 30 min using a modification of the technique developed by Selye $et\ al.$ (1960). The validity of this technique for temporary coronary occlusion and reperfusion was affirmed by adding a small amount of methylene blue in the preliminary experiments. ^{45}Ca was added to the perfusate 10 min before LAD

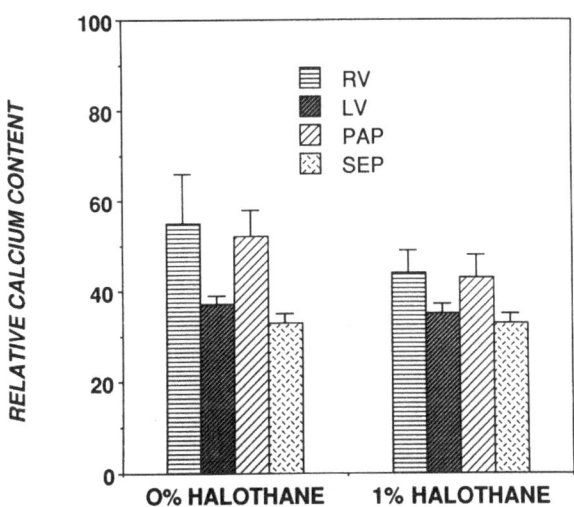

FIGURE 2. Relative calcium content in different parts of the heart following 40 min ^{45}Ca loading, without and in the presence of 1% halothane. RV (right ventricle), LV (left ventricle), PAP (papillary muscle), and SEP (septum). No significant difference was observed in the ^{45}Ca uptake. N = 6. Values are mean ± SE.

occlusion and was continued through the reperfusion period of 20 min for a total loading time of 60 min. Halothane, when present, was added 10 min before loading with ^{45}Ca and was present throughout reperfusion. The perfusate was then changed to ^{45}Ca and halothane-free Krebs for a 20-min washout period. Two tissue samples were removed from the anterior wall of the left ventricle, representing the ischemic region, and one from the posterior wall of the left ventricle, representing a nonischemic region.

Baseline heart rates were 177 ± 5 bpm ($N = 9$) in the control group and 174 ± 4 bpm ($N = 8$) in the halothane group. Halothane alone caused a significant decrease in the heart rate (to 131 ± 7 bpm) relative to the baseline value and on comparison to the preocclusion rate (170 ± 5 bpm) in the control group. The rates observed during LAD occlusion and after release in the presence of halothane were also less ($P < 0.05$) than the rates obtained at the same times in the control group. The baseline coronary flows were 1.78 ± 0.12 and 1.60 ± 0.04 ml/min/g tissue in the control and halothane groups, respectively. Halothane alone significantly increased coronary flow to 2.12 ± 0.14 ml/min/g tissue relative to the baseline value. During occlusion, coronary flow decreased in both groups ($P < 0.05$) while flow was greater ($P < 0.05$) in the halothane group (1.60 ± 0.08 ml/min/g) than the control group (1.36 ± 0.06 ml/min/g) just before release of occlusion and during the reperfusion period. Heart rate and coronary flow at the end of the washout period were not significantly different between the two groups. Figure 3 shows the relative tissue calcium contents of the nonischemic zone and the two ischemic zones in the absence and presence of halothane. In the control group, ischemia increased the relative calcium content to an average of 202% ($P < 0.05$) of the calcium content in the nonischemic region. The average percent increase in the relative calcium content in the ischemic zones (143%) was significantly less ($P < 0.05$) in the presence of halothane compared to that in the control group. There was no significant difference between the relative calcium contents of the nonischemic regions between the

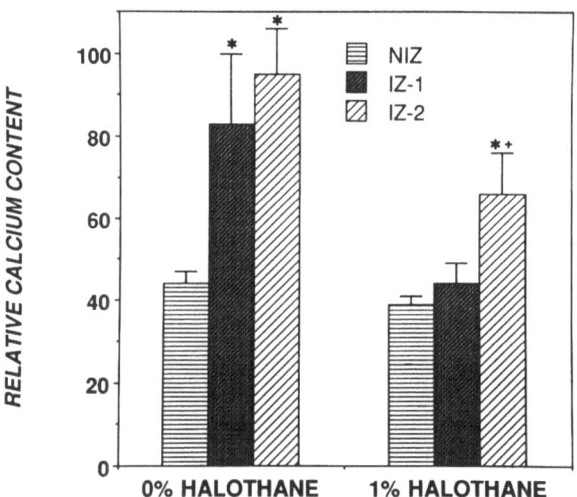

FIGURE 3. Relative calcium content from a nonischemic zone (NIZ, $N = 9$) and two ischemic zones (IZ-1 and IZ-2, $N = 8$) under control conditions (0% halothane) and in the presence of 1% halothane. *$P < 0.05$ versus NIZ. +$P < 0.05$ versus 0% halothane IZ-2. Values are mean ± SE. (Modified from Hoka et al., 1987.)

control and halothane groups. The results indicate that halothane inhibits calcium accumulation associated with regional myocardial ischemia.

4. CALCIUM PARADOX

The hearts were subjected to 10 min of calcium-free perfusion followed by calcium repletion (Hoka *et al.*, 1987). Loading of ^{45}Ca was started with calcium repletion for 40 min followed by a 20-min washout period. Halothane was added 10 min before the calcium-free perfusion and throughout ^{45}Ca loading. Tissue samples were obtained from right ventricle, left ventricle, and left ventricular papillary muscle.

Three groups were studied including a control group not subjected to calcium paradox (control), a calcium paradox group (CP), and a calcium paradox group in the presence of 1% halothane (CP + Hal). The baseline heart rates which ranged from 172 to 179 bpm and coronary flows which ranged from 1.66 to 1.76 ml/min/g tissue did not differ between the three groups. In the group pretreated with halothane, the heart rate (140 ± 8 bpm) was lower and the coronary flow (2.60 ± 0.26 ml/min/g) was higher than in the control group just before calcium-free perfusion. During calcium-free perfusion, the hearts ceased beating, and on repletion of calcium the heart rates gradually increased. Calcium-free perfusion tended to increase coronary flow in the CP group but did not alter the relatively high coronary flow already present in the CP + Hal group. Reperfusion with normal calcium medium caused a significant decrease ($P < 0.01$) in coronary flow in both the CP and CP + Hal groups (1.00 ± 0.34 and 1.58 ± 0.30 ml/min/g, respectively) while the coronary flow during repletion in the CP + Hal group was greater than in the CP group without halothane. Figure 4 shows the relative calcium content in all three groups. The relative calcium contents in the CP and CP + Hal groups were significantly greater ($P < 0.05$) than in the control group. In addition, the relative calcium content in the CP + Hal group was significantly less ($P < 0.05$) for the RV and LV myocardium (186 and 212% of the control, respectively) than that found in the CP group (RV = 635%, LV = 526%).

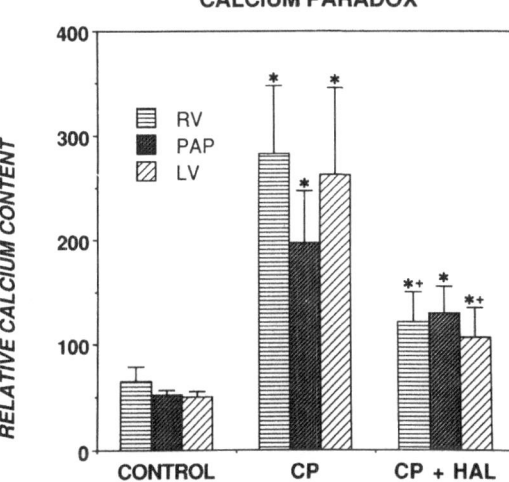

FIGURE 4. Relative calcium content in the control hearts (N = 5), calcium paradox experiments (CP, N = 8), and calcium paradox in the presence of 1% halothane (CP + HAL, N = 6). RV (right ventricle), PAP (papillary muscle), and LV (left ventricle). Halothane depressed the increase in calcium content caused by CP. *$P < 0.05$ versus control. +$P < 0.05$ versus CP. Values are mean ± SE. (Modified from Hoka *et al.*, 1987.)

5. CALCIUM EFFLUX

Calcium efflux was assessed by measuring the washout of ^{45}Ca after a loading period of 60 min in five guinea pig papillary muscle preparations. The papillary muscle was superfused at a constant flow rate of 10 ml/min with Krebs solution as described previously (Bosnjak and Kampine, 1986). The muscle was stimulated using a bipolar electrode at 0.5-sec intervals. The effluents were collected in 1-min periods in a fraction collector during the 60 min after initiating washout. Each outflow fraction was weighed and radioactivity per 1 ml/min measured in liquid scintillation counter (0.5 ml of superfusate was added to 10 ml of Safty-Solve by RPI), and an efflux curve of the ^{45}Ca was made. The half-time of the final phase of the efflux curve was approximately 50 min. Efflux of ^{45}Ca was not significantly altered by halothane.

6. SUMMARY

The present study was designed to examine the effects of halothane on the abnormal calcium ion movements associated with regional myocardial ischemia and the calcium paradox in the isolated perfused guinea pig heart. This is potentially important since alterations of calcium ion homeostasis play an important role in the cascade of events associated with irreversible myocardial injury in response to ischemia. Halothane was studied since it is often utilized during open-heart surgery and its direct negative inotropic actions are in part related to inhibition of calcium ion movements. The results in this model indicate that halothane reduces calcium accumulation associated with myocardial ischemia and the calcium paradox, and suggest that halothane may have a protective effect on myocardial injury under certain experimental conditions.

The actions of halothane were assessed in a nonworking normothermic Langendorff preparation of guinea pig heart at a constant coronary perfusion pressure of 40 mm Hg. A perfusion pressure of 40 mm Hg was selected because it has been shown that the severity of myocardial damage in the calcium paradox increases with the coronary perfusion pressure and flow rate (Koomen et al., 1980). Ohhara et al. (1982) showed that pretreatment with verapamil had a protective effect only on mild injury occurring at a perfusion pressure of 30 mm Hg but not on severe injury associated with calcium-free perfusion at a perfusion pressure of 60 mm Hg. The intermediate coronary perfusion pressure of 40 mm Hg was selected to produce a moderate degree of myocardial injury instead of a complete and unprotectable damage. Therefore, the observed protective effects of halothane may not apply under conditions of higher perfusion pressure. Calcium ion movements were assessed by equilibrating the tissue with ^{45}Ca perfusate for either 40 min (calcium paradox experiments) or 60 min (regional ischemia experiments). Equilibration appeared to be complete after 40 min since the control preparations did not exhibit further increase in ^{45}Ca content after 60 min of exposure. Therefore, changes in the relative tissue ^{45}Ca content associated with the experimental interventions should reflect changes in the total calcium content. Following equilibration with isotope, a single measurement of tissue ^{45}Ca was obtained after 20 min of washout with tracer-free 2.5-mmol Ca^{2+} perfusate with the assumption that this part of the calcium efflux originates primarily from an intracellular compartment. This assumption is supported by the kinetic studies of Pytkowski et al. (1983), Lewartowski et al. (1984), and Shine et al. (1971), who demonstrated that Ca^{2+} originating from the interstitial space and rapidly exchangeable intracellular sources is nearly eliminated after 20 min of washout, leaving a more permanently bound intracellular calcium pool. Therefore, it is probable that the relative calcium content measured represents a slowly exchangeable calcium compartment originating from the intracellular space. It has been shown that calcium accumulates in the mitochondria after myocardial ischemia (Henry *et al.*, 1977), on reoxygenation after hypoxia (Nakanishi *et al.*, 1982), and as the result of calcium paradox (Ruigrok, 1985).

The decrease in calcium accumulation in the presence of halothane on reperfusion of the ischemic myocardium could be due to an inhibition of calcium influx or enhancement of calcium efflux. In the ischemia experiments, total coronary flow was higher in hearts pretreated with halothane than control hearts during both the period of LAD occlusion and the reperfusion phase. It might be hypothesized that this greater coronary flow may increase calcium efflux. However, it seems unlikely that electrical and ionic gradients for calcium, even after the injury, would favor substantial calcium efflux, which is mostly an energy-dependent process. We also found that halothane did not change the rate of calcium efflux from the nonischemic papillary muscle preparation under constant-flow conditions. It has been reported that calcium efflux is not increased following ischemia and the calcium accumulation on reperfusion is due to enhanced calcium influx (Bourdillon and Poole-Wilson, 1981). Therefore, it seems unlikely that halothane enhances calcium efflux during the reperfusion phase at the time of net calcium accumulation.

On the other hand, the decrease in calcium accumulation by halothane may be due to the inhibition of calcium influx. It has been shown that halothane can inhibit slow-channel calcium influx (Lynch *et al.*, 1981; Ikemoto *et al.*, 1985; Bosnjak and Kampine, 1983, 1986). This may contribute in part to the depression of calcium accumulation, since the slow calcium channel may be one of several potential pathways for calcium accumulation in myocardial injury (Henry *et al.*, 1977; Lefer *et al.*, 1979). Despite the actions of halothane which could inhibit slow calcium inward current, calcium accumulation in nonischemic tissue was not different between halothane and nonhalothane groups (Fig. 3), indicating that total calcium uptake over time is not altered by halothane in a normal beating heart. This finding is consistent with the results of Porsius and van Zwieten (1975), who found that halothane changed the rate of calcium uptake but did not alter total calcium uptake in guinea pig atria. Verapamil has also been found to have no influence on net calcium influx or efflux in nonischemic myocardium by similar exchange studies (Bourdillon and Poole-Wilson, 1982).

Changes in coronary flow produced by halothane may favor the inhibition of calcium accumulation with myocardial ischemia due to an increased oxygen supply or a more favorable flow distribution. It has been shown that halothane improved myocardial perfusion and oxygenation (Bland and Lowenstein, 1976; Verrier *et al.*, 1980; Davis *et al.*, 1983). Verrier *et al.* (1980) showed that halothane increased the coronary vascular reserve, suggesting a possible increase in myocardial oxygen supply in the ischemic heart. Therefore, the decrease in calcium accumulation by halothane might be due in part to an increase in oxygen supply during coronary occlusion with either less severe ischemia or a smaller area of ischemic myocardium.

On the other hand, an increase in coronary flow does not necessarily favor decreased calcium accumulation with the calcium paradox. It has been shown that higher coronary flow during calcium-free perfusion produces more severe myocardial damage and greater calcium accumulation (Koomen *et al.*, 1980; Ohhara *et al.*, 1982). Accordingly, it seems unlikely that the protective effect of halothane on calcium accumulation associated with calcium paradox is due to the increased coronary flow by halothane.

Calcium accumulation may be a primary event leading to cell necrosis in myocardial ischemia and calcium paradox, the result of damage during these interventions, or one of several consequences of depletion of tissue ATP resulting in inability to maintain normal ionic homeostasis. Halothane may be beneficial to an ischemic heart not only because of decreased calcium influx and increased oxygen supply, but also because of its cardiodepressant effect. The negative inotropic (Prys-Roberts *et al.*, 1972; Filner and Karliner, 1976) and chronotropic (Bosnjak and Kampine, 1983) actions of halothane may limit the rate of decline of ATP at the onset of ischemia such that relatively more ATP is available to maintain cellular integrity. Finally, a specific inhibitory effect of halothane or superoxide production (Nakagawara *et al.*, 1986), which may lead ultimately to a loss of membrane integrity (Ferrari *et al.*, 1985), may also favor the preservation of the ischemic heart by halothane.

In conclusion, recent studies indicate that clinically relevant concentrations of the volatile anesthetic halothane reversibly inhibits the slow calcium influx during the generation of cardiac action potential. In addition, halothane has been shown to inhibit the abnormal calcium accumulation associated with postischemic reperfusion and the calcium paradox. The results from these studies do not directly implicate the clinical benefit of halothane anesthesia to patients who have myocardial injuries.

ACKNOWLEDGMENTS. Supported in part by NIH grant HL34708 and the Medical Research Service of the Veterans Administration. The authors thank Ms. Miriam Mick for her assistance in preparing this manuscript.

REFERENCES

Bland, G. H. L., and Lowenstein, E., 1976, Halothane-induced decrease in experimental myocardial ischemia in the non-failing canine heart, *Anesthesiology* **45**:287–293.

Bourdillon, P. D., and Poole-Wilson, P. A., 1981, Effects of ischemia and reperfusion on calcium exchange and mechanical function in isolated rabbit myocardium, *Cardiovasc. Res.* **15**:121–130.

Bourdillon, P. D., and Poole-Wilson, P. A., 1982, The effects of verapamil, quiescence, and cardioplegia on calcium exchange and mechanical function in ischemic rabbit myocardium, *Circ. Res.* **50**: 360–368.

Bosnjak, Z. J., and Kampine, J. P., 1983, Effects of halothane, enflurane, and isoflurane on the SA node, *Anesthesiology* **58**:314–321.

Bosnjak, Z. J., and Kampine, J. P., 1986, Effects of halothane on transmembrane potentials, Ca^{++} transients, and papillary muscle tension in the cat, *Am. J. Physiol.* **251**:H374–H381.

Davis, R. F., Deboer, L. W. V., Rude, R. E., Lowenstein, E., and Maroko, P. R., 1983, The effect of halothane anesthesia on myocardial necrosis, hemodynamic performance, and regional myocardial blood flow in dogs following coronary artery occlusion, *Anesthesiology* **59**:402–411.

Ferrari, R., Ceconi, C., Curello, S., Guarnieri, C., Caldarera, C. M., Albertini, A., and Visioli, O., 1985, Oxygen-mediated damage during ischemia and reperfusion: Role of the cellular defenses against oxygen toxicity, *J. Mol. Cell. Cardiol.* **17**:937–945.

Filner, B. E., Karliner, J. S., 1976, Alterations in normal left ventricular performance by general anesthetic, *Anesthesiology* **45**:610–621.

Henry, P. D., Schuchleib, R., Davis, J., Weiss E. S., and Sobel B. E., 1977, Myocardial contracture and accumulation of calcium in ischemic rabbit heart, *Am. J. Physiol.* **233**:H677–H684.

Hoka, S., Bosnjak, Z. J., and Kampine, J. P., 1987, Halothane inhibits calcium accumulation following myocardial ischemia and calcium paradox in guinea pig hearts, *Anesthesiology* **67**:197–202.

Hug, C. C., 1979, Pharmacology-anesthetic drugs, in: *Cardiac Anesthesia* (J. A. Kaplan, ed.), Grune and Stratton, Orlando, pp. 3–37.

Ikemoto, Y., Yatani, A., Arimura, H., Yoshitake, J., 1985, Reduction of the slow inward current of isolated rat ventricular cells by thiamylal and halothane, *Acta Anesthesiol. Scand.* **29**:583–586.

Koomen, J. M., Jager, L. P., and van Noordwijk, J., 1980, Effects of perfusion pressure on coronary flow, myocardial Ca^{2+}-washout, and the occurrence of calcium paradox in isolated perfused rat heart ventricles, *Basic Res. Cardiol.* **75**:318–327.

Kroll, D. A., and Knight, P. R., 1984, Antifibrillatory effects of volatile anesthetics in acute occlusion/reperfusion arrhythmias, *Anesthesiology* **61**:657–661.

Lefer, A. M., Polansky, E. W., Bianchi, C. P., and Narayan, S., 1979, Influence of verapamil on cellular integrity and electrolyte concentrations of ischemic myocardial tissue in the cat, *Basic Res. Cardiol.* **74**:555–567.

Lewartowski, B., Pytowski, B., and Janczewski, A., 1984, Calcium fraction correlating with contractile force of ventricular muscle of guinea-pig heart. *Pflugers Arch.* **401**:198–203.

Lynch, III, C., Vogel, S., and Sperelakis, N., 1981, Halothane depression of myocardial slow action potentials, *Anesthesiology* **55**:360–368.

Nakagawara, M., Takeshige, K., Takamatsu, J., Takahashi, S., Yoshitake, J., and Minakami, S., 1986, Inhibition of superoxide production and Ca^{++} mobilization in human neutrophils by halothane, enflurane, and isoflurane, *Anesthesiology* **64**:4–12.

Nakanishi, T., Hishioka, K., and Jarmakani, J. M., 1982, Mechanism of tissue Ca^{2+} gain during reoxygenation after hypoxia in rabbit myocardium, *Am. J. Physiol.* **242**:H437–H449.

Nayler, W. G., Sturrock, W. J., and Panagiotopoulos, S., 1985, Calcium and myocardial ischaemia, in: *Control and Manipulation of Calcium Movement* (J. R. Parratt, ed.), Raven Press, New York, pp. 303–324.

Ohhara, H., Kanaide, H., and Nakamura, M., 1982, A protective effect of verapamil on the calcium paradox in the isolated perfused rat heart, *J. Mol. Cell. Cardiol.* **14**:13–20.

Porsius, A. J., and van Zwieten, P. A., 1975, Influence of halothane on calcium movements in isolated heart muscle and in isolated plasma membrane, *Arch. Int. Pharmacodyn. Ther.* **218**:29–39.

Prys-Roberts, C., Gersh, B. J., Baker, A. B., and Reuben, S. R., 1972, The effects of halothane on the interactions between myocardial contractility, aortic impedance and left ventricular performance. Theoretical considerations and results, *Br. J. Anaesthesiol.* **44**:634–649.

Pytkowski, B., Lewartowski, B., Prokopczuk, A., Zdanowski, K., and Lewandowska, K., 1983, Excitation- and rest-dependent shifts in guinea-pig ventricular myocardium, *Pflugers Arch.* **398**: 103–113.

Ruigrok, T. J. C., 1985, The calcium paradox and the heart, in: *Control and Manipulation of Calcium Movement* (J. R. Parratt, ed.), Raven Press, New York, pp. 341–365.

Selye, H., Bausz, E., Grasso, S., and Mendell, P., 1960, Simple techniques for the surgical occlusion of coronary vessels in the rat, *Angiology* **11**:398–407.

Shen, A. C., and Jennings, R. B., 1972, Myocardial calcium and magnesium in acute ischemic injury, *Am. J. Pathol.* **67**:417–440.

Shine, K. I., Serena, S. D., and Langer, G. A., 1971, Kinetic localization of contractile calcium in the rabbit myocardium, *Am. J. Physiol.* **221**:1408–1417.

Verrier, E. D., Edelist, G., Consigny, P. M., Robinson, S., and Hoffman, J. I. E., 1980, Greater coronary vascular reserve in dogs anesthetized with halothane, *Anesthesiology* **53**:445–459.

Watts, J. A., Koch, C. D., and LaNoue, K. F., 1980, Effects of Ca^{2+} antagonism on energy metabolism: Ca^{2+} and heart function after ischemia, *Am. J. Physiol.* **238**:H909–H916.

62

Defective Ca^{2+} Functions in Protein (47-kDa) Phosphorylation and in the Coupling to Physiological Responses in Platelets from Stroke-Prone Spontaneously Hypertensive Rats

TAKAKO TOMITA, KEIZOU UMEGAKI, MASAHIKO IKEDA,
NOBUAKI TAKESHITA, KAZUKI NAKAMURA, and
YASUHIDE INOUE

1. INTRODUCTION

Although hypertension is assumed to be one of the important factors in thrombotic disease, inconsistent reports have been presented concerning changes in platelet function due to hypertension. Stroke-prone spontaneously hypertensive rats (SHRSP), a substrain of spontaneously hypertensive rats (SHR), which was established in 1974 (Okamoto *et al.*, 1974), develops more severe hypertension spontaneously than SHR, and dies of massive cerebral hemorrhage or infarction between 100 and 300 days after birth. Changes in platelet functions were investigated using these strains of spontaneous hypertension which are considered to represent the closest model to human hypertension. In addition, platelets are a tissue in which pathophysiological changes are intimately related to those of blood vessels (Mustard *et al.*, 1964). Both contain contractile proteins and Ca^{2+} plays important roles in both tissues (Weiss, 1975). Thus, the study of abnormalities in platelets from hypertensive animals may shed light on the etiology of hypertension as well as of diseases consequent to hypertension.

We demonstrated that aggregation, secretion, and 47-kDa protein phosphorylation in response to various agents are significantly reduced in SHRSP platelets compared with those in platelets from age-matched, normotensive Wistar Kyoto rats (WKY). The abnormalities in SHRSP platelets are not secondary to hypertension but result from defective Ca^{2+} functions

TAKAKO TOMITA, KEIZOU UMEGAKI, MASAHIKO IKEDA, NOBUAKI TAKESHITA, KAZUKI NAKAMURA, and YASUHIDE INOUE ● Department of Pharmacology, University of Shizuoka School of Pharmaceutical Sciences, Shizuoka, Japan.

in phosphorylation of 47-kDa protein and in the coupling of phosphorylation to physiological responses. This study presents the first evidence to link defective protein phosphorylation to the pathophysiology of platelets.

2. MATERIALS AND METHODS

SHRSP and WKY were provided by Professor K. Okamoto of Kinki University Medical School and maintained by brother–sister breeding in our laboratory. Male SHRSP at the ages of 10–17 weeks and age-matched WKY were used throughout the study. Blood was removed from the abdominal aorta and washed platelets were prepared as described elsewhere (Tomita *et al.*, 1983). Platelets prepared from rats at these ages showed constant and characteristic properties for each strain. Platelet aggregation was measured by a turbidometric method. [^{14}C]Serotonin release assay was performed by a modification of the two methods (Holmsen and Dangelmaier, 1981; Hofmann *et al.*, 1982). For analysis of platelet-protein phosphorylation, ^{32}P-prelabeled platelets which were prepared according to Feinstein *et al.* (1983) were incubated in a Tris buffer (pH 7.4) with aggregating agents; the reaction was terminated by adding a sodium dodecyl sulfate (SDS) stopping buffer. The protein was subjected to SDS–PAGE and phosphorylated protein was analyzed by radio autography. Changes in intracellular Ca^{2+} concentration ($[Ca^{2+}]i$) were monitored using Quin2 (Rink *et al.*, 1982).

3. ABNORMAL FUNCTIONS OF PLATELETS FROM SHRSP

3.1. Blood Pressure and Hematological Characteristics

The blood pressure of male SHRSP was not different from that of WKY at 4 weeks whereas it rose to 210–220 mm Hg at 11 and 17 weeks. The blood pressure of WKY was always around 130 mm Hg. The number of platelets at prehypertensive ages was equal in the two strains whereas it decreased in SHRSP with the development of hypertension. The number of platelets in SHRSP was reduced to almost one-half of WKY at age of 11–17 weeks and the size increased with age (Tomita *et al.*, 1984). These changes in the number and the size of SHRSP platelets seem unlikely to result from the exhaustion of platelets due to vascular lesion because these changes occurred even before the appearance of the lesion (Umegaki *et al.*, 1985).

3.2. Hypoaggregability of SHRSP Platelets

To examine platelet functions without humoral factors, washed platelets were prepared. Aggregation of washed platelets from SHRSP by thrombin, collagen, ADP, and ionophore A23187 was markedly reduced with the development of hyptertension in comparison with age-matched WKY platelets (Fig. 1). The degree of aggregation in SHRSP platelets was similar to that of WKY platelets at the age of 4 weeks, but the responses in SHRSP gradually diminished with the increasing blood pressure. The study with substrains of SHR with different degrees of hypertension revealed that the intensity of hypoaggregability was inversely correlated with the severity of hypertension (Tomita *et al.*, 1985). In addition, this abnormality in SHRSP platelets was not recovered by a long-term hypotensive treatment

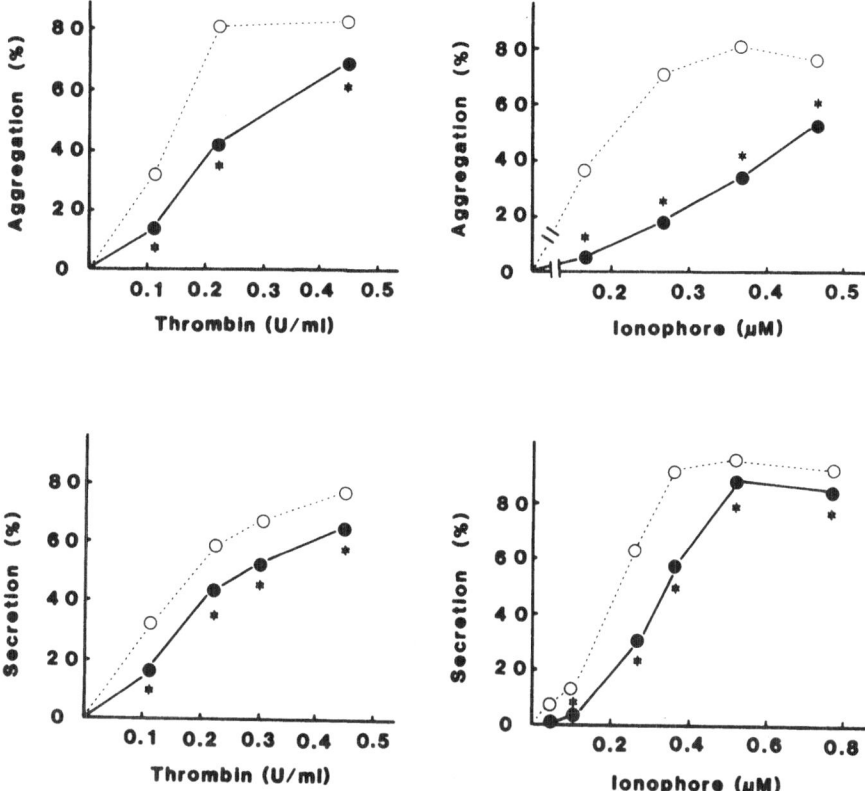

FIGURE 1. Thrombin- and Ionophore A23187-induced aggregaton and [^{14}C]serotonin release in platelets from SHRSP (—o—) and WKY (--o--). Washed platelets (4 × 10^8 cells/ml) were stimulated at 37°C either with thrombin for 3 min or with Ionophore A23187 for 5 min (aggregation) and 3 min (secretion) in the presence of 1.5 mM Ca^{2+}. Results are expressed as mean ± SD for five determinations.

(Umegaki *et al.*, 1983), implying that it was genetically determined. Reduced responses of SHRSP platelets were also observed with platelet-rich plasma which contained various humoral factors, and whole blood in the presence of other blood cells. Since the hypoaggregability was demonstrated with various stimulants including Ionophore A23187, it seems likely that defects in SHRSP platelets are at a stage distal to receptor occupancy.

3.3. Decrease in Release Reaction

When [^{14}C]-prelabeled platelets were stimulated with varying amounts of thrombin, [^{14}C]serotonin was dose-dependently released into the medium as shown in Fig. 1. The release reached a maximum at 0.5–1 min. Accompanying hypoaggregability, [^{14}C]serotonin release from SHRSP platelets was reduced with the increasing of blood pressure compared with that of age-matched WKY. The release was affected less than the aggregation in SHRSP platelets. The release of ADP and ATP in response to thrombin was also decreased in SHRSP platelets (Umegaki *et al.*, 1986a). Ionophore A23187-induced [^{14}C]serotonin release as well as aggregation was significantly attenuated in SHRSP platelets. However, there was no difference between the contents of dense granules in SHRSP and WKY at 5–16 weeks of age (Umegaki *et al.*, 1986a), whereas they were significantly lowered in SHRSP platelets at higher ages (Umegaki *et al.*, 1985). It is therefore suggested that the reduced platelet aggregation and secretion observed in SHRSP platelets at the ages of 10–17 weeks are not secondary phenomena to the circulation of degranulated platelets due to vascular injuries caused by hypertension.

3.4. Defective Protein Phosphorylation in Platelets

In human platelets stimulated with thrombin or other physiological stimuli, there was a rapid and transient rise in 1,2-diacylglycerols (DG) and intracellular Ca^{2+} concentration ($[Ca^{2+}]_i$). On the other hand, two endogenous proteins with approximate molecular masses of 47 and 20 kDa (substrates of protein kinase C and myosin light-chain kinase, respectively) are rapidly and heavily phosphorylated in parallel with physiological responses (Nishizuka, 1984). Diacylglycerols produced by the hydrolysis of phosphatidylinositol acts as as a signal molecule by enhancing the Ca^{2+} sensitivity of protein kinase C. Several phorbol esters such as 12-O-tetradecanoylphorbol-13-acetate (TPA) (a potent tumor promoter) mimic endogenously produced DG in activating protein kinase C, thus bypassing the receptor signal (Nishizuka, 1984).

Figure 2 shows the time course of 47-kDa protein phosphorylation in ^{32}P-prelabeled SHRSP and WKY platelets stimulated by thrombin (0.15 U/ml), Ionophore A23187 (0.5 μM), and TPA (80 nM). Complete phosphorylation occurred within 30 sec with these three stimulants. Thrombin (0.2 U/ml) and TPA (10 nM) induced full response of 47-kDA protein phosphorylation. As clearly shown in Fig. 2, there was a significant decrease in phosphorylation of 47-kDa protein in SHRSP platelets in comparison with that of WKY platelets. The phosphorylation of the 20-kDa protein occurred in similar degree in the platelets from the two strains. By simultaneous measurement of aggregation in ^{32}P-prelabeled platelet preparation, a concomitant decrease in aggregation of SHRSP platelets was confirmed.

The TPA effect on phosphorylation was seen at a concentration as low as 0.5 nM, peaking at 30 sec to 1 min. In contrast to reduction of thrombin-induced protein phosphorylation in SHRSP platelets, there was no difference in the magnitude of 47-kDa protein phosphorylation between SHRSP and WKY platelets when they were stimulated with 80 nM TPA. The TPA at concentrations less than 20 nM did not induce aggregation or [^{14}C]serotonin release, whereas at the higher concentrations aggregation was induced slowly with time. Unlike thrombin stimulation, the aggregation and release response to TPA (80 nM) were identical in both platelets.

Protein phosphorylation in response to Ionophore A23187 which bypassed Ca^{2+} channels is shown on the right-hand side of Fig. 3. In parallel with the attenuated responses in aggregation and secretion, Ionophore A23187 (0.5 μM)-induced protein phosphorylation (47-kDa) was reduced in SHRSP platelets compared with WKY platelets. However, the degree of phosphorylation of 20-kDa protein of both platelets was similar.

These results indicated that protein kinase C and its substrate (47-kDa protein) were normally present in SHRSP platelets, and the defective protein phosphorylation would be ascribed to incomplete receptor-mediated activation of protein kinase C.

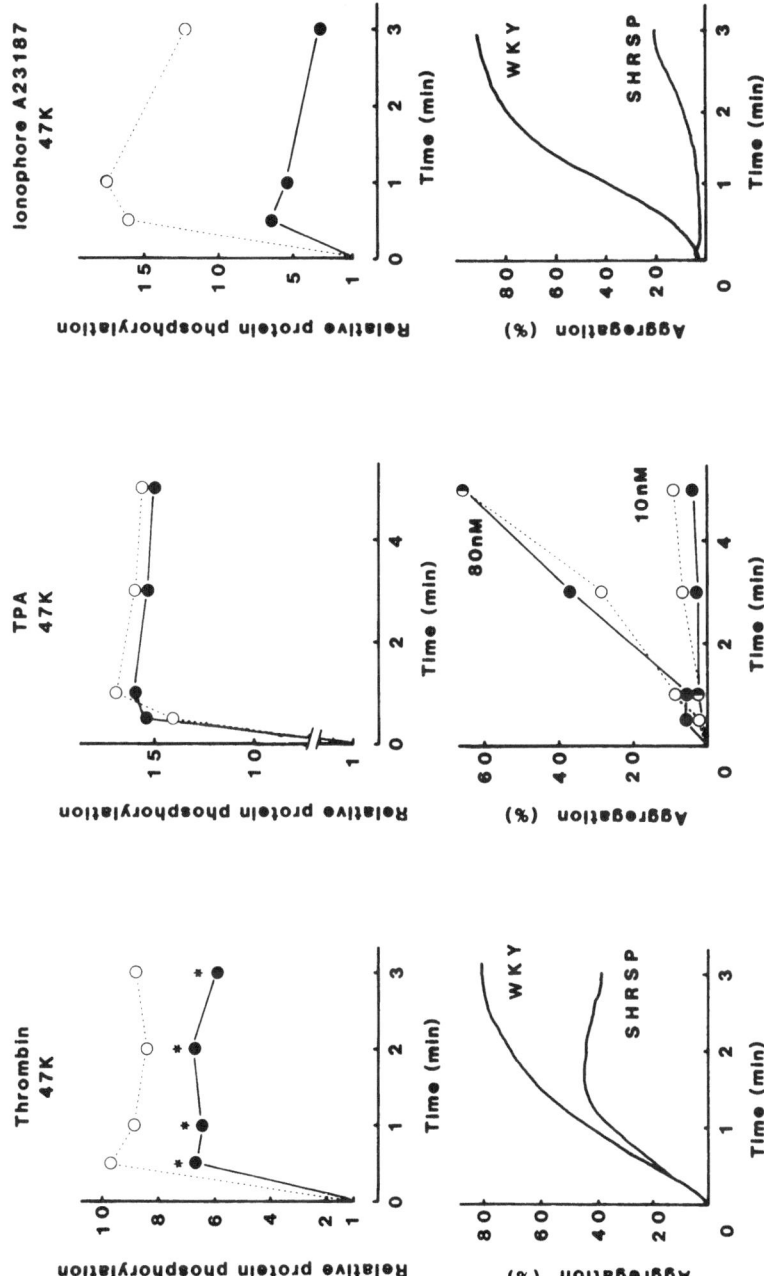

FIGURE 2. Thrombin-, TPA-, and ionophore-induced protein phosphorylation and aggregation responses simultaneously measured with ^{32}P-prelabeled platelets from SHRSP (—o—) and WKY (--o--). ^{32}P-prelabeled platelets (5×10^8 cells/ml) in 25 mM Tris-HCl buffer (pH 7.4) were stimulated with thrombin (0.15 U/ml), with TPA (80 and 10 nM), or with ionophore A23187 (0.5 μM). Phosphorylated protein was subjected to SDS–PAGE and analyzed by autoradiography.

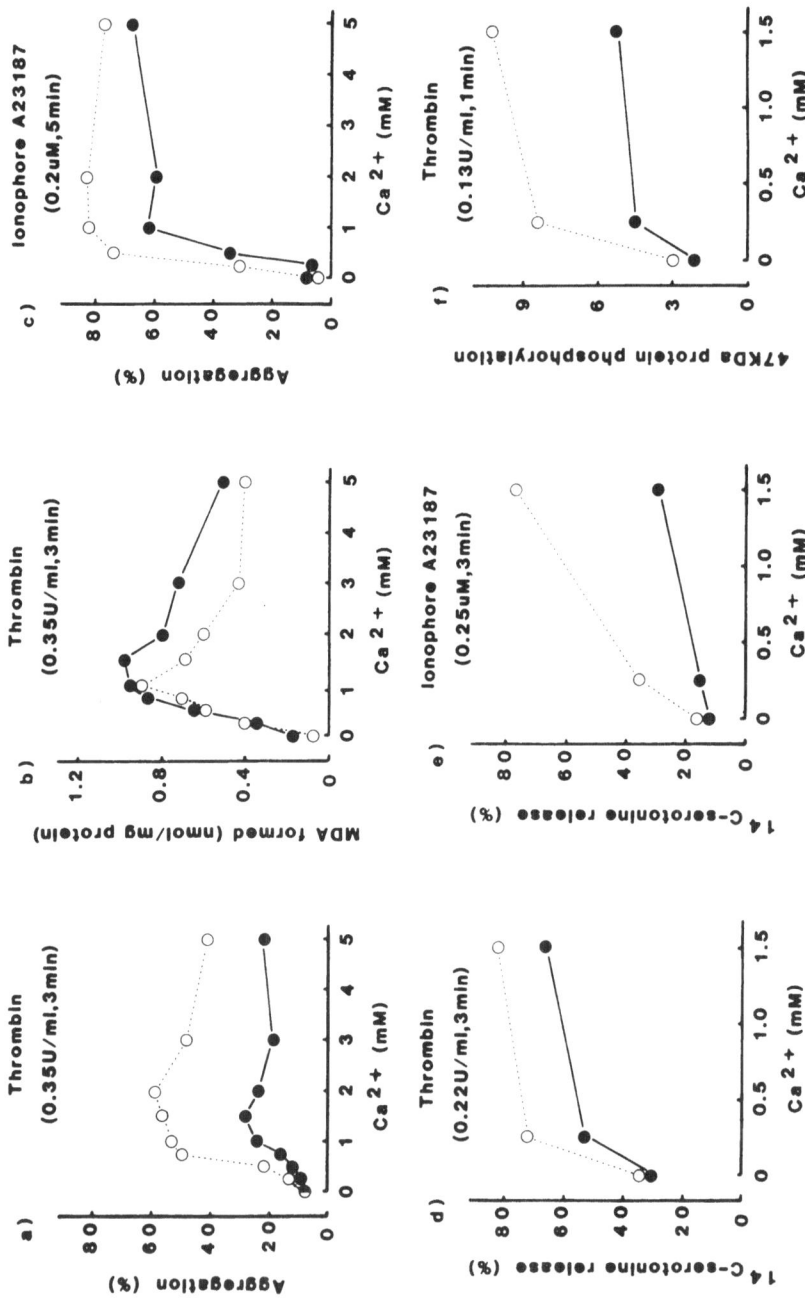

FIGURE 3. Effects of extracellular Ca²⁺ concentrations on thrombin- and Ionophore A23187-induced aggregation (a, c), [¹⁴C]serotonin release (d, e), malondialdehyde formation (b), and 47-kDa protein phosphorylation (f) in SHRSP and WKY platelets.

4. CHANGES IN Ca²⁺ DEPENDENCY IN AGGREGATION, SECRETION AND PROTEIN PHOSPHORYLATION

4.1. Aggregation and Malondialdehyde (MDA) Formation

Thrombin (0.35 U/ml)- and Ionophore (0.2 μM)-induced aggregation as a function of the exogenous Ca^{2+} concentration were compared in SHRSP and WKY platelets (Fig. 3). The SHRSP and WKY platelets aggregated to the same degree in the absence of Ca^{2+} in all doses examined. However, the enhancement of aggregation by extracellular Ca^{2+} was remarkably higher in WKY platelets than in SHRSP platelets regardless of the type of stimulants, including ADP. At prehypertensive ages (less than 4 weeks), there was no difference in the degree of Ca^{2+} enhancement between platelets from SHRSP and WKY. Platelet MDA formation is regarded as an indicator of cyclooxygenase and thromboxane synthase activities and also of the release of arachidonic acid (AA) from membrane phospholipids (Smith et al., 1976). Ca^{2+} dependence of MDA formation was simultaneously examined (Fig. 3, upper center). In contrast to aggregation, the magnitude of enhancement in MDA formation by Ca^{2+} was almost the same in SHRSP as in WKY below 1 mM Ca^{2+} and became greater in SHRSP than in WKY over 1.5 mM Ca^{2+}. These data indicated that abnormalities of SHRSP platelets appear to exist in Ca^{2+} functions involved in aggregation and secretion but not in AA metabolism.

4.2. [¹⁴C]Serotonin Release

The lower left and center figures in Fig. 3 show the effect of extracellular Ca^{2+} on [¹⁴C]serotonin release. Thrombin (0.22 U/ml)- and Ionophore A23187 (0.25 μM)-induced release from SHRSP platelets was identical to that of WKY platelets in the absence of Ca^{2+}. In the presence of 0.25 and 1.5 mM Ca^{2+}, the release was enhanced in smaller degree in SHRSP platelets than in WKY platelets. 12-O-Tetradecanoylphorbol-13-acetate (100 nM), which activated platelets without elevating $[Ca^{2+}]_i$, released [¹⁴C]serotonin from SHRSP and WKY platelets at similar degrees regardless of extracellular Ca^{2+} concentration.

4.3. 47-kDa Protein Phosphorylation

Thrombin-induced protein (47-kDa) phosphorylation of ³²P-prelabeled platelets from SHRSP was not significantly different from WKY in the absence of Ca^{2+} (Fig. 3, lower right). As was seen in aggregation and release response, the enhancement by the presence of extracellular Ca^{2+} was less in protein (47-kDa) phosphorylation of SHRSP platelets than in that of WKY platelets. The results obtained here suggest that impaired are Ca^{2+} functions involved in phosphorylation of 47-kDa protein, aggregation, and secretion, but not in phospholipase A_2, cyclooxygenase, and thromboxane synthase pathways.

5. MECHANISMS OF DEFECTIVE PROTEIN PHOSPHORYLATION

5.1. Contents of Phosphoinositides in Platelets

Phosphatidylinositol (PI), phosphatidylinositol-4-phosphate (PIP), and phosphatidyli-nositol-4,5-diphosphate (PIP_2) are minor constituents of mammalian membranes. Agonist-induced phosphoinositide breakdown functions as a signal-generating system by yielding two second messengers, DG and inositol trisphosphate (IP_3) (Nishizuka, 1984). Diacylglycerols stimulate protein kinase C synergistically with Ca^{2+} while IP_3 is proposed as a second messenger for Ca^{2+} mobilization (Berridge and Irvine, 1984). It is therefore interesting to investigate whether there are changes of membrane lipids, especially inositol phospholipids in SHRSP platelets.

Phosphatidylinositol was 20% less in SHRSP platelets [nmol/mg protein, mean ± SD(n): 8.27 ± 1.33(9) in SHRSP versus 10.60 ± 1.59(9) in WKY] while the content of PIP and PIP_2 were not significantly different between the two strains [nmol/mg protein, mean ± SD(n): PIP 1.45 ± 0.30(14) in SHRSP versus 1.45 ± 0.29(15) in WKY, PIP_2 1.00 ± 0.15(18) in SHRSP versus 0.93 ± 0.16(18) in WKY]. No changes in the contents of PC and PE were observed in SHRSP platelets compared with WKY platelets (Ikeda *et al.*, 1987). This result coincided with results observed in erythrocyte membranes from SHR and WKY (Koutouzov *et al.*, 1983).

Formation of DG following the receptor-mediated hydrolysis of phosphatidylinositols was investigated in [³H]AA-labeled platelets from SHRSP and WKY. Diacylglycerol formation peaked at 15 sec after a thrombin stimulation. Aggregation responses of SHRSP platelets to lower doses of thrombin (<0.15 U/ml) were smaller than those in WKY platelets. With these concentrations of thrombin, no significant formation of DG was observed in either platelets type. In supramaximal doses of thrombin, there was no difference in aggregation of SHRSP and WKY platelets; DG formation was apparently observed, and it was lower in SHRSP platelets than in WKY platelets, which is assumed to be reflected by decreased PI content in SHRSP platelets.

Thrombin stimulation at low concentrations hydrolyzes PIP_2, and PI hydrolysis occurs only at supramaximal concentrations of thrombin. Therefore, it indicates that attenuated responses of aggregation, secretion, and protein phosphorylation in SHRSP are not related to the degree of DG formation. This concept is supported by phospholipase C activity following thrombin stimulation. Phospholipase C activity was estimated through phosphatidic acid formation from ³²P-prelabeled platelets (Lapetina and Cuatrecasas, 1979). There was no difference of phosphatidic acid formation between SHRSP and WKY platelets which were stimulated with thrombin (0.13–1 U/ml) for 30 sec.

5.2. Changes in Cytoplasmic Ca^{2+} Concentration

Changes of cytoplasmic $[Ca^{2+}]_i$ were examined by using Quin2 (Rink *et al.*, 1982). $[Ca^{2+}]_i$ increase in response to thrombin (0.03–0.1 U/ml) was significantly delayed in SHRSP platelets compared with WKY platelets. The time (sec) to peak in $[Ca^{2+}]_i$ was about two times longer in SHRSP platelets. $[Ca^{2+}]_i$ levels at resting state were significantly lower in SHRSP platelets while there was no difference in maximal $[Ca^{2+}]_i$ level in response to thrombin between two strains (Umegaki *et al.*, 1986b). This result on resting $[Ca^{2+}]_i$ level in platelets from spontaneously hypertensive rats contradicts the results from human platelets of patients with essential hypertension (Sang and Devynck, 1986). In addition, thrombin-induced $^{45}Ca^{2+}$ uptake was noticeably delayed in SHRSP platelets. The cause of the delay in $[Ca^{2+}]_i$ increase and the influence of the delay on platelet activation remains to be explored.

5.3. Effect of TPA and Ionophore A23187 on Platelet Aggregation and Protein Phosphorylation

Finally, Fig. 4 shows independent and synergistic effects of subminimum doses of Ionophore A23187 and TPA on platelet functions. Ionophore A23187 (0.1–0.2 μM) induced neither aggregation nor phosphorylation whereas TPA (10–20 nM) induced full phosphorylation of 47-kDa protein equally in platelets from SHRSP and WKY without inducing aggregation. However, TPA (20 nM) and Ionophore A23187 (0.2 μM) induced both platelet aggregation and phosphorylation of 47-kDa protein: aggregation was less in SHRSP than in WKY while phosphorylation was similar in both SHRSP and WKY. Thus, it is suggested that in SHRSP platelets, Ca^{2+} does not function normally, first in activating protein kinase C and second in the coupling of protein phosphorylation to physiological responses. The result implies that senstivity to Ca^{2+} of protein kinase C must be decreased in SHRSP platelets.

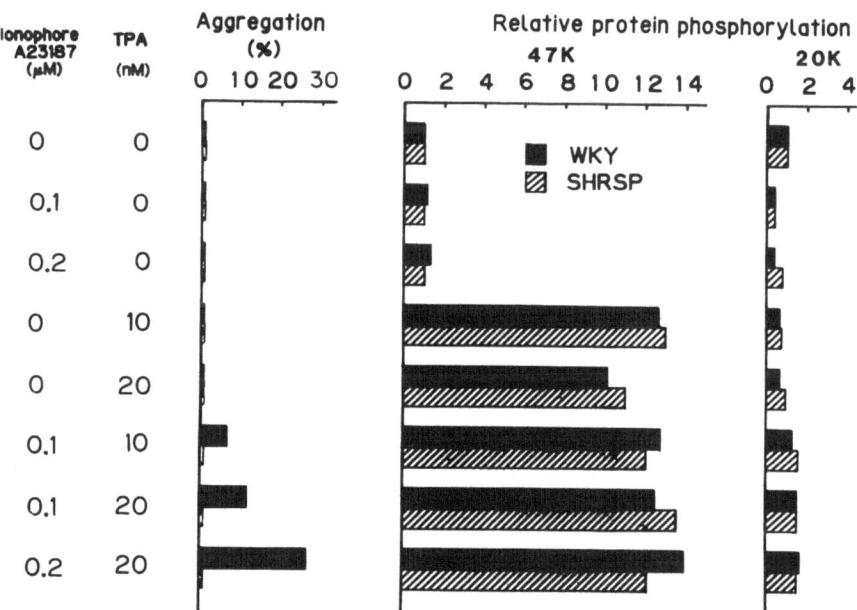

FIGURE 4. Effects of TPA and Ionophore A23187 on aggregation and protein phosphorylation in platelets from SHRSP and WKY. ^{32}P-labeled platelets (5 $\times$ 10^8 cells/ml) from SHRSP and WKY were stimulated with TPA and/or Ionophore A23187 for 1 min in the presence of 1.5 mM Ca^{2+}.

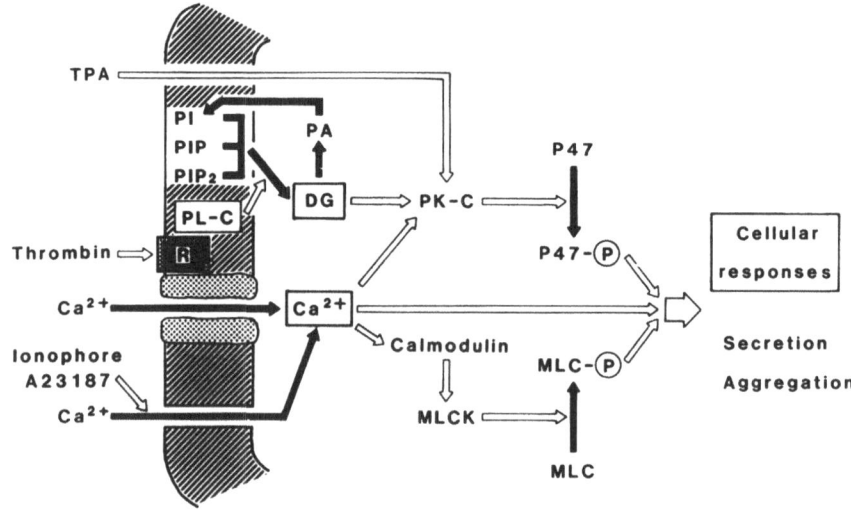

Stimulus-Response Coupling

FIGURE 5. A schematic representation of signal pathways in platelets. R (receptor), TPA (12-O-tetradecanoylphorbor-13-acetate), PI (phosphatidylinositol), PIP (PI-4-phosphate), PIP_2 (PI-4,5-diphosphate), PA (phosphatidic acid), DG (diacylglycerols), P47 (47-kDa protein), P47-P (phosphorylated P47), MLC (myosin light chain), MLC-P (phosphorylated myosin light chain), MLCK (MLC kinase), PK-C (protein kinase), PL-C (phospholipase C).

6. DISCUSSION

A calcium–calmodulin complex activates myosin light-chain kinase (Hathaway and Adelstein, 1979). The 20-kDa light chains of myosin are consequently phosphorylated to form an actin–myosin complex that can hydrolyze ATP. Phospholipid-dependent and Ca^{2+}-activated protein kinase, protein kinase C, phosphorylates a 47-kDa protein in platelets prior to the secretion reaction (Nishizuka, 1984). Our study showed that thrombin- and Ionophore A23187-induced aggregation, secretion, and 47-kDa protein phosphorylation in SHRSP platelets were reduced compared with WKY platelets in the presence of extracellular Ca^{2+}. With other compiled evidence, it was assumed that defective calcium functions in activating protein kinase C and in coupling the protein phosphorylation to physiological responses were underlying mechanisms for the dysfunction of SHRSP.

The nature of the 47-kDa protein is not clearly known. It was recently reported that this protein is polyphosphoinositide-specific phosphatase which is activated by phosphorylation. (Connolly *et al.*, 1986). Touqui *et al.* (1986) provided evidence that the 47-kDa protein was lipocortin, by whose phosphorylation the inhibitory action on phospholipase A_2 was lost. The most plausible report to explain our data is that 47-kDa protein is an inhibitor of actin polymerization. Phosphorylation switches on actin polymerization (Imaoka, 1987).

Several lines of evidence showed that there is an alternation in calcium regulation in excitable cells of SHR. It was reported that Ca^{2+} binding in the erythrocyte membranes, the cardiac sarcoplasmic reticulum, the brain synaptosomal membrane, and the calcium uptake in aortic microsomes are reduced in SHR compared with WKY. There are reports on the decrease in calmodulin in SHR. It remains unclear whether these phenomena in SHR are related to the delay of $[Ca^{2+}]_i$ increase in SHRSP platelets. In human platelets from genetic hypertensive patients, platelet functions are enhanced (Valtier *et al.*, 1986) and the resting $[Ca^{2+}]_i$ is higher than those of platelets from normal persons (Sang and Devynck, 1986). It would be difficult to explain the etiology of hypertension and of the consequent diseases in hypertensive rats from our present observation in SHRSP platelets. Further detailed investigation on the defects found in this study are expected to explain the discrepancy between human and rats.

7. SUMMARY

1. Thrombin-, collagen-, ADP-, and Ionophore A23187-induced aggregation, as well as serotonin release, were greatly reduced in washed platelets from stroke-prone spontaneously hypertensive rats (SHRSP) compared with those of platelets from age-matched normotensive Wistar Kyoto rats (WKY).

2. Enhancement of platelet functions by extracellular Ca^{2+} was much less in SHRSP platelets.

3. Consistent with the hypofunctions, thrombin-induced phosphorylation of 47-kDa protein in SHRSP platelets was significantly reduced compared with that of age-matched WKY platelets.

4. However, TPA-induced aggregation, secretion, and 47-kDa protein phosphorylation in SHRSP platelets were similar to those in WKY platelets.

5. On the other hand, Ionophore A23187 (0.2–0.5 μM)-induced responses were greatly reduced in SHRSP platelets.

6. Ionophore A23187 (<0.2 μM) induced neither aggregation nor phosphorylation whereas TPA (10 nM) fully phosphorylated 47-kDa protein in both SHRSP and WKY platelets without inducing aggregation. TPA and Ionophore A23187, together in these concentrations, induced greater aggregation responses in WKY platelets than those in SHRSP platelets in spite of similar degrees of phosphorylation of 47-kDa proteins.

7. Mass content of PI in SHRSP platelets was 20% lower than that of WKY platelets, while that of PIP₂ and PIP was identical with that in WKY platelets.

8. Thrombin-induced formation of [³H]AA-diacylglycerols and [³²P]phosphatidic acid in SHRSP platelets, was unaltered from that of WKY platelets in the submaximal concentrations.

9. $[Ca^{2+}]_i$ increase following thrombin stimulation was delayed in SHRSP platelets, but the maximal $[Ca^{2+}]_i$ was similar in both platelets.

10. It is concluded that Ca^{2+} functions in SHRSP platelets are impaired first in phosphorylation of 47-kDa protein and second in coupling of protein phosphorylation to physiological responses. This is the first evidence to link defective protein phosphorylation to the pathophysiology of platelets.

AKNOWLEDGMENT. The authors wish to express their sincere thanks to Professors K. Okamoto of Kinki University Medical School and Y. Nishizuka of Kobe University School of Medicine for their encouragement and pertinent advice throughout this work.

REFERENCES

Berridge, M. J., and Irvine, R. F., 1984, Inositol trisphosphate, a novel second messenger in cellular signal transduction, *Nature* **312:**315–321.

Connolly, T. M., Lawing, W. J., and Majerus, P. W., 1986, Protein kinase C phosphorylates platelet inositol trisphosphate 5′-phosphomonoesterase, increasing the phosphatase activity, *Cell* **46:**951–958.

Feinstein, M. B., Egan, J. J., and Opas, E. E., 1983, Reversal of thrombin-induced myosin phosphorylation and the assembly of cytoskeletal structures in platelets by the adenylate cyclase stimulants prostaglandin D_2 and forskolin, *J. Biol. Chem.* **258:**1260–1267.

Hathaway, D. R., and Adelstein, R. S., 1979, Human platelet myosin light chain kinase requires the calcium binding protein calmodulin for activity, *Proc. Natl. Acad. Sci. USA* **76:**1653–1657.

Hofmann, S. L., Prescott, S. M., and Majerus, P. W., 1982, The effects of mepacrine and p-bromophenacyl bromide on arachidonic acid release in human platelets, *Arch. Biochem. Biophys.* **215:**237–244.

Holmsen, H., and Dangelmaier, C. A., 1981, Evidence that the platelet plasma membrane is impermeable to calcium and magnesium complex of A23187, *J. Biol. Chem.* **256:**10449–10452.

Ikeda, M., Umegaki, K., Takeshita, N., Nakamura, K., Inoue, Y., and Tomita, T., 1987, Changes of platelet functions due to spontaneous hypertension: Reduced response to ionophore A23187 in SHRSP platelets, *Jpn. Heart J.* **28:**577.

Imaoka, T., 1987, Unphosphorylated P47 had an inhibitory activity in platelet actin polymerization. Purified phosphorylated P47 looses the binding activity with membrane, also the inhibitory activity in actin polymerization, *Fed. Proc. of American Society of Biochemistry.* **46:**2064.

Koutouzov, S., Marche, P., Girard, A., and Meyer, P., 1983, Altered turnover of polyphosphoinositides in the erythrocyte membrane of the spontaneously hypertensive rat, *Hypertension* **5:**409–414.

Lapetina, E. G., and Cuatrecasa, P., 1979, Stimulation of phosphatidic acid production in platelets precedes the formation of arachidonate and parallels the release of secretion, *Biochim. Biophys. Acta* **573:**394–402.

Mustard, J. F., Murphy, E. A., Rowsell, H. C., and Downie, H. G., 1964, Platelets and atherosclerosis, *J. Atheroscler. Res.* **4:**1–28.

Nishizuka, Y., 1984, The role of protein kinase C in cell surface signal transduction and tumour promotion, *Nature* **308:**693–698.

Okamoto, K., Yamori, Y., and Nagakoa, A., 1974, Establishment of the stroke-prone SHR, *Circ. Res.* **34, 35,** Suppl. 1:143–153.

Rink, T. J., Smith, S. W., and Tsien, R. V., 1982, Cytoplasmic free Ca^{2+} in human platelets: Ca^{2+} thresholds and Ca^{2+}-independent activation for shape-change and secretion, *FEBS Lett.* **148:**21–26.

Sang, K. H. L. Q., and Devynck, M., 1986, Increased platelet cytosolic free calcium concentration in essential hypertension, *J. Hypertension* **4:**567–574.

Smith, J. B., Ingerman, C. M., and Silver, M. J., 1976, Malondialdehyde formation as an indicator of prostaglandin production by human platelets, *J. Lab. Clin. Med.* **88:**167–172.

Tomita, T., Umegaki, K., and Hayashi, E., 1983, Basic aggregation properties of washed rat platelets: Correlation between aggregation, phospholipid degradation, malondialdehyde, and thromboxane formation, *J. Pharmacol. Meth.* **10:**31–44.

Tomita, T., Umegaki, K., and Hayashi, E., 1984, Hypoaggregability of washed platelets from stroke-prone spontaneously hypertensive rats (SHRSP), *Stroke* **15:**70–75.

Tomita, T., Umegaki, K., and Hayashi, E., 1985, Changes in platelet functions due to hypertension: Defective Ca^{2+} transport and functions in SHRSP platelets, *Jpn. Heart J.* **26:**664.

Touqui, L., Rothhut, B., Shaw, A. M., Fradin, A., Vargaftig, B. B., and Russo-Marie, F., 1986, Platelet activation: role for a 40K antiphospholipase A_2 protein indistinguishable from lipocortin, *Nature* **321:**177–180.

Umegaki, K., Tomita, T., and Hayashi, E., 1983, Effects of hypotensive treatement on platelet aggregation in SHRSP, *Jpn. Heart J.* **24:**831.

Umegaki, K., Inoue, Y., and Tomita, T., 1985, The appearance of "exhausted" platelets at the time of stroke-prone spontaneously hypertensive rats, *Throm. Haem.* **54:**764–767.

Umegaki, K., Nakamura, K., and Tomita, T., 1986a, Primary dysfunction in aggregation and secretion of SHRSP platelets: Not secondary to the circulation of "exhausted" platelets, *Blut* **52:**17–27.

Umegaki, K., Ikeda, M., and Tomita, T., 1986b, Defective calcium transport in platelets from stroke-prone spontaneously hypertensive rats, *Throm. Res.* **41:**415–423.

Valtier, D., Guicheney, P., Baudouin-Legros, M., and Meyer, P., 1986, Platelets in human essential hypertension: In vitro hyperreactivity to thrombin, *J. Hypertension* **4:**551–555.

Weiss, H. G., 1975, Platelet physiology and abnormalities of platelet function, *N. Engl. J. Med.* **293:** 531–541.

63

Effects of Elastin Peptides on Ion Fluxes

T. FÜLÖP, Jr., M. P. JACOB, G. FÓRIS, Zs. VARGA,
and L. ROBERT

1. INTRODUCTION

The degradation of elastin by elastase-type enzymes was shown to play an important role in several pathological processes such as the development of emphysema (Crystal, 1976; Bignon et al., 1978), arteriosclerosis (Robert et al., 1980, 1984), and a variety of skin diseases (Frances et al., 1984). All these enzymes, although of different natures (Bourdillon et al., 1984), are able to hydrolyze elastin and release soluble peptides. The released peptides (mainly kappa elastin, KE) were shown to have a variety of biological properties (Robert et al., 1970; Jacob et al., 1984). One of these was the chemotactic effect on monocyte and fibroblasts (Senior et al., 1980). Robert et al. (1967, 1970) demonstrated their antigenic nature. Rabbits immunized with elastin peptides were shown to develop severe arteriosclerosis (Robert et al., 1971; Jacob et al., 1984) and also lesions of pulmonary arteries. Transmembrane cation fluxes (Na^+, K^+, Ca^{2+}) were shown to play an important role in cell activity regulation (Scully et al., 1984; Blaustein et al., 1984). The chemotactic peptide receptors were shown to be coupled to the phosphoinositide-specific phospholipase C through a guanine nucleotide regulatory protein (Berridge et al., 1984). This receptor activation involves hydrolysis of phosphoinositides followed by generation of inositol trisphosphate and diacylglycerol. The inositol trisphosphate is believed to induce the release of Ca^{2+} from an intracellular storage and as a consequence the level of intracellular free Ca^{2+} is increased (Reynolds, 1985). Furthermore, the formation of inositol tetrakisphosphate from inositol trisphosphate seems to have ionophore effects (Trimble et al., 1987). Our aims were to investigate the effects of elastin peptides on ion fluxes and to elucidate their mechanism of action at the cellular and intracellular levels.

T. FÜLÖP Jr., G. FÒRIS, and Zs. VARGA • First Department of Medicine, University Medical School of Debrecen, H-4012 Debrecen, Hungary.   M. P. JACOB and L. ROBERT • Laboratory for Biochemistry of Connective Tissue, University of Paris XII, Faculty of Medicine, 94010 Créteil, France.

2. EFFECTS OF ELASTIN PEPTIDES ON ION FLUXES

It was previously demonstrated (Robert et al., 1971; Jacob et al., 1984) that immunization of rabbits with highly purified elastin peptides in complete Freund's adjuvant induced calcified arterial lesions accompanied by pronounced fragmentation of the elastic fibers in the aorta and lung vessels. Thus, a calcification-inducing effect of elastin peptides was supposed. Jacob et al. (1987b) first carried out experiments on ion fluxes induced by KE using smooth-muscle cells. They found that KE induced a Ca^{2+} influx in cells which was prevented by calcitonin.

The effect of KE on ion fluxes was further investigated using monocytes, smooth-muscle cells, and fibroblasts, all of which play an important role in the development of arteriosclerosis. Our results confirmed that KE markedly stimulated the Ca^{2+} influx but at the same time inhibited the Ca^{2+} extrusion by an apparently calmodulin-dependent mechanism (Jacob et al., 1987a). It seems, then, that KE acts on ion fluxes in a way opposite to that of formyl peptides concerning the Ca^{2+} transport (Fülöp et al., 1987; Lagast et al., 1984). The stimulation of Ca^{2+} influx is an important trigger for the activation of monocytes (Scully et al., 1984). As the extrusion of Ca^{2+} is equally important both for cell activation and for regulation of intracellular Ca^{2+} concentration, its inhibition by elastin peptides may well result in an accumulation of intracellular Ca^{2+}, as it is suggested by the sustained rise of intracellular free Ca^{2+} measured by Quin2 (Fülöp et al., 1986). These investigations did not permit the determination of whether the increase of intracellular free Ca^{2+} is due entirely to external Ca^{2+} influx or if the liberation of Ca^{2+} from intracellular storage sites also participates. After our recent studies in Ca^{2+} free medium, both of these Ca^{2+} mobilizations play an important role; however, the extracellular Ca^{2+} influx seems to be more important in the rise of intracellular free Ca^{2+}.

The results concerning the $^{22}Na^+$ and $^{86}Rb^+$ fluxes indicate an ouabain-like activity of elastin peptides. The $^{22}Na^+$ influx was stimulated while the $^{86}Rb^+$ influx was inhibited. These findings further suggest the accumulation of intracellular Ca^{2+} because of the reduced Ca^{2+} efflux via the Na^+-Ca^{2+} exchange mechanisms. This will decrease because of the rise in intracellular Na^+, possibly as a consequence of an inhibited Na^+,K^+-ATPase. The above effects of elastin peptides are similar to those of the so-called natriuretic hormones (Blaustein et al., 1984). On the other hand, the alterations of ion fluxes produced by kappa elastin appear to be similar to those observed in hypertension (Sprenger, 1985).

3. ELASTIN PEPTIDE-INDUCED CHANGES IN $[Ca^{2+}]_i$ USING HUMAN PMNLs AS A MODEL

As the KE stimulates the Ca^{2+} influx and inhibits the Ca^{2+} efflux, the result should be the increase of intracellular Ca^{2+}. As this effect of KE is different from that of other formyl peptides, the effect of FMLP (N-formyl-methionyl-leucyl-phenylalanine) was also investigated on intracellular free-Ca^{2+} changes. The intracellular Ca^{2+} changes were also studied in PMNLs (polymorphonuclear leukocytes) of healthy and arteriosclerotic elderly, since earlier experiments showed that Ca^{2+} transport is altered with aging (Fülöp et al., 1987).

The basic $[Ca^{2+}]_i$ in PMNLs increases with aging and a further increase was detected in PMNLs of aged patients with clinically manifest arteriosclerosis (Varga et al., 1988b). Addition of KE to PMNLs of healthy young subjects very rapidly enhanced the cytosolic

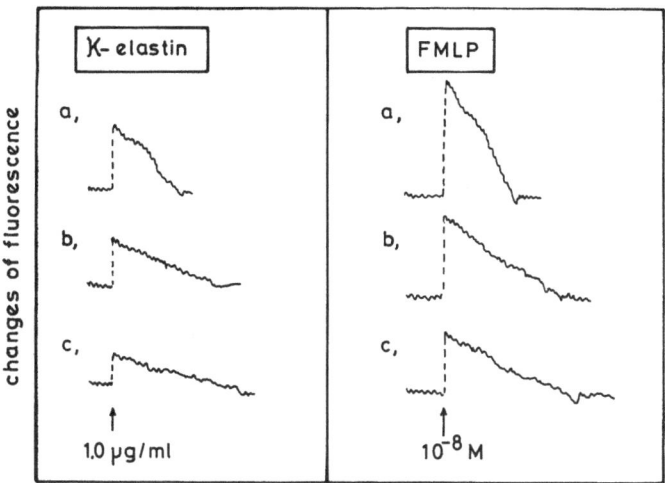

FIGURE 1.   Effect of KE (1.0 μg/ml) and FMLP (10^{-8} M) on [Ca^{2+}]$_i$ in PMNLs of (a) healthy middle-aged, (b) healthy aged, and (c) aged arteriosclerotic subjects.

free-Ca^{2+} concentration, measured by Quin2 (Fig. 1). The most important increase was obtained by 1 μg/ml of KE. The FMLP also induced an increase in [Ca^{2+}]$_i$.

With aging, the mobilization of [Ca^{2+}]$_i$ decreased with either KE or FMLP stimulation. It should be noted that the FMLP-induced raising of [Ca^{2+}]$_i$ was higher than the KE-induced raising in all experimental groups.

Taking into consideration all of the above presented results on KE biological actions, it seems that (1) KE possesses a receptor site on either monocytes or PMNL membranes; (2) even if the KEs act differently on Ca^{2+} ion fluxes than FMLP, the common result is the increase of [Ca^{2+}]$_i$; (3) by raising the [Ca^{2+}]$_i$, KE seems to activate the cells through phosphoinositide hydrolysis; and (4) with aging, the effects of these peptides change.

In the following we focused our attention on the elucidation of the postreceptorial signal transduction of KE.

4. STUDIES ON POSTRECEPTORIAL SIGNAL TRANSDUCTION MECHANISM OF KE

Previous data obtained suggested that the phosphoinositide breakdown could play an important role in the postreceptorial signal transduction. This hypothesis is based on findings that the KE-triggered [Ca]$_i$ mobilization can be inhibited by pertussis toxin (PT) in a similar way to the well-known effect of PT on FMLP-induced transduction mechanism (Fig. 2). The [Ca^{2+}]$_i$ elevation in PMNLs of aged patients was only partially inhibited and not prevented by PT. From these results it seems that KE acts by the hydrolysis of phospho-inositide (Varga et al., 1988a). To further investigate the mechanism of action of KE, we measured the respiratory burst and the [Ca^{2+}]$_i$ in PMNLs in the presence of Li, a well-known inhibitor of the monophosphoesterase enzyme which is responsible for the formation

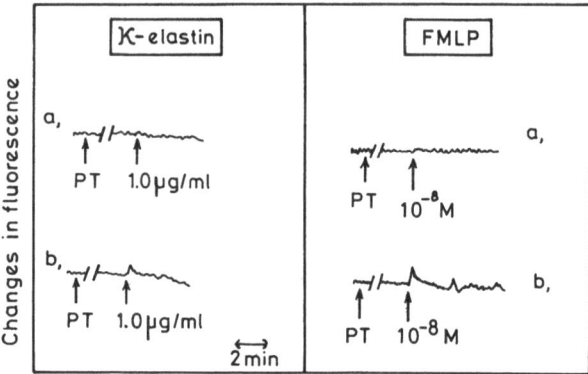

FIGURE 2. Effect of pertussis toxin (PT) on KE- and FMLP-induced $[Ca^{2+}]_i$ elevation in PMNLs of (a) healthy middle-aged and (b) healthy aged subjects.

of myosinositol in the PtdIns cycle (Berridge, 1986). The FMLP served as another known stimulant.

It was found that KE and FMLP stimulated the O_2^- production in PMNLs of healthy, middle-aged subjects while this stimulation was much less in healthy elderly. The Li pre-treating process decreased less the stimulating effect of FMLP on O_2^- production in all subjects compared to that of KE. This would suggest that in the postreceptorial signal transduction of FMLP leading to O_2^- production, other phenomena than the phospholipase C (PLC)-dependent phosphoinositide breakdown could play an important role, such as the phospholipase A_2 (PLA_2)–arachidonic acid (AA) cascade (Irvine, 1982).

When Li was applied before KE stimulation, $[Ca^{2+}]_i$ mobilization diminished significantly but the shape of the curve did not change. In contrast, the FMLP-induced stimulation of $[Ca^{2+}]_i$ mobilization was much less decreased by Li, but the shape of the curve changed.

In summary, it can be said that KE seems to act mostly through the phosphoinositide breakdown, while FMLP seems to have several pathways for postreceptorial signal trans-duction. As a consequence, Li has a greater effect on the biological actions of KE than on those of FMLP. Furthermore, the reactions to KE of the healthy elderly and the arteriosclerotic subjects are different from each other as well as from those of middle-aged subjects. Thus it can be suggested that in the age-dependent KE postreceptorial signal transduction the G_i protein alteration seems to play a determining role. It is evident that several postreceptorial signal transduction mechanisms exist when the cell is stimulated through a specific receptor to achieve a cellular response (Fig. 3).

To approach more closely the KE postreceptorial signal transduction pathway, supposed to be the PLC-dependent phosphoinositide breakdown, several inhibitors were used (PLA_2 inhibitor: mepacrin; PLC inhibitor: neomycin; GTP-binding regulatory protein inhibitor: PT; and PKC inhibitor: retinal) in the study of O_2^- production, AA release, and cGMP formation (Fig. 4). The results obtained show that the only inhibitors that affect KE action on O_2^- production and cGMP formation are those which block some part of the phosphoinositide breakdown. The AA was not affected at all by KE.

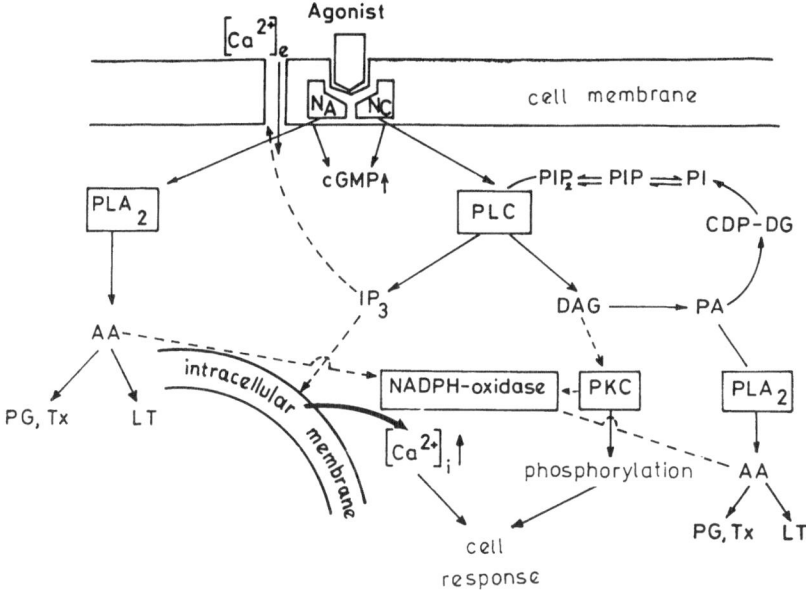

FIGURE 3. Different postreceptorial signal transduction pathways, during receptor stimulation, leading to cell response. PLA$_2$, phospholipase A$_2$; PLC, phospholipase C; DAG, diacylglycerol; PA, phosphatidic acid; AA, arachidonic acid; PG, prostaglandin; LT, leukotrienes; PKC, protein kinase C; IP$_3$, inositol trisphosphate.

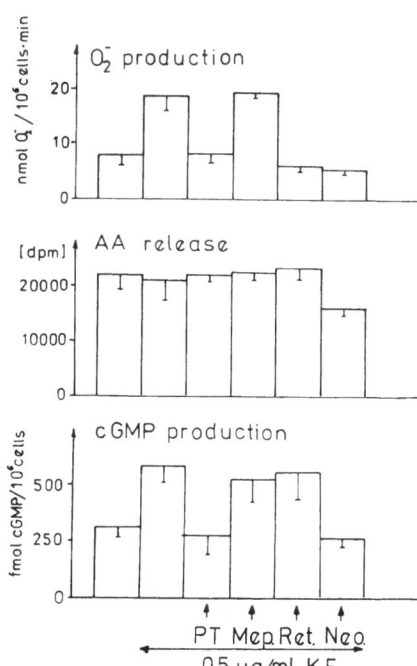

FIGURE 4. Modulation of KE-induced O$_2^-$ production, AA release, and cGMP production by various agents. PT, pertussis toxin; Mep, mepacrin; Ret, retinal; Neo, neomycin.

5. CONCLUSION

We can say that KE, which is a degradation product of elastin in several pathological processes that exists in the sera of patients suffering from various diseases (e.g., arteriosclerosis and experimentally induced severe calcifying arterial lesions), has several important biological actions. The action of KE on ion fluxes, oxygen metabolism, cyclic nucleotides and $[Ca^{2+}]_i$ mobilization suggests that there exists a receptor site on monocytes and PMNL membranes. The elucidation of the postreceptorial signal transduction mechanism using various inhibitors revealed that KE acts through phosphoinositide breakdown. The direct verification of this indirect evidence is currently under investigation in our laboratory, with the concomitant separation of the KE receptor site. Hopefully, the elucidation of these mechanisms might give us useful pharmacological tools which could be efficiently used in such diseases as arteriosclerosis and emphysema.

REFERENCES

Berridge, M. J., and Irvine, R. F., 1984, *Nature* **312**:315–321.
Berridge, M. D., 1986, *Biol. Chem. Hoppe-Seyler* **367**:447–456.
Bignon, J., and Robert, L., 1978, *La revue du Médecin* **19**:764–778.
Blaustein, M. D., and Hamlyn, J. M., 1984, *Am. J. Med.* **77**:45–59.
Bourdillon, M. C., Hornebeck, W., Soleilhac, J. M., and Robert, L., 1984, *Biochem. Soc. Trans.* **12**: 876–887.
Crystal, R. G., 1976, *The Biochemical Basis of Pulmonary Function*, Marcel Dekker, New York.
Fülöp, T., Jr., Jacobs, M. P., Varga, Zs., Fóris, G., Leövey, A., and Robert, L., 1986, *Biochem. Biophys. Res. Commun.* **141**:92–98.
Fülöp, T., Jr., Hauck, M., Wórum, I., Föris, G., and Leövey, A., 1987, *Immunol. Lett.* **14**:283–286.
Frances, C., and Robert, L., 1984, *Int. J. Dermatol.* **23**:166–179.
Irvine, R. F., 1982, *Biochemistry* **204**:3–10.
Jacob, M. P., Hornebeck, W., Lafuma, C., Bernaudin, J. F., Robert, L., and Godean, G., 1984, *Exp. Mol. Pathol.* **41**:171–190.
Jacob, M. P., Fülöp, T., Jr., Fóris, G., and Robert, L., 1987a, *Proc. Natl. Acad. Sci. USA* **84**: 995–1000.
Jacob, M. P., Hornebeck, W., and Robert, L., 1987b, *Exp. Mol. Pathol.* **46**:345–356.
Lagast, H., Lew, P. D., and Waldvogel, A., 1984, *J. Clin. Invest.* **73**:107–115.
Reynolds, E. E., and Dubyak, G. R., 1985, *Biochem. Biophys. Res. Commun.* **130**:627–632.
Robert, L., Stein, F., Pezess, M. P., and Poullain, N., 1967, *Arch. Mal. Coeur* **60**:233–241.
Robert, L., Robert, B., and Robert, A. M., 1970, *Exp. Gerontol.* **5**:339–356.
Robert, A. M., Grosgogeat, Y., Reverdy, V., Robert, B., and Robert, L., 1971, *Atherosclerosis* **13**: 427–449.
Robert, L., and Robert, A. M., 1980, in *Frontiers of Matrix Biology: Biology and Pathology of Elastic Tissues*, Vol. 8 (A. M. Robert and L. Robert, eds.), Karger, Basel, pp. 130–173.
Robert, L., Chaudiere, J., and Jacotot, B., 1984, in: *Regression of Atherosclerotic Lesions: Experimental Studies and Observations in Humans* (M. R. Malinow and V. H. Blaton, eds.), NATO ASI Series, Life Sciences, vol. 79, Plenum Press, New York, pp. 145–173.
Scully, S. P., Segel, G. B., and Lichtman, M. A., 1984, *J. Clin. Invest.* **74**:589–599.
Senior, R. M., Griffin, G. L., and Mechan, R. P., 1980, *J. Clin. Invest.* **66**:859–862.
Sprenger, K. B. G., 1985, *Clin. Physiol. Biochem.* **3**:208–220.
Trimble, E. R., Bruzzone, R., Meehan, C. J., and Biden, T. J., 1987, *Biochem. J.* **242**:289–292.
Varga, Zs., Jacobs, M. P., Robert, L., and Fülöp, T., Jr., 1988a, *Proc. Natl. Acad. Sci.* (in press).
Varga, Zs., Kovács, É. M., Paragh, G., Fülöp, T., Jr., Jacobs, M. P., and Robert, L., 1987b, *Clin. Biochem.* **21**:127–130.

Participants

SAMUEL EVANS ADUNYAH
Department of Biochemistry
University of Louisville School of Medicine
Louisville, Kentucky 40292

PER ARKHAMMER
Department of Medical Cell Biology
Biomedicum
University of Uppsala
S 751 23 Uppsala, Sweden

TAMAS BALLA
Endocrinology and Reproduction Research
 Branch
National Institute of Child Health and Human
 Development
National Institutes of Health
Bethesda, Maryland 20892

JACQUES BARHANIN
Center for Biochemistry of the CNRS
Parc Valrose
06034 Nice Cedex, France

ALBERT J. BAUKAL
Endocrinology and Reproduction Research
 Branch
National Institute of Child Health and Human
 Development
National Institutes of Health
Bethesda, Maryland 20892

ZAFARUL H. BEG
Molecular Disease Branch
National Heart, Lung, and Blood Institute
National Institutes of Health
Bethesda, Maryland 20892

IRENE K. BEREZESKY
Department of Pathology
University of Maryland School of Medicine,
 and
The Maryland Institute for Emergency Medical
 Services Systems
University of Maryland Medical System
Baltimore, Maryland 21201

PER-OLOF BERGGREN
Department of Medical Cell Biology
Biomedicum
University of Uppsala
S-751 23 Uppsala, Sweden

CELENE F. BERNARDES
Department of Biochemistry
I.B. The State University of Campinas
C.P. 6109
CEP. 13081, Brazil

MICHAEL JOHN BERRIDGE
Unit of Insect Neurophysiology and
 Pharmacology
Department of Zoology
University of Cambridge
Cambridge CB2 3EJ, United Kingdom

SYAMAL K. BHATTACHARYA
Edward Dana Mitchell Surgical Research
 Laboratories
Departments of Surgery, Anatomy, and
 Neurobiology
University of Tennessee Medical Center
Memphis, Tennessee 38163

PETER F. BLACKMORE
Howard Hughes Medical Institute Laboratories
Department of Molecular Physiology and
  Biophysics
Vanderbilt University School of Medicine
Nashville, Tennessee 37232

T. J. J. BLANCK
Division of Cardiac Anesthesia
Department of Anesthesiology and Critical
  Care Medicine
Johns Hopkins University
Baltimore, Maryland 21205

STEPHEN B. BOCCKINO
Howard Hughes Medical Institute Laboratories
Department of Molecular Physiology and
  Biophysics
Vanderbilt University School of Medicine
Nashville, Tennessee 37232

DEJAN BOJANIC
Department of Biochemistry
College of Medicine
University of Tennessee
Memphis, Tennessee 38163

ZELJKO J. BOSNJAK
Department of Anesthesiology
Medical College of Wisconsin and USVA
  Medical Center
Milwaukee, Wisconsin 53295

ALTON L. BOYNTON
Cancer Research Center of Hawaii
University of Hawaii
Honolulu, Hawaii 96813

H. BRYAN BREWER, JR.
Molecular Disease Branch
National Heart, Lung, and Blood Institute
National Institutes of Health
Bethesda, Maryland 20892

L. MAXIMILIAN BUJA
Department of Pathology
University of Texas Southwestern Medical
  Center at Dallas
Dallas, Texas 75235-9072

STUART K. CALDERWOOD
Joint Center for Radiation Therapy and
Dana Farber Cancer Institute
Harvard Medical School
Boston, Massachusetts 02115

MAURIZIO C. CAPOGROSSI
Laboratory of Cardiovascular Science
Gerontology Research Center
National Institute on Aging
National Institutes of Health, and
  Johns Hopkins Medical Institutions
Baltimore, Maryland 21224

ERNESTO CARAFOLI
Laboratory of Biochemistry
Swiss Federal Institute of Technology (ETH)
8092 Zurich, Switzerland

EVA G. S. CARNIERI
Department of Biochemistry
I.B. The State University of Campinas
C.P. 6109
CEP. 13081, Brazil

E. S. CASELLA
Department of Cardiac Anesthesia
Department of Anesthesiology and Critical
  Care Medicine
Johns Hopkins University
Baltimore, Maryland 21205

KEVIN J. CATT
Endocrinology and Reproduction Research
  Branch
National Institute of Child Health and Human
  Development
National Institutes of Health
Bethesda, Maryland 20892

WILLIAM A. CATTERALL
Department of Pharmacology
University of Washington
Seattle, Washington 98195

JOO CHEON
Roche Institute of Molecular Biology
Roche Research Center
Nutley, New Jersey 07110

ROY K. CHEUNG
Divisions of Cell Biology and Immunology
    and Rheumatology
Research Institute
The Hospital for Sick Children, and
    Department of Biochemistry
University of Toronto
Toronto, Ontario M5G 1X8, Canada

SHEAU-HUEI CHUEH
Department of Biological Chemistry
University of Maryland School of Medicine
Baltimore, Maryland 21202

MICHAEL G. CLARK
Department of Biochemistry
University of Tasmania
Hobart, Australia

PERRY J. F. CLELAND
Department of Biochemistry
University of Tasmania
Hobart, 7001, Tasmania, Australia

ANNA COCO
Environmental Health Sciences Center
Department of Biophysics
University of Rochester Medical Center
Rochester, New York 14642

ISAAC COHEN
Atherosclerosis Research Laboratory
Northwestern University
Chicago, Illinois 60611

JERRY COLCA
The Upjohn Company
Kalamazoo, Michigan 49001

ERIC Q. COLQUHOUN
Department of Biochemistry
University of Tasmania
Hobart, Australia

WILSON C. COLUCCI
Cardiovascular Division
Department of Medicine
Brigham and Women's Hospital, and
    Harvard Medical School
Boston, Massachusetts 02115

CHRISTY L. COOPER
Department of Biological Chemistry and
    Structure
University of Health Sciences/The Chicago
    Medical School
North Chicago, Illinois 60064

BARBARA E. CORKEY
Department of Biochemistry and
    Division of Diabetes and Metabolism
Evans Department of Medicine
Boston University School of Medicine
Boston, Massachusetts 02118

ALICE J. CRAWFORD
Edward Dana Mitchell Surgical Research
    Laboratories
Departments of Surgery, Anatomy, and
    Neurobiology
University of Tennessee Medical Center
Memphis, Tennessee 38163

MICHAEL F. CROUCH
Department of Pharmacology
Australian National University
Canberra, A.C.T. 2601, Australia

TOM CURRAN
Department of Molecular Oncology
Roche Institute of Molecular Biology
Roche Research Center
Nutley, New Jersey 07110

BENSON M. CURTIS
Department of Pharmacology
University of Washington
Seattle, Washington 98195
Present address:
Division of Biology
California Institute of Technology
Pasadena, California 91125

ANTHONY F. CUTRY
Department of Biochemistry
Roswell Park Memorial Institute
Buffalo, New York 14263

WILLIAM L. DEAN
Department of Biochemistry
University of Louisville School of Medicine
Louisville, Kentucky 40292

JUDE T. DEENEY
Division of Diabetes and Metabolism
Evans Department of Medicine
Boston University School of Medicine
Boston, Massachusetts 02118

RICHARD M. DENTON
Department of Biochemistry
University of Bristol Medical School
Bristol B58 1TD, United Kingdom

GWYNETH DE VRIES
Department of Medical Biochemistry
University of Calgary
Calgary, Alberta T2N 4N1, Canada

JOHN H. EXTON
Howard Hughes Medical Institute Laboratories
Department of Molecular Physiology and
  Biophysics
Vanderbilt University School of Medicine
Nashville, Tennessee 37232

MARCIA M. FAGIAN
Department of Biochemistry
I.B. The State University of Campinas
C.P. 6109, CEP. 13081, Brazil

JOHN N. FAIN
Department of Biochemistry
College of Medicine
University of Tennessee
Memphis, Tennessee 38163

DONNA FALETTO
Environmental Health Sciences Center
Department of Biophysics
University of Rochester Medical Center
Rochester, New York 14642

RORY A. FISHER
Department of Biochemistry
University of Texas Health Science Center
San Antonio, Texas 78284

GARY FISKUM
Departments of Biochemistry and Emergency
  Medicine
The George Washington University School of
  Medicine and Health Sciences
Washington, D.C. 20037

G. FÓRIS
First Department of Medicine
University Medical School of Debrecen
H-4012 Debrecen, Hungary

MICHEL FOSSET
Center for Biochemistry of the CNRS
Parc Valrose
06034 Nice Cedex, France

JUNICHI FUJII
Division of Cardiology
First Department of Medicine, and
  Department of Pathophysiology
Osaka University School of Medicine
Fukushima-ku, Osaka 553, Japan

T. FÜLÖP, JR.
First Department of Medicine
University Medical School of Debrecen
H-4012 Debrecen, Hungary

JEAN-PIERRE GALIZZI
Center for Biochemistry of the CNRS
Parc Valrose
06034 Nice Cedex, France

JAMES C. GARRISON
Department of Pharmacology
University of Virginia School of Medicine
Charlottesville, Virginia 22908

JAMES GAUT
Environmental Health Sciences Center
Department of Biophysics
University of Rochester Medical Center
Rochester, New York, 14642

ERWIN W. GELFAND
Divisions of Cell Biology and Immunology
   and Rheumatology
Research Institute
The Hospital for Sick Children, and
   Department of Biochemistry
University of Toronto, Ontario M5G 1X8,
   Canada

JONATHAN M. GERRARD
Department of Pediatrics
University of Manitoba
Winnipeg, Canada R3E 0Z3

TARUN K. GHOSH
Department of Biological Chemistry
University of Maryland School of Medicine
Baltimore, Maryland 21202

DONALD L. GILL
Department of Biological Chemistry
University of Maryland School of Medicine
Baltimore, Maryland 21202

M. CLAY GLENNON
Department of Biochemistry and Biophysics
University of Pennsylvania School of Medicine
Philadelphia, Pennsylvania 19104

GREGORY J. GORES
Laboratories for Cell Biology
Department of Cell Biology and Anatomy
School of Medicine
University of North Carolina at Chapel Hill
Chapel Hill, North Carolina 27599-7090

GEORGE GRAY
Environmental Health Sciences Center
Department of Biophysics University of
   Rochester Medical Center
Rochester, New York 14642

SERGIO GRINSTEIN
Divisions of Cell Biology and Immunology
   and Rheumatology Research Institute
Hospital for Sick Children, and
Department of Biochemistry University of
   Toronto, Ontario M5G 1X8, Canada

FERENC GUBA, JR.
Department of Pharmacology and Cell
   Biophysics
University of Cincinnati College of Medicine
Cincinnati, Ohio 45267-0575

GAETAN GUILLEMETTE
Endocrinology and Reproduction Research
   Branch
National Institute of Child Health and Human
   Development
National Institutes of Health Bethesda,
   Maryland 20892, and
Department of Pharmacology
University of Sherbrooke School of Medicine
Sherbrooke, Quebec JIH 5N4, Canada

FOZIA HAMUD
Department of Biochemistry
George Washington University School of
   Medicine
Washington, DC 20037

CARL A. HANSEN
Department of Biochemistry and Biophysics
University of Pennsylvania School of Medicine
Philadelphia, Pennsylvania 19104-6089

RICHARD G. HANSFORD
Energy Metabolism and Bioenergetics Section
Laboratory of Cardiovascular Science
Gerontology Research Center
National Institute on Aging
National Institutes of Health
Baltimore, Maryland 21224

ROBERT C. HAYNES, JR.
Department of Pharmacology
University of Virginia
Charlottesville, Virginia 22908

BRIAN HERMAN
Laboratories for Cell Biology
Department of Cell Biology and Anatomy
School of Medicine
University of North Carolina at Chapel Hill
Chapel Hill, North Carolina 27599-7090

TIMOTHY D. HILL
Cancer Research Center of Hawaii
University of Hawaii
Honolulu, Hawaii 96813

JAN B. HOEK
Department of Pathology and Cell Biology
Thomas Jefferson University
Philadelphia, Pennsylvania 19107

SUMIO HOKA
Department of Anesthesiology
Medical College of Wisconsin and
    USVA Medical Center
Milwaukee, Wisconsin 53295

M. MARLENE HOSEY
Department of Biological Chemistry and
    Structure
University of Health Sciences/The Chicago
    Medical School
North Chicago, Illinois 60064

STAN HRENIUK
Department of Physiology
Milton S. Hershey Medical Center
Pennsylvania State University
Hershey, Pennsylvania 17033

MASAHIKO IKEDA
Department of Pharmacology
University of Shizuoka School of
    Pharmaceutical Sciences
Shizuoka, Japan

PETER INGRAM
Research Triangle Institute
Research Triangle Park,
    North Carolina 27709

YASUHIDE INOUE
Department of Pharmacology
University of Shizuoka School of
    Pharmaceutical Sciences
Shizuoka, Japan

PAUL INSEL
INSERUM U-99
Hôpital Henri Mondor
94010 Créteil, France

R. F. IRVINE
Department of Biochemistry
AFRC Institute of Animal Physiology
Babraham, Cambridge CB2 4AT
    United Kingdom

KIYOSHI ITAGAKI
Department of Pharmacology and Cell
    Biophysics
University of Cincinnati College of Medicine
Cincinnati, Ohio 45267-0575

NICHOLAS J. IZZO, JR.
Cardiovascular Division
Brigham and Women's Hospital
Boston, Massachusetts 02115

M. P. JACOB
Laboratory for Biochemistry of Connective
    Tissue
University of Paris XII
Faculty of Medicine
94010 Créteil, France

WILLIAM R. JACOBS
Duke University Medical Center
Durham, North Carolina 27710

LIONEL F. JAFFE
Marine Biological Laboratory
Woods Hole, Massachusetts 02543

ROY A. JOHANSON
Department of Biochemistry and Biophysics
University of Pennsylvania School of Medicine
Philadelphia, Pennsylvania 19104-6089

PATTI L. JOHNSON
Edward Dana Mitchell Surgical Research
    Laboratories
Departments of Surgery, Anatomy, and
    Neurobiology
University of Tennessee Medical Center
Memphis, Tennessee 38163

RANDOLPH M. JOHNSON
Department of Biomolecular Pharmacology
Genentech, Inc.
South San Francisco, California 94080

MASAAKI KADOMA
Division of Cardiology
First Department of Medicine, and
    Department of Pathophysiology
Osaka University School of Medicine
Fukushima-ku, Osaka 553, Japan

KIMINORI KAJIYAMA
Department of Medicine Division of
    Cardiovascular Disease
University of Alabama at Birmingham
Birmingham, Alabama 35294, and
Department of Medicine
Section of Cardiovascular Sciences
Baylor College of Medicine
Houston, Texas 77030

JOHN P. KAMPINE
Department of Anesthesiology
Medical College of Wisconsin and
    USVA Medical Center
Milwaukee, Wisconsin 53295

ALAN J. KINNIBURGH
Department of Human Genetics
Roswell Park Memorial Institute
Buffalo, New York 14263

C. J. KIRK
Department of Biochemistry
University of Birmingham
Birmingham B15 2TT, United Kingdom

ITARU KOJIMA
Cell Biology Research Unit
Fourth Department of Internal Medicine
University of Tokyo School of Medicine
3-28-6 Mejirodai
Bunkyo-ku, Tokyo 112, Japan

EDWARD G. LAKATTA
Laboratory of Cardiovascular Science
Gerontology Research Center
National Institute on Aging
National Institutes of Health, and
    Johns Hopkins Medical Institutions
Baltimore, Maryland 21224

G. A. LANGER
Cardiovascular Research Laboratories
University of California
Los Angeles School of Medicine
Center for the Health Sciences
Los Angeles, California 90024-1760

KATHRYN F. LANOUE
Department of Physiology
Milton S. Hershey Medical Center
Pennsylvania State University
Hershey, Pennsylvania 17033

EDUARDO G. LAPETINA
Department of Cell Biology
Burroughs Wellcome Co.
Research Triangle Park, North Carolina 27709

JOSEPH P. LAURINO
Departments of Pathology and Medicine
Washington University School of Medicine
St. Louis, Missouri 63110

MICHEL LAZDUNSKI
Center for Biochemistry of the CNRS
Parc Valrose
06034 Nice Cedex, France

ANN LEFURGEY
Department of Physiology
Duke University Medical Center
Durham, North Carolin· ·7710

KIRK J. LEISTER
Department of Biochemistry
Roswell Park Memorial Institute
Buffalo, New York 14263

JOHN J. LEMASTERS
Laboratories for Cell Biology
Department of Cell Biology and Anatomy
School of Medicine
University of North Carolina at Chapel Hill
Chapel Hill, North Carolina 27599-7090

MELVYN LIEBERMAN
Department of Physiology
Duke University Medical Center
Durham, North Carolina 27710

SUE-HWA LIN
Department of Biochemistry and Molecular
  Biology
Harvard University
Cambridge, Massachusetts 02138

H. LOATS
Loats Associates, Inc.
Westminster, Maryland 21157

ALEX L. LOEB
Department of Anesthesia
University of Pennsylvania School of Medicine
Philadelphia, Pennsylvania 19104

RICHARD M. LOPACHIN, JR.
Department of Anesthesiology
State University of New York Health Sciences
  Center
Stony Brook, New York 11794-8480

VICKI R. LOPACHIN
Department of Anesthesiology
State University of New York Health Sciences
  Center
Stony Brook, New York 11794-8480

CHRISTOPHER J. LYNCH
Howard Hughes Medical Institute Laboratories
Department of Molecular Physiology and
  Biophysics
Vanderbilt University School of Medicine
Nashville, Tennessee 37232

SOPHIE LOTERSZTAJN
INSERM U-99
Henri Mondor Hospital
94010 Créteil, France

IAN G. MACARA
Environmental Health Sciences Center
Department of Biophysics
University of Rochester Medical Center
Rochester, New York 14642

STEPHEN MACDOUGALL
Divisions of Cell Biology and Immunology
  and Rheumatology
Research Institute, Hospital for Sick Children,
  and
Department of Biochemistry
University of Toronto, Ontario M5G 1X8,
  Canada

SAMIRA MAHMOUD
Department of Biochemistry
George Washington University School of
  Medicine and Health Sciences
Washington, D.C. 20037

ARIANE MALLAT
INSERM U-99
Hôpital Henri Mondor
94010 Crétein, France

LAZARO J. MANDEL
Duke University Medical Center
Durham, North Carolina 27710

IONE S. MARTINS
Department of Biochemistry
I.B. The State University of Campinas
C.P. 6109
CEP. 13081, Brazil

FRANZ M. MATSCHINSKY
Department of Biochemistry and Biophysics
University of Pennsylvania School of Medicine
Philadelphia, Pennsylvania 19104

DAVID J. MCCONKEY
Department of Toxicology
Karolinska Institutet
S-104 01 Stockholm, Sweden

JAMES G. MCCORMACK
Department of Biochemistry
University of Leeds
Leeds LS2 9JT, United Kingdom

JAY M. MCDONALD
Departments of Pathology and Medicine
Washington University School of Medicine
St. Louis, Missouri 63110

JOHN R. MCDONALD
Department of Medical Biochemistry
University of Calgary
Calgary, Alberta T2N 4N1, Canada

EDWARD MCKENNA
Department of Pharmacology and Cell
    Biophysics
University of Cincinnati College of Medicine
Cincinnati, Ohio 45267-0575

JEANIE B. MCMILLIN
Department of Medicine
Division of Cardiovascular Disease
University of Alabama at Birmingham
University Station
Birmingham, Alabama 35294, and
Department of Medicine
Section of Cardiovascular Sciences
Baylor College of Medicine
Houston, Texas 77030

K. MEADE-COBUN
Johns Hopkins Oncology Center
Baltimore, Maryland 21205

R. H. MICHELL
Department of Biochemistry
University of Birmingham
Birmingham B15 2TT, United Kingdom

KUNIHISA MIWA
Department of Pharmacology and Cell
    Biophysics
University of Cincinnati College of Medicine
Cincinnati, Ohio 45267-0575

JONATHAN R. MONCK
Department of Biochemistry and Biophysics
University of Pennsylvania School of Medicine
Philadelphia, Pennsylvania 19104-6089

TERRY W. MOODY
Department of Biochemistry
George Washington University School of
    Medicine and Health Sciences
Washington, D.C. 20037

RAFAEL MORENO-SÁNCHEZ
Energy Metabolism and Bioenergetics Section
Laboratory of Cardiovascular Science
Gerontology Research Center
National Institute on Aging
National Institutes of Health
Baltimore, Maryland 21224

JAMES I. MORGAN
Department of Neurosciences
Roche Institute of Molecular Biology
Roche Research Center
Nutley, New Jersey 07110

JULIENNE M. MULLANEY
Department of Biological Chemistry
University of Maryland School of Medicine
Baltimore, Maryland 21201

ANNE N. MURPHY
Department of Biochemistry
George Washington University School of
    Medicine and Health Sciences
Washington, D.C. 20037

ELIZABETH MURPHY
Laboratory of Molecular Biophysics
National Institute of Environmental Health
    Sciences
Research Triangle Park, North Carolina 27709

KOH-ICHI NAGATA
Department of Biochemistry
Gifu University School of Medicine
Tsukasamachi 40
Gifu, 500 Japan

KAZUKI NAKAMURA
Department of Pharmacology
University of Shizuoka School of
    Pharmaceutical Sciences
Shizuoka, Japan

SHIGERU NAKASHIMA
Department of Biochemistry
Gifu University School of Medicine
Tsukasamachi 40
Gifu, 500 Japan

T. E. NELSON
Department of Anesthesiology
University of Texas Health Science Center at
    Houston
Houston, Texas 77030

PIERLUIGI NICOTERA
Department of Toxicology
Karolinska Institutet
S-104 01 Stockholm, Sweden

ANNA-LIISA NIEMINEN
Laboratories for Cell Biology
Department of Cell Biology and Anatomy
School of Medicine
University of North Carolina at Chapel Hill
Chapel Hill, North Carolina 27599-7090

THOMAS NILSSON
Department of Medical Cell Biology
Biomedicum
University of Uppsala
S 751 23 Uppsala, Sweden

IKUO NISHIMOTO
Cell Biology Research Unit
Fourth Department of Internal Medicine
University of Tokyo School of Medicine
3-26-6 Mejirodai
Bunkyo-ku, Tokyo 112, Japan

YOSHINORI NOZAWA
Department of Biochemistry
Gifu University School of Medicine
Tsukasamachi 40
Gifu, 500 Japan

CLIFF M. O'CALLAHAN
Department of Biological Chemistry and
    Structure
University of Health Sciences/The Chicago
    Medical School
North Chicago, Illinois 60064

ETSURO OGATA
Cell Biology Research Unit
Fourth Department of Internal Medicine
University of Tokyo School of Medicine
3-28-6-Mejirodai
Bunkyo-ku, Tokyo 112, Japan

YUKIO OKANO
Department of Biochemistry
Gifu University School of Medicine
Tsukasamachi 40
Gifu, 500 Japan

MERLE S. OLSON
Department of Biochemistry
University of Texas Health Science Center
San Antonio, Texas 78284

STEN ORRENIUS
Department of Toxicology
Karolinska Institutet
S-104 01 Stockholm, Sweden

JANICE C. PARKER
Department of Biochemistry and Biophysics
University of Pennsylvania School of Medicine
Philadelphia, Pennsylvania 19104

J. B. PARRY
Department of Biochemistry
University of Birmingham
Birmingham B15 2TT, United Kingdom

DANIEL F. PAULY
Department of Medicine
Division of Cardiovascular Disease
University of Alabama at Birmingham
University Station
Birmingham, Alabama 35294, and
Department of Medicine
Section of Cardiovascular Sciences
Baylor College of Medicine
Houston, Texas 77030

CATHERINE PAVOINE
INSERM U-99
Hôpital Henri Mondor
94010 Créteil, France

MICHAEL J. PEACH
Department of Pharmacology
University of Virginia School of Medicine
Charlottesville, Virginia 22908

FRANÇOISE PECKER
INSERM U-99
Hôpital Henri Mondor
94010 Créteil, France

LUCIA PEREIRA-DA-SILVA
Department of Biochemistry
I.B. The State University of Campinas
C.P. 6109,
CEP. 13081, Brazil

J. POGGIOLI
Cellular Physiology and Pharmacology
  Research Unit
INSERM U-274
Universite Paris-Sud
F-91405 Orsay Cedex, France

MARC PRENTKI
Department of Biochemistry and Biophysics
University of Pennsylvania School of Medicine
Philadelphia, Pennsylvania 19104

GUNDU H. R. RAO
Department of Laboratory Medicine and
  Pathology
University of Minnesota
Minneapolis, Minnesota 55455

STEPHEN RATTIGAN
Department of Biochemistry
University of Tasmania
Hobart, Australia

JOHN P. REEVES
Roche Institute of Molecular Biology
Roche Research Center
Nutley, New Jersey 07110

D. RENARD
Cellular Physiology and Pharmacology
  Research Unit
INSERM U-274
Universite Paris-Sud
F-91405 Orsay Cedex, France

ERIK A. RICHTER
August Krogh Institute
University of Copenhagen
Denmark

L. ROBERT
Laboratory for Biochemistry of Connective
  Tissue
University of Paris XII
Faculty of Medicine
94010 Créteil, France

ROBERT E. ROSENTHAL
Department of Emergency Medicine
George Washington University School of
  Medicine
Washington, D.C. 20037

EMANUEL RUBIN
Department of Pathology and Cell Biology
Thomas Jefferson University
Philadelphia, Pennsylvania 19107

RAPHAEL RUBIN
Department of Pathology annd Cell Biology
Thomas Jefferson University
Philadelphia, Pennsylvania 19107

NEIL B. RUDERMAN
Division of Diabetes and Metabolism
Evans Department of Medicine
Boston University School of Medicine
Boston, Massachusetts 02118

ALBERT J. SAUBERMANN
Department of Anesthesiology
State University of New York Health Sciences
    Center
Stony Brook, New York 11794-8480

RUSSELL SCADUTO, JR.
Department of Physiology
Milton S. Hershey Medical Center
Pennsylvania State University
Hershey, Pennsylvania 17033

RICK SCHNELLMANN
Duke University Medical Center
Durham, North Carolina 27710

ARNOLD SCHWARTZ
Department of Pharmacology and Cell
    Biophysics
University of Cincinnati College of Medicine
Cincinnati, Ohio 45267-0575

MICHAEL J. SEAGAR
Department of Pharmacology
University of Washington
Seattle, Washington 98195

MARIA SGAMBATI
Duke University Medical Center
Durham, North Carolina 27710

S. B. SHEARS
Department of Biochemistry
University of Birmingham
Birmingham B15 2TT, United Kingdom

HAROLD A. SPURGEON
Laboratory of Cardiovascular Science
Gerontology Research Center
National Institute on Aging
National Institutes of Health and
Johns Hopkins Medical Institutions
Baltimore, Maryland 21224

GIOVANNI SPERTI
Cardiovascular Division
Department of Medicine
Brigham and Women's Hospital, and
Harvard Medical School
Boston, Massachusetts 02115

JAMES M. STADDON
Energy Metabolism and Bioenergetics Section
Laboratory of Cardiovascular Science
Gerontology Research Center
National Institute on Aging
National Institutes of Health
Baltimore, Maryland 21224

MARK E. STEINHELPER
Department of Biochemistry
University of Texas Health Science Center
San Antonio, Texas 78284

DOMINIQUE STENGEL
INSERM U-99
Hôpital Henri Mondor
94010 Créteil, France

MICHAEL D. STERN
Laboratory of Cardiovascular Science
Gerontology Research Center
National Institute on Aging
National Institutes of Health, and
Johns Hopkins Medical Institutions
Baltimore, Maryland 21224

ANNA STERNICZUK
Department of Physiology
Milton S. Hershey Medical Center
Pennsylvania State University
Hershey, Pennsylvania 17033

MARY ANN STEVENSON
Joint Center for Radiation Therapy and
    Dana Farber Cancer Institute
Harvard Medical School
Boston, Massachusetts 02115

JOHN A. STONIK
Molecular Disease Branch
National Heart, Lung, and Blood Institute
National Institutes of Health
Bethesda, Maryland 20892

MICHIHIKO TADA
Division of Cardiology
First Department of Medicine, and
Department of Pathophysiology
Osaka University School of Medicine
Fukushima-ku, Osaka 553, Japan

MASAMI TAKAHASHI
Department of Pharmacology
University of Washington
Seattle, Washington 98195, and
Mitsubishi, Kasei
    Institute of Life Sciences
Machida-Shi, Japan

NOBUAKI TAKESHITA
Department of Pharmacology
University of Shizuoka School of
    Pharmaceutical Sciences
Shizuoka, Japan

JAY H. THAKAR
Edward Dana Mitchell Surgical Research
    Laboratories
Departments of Surgery, Anatomy, and
    Neurobiology
University of Tennessee Medical Center
Memphis, Tennessee 38163

ANDREW P. THOMAS
Department of Pathology and Cell Biology
Thomas Jefferson University
Philadelphia, Pennsylvania 19107

TAKAKO TOMITA
Department of Pharmacology
University of Shizuoka School of
    Pharmaceutical Sciences
Shizuoka, Japan

TOYOHIKO TOHMATSU
Department of Biochemistry
Gifu University School of Medicine
Tsukasamachi 40
Gifu, 500 Japan

KEITH TORNHEIM
Department of Biochemistry and Division of
    Diabetes and Metabolism
Evans Department of Medicine
Boston University School of Medicine
Boston, Massachusetts 02118

BENJAMIN F. TRUMP
Department of Pathology
University of Maryland School of Medicine,
    and
The Maryland Institute for Emergency Medical
    Services Systems
Baltimore, Maryland 21201

R. W. TUCKER
Johns Hopkins Oncology Center
Baltimore, Maryland 21205

LAWRENCE TURNER
Department of Anesthesiology
Medical College of Wisconsin and
    USVA Medical Center
Milwaukee, Wisconsin 53295

KEIZOU UMEGAKI
Department of Pharmacology
University of Shizuoka School of
    Pharmaceutical Sciences
Shizuoka, Japan

PAL L. VAGHY
Department of Pharmacology and Cell
    Biophysics
University of Cincinnati College of Medicine
Cincinnati, Ohio 45267-0575

SYLVIE VANDAELE
Center for Biochemistry of the CNRS
Parc Valrose
06034 Nice Cedex, France

JACK Y. VANDERHOEK
Department of Biochemistry
The George Washington University School of
  Medicine and Health Sciences
Washington, D.C. 20037

ZS. VARGA
First Department of Medicine
University Medical School of Debrecen
H-4012 Debrecen, Hungary

ANIBAL E. VERCESI
Department of Biochemistry
I.B. The State University of Campinas
C.P. 6109
CEP. 13081, Brazil

BECKY M. VONAKIS
Department of Biochemistry
The George Washington University School of
  Medicine and Health Sciences
Washington, D.C. 20037

BERNHARD WAGENKNECHT
Department of Physiology
Duke University Medical Center
Durham, North Carolina 27710

MICHAEL A. WALLACE
Department of Biochemistry
University of Tennessee College of Medicine
Memphis, Tennessee 38163

MICHAEL P. WALSH
Department of Medical Biochemistry
University of Calgary
Calgary, Alberta T2N 4N1, Canada

C. E. WENNER
Department of Biochemistry
Roswell Park Memorial Institute
Buffalo, New York 14263

JAMES G. WHITE
Department of Laboratory Medicine and
  Pathology
University of Minnesota
Minneapolis, Minnesota 55455

JAMES T. WILLERSON
Department of Pathology
University of Texas Southwestern Medical
  Center at Dallas
Dallas, Texas 75235-9072

JOHN R. WILLIAMSON
Department of Biochemistry and Biophysics
University of Pennsylvania School of Medicine
Philadelphia, Pennsylvania 19104-6089

THERESA WINGROVE
Environmental Health Sciences Center
Department of Biophysics
University of Rochester Medical Center
Rochester, New York 14642

CARL J. WITKOP, JR.
Department of Human and Oral Genetics
University of Minnesota School of Dentistry
Minneapolis, Minnesota 55455

RICHARD J. H. WOJCIKIEWICZ
Department of Biochemistry
University of Tennessee College of Medicine
Memphis, Tennessee 38163

ALAN WOLFMAN
Environmental Health Sciences Center
Department of Biophysics
University of Rochester Medical Center
Rochester, New York 14642

BARNABY E. WRAY
Laboratories for Cell Biology
Department of Cell Biology and Anatomy
School of Medicine
University of North Carolina at Chapel Hill
Chapel Hill, North Carolina 27599-7090

KOUJI YAMADA
Department of Biochemistry
Gifu University School of Medicine
Tsukasamachi 40
Gifu, 500 Japan

JEAN ZWILLER
Cancer Research Center of Hawaii
University of Hawaii
Honolulu, Hawaii 96813

Index